Interstitial Intermetallic Alloys

NATO ASI Series

Advanced Science Institutes Series

A Series presenting the results of activities sponsored by the NATO Science Committee, which aims at the dissemination of advanced scientific and technological knowledge, with a view to strengthening links between scientific communities.

The Series is published by an international board of publishers in conjunction with the NATO Scientific Affairs Division

A	**Life Sciences**	Plenum Publishing Corporation
B	**Physics**	London and New York
C	**Mathematical and Physical Sciences**	Kluwer Academic Publishers
D	**Behavioural and Social Sciences**	Dordrecht, Boston and London
E	**Applied Sciences**	
F	**Computer and Systems Sciences**	Springer-Verlag
G	**Ecological Sciences**	Berlin, Heidelberg, New York, London,
H	**Cell Biology**	Paris and Tokyo
I	**Global Environmental Change**	

PARTNERSHIP SUB-SERIES

1. Disarmament Technologies	Kluwer Academic Publishers
2. Environment	Springer-Verlag
3. High Technology	Kluwer Academic Publishers
4. Science and Technology Policy	Kluwer Academic Publishers
5. Computer Networking	Kluwer Academic Publishers

The Partnership Sub-Series incorporates activities undertaken in collaboration with NATO's Cooperation Partners, the countries of the CIS and Central and Eastern Europe, in Priority Areas of concern to those countries.

NATO-PCO-DATA BASE

The electronic index to the NATO ASI Series provides full bibliographical references (with keywords and/or abstracts) to more than 30000 contributions from international scientists published in all sections of the NATO ASI Series.
Access to the NATO-PCO-DATA BASE is possible in two ways:

– via online FILE 128 (NATO-PCO-DATA BASE) hosted by ESRIN, Via Galileo Galilei, I-00044 Frascati, Italy.

– via CD-ROM "NATO-PCO-DATA BASE" with user-friendly retrieval software in English, French and German (© WTV GmbH and DATAWARE Technologies Inc. 1989).

The CD-ROM can be ordered through any member of the Board of Publishers or through NATO-PCO, Overijse, Belgium.

Series E: Applied Sciences - Vol. 281

Interstitial Intermetallic Alloys

edited by

Fernande Grandjean

Institute of Physics,
University of Liège,
Sart-Tilman, Belgium

Gary J. Long

Department of Chemistry,
University of Missouri-Rolla,
Rolla, Missouri, USA

and

K. H. J. Buschow

Van der Waals-Zeeman Laboratory,
University of Amsterdam,
Amsterdam, The Netherlands

Springer Science+Business Media, B.V.

Proceedings of the NATO Advanced Study Institute on
Interstitial Alloys for Reduced Energy Consumption and Pollution
Il Ciocco, Italy
12–24 June 1994

Library of Congress Cataloging-in-Publication Data

DOI 10.1007/978-94-011-0295-7

Printed on acid-free paper

TABLE OF CONTENTS

Preface

It is well known that the density of molecular hydrogen can be increased by compression and/or cooling, the ultimate limit in density being that of liquid hydrogen. It is less well known that hydrogen densities of twice that of liquid hydrogen can be obtained by intercalating hydrogen gas into metals. The explanation of this unusual paradox is that the absorption of molecular hydrogen, which in TiFe and $LaNi_5$ is reversible and occurs at ambient temperature and pressure, involves the formation of hydrogen atoms at the surface of a metal. The adsorbed hydrogen atom then donates its electron to the metal conduction band and migrates into the metal as the much smaller proton. These protons are easily accomodated in interstitial sites in the metal lattice, and the resulting metal hydrides can be thought of as compounds formed by the reaction of hydrogen with metals, alloys, and intermetallic compounds.

The practical applications of metal hydrides span a wide range of technologies, a range which may be subdivided on the basis of the hydride property on which the application is based. The capacity of the metal hydrides for hydrogen absorption is the basis for batteries as well as for hydrogen storage, gettering, and purification. The temperature-pressure characteristics of metal hydrides are the basis for hydrogen compressors, sensors, and actuators. The latent heat of the hydride formation is the basis for heat storage, heat pumps, and refrigerators. Although metal hydride applications often utilize all three of these features, usually one property is central to a specific application.

Apparently unrelated to the study of metal hydrides is the study of permanent magnetic materials. There has been considerable progress in this field over the past two decades and the discovery of $Nd_2Fe_{14}B$ in 1983 led to the extensive replacement of $SmCo_5$ magnets by the much cheaper and stronger $Nd_2Fe_{14}B$-based magnets. Several hundred research scientists are involved in a worldwide research program trying to improve the high-temperature properties of the $Nd_2Fe_{14}B$ magnets so that they can be used in small, powerful, and efficient motors. The demand for small strong magnets with good high-temperature properties led to the recent discovey that Sm_2Fe_{17} can absorb large quantities of nitrogen. The Curie temperature of the corresponding interstitial nitride, $Sm_2Fe_{17}N_{2.7}$, is about 400° higher than that of Sm_2Fe_{17} and its magnetocrystalline anisotropy is axial. This dramatic improvement of the magnetic properties of Sm_2Fe_{17} has led many scientists, working in the field of permanent magnets, to investigate the absorption of molecular nitrogen by many other intermetallic compounds. Many of these experiments are similar to the early experiments on hydrogen absorption. It is unfortunate that the scientists working on intermetallic nitrides have not profited from the enormous experience generated in the study of intermetallic hydrides. One of the goals of this Advanced Study Institute was to bring these two groups of scientists together to induce scientific cross-fertilization. The second goal of this Advanced Study Institute was to emphasize the environmental aspects of current hydride research and to demonstrate to young scientists and engineers the exciting challenges available in this newly emerging field of research.

Virtually all of the lecturers honored the request of the Advanced Study Institute organizing committee to present their lectures at a tutorial level which would be appropriate for typical graduate students. The resulting lectures and the informal and friendly atmosphere of Il Ciocco encouraged the students to participate actively in the various discussions. During the lectures there were many requests for clarification by the students. The Advanced Study Institute ended with a pannel discussion in which the main issues of the lectures and possible future developments in the field were discussed. Students were also invited to present their research in the field of intermetallic hydrides, either as short oral presentations or as posters. The posters were on display during the entire Advanced Study Institute and served as nucleation centers for numerous discussions. A group of student participants who did not have posters on display selected the posters by Dave P. Middleton and by Fermin Cuevas and Jose F. Fernandez, as the best poster presentations. The directors selected the talk by Luis Miguel Garcia as the best oral research presentation and the talks by Peter H. L. Notten as the best received lectures.

The success of an Advanced Study Institute is very dependent upon the venue and its meeting facilities. In this respect Il Ciocco, located in Castelvechio-Pascoli, a few kilometers from the beautiful village of Barga, in the Serchio valley of Tuscany, Italy, is ideal. Il Ciocco is a conference and sports center with accomodation and meeting facilities which are perfect for an Advanced Study Institute. Indeed, one of the advantages is its isolation which ensures excellent attendance by the participants throughout the scientific sessions as well as at the coffee breaks, meals, and other social functions. The organizers are convinced that as many personal contacts were established in the dining room and during the many frequent games which occured in the evenings as during the formal lecture sessions.

Finally, the editors would like to thank the organizing committee and especially Peter Notten and Umberto Russo, for many helpful discussions during the planning of the Advanced Study Institute and each of the lecturers for their timely and informative chapters. The editors, as well as all of the participants in the ASI, extend their warm thanks to Mr. Bruno Giannasi and the gracious staff of Il Ciocco, for their help in making their visit so pleasant. Of course, this NATO-Advanced Study Institute could not have taken place without the financial support provided by the Scientific Affairs Division of NATO, support which is gratefully appreciated. The support provided by the Italian Consiglio Nazionale delle Ricerche, the US National Science Foundation, the Société Générale de Banque, the University of Liège, and the University of Missouri-Rolla is also gratefully appreciated.

August 1994
Liège, Belgium

Fernande Grandjean
Gary J. Long
K. H. J. Buschow

Chapter 1

AN INTRODUCTION TO THE FORMATION OF INTERSTITIAL ALLOYS THROUGH GAS-SOLID REACTIONS

K. H. J. Buschow
Van der Waals-Zeeman Laboratory
University of Amsterdam
Valckenierstraat 65
NL-1018 XE Amsterdam
The Netherlands

Gary J. Long
Department of Chemistry
University of Missouri-Rolla
Rolla, MO 65401-0249
USA

Fernande Grandjean
Institute of Physics, B5
University of Liège
B-4000 Sart-Tilman
Belgium

1. Historical Introduction

Gas-solid reactions date back several centuries B.C. to when Aristotle postulated his famous principle of the four elements. He postulated that all materials could be characterized by combinations of at least two of the four elements, *earth*, which is cold and dry, *water*, which is cold and humid, *air*, which is warm and humid, and *fire*, which is warm and dry. The two properties associated with each element, the so-called 'primae qualitates' are as important as the four elements themselves. These two properties allow the four elements to transform into each other and make it possible to specify the composition of a given material. *Earth* predominates in heavy solids such as metals, although the presence of *water* is also required to explain why metals melt upon heating. Gas-solid reactions take the form of *earth-fire* combinations within this postulate.

Much later, in the seventeenth century, the English chemist and physicist, Robert Boyle, reported that the model of Aristotle was unsatisfactory for explaining his observation that combustion occurs only in the presence of air and that part of the air disappears with a concomitant increase in weight of the combustion products.

In the eighteenth century the English experimentalist and philosopher, Henry Cavendish, discovered a new type of *air* that escaped from certain metals. This air was flammable and led to explosive combustion in combination with normal air. However, his discovery did not pertain to metal hydrides, because the liberation of the flammable air from the metal required that it be placed in sulfuric acid.

F. Grandjean et al. (eds.), Interstitial Intermetallic Alloys, 1–6.

The first metal hydride was discovered, almost a hundred years later, by the Scottish chemist and physicist, T. Graham, while studying the penetration of hydrogen into heated palladium. It is interesting to note that his experimental results led him to assume that hydrogen is the vapor of a volatile metal.

2. Metal Hydrides

The hydrogen atom is by far the smallest and simplest atom in the periodic table. It is capable of forming compounds with virtually all elements, including metals and non-metals. When combined with the halogens, hydrogen forms strong acids, such as HF, HCl, and HI. When combined with elements located to the left of the halogens, it forms covalently bonded compounds, such as H_2O, NH_3, CH_4, B_2H_6, and GeH_4. There are numerous hydrides in which hydrogen is combined with the metals. These include the hydrides of the d-block transition metals, the lanthanide and actinide metals, and the groups I, II, IIIa, IVa, and Va metals.

A closer examination of the electronic properties of compounds such as HCl and CsH indicates that, in the former, the hydrogen atoms are positively charged, whereas in the latter they are negatively charged. The difference in charge is reflected by a distinct difference in the radii of the hydrogen which are typically 0.32Å for hydrogen cations and 2.08Å for hydride anions. Large differences in the hydrogen charge exist among the metal hydrides. Thus it has become customary to classify [1] the metal hydrides according to the nature of their metal-hydrogen bonds. The groups Ia and IIa elements form the so-called ionic or saline hydrides, whereas GeH_4, AsH_3, BeH_2, CuH, and $LiAlH_3$ are representative of the covalent hydrides. By far the largest class of hydrides is the metallic hydrides, which are discussed in detail in this volume. These metallic hydrides include those of the lanthanide and actinide metals, those formed with the group IIIa, IVa, and Va metals, and those formed with some of the d-block transition metals, such as chromium, nickel, and palladium. This class also contains the many hydrides formed through hydrogen absorption into intermetallic compounds.

The determination of the valence state of hydrogen in metallic hydrides is difficult and many experimental efforts have been devoted to distinguishing [2] betweeen the so-called hydridic or anionic model and the protonic model. In the hydridic model the two 1s states of hydrogen are found below the Fermi energy of the metal. In the anionic hydrides both of these states are occupied because an additional electron is provided by the conduction band and thus hydrogen is present as the hydride ion, H^-. In contrast, the two 1s states of hydrogen are found above the Fermi level in the protonic hydrides. In this case the hydrogen donates one electron to the conduction band upon hydride formation and is present as H^+. The difficulties associated with both models result because, for a given hydride, some experimental results favor the anionic model and other results favor the protonic model.

Modern electronic band structure calculations, performed on a large number of metallic hydrides, have removed much of the difficulty in distinguishing between the two models. These calculations indicate that the hydrogen has to be regarded as a proton screened by varying numbers of valence electrons. A comprehensive description of the results of electronic band structure calculations can be found in the reviews by Schlapbach et al. [3] and Yamaguchi and Akiba. [4] In these reviews, a description of the changes in the physical properties upon hydrogen absorption, such as the magnetic properties and the electrical transport properties, can be found.

The reversible nature of hydrogen absorption in metallic systems leads to a wide gamut of practical applications, applications which are often based on a specific property of the metallic hydride. The hydrogen sorption capacity is most important both in applications involving batteries and in hydrogen storage, gettering, and purification devices. The hydrogen pressure-temperature characteristics are most important in compressors, sensors, and actuators. Finally, the latent heat of metal hydride formation is most important in heat storage, heat pumps, and refrigerators. Many of these applications and especially hydride batteries and heat pumps, will have a strong impact upon reducing environmental pollution and lowering energy consumption. Several of these applications are discussed in detail in various chapters in this book. Over the past few decades the development of these metal hydride applications has proceeded hand in hand with fundamental research on the nature of hydrogen sorption reactions and their thermodynamic properties. This fundamental research is also discussed in detail in various chapters in this book.

3. Interstitially Modified Hard Magnetic Materials

Permanent magnetic materials, and their modification through the formation of interstitial hydrides, nitrides, and carbides, form a separate field of metal-gas reaction studies. In these studies the gases involved are typically H_2, N_2, NH_3, and CH_4, and occasionally heavier hydrocarbon gases. These investigations have become a field of research of their own and, as will be described below, there has been comparatively little contact between research scientists working in this field and the field of metal hydrides discussed above.

It is interesting to note that the most important metal hydride, $LaNi_5H_x$, was discovered during an investigation of the equally important permanent magnetic material, $SmCo_5$. Intermetallic compounds capable of absorbing large quantities of hydrogen gas were already known by the 1950s. These early studies were carried out on intermetallic compounds, compounds in which the early transition metals were usually predominant. $LaNi_5$ probably escaped early detection because compounds with a high concentration of a late transition metal were generally considered as incapable of absorbing large quantities of hydrogen gas.

As discussed in detail by Willems and Buschow, [5] $LaNi_5$ was discovered by chance in the 1960s when the attention of scientists working in the field of permanent magnets was focused on $SmCo_5$ and attempts to improve its hard magnetic properties. Milling and grinding was a common method for increasing its coercivity, but, unfortunately, the coercivity decreased when the powdered $SmCo_5$ was kept in air for some time. The explanation offered for this decrease was based on the reaction of $SmCo_5$ with the water in moist air,

$$2SmCo_5 + 3H_2O \rightarrow Sm_2O_3 + 3H_2 + 10Co.$$

The samarium oxide and the cobalt were assumed to remain on the surface of the powder, while the hydrogen gas was absorbed by the unreacted $SmCo_5$ and caused a decrease in its coercivity. In order to test this hypothesis, experiments were performed in which $SmCo_5$ was deliberately charged with hydrogen gas. Figure 1 shows the resulting concentration dependence of the coercivity and clearly demonstrates the detrimental effect of the hydrogen on the coercivity of $SmCo_5$. It is interesting that this curve mimics the concentration dependence of the hydrogen content, a dependence which was not measured until many years later.

Experimental attempts to determine the location of the hydrogen atoms by proton NMR studies failed because of the strong ferromagnetic character of the hydride of $SmCo_5$. The search for an isotypic paramagnetic compound eventually led to $LaNi_5$ and its favorable hydrogen sorption properties. This discovery led to a strong increase in research, especially by the hydride research scientists. These studies dealt with the hydrogen sorption properties, as well as surface effects, reaction kinetics, and reaction

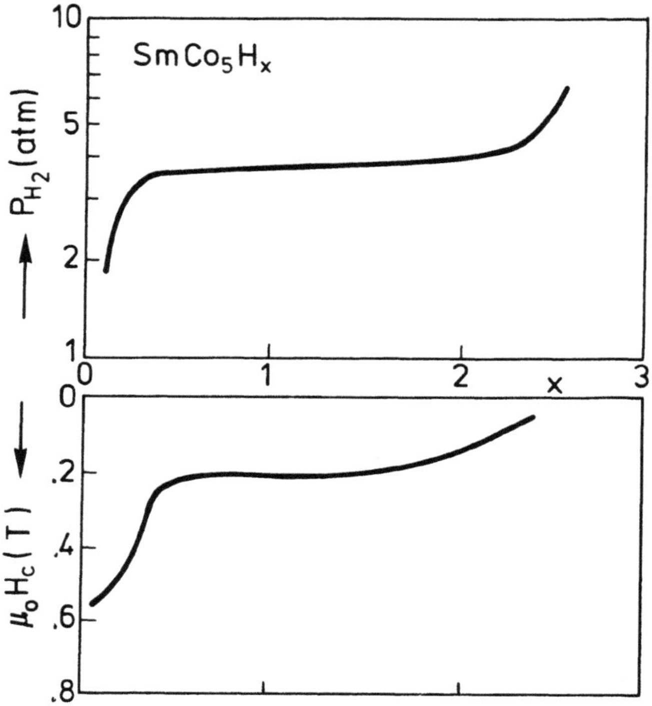

Figure 1. A comparison of the concentration dependence of the hydrogen pressure, top, and the coercivity, bottom, in the permanent magnet material, $SmCo_5$.

thermodynamics of many new and novel ternary metallic hydrides. The interests of hydride research scientists and permanent magnet research scientists, having enjoyed a common basis for a very short period, then began to diverge. However, contacts between the two groups were not completely broken, because the continued research on $SmCo_5$ had led to the discovery of a novel manufacturing route to permanent magnets, a route involving liquid phase sintering. As is discussed in more detail in a chapter [6] in this book, liquid phase sintering leads to permanent magnet materials which have both a density close to that of a single crystal of $SmCo_5$ and a very high coercivity. In this sintering approach, coercivity decreases resulting from the reaction given above are no longer a problem. However, liquid phase sintering has to be performed on magnetically aligned pressed powders. In 1972, Harris [7] realized that the strong hydrogen absorption capacity and the corresponding volume increase in $SmCo_5$, could be advantageously used for the preparation of rare earth based permanent magnets. The so-called hydrogen decrepitation process is described [6] in this book.

However, the introduction of hydrogen decrepitation did not prevent permanent magnet research scientists from looking for materials with hard magnetic properties even better than those of $SmCo_5$. One of the obvious candidates was Sm_2Co_{17}, a compound which has a higher saturation magnetization and less of the expensive samarium than $SmCo_5$. Unfortunately, in Sm_2Co_{17} the cobalt sublattice magnetic anisotropy is basal rather than uniaxial, as is required for a permanent magnet material. In a way this basal magnetic anisotropy was surprising because the crystal structure of Sm_2Co_{17} is derived from that of $SmCo_5$ which does have uniaxial magnetic anisotropy. The former can be thought of as built from the latter by the substitution of a pair of cobalt atoms onto the rare earth site of every third unit cell in $SmCo_5$. These so-called dumbbell cobalt atom pairs were soon found to be responsible for the unfavorable magnetic anisotropy of the cobalt sublattice in Sm_2Co_{17}.

Eventually a compromise was reached between $SmCo_5$ and Sm_2Co_{17} in the form of a composite material, with the approximate composition of $SmCo_{7.7}$, in which part of the cobalt was replaced by iron and relatively small amounts of copper and zirconium. The manufacturing process is fairly complex, involves carefully controlled precipitation kinetics, and results in the formation of microstructures in which large grains of a cobalt rich phase are separated by, and often completely surrounded by, a thin intergranular layer of $SmCo_5$. A third phase, the z-phase, is present within the heavily twinned cobalt rich grains and its nature is not yet completely understood. [8]

Both of the samarium-cobalt magnetic materials described above have been commercially available for some time. Because of their high Curie temperatures and their excellent resistances to corrosion, they are mainly used in high-temperature applications. In many electromotive applications, a considerable reduction in energy consumption can be achieved by using these materials. Unfortunately, samarium and cobalt are both fairly expensive and, for this reason, many investigations were carried out in the 1970s with the goal of finding iron-based compounds suitable for permanent magnets. Unfortunately, $SmFe_5$ does not exist and Sm_2Fe_{17} does not have the uniaxial magnetic anisotropy and high Curie temperature required of permanent magnets. Only after more than a decade of research was a remedy found to cope with the deficiencies of Sm_2Fe_{17}. The introduction of interstitial carbon atoms [9] by alloying Sm_2Fe_{17} with carbon during melting changes the magnetic anisotropy and increases the Curie temperature, such that these materials become candidates for permanent magnets. Subsequent studies by Sun et al. [10] indicated that higher interstitial concentrations could be reached through the solid-gas reactions of R_2Fe_{17} compounds with molecular nitrogen. These discoveries have given new impetus to the development [11] of permanent magnetic materials based on Sm_2Fe_{17}.

Before the discovery of interstitially modified Sm_2Fe_{17}, an event occured which had already changed the permanent magnet industry. In 1984 two independent reports were published [12,13] which described the preparation and properties of a new permanent magnet material based on $Nd_2Fe_{14}B$. A full account of the properties of $Nd_2Fe_{14}B$ and its use in hard magnetic materials can be found elsewhere. [14] Permanent magnets based on $Nd_2Fe_{14}B$ became commercially available in the form of sintered magnets soon after this discovery. Unfortunately, no coercive powders could be obtained through a simple milling of the $Nd_2Fe_{14}B$ ingots and this limitation hampered the manufacture of resin bonded magnets. The solution [15,16] to this problem was eventually found in the form of a solid-gas reaction in which absorbed hydrogen was employed to form a transient medium which completely modified the microstructure of the original cast $Nd_2Fe_{14}B$ grains. This process is called hydrogen decrepitation desorption recombination and is discussed in detail in reference 6.

It is interesting that in both hydrogen decrepitation and hydrogen decrepitation desorption recombination, no hydrogen is left in the final magnet material and that the metal hydride, $Nd_2Fe_{14}BH_x$, plays only a transient role. This is probably the reason why $Nd_2Fe_{14}BH_x$ has attracted little interest among reseach scientists, and why its sorption isotherms and related thermodynamic quantities have not as yet been measured. A similar lack of information applies to $R_2Fe_{17}N_x$ and other types of interstitial nitrides, although, in this case, experimental difficulties may prove an obstacle to a thorough thermodynamic study of the underlying solid-gas reactions. In contrast, in order to better understand the changes in magnetic properties, many research scientists working in the field of magnetism have devoted considerable effort to determining the number of interstitial atoms absorbed per formula unit and in determining their location in the crystal structure of the R_2Fe_{17} compounds.

It has been the goal of this Advanced Study Institute to bring together these two more or less separate groups of scientists in order to induce a cross-fertilization of their scientific work. It is important to note that the main goal of both groups of scientists is the protection of the environment through reduced energy consumption and pollution.

References

[1] W. M. Müller, J. P. Blackledge, and G. G. Libowitz, *Metal Hydrides*, Academic Press, New York, 1968.
[2] G. G. Libowitz, Inorg. Chem., **10**, 79 (1972).
[3] L. Schlapbach, I. Anderson, and J. P. Burger, in *Materials Science and Technology,* Vol. 3B, K. H. J. Buschow, ed., VCH-Verlag, Weinheim, Germany, 1994.
[4] M. Yamaguchi and E. Akiba, in *Materials Science and Technology*, Vol. 3B, K. H. J. Buschow, ed., VCH-Verlag, Weinheim, Germany, 1994.
[5] J. J. G. Willems and K. H. J. Buschow, J. Less-Common Met., **129**, 13 (1987).
[6] K. H. J. Buschow, in *Interstitial Intermetallic Alloys*, F. Grandjean, G. J. Long, and K. H. J. Buschow, eds., Kluwer Academic Publishers, Dordrecht, 1995, p. 633.
[7] I. R. Harris, U. K. Patent 1, 262,594 (1972).
[8] K. Strnat, in *Ferromagnetic Materials*, Vol. 4, K. H. J. Buschow, ed., Elsevier Science Publishers, Amsterdam, 1988, p. 131.
[9] D. B. de Mooij and K. H. J. Buschow, J. Less-Common Met., **142**, 349 (1988).
[10] H. Sun, J. M. D. Coey, Y. Otani, and D. P. F. Hurley, J. Phys.: Condens. Matter, **2**, 6465 (1990).
[11] K. H. J. Buschow, in *Interstitial Intermetallic Alloys*, F. Grandjean, G. J. Long, and K. H. J. Buschow, eds., Kluwer Academic Publishers, Dordrecht, 1995, p. 617.
[12] J. J. Croat, J. F. Herbst, R. W. Lee, and F. E. Pinkerton, J. Appl. Phys., **55**, 2078 (1984).
[13] M. Sagawa, S. Fujimura, M. Togawa, and Y. Matsuura, J. Appl. Phys., **55**, 2083 (1984).
[14] G. J. Long and F. Grandjean, eds., *Supermagnets, Hard Magnetic Materials*, Kluwer Academic Publishers, Dordrecht, 1991.
[15] P. J. Mc Guiness, X. J. Zhang, X. J. Yin, and I. R. Harris, J. Less-Common Met., **158**, 359 (1990).
[16] T. Takeshita and R. Nakayama, *Proceedings of the Eleventh International Workshop on Rare-Earth Magnets and their Application*, Vol. I, S. G. Sankar, ed., Carnegie Mellon University, Pittsburg, PA, 1990, p. 49.

Chapter 2

FUNDAMENTAL CONCEPTS AND UNITS IN MAGNETISM AND THERMODYNAMICS

Fernande Grandjean
Institut de Physique, B5
Université de Liège
B-4000 Sart-Tilman
Belgium

and

Gary J. Long
Department of Chemistry
University of Missouri-Rolla
Rolla, MO 65401-0249, USA

1. Introduction

In the second half of the twentieth century, interstitial alloys have attracted extensive interest because of their potential applications as either hard magnetic materials or hydrogen storage materials. Hard magnetic materials are used in numerous devices [1] such as electromechanichal machines, automobiles, data storage equipment, laboratory and scientific equipment. Hydrogen storage materials are used in batteries [2] and in thermodynamic engines.[3] The improvement of the magnetic and thermodynamic properties of interstitial alloys leads to materials which are more environmentally friendly and reduce the worldwide energy consumption. Indeed, electric motors based on magnets with a large energy product are smaller and hence lighter and cheaper to move in an automobile, for instance. Batteries based on hydrides are more environmentally friendly than the Ni-Cd batteries. Finally, hydride based thermodynamic engines do not pollute the atmosphere. This book is devoted to these interstitial alloys, and particularly to the study of their magnetiic and thermodynamic properties. To help the reader in understanding the following chapters, this chapter introduces the different concepts and their units in magnetism and thermodynamics.

2. Concepts in Magnetism

2.1. SI Units in Magnetism

SI is the abbreviation for "Système International d'Unités." The SI units are based on the usual fundamental units, the meter, the kilogram, and the second, as well as the ampere, the unit for electric current intensity. A more detailed description of all SI units may be found in reference 4. Because of the difficult problem of consistent units in magnetism, it is useful to begin by discussing how they are defined starting from the ampere. The use of the SI units in magnetism is rather recent and many textbooks and important reference works [5 - 7] use the cgs-emu units, and hence we will show how these units can be

F. Grandjean et al. (eds.), Interstitial Intermetallic Alloys, 7–21.

can be converted to the SI units. In addition we will show how the unit system affects the expressions defining the magnetic properties. Tables I and II summarize the unit conversion and give the relationships between the two systems.

2.2. Definition of the Ampere

When two infinitely long straight parallel wires, separated by a distance d, as shown in Figure 1, carry an electric current, I, one of these wires attracts a unit length of the other wire with a force which is given by

$$F/l = (\mu_0 /2\pi)\, I^2/d, \tag{1}$$

where μ_o is the *permeability of free space*. Equation 1 defines the ampere, the unit of electric current intensity. If the constant μ_o, is taken to be $4\pi\times10^{-7}$ $kgms^{-2}A^{-2}$, which is also referred to as a henry/m, or H/m, then the ampere is that constant current which, if maintained in two straight parallel conductors of infinite length and of negligible cross section, and placed one meter apart in free space, would produce between these conductors a force per unit length equal to 2×10^{-7} newton per meter. [4]

2.3. Definition of Magnetic Induction

The presence of a *magnetic induction*, generated in space by an electric current in a conductor or a magnet, may be detected by the force exerted on a moving electric charge, just as an electric field may be detected by the force exerted on a static electric charge. The magnetic induction, **B**, of a bar magnet may be represented, as shown in Figure 2, by its magnetic field lines which are always closed. When an electric charge, q, moves with a velocity, **v**, the force, **F**, acting on the charge is given by,

$$\mathbf{F} = q\mathbf{v}\times \mathbf{B}, \tag{2}$$

which defines the tesla, T, the unit of magnetic induction. One tesla is the magnetic induction at a point which exerts a force of one newton on an electric charge of one coulomb moving with a velocity of one m/s perpendicular to the magnetic induction. Table III gives some typical magnitudes for the magnetic induction.

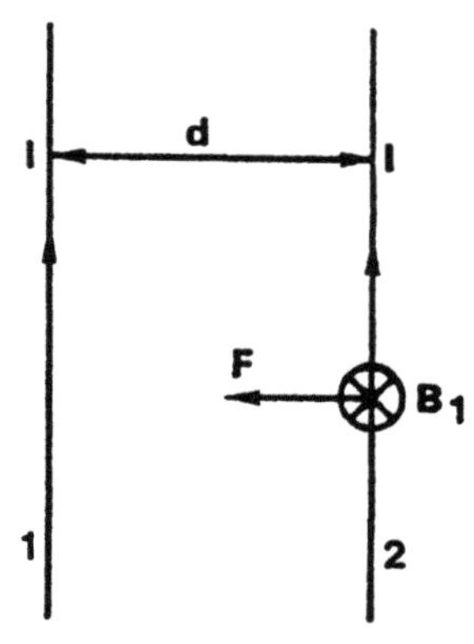

Figure 1. The definition of the ampere

Table I. A comparison of SI and cgs-emu units in magnetism

Symbol	Quantity	SI Unit	cgs-emu Unit	Dimension	Conversion Factor SI/cgs-emu
B	magnetic induction	tesla, T, or Wb/m^2	gauss, G	$MT^{-2}A^{-1}$	10^{-4} T/G
E	energy	joule, J	erg	ML^2T^{-2}	10^{-7} J/erg
F	force	newton, N	dyn	MLT^{-2}	10^{-5} N/dyn
H	magnetic field strength	ampere/m A/m	oersted, Oe	$L^{-1}A$	$10^3/4\pi$ (A/m)/Oe
I	electric current	ampere, A	emu	A	10 A/emu
k	Boltzmann constant	joule/kelvin J/K	erg/kelvin erg/K	$ML^2T^{-2}K^{-1}$	10^{-7} J/erg
K	magnetic anisotropy	joule/meter^3 J/m^3	erg/centimeter^3 erg/cm^3	$ML^{-1}T^{-2}$	$10^{-1}\ \frac{J/m^3}{erg/cm^3}$
m	magnetic moment	ampere meter^2 Am^2	erg/gauss erg/G	L^2A	10^{-3} Am^2/(erg/G)
M	magnetization	ampere/m, A/m	emu	$L^{-1}A$	10^3 Am^{-1}/emu
M	molar mass	kilogram/mole k mol^{-1}	gram/mole g mol^{-1}	$Mmol^{-1}$	10^{-3} kg/g
N_A	Avogadro no.	$mole^{-1}$, mol^{-1}	$mole^{-1}$, mol^{-1}	mol^{-1}	-
T	temperature	kelvin, K	kelvin, K	K	-
W	work	joule, J	erg	ML^2T^{-2}	10^{-7} J/erg
μ	permeability	henry/meter H/m	emu H/m	$MLT^{-2}A^{-2}$	$4\pi\times10^{-7}$(H/m)/emu
μ_o	permeability of vacuum	henry/meter H/m	emu	$MLT^{-2}A^{-2}$	$4\pi\times10^{-7}$(H/m)/emu
μ_B	Bohr magneton	ampere meter^2 Am^2	erg/gauss erg/G	L^2A	10^{-3} Am^2/(erg/G)
μ_{eff}	magnetic moment of an atom in μ_B,	unitless	unitless	1	-
μ_r	relative permeability	unitless	unitless	1	-
ϕ	flux	weber, Wb	maxwell	$ML^2T^{-2}A^{-1}$	10^{-8}Wb/maxwell
ρ_E	energy density	joule/meter^3 J/m^3	erg/centimeter^3	$ML^{-1}T^{-2}$	$10^{-1}\ \frac{J/m^3}{erg/cm^3}$
χ	volume magnetic susceptibility	unitless	unitless	1	-
χ_M	molar magnetic susceptibility	meter^3/mole m^3/mol	centim.^{3}/mole cm^3/mol	L^3mol^{-1}	$4\pi\times10^{-6}$ m^3/cm^3
χ_m	mass magnetic susceptibility	meter^3/kilogram m^3/kg	centim.^{3}/gram cm^3/g	$M^{-1}L^3$	$4\pi\times10^{-3}\ \frac{m^3/kg}{cm^3/g}$
BH	energy product	joule/meter^3 J/m^3	gauss oersted GOe	$ML^{-1}T^{-2}$	$\frac{10^{-1}}{4\pi}\ \frac{J/m^3}{GOe}$

The magnetic induction is often called the *magnetic flux density* [8] or, unfortunately, the magnetic field.[9] Magnetic flux density is correct because the *magnetic flux*, ϕ, penetrating a surface is the product of the magnetic induction, B, and the area of the surface, when **B** is normal to the surface. The magnetic flux unit is the weber, Wb,

Table II. Magnetic relationships in rationalized SI and in unrationalized cgs-emu units

Relationship	SI	cgs-emu
Definition of the ampere	$F/l = (\mu_0/2\pi)I^2/d$	$F/l = I^2/d$
Magnetic induction in a material	$\mathbf{B} = \mu_0(\mathbf{H} + \mathbf{M})$	$\mathbf{B} = \mathbf{H} + 4\pi\mathbf{M}$
Volume magnetic susceptibility	$\chi = M/H$	$\kappa = M/H$
Relative permeability	$\mu_r = \chi + 1$	$\mu_r = 4\pi\kappa + 1$
Mass magnetic susceptibility	$\chi_m = \chi/\rho$	$\chi_m = \kappa/\rho$
Molar magnetic susceptibility	$\chi_M = M\chi/\rho$	$\chi_M = M\kappa/\rho$
Magnetic moment of a paramagnetic species in Am^2	$\mu_{eff} = (3k\chi_M T/N_A\mu_0)^{1/2}$	

which is a tesla m^2, Tm^2. Magnetic field is an unfortunate abuse of language, [9] as will be discussed in the following section.

2.4. Definition of the Magnetic Field Strength, H

In free space, the magnetic induction, **B**, is related to the *magnetic field strength*, **H**, by

$$\mathbf{B} = \mu_o\mathbf{H} = 4\pi\times10^{-7}\,\mathbf{H}, \qquad (3)$$

and thus the unit for magnetic field strength, **H**, is A/m. The magnetic field strength is a

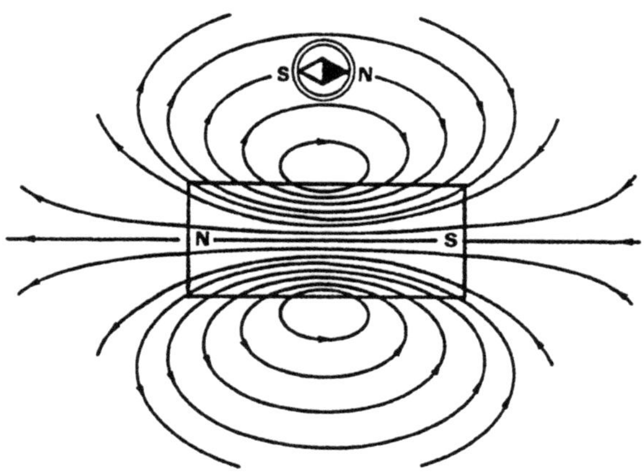

Figure 2. Magnetic induction lines of bar magnet.

Table III. Typical magnitudes of magnetic induction

Source	**B**, T
Magnetic induction at the surface of the earth	5×10^{-5}
Permanent magnets	10^{-2} to 1
Iron-core electromagnets	up to 3
Superconducting magnets	up to 20

quantity which is related to the macroscopic origin of the magnetic induction, i.e. either the electric current generating the magnetic induction, or a magnetic inhomogeneity in space. However, it is not related to the material in which the magnetic induction occurs. In a material, the permeability of free space, μ_o, in Equation 3, must be replaced by the permeability of the material, as will be discussed below. The confusion [9] between the magnetic induction, **B**, and the magnetic field strength, **H**, results from the use of cgs-emu units. Indeed, as shown in Table I, in this unit system, the constant μ_o is taken as equal to one gauss/oersted, and as a result in cgs-emu units, the magnetic induction, **B**, and the magnetic field strength, **H**, are given by the same number, expressed in gauss or oersted, respectively.

2.5. Magnetic Field Strength and Magnetic Induction in a Material

If a material is placed in an external magnetic induction, $\mathbf{B}_o$, three types of magnetic behavior, *diamagnetism, paramagnetism, or ferromagnetism,* are observed. In a diamagnetic material, the internal magnetic induction, $\mathbf{B}_{int}$, is somewhat smaller than the external magnetic induction, $\mathbf{B}_o$. In a paramagnetic material, the internal magnetic induction is somewhat larger than the external magnetic induction. In a ferromagnetic material, the internal magnetic induction is much larger than the external magnetic induction. Hence, the magnetic induction lines are diluted by a diamagnetic material, concentrated by a paramagnetic material, and strongly concentrated by a ferromagnetic material. In diamagnetic and paramagnetic materials, small applied fields produce an internal magnetic induction, $\mathbf{B}_{int}$, which is directly proportional to the applied field strength,

$$\mathbf{B}_{int} = \mu_r \mu_o \mathbf{H}, \tag{4}$$

where μ_r, the *relative permeability* of the magnetic material, is a unitless constant.

The *permeability* of the magnetic material, μ, is defined as

$$\mu = \mu_r \, \mu_o, \tag{5}$$

and has the unit kg m $s^{-2}A^{-2}$, H/m, or Tm/A. The relative permeability measures the ease with which the lines of magnetic induction penetrate a material and is smaller than unity for a diamagnetic material and larger than unity for a paramagnetic material. In a ferromagnetic material, the relationship between $\mathbf{B}_{int}$ and **H** is not as simple and will be discussed in detail in Section 2.8.

2.6. Definition of the Magnetic Moment

A planar loop of electric current, I, of area a, see Figure 3, has a *magnetic dipole moment,*

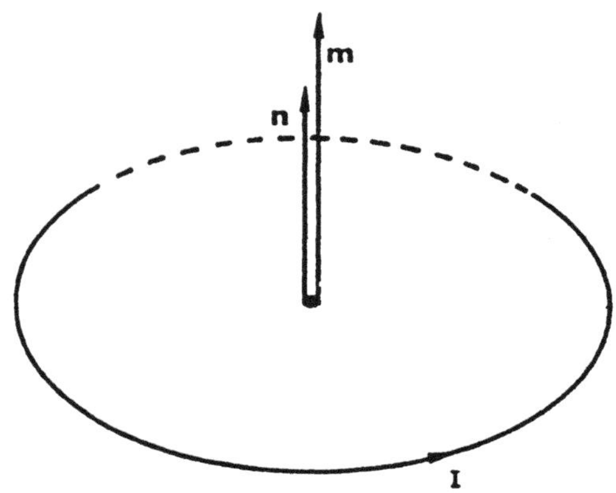

Figure 3. A loop of current and its magnetic dipole moment.

m, given by the expression

$$\mathbf{m} = Ia\mathbf{n}, \qquad (6)$$

where **n** is the unit vector perpendicular to the plane of the loop, with its direction given by the right hand rule. The unit for the magnetic dipole moment is then Am^2.

When a current loop experiences an external magnetic induction, $\mathbf{B}_o$, it tends to orient its magnetic dipole moment, **m**, parallel to the magnetic induction through the action of the torque, **T**, given by

$$\mathbf{T} = \mathbf{m} \times \mathbf{B}_o. \qquad (7)$$

The significance of the magnetic dipole moment of a bar magnet may be explained as follows. If the poles of a bar magnet could be separated, they would experience forces, $\mathbf{F}_N$ and $\mathbf{F}_S$, as shown in Figure 4, when the magnet is placed in an external magnetic induction, $\mathbf{B}_o$. These forces are proportional to the external magnetic induction, such that

$$\mathbf{F} = \pm\, |\varphi|\mathbf{B}_o. \qquad (8)$$

The proportionality constant, φ, is the magnetic pole strength and is positive for the north pole and negative for the south pole and has the unit newton/tesla, N/T, which is equivalent to Am. Then the magnetic dipole moment, **m**, of the bar magnet is defined as,

$$\mathbf{m} = 2|\varphi|\mathbf{d}_N, \qquad (9)$$

where $\mathbf{d}_N$, is the vector extending from the center of the bar magnet to its north pole. The units for the magnetic moment, **m**, for a bar magnet is thus Am^2, the same as for a current loop.

The torque, given by Equation 7, rotates the plane of the current loop until it is perpendicular to the lines of magnetic induction. In a similar fashion, the torque rotates the bar magnet until it is parallel to the lines of magnetic induction. In other words, the torque rotates the loop until its magnetic moment is parallel to the lines of magnetic induction. At this orientation, the magnetic energy of the current loop or the bar magnet, E_p, defined as,

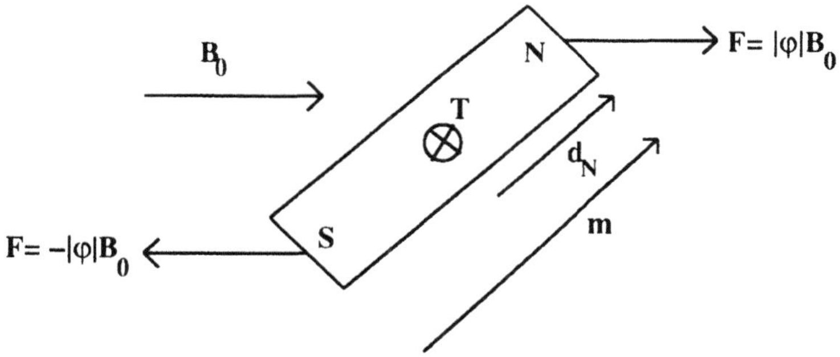

Figure 4. A bar magnet in an external magnetic induction and its magnetic dipole moment.

$$E_p = -\,\mathbf{m}\cdot\mathbf{B}_o, \tag{10}$$

is minimized, and the magnetic flux through the loop or through the cross section of the bar magnet is maximized.

2.7. Definition of the Magnetization and the Magnetic Susceptibility

The *magnetization*, **M**, is the magnetic moment per unit volume,

$$\mathbf{M} = n\mathbf{m}, \tag{11}$$

where n is the number of magnetic dipole moments, **m**, per cubic meter. Thus the unit for magnetization is A/m, the same as the magnetic field strength, **H**.

If a material is placed in an external magnetic induction, $\mathbf{B}_o$, or an external magnetic field of strength, **H**, the internal magnetic induction, $\mathbf{B}_{int}$, is

$$\mathbf{B}_{int} = \mathbf{B}_o + \mu_o\mathbf{M} = \mu_o(\mathbf{H}+\mathbf{M}), \tag{12}$$

where demagnetization effects have been neglected and thus $\mathbf{H}_{int}$ has been approximated by **H**. For a diamagnetic or paramagnetic material, the elimination of $\mathbf{B}_{int}$ by a combination of Equations 4 and 12 yields,

$$\mathbf{M} = (\mu_r-1)\mathbf{H} = \chi\mathbf{H}, \tag{13}$$

for the magnetization. In Equation 13 the proportionality constant, μ_r-1, defines the *magnetic susceptibility*, χ, which is unitless and is also known as the ***volume*** *susceptibility*. It measures the ease with which a given material may be magnetized. Various other expressions of the magnetic susceptibility are given in Table II and discussed in more detail in reference 3.

For paramagnetic materials, the temperature dependence of the magnetic susceptibility follows the Curie law,

$$\chi = C/T, \tag{14}$$

where C is the Curie constant and T is the absolute temperature.

2.8. The Hysteresis Loop

The characteristic properties of any magnetic material are best described in terms of the material's hysteresis (B,H) loop. A ferromagnet has, as the origin of its magnetism, a net alignment of its atomic magnetic moments in many domains. The magnetic moments of the different individual domains are not necessarily aligned in the magnet as a whole. Above the *Curie temperature*, T_C, the atomic magnetic moments are dynamically and randomly oriented because of thermal agitation, and the ferromagnet has become a paramagnet. The susceptibility of a ferromagnet follows [6, 7] a modified Curie law, the Curie-Weiss law,

$$\chi = C/(T - \theta), \tag{15}$$

where θ is the paramagnetic Curie temperature. Experimentally it is found that θ is somewhat greater than T_C in ferromagnetic materials. [6, 7] Below the Curie temperature, the magnetic moments in a single domain are aligned parallel to each other because of the quantized magnetic exchange interaction. The non-zero magnetic moments of the various single domains can be aligned by placing the ferromagnet in an external magnetic field of strength, **H**. Then the magnetic induction, $\mathbf{B}_{int}$, in the ferromagnetic material is given by

$$\mathbf{B}_{int} = \mu_o(\mathbf{H}_{int} + \mathbf{M}), \tag{16}$$

where the internal magnetic field, $\mathbf{H}_{int}$, is the sum of the external magnetic field, **H**, and the demagnetizing field arising from inhomogeneities in the magnetization, **M**. The demagnetizing field will be discussed in more detail below.

A permanent magnetic material which is subjected to an external magnetic field strength, **H**, follows a hysteresis loop, as shown in Figure 5. Consider a sample of a ferromagnetic material which is placed between the poles of an electromagnet, or better, inside an infinitely long solenoid with n loops per unit length. When the electromagnet is turned on, or a current, I, is passed through the solenoid, it subjects the ferromagnetic material to a magnetizing field strength, H = nI. A magnetic induction, $\mathbf{B}_{int}$, arises in the ferromagnetic material in response to the presence of the external magnetic field, **H**. If the ferromagnet is initially unmagnetized, i.e., the magnetic moments of the domains are randomly oriented, the magnetic induction, $\mathbf{B}_{int}$, follows the curve Oa in Figure 5, when the current in the solenoid or electromagnet is slowly increased. The rate of increase in the magnitude of $\mathbf{B}_{int}$, at zero applied field, defines the *initial permeability*, μ_i, of the material,

$$\mu_i = dB_{int}/dH_{int} = \mu_o + dM/dH_{int}. \tag{17}$$

The *initial relative permeability* is thus μ_i/μ_o. Eventually, at point b in Figure 5, the ferromagnet is saturated and all the magnetic moments of the domains are aligned; a further increase in the applied magnetic field, **H**, yields no further increase in the magnetization. The increase in $\mathbf{B}_{int}$ beyond point b in Figure 5 is due to the $\mu_o\mathbf{H}_{int}$ term

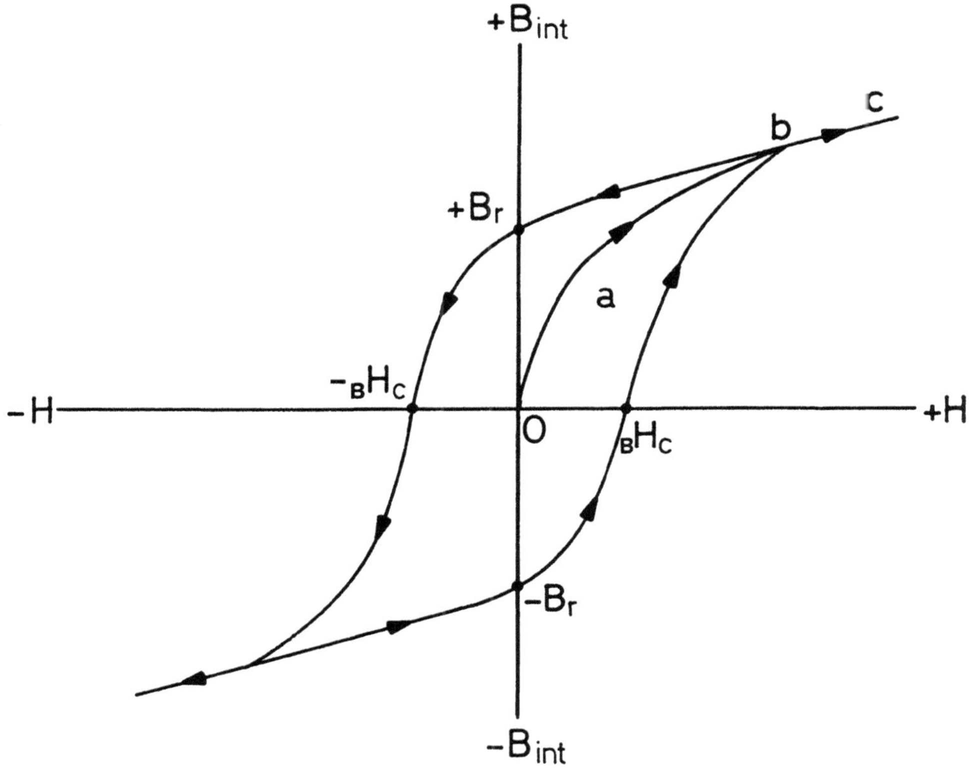

Figure 5. Magnetic induction, B_{int}, versus an applied magnetic field of strength H, in a ferromagnetic material, where the demagnetizing field has been neglected.

in Equation 16. The linear portion, bc, of the curve in Figure 5 has a positive slope equal to the permeability of free space, μ_o. In a hard magnetic material, a large magnetic field strength, **H**, is required to produce saturation magnetization. In a soft magnetic material, a much smaller magnetic field strength, H, is required to produce saturation magnetization. Soft and hard magnetic materials characteristically have comparable saturation inductions.

When the applied field of strength, **H**, is slowly reduced, the magnetic induction, $\mathbf{B}_{int}$, also decreases, but does not follow the curve baO, in Figure 5, which was generated during the initial magnetization of the material. If a ferromagnetic material has been magnetized to saturation, when the applied field is subsequently decreased to zero, the magnetic induction remains positive. The value $\mathbf{B}_r$, achieved after a ferromagnet has first been magnetized to saturation is called the *remanent magnetic induction*, see Figure 5. The ferromagnet has become a permanent magnet. If subsequently the applied field, **H**, is increased with its polarity reversed, the magnetic induction, $\mathbf{B}_{int}$, continues to decrease, reaching zero at the *coercive field*, $-_BH_C$. The $\mathbf{B}_{int}$ vs. **H** curve, in the third and fourth quadrants of Figure 5a and b, is a centrosymmetric image of the curve in the first and second quadrants. The resulting loop is called a hysteresis loop. Soft and hard magnetic materials differ by their vastly different coercivities, which typically have values between 0.2 to 100A/m for a soft magnetic material and between 200 to 2000kA/m for a hard magnetic material.

A hysteresis loop of the magnetization, **M**, versus the applied field strength, **H**, can also be generated, as shown in Figure 6a. The shape and properties of this loop are similar to those shown in Figure 5. In such a loop, the differential magnetic susceptibility, dM/dH, or χ, is the slope of M vs. H along Oa during the initial magnetization. To a first approximation, at small applied fields, the susceptibility is independent of H, whereas at larger applied fields it depends on H. It follows that the relative permeability, $\mu_r = 1 + \chi$, also depends on H at large applied fields. In the **M** vs. **H** loop shown in Figure 6a, **M** reaches a maximum at b. The portion bc of the curve has a zero slope unlike the portion bc of the magnetic induction curve shown in Figure 5. When the magnetized material is removed from the external applied magnetic field, it undergoes spontaneous partial demagnetization to a value, $+M_r$, as shown in Figure 6a. The coercive field, $-_M H_C$, is not necessarily identical to the coercive field, $-_B H_C$, in Figure 5.

The demagnetizing effect can best be understood by considering a bar magnet in zero external field, as shown in Figure 7. The magnet has north and south poles, and the lines of its magnetic induction, **B**, go from the north to the south pole outside the magnet. At the north and south poles of the magnet, these lines follow the direction of the applied field that was used to magnetize the bar. Therefore, the magnetization, **M**, inside the bar magnet is directed from the south to the north pole. Because of the discontinuity of the normal component of the magnetization at the surface of the magnet, as compared to that at the poles of the magnet, there is a demagnetizing field, $\mathbf{H}_d$, in a direction opposite to the magnetization direction inside the bar magnet. Thus the magnetic induction, **B**, produced by the magnet is given by the equation

$$\mathbf{B} = \mu_o(\mathbf{M} + \mathbf{H}_d). \tag{18}$$

The demagnetizing field, $\mathbf{H}_d$, depends on the shape of the magnet and is, for a homogeneously magnetized ellipsoid, expressed by

$$\mathbf{H}_d = -N_d\mathbf{M}, \tag{19}$$

where N_d, which ranges between zero and one, is the demagnetizing factor and is a sensitive function of the geometry of the magnet. The demagnetizing field is also present if the magnet is placed in an external field, **H**. We thus have to correct the magnetization hysteresis loop shown in Figure 6a, so that it is independent of the shape of the magnet. This correction is shown in Figure 6b, where Equation 19 is plotted as the straight line Oα. For each value of the magnetization, the demagnetizing field is given as the horizontal line segment PQ, which is then subtracted from the external applied field, to obtain the corrected value of H_{int} at the point P'. The magnetization hysteresis loop can thus be replotted as a function of the internal magnetic field, H_{int}, as shown by the solid line in Figure 6b.

When a permanent magnet is removed from the external field, the only field acting on the magnet is the demagnetizing field, $\mathbf{H}_d$. Thus upon removal, the magnetic induction, **B**, proceeds down the hysteresis loop, as given by Equation 18, in the second quadrant of Figure 5 until it reaches a point which is determined by the shape of the magnet. This portion of the hysteresis loop is known as the *demagnetization curve.* A permanent magnet, in a practical application, will operate at some point P in this second quadrant, as

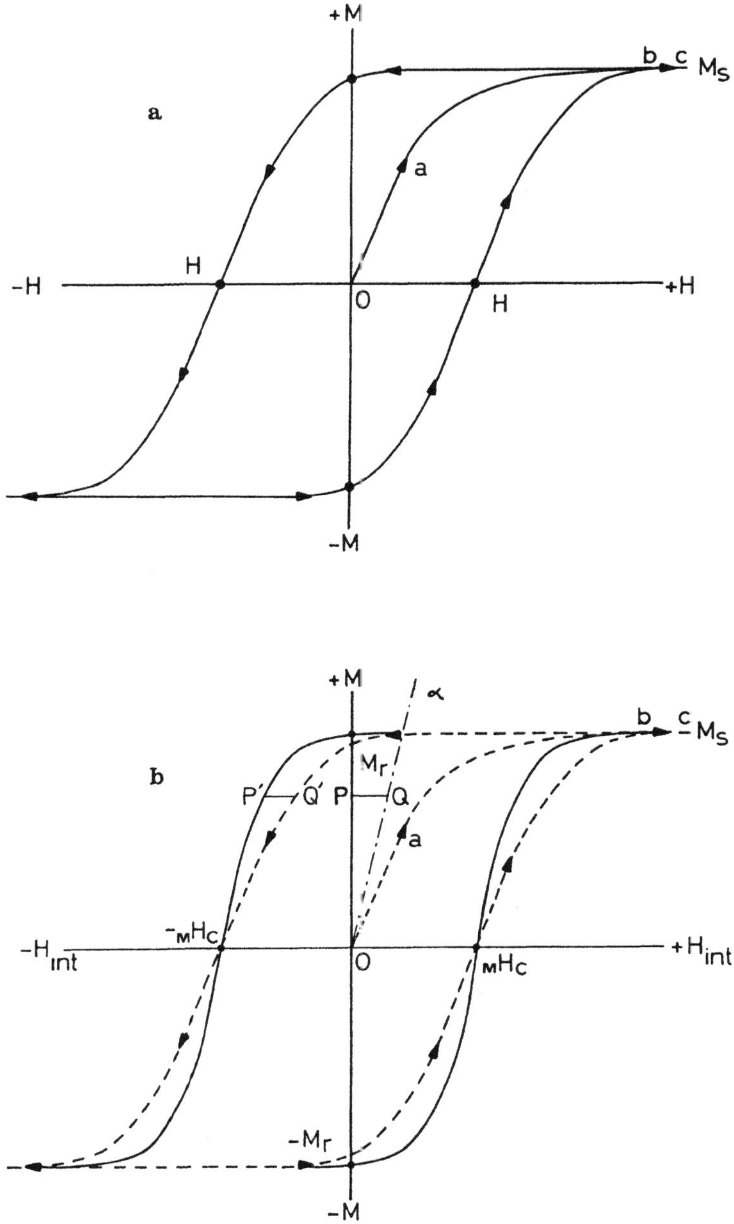

Figure 6. The magnetization, M, in a ferromagnetic material, versus an applied magnetic field of strength, H, a, and the internal magnetic field of strength H_{int}, b. In this plot $1/N_d$ is the slope of the line Oα, Equation 19, as discussed in the text.

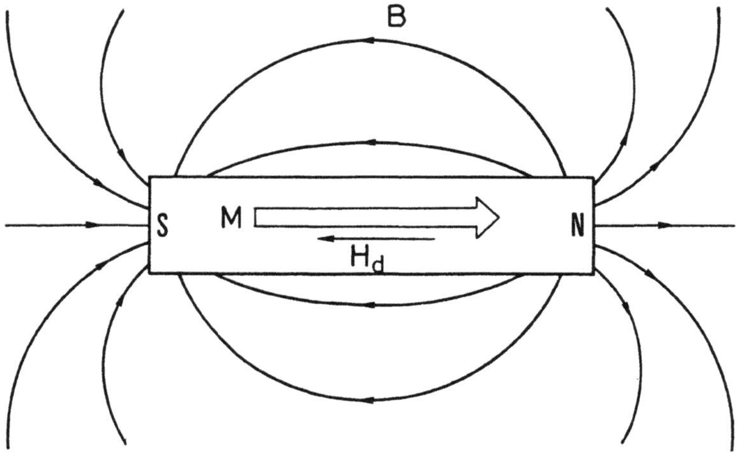

Figure 7. Magnetic induction, magnetization, and the demagnetizing field for a bar magnet.

shown in Figure 8. This point is at the intersection of the demagnetization curve, expressed by Equation 18, and the straight line OP, given by the equation,

$$\mathbf{B} = \mu_o(1 - N_d)\mathbf{M} = \mu_o[(1/N_d) - 1]\mathbf{H}_{int}. \tag{20}$$

The area enclosed by the (B,H) hysteresis loop is an important quantity, which is directly related to an energy. When a hard magnetic material is removed from the magnetizing field, energy is stored in the magnet, which then has a magnetic induction, **B**. The magnetic energy density, ρ_M in J/m^3, stored in the magnetic induction is given by

$$\rho_M = B^2/2\mu_o = BH/2. \tag{21}$$

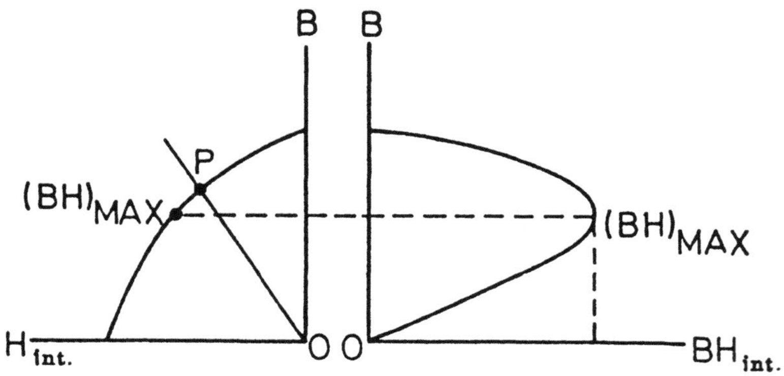

Figure 8. The demagnetization and B versus BH curves for a magnet.

Equation 21 shows that the product, BH, has the dimension of an energy per unit volume. The area of the hysteresis loop, shown in Figure 5, is proportional to the amount of energy stored in the magnet, and the product, BH, at any point on the loop in the second quadrant, is known as the *energy product* of the magnet. The energy product has a maximum value for a given point on the loop in the second quadrant, see Figure 8. This maximum value corresponds to the rectangle of maximum area which can be inscribed in the hysteresis loop. The maximum energy product is a figure of merit for a permanent magnet.

2.9. Magnetic Anisotropy Constants

The existence of preferred directions of magnetization in a given material is due to its magnetocrystalline anisotropy. The origin of magnetocrystalline anisotropy lies both in the dipolar interaction between the magnetic moments in non-cubic crystals and in a complex coupling between electronic spins and crystal symmetry due to spin-orbit and exchange interactions.

The interaction energy between two magnetic dipole moments is a function of their separation and the angle between the moment direction and a line connecting the moments. In a non-cubic crystal, the magnetic dipolar energy, which represents the sum of all dipolar interactions in the crystal, is thus a function of the orientation of the magnetization in the crystal. The dipolar magnetocrystalline anisotropy results from the tendency of the magnetic system to minimize its dipolar energy.

In a cubic or uniaxial crystal, the orbital angular momentum of an electron, or an ensemble of electrons, may have a strongly directional character which is determined by the particular symmetry of the lattice site. There is also a coupling between the spin and orbital angular momentum of an electron or an ensemble of electrons, and an exchange coupling between spin moments. Hence, the energy of the crystal depends on the orientation of the spin moments with respect to the total angular momenta of the electrons, and thus with respect to the crystal axes.

In a ferromagnetic material, in the absence of an external magnetic field, the magnetic moment is directed in the crystallographic direction which yields the minimum free energy. [5] This direction is referred to as the *easy axis of magnetization*. Hence, the total free energy of a magnetic material depends on the direction which the magnetization takes with respect to the crystallographic axes of the material. This magnetocrystalline free energy is usually refered to as the magnetic anisotropy and is expressed in terms of simple sine and cosine functions. For a tetragonal or hexagonal crystal, this magnetocrystalline anisotropy energy, limited to the first and second order terms, is given by

$$E_a = V(K_1 \sin^2\theta + K_2 \sin^4\theta), \qquad (22)$$

where V is the volume of the crystal and θ is the angle between the magnetization direction and the principal crystallographic axis of the crystal. Further terms in the series are all even powers because the energy terms must be symmetric with respect to θ. In general, only the first two terms of the series are significant or can be measured with much accuracy. The SI unit of the anisotropy constants, K_1 and K_2, is J/m^3, whereas the cgs-emu unit is GOe, where one MGOe is $7.9577 kJ/m^3$. These anisotropy constants reflect the macroscopic magnetic anisotropy and have no direct relationship to the physical origin of the magnetic anisotropy.

If the magnetization is close to the easy magnetization direction, which must be along an axis of at least three-fold symmetry, the magnetic anisotropy may be described as due to an apparent magnetic field strength, H_a, called the *anisotropy field*, and given by

$$H_a = 2K_1/(\mu_0 M_s). \tag{23}$$

The anisotropy field is the field required to rotate the magnetization in presence of an anisotropy constant K_1.

3. Thermodynamic and Electrochemical Units

In general there is little confusion over the units for thermodynamic quantities, but one must be very careful when specifying the thermodynamic standard state and using terms such as 'specific' and 'molar' as a part of the name of a thermodynamic quantity.[10] *Specific* before the name of an extensive thermodynamic quantity is used to mean that this quantity is divided by mass. In contrast, *molar* before the name of an extensive thermodynamic quantity is used to mean that the quantity is divided by the amount of substance, given in moles. Some of the thermodynamic quantities used in this book are given in Table IV along with their standard symbol, and SI and non-SI units. The conversion factor between the calorie and joule is 4184 J/cal.

Table IV. Thermodynamic and Electrochemical Quantities and Units

Quantity	Symbol	SI Unit	Non-SI Unit
Volume	V	m^3	liter
Pressure	p, P	Pa, N/m^2	Torr, bar, atm
Heat	q, Q	J	kcal
Work	w, W	J	kcal
Internal energy	U	J	kcal
Enthalpy	H	J	kcal
Reaction enthalpy	ΔH_r	J/mol	kcal/mol
Entropy	S	J/K	kcal/K
Reaction entropy	ΔS_r	J/mol K	kcal/mol K
Helmholtz energy	F, A	J	kcal
Gibbs energy	G	J	kcal
Reaction Gibbs energy	ΔG_r	J/mol	kcal/mol
Chemical potential	μ	J/mol	kcal/mol
Compressibility	κ	Pa^{-1}	-
Heat capacity	C	J/K	kcal/K
Specific heat capacity	c	J/kg K	kcal/kg K
Thermal conductivity	λ, k	W/m K	kcal/s m K
Elementary charge	e	C	-
Faraday constant	F	C/mol	-
Electrode potential	E	V	-

4. Pressure Units

The SI unit for pressure is the pascal, where one pascal, Pa or N/m^2, is the pressure exerted by a force of one newton normal to an area of one square meter. In practice several other pressure units are often used and Table V gives the conversion factor between the most common pressure units.

Table V. Conversion Factors for Pressure Units

	Pa	bar	atm	Torr[a]
1 pascal	1	10^{-5}	9.869×10^{-6}	7.501×10^{-3}
1 bar	10^5	1	0.987	750.06
1 atmosphere	1.013×10^5	1.013	1	760.00
1 Torr[a]	133.3	1.333×10^{-3}	1.316×10^{-3}	1

[a]The equivalent of 1 mm of Hg conventional.[10]

References

[1] D. Howe, Review of Permanent Magnet Applications and the Potential for High Energy Magnets, in *Supermagnets, Hard Magnetic Materials*, G. J. Long and F. Grandjean, eds., Kluwer Academic Publishers, Dordrecht (1991) p. 585-616.

[2] P. H. L. Notten, Rechargeable Nickel-metalhydride Batteries: a Successful New Concept, in *Interstitial Intermetallic Alloys*, F. Grandjean, G. J. Long, and K. H. J. Buschow, Kluwer Academic Publishers, Dordrecht (1994) p.

[3] P. Dantzer, Progress in Metal Hydride Technology, in *Interstitial Intermetallic Alloys*, F. Grandjean, G. J. Long, and K. H. J. Buschow, Kluwer Academic Publishers, Dordrecht (1994) p.

[4] *Symbols, Units, and Nomenclature in Physics*, International Union of Pure and Applied Physics (1978); *Quantities, Units and Symbols in Physical Chemistry*, International Union of Pure and Applied Chemistry, Blackwell Scientific Publications, Oxford (1988).

[5] B. D. Cullity, *Introduction to Magnetic Materials*, Addison-Wesley, Reading, MA (1972) p. 491-555.

[6] H. J. Zeiger and G. W. Pratt, *Magnetic Interactions in Solids*, Clarendon Press, Oxford (1973).

[7] R. M. Bozorth, *Ferromagnetism*, D. Van Nostrand, New York (1951).

[8] T. I. Quickenden and R. C. Marshall, J. Chem. Ed. **49**, 114 (1972).

[9] J. E. Crooks, J. Chem. Ed. **56**, 301 (1979).

[10] I. Mills, T. Cvitas, K. Homann, N. Kallay, and K. Kuchitsu, *Quantities, Units and Symbols in Physical Chemistry*, Blackwell Scientific Publications, Oxford (1988).

Chapter 3

MATTER AND MAGNETISM: THE ORIGIN OF THE PROBLEM

Umberto Russo
Dipartimento di Chimica Inorganica, Metallorganica ed Analitica
Via Loredan 4
I-35131 Padova
Italy

Francesca Capolongo
Istituto Zooprofilattico Sperimentale delle Venezie
Via Orrus 2
I-35100 Padova
Italy

INTRODUCTION

A few physical questions arise in the history of science from its very beginning, when they were mixed with philosophy, religion, and cosmology. Several of these questions received definitive answers only when physics became a science in the modern sense, after centuries of difficult, and sometimes dangerous, work by many scientists and philosophers, now too often forgotten. Some other questions instead still remain open, even if they have been formulated in a clearer and more logical fashion. In every case progress occurred only after Galileo developed the scientific method at the beginning of the XVII century.

Among these questions two seem to be particularly interesting for this chapter, what the structure of matter? and what is magnetism? They are actually closely related topics and, in fact, their origin and development through the centuries and their, at least partial, solution are strictly interconnected; the answer to the first question must also contain an answer to the second.

The first question may be formulated in modern terms in a very simple way, Is matter continuous or is it formed by small, discrete particles? The question on magnetism can be formulated as, What is the force attracting or repelling two particular pieces of matter?

These questions have surely arisen in several, if not all cultures at a certain point in their development. Thus similar histories can be written on the different paths followed by different civilizations in answering these questions. We will mainly be concerned with the western aspects of the problem, even though we will also be interested in what happened in distant countries, such as China, because, after Marco Polo's travels, they had a strong influence on the development of our culture.

When these questions were first formulated some 3000 years ago, they involved the origin and the structure of the universe, and the presence and the role of gods in the development of the world and of the human race.

F. Grandjean et al. (eds.), Interstitial Intermetallic Alloys, 23–41.

The word 'magnetism', probably more than any other scientific term, has widened its meaning to such an extent that it has invaded fields that do not have anything in common with science. People easily admit the existence of a magnetic fluid, an unearthly and hypothetical power, emitted by someone able to transmit his thoughts and will to other people, or a magnetic sensibility considered as a particular reaction of a person affected with morbid nervousness when in contact with a magnetic personality. And, moreover, we can speak of magnetic insight and magnetism as synonymous with hypnotism, suggestion, and telepathy, and we can even say to magnetize instead of to charm or to ravish with a magnetic insight. But other forms of attraction are also called magnetic, such as the mysterious and beautiful force which links two human beings together, the obscure energy by which santons and faith-healers cure people from some kinds of sickness, hypnotism, love potions, the enormous ability to charm people of illusionists, mob-leaders, dictators, and the astonishing power of rings and bracelets. All this derives from the mysterious attractive force which is able to work from a distance, through other materials, a force that can pass from one object to another, and yet remain unchanged in every part, resulting from a broken stone.

Who first handed down the knowledge of this phenomenon? And where did the word 'magnetism' come from?

MAGNETISM IN CHINA

Since the second half of the first millennium B.C., magnetism has been one of the most frequently appearing concepts in the Chinese literature. However, even today, the date when magnets were first discovered and the region in which this discovery took place, are not completely clear. The hypothesis that magnets were discovered and used in China, even if with purposes completely different from today's uses, and that they spread from China all over the world, has been accepted. For instance, the south pointing carriage, which is the first object known today connected with a magnet, dates back to the Han period in the III century B.C.. Strangely, it had nothing in common with the problem of finding direction, but rather was simply a mechanical device, which, by means of a ring nut, always pointed to the south, independent of the movement of the carriage. In a book by Lung Heng, "Discourses Weighed in the Balance", 83 A.D., an interesting connection between a magnetic compass and the 'diviner's board' of the same Han period can be found. This board has a precisely shaped spoon made from a block of loadstone. The spoon is free to rotate over the smooth surface of a board called the "diviner's board" because of its use. It dates back to the first century B.C., but probably it had been used for the previous two centuries as a secret of the court wizards. The first magnetic needles appeared in China much later, in 1080, about a century before they appeared in Europe. The discovery of magnetic declination dates back to an uncertain period between the VII and the X century A.D. in China.

The compass was used as a navigational aid in China, starting in about the eleventh century, about a hundred years before it was used in Europe. In China, as well as in Europe, magnets received various names, the most common of which was 'Tzhu Shih', or Loving Stone, or 'Tzhu', which means breeding or copulation. The second name is particularly interesting because it shows that this phenomenon was described in China as it was in ancient Greece by many philosophers such as Tales.

Another name attributed to the magnet was 'Hsuan Shih', or mysterious stone; this term was later also used for non-magnetic iron ores.

Chinese literature, like European literature, between the III century B.C. and the VI A.D., reveals a great interest in the attractive power of magnets as shown in Table 1. No Chinese writer more ancient than Tales is known, and "Master Lu's Spring and Autumn Annals", a summary of science, is the oldest book dealing with magnetism. It was written at the end of the III century B.C. and hence is contemporary to Archimedes. In 83 A.D. the "Discourses Weighed in the Balance" was published; in this book the author describes both the magnetic and the amber related attraction, as examples of attractions of 'sympathy'. This attraction is also extended to materials that are able to experience the effect of magnets or of electrified amber; an extension which stresses some concepts characteristic of the Chinese culture, such as 'resonance' and action from a distance.

The V century A.D. saw the first attempts to make quantitative measurements of the magnetic force, measurements which were important in the medical applications of magnets. In order to achieve these applications, it was necessary to separate true magnets from normal stones that were often considered toxic. In "Lei Kung Phao Chih," or "[Handbook Based on the] Venerable Master Lei's [Treatise on] the Preparation [of Drugs]", the system used to test the medical-magnetic properties of stones was described. About 600g of stone were used. If the same amount of iron was attracted from four different directions, the material was considered to be a high quality magnet and was named 'Yen Nien Sha'. If the amount of iron attracted from each side was between 200 and 600g, the stone was an average quality magnet and named 'Hsu Tshai Shih'. If it could move less than 200g of iron, the stone was a low quality magnet and simply named loadstone. Stones which attracted less than 100g were considered non-magnetic ores.

Many legends occur dealing with the magic powers attributed to magnets, powers of which the court fortune-tellers and magicians took great advantage in their prognostications. One of these legends tells about islands in the sea that did not allow the passage of wooden ships because nails were extracted from the boards which then disappeared into the sea. It is interesting to note the tale about magnetic islands between Sri Lanka and Malay is reported by Ptolomeus, in the II century A.D., and is also common in Europe and Arabia in addition to China. In other legends there are gates which, due to the attractive power of magnets, forbid the passage of people bearing iron weapons; others tell of iron statues floating between heaven and earth.

As far as the practical applications of magnets are concerned, alchemy and medicine take precedence. For instance, in the "Sung Medical Books", V century A.D., the extraction of extraneous matter, such as arrow tips from the body or the opening of blocked canals by means of magnets, is described.

To summarize, as reported in Table 2, there were only minor differences between the ancient knowledge of magnets in China and Europe, although theory was more developed in China, probably because the concept of action from a distance was more acceptable in the Chinese culture than in the Greek culture. This was because, in Greek philosophy, everything had its natural place in the universe and this led to the Aristotelian ideas of violent motion and of the necessity of physical contact between cause and effect. In this reference frame, it is extremely difficult to accept

Table 1. Principal written sources on magnets in ancient China

CANON OF THE TAO AND ITS VIRTUE		300 B.C.
MASTER LU'S SPRING AND AUTUMN ANNALS	LUSHIH CHHUN CHHIU	300 B.C.
DISCOURSES WEIGHED IN THE BALANCE	LUNG HENG	83 A.D.
COMMENTARY ON THINGS OLD AND NEW	KU CHIN CHU	IV A.D.
SUNG MEDICAL BOOKS		V A.D.
HANDBOOK BASED ON THE VENERABLE MASTER LEI	LEI KUNG PHAO CHIH	V A.D.
LEI'S TREATISE ON THE PREPARATION OF DRUGS		
DREAM POOL ESSAY	MENG CHHI PI THAN	1088
PHING-CHOU KHO THAH	CHU YU	1111-17

the behavior of magnets. Nevertheless Ermogens, a Greek heretic who lived in the II century A.D., proposed the Chinese hypothesis according to which God created the universe from nothing and then organized it with the help of a magnet. Clearly these ideas were not acceptable to the western culture, but it is important to underline their close similarity with the Chinese doctrine of Tao.

The magnetic compass is, together with the sun dial and the weather vane, among the ancient instruments, the one that played the most important role in the development of modern science. The sun-dial was obviously the most ancient, but no part of it moved; the weather vane, which indicated the wind direction, lacked precision, because a graduated circular scale was absent; the compass-rose was indeed introduced much later. So their roles were marginal compared to the compass, which was clearly the most complex and elaborate instrument of its time. In its most ancient form, it was self-regulating and thus subject to large errors; only much later did the introduction of a magnetic needle allow a higher level of precision and finally the introduction of the graduated scale brought the instrument to near perfection. In the beginning however, the compass was used in the magic rites typical of the Chinese imperial courts. Later it found new applications in agriculture and in the navy. But its diffusion was extremely slow because of its connections with the secret arts of geomancy and because, throughout the Middle Ages, navigation was limited to rivers and there was no need for compasses. So, for a very long time, geomancy

Table 2. A comparison of the development of magnetism in China and Europe

	CHINA		EUROPE	
	experimental period	actual period	experimental period	actual period
MAGNETIC SPOON ROTATING ON A BRONZE PLATE	I B.C.	83	-	-
FLOATING MAGNET	-	1020	-	1190
MAGNETIC DECLINATION	-	1030	-	1450
SUSPENDED MAGNETIC NEEDLE	X B.C.	1086	-	-
NAVIGATION COMPASS	XI B.C.	1117	-	-
MAGNETIC DECLINATION	-	1174	-	1600

conditioned the development of the magnetic compass. Its diffusion began in the IV century B.C., in the Warring State period, with the naturalist philosopher Tsou Yen, it continued during the Han period, and became a 'common' tool in the San Kuo period, 221-65 A.D.. After the VII century the magnetic compass divided geomancy into two schools. One, the most progressive and open to the future, formed by people from the coast, supported the technical development of the magnetic compass without the limits imposed on it by astrology and geomancy.

THE COMPASS: ITS ORIGIN

The 'South Pointing Spoon' was the starting point for instruments which help people find particular directions. It was described, as a part of the 'Diviner's Board' in the "Discourses weighed in the Balance" written in 83 A.D.. This spoon is a piece of loadstone shaped to imitate the constellation of the Great Bear. The Diviner's Board, of which many fragments have been found in Han period tombs, is formed by two plates, the lower one is square and symbolizes the earth, the upper one is circular and represents the sky. The upper plate, on which the 24 Chinese cardinal points were carved at the edge, had a Great Bear in its center and the spoon rotated on a central pivot. On the edge of the lower plate there were the 28 Lunar Mansions with the 24 cardinal points inside and the Trigram of the Eight-Chief of the Book of Changes disposed in such a way that the symbol of Chhien occupied the northwest position and the Khun symbol the south east position. Interestingly, this arrangement is the same as that found in the computing device described in the mathematics text by Shu Shu Chi I. Normally, the lower plate was made of maple, while the upper one, on which the spoon rotated, was made of the much harder jujube wood, sometimes covered with bronze. So it is clear that the handle of the plough of the constellation can be considered the most ancient pointer reading, and this system can be viewed as the first step towards the compass-rose.

The Diviner's Board has been known since the Han period, the first written reference to this instrument was the "Tao Te Ching" or "Canon of the Tao and its Virtue", which appeared in 300 B.C. and it was reported in many other books by the VI century. In 1150 it could be found connected with alchemy, in which it was used to foresee the results of alchémy reactions. As far as the pointer spoon is concerned, the first reference can be found in the "Han Fei Tzu" or the "Book of Master Han Fei", where, surprisingly, this instrument does not appear grouped together with other astronomical tools, probably because it had been developed by geomantics, a group of technicians rather far from astrological and astronomical interests. No written reports can be found for the period from the II century B.C. to the IX century A.D., probably because this instrument was being substituted with something else, halfway between the spoon and the needle, or because of the secrecy imposed on it by geomancy. The journals of Arab travelers to China do not even report the first examples of the compass, which surely had appeared by this time. Only a painting of the 28 Lunar Mansions was found, and the figure was reported in the compasses of the Thang period. But only the southern sky region, between 15 E and 45 E is described and this means that no progress was made in the development of the compass card after the Diviner's Board.

THE COMPASS: FROM THE SPOON TO THE NEEDLE

According to some western sources, the most ancient word for the compass was 'calamita' which seems to be derived from the Greek word 'kalamos' for the small tube which helped a needle to float. According to other hypotheses, 'calamita' indicated a small frog or a tadpole. In "Ku Chin Chu" or the "Commentary on Things Old and New", a Chinese dictionary of the IV century A.D., a mysterious needle is described which looked like a stick with a round head and a thin tail, like a tadpole. This object may well represent the connection between the spoon and the needle and so the lack of written information between the IV and the IX centuries may be explained by the evolution of this south pointing iron fish. In the same period Chinese people used floating wooden objects shaped like fish or turtles which concealed a magnet or a pieced of magnetized iron inside.

A summery of military technology of 1044 describes how a floating compass was made by the Sung technicians. They prepared a thin iron leaf which was shaped like a slightly concave fish, such that it could float on water. Then this leaf was heated to 600 or 700° C and quickly cooled, so that it remained magnetized along the earth's magnetic field. In this way magnets were no longer necessary in the preparation of a compass. A clear description of the compass, together with a definition of magnetic declination, was reported in "Meng Chhi Pi Than" or the "Dream Pool Essay" in 1088. This description indicates that a lot of experimental work had already been carried out. For instance, a single silk thread had been substituted for hemp with a great improvement in the sensitivity and accuracy of the instrument. Sadly a large part of the literature of this period was destroyed, either by the Jesuits or by the Emperor Chhin Shih Huang Ti. It is in fact still normal in China for a new dynasty to destroy whatever the former dynasty had accomplished.

THE USE OF THE COMPASS IN NAVIGATION

The first description of the use of the compass as an aid to navigation can be found in a work by Chu Yu, "Phung Chan Kho Thah", written between 1111 and 1117. It reports on merchant activity in the port of Canton after the year 1086 and seems to indicate that a long time had passed between the appearance of the magnetic needle in geomancy and its use in navigation. The first attempt to magnetize an iron needle by means of a magnet probably dates back to the V century, but application of the compass to the solution of navigational problems surely occurred later than the X century. The most probable period for the development of the magnetic needle was the Sung period between 850 and 1050. However, the use of the magnetic needle in navigation was also delayed because of the particular material used for the needle. Unfortunately, mild iron, which was used for the needle, lost its magnetization far too quickly for use on long trips. So it was necessary to use steel instead of iron, but its preparation was rather difficult and complex. Steel was probably imported into China from Hyderabad, India, or from Turkey, at very high costs.

The help given by the compass to navigation is a very common argument in the Chinese literature after the XII century. For instance, a geographer of the Sung dynasty in 1225 described the difficulties found by sailors near the island of Hainan, due to the presence of reefs, underlying the necessity of the use of a compass. Chinese sailors soon became very skillful in the use of the compass and proved so through their ability to cross the Singapore Main Strait, a crossing that was even avoided by the Portuguese in the XVII century.

One of the continuing problems was the difficulty the helmsmen had in following the direction of the compass. He had to maintain an imaginary axis, connecting the bow and the stern of the ship, parallel to the needle of the compass. To overcome this problem, a compass card was added to the compass and the entire instrument was placed in a round box. The compass card was invented in Amalfi, Italy, shortly after 1300, and only in the XVI century did it arrive in China, carried by Portuguese and Danish sailors. It is interesting to note that Chinese ships not only carried loadstone to remagnetize their iron needles, but also special types of water which were used in the floating compasses to maintain the ancient and strange tradition of the manner in which the needle had to float. They also used a ceremonial libation on the occasion of the installation of a compass on a ship.

To summarize, in China as well as in Europe, during the birth and the preliminary development of the compass, magnetism was deeply mixed with magic and superstition. The transition to science and technology required a long period and was successful only after the separation of magnetism from irrational beliefs and after numerous contacts with other civilizations, first with Greece, and then with all the different western cultures.

ANCIENT TIMES IN THE WEST

As far as western culture is concerned, history begins around 600 B.C. as usual in that large part of the Mediterranean area dominated by Greek culture. It seems the Greek people had a speculative mental attitude that pushed them to investigate the fundamental questions proposed by nature. As a consequence they founded, in Greece, Asia Minor, and southern Italy, Magna Grecia, the first scientific and philosophic schools with the goal of solving these problems. The Greek philosophers benefited from the practical experience of the craftsmen of the early empires of the Middle East. They were able to work many natural materials, such as rocks, bones, textiles, and leather, a few metals, such as iron, gold, lead, tin, and mercury, as well as then pottery and glass. So they knew the properties and characteristics of all these materials and were able to carry out various chemical processes, such as glass manufacturing and iron ore reduction.

A broad and deep interest was displayed by Greek people in magnetism. In their world, magnetism was very important both in medicine, for curing the eyes and various burns, and in magic. Actually, all the great philosophers and naturalists, from Tales to Aristotle, were engaged in its study, and tried to give explanations, most of which are bizarre, as to its peculiar characteristics. According to Lucretius, Rome, 96-53 B.C., in "On the Nature of Things", the term magnetism derives from the city of Magnesia, near the border with Macedonia. In contrast, Nicander of Colophon, Greece, ca. 150 BC, says that magnetism comes from Magnes, a Greek shepherd who first found a magnet on Mount Ida. Legend says that the nails of Magnes' shoes and the point of his stick were attracted by a magnet while he was tending his flock. As reported in Table 3, magnets were classified into different varieties according to their origin in Ethiopia, the city of Magnesia, Beotia, Troad, or Asia. They were also given a sex, the strong magnets were masculine, obviously in a society still based on the principle of the king warrior, and the weak ones were feminine. Finally they were distinguished by their color, which ranged from deep red, the strongest and most masculine magnets, to white, the weakest and most feminine magnets.

The development of theories of the structure of matter is summarized in Table 4. Thales of Miletus, VII-VI century B.C., was the first great philosopher and scientist in the Greek world. He founded an important school in Miletus that strongly conditioned the future of Greek culture. He considered matter as having a unique origin in water, a ubiquitous substance existing in three different states or 'phases', vapor, liquid, and solid. It is evident that, from the convergence of religious and magic influences, as did the animists, he gave magnets a divine origin. But nevertheless he posed the question of the nature of matter and gave a first answer, a simple monist theory halfway between a continuous and atomistic theory.

Thales disciple, Anaximander of Miletus, 610-547 B.C., extended his theory by postulating the existence of a single, anonymous, and indeterminate substance present in four forms, earth, mist, fire, and water. Anaximenes of Miletus, 586-528 B.C., proposed the existence of a fundamental substance, mist or air (πνευμα) that is transformed into the different forms of matter by the process of rarefaction and condensation. This is the first mechanism ever proposed to explain the transformations occurring in the universe. One hundred years later, Empedocles, in the V century, abandoned the monistic theory in favor of four basic types of matter, earth, air, fire, and water, forming all common things by means of two universal powers, love and strife. These powers are today called interatomic forces.

Up to this point, philosophers had not clearly faced the dichotomy between the continuum, that assumes that matter can be divided into smaller and smaller parts and still retain its original characteristics, and atomism, that assumes that matter consists of discrete, indivisible particles. Clearly, magnetism can be explained only in the second framework, and in fact Anaxagoras, from Asia Minor, 496-428

Table 3. Classification of magnets.

ORIGIN	ETHIOPIA MAGNESIA BEOTHIA TROAD ASIA
SEX	STRONG AND MALE WEAK AND FEMALE
COLOR	DEEP RED AND STRONG WHITE AND WEAK

B.C., was still strictly bound to divine and magic power when he explained the attractive power of magnets. Only at this time, did the first atomic theory appear based on the work of Leucippus and Democrats between 450 and 420 B.C.. The ideas proposed by these two philosophers are rather complex and many physical processes, such as expansion and contraction or solution and precipitation, could immediately be explained. Everything is made up of small, invisible, and indivisible particles, called atoms, existing in an absolute and ubiquitous void. They are in continuous motion and yield various species characterized by different shapes. The number of species is finite, but there are an infinite number of atoms in each species. The atoms can combine with each other to form the different compounds. They stick together because of the presence of hooks on their surface and this can explain the hardness of some substances and the softness of others. So the theory was nearly complete from its birth, and this was exploited by Epicures of Samo, 341-271 B.C., who widened it to include an explanation of the behavior of magnets as reported in Table 5. According to Epicurus, the void existing between a magnet and any object under its influence is the cause of a movement of atoms from the iron toward the magnet in order to fill the vacuum. As a consequence of this atomic movement, the entire object moves toward the vacuum and finally falls onto the magnet. In the meantime Plato, Athens, 427-347 B.C., had given an accurate description of the properties of magnets. "This stone not only attracts iron rings, but also gives them a new power, by which they are themselves able to do the same things that a magnet does, and so now they can attract other rings. For this reason, long chains of iron pieces and rings, suspended one from another, are sometimes formed. All this is only due to the power of the first magnet" (Ion, 533, D-E). In other words, he discovered or at least first recorded a phenomenon that today is called magnetic induction.

Diogenes explain the magnetic attraction in a strange way. He claimed that the humidity of iron was attracted to the dryness of the magnet. Aristotle, Stagira, 384, Calcide, 322 B.C., faithful to his principle of the unity of the mover with the moved, affirmed that a magnet was not physically able to attract iron, but that it induced a magnetic virtue inside the iron object. This magnetic virtue was thus responsible for any actual movement. Averroes, Cordoba, 1126-1198, comments on the "Physics" of Aristotle, confirmed this supposition. He also solved the enormous problem of the lack of unity between the magnet and the iron in an interesting way.

Table 4. The development of the structure of matter in the Greek culture

THALES	→	WATER	→	VAPOR, LIQUID, SOLID
ANAXIMANDER	→	SUBSTANCE	→	EARTH, MIST, FIRE, WATER
ANAXIMENES	→	MIST	→	MATTER
EMPEDOCLES	→	EARTH, AIR, FIRE, WATER	→	MATTER
LEUCIPPUS AND DEMOCRITUS	→	ATOMISM → VOID	→	MOTION → SHAPE

He suggested that a magnet had the ability to modify the property of any object with which it came into physical contact, this modification, species magnetica, could pass from one object to another until it arrived at a piece of iron.

EARLY MIDDLE AGES

During the following millennium or so, no further progress in the field of magnetism was achieved, at least in Europe, In fact exhaustive and detailed explanations about the nature and the properties of matter and of magnets were carried out neither by the Greeks, who were too descriptive and too speculative, or by the Romans, who were more interested in the practical and utilitarian aspects of magnetism rather than in the interpretation of physical events.

During the Middle Ages, magnetic properties found their first true and practical application, even in the absence of any logical explanation. Probably introduced from China at the turn of the tenth century, the compass appeared in the Mediterranean basin and very shortly became an indispensable aid to navigation. It took the place of the birds, which, for at least four thousands years, had helped sailors to find the mainland. Thanks to the birds, the Viking sailor, Floki, was able to land in Ice

Table 5. Magnetism and some Greek philosophers and philosophy schools

PLATO	MAGNETIC INDUCTION
EPICUREANS	MOVEMENT OF MAGNETS
ANIMISTS	DIVINE ORIGIN OF MAGNETISM
DIOGENES	HUMIDITY OF IRON AND DRYNESS OF MAGNETS
ARISTOTLES	MAGNETIC VIRTUE

land in about 800. The compass was surely introduced after this event, but before the middle of the XIII century, because its use is recorded in many books of this period.

The first approximate description of a compass is reported in "De Naturis Rerum" or "About the Nature of Things", by Alexander Neckam, written at the beginning of the XIII century. The interest shown in this strange object spread very quickly, and on 8 August 1269, Petrus Peregrinus from Maricourt wrote the "Epistola de Magnete" or "Letter on the Magnet". This letter was the first accurate description of a magnet and its properties, and of a compass. It may well be considered the best example of the scientific method in the Middle Ages and it reports numerous experiments that led to the beginning of true scientific research on magnetism.

At the beginning of his letter, Petrus Peregrinus deals with the possibility of distinguishing a magnet from a common stone. Then he describes two different ways to determine the position of the poles on a spherical magnet; the intersection of two meridians and the point on which a magnetic needle remains vertical. He also found the way to distinguish between the two poles and realized that the polarity could be inverted by the action of a second stronger magnet on the first magnet. He then gives a long detailed description of a floating magnet. Petrus Peregrinus was really fascinated by its ability, when free to move, to always orientate itself in the same direction. He also explained with great elegance why a magnet, once it had been broken up, could only be put back together in one way. "Nature has a tendency to act or behave to its best ability" he said. "The initial course of action is always the one in which the identity is best preserved." He correctly observed that the alignment of magnets in the same direction could not be due to the presence of iron at the poles because iron is found everywhere. But then he was wrong when, from that observation, he presumed that the attraction was due to the celestial poles, even though he realized that the needle did not point toward the crossing of the earth's meridians.

At the end of this long, but accurate theoretical description of magnetism and of its origin and effects, Petrus Peregrinus, described some clever instruments that resulted from the practical application of his theoretical knowledge. One of these instruments was made by applying a floating magnet to an astrolabium. In this way the measurement of the azimuth of the stars was easier and, as longitude and latitude were known, navigation was possible and sure. A second instrument, described in the letter, consisted of a wheel that, by means of magnets, one fixed and several others inserted in the wheel, should have undergone perpetual motion. Three centuries later, in 1558 and again in 1562, the "Epistola de Magnete" was printed, demonstrating its importance and its great and lasting value. Indeed, with this work, magnetism for the first time left the fold of magic and entered a new field that was just beginning to take shape. The scientific method, experimentation, and accuracy of measurement were then new ideas that had just began their long and hard journey to the definitive assessment given to them by Galileo

Contemporary with Petrus Peregrinus and, at the same time his teacher and disciple, Roger Bacon (1214-1292) spent his cultural life between science and magic. He believed that stones and grass had occult virtues, but he also realized, for the first time, quite clearly the practical character of experiments and of scientific research. He gave his support to the practical goals of science, describing in his work "Epistola de Secretis Operibus" or " Letter About Secret Works" the realization of ships, submarines, tanks, aircraft, and automobiles. His ideas on magnets and their uses were very coherent and he reported them very accurately in his work "Opus Minus" or "Minor Work".

The work of Jean de Saint Amand (1261-1298), a French physician, is also closely related to that of Averroe. His work "Super Antidotarum Nicolai" reports a complex and elabo-rate explanation both of the passage of magnetic virtue from a magnet to an iron needle, even if a dish of water separates them, and of the reciprocal positions of the north and south poles. This explanation was restated and confirmed by the French Averroist, Jean de Jandum, (?-1328) in his comments on the work of Aristotle. After nearly seventeen centuries, Aristotle's clear statement of the impossibility of any action without physical contact between the mover and the moved still compelled the scientists of the Middle Ages to hypothesize the existence of some mysterious fluid which could guarantee the physical contact among the various bodies experiencing magnetic attraction. This idea also led to the medieval concept of impetus and of ether, an obscure substance that permeated the entire universe. In this way the existence of the force of gravity and the illumination of the earth by the sun became possible.

THE REBELLION AGAINST ARISTOTLE

The first opposition to Aristotle was proposed, with great courage, by William of Ockham (1297-1350) on the simple basis of his working method. In fact, he clearly stated in his remarkable work that no methodological or philosophical obstacle could prevent the magnetic attraction even in the absence of a physical contact. He did not consider it important to keep up appearances and to respect, even formally, the authority of Aristotle. He refused to study intermediate species in order to justify the distant action of magnetism and of the light of the sun. The door of the rebellion against Aristotle was thus half open, but additional centuries had to pass, and the harsh sentence against Galileo had to be proclaimed, before it was completely opened. Even today, for many, this door is not completely open.

LATE MIDDLE AGES

By the beginning of the XIV century, technological research became autonomous and original, and moved away from the main trend of physics that remained with philosophers and theologists of the universities. In contrast the interest in the properties of materials and magnetism was advanced by technicians and artisans, people not officially cultural, probably because research was closely connected with practical interests. Among these, navigation was particularly important and during the XV and XVI centuries, it began to have economic and political importance, at least in England. Seeing that the most essential instrument for the extensive development of navigation and world exploration was the compass, its theoretical and practical study acquired a great importance. These studies were mainly carried out by people who, even if normally without a wide cultural and academic background, had the manual skills necessary to make accurate compasses. Among the greatest of these technicians, Robert Normann stands out prominently. He was among the first to claim the technicians right to be the owner of their own work and to share its destiny with an assured and sharp tone. He wanted to free himself of the abuse of power of the academic scholars who considered themselves the only repositories of science and of the language by which it could be propagated. However his greatest accomplishment was the solution of a problem which the makers of compasses had faced for about three centuries without solution. These craftsmen soon realized that a magnetized needle suspended on its center of gravity inclined downwards while pointing to the north. They found a remedy to this 'defect' by putting a small counterweight on the south end of the needle. Normann, in contrast, solved the problem by simply balancing the needle in such a way that it could move not only in the horizontal plane, but also in the vertical plane. What he defined as "Newe Attractive" is simply the magnetic inclination.

The study of the fundamental nature of matter, was much slower, probably because theoretical explanations and accurate experiments were not required for the simple materials employed at that time. A main problem was the resistance of wires to different loads. Leonardo da Vinci, Florence, 1452-1519, devised a system "to find the load an iron wire can carry" that represents the first experimental approach to the study of the properties of matter. There was no explicit reference to the structure of matter, even if atomism seems to have been the accepted theory, even if unconsciously, during the Renaissance. The problem of the structure of matter was later reproposed at the same level by Galileo, who, like Leonardo studied the resistance of different materials with different shapes and dimensions along different directions. In agreement with his geometric background, Galileo treated the problem from a quantitative point of view. He raised the question of the tensile strength of ropes, and in this way, dealt with the question of the composition of matter. In a rather confused way he accepted the atomistic theory. Ropes are made up of small fibers which give them resistance. In a similar way, solid materials are made up of small particles that are held together by an undefined and mysterious glue. However he never realized that the 'atoms' may move with respect to one another under the influence of an applied force.

A few years before Galileo became famous in Italy and throughout Europe, an English physician was occupying an important position in the history of science. The importance of this individual increased with the years until he was considered the father on magnetism and, according to some historians, a competitor to Galileo in the definition of the modern scientific method.

WILLIAM GILBERT, FATHER OF MAGNETISM

William Gilbert, as reported in Table 6, was born in Colchester, Essex, in 1544 to a middle class family of five children. In 1588 he entered St.John's College, Cambridge, where he obtained his A.B. degree in 1561, his M.A. in 1564, and his M.D. in 1569. Sometime around 1575 he moved to London where he practiced medicine and became a member of the Royal College of Physicians. By the beginning of the 1580's he was already a famous physician and soon became the physician to Queen Elisabeth I and then to King James I. He was also responsible for the health of the men in the Royal Navy. At the same time he served the College in many positions and finally as its President in 1600. Not much is known about his life, as, upon his death in 1603, probably from the plague, he left all his books, instruments, and notes to the library of the Royal College of Physicians, which was destroyed in the fire of London. The results of his studies were published in 1600 under the title "De magnete, magneticisque corporibus, et de magno magnete tellure: physiologia nova, plurimi et argumentis et experimentis demonstrata" or "Concerning the Magnet, Magnetic Bodies, and the Great Magnet the Earth: a New Physiology Demonstrated by Many Arguments and by Many Experiments". The notes on his other studies were published in 1651 by his brother under the title "De mundo nostro sublunari philosophia nova" or "About a New Philosophy on Our Sublunary World".

Gilbert occupied a position that was completely original in the scientific and cultural world of the seventeenth century. His masterpiece, recognized worldwide as the basis of the modern science of magnetism, is, at the same time, a grandiose summary of occultism and magic, and the first treatise about a scientific problem in which a modern scientific approach and terminology were used. It was the first book even written by a cultivated person, someone who graduated from the Cambridge

Table 6. A short outline of the life of William Gilbert.

1544	Born in Colchester, Essex, England
1558	Matriculated as a member of St. John's College, Cambridge
1561	received B.A.
1564	received M.A.
1565-66	Mathematical examiner of St. John's College
1569	received M.D.
About 1575	Settled in London
About 1580	Member of the Royal College of Physicians
1588	Requested by the Privy Council to provide care for the health of the men of the Royal Navy
1600	Physician to the Queen Elisabeth I President of the Royal College of Physicians Published "De Magnete"
1603	Physician to King James I
1603	Died in London on November 30th
1628-1633	"De Magnete" republished
1651	"De Mundo" published

University, which dealt with technical arguments, studied by means of empirical observations and practical experiments, and was addressed to technicians, artisans, and sailors.

Gilbert tends to hesitate between the position of a modern scientist, attentive to natural phenomena, to the precision of measurements, to the accuracy and reproducibility of experiments, and that of an occultist or an animist, sure in his mind of the active role of natural things, confident in horoscopes, and in the influence of stars on a new born baby. However, it is important to remember that he accepted animistic arguments only if they did not contradict the results and the achievements of his research; so he absolutely rejected any sort of constituted authority of ancient scholars, he mocked the old beliefs according to which diamond and garlic destroy magnetism, and to prove this he destroyed seventy-five diamonds! or that magnets put between the sheets pull away an adulteress, but not an adulter! In contrast he accepted the Aristotelian theory of form and matter, but, also in this case, he adopted it to rationalize the phenomena that otherwise would have been completely incomprehensible. Even the title of his masterpiece promises a lot more than we are used to looking for today in a scientific book. Besides the discussion on magnetic and electric phenomena and numerous applications mainly related to solving navigation problems, his work often trespasses on philosophy.

De Magnete is divided into six chapters, as shown in Table 7, the first of which covers the history of magnetism from the ancient Greek tales to the contemporary ones. At this time, the subject of magnetism was still principally composed of

Table 7. Index of Gilbert's "De Magnete"

1. HISTORY OF MAGNETISM
2. ATTRACTION
3. DIRECTION
4. VARIATION
5. INCLINATION
6. REVOLUTION

tales, and the magnet itself was still considered to be full of obscure, mysterious, and magic forces. Magnetic mountains rose out of the sea to pull the nails out of ships; magnets were also a protection against witches and a cure for many sicknesses. The first chapter ends with an assumption that was to be fundamental in the development of Gilbert's cosmological theories. He claimed that the earth was a giant magnet and, as such, it was endowed with all the magnet's properties and rules. This hypothesis, as was customary for Gilbert, was supported by an accurate comparison between the shape and the magnetic properties of the earth and those of a small spherical magnet which he called a "Terrella". Backed by this assumption, he did not hesitate to strongly oppose Aristotle. Without any reverential awe, he was able to draw all his own conclusions. What is proven true for the small "Terrella", that is clearly easier to manipulate and investigate, also must be true for the earth.

Each of the subsequent five chapters deals with one of the movements associated with a magnet, attraction, direction, variation, declination, and revolution. In the second chapter, in which attraction is described, the difference between magnetic and electrostatic attraction is, for the first time, analyzed and properly defined. The first kind of attraction is due to the shape of the body, and can pass through certain materials. For this argument, his experiments with water and wood are rightfully famous. The second kind of attraction, the electrostatic attraction, is due to the substance, it cannot pass through another material, and can be revealed by an electroscope or Versorium. The first version of this instrument, invented by Gilbert, consisted of a small metallic needle, balanced on its pivot in such a way that it could rotate if an electrified object was brought close to its tip. In this book the endless uncertainty of Gilbert between animism and science is very evident. He follows the ancient concept of a difference between shape and matter to differentiate the origin of magnetism from that of electricity. Magnetism now becomes the first essence of the earth, its 'soul', an essence that remains intact only in magnets and in pure iron. As a consequence of this, every magnet, including the earth, endows a portion of the surrounding space with its own properties, its Orb of Virtue. Every other magnetic body in this part of space feels the existence of the magnet. In other words, magnetic attraction is no longer a force emitted by a magnet that compels a second body to fall onto the magnet. Now it has become a reciprocal attraction that harmonically moves the two bodies, one towards the other. The intensity of this attraction is a function of the size and the purity of the magnets, and its nature is comparable to the instinctive attraction existing between the two sexes, that is coition. Except for this last statement, this chapter is remarkable for the accuracy of the concepts and the words used to describe magnets. In fact there is only a small difference, at least on a physical level, between the 'orb of virtue' and the current concept of the magnetic field. The differences are apparent only if the ideas of Gilbert are considered to be born simply from the necessity of a philosophical answer to the problem of the action of distant magnets. In any event, the consequences of this idea spread out to affect the structure and organization of the whole universe. What today is the electric, magnetic, and gravitational field can be found in the 'orb of virtue'. In this way Gilbert tried to offer an overall explanation to the current problems; but this synthesis remains typically animistic and magnetism is still considered the 'finger of God' that builds up the universe.

The content of the third chapter is more technical and covers the behavior of the compass. It was the most important contribution of the book because the compass was closely related to navigation and so to the main political and economic interests of Great Britain. In fact in those years the art of navigation was developing very quickly because England planned to expand its empire thanks to navigation. The ideas of the 'orb of virtue' has been useful to explain the various movements of the magnetic needle and, in fact, each of them found in Gilbert's ideas a complete rationalization.

After the orientation of the magnetic needle, in the fourth chapter Gilbert deals with the problem of the variation of the orientation, a variation that today is called the declination. He uses his nautical knowledge and his friendship with sailors and famous scholars extensively. In this way, Gilbert proves to be a new man, headed towards the future, who has forgotten the awe of the past cultural basis of his time. In fact it was not then easily accepted that a scholar, who had graduated from a university, was familiar with, or even worse, discussed scientific problems with artisans, technicians, and other people who were obliged to use their hands in their work. Gilbert already knew that declination was different from place to place and had published a table in which he reported the declination values measured in many different locations by sailors from England, Spain, Portugal, and Holland. He demonstrated that the declination was not influenced by large iron deposits such those in the isle of Elba in the Tirrenic Sea, thus overturning many ancient tales. He also showed that the declination did not decrease with latitude. He still did not realize that declination also varies with time.

The fifth chapter deals with the inclination of the magnetic needle with respect to the horizontal plane. This phenomenon, previously studied by Normann, was illustrated by Gilbert with great clarity and simplicity by means of an experiment with the 'Terrella'. The current interest in this phenomenon was limited to its use in mines, but Gilbert, aware that the inclination increases on going from the equator to the poles, found an application that, at his time, was very important. By knowing a sufficiently large number of values of inclination and declination throughout the world, it should have been possible to establish one's position even in the absence of the sun or the stars. With this aim, he developed an instrument, the inclinometer, that allowed the determination of both parameters simultaneously. Unfortunately, this instrument was too complex and difficult to use; the values required needed too high a precision and, moreover, were not constant with time. As a consequence, the instrument failed. But what remained was the importance of the new link between science, technology, and political, economic, and military interests.

De Magnete ends with the description of the cosmological theories of Gilbert. This chapter probably represents the highest price he had to pay to the culture of his time In this chapter numerous and far reaching concepts are presented. In fact, he hypothesized that the earth was formed with an iron core in which magnets maintain the purity and intimate structure, i.e., the earth's soul. All the materials that could be found on the earth surface derive from the decomposition and decay of this extremely pure soul. He also established that, by inclining the magnetic axis with respect to the geographic axis, creating thus the declination, the 'Terrella', and thus the earth, would begin to rotate. He then extended the structure of the earth to the moon, sun, and the stars, such that, for instance, the moon is inside the 'orb of virtue' of the earth and that this is the origin of its rotation and of the tides on the earth. It is important to note that this explanation of the tides is much closer to the present day ideas than Galileo's theories which were nearly contemporary. In a similar way, the earth is inside the 'orb of virtue' of the sun. Unfortunately, Gilbert lacked the courage to give the earth a second, orbital movement. Moreover, he was able to extend his theory to the celestial spheres, and he proved that the fixed stars were at different distances from the earth. To summarize, the solar system and the universe, in agreement with Giordano Bruno, did not have boundaries; they were a huge magnetic machine in which all the bodies, planets, comets, and stars were held together by magnetic forces. With this hypothesis, Gilbert was far ahead of his time, his universe was extremely rational and reasonable, and quite close to the universe of today. It is again important to remember that De Magnete was written in 1600, before Kepler and Galileo began to publish their works.

With De Magnete, magnetism reached a high point, even though it was still considered magical and superstitious, with occult and medical powers that could provide power and wealth. But it also became much more than this, it even became much more than what it is today, a simple natural phenomenon. For Gilbert magnetism was the key to understanding and rationalizing everything. It was the motor, the active principle of the universe, the way by which God orders and structures nature; it is the soul of the earth and the universe. But this is an apology only for its occult, animistic, and naturalistic aspects, aspects which are typical of the Renaissance culture. De Magnete is the record of the birth of the scientific and technological aspects of magnetism. It is magnetism first rationalization, an experimental base comparable in its philosophical basis with what Galileo was going to do for dynamics. What is completely missing in the work of Gilbert is a quantitative description of the magnetic phenomena. Skillful, accurate, and precise in the descriptions of his experimental observations, he completely lacked any interest in the quantitative mathematical relationships between the various physical variables. Gilbert never tried any calculation or searched for any clear and precise rule by which to fit his theories into an organic framework. On the other hand, the intrinsic difficulty of this science must be stressed. In fact, magnetism remained at this level for a long time after the death of Gilbert. The next progress did not occur until the arrival of Coulomb (1736-1806) who, with the help of more than 150 years of scientific research and progress in mathematics, established the quantitative basis of both electricity and magnetism.

With the publication of Gilbert's masterpiece, the prehistory of magnetism came to an end. From then on, magnetism was clearly defined, and within this framework, there was no room for either magical or animistic tales, medieval superstitions or fears, or for phenomena that, even if perfectly scientific, followed different rules, such as electricity. Some aspects of magnetism still remained mysterious. In fact the origin and the role of magnetism in the universe were still bound to the past; the limits of its action were still completely uncertain; under magnetism, phenomenon that are completely foreign can still be found. The quantitative aspects had yet to be developed, but, in any case, the seed had been scattered and the

ground was fertile; only time, mathematical knowledge, and experimental techniques were required for magnetism to be fully developed. De Magnete is however a fixed point, a
gateway, the record of the death of magic in magnetism, and, at the same time, the door through which one enters an extremely wide and fascinatıng scientific field, a field that today, more than ever before, shows all its validity, vitality, and importance.

THE BEGINNING OF THE HISTORY OF MAGNETISM

The work by Gilbert strongly influenced the opinion of another English scientist, who, however, is today better remembered as the founder of economics. In 1674 Sir William Petty presented a lecture before the Royal Society of London in which he suggested that atoms are tiny magnets. As summarized ın Table 8, he affirmed that matter is made up of corpuscles which are the smallest visible bodies, and that corpuscles are made up of atoms, the smallest bodies in the universe. At least one million atoms are necessary to make a corpuscle. Atoms were immutable, but differ from each other in shape and size. The novelty is that atoms have two magnetic poles and a center of gravity, so they can rotate on their own axis and revolve around one another. In this way matter became a small universe with earths and moons; the mysterious glue of Galileo becomes gravity. Moreover atoms, possessing a magnetic dipole, tend to align with the earth's magnetic field, but are prevented from doing so because of their motion. This theory is very up to date, espe-

Table 8. A schematic summary of the structure of matter according to Sir William Petty

MATTER
↓
BODIES
↓
CORPUSCLES
↓
ATOMS = MAGNETS

cially because it accepts the same law for the universe and the small atoms. But some links with the Middle Ages were still alive, atoms still could be male or female. However at this stage magnetism and the structure of matter have really become a unique science and the world was ready to accept the modern theories of electricity and magnetism.

SUGGESTED REFERENCES

1. C.A.Ronan, "The Shorter Science and Civilisation in China", Cambridge University Press, Cambridge, USA, 1986.
2. C.Messina, "Il Magnetismo ed i suoi Misteri", MEB, Padova, Italy, 1983.
3. G.Asti, "Il Magnetismo", Editori Riuniti, Roma, Italy, 1985.
4. M.R.Cohen and I.E.Drabkin, "A Source Book in Greek Science", Harvard University Press, Cambridge, USA, 1948.

5. R.S.Westfall, "La Rivoluzione Scientifica del XVII Secolo", Il Mulino, Bologna, Italy, 1984.
6. J.D.Bernal, "Storia della Fisica", Editori Riuniti, Roma, Italy, 1983.
7. A.C.Cromble, "Da S.Agostino a Galileo", Feltrinelli, Milano, Italy, 1970.
8. M.Boas, "The Scientific Renaissance 1450-1630", R.Clark, Edinburgh, England, 1962.
9. A.C.Crombie, "Roberto Grosseteste and the Origin of the Experimental Science", Clarendon Press, Oxford, Great Britain, 1953.
10. "Le Radici del Pensiero Scientifico", P.P.Wiener and A.Noland ed., Feltrinelli, Milano, Italy, 1977.
11. L.Hollyday, Early Views on Forces between Atoms, Scientific American, May 1970, p.116.
12. F.Bonaudi, "Groping in the Dark: Magnetism and Electricity from Prehistory to (almost) Maxwell", Nucl.Phys.B, 33C, 8, 1993.

Chapter 4

THERMODYNAMICS OF METAL-GAS REACTIONS

TED B. FLANAGAN
Chemistry Department, University of Vermont,
Burlington VT 05405

1. Dilute Phases

1.1. INTRODUCTION

Since this conference concerns interstitial alloys, this thermodynamic review will be limited to that type of gas-metal system. The review will be divided into three sections:

i. dilute solutions of gases in metals and terminal solubilities, i.e., the limiting solid solution concentrations which co-exist with the more concentrated solute phases, e.g., oxides,
ii. the formation and decomposition of metal hydrides, nitrides and oxides,
iii. experimental methods for obtaining thermodynamic data for gas-metal systems.

1.2. THERMODYNAMIC REVIEW

Metals dissolve diatomic gases in the form of dissociated species. Thus the equation corresponding to equilibrium can be represented somewhat schematically by

$$\frac{1}{2}X_2(g) = [X] \tag{1}$$

where [X] refers to the dissolved species and (g) the gas phase. Examples of such reactions are:

$$\frac{1}{2}H_2(g) = [H]$$
$$\frac{1}{2}N_2(g) = [N]$$

F. Grandjean et al. (eds.), Interstitial Intermetallic Alloys, 43–76.

$$\begin{aligned}\frac{1}{2}O_2(g) &= [O]\\ CO(g) &= [C]+[O]\end{aligned}$$

From equation (1)

$$K = \frac{a_X}{\sqrt{p_{X_2}}} \tag{2}$$

where K is an equilibrium constant and at small concentrations, the activity of [X], a_X, can be replaced by its concentration which we will express as the [X]-to-metal, atom ratio, r, and therefore from equation (2),

$$r = K_s\sqrt{p_{X_2}}. \tag{3}$$

This is Henry's law for a dissociating gas and was discovered to apply for the solution of hydrogen in palladium in the last century [1] and not long after that, Sieverts and Hagenacker [2] found in 1909 that it also holds for oxygen in silver. Equation (3) is commonly referred to as Sieverts' law and the equilibrium constant, K_s, is referred to as Sieverts' constant.

The following equality must hold for phase equilibrium to obtain between the diatomic gas, X_2, and the dissolved dissociated species, [X],

$$\mu_X = \frac{1}{2}\mu_{X_2}(g) = \frac{1}{2}\mu^\circ_{X_2} + \frac{1}{2}RT\ln p_{X_2} \tag{4}$$

where the standard superscript designation on μ_{X_2} indicates 1 bar pressure of $X_2(g)$. The physical situation, which has been confirmed by neutron diffraction, is that the dissolved species occupy interstitial positions singly; there are β interstices per metal atom. At low concentrations of X, μ_X is given by

$$\mu_X = \mu^\circ_X + RT\ln\left(\frac{r}{\beta - r}\right) \tag{5}$$

where the superscript standard designation on μ_X refers to infinite dilution of X without the contribution of the configurational term. The last term in equation (5) is obtained from differentiation of the integral molar entropy obtained from the distribution of identical objects (X) into identical boxes (interstices) using the Boltzmann equation, i.e.,

$$S^{id.conf}_m = k\ln\ W = k\ln\frac{(\beta N_M)!}{N_X!(\beta N_M - N_X)!} \tag{6}$$

where $S^{id,conf}_m$ is the ideal integral configurational entropy per mole of metal, and N_M, βN_M, N_X are the numbers of metal atoms, interstices and occupied interstices, respectively. When Stirling's approximation is applied to this equation, an expression for $S_m^{id,conf}$ can obtained in terms of r

$$-\frac{S_m^{id,conf}}{R} = r\ln\frac{r}{\beta} + (\beta - r)\ln(1-\frac{r}{\beta}) \tag{7}$$

where $r = n_X/n_M$ where n_X and n_M are the moles of X and M, respectively. We are interested in the partial molar ideal configurational entropy of X which can be obtained from $S_m = S/n_M$ as

$$S_X = \left(\frac{\partial S}{\partial n_X}\right)_{p,T,n_M} = \left(\frac{\partial S}{\partial r}\right)\left(\frac{\partial r}{\partial n_X}\right) = \left(\frac{\partial S_m}{\partial r}\right) \tag{8}$$

Therefore differentiation of equation (7) with respect to r gives S_X, the partial molar entropy. If this differentiation is carried out, we obtain the ideal partial molar configurational entropy,

$$S_X^{id,conf} = -R\ln\left(\frac{r}{\beta - r}\right). \tag{9}$$

This can be used to obtain the ideal partial molar configurational chemical potential of X,

$$\mu_X^{id,conf} = RT\ln\left(\frac{r}{\beta - r}\right), \tag{10}$$

and substitution of this into equation (4) gives

$$\begin{aligned} \mu_X - \frac{1}{2}\mu_{X_2}^\circ &= \frac{1}{2}RT\ln\ p_{X_2} = \Delta\mu_X \\ \frac{1}{2}RT\ln\ p_{X_2} &= \mu_X^\circ - \frac{1}{2}\mu_{X_2}^\circ + RT\ln\left(\frac{r}{\beta - r}\right) \end{aligned} \tag{11}$$

where μ_X° is the value of μ_X without the configurational term. As $r \to 0$, equation (11) gives

$$\begin{aligned} RT\ln\left(\frac{r}{p_{X_2}^{1/2}}\right) - RT\ln\beta &= -\left(\mu_X^\circ - \frac{1}{2}\mu_{X_2}^\circ\right) = RT\ln K_s \\ RT\ln\ K_s &= -(\Delta H_X^\circ - T\Delta S_X^\circ) \end{aligned} \tag{12}$$

A plot of $\ln K_s$ against $1/T$ gives values for $-\Delta H_X^\circ/R$ from the slope and $-\Delta S_X^\circ$ from the intercept.

Examples of the solution of oxygen, nitrogen and hydrogen in specific pure metals will now be discussed. These will be employed to illustrate examples of equation (1) which correspond to: (*i*) an endothermic solution reaction, (*ii*) a very exothermic solution reaction and, (*iii*) a slightly exothermic solution reaction.

TABLE 1. Solubility of Oxygen in Silver after reference 3

T/°C	(mol O)/(mol Ag) at 1 atm	$p_{O_2}(Ag_2O)$/atm	terminal O solubility
200	2.2×10^{-7}	1.74	2.9×10^{-7}
300	2.0×10^{-6}	19.8	9.0×10^{-6}
400	9.4×10^{-6}	109	9.85×10^{-5}
500	3.0×10^{-5}	386	5.8×10^{-4}

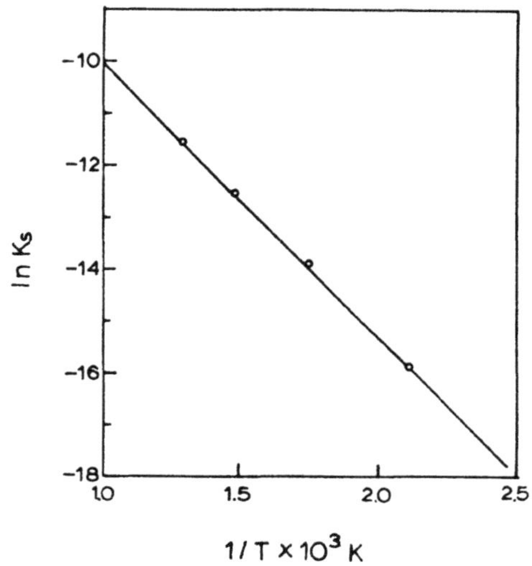

Figure 1. Solubility of oxygen in silver plotted as ln K_s against 1/T.

1.3. SPECIFIC GAS-METAL SYSTEMS

1.3.1. *Oxygen Solution in Silver*

Data for the solubility of oxygen in solid silver are given in Table 1 from the book *Gases in Metals* by Fast [3] which is an excellent introduction to this topic. The collection of data by Fromm and Gebhardt [4] should also be mentioned as a very useful reference. These solubilities are so small that Sieverts' law of ideal solubility undoubtedly holds. From each solubility determination at 1 atm., a corresponding value of K_s can be calculated from equation (3), e.g., 473 K, $K_s = r/\sqrt{p_{H_2}} = 2.2 \times 10^{-7}/(\text{atm})^{1/2}$ where r is the O-to-Ag, atom ratio, and the pressure is in atm. Such a plot for the data shown in Table 1 appears in Figure 1. The solubility increases with temperature increase and therefore the solubility, represented by reaction (1), must be endothermic. This can be appreciated from LeChatelier's principle because as the temperature is raised and "heat transferred to the system", it reacts so as to "remove the heat" which it can do for an endothermic reaction by shifting in the forward direction of reaction (1) giving a greater solubility. Since the solubility behavior is probably ideal because of the small solubilities, it can be calculated for any pressure from the K_s for that temperature. The solubility of gases in metals reaches a limit whenever an interstitial compound forms, e.g., an oxide, hydride, etc. In the case of the $Ag–O_2$ system the oxide which forms is Ag_2O. The dissociation pressure of Ag_2O is known as a function of temperature and therefore the terminal solubility, a, of oxygen in silver can be obtained at a given temperature from the above values of K_s and the corresponding dissociation pressure of Ag_2O. The terminal oxygen solubility is the solubility at a given temperature where the oxide phase first appears. Thus at 473 K where the dissociation pressure is 1.74 atm (Table 1), $a=K_s\sqrt{p_{O_2}}=2.2\times10^{-7}\times\sqrt{1.74}=2.9\times10^{-7}$. Thus if the solubility exceeds $r=2.9\times10^{-7}$ (473 K) the oxide phase, Ag_2O, precipitates. A similar calculation of a values can, of course, be made for any gas-metal system provided that the solubilities are ideal over the range of interest.

1.3.2. *Nitrogen in Tantalum*

Generally the equilibrium of $N_2(g)$ with metals is established slowly compared with the equilibrium with H_2 because of the slow surface reaction: $N_2 \rightarrow 2N_*$ where N_* indicates chemisorbed N. The equilibria are consequently restricted to rather high temperatures. For the case of nitrogen in tantalum, high temperatures are needed for the determination of thermodynamic data from equilibrium pressures because the equilibrium pressures are very small, e.g., at 1673 K less than 10^{-4} torr at the terminal solubility and they would be unmeasureable much below this temperature. An example of an exothermic solution is nitrogen in tantalum. This equilibrium has been studied at elevated temperatures by several different groups. Figure 2 shows a plot of isotherms at various temperatures for the dilute region and for the two phase region consisting of the dilute N–saturated metal phase and a nitride phase of approximate composition Ta_2N.

The dilute phase solubility isotherms appear to have constant slopes on the log–log scale (Fig. 2) which indicates that the solution closely obeys Sieverts' law of ideal solubility up to the terminal solubility. An equation which describes the solubility of nitrogen in tantalum is given in Fromm and Gebhardt's review [4], i.e.,

$$\ln r = \ln (p^{1/2}/\mathrm{bar}^{1/2}) + \frac{21880}{T} - 7.277 \quad (13)$$

where, in this case, r=N-to-Ta, atom ratio. This solubility equation gives

$$\Delta H^\circ_N = -182.0\ \mathrm{kJ/mol}\ \frac{1}{2}N_2,\ \text{and}\ \Delta S^\circ_N = -60.25\ \mathrm{J/K\ mol}\ \frac{1}{2}N_2; \quad (14)$$

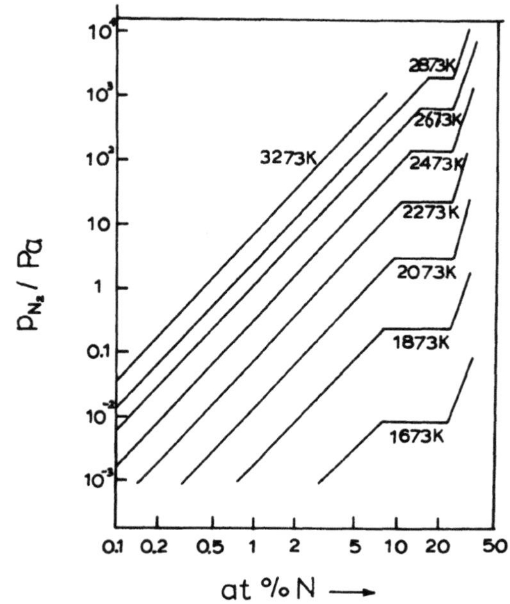

Figure 2. Isotherms for nitrogen in tantalum from reference 4

thus it can be seen that the solution of nitrogen gas in Ta is very exothermic. The terminal solubilities can be determined from this solubility equation and the dissociation pressure of the Ta_2N compound. The difference between ΔH°_N and $\Delta H^\circ(Ta_2N, \text{dissoc.})$ is rather small and therefore the lower phase boundary is rather steep, i.e., it only changes from about a=0.07 to 0.15 as the temperature increases from 1400 to 2600°C. The relationship between the enthalpy for the solution of $N_2(g)$ and that for the dissociation of Ta_2N follows from Hess's law, i.e.,

$$\frac{1}{2}N_2(g) \rightarrow [N](\text{at } a), \; ; \Delta H^\circ_N \quad = \quad -182 \text{ kJ} \qquad (15)$$

$$Ta_2N(s) \rightarrow 2Ta(\text{at } a) + (1/2)N_2(g) \; ; \Delta H^\circ \quad = \quad 204 \text{ kJ} \qquad (16)$$

$$Ta_2N(s) \rightarrow 2Ta + [N](\text{at } a) \; ; \Delta H \quad = \quad 22 \text{ kJ} \qquad (17)$$

where a refers to the N-to-Ta atom ratio of the solution which co–exists with the nitride phase, Ta_2N, and the corresponding process is the transfer of one mole of N in the nitride phase to the solution phase. It is sometimes referred to as the solvus enthalpy. Plots of ln a against 1/T should give ΔH= 22 kJ/ mol for this system.

1.3.3. *Hydrogen in Palladium*

This system has been selected as an example where there is an extensive region of dilute phase solubility at relatively low temperatures and hence

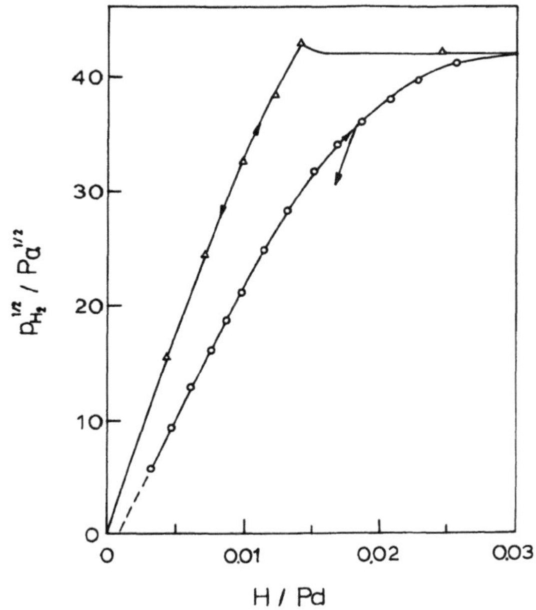

Figure 3. Dilute phase isotherms for hydrogen in palladium 298 K. △, well-annealed Pd; ○, "cycled" Pd (after reference 6).

non–ideality, i.e., deviations from Sieverts' law, is a factor because the interaction energy is comparable to kT. Experimental data for Pd–H are shown in Figure 3 where it can be seen that the data no longer follow Sieverts' law of ideal dilute solubility in the higher H content region where the equilibrium hydrogen pressures are lower than predicted by Equation (3). For systems exhibiting non–ideality, equation (11) can be modified by including an excess term and specificaly for X=H we have:

$$\frac{1}{2}\mathrm{RT}\ln\ p_{\mathrm{H}_2} = \mu_{\mathrm{H}}^{\circ} - \frac{1}{2}\mu_{\mathrm{H}_2}^{\circ} + \mathrm{RT}\ln\left(\frac{\mathrm{r}}{\beta - \mathrm{r}}\right) + \mu_{\mathrm{H}}^{\mathrm{E}}(\mathrm{r}) \qquad (18)$$

where $\mu_{\mathrm{H}}^{\mathrm{E}}(\mathrm{r})$ is the excess hydrogen chemical potential which accounts for the non–ideality. The non-ideality arises from several sources such as the H–H interaction and the effect of the H on the electronic band structure of the palladium. The H–H interaction is attractive at small contents and then changes to a repulsive interaction at larger H contents. It can be expressed as a polynomial in r where the coefficients are temperature dependent. The coefficients have been evaluated from experiment for Pd–H [5].

For Pd–H the values of $\Delta \mathrm{H}_{\mathrm{H}}^{\circ}$ and $\Delta \mathrm{S}_{\mathrm{H}}^{\circ}$ are found to be temperature dependent and, for example, Lässer and Powell [7] have measured values over a very large temperature range for all three hydrogen isotopes. At 300 K, $\Delta \mathrm{H}_{\mathrm{H}}^{\circ} = -10.0$ kJ/mol($\frac{1}{2}\mathrm{H}_2$) and $\Delta \mathrm{S}_{\mathrm{H}}^{\circ} = -54.5$ J/ K mol($\frac{1}{2}\mathrm{H}_2$) where

TABLE 2. Thermodynamic Parameters for Hydrogen Solution at Infinite Dilution for Hydrogen in Some Pure Metals where ΔS_H° is for β=1.

Metal	average T/K	ΔH_H°/kJ/ mol $\frac{1}{2}H_2$	ΔS_H°/J/ K mol $\frac{1}{2}H_2$	ref.
Ni	1150	16.6	–48.6	[8]
FeTi	600	10.6	–35.6	[9]
Pd	300	-10.0	–54.5	[7]
Nb	400	–36.2	–71.3	[10, 11]
V	675	–29.9	–72.6	[12]
α-Ti	950	–46.0	–49.3	[13]
α-Zr	928	–49.3	–55.5	[14]

these negative values are for absorption. The entropy of $\frac{1}{2}H_2(g)$ is 65.3 J/ K mol at this temperature and therefore ΔS_H° for reaction (1) is somewhat less negative than given by the loss of the gaseous entropy alone. The dissolved hydrogen must have some vibrational entropy.

1.4. SOME GENERAL RESULTS

The enthalpy of solution at infinite dilution for hydrogen is exothermic in palladium but not nearly as exothermic as for some other pure metals as shown in Table 2. The values shown in Table 2 are representative of a few of the values available and they illustrate the wide range of behavior from endothermic to exothermic absorbers. The temperatures should not influence the thermodynamic values shown in Table 2 very much especially for the non–fcc metals.

The entropy of solution at infinite dilution shown in the Table does not include the configurational contribution which goes to infinity as $r \rightarrow 0$. These ΔS_H° values have all been determined using β=1 (equation 9) which, in some cases, required changing the values given in the references. With the exception of the FeTi-H system, the values of the entropy change for reaction (1) with $\frac{1}{2}H_2(g)$ range from about –48 to about –72 J/ K mol H where the latter value is for Nb or V; it is more negative than the others because there are 6 tetrahedral interstitial positions available per metal atom in these bcc metals and this has not been allowed for. If these values are corrected for the possibility of 6 interstices per metal atom by adding $R\ln 6$ to their values, then they become close to the other values shown in Table 2.

Few reliable thermodynamic parameters at infinite dilution are available for intermetallic compounds because normally these are activated before use, i.e., hydrided and dehydrided several times and this introduces defects

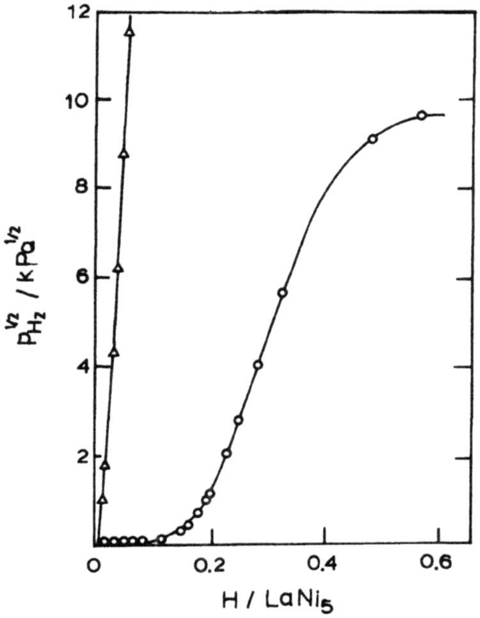

Figure 4. Comparison of dilute phase isotherms at 273 K for activated and for $LaNi_5-H$ annealed at 1023 K. △, annealed; ○, activated (after reference 14)

which cause deviations from Sieverts' law (Fig. 4). Even if the material has not been activated, an intermetallic compound may contain structural defects such as interchanged metal atoms which may affect the solubility at very small solubilities. Welter *et al* [9] determined thermodynamic parameters at infinite dilution for well characterized, non-activated FeTi. Although the plateau enthalpy for this compound is exothermic, the enthalpy at infinite dilution is seen in Table 2 to be endothermic. The entropy appears to be somewhat small in magnitude and this may be due to experimental error. Flanagan *et al* [15] employed unactivated $LaNi_5$ and obtained values of $\Delta H^{\circ}=-5.4$ kJ/ mol H and $\Delta S^{\circ}=-61.7$ J/K mol H using $\beta=1$ or -52.6 J/ K mol H using $\beta=3$. These values seem to be of the correct magnitude.

1.4.1. *Effect of Defects on Dilute Phase Solubility*

This topic has been discussed by Kirchheim [16] and therefore only material not covered in his review will be discussed here. Attention is drawn to Figure 3 which shows the dilute phase solubilities of hydrogen in two different pure palladium samples; one has been well annealed and the other has been hydrided completely and then dehydrided at the same temperature before determining the dilute phase solubility. It can be seen that there are large differences which are due to the high dislocation densities created by the phase change. In addition to the larger solubility in the "cycled" sample, it can be seen that as the equilibrium pressures approach the plateau pressure, the isotherm is no longer reversible whereas at similar pressures, the annealed sample exhibits completely reversible behavior. On the curved part of the isotherm for the "cycled" sample in Figure 3 two arrows are shown indicating absorption of hydrogen and its subsequent desorption; the fact that the directions of these arrows do not coincide, as they do for annealed Pd (Fig. 3), demonstrates the irreversibility in this region. This behavior is attributed to hydride formation in the "cycled" sample at regions having some long range stresses.

A similar but more pronounced difference in dilute phase solubilities is seen for the intermetallic compound–H system, $LaNi_5$ – H [15] in Figure 4. The large difference is attributed to defects such as structural defects, introduced by the activation procedure. Differences in hydrogen solubilities between cycled and annealed materials can be expected generally and this affects their thermodynamic behavior in the dilute phase.

2. Concentrated Phases

2.1. INTRODUCTION

This section concerns concentrated solute phases in metals such as hydride phases rather than dilute solutions which have been discussed in the first section. It was noted there that if the terminal solubility of the dissolved species is exceeded, a solute-rich phase precipitates. For the case of metal-hydrogen solutions a hydride phase precipitates. This review will employ primarily hydrogen as the representative interstitial solute but nitrogen or oxygen could also have been employed. There are, however, more data available for hydrogen and so it seems to be the natural choice.

The sub-lattice of the metal phase of the precipitating hydride phase may have the same structure as the parent metal or else, and more commonly, a different structure. In the former type of hydride the hydrogen atoms are in a disordered array unless a completely stoichiometric hydride forms, and in the latter type, the hydrogen atoms are ordered. In the case of the former, the fact that a hydride phase actually forms has been questioned because its structure is the same as the parent phase. Such miscibility gap hydride phases are, however, phases, because they satisfy Gibbs' phase rule, i.e., when such a hydride phase is present together with the dilute and gas phases, the system is univariant. X–ray diffraction also reveals that a true phase is present because two sets of x-ray reflections are found corresponding to two lattice parameters where both sets originate from the metal sub-lattice of the same symmetry. Alefeld [17] has compared a

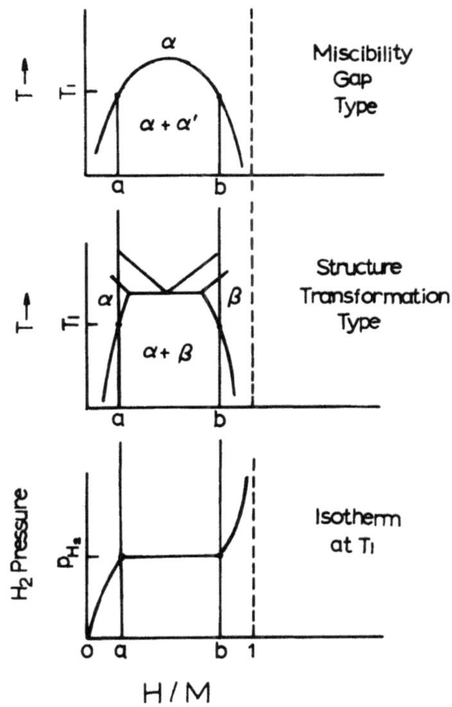

Figure 5. Miscibility gap and structural transformation–type phase diagrams for metal hydrogen systems with the accompanying Isotherm. The structural limiting composition is assumed to be MH (the figure is drawn after reference 18).

miscibility gap metal–hydrogen system to a unary gas/liquid where the formation of the hydride phase from the dilute solution is then analogous to the liquid precipitating from the gaseous phase. The dilute and hydride phases are both disordered just as gas and liquid phases are. In this analogy the metal sublattice acts simply as an inert matrix of interstices In miscibility gap systems, e.g., Pd–H, Nb–H, there is a critical point above which the miscibility gap disappears. A representative phase diagram for a miscibility gap metal–hydrogen system is shown in Figure 5. The hydride phase is designated as the α' phase. At temperature T_1 the two phase boundary compositions a and b, expressed as H/metal=r, co-exist according to the phase diagram shown in Figure 5. In contrast to miscibility gap systems, structural transformation systems do not have critical points and their phase diagrams may appear as shown in Figure 5 (after Rudman [18]). When there is only one hydride phase in a structure transformation system or when referring to the lowest hydrogen content phase, the designation β–phase is usually employed for the hydride phase.

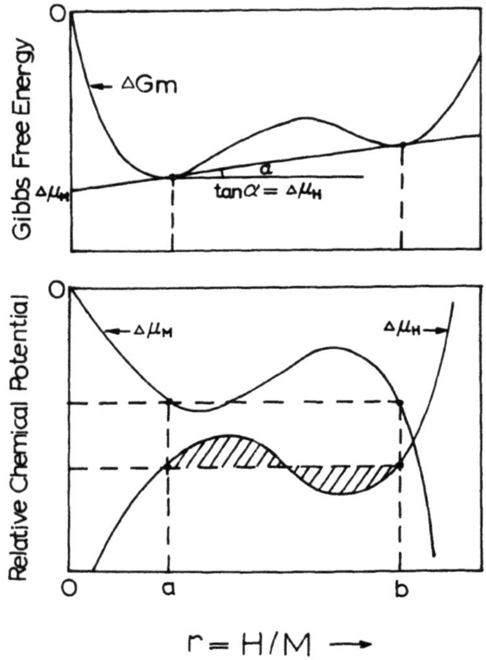

Figure 6. A schematic plot of ΔG_m against r for a miscibility gap metal hydrogen system (top). Plots of $\Delta\mu_H$ and $\Delta\mu_M$ as a function of r are also shown (bottom).

Despite the difference in their phase diagrams, a miscibility gap and a structural transformation system exhibit many common features. They are both univariant when the dilute and hydride phases co-exist together with the gas phase. The univariance leads to a constant equilibrium pressure when the dilute and hydride phases co-exist (Fig. 5); this univariant pressure region is known as the plateau pressure. A plateau pressure is exhibited for both types of systems as shown by the isotherm below the phase diagrams in Figure 5. Thus the isotherms at T_1 may have the same appearance despite the different phase diagrams. (X-ray diffraction studies are needed to distinguish the two types of phase diagrams or else a determination of the isotherms over an extended temperature range in order to construct a phase diagram).

2.2. THERMODYNAMICS OF HYDRIDE PHASE FORMATION

2.2.1. *General*

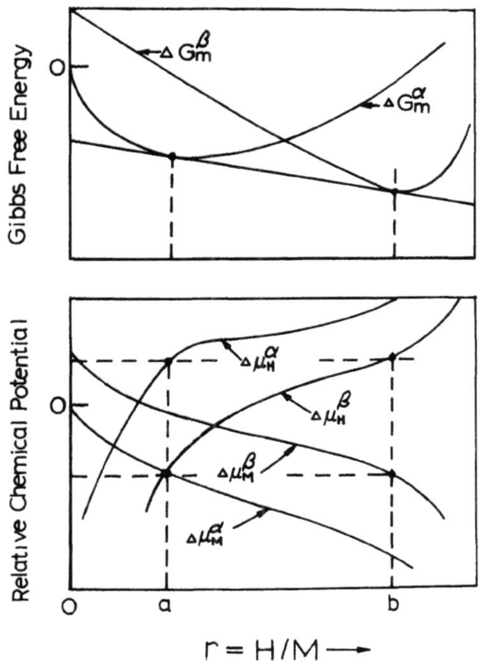

Figure 7. A schematic plot of ΔG_m against r for a structural transformation metal hydrogen system (top). Plots of $\Delta\mu_H$ and $\Delta\mu_M$ as a function of r are also shown (bottom).

Figures 6 and 7 show schematic plots of ΔG_m against r for a miscibility gap and a structural transformation system, respectively, where $\Delta G_m = \Delta G/n_M$ and Δ means that μ_H is relative to $\frac{1}{2}\mu_{H_2}^{\circ}$ and μ_M to its value at r=0. [1] The accompanying plots of $\Delta\mu_H$ and $\Delta\mu_M$ are also shown. These plots have been made with r as the composition variable rather than atom fraction H, X_H, which would be the usual composition variable employed if this were a substitutional alloy system, and this means that the slope, $(\partial\Delta G_m/\partial r)$, $= \Delta\mu_H$. In the more usual case of substitutional alloys, the intercept at X_H=1 would be $\Delta\mu_H$. For interstitial solutions, r is a more convenient and natural variable than X_H because it is not possible to attain X_H=1 experimentally.

The possibility of two phase formation for a miscibility gap system is shown by the maximum located between the two local minima shown at

[1] Upper case and lower case subscripts on the thermodynamic functions indicate partial molar and integral molar properties, respectively, according to the recommendations of the IUPAC Commission 1.2 on Thermodynamics.

the top of Figure 6 because, if the system lies on the ΔG_m curve at a composition lying between the two local minima, its Gibbs free energy can decrease if it separates into two phases of compositions a and b. The compositions of the two phases, a and b, can be determined using a line of common tangents drawn touching the ΔG_m curve on either side of the maximum as illustrated in Figure 6. The slope of the common tangent line corresponds to the relative chemical potential of hydrogen for the co-existing phases, $\Delta\mu_H(\text{plat})$, and its intersection at r=0 gives $\Delta\mu_M(a) = \Delta\mu_M(b)$ where a and b are the r values of the co-existing phases (Figs. 5 and 6).

The reason why the slope of the equi-tangent line is $\Delta\mu_H(\text{plat})$ and its intersection at r=0 is equal to $\Delta\mu_M$ arises from the condition of phase equilibria requiring that when phases are in equilibrium the chemical potentials of the components in each phase must be equal and from some results of the thermodynamics of interstitial solutions given below. For the specific case of a metal hydride system phase equilibria requires that $\mu_H(a) = \mu_H(b)$ and $\mu_M(a) = \mu_M(b)$.

Using H as the solute for purposes of illustration, the additivity relation is given for interstitial solutions as

$$G/n_M = G_m = r\mu_H + \mu_M. \tag{19}$$

The equations for metal-gas systems expressed in terms of r differ somewhat from those for a binary substitutional alloy A,B where X_B is the variable; the appropriate equations for interstitial systems have been given by Oates and Flanagan [19]. The Gibbs-Duhem equation in terms of r is given as

$$r d\mu_H + d\mu_M = 0 \tag{20}$$

and from this, the following expression can be obtained for $\Delta\mu_M = \mu_M - \mu_M^\circ$,

$$\mu_M - \mu_M^\circ = \Delta\mu_M = -r\mu_H + \int_0^r \mu_H dr. \tag{21}$$

It follows by substitution of equation (21) into (19) that μ_H is given by

$$\mu_H = \left(\frac{\partial G_m}{\partial r}\right)_T. \tag{22}$$

This equation could equally well be given in terms of $\Delta\mu_H$ and ΔG_m. Equation (22) means that the slope of the $G_m - r$ curve at any value of r gives μ_H. It follows from equation (19) that at the same value of r where the slope is evaluated as $\Delta\mu_H$, the intercept at r=0 is equal to μ_M. It follows from the integration of equation (22) that

$$\Delta G_m = \int_0^r \mu_H dr. \tag{23}$$

In the special case where the two phases coexist, the same considerations hold except that now the slope and the intercept are equal for the equi-tangent line as shown in Figure 6. In the lower part of Figure 6, plots of $\Delta\mu_H$ and $\Delta\mu_M$ against r are shown. For $\Delta\mu_H$, Maxwell's "rule-of-equal-areas" holds as seen by the dashed regions of the $\Delta\mu_H - r$ curve and this locates the equilibrium state. This is a consequence of the equi-tangent line shown in the upper plot, i.e., the two requirements are equivalent because the right-hand-side of equation (21) will vanish because of the "rule-of-equal-areas" and therefore the left-hand-side of equation (21) must also vanish, i.e., $\mu_M(b) - \mu_M(a)=0$. The $\Delta\mu_M - r$ curve is also shown which has been calculated from the Gibbs–Duhem equation, i.e., equation (21). Although it does not follow a "rule of equal areas", it is seen in Figure 6 to also have a loop.

Figure 7 shows analogous plots for a structure transformation system which has been drawn after a figure given by Rudman [18]. In this case the integral free energy curves for the dilute and hydride phases differ because they no longer have common sub-lattices as for the miscibility gap system shown in Figure 6. The β–metal sublattice of the hydride phase is metastable as shown in Figure 7 because it lies above the stable α–sublattice as $r \rightarrow 0$. The two phase equilbrium is established by drawing an equi-tangent line to the two curves; the slope of this line is $\Delta\mu_H$(plat) and, its intercept at r=0, gives the metal atom chemical potentials. The lower drawing in Figure 7 shows the corresponding $\Delta\mu_H$ and $\Delta\mu_M$ relations. The "rule of equal areas" is not applicable because of the two separate $\Delta\mu_H$ relationships. The chemical potentials and metal potentials must, however, be equal at equilibrium as shown in the lower drawing by the horizontal dashed lines intersecting the ΔG_m plots at equal values of $\Delta\mu_H$ and $\Delta\mu_M$.

The plateau pressures and their temperature dependences are most frequently employed to determine thermodynamic properties for metal hydride systems and indeed similar plots are employed for the analogous nitride and oxide systems. Plots of ln p_f or ln p_d against 1/T are called van't Hoff plots and these are generally linear. Such a plot for a miscibility gap system is illustrated schematically in Figure 8 where the isotherms are also shown. The slope of the linear plot is $\Delta H_{plat}/R$ for hydride formation and from the slope and intercept $\Delta\mu_{plat}$, ΔS_{plat} can be calculated.

Analogously to metal/hydrogen systems, nitride phases precipitate from nitrogen solutions and oxide phases from oxygen solutions when their terminal solubilities are exceeded. These are invariably structure transformation systems where the N and O are ordered within the metal sub-lattice. The degree of non-stoichiometry is usually not as great for these as for hydrides although they can be also be very significant, e..g, CeO_{2-x}. Much research has been carried out on the characterization of the rare earth oxide/O_2(g)

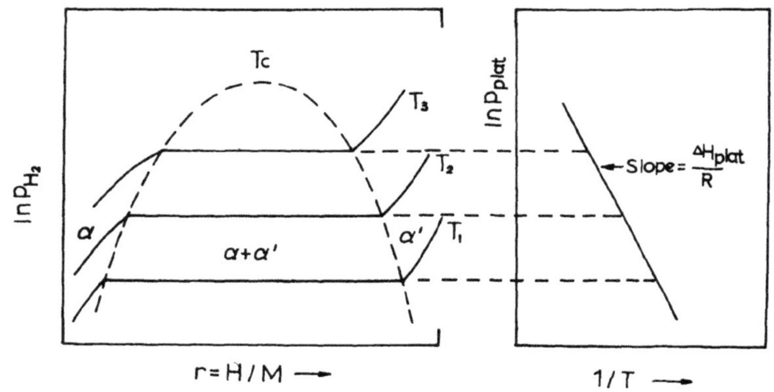

Figure 8. A schematic plot of isotherms for a miscibility gap system and the accompanying van't Hoff plot.

equilibria. A useful review is given by Sørenson [20]. Ag_2O formation has been referred to in the first section (Table 1 and Figure 1) as has Ta_2N, whose isotherms are shown in Figure 2.

2.2.2. *Regular Interstitial Solution*

The dilute solution was reviewed in the first section and here we will again employ metal-hydrogen systems for the purpose of illustration of the phenomenon of formation of a concentrated phase from a dilute solution. In the case of a metal-hydrogen system a hydride precipitates when the hydrogen content reaches the terminal hydrogen solubility. The simplest model which leads to hydride formation is the regular interstitial model [19] given by the following equation

$$\mu_H = \mu_H^\circ + RT \ln\left(\frac{r}{\beta - r}\right) + h_1 r \tag{24}$$

where $h_1 r$ is an H–H interaction enthalpy which will be assumed to be independent of temperature. The integral molar free energy can be obtained from equation (23) and $\Delta\mu_M$ for a miscibility gap system can be obtained from equation (21). In practice, $\Delta\mu_H = \frac{1}{2}RT \ln p_{H_2}$ is substituted for μ_H in the equation above where the Δ refers to 1 bar $H_2(g)$ in $\Delta\mu_H$ and to μ° which is μ_M at $r=0$ for $\Delta\mu_M$.

The reaction which takes place when the hydride phase forms along a plateau is

$$\frac{1}{2}H_2(g, 1\text{ bar}) + \frac{1}{(b-a)}MH_a \rightarrow \frac{1}{(b-a)}MH_b, \tag{25}$$

which corresponds to the difference of two relative integral molar properties where a and b are the lower and upper phase boundary compositions, respectively, expressed as r=H/M. This is true for both miscibility and structure transformation systems. Only for the former, however, can the difference of the relative integral molar properties be written from equation (23) as

$$\Delta H_{plat} = \frac{1}{(b-a)} \int_a^b \Delta H_H dr. \tag{26}$$

Thus if the relative partial molar enthalpy is known over the whole range of r, the plateau enthalpy can be calculated for a miscibility gap system. It is also clear from equation (26) that for miscibility gap systems a plateau property represents an average of the relative partial property value over the (b–a) interval.

For the regular interstitial solution, equation (24), it is easy to show [19], using equation (26) that

$$\begin{aligned} \Delta H_{plat} &= \frac{1}{(b-a)} \int_a^b (\Delta H_H^\circ + h_1 r) dr \\ \Delta H_{plat} &= \Delta H_H^\circ + \frac{1}{2} h_1. \end{aligned} \tag{27}$$

Equation (27) results from the fact that the regular interstitial solution model leads to a phase diagram which must be symmetrical about r=0.5, assuming that there is one interstice per metal atom, and this leads to the result that (b + a) =1.

Pursuing the same approach using the regular interstitial solution model for the plateau entropy change leads to $\Delta S_{plat}=\Delta S_H^\circ$, i.e., according to this model the plateau entropy change should be similar for most systems which is indeed found to be the case. Although the regular interstitial solution model is physically unrealistic for hydrogen-metal systems, it is of heuristic value because its qualitative predictions are valid and this helps to understand these systems.

For structure transformation systems there is no analytical model which may be employed for both phases but models can be applied to each of the two metal sub-lattices separately, i.e., the regular interstitial solution can be applied separately to the two sub-lattices where the β–phase metal sub-lattice is metastable to $r \to 0$ as shown in Figure 7.

2.3. REAL METAL-GAS SYSTEMS

2.3.1. *Thermodynamic Data for Representative Systems*

Figure 9 shows a compilation of plateau pressure plots gathered by Penzhorn *et al* [21]. The plots show an amazing range of plateau pressures for

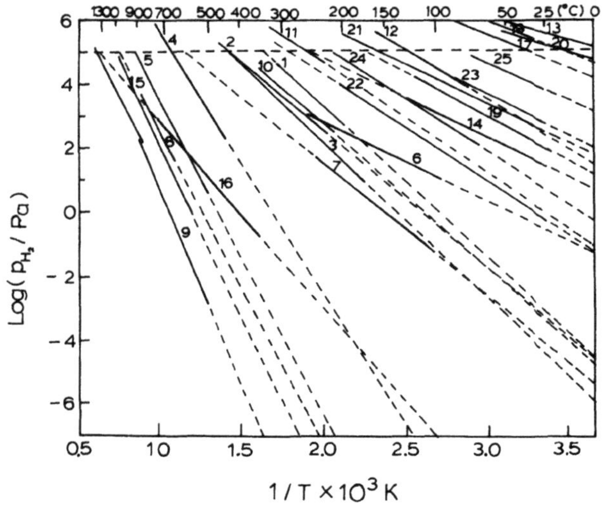

Figure 9. Van't Hoff plots for a number of different hydrogen-metal systems adapted from reference 21. The numbers identify the specific systems and they can be identified by referring to the paper of Penzhorn *et al.* A few representative ones are identified here, i.e., #3≡U–H [22], #13≡FeTi–H [23],#20≡$LaNi_5$–H [24].

any given temperature. This plot is useful in enabling the choice of a metal hydride to be made for technological purposes requiring specific plateau pressures over a specific temperature range. Penzhorn *et al* pointed out the need for a metal hydride system with a low plateau pressure, $\approx 10^{-3}$ Pa, near room temperature but which reaches a pressure of 1 bar below 773 K; this is required as a replacement system for uranium as a tritium storage material. The ZrCo–H(T) (van't Hoff plot #1, Fig. 9) appeared to be a suitable choice and worthy of examination of its other properties for its possible suitability as a tritium storage system.

At a given temperature, log p_{H_2} is proportional to the free energy of formation of the hydride phase from $\frac{1}{2}H_2$(g, 1 bar). Thus when a specific van't Hoff line is lower than another one, the former metal will dehydride the latter one. As $(1/T) \to 0$ the van't Hoff plots for the metal–gas systems should all intersect the same value of the ordinate, i.e., $\frac{1}{2}\Delta S_{plat}/(2.303\ R) + \log 10^5$; this neglects any temperature dependence of ΔS_{plat}. Except for one or two systems, this appears to be the case in Figure 9.

Fukai [25] has recently plotted plateau pressure data as $\Delta\mu_{plat} = RT \ln p_{H_2}^{1/2}$ against T which he points out is an "Ellingham diagram" for metal hydride systems. On the other hand, Figure 9 is perhaps more convenient than the

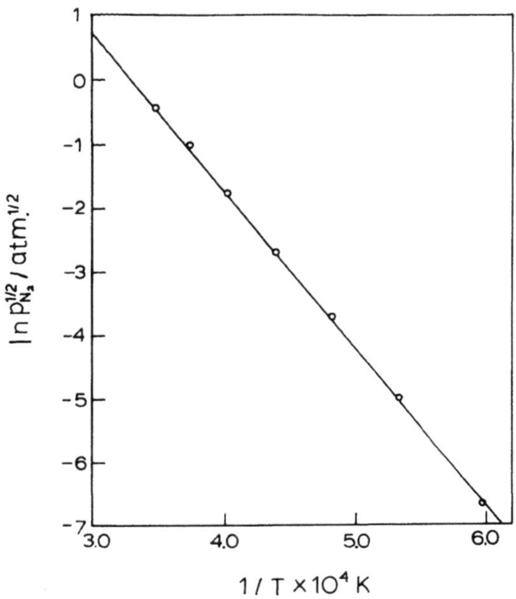

Figure 10. A van't Hoff plot for the nitrogen-tantalum system from the data shown in Figure 2. The plot is surprisingly linear for such an extended temperature range.

Ellingham presentation for the purpose of selecting systems for technological uses.

We saw in Figure 2 that there is a plateau in the Ta–N_2 system near the composition Ta_2N. Plateau pressures for Ta–N_2 have been plotted as a van't Hoff plot in Figure 10 and the plot is seen to be quite linear over a very large temperature range. The slope yields a very exothermic enthalpy, ΔH_{plat} =−205 kJ/ mol N. The plateau pressures of N_2 are consequently very low even at 1673 K.

2.3.2. *Sloping of Plateaux*

For both types of metal-hydride systems, i.e., the miscibility gap and structure transformation type, the plateau pressures frequently increase as the fraction of hydride phase increases during hydride formation and decrease as the fraction of the dilute phase increases during decomposition, i.e., there are "sloping plateaux" as illustrated by Figure 11. The sloping plateaux shown in Figure 11 are more characteristic of pure metals than some intermetallic compound-H systems which slope even at small fractions of phase conversion and the degree of the sloping is greater. The origin of the sloping may be different for the two types of systems. The reasons for the sloping are not always clear but in some cases of alloys or intermetallic compounds

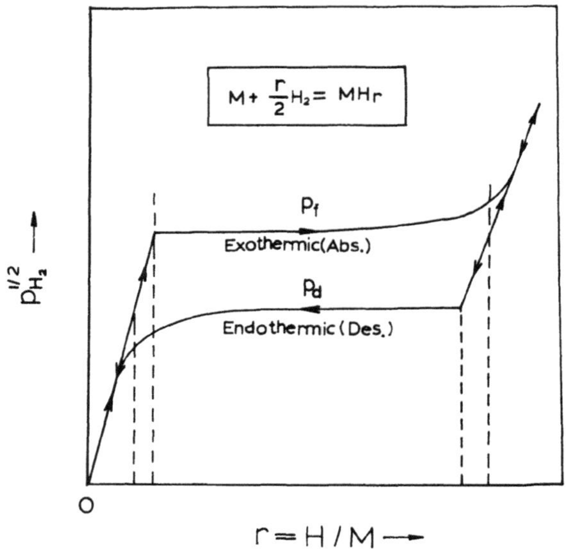

Figure 11. A schematic isotherm showing sloping of the isotherms especially after about 50 % conversion during hydriding and dehydriding. Hysteresis is also illustrated.

it is due to compositional inhomogeneities. Evidence for this is that the plateaux of "as-cast" intermetallic compounds usually slope significantly more than the ones of the same compounds after annealing [26]. Obviously inhomogeneities cannot be the sole reason for plateau sloping because sloping is also found in pure metals. In pure metals the sloping may be to due to the stress which originates from the inward penetration of the hydride/dilute interface during the hydriding and the inward penetration of the dilute/hydride interface during dehydriding.

2.3.3. *Hysteresis*

Both miscibility gap and structural transformation systems exhibit hysteresis, i.e., different pathways are followed during the forward process of hydriding and the reverse process of dehydriding, despite the fact that both processes may be carried out extremely slowly, i.e., quasi-statically. The plateau pressure for hydride formation, p_f, is then greater than that for decomposition, p_d, i.e., $p_f > p_d$ (Fig. 11).

The most fundamental thermodynamic expression for pressure hysteresis is given by

$$\text{free energy dissipated} = \frac{1}{2}\text{RT}\ln\left(\text{p}_\text{f}/\text{p}_\text{d}\right), \tag{28}$$

where p_f and p_d are the formation and decomposition plateau pressures, respectively. Expression (28) also gives the minimum work done on the system by the surroundings and this work is subsequently lost to the surroundings as heat. The effect of hysteresis on the van't Hoff plots is shown in Figure 12 for ZrNi–H [27] where the van't Hoff slope for the decomposition pressures is greater than for the formation pressures, as expected. The values of ΔH_{plat} for hydride formation and decomposition are −34.1 and 36.8 kJ/ mol $\frac{1}{2}H_2$, respectively. The calorimetric value determined for hydrogen absorption at a lower temperature, 298 K, is −34.0±0.5 kJ/ mol $\frac{1}{2}H_2$ [27]. The fact that the calorimetric value is closer to the value derived from the van't Hoff plot for p_f rather than for p_d is believed not to be significant because of experimental error. The inset shows more recent data from our laboratory for plateau pressures for this system starting at the high temperature end of the main plot and extending it to higher temperatures. This inset shows the near disappearance of hysteresis at the highest temperature. These data give ΔH_{plat} for hydride formation=−32.2 and for decomposition=39.8/ mol $\frac{1}{2}H_2$. This illustrates that in the range where the hysteresis is disappearing, it affects the van't Hoff plots more significantly than in the range where it is more nearly constant.

Often the plateau pressures, and consequently hysteresis, vary with cycling, i.e., with the number of times the hydride phase has been formed and decomposed without an intermediate annealing treatment. This is shown for $LaNi_5 - H$ and a $Pd_{0.9}Ni_{0.1}$ alloy in Figures 13 and 14.

It can be seen for the former that the formation plateau pressure decreases markedly during the first two cycles of hydriding and then remains constant during further cycles. For the $Pd_{0.9}Ni_{0.1}$ alloy both the formation and decomposition isotherms change but the former changes more than the latter one. The hysteresis decreases markedly for both of them.

Hysteresis varies from system to system and it is generally larger for those where $(b-a)$ is large. It tends to decrease with increase of temperature but not linearly. It is also present in rare earth metal oxide systems even though these are measured at quite high temperatures [29].

2.3.4. *Other Complications Found in Real Systems*

Frequently several hydride phases exist within a given system; they form sequentially as the hydrogen content is increased isothermally. For example, the $ErFe_2 - H$ system appears to have at least four different hydride phases as the hydrogen content increases from $r=0$ to 3.2 [31, 32]. The

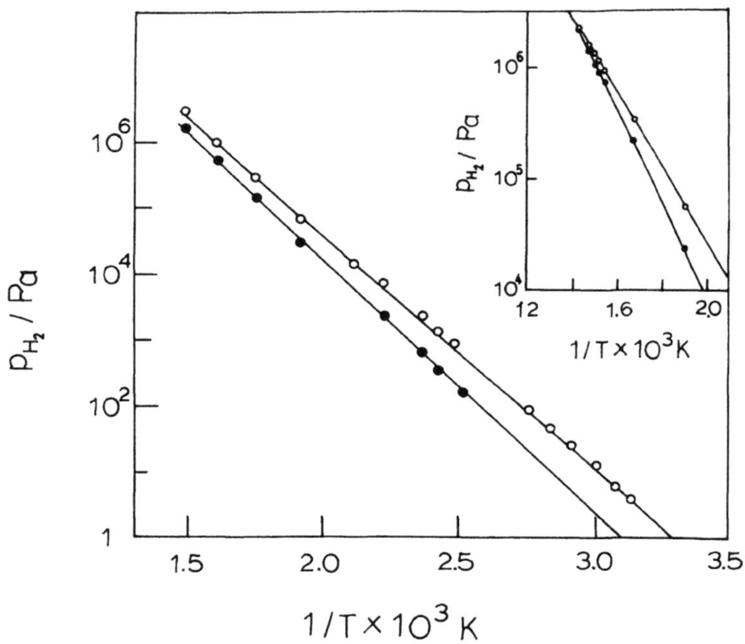

Figure 12. A van't Hoff plot of plateau pressures for the co-existence of the $\beta + \gamma$ phases in the ZrNi/H system [27]. The inset shows some of our recent data in the higher temperature region [28]. The open and closed symbols refer to absorption and desorption data, respectively.

chemical potential of hydrogen must increase with increase of r within a single phase region. This means that in a two phase region the chemical potential of hydrogen will remain constant because the hydrogen contents in both phases do not change as the fraction of the two phases change with the overall hydrogen content. This follows from thermodynamics [34], however, the values of ΔH_H and ΔS_H need not change with increase of r in any predictable manner as long as $\Delta\mu_H = \Delta H_H - T\ \Delta S_H$ increases with r. Sometimes it has been observed that a second or third plateau region might have a more exothermic ΔH_{plat} than the initial one.

3. Experimental Methods

3.1. INTRODUCTION

This section concerns experimental methods for obtaining thermodynamic data for gas-metal systems. Many of the techniques have been alluded to earlier but they will be discussed here in more detail. The experimental methods will be illustrated largely with hydrogen-metal systems but

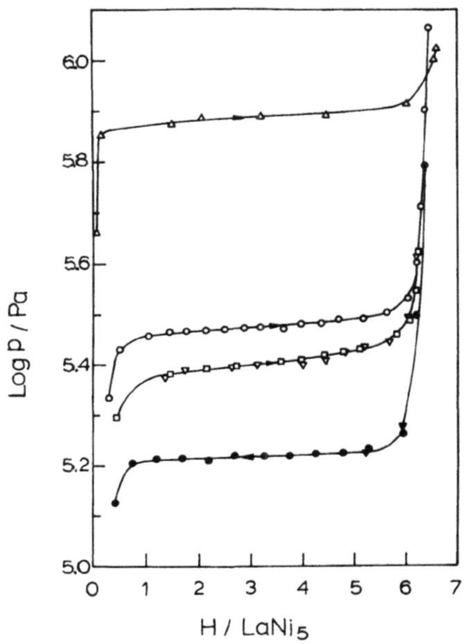

Figure 13. Effect of repeated hydriding and dehydriding of $LaNi_5$ at 298 K (after reference [28]. △, first hydriding; ○, second hydriding; ▽, third hydriding; □, fourth hydriding. Closed symbols represent decomposition which appears to be unaffected by the number of cycles.

most of the methods apply to other gas-metal systems except that higher temperatures must be employed for other gas–metal systems. The reason for this is clear from a simple consideration of the diffusion constants of H and O in iron at 298 K where $D(H) = 5.2 \times 10^{-9}$ m^2 s^{-1} and $D(O) = 4.2 \times 10^{-29}$ m^2 s^{-1} [4]; this startling difference mandates different temperature regimes. The higher temperatures needed for interstitial species other than H may require special reaction vessel material, etc.

3.2. ISOTHERM MEASUREMENTS

The p-c-T method is based on measurements of the free energy of reactions (25) and (29) and their temperature dependences, i.e., in a single phase region

$$\frac{1}{2}H_2(1\text{bar}, g) \rightarrow [H] \tag{29}$$

and, in two phase regions, the reaction is given by (25); the reverse reactions can, of course, also be determined. The determination of the free energy

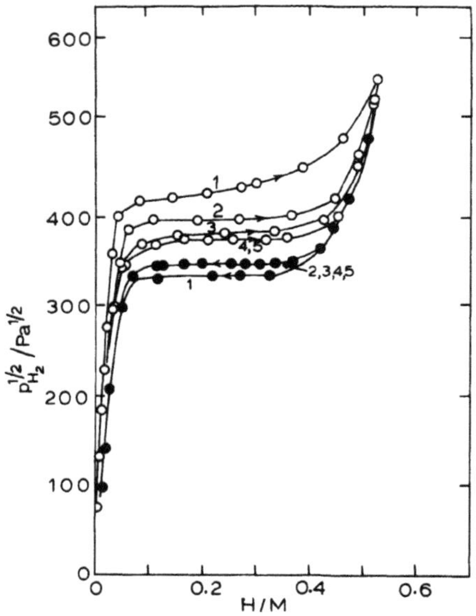

Figure 14. Effect of repeated hydriding and dehydriding of $Pd_{0.9}Ni_{0.1}$ at 303 K. The numbers on the plateaux indicate the number of the cycle. The open and closed symbols represent hydride formation and decomposition, respectively (after reference [33].

change at a given (H/M)=r, $\Delta\mu_H = \mu_H - \frac{1}{2}\mu^{\circ}_{H_2}$, is made *via* measurements of the equilibrium hydrogen pressure since

$$\Delta\mu_H = RT \ln\ p_{H_2}^{1/2}, \tag{30}$$

where p_{H_2} is the equilibrium pressure in bar and $\Delta\mu_H$ is the hydrogen chemical potential at r relative to $\frac{1}{2}H_2$(g, 1 bar). The temperature dependence of $\Delta\mu_H$ gives ΔH_H and ΔS_H.

It should be emphasized that the p-c-T method is an equilibrium method. It requires that equilibrium be established, i.e., the gaseous pressure must have reached equilibrium and is therefore a true indication of the chemical potential of the dissolved hydrogen.

Reaction (25) is often referred to as the plateau reaction and refers to hydride formation and decomposition. The thermodynamic parameters for reaction (25) are then referred to as plateau values, e.g., ΔH_{plat}. These are obtained from van't Hoff plots of ln p_{plat}. These van't Hoff plots are invariably linear for metal hydride systems, i.e., ΔH_{plat} and ΔS_{plat} are independent of temperature. This follows according to Rudman [18] if the dependence of ΔH_H on r does not show any curvature or else if the phase boundary is symmetrical. Some possible errors connected with obtaining thermodynamic parameters from p-c-T data, aside from the obvious one of not employing equilibrium pressure data, are discussed below.

3.2.1. *The Sieverts' Method*

The conventional method for measuring isotherms, ($p_{H_2} - r$ relationships), employs a so-called Sieverts' apparatus. From isotherms, the temperature dependence of $\Delta\mu_H$ at various r values, and thus all of the thermodynamic parameters, can be obtained. In this method known pressures of hydrogen are added to a known dosing volume which is then opened to the known reaction volume containing the sample. From the magnitudes of the equilibrium pressure decreases, the amounts of hydrogen absorbed can be determined for each value of r. The reverse procedure is employed to determine the desorption data. It is useful to employ a strip-chart recorder to monitor the pressure-time after the introduction or removal of each dose of hydrogen in order to establish with some certainty when equilibrium is established. The Sieverts' method does not work very well when the hydrogen pressures are large and the uptake of hydrogen relatively small. For example, in the single hydride phase of Pd–H where the hydrogen chemical potential increases markedly with r the method is rather poor because it is based on differences of large pressures in a range where the pressure measurements are generally not very precise.

This method obviously does not work very well if the pressures are so low that they cannot be measured accurately. The electrochemical method as shown by Kirchheim [16] and also Boes and Züchner [35] may be a viable method in this case because it is easy to measure electrode potentials corresponding to extremely low pressures (see below).

Before about 1960, glass (pyrex) systems were generally employed for metal-gas thermodynamic studies; these contained mercury and stopcock grease as contaminants. Contamination of the surface of the hydrogen-absorbing metal by mercury and grease in such vacuum systems is shown by the observation that equilibrium cannot be established between bulk Pd and H_2(g) at temperatures lower than about 400 K [36] in these glass systems but it is readily established in grease and mercury free all-metal systems at temperatures well below even 273 K [37]. Some workers continue to employ glass-mercury systems [38] and such apparatuses can be quite successful but only at elevated temperatures.

Most workers now employ all–metal systems which is, of course, a necessity when operating much above 1 bar pressure. Limiting measurements to $p_{H_2} \leq 1$ bar often constitutes a serious restriction in the study of typical metal-hydrogen systems. It is necessary to correct for the non–ideality of the hydrogen even at hydrogen pressures of <10 bar because errors are cumulative during isotherm determinations.

For accurate pressure measurements two types of electronic diaphragm gauges are now generally employed in all metal systems. In one type, the capacitance manometer, pressure changes on the diaphragm cause a change in a capacitance bridge in the manometer which is converted to an electrical output proportional to the total gas pressue. In the second type, the strain gauge manometer, a strain gauge mounted on the diaphragm produces an electrical output proportional to the total gas pressure. The latter type is generally less expensive but has a greater error. All of these electronic gauges have the great advantage that they are independent of the type of gas which is measured and produce an electric signal which can be easily interfaced with data acquisition equipment. Another advantage of electronic gauges is their fast response times as compared to manometers or McLeod gauges which are generally employed in glass systems. It should be noted that being "black boxes", the electronic gauges must be calibrated occassionally against a more fundamental gauge or by the use of known pressures from a well-characterized metal-gas system.

Another gauge which is suitable for all-metal systems, and which is independent of the type of gas and which can be used in an all-metal apparatus, is the Bourdon gauge. This is a coiled bimetallic tube which distorts according to the pressure of the gas in the tube. These Bourdon gauges have the disadvantage that they have larger internal (dead) volumes than the electronic diaphragm gauges; a small "dead volume" is especially desirable at elevated pressures.

Instead of determining isotherms at separate temperatures, a variant of the usual p-c-T method is to add a certain amount of hydrogen to the sample and then to decrease the temperture, stopping at several different temperatures where the equilibrium pressures are recorded. The sample is then degassed and another, different dose of hydrogen is added to the sample and the procedure is repeated. This method can be employed to determine a series of isotherms. It has been used by Weiss and his coworkers, e.g., [38], and has been described by them as the manometric method to distinguish it from the volumetric method where the pressure is maintained constant and volume changes are measured [39]. They have employed both methods. This manometric method is also the basis of the automatic apparatus employed by Lässer and his coworkers [40]. It may lead to difficulties in two phase regions due to hysteresis effects.

3.2.2. *Automatic Methods*

Automatic volumetric p-c-T apparatuses are now available commercially. Isotherms are determined by automatic dosing using valves powered by compressed air or a compressed gas.

3.2.3. *Gravimetric Method*

Gravimetric methods are superior to p-c-T methods in regions where the pressure increases steeply with r. Corrections must be made for bouyancy unless twin suspensions are employed. A newly developed commercial gravimetric system is fully automated and is, reportedly, "tailor made" for gas–metal systems having been developed by workers in this field [41].

3.2.4. *Electrochemical Method*

As mentioned above, the electrochemical method is useful in regions of low hydrogen chemical potential where pressure measurements are difficult, if not impossible. Lewis and his coworkers have employed a mixed gas phase electrochemical method where the hydrogen is absorbed from $H_2(g)$ saturated solutions but the chemical potential of the dissolved hydrogen is measured by the electrode potential. Some of this earlier work is described in Lewis' book [42] and the later work is reviewed in the "Platinum Metals Review" by Lewis [43]. Reactions (25) or (29) can be carried out electrochemically, i.e.,

$$H^+(aq) + \bar{e} \;=\; [H];\ \mathcal{E} \qquad (31)$$

$$\frac{1}{2}H_2(g) \;=\; H^+(aq) + \bar{e};\ \ -\mathcal{E}^{\circ}_{H^+/\frac{1}{2}H_2(g)} \qquad (32)$$

where $\mathcal{E}$ is the electrode potential for the half cell reactions. The sum of these two electrode half cell reactions corresponds with the gas/metal reaction given by reactions (25) or (29). The relationship between $\mathcal{E}$ and p_{H_2} for hydrogen absorption is

$$(\mathcal{E} - \mathcal{E}^{\circ}) = -\frac{RT}{\mathcal{F}} \ln \left(p_{H_2}/1\ \mathrm{bar}\right)^{1/2} \qquad (33)$$

where $\mathcal{F}$ is the Faraday. A positive voltage indicates a spontaneous change with respect to the standard hydrogen electrode (S.H.E.), i.e., reaction (32). Of course, it is often more convenient to use a reference electrode other than the hydrogen electrode.

As Kirchheim has pointed out [16], the use of relatively viscous phosphoric acid electrolyte or else scrupulously O_2-free H_2SO_4 can be employed for the measurement of the equivalent of very small hydrogen pressures. For example, the equilibrium pressure of the plateau region, $(\alpha+\beta)$, of Ta/H_2 is 10^{-5} torr at 313 K which is difficult to measure accurately by pressure gauges but has been readily measured electrochemically as the equivalent of 0.245 V against S.H.E. by Boes and Züchner [35]. Kirchheim has measured $\mathcal{E}$ values for hydrogen in palladium corresponding to a few parts per million with no difficulties although the equilibrium pressures would be difficult to measure, e.g., $\approx 10^{-7}$ torr.

3 3. CALORIMETRIC METHODS

These remarks hold for any of the interstitial gas/metal systems. Enthalpies of reaction can be obtained using a quasi-isothermal calorimeter; heat capacities can be obtained in a non-isothermal calorimeter, such as a differential scanning calorimeter. Calorimetric methods do not require chemical equilibrium to be achieved as in the p-c-T method although the system should not deviate grossly from equilibrium, i.e., the bulk hydrogen content should be nearly uniform or, if two phases are present, each phase must be nearly homogeneous in order for r in the derived ΔH_H against r relationships to have significance. The most powerful method is the simultaneous measurement of the enthalpy *and* the equilibrium hydrogen pressure because in this way isothermal measurements will provide all three thermodynamic parameters, $\Delta\mu_H$, ΔH_H, and ΔS_H, needed to characterize the system from the reaction point-of-view.

3.3.1. *Reaction Calorimetry*

Reaction calorimetry refers to the technique where a calorimeter is employed to measure heats of gas/metal reactions directly, e.g., reaction (1). In practice, the H_2(g) does not have to be at 1 bar because the enthalpy change will not depend on the pressure except at high pressures. The heat for the reverse reaction can also be measured. In the two, solid-phase region the reaction whose heat is measured corresponds to reaction (25) for metal–hydrogen systems.

This calorimetric method was first used for Pd–H in its plateau region in 1874 by Favre [44] and he obtained an enthalpy of magnitude of 20.2 kJ/ mol H for the plateau value for Pd black, i.e., finely divided Pd. In an earlier set of experiments [45] he found a value of 17.3 kJ/ mol H. Mond, *et al* [46] found for the plateau reaction with Pd black at 0°C a value of 19.3 kJ/ mol H using an ice calorimeter. This value is amazingly close to recently measured values for the reaction of hydrogen with bulk Pd found by the author and his co–workers [47] who reported for hydride formation −19.09 kJ/ mol H and, 19.28 kJ/ mol H for decomposition at 298 K. A difference between the older calorimetric results and the newer ones, is that in the latter relative partial molar values for the single phase regions are determined rather than just values for the plateau reaction.

Flanagan *et al* [47] employed a dual cell heat leak calorimeter for the study of the Pd–H(D) system which was the type as employed by Murray *et al* for their studies of the $LaNi_5$ – H system [48]. In fact, the most versatile and simplest calorimeter to employ for reaction calorimetry of gas/metal systems is the twin-cell heat leak type. It has been employed extensively in recent years for the study of interstital gas/metal systems. It should be noted that frequently the largest error in this type of calorimetry arises from uncertainties in δn_H rather than δq where δn_H is the mols of H added or removed from the system and

$$\Delta H_H = \left(\frac{\delta q}{\delta n_H}\right)_r, \text{ as } \delta n_H \rightarrow 0. \tag{34}$$

In certain ranges the evaluation of δn_H is subject to considerable error, e.g., in the hydride single phases where the equilibrium pressures are large and increase steeply with r. This means that for these regions very precise measurements of δq employing elaborate calorimeters will not necessarily lead to accurate values of ΔH_H unless the pressure and volume measurements are also very precise and it is nearly impossible for the pressure measurements to be sufficiently precise in the range of pressures above about 10 bar.

A drawing of this type of Tian-Calvet calorimeter is shown schematically in Figure 15. Its *modus operandi* is that the two cells, which are matched as closely as possible thermally, are situated in a large heat sink. Many thermocouples are arranged in series about the two cells such that the resulting thermopiles monitor the temperature difference between the two cells. Both cells are at the same temperature until reaction takes place in the reaction cell. The resulting imbalance of the two cells is then recorded as the integral of the voltage–time signal. The heat leak is arranged to be optimal, i.e., so that the cells return to the same temperature after a time interval which is neither too fast or too slow. Most of the heat leak takes place through the thermocouple wires themselves. The twin cell design eliminates the need for corrections due to the work of gas expansion [49, 50] according to Murray *et al* [48] since the work effect cancels. Calibration of the calorimeter may be accomplished either by Joule heating using a known electrical resistance either in the reaction vessel or in the blank vessel. An alternate procedure is to employ the heat of a known reaction; this procedure is perhaps the best calibration method if the known reaction is comparable to the one being investigated. A Tian-Calvet heat leak calorimeter is often referred to as a micro-calorimeter but it need not be "micro", e.g., the one illustrated in Figure 15 is not "micro".

Dantzer and Guillot [51] have described a high-temperature heat flow calorimeter capable of operating up to 1500 K. It consists of a massive ceramic furnace containing 180 Pt-Pt (10%Rh) thermocouples to detect the signal. This type of calorimeter has been automated for metal–hydrogen systems by Dantzer and his coworkers.

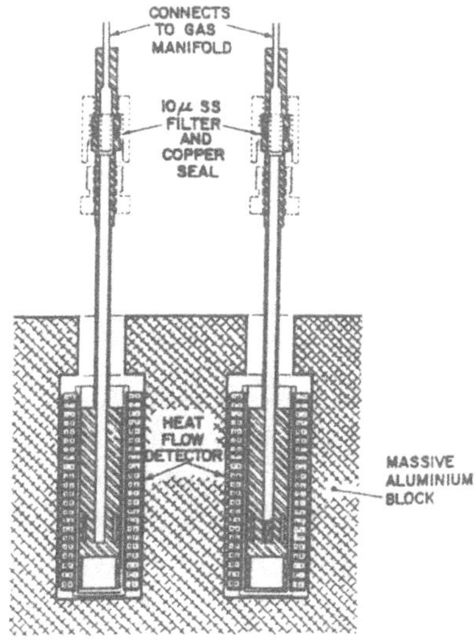

Figure 15. A typical dual-cell, heat leak calorimeter of the Tian-Calvet type (reference 48 and with permission of the publisher)

A similar twin cell (Tian–Calvet type) calorimeter has been employed by Inaba, Navrotsky, and Eyring [52] to study the reaction

$$\frac{1}{7}Pr_7O_{12} + (\frac{1}{7} - \frac{x}{2})O_2(g) \rightarrow PrO_{2-x} \quad (35)$$

at ≈973 K.

3.3.2. *Solution Calorimetry*

This type of calorimetry works quite well for very stable hydrides, nitrides, etc. It will be illustrated by Turnbull's [53] determination of the heat of formation of ZrH_2 from the heats of solution of a series of ZrH_{2-x} compounds in HF(aq). The difference between the heats of solution of the non-stoichiometric hydride and Zr(s) gives the enthalpy of formation of the former and determination of the enthalpies of formation of several ZrH_{2-x} compounds and extrapolation to a stoichiometry of ZrH_2 gives that for the stoichiometric compound where (s) indicates solid Zr. This method of calorimetry is relatively straight-forward because solution calorimeters are available commercially and high precision pressure gauges are not needed. The disadvantages are that the metal hydride must be quite stable, i.e., can be handled without loosing hydrogen, and it is laborious because a given sample composition MH_x must be prepared for each individual determination.

3.3.3. *Thermal Analysis (DSC and DTA)*

Since the advent of modern thermal analysis equipment, this technique has been employed for studies of gas–metal systems. It can be employed by using samples of fixed hydrogen content or else by either hydriding or dehydriding the sample *in situ*. The former is restricted to samples and temperatures where the hydrogen loss is nil. This method has been extensively employed for phase diagram determination, e.g., in a recent study Kronski and Schober [54] determined part of the phase diagram of ZrNi–H using DTA (Differential Thermal Analysis). Heat capacities due to the interstitial species can be conveniently determined using thermal analytic methods. Thus the heat capacity of VH_r and VD_r was determined in this way by Meuffels [55]. Thermal analysis is also well-suited for the determination of the enthalpy and entropy of phase changes [56]. Craft *et al* [57] have employed DSC (Differential Scanning Calorimetry) to determine the enthalpy and entropy changes for the triple point change in NbH_r and reconciled the values as a function of r with the thermodynamic parameters for the related gas phase processes.

It seems that the utilization of thermal analysis to measure reaction calorimetric changes upon the absorption or desorption of hydrogen have not been very successful; relative integral enthalpies are measured which, in themselves, are not very useful unless many values are measured as a function of r. In any case, the H contents are not directly measureable after the loss or gain of hydrogen due to reaction within the thermal analytical cell. Since both relative partial molar and plateau enthalpies can be measured quite accurately with twin-cell heat leak calorimeters under conditions where the hydrogen content is known, thermal analytical methods are not recommended for this purpose.

3.4. SOURCES OF ERROR IN P-C-T STUDIES

3.4.1. *Thermal Transpiration*

If temperatures of the sample-containing reaction vessel and the pressure measuring gauge differ, which is the usual experimental situation, and if the pressure is below about 15 Pa and the connecting tubing between the two different regions is of small diameter, corrections must be made for thermal transpiration. The corrections can be made by employing various equations which are not completely accurate in regions where the mean free path of the molecules becomes comparable to the diameter of the tubing. Wallbank and McQuillan [58] have discussed earlier equations and have proposed a semi-empirical equation which should be applicable even in the transition region between Knudsen and viscous flow. Another approach is to carry out the measurements at various diameters of the connecting tubing and to extrapolate to a large diameter where the correction vanishes. The errors introduced by thermal transpiration have often been neglected when the reaction vessel containing the sample is at an elevated, rather than a low, temperature.

3.4.2. *van't Hoff Plots*

In the dilute phase region of activated intermetallic compounds there is non–Sieverts' behavior due to trapping of hydrogen by defects introduced by the activation process [59] (Fig. 13). This makes the p-c-T method difficult in this region of hydrogen contents and consequently, reaction calorimetry is the preferred method.

In the plateau region the pressures frequently increase as the fraction of hydride phase increases during hydride formation and decrease as the fraction of dilute phase increases during decomposition, i.e., the plateaux slope. For the determination of the thermodynamic parameters, the usual method of dealing with this sloping is to evaluate the pressures at a given r value at the various temperatures; in this way it is believed that the effect of the sloping is minimized.

3.5. HYSTERESIS

The plateau pressure for hydride formation, p_f, is greater than that for decomposition, p_d, i.e., $p_f > p_d$, because of hysteresis. This introduces some ambiguities into the determination of thermodynamic properties for the reaction which occurs in the plateau region. Hysteresis mandates irreversibility whereas the p-c-T method requires reversibilty, i.e., equilibrium.

Either $\ln p_f$ and/or $\ln p_d$ may be plotted against 1/T. Because of hysteresis, the two van't Hoff plots will have different slopes and the decomposition pressure plot is generally steeper than the formation plot. For systems where the hysteresis changes markedly over the temperature interval comprising the van't Hoff plots, it is clear that considerable error may be introduced into the thermodynamic parameters. It seems that generally the magnitudes of the true thermodynamic parameters must lie between the magnitudes of the enthalpies obtained from the p_f and p_d van't Hoff plots.

Often the plateau pressures, and consequently hysteresis, vary with cycling, i.e., with the number of times the hydride phase has been formed and decomposed without an intermediate annealing treatment. This must be allowed for and data must not be obtained until reproducible behavior obtains during the cycling.

Thermodynamic parameters for the plateau region obtained from data over a narrow temperature range, cannot be assumed to be reliable. This is true for the van't Hoff plot for any chemical reaction but it is especially true for gas-metal systems where complicating factors such as hysteresis and cycling occur.

References

1. C. Hoitsema, *Z. phys. Chem.*, **17** (1895) 1.
2. A. Sieverts, J. Hagenacker, *Z. phys. Chem.*, **68** (1909) 115.
3. J.D. Fast, *Interaction of Metals and Gases* Vol. 1, Academic Press, N. Y. 1965.
4. E. Fromm, E. Gebhardt, *Gase und Kohlenstoff in Metallen*, Springer–Verlag, Berlin, 1976.
5. T. Kuji, W. Oates, B. Bowerman, T. Flanagan, J..Phys. F: Met. Phys., **13** (1983) 1785.
6. S. Kishimoto, T. Flanagan, G. Biehl, in *Chemical Metallurgy–A Tribute to Carl Wagner*, N. Gokcen, ed., The Metallurgical Society, AIME, Warrendale, PA, 1981, p. 481.
7. R. Lässer, L. Powell, *Phys. Rev.*, **B34** (1986) 578.
8. R. McLellan, W. Oates, *Acta. Met.*, **21** (1973) 181.
9. J.-M. Welter, G. Arnold, H. Wenzl, *J. Phys.F: Met. Phys.*,**13** (1983) 1773.
10. T. Kuji, W. Oates, *J. Less-Common Mets.*, **102** (1984) 251.
11. W. Luo, T. Kuji, J. Clewley, T. Flanagan, *J. Chem. Phys.*, **94** (1991) 6179.
12. E. Veleckis, R. Edwards, *J. Phys. Chem.*, **73** (1969) 683.
13. M. Mrowietz, A. Weiss, *Ber. Bunsenges Physik. Chem.*, **89** (1985) 62.
14. P. Dantzer, W. Luo, T. Flanagan, J. Clewley, *Met. Trans. A.*, **24A** (1993) 1471.
15. T. Flanagan, N. Mason, B. Bowerman, *J. Less-Common Mets.*, **91** (1983) 107.
16. R. Kirchheim, *Prog. in Materials Sci.*, **32** (1988) 261.
17. G. Alefeld, *Ber. Bunsenges. Physik. Chem.*, **76** (1972) 746.
18. P. Rudman, *Int. J. Hydrogen Energy*, **3** (1978) 431.
19. W. Oates, T. Flanagan, *J. Mat. Sci.*, **16** (1981) 3235.
20. O. T. Sørenson, ed., *Non-Stoichiometric Oxides*, Academic Press, N.Y. (1981).
21. R. Penzhorn, M. Devillers, M. Sirch, *J. Nucl. Materals.*, **170** (1990) 217.
22. J. Condon, *J. Chem. Thermodyn.*, (1980) 1069.
23. J. Reilly, R. Wiswall, *Inorg. Chem.*, **13** (1974) 218.
24. H. van Mal, K. Buschow, A. Miedema, *J. Less-Common Mets.*, **35** (1975) 65.
25. Y. Fukai, *The Metal-Hydrogen System*, Springer-Verlag, Berlin, 1993.
26. R. van Essen, K. Buschow, *Mat. Res. Bull.*, **15** (1980) 1149.
27. W. Luo, A. Craft, T. Kuji, H. Chung, T. Flanagan, *J. Less-Common Mets.*, **162** (1990) 251.
28. S. Luo, J. Clewley, T. Flanagan, *unpublished data.*
29. B. Hyde, D. Bevan, L. Eyring, *Phil. Trans. Roy. Soc. A, London*, **259** (1966) 583.
30. W. Luo, J. Clewley, T. Flanagan, *Zeit. Physik. Chem.*, **179** (1993) 77.
31. H. Kierstead, *J. Less-Common Mets.*, **70** (1980) 199.
32. T. Flanagan, J. Clewley, N. Mason, H. Chung, *J. Less-Common Mets.*, **130** (1987) 309.
33. H. Noh, T. Flanagan, *to be published.*
34. J. Kirkwood, I. Oppenheim, *Chemical Thermodynamics*, McGraw-Hill, New York, 1961.
35. N. Boes, H. Züchner, *Ber. Bunsenges Physik. Chem.*, **80** (1976) 22.
36. D. Everett, P. Nordon, *Proc. Roy. Soc. London, Ser. A*, **259** (1960) 341.

37. J. Clewley, T. Curran, T. Flanagan, W. Oates, *J.C.S. Faraday Trans.I,* **69** (1973) 449.
38. S. Ramaprabhu, R. Leiberich, A. Weiss, *Zeit. Physik. Chem. N.F.,* **161** (1989) 83.
39. R. Kadel, A. Weiss, *Ber. Bunsenges Physik. Chem.,* **82** (1978) 1290.
40. R. Lässer, K.-H. Klatt, *Phys. Rev.,* **B28** (1983) 748.
41. M. Benham, K. Grose, D. Ross, *J. Less-Common Mets.,* **130** (1987) 541.
42. F. Lewis, *The Palladium/Hydrogen System*, Academic Press, New York, 1967.
43. F. Lewis, *Plat. Met. Rev.,* **26** (1982) 20,70, 121.
44. P. Favre, *Comtes Rendes.,* **78** (1874) 1257
45. P. Favre, *Comtes Rendus.,* **68** (1869) 1520.
46. L. Mond, W. Ramsay, J. Shields, *Trans. Roy. Soc. London,* **A191** (1898) 105.
47. T. Flanagan, W. Luo, J. Clewley, *J. Less–Common Mets.,* **172-174** (1991) 42.
48. J. Murray, M. Post, J. Taylor, *J. Less-Common Mets.,* **73** (1980) 33.
49. D. Nace, J. Aston, *J. Am. Chem. Soc.,* **79** (1957) 3619.
50. G. Boureau, O. Kleppa, *J. Chem. Thermodynamics,* **5** (1977) 543.
51. P. Dantzer, A. Guillot, *J. Phys. E: Sci. Instrum.,* **15** (1982) 1373.
52. H. Inaba, A. Navrotsky, L. Eyring, *J. Solid State Chemistry* **37** (1981) 67.
53. A. Turnbull, *Aust. J. Chem.,* **17** (1964) 1063.
54. R. Kronski, T. Schober, *J. Alloys and Compounds.,* **205** (1994) 175.
55. P. Meuffels. *PhD dissertation*, Aachen, 1982.
56. K. Watanabe, Y. Fukai, *J. Phys. Soc. Japn.,* **54** (1985) 3415.
57. A. Craft, T. Kuji, T. Flanagan, *J. Phys. F: Met. Phys.,* **18** (1988) 1149.
58. A. Wallbank, A. McQuillan, *J.C.S. Faraday Trans. I.,* **71** (1975) 685.
59. T. Flanagan, C. Wulff, B. Bowerman, *J. Solid State Chem.,* **34** (1980) 215.

Chapter 5

A SURVEY OF BINARY AND TERNARY METAL HYDRIDE SYSTEMS

A. PERCHERON GUEGAN
Laboratoire de Chimie Métallurgique et Spectroscopie des Terres Rares
Centre National de la Recherche Scientifique
1 Place Aristide Briand
F-92195 Meudon Cedex
France

1. Introduction

The aim of this chapter is to provide a brief review of the fundamental properties of the major families of metal hydrides, hydrides whose potential applications have been studied worldwide for the past thirty years. After the publication of the classic book *Metal Hydrides* in 1968 [1], a large portion of metal hydride research has been presented, since 1972, at the "International Conference on Hydrogen in Metals" and, since 1977, at the "International Symposium on the Properties and Applications of Metal Hydrides." These two conferences merged in 1988 when they became the conference on "Metal-Hydrogen Systems—Fundamentals and Applications." The proceedings are published as special issues of the *Journal of Less-Common Metals* [2] and *Zeitschrift für Physikalische Chemie* [3, 4].

The phenomena related to the physics of hydrogen in metals, ranging from dilute solid solutions to the binary hydrides, AH_x, are reviewed in *Hydrogen in Metals* [5]. More recently, the different properties of ternary metal hydrides, ABH_y, are presented in *Hydrogen in Intermetallic Compounds* [6]. These books give a thorough description of the various ternary hydrides, whose fascinating properties have been known for the last thirty years. Finally, *The Metal Hydrogen System* [7] provides a coherent and consistent description of the properties of the hydrides, ranging from macroscopic properties, such as solubilities and phase diagrams, to microscopic properties, such as atomic states and diffusion.

In view of the coverage provided in the above books, this chapter will not attempt to survey all aspects of binary and ternary metal hydrides but, rather will concentrate on the thermodynamic and structural properties of the most representative systems, especially those with potential applications. This chapter will thus, first briefly classify the hydrides and discuss their formation and structures. Second, this chapter will discuss the properties of the palladium-hydrogen system as an example of a binary metal hydride. This will be followed by a description of some intermetallic ternary hydrides. The changes in the stability and hydrogen content of the binary hydrides after partial substitution of the metal is very useful in tuning the characteristics of a system to the particular requirements of a given application, as will be illustrated for palladium and the AB_5 intermetallic compounds. Finally, semiempirical models which relate the hydriding characteristics, such as stability and hydrogen content, to other physical properties of the parent metal, such as the enthalpy of formation, unit cell volume, interstitial site size, and electronic structure, will be briefly reviewed.

F. Grandjean et al. (eds.), Interstitial Intermetallic Alloys, 77–105.

2. Formation of Hydrides

2. 1. CLASSIFICATION OF HYDRIDES

Metals which dissolve hydrogen can be divided into two classes. The first class contains those metals which react exothermically to form a hydride phase by the direct combination of metal and hydrogen. This class includes the alkali and alkaline-earth metals, the titanium and vanadium subgroups, the rare earths, the actinides, and palladium. The second class contains metals, such as iron, cobalt, nickel, copper, silver, and platinum, which dissolve hydrogen endothermically. The solubility of hydrogen in this second class is considerably smaller at room temperature than that in the exothermic class. Within the two classes, metal hydrides are classified by the nature of the metal hydrogen bond [6] as either covalent, ionic, or metallic. The binary hydrides formed by metals are shown in the periodic table of Figure 1. In this chapter we will only consider the metal hydrides formed by transition metals in the first class.

2. 2. THERMODYNAMIC PROPERTIES

An exothermic and reversible hydrogenation reaction may be written as

$$2A + xH_2 \leftrightarrow 2AH_x.$$

Saline: IA	Saline: IIA	Metallic: IIIA	Metallic: IVA	Metallic: VA	Metallic: VIA	Metallic, Complex Covalent: VIIA	Metallic, Complex Covalent: VIIIA	Metallic, Complex Covalent: VIIIA	Metallic: VIIIA	Covalent: IB	Covalent: IIB	Covalent: IIIB	Covalent: IVB
LiH	$(BeH_2)_n$ Covalent											Series of Boron Hydrides	
NaH	MgH_2											$(AlH_3)_n$	Series of Si Hydrides
KH	CaH_2	ScH_2	TiH_2 (cubic and tetrag.)	VH VH_2	CrH CrH_2	Mn -	Fe -	Co -	NiH	CuH	$(ZnH_2)_n$	$(GaH_3)_n$	Series of Ge Hydrides
RbH	SrH_2	YH_2 YH_3	ZrH_2 (cubic and tetrag.)	NbH NbH_2	Mo -	Tc -	Ru -	Rh -	PdH	Ag -	$(CdH_2)_n$	$(InH)_n$ $(InH_3)_n$	SnH_4 Sn_2H_6
CsH	BaH_2	See Rare Earth Series	HfH_2 (cubic and tetrag.)	TaH	W -	Re -	Os -	Ir -	Pt -	Au -	$(HgH_2)_n$	$(TlH)_n$ $(TlH_3)_n$	PbH_4

LaH_{2-3}	CeH_{2-3}	PrH_{2-3}	NdH_{2-3}	Pm ?	SmH_2 SmH_3	EuH_2	GdH_2 GdH_3	TmH_2 TmH_3	DyH_2 DyH_3	HoH_2 HoH_3	ErH_2 ErH_3	TmH_2 TmH_3	YbH_2 YbH_3 (?)	LuH_2 LuH_3

AcH_2	ThH_2 Th_4H_{15}	PaH_3	UH_3	NpH_2 NpH_3	PuH_2 PuH_3	AmH_2 AmH_3 (?)

Figure 1. Classification of the binary hydrides formed by the metals [1].

To obtain the same exothermic reversible reaction with an intermetallic compound,

$$2AB + xH_2 \leftrightarrow 2ABH_x,$$

it is necessary that at least one of the metal forms a stable metal hydride.

A measurement of the equilibrium hydrogen pressure as a function of hydrogen content and temperature is the most useful method for determining the thermodynamic properties of a metal hydride. An ideal pressure-composition-temperature diagram is shown in Figure 2 for a single hydride phase. In this diagram, three different regions may be distinguished. Two of these regions, in which the pressure changes with the hydrogen content, because of the Gibbs' phase rule, correspond to the solution of hydrogen in the metal α-phase and in the metal hydride phase β-phase. The third region, with the constant pressure plateaus, corresponds to the coexistence of the hydrogen saturated metal α-phase and the hydrogen deficient metal hydride β-phase. The plateau pressure and the hydrogen content at this pressure, the so-called capacity, are the two essential characteristics of this system. The plateau pressure changes rapidly with temperature, following the Van't Hoff rule which may be written as,

$$\log p_{H_2} = (\Delta H/RT) - (\Delta S/R),$$

where ΔH and ΔS are the changes in enthalpy and entropy for the hydrogenation reaction. By assuming that ΔH and ΔS do not change over a small range of temperature, it is possible to determine their values by measuring the isotherms at different temperatures and plotting the log p_{H_2} versus the inverse temperature, as is shown to the right of Figure 2. With increasing temperature, the width of the plateau decreases with a concurrent increase in the regions of the α and β phases. Above a critical temperature, T_c, there is a continuous solution of hydrogen in the metal and a disappearance of the two phases.

The actual behavior of most hydride forming materials deviates from the ideal description [8] because of two phenomena, the plateau slope and hysteresis, as is shown in Figure 3. The plateau slope is essentially due to compositional inhomogeneities in the parent compounds and can be minimized through an appropriate annealing treatment [9]. The hysteresis results because the desorption plateau pressure, P_d, is smaller than the absorption plateau pressure, P_a. The hysteresis is given by the ratio of the two pressures, as is indicated in Figure 3. The thermodynamic origins of the hysteresis pressures are discussed in detail in reference 10.

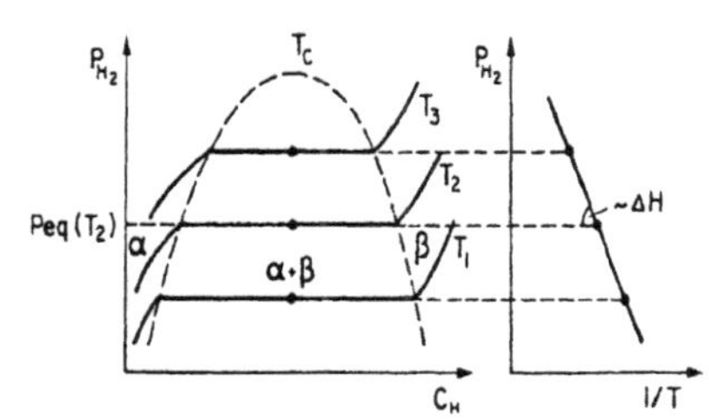

Figure 2. The pressure-composition isotherms [6] for the solid solution of hydrogen, the metal α-phase and the metal hydride β-phase. The region of coexistence of the two phases, characterized by the plateau equilibrium pressures, p_{eq}, terminates at the critical temperature, T_c.

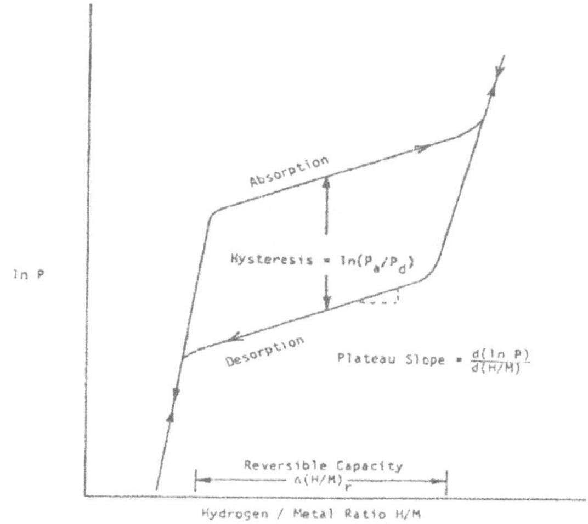

Figure 3. A hypothetical pressure-composition loop [8] with definitions of the hysteresis, plateau slope, and reversible capacity.

The pressure-composition-temperature diagram can be measured by several different methods. The *volumetric* method involves the measurement of the hydrogen pressure change as a function of time in a calibrated volume maintained at a constant temperature. The *gravimetric* method involves the measurement of the change in weight of a sample, maintained at constant temperature, as a function of time. Physical methods involve the measurement of particular properties, such as the magnetic susceptibility or the electrical resistivity, as a function of both hydrogen absorption and temperature.

2. 3. STRUCTURAL PROPERTIES

Binary and ternary metal hydrides form by filling the interstices of the host metallic lattice with hydrogen atoms. The amount of hydrogen is, approximately one hydrogen atom per metal atom, a hydrogen concentration per unit volume which is often higher than that of liquid hydrogen. The major factor limiting hydrogen capacity is the filling of all available interstices in the metal framework. Intersite spacing is an important factor in determining the maximum hydrogen concentration in crystalline metals and intermetallic compounds. Switendick [11] has stated that, when the addition of hydrogen to a metal introduces new states below the Fermi level, the electronic properties are very sensitive to the hydrogen-hydrogen distance. From a survey of the stable transition metal hydrides, it seems that the minimum hydrogen-hydrogen distance is about 2.1Å. Westlake [12], in a critical review of the qualitative and quantitative models used to rationalize the observed hydrogen atom site occupancies, has developed an approach, based on geometric considerations, which predicts the stoichiometry and the preferred

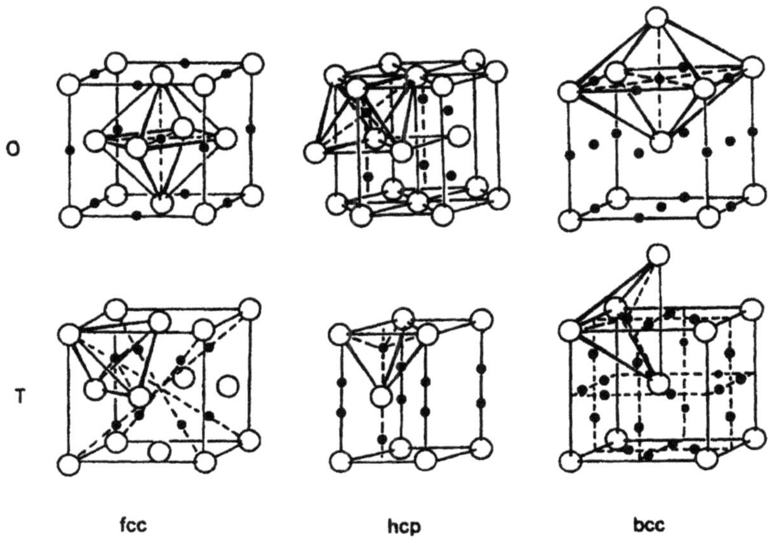

Figure 4. The interstitial octahedral sites, O, and tetrahedral sites, T, in the fcc, hcp, and bcc metallic lattices [7].

occupation of the interstitial sites by hydrogen. This approach uses two criteria, (i) the minimum radius of an interstitial site to be filled is 0.4Å, and (ii) the minimum distance between two hydrogen atoms is 2.1Å. On the atomic scale, it is schematically possible to illustrate the typical structure of binary hydride by showing the three principal metal crystal structures, fcc, hcp, and bcc. In practice, only two types of interstitial sites, octahedral, O, and tetrahedral, T, are occupied by hydrogen in the binary and ternary hydrides, as is shown in Figure 4. The size and number of interstitial sites per metal atom are given in Table 1.

In the fcc lattice, a T or O interstitial site is surrounded by a regular tetrahedron or octahedron of metal atoms. In the hcp lattice, the polyhedra formed by the metal atoms distort as the c/a ratio deviates from the ideal value of 1.633. In the bcc lattice, an O site is surrounded by a highly distorted octahedron which has two metal atoms at a much shorter distance than the remaining four metal atoms. Thus, the O sites in the bcc lattice are further divided into O_x, O_y, and O_z sites, depending upon the direction of the four-fold symmetry axis of the polyhedron. Similarly, the T sites are divided into T_x, T_y, and T_z sites.

Table 1. The number and size of interstitial sites in different metallic lattices.

Structure	fcc	fcc	hcp[a]	hcp[a]	bcc	bcc
Site geometry	O	T	O	T	O	T
Number per metal atom	1	2	1	2	3	6
Size[b]	0.414	0.225	0.414	0.225	0.155	0.291

[a]For a c/a ratio of 1.633. [b]The maximum radius of a sphere accomodated in an interstitial space in units of the metal atom radius.

Macroscopically, the absorption of hydrogen gas leads to a large expansion of the metal hydride unit cell volume. Consequently, intermetallic compounds which are brittle break into small particles, upn hydrogenation, a phenomenon which is often called decrepitation. The resulting fine powders exhibit a surface area in the range of 0.2 to $2m^2/g$ depending upon the nature of the intermetallic compound.

The structure of metal hydrides and the positions of the hydrogen atoms are determined by X-ray and neutron diffraction experiments. In most cases, neutron diffraction experiments are performed on deuterides, rather than on hydrides, because the coherent scattering cross section of deuterium is much larger than that of hydrogen and because the scattering of hydrogen is incoherent.

3. The Palladium-Hydrogen System: An Example of a Binary Metal Hydride

Since 1866, when T. Graham [13] discovered that palladium could absorb large amounts of hydrogen, the resulting palladium hydride has been extensively studied and is now one of the most interesting model systems from the structural and thermodynamic point of view [14,15].

The most complete phase diagram of the palladium-hydrogen system, shown in Figure 5, has been determined both by gravimetric methods and by magnetic susceptibility measurements [16]. At room temperature the limit of the saturated α-phase is ca. 0.008 hydrogen atoms per palladium. Above this limit, the β-phase forms with a minimum of 0.61 hydrogen atoms per palladium. The plateau pressure in the two phase region increases with increasing temperature according to the Van't Hoff rule. The enthalpy of desorption is 41,000±400 kJ/mol of H_2 and the entropy of desorption is 97.5±0.8 kJ/K mol of H_2, between 20 and 300°C. The two phase region disappears above a critical temperature of 298°C.

At room temperature, upon the absorption of hydrogen, the fcc palladium lattice parameter increases from 3.891 to 3.894Å for the saturated $PdH_{0.008}$ α-phase. Upon further absorption of hydrogen, the fcc structure of palladium is retained with a lattice parameter of 4.025Å for $PdH_{0.61}$. This corresponds to a 10.7 percent increase in volume. The hydrogen atoms occupy octahedral sites, see Figure 6, as has been shown by neutron diffraction studies [17] on the $PdH_{0.706}$ and $PdD_{0.658}$ β-phases.

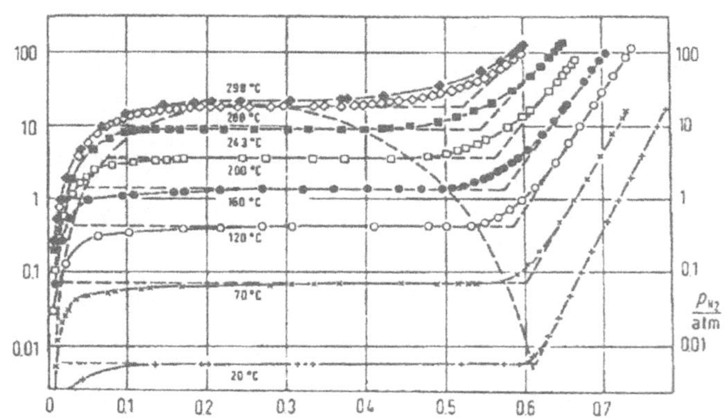

Figure 5. The palladium-hydrogen isotherms for bulk palladium at several different temperatures [14].

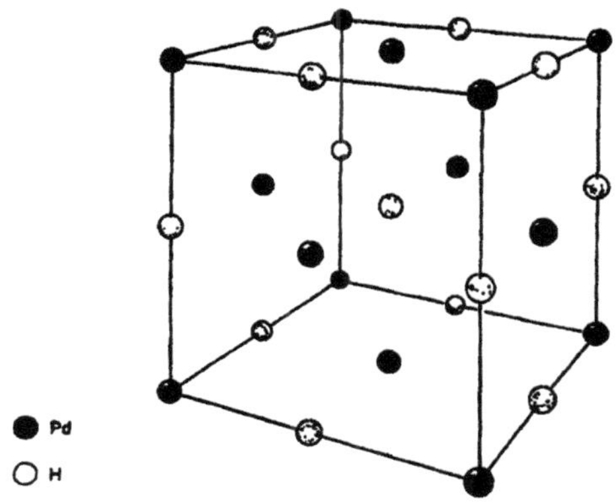

Figure 6. A schematic diagram showing the positions of hydrogen and deuterium in the Pd-H and Pd-D β-phase.

4. Ternary Hydrides

The first true ternary hydride was discovered by Libowitz et al. [18] in 1958. They showed that the intermetallic compound, ZrNi, forms exothermically and reversibly $ZrNiH_3$. They found that the dissociation pressure was intermediate between that of the respective binary hydrides, the very stable ZrH_2 and the very unstable NiH. Although this discovery considerably expanded the world of binary hydrides, the potential of intermetallic hydrides was not fully appreciated until the 1970's when groups at the Philips Research Laboratories in the Netherlands and at Brookhaven National Laboratory in the US discovered the first examples of several families of intermetallic hydrides based on $LaNi_5$ [19], TiFe [20], and Mg_2Ni [21]. These compounds are readily and reversibly formed with a high hydrogen concentration at room temperature under a modest hydrogen pressure. Several years later, hydrides of the AB_2-type Laves phases were independently discovered by Shaltiel et al. [22] and Ishido et al. [23].

Over the past two decades, extensive experimental work on the properties of these and isostructural compounds has been carried out by changing the elemental constituents through partial metal substitution. The following sections will concentrate on crystalline materials and be restricted to a discussion of four major families of intermetallic compounds, namely the AB_5, AB_2, AB, and A_2B families of compounds, compounds which may be represented by $LaNi_5$, $TiCr_2$, TiFe, and Mg_2Ni. In this notation, A may be an alkaline earth, an early transition metal, a rare-earth metal, or an actinide which forms a stable hydride and B may be a late transition metal which forms an unstable hydride.

4. 1. THERMODYNAMIC PROPERTIES

Isotherms with plateau pressures in the range of 0.01 to 1MPa are shown in Figure 7 for the representative compounds. It should be noted that there is a wide temperature range, between -90 and 300°C, with similar dissociation pressures. TiFe and $TiCr_{1.8}$ show a second plateau corresponding to the formation of a second hydride, $TiFeH_2$ and $TiCr_{1.8}H_{3.6}$, respectively. Thermodynamic properties [24-27] are summarized in Table 2.

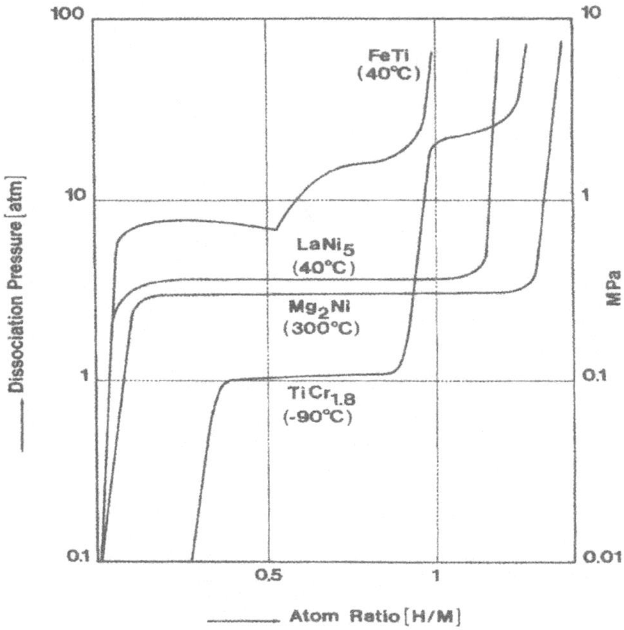

Figure 7. A comparison of the desorption isotherms of representative intermetallic compounds at different temperatures.

Table 2. A comparison of the characteristics of several representative ternary hydrides.

Ternary hydride	$LaNi_5H_{6.7}$	$FeTiH_{1.8}$	$TiCr_{1.9}H_{3.6}$	Mg_2NiH_4
Hydrogen to metal ratio	1.1	0.9	1.28	1.3
Weight capacity, %	1.5	1.8	2.5	3.8
Hydrogen volume capacity, g/dm^3	140	95	126	96
Equilibrium pressure at 25°C, MPa	1.07	1	7	10^{-4}
Temperature for 0.1MPa H_2, °C	10	0	-60	250
ΔH of plateau desorption, kJ/mol	-31.8	-33.8	-26.2[a]	-64
Reference	24	25	26	27

[a]Value measured for $TiCr_{1.9}H_{2.5}$.

Van't Hoff plots for several metallic and intermetallic hydrides are shown in Figure 8. A line connecting 25°C and the dissociation pressure of 1atm or 0.1MPa may be used to classify the hydrides as unstable or stable above this pressure. It is noteworthy that the stable compounds on the left side of this line in Figure 8 exhibit a high hydrogen weight capacity, whereas the unstable compounds on the right side of this line have a small hydrogen weight capacity. In contrast, the volume capacity shows the opposite dependence. A schematic diagram, see Figure 9, gives a comparison of the density of hydrogen in different intermetallic compounds, with that of compressed hydrogen and liquid hydrogen.

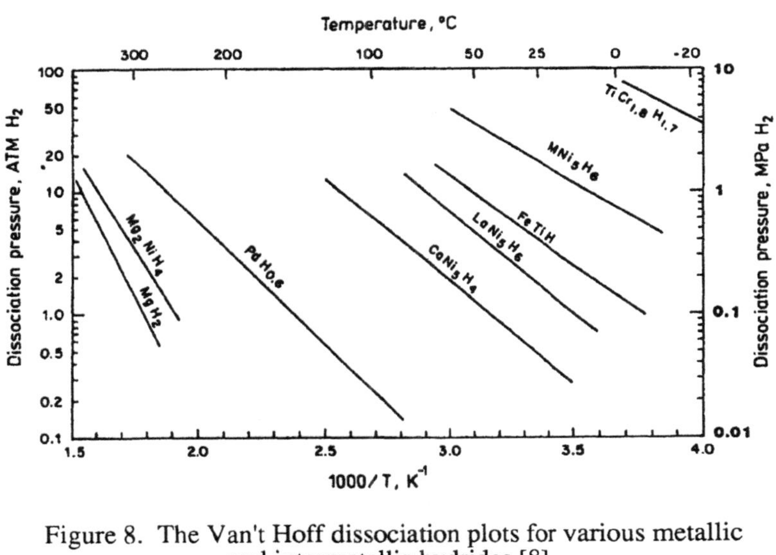

Figure 8. The Van't Hoff dissociation plots for various metallic and intermetallic hydrides [8].

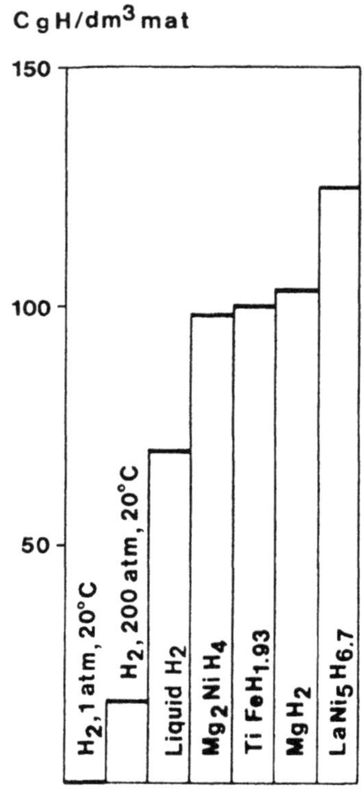

Figure 9. A comparison of the hydrogen density in several representative materials.

4. 2. STRUCTURAL PROPERTIES

As in the case of the binary hydrides, ternary intermetallic hydrides form by filling the interstitial sites in the metallic lattice according to the general rules given in Section 2.3. Preferential hydrogen site occupancies may be interpreted by assuming both that hydrogen prefers to occupy large interstitial sites and that there is a correlation between hydrogens on the occupied sites. As in the binary hydrides, only octahedral and tetrahedral interstitial sites are occupied by hydrogen. Ternary metal hydrides show a variety of complex structures when compared with metal hydrides which are characterized by highly symmetric metal arrangements. With the exception of the usual expansion and distortion, most structures are similar to those of the corresponding parent intermetallic compound.

An extensive review of ternary metal hydrides has been published by Yvon et al. [28] and the structure of some representative compounds are listed in Table 3. Most structural studies used a Rietveld line profile analysis of powder neutron diffraction results [29] to obtain the lattice parameters, atomic positional parameters for the metal and hydrogen or deuterium atoms, hydrogen or deuterium site occupancies, and the amplitude of thermal vibration. The complexity of the crystal structure, the large number

Table 3. Ternary metal deuterides studied by neutron diffraction.

Compound	Structure type	Space group	Lattice parameters	Occupied interstices	Ref.
β-$LaNi_5D_{5.5}$, 4bar	$CaCu_5$	*P31m*	a=5.336Å c=4.259Å	$LaNi_3$, La_2Ni_4, La_2Ni_2	30
β-$LaNi_{5.12}D_{6.7}$, 5atm	$CaCu_5$	*P6/mmm*	a=5.399Å c=4.290Å	$LaNi_3$, La_2Ni_4, La_2Ni_2, Ni_4	31
β-$LaNi_5D_{5.0}$, 363K, 24bar	$CaCu_5$	*P6₃mc*	a=5.395Å c=8.484Å	$LaNi_3$, La_2Ni_2	32
β-$LaNi_5D_{6.7}$, 19.5bar	$CaCu_5$	*P6₃mc*	a=5.412Å c=8.599Å	$LaNi_3$, La_2Ni_2, Ni_4	32
β-$LaNi_5D_6$, 276K	$CaCu_5$	*P6₃mc*	a=5.388Å c=8.559Å	$LaNi_3$, La_2Ni_4, La_2Ni_2, Ni_4	33
γ-$TiFeD_{1.9}$	CsCl	*P2/m*	a=4.704Å b=2.830Å c=4.704Å β=96.97°	Ti_4Fe_2, Ti_2Fe_4	34
γ-$TiFeD_{1.8}$, 300K	CsCl	*P2/m*	a=4.708Å b=4.697Å c=2.835Å γ=97.05°	Ti_4Fe_2, Ti_2Fe_4	35
γ-$TiFeD_{1.94}$, 113bar	CsCl	*Cmmm*	a=7.029Å b=6.233Å c=2.835Å	Ti_4Fe_2, Ti_2Fe_4	36
β-Mg_2NiD_4	CaF_2	*C2/m*	a=6.496Å b=6.412Å c=6.602Å β=93.2°	Mg_4Ni	37
β-Mg_2NiD_4	CaF_2	*Ia*	a=13.197Å b=6.403Å c=6.489Å β=93.23°	Not specified	38
β-Mg_2NiD_4	CaF_2	*C2/c*	a=14.342Å b=6.404Å c=6.483Å β=113.52°	Mg_4Ni	39
$ZrMn_2D_{3.0}$	$MgZn_2$	*P6₃/mmc*	a=5.391Å c=8.748Å	Zr_2Mn_2	40
$ZrV_2D_{4.9}$	$MgCu_2$	*Fd3m*	a=7.913Å	Zr_2V_2, ZrV_3	41

of atoms, the poor crystallinity of the materials, and anisotropic line broadening due to microstrains, are the limiting factors in the determination of the structures of ternary intermetallic hydrides [30-41]. These complexities account for the different results obtained for the structure of β-Mg_2NiD_4 and given in Table 3. To the author's knowledge, there is no structural determination on a $TiCr_2$ hydride and hence the rather unstable $ZrMn_2D_{3.0}$ and $ZrV_2D_{4.9}$ compounds have been used to illustrate the structure of the hexagonal and cubic AB_2 Laves phase hydrides.

Without going into full detail, the structures of the representative compounds will be briefly discussed. $LaNi_5$ crystallizes in the hexagonal $CaCu_5$ structure with the space group *P6/mmm*. Upon hydrogenation $LaNi_5$ forms the expanded β-hydride, $LaNi_5H_x$, with x greater than 5.8. Two different space groups have been proposed for this hydride, *P31mm* and *P6/mmm*. In the first structure, the deuterium is located in two different sites, as is shown in Figure 10. In the second structure, one of the deuterium atom is distributed between five different interstitial sites, which can be divided into two groups. Further studies, as a function of deuterium content indicated that a structural phase transformation resulting from the ordering of the deuterium atoms occurs [32] in β-$LaNi_5D_x$, when x is greater than 5. This structural transformation leads to the *$P6_3mc$* structure with a total of seven interstices, interstices which are partially occupied by deuterium.

Two different space groups, monoclinic ***P2m*** and orthorhombic ***Cmm***, have been proposed [34-36] for the structure of γ-$TiFeD_{1.94}$. As is shown in Figure 11, the structure of γ-$TiFeD_{1.94}$ has three types of octahedral metal interstices, two composed of Ti_4Fe_2 and occupied by deuterium atoms, D1 and D3, and one composed of Ti_2Fe_4 and occupied by deuterium atom, D2. The first two sites are 100 percent occupied whereas the third site is only 91 percent occupied. The shortest deuterium-deuterium interatomic distance is 2.35Å.

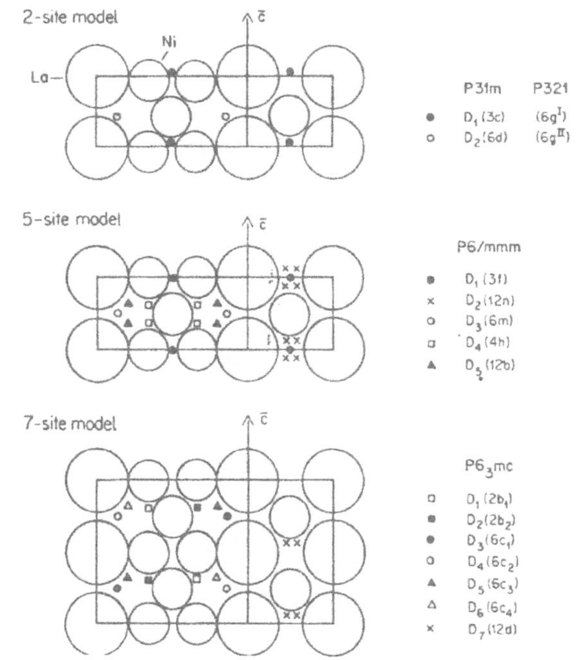

Figure 10. The deuterium atom sites for the three multisite models proposed [32] for the structure of β-$LaNi_5D_x$.

Several space groups have also been proposed for β-Mg_2NiD_4. In the monoclinic *C2m* structure, four deuterium atoms form a planar array around the nickel atoms [37]. However, high resolution neutron diffraction results indicated, in constrast with previous work, that the deuterium atoms form a tetrahedron around the nickel atoms, as is shown in Figure 11.

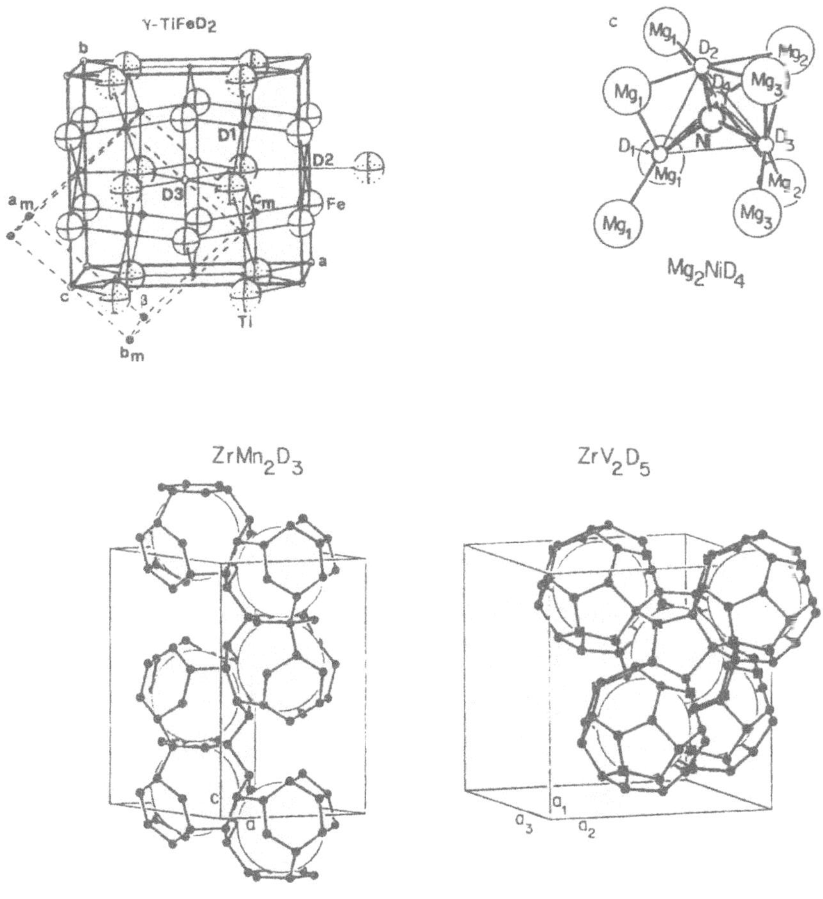

Figure 11. Upper left. The orthorhombic unit cell, thick lines, of γ-$TiFeD_{1.94}$ as compared with the monoclinic unit cell, broken lines [36]. The octahedral coordination bonds of the deuterium atoms are shown. Upper right. The transition metal environment in β-Mg_2NiD_4[39]. Bottom. The distribution of deuterium atom sites in hexagonal $ZrMn_2D_3$ and cubic ZrV_2D_5 [42]. The large points represent zirconium, the small points and square points represent deuterium. The vanadium and manganese atoms are omitted. The solid lines indicate possible diffusion paths in the structure.

At room temperature, the hexagonal C14 Laves phase, $ZrMn_2$, forms $ZrMn_2D_3$ in which the deuterium atoms occupy only one type of tetrahedral interstice, an interstice which is formed by two zirconium and two manganese atoms [40]. This interstice is partially occupied and provides a three dimensional infinite network of diffusion paths for the deuterium atoms, as is shown at the bottom of Figure 11. Each deuterium atom site in this network is approximately 1.3Å from at least two other deuterium atom sites. In contrast, in the cubic C15 Laves phase, $ZrV_2D_{4.5}$, the deuterium atoms occupy two different types of tetrahedral interstices; ZrV_3 interstices which have a deuterium occupancy of 39 percent and Zr_2V_2 interstices which have a deuterium occupancy of 25 percent. The adjacent deuterium atom sites in this network are approximately 1.3Å apart and the average zirconium-deuterium and vanadium-deuterium distances are 2.10 and 1.77Å, respectively [41, 42].

5. The Influence of Substitution in Palladium and Intermetallic Compounds

5.1. PALLADIUM ALLOYS

Numerous studies have shown the influence upon the stability and hydrogen content resulting from the alloying of metals with palladium [14, 43]. Figure 12 shows the 50°C isotherm of palladium [44] as compared with the isotherms of palladium alloyed with silver [45], lead [46] and rhodium [46]. With increasing silver content, the plateau pressure decreases with a concomitant decrease of the hydrogen content and above 40 percent silver disappears. The same decrease is observed for 5 percent of lead in palladium, whereas an increase is observed for 5 percent of rhodium in palladium.

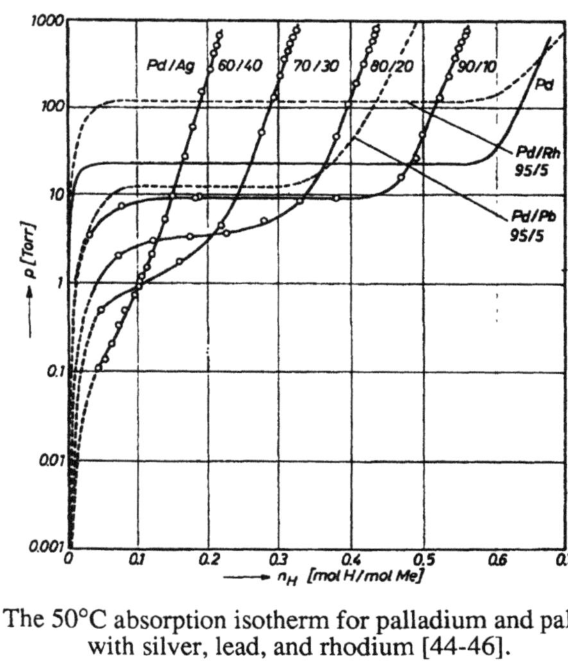

Figure 12. The 50°C absorption isotherm for palladium and palladium alloys with silver, lead, and rhodium [44-46].

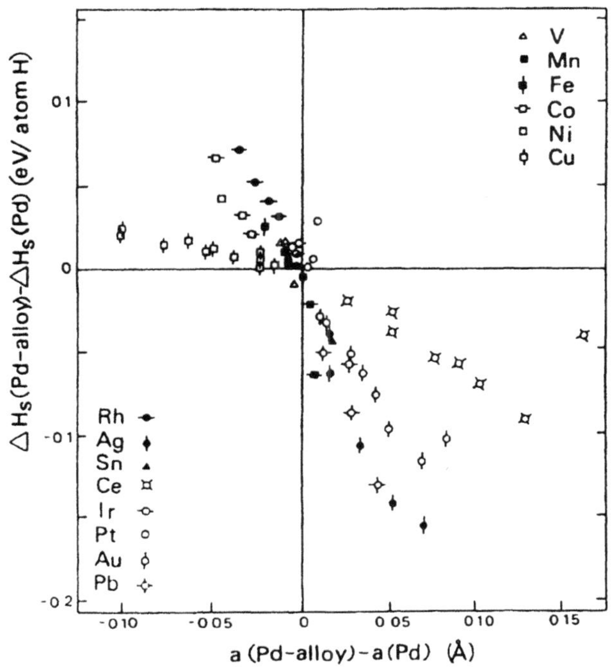

Figure 13. The heat of solution of hydrogen at infinite dilution in palladium alloys as a function of their lattice parameter [49].

Fukai [7] provided a geometric and electronic explanation for the above palladium alloy variations, a summary of which is given here. Good correlation exists between the heat of solution at infinite dilution, ΔH°_s, and the lattice parameters of the alloys. As is shown in Figure 13, the heat of solution becomes more negative with an increase in the alloy lattice parameter. In contrast, if the heat of solution at infinite dilution is plotted as a function of the average electron to atom ratio, e/a, for a series of palladium alloys [45-59], see Figure 14, no systematic variation is observed, except that several of the curves show a minimum for an e/a ratio in the range of 0.4 to 0.6. This corresponds to the point at which, according to the band filling model, the Fermi level is expected to move from the 4d band to the 5d band, a change which is excepted to lead to a more rapid increase of the electronic energy with the increasing addition of alloying elements. Thus, in the case of palladium based alloys, the volume is more important that the average electronic energy in determining the heat of solution. In addition, it is impossible to explain the observed variation in the heat of solution by adding the elastic and electronic contribution, i.e. for instance as $\Delta H^\circ_s = c_1(1/a)(da/dcs) + c_2(e/a)$, where c_1 and c_2 are arbitrary coefficients for a given host metal.

In conclusion, the attempt to correlate the change in the heat of solution upon alloying with the average of the elastic and electronic energies is only partially successful. Other mechanisms can be proposed by studying the entropy of solution at infinite dilution, ΔS°_s. Figure 15 indicates that, for a variety of alloying elements, ΔS°_s decreases linearly with increasing solute concentration. This suggests that the number of available interstitial sites decreases with increasing solute concentration because of the repulsive interactions between hydrogen and the alloying elements.

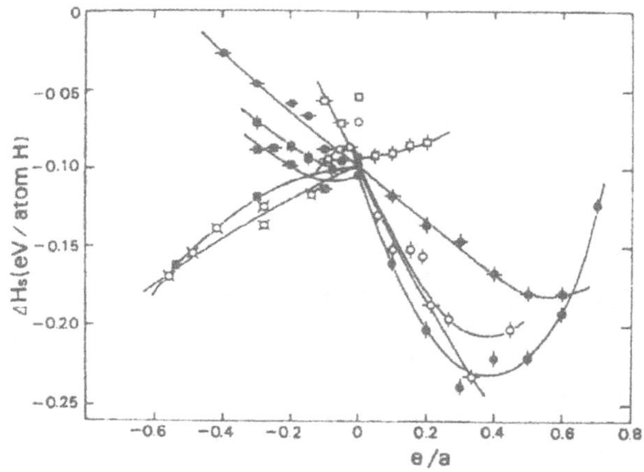

Figure 14. The heat of solution of hydrogen at infinite dilution in palladium alloys as a function of the electron to atom ratio, e/a, for a variety of alloying metals [45-58].

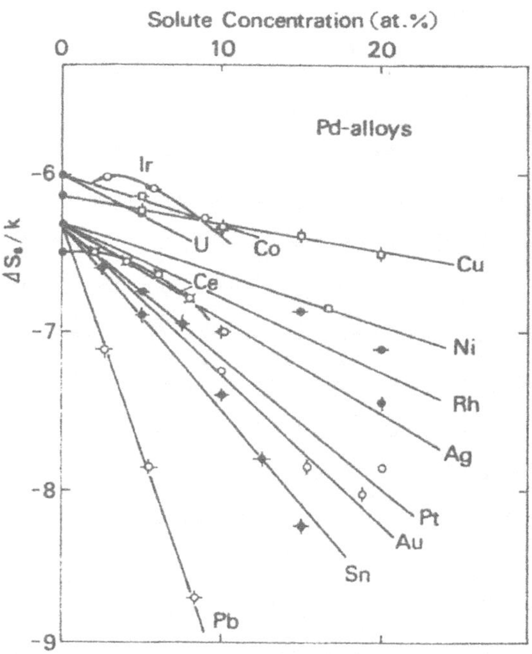

Figure 15. The entropy of solution of hydrogen at infinite dilution in palladium alloys as a function of solute concentration [45-59].

5. 2. $LaNi_5$ COMPOUNDS

The exceptional hydriding properties of $LaNi_5$ were first reported by Van Vucht et al. [19], who found at room temperature that $LaNi_5$ rapidly and reversibly absorbed more than six atoms of hydrogen per formula unit at a hydrogen pressure of approximately 2.5atm. Many pseudobinary compounds of the formula $La_{1-x}R_xNi_5$ and $LaNi_{5-x}M_x$ have subsequently been studied [60-64] with the double goal of both changing the stability of their hydrides and understanding which parameters play a role in the stability and hydrogen content of their hydrides.

5. 2. 1. *Parent Intermetallic Compounds.* Although the $La_{1-x}R_xNi_5$ solid solutions, where R represents a rare-earth, exist over an entire range of x values, the $LaNi_{5-x}M_x$ solid solutions do not necessarily all exist. In the latter case, the maximum value of x varies according to the substitutional element, M, and is 0.6 for silicon, 1.2 for iron, 1.3 for aluminum, 2.2 for manganese, and 5 for copper and cobalt. This maximum value depends on the heat treatment of the sample which has to be optimized for each family of alloys. For example, this maximum can be increased by lowering the annealing temperature in the case of $LaNi_{5-x}Mn_x$ [64] or by increasing it in the case of $LaNi_{5-x}Fe_x$ [62]. However, it seems that the atomic size and the electronic structure of the M atom are the main factors limiting the value of x.

All of the $La_{1-x}R_xNi_5$ and $LaNi_{5-x}M_x$ solid solutions crystallize in the hexagonal $CaCu_5$ structure with unit cell dimensions which vary with R and M. For a given solid solution, the unit cell volume varies linearly with x, and specifically it decreases in the $La_{1-x}R_xNi_5$ solid solutions, whereas it increases in the $LaNi_{5-x}M_x$ solid solutions. Figure 16 indicates that the change in unit cell volume induced by silicon in $LaNi_{5-x}Si_x$ is very small in contrast to the change induced by manganese in $LaNi_{5-x}Mn_x$.

The distribution of the substitutional atoms within the $LaNi_{5-x}M_x$ solid solutions has been studied by neutron diffraction and for the $LaNi_{5-x}Fe_x$ solid solutions, by Mössbauer spectroscopy. The partial replacement of nickel has been found to occur randomly, but mainly at z=1/2, the least dense atomic plane, as is shown in Figure 17. In the $LaNi_{5-x}Al_x$ solid solutions, for $x \leq 1$, and in $LaNi_{4.5}Si_{0.5}$, aluminum and silicon occupy only the 3g crystallographic sites [65, 66]. In contrast, some of the manganese, iron, and cobalt atoms also occupy the 2c basal plane crystallographic sites in $LaNi_{5-x}Mn_x$, with $x \leq 2$ [64], $LaNi_4Fe$ [62], and $LaNi_4Co$ [67]. An exception occurs for $LaNi_4Cu$ in which copper is found in both atomic planes, with a slight preference for the 2c site [66].

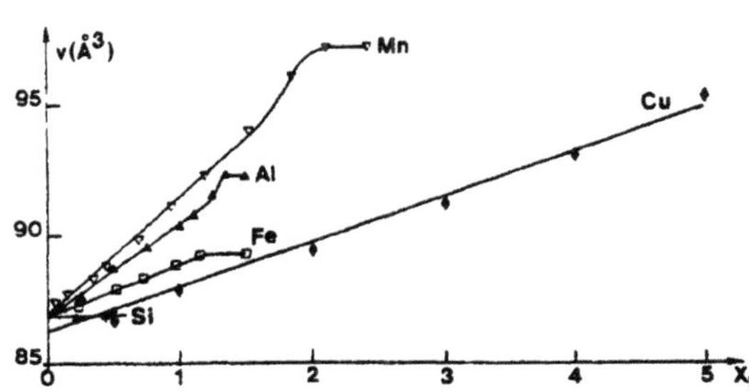

Figure 16. The change in unit cell volume of several $LaNi_{5-x}M_x$ solid solutions as a function of the M concentration.

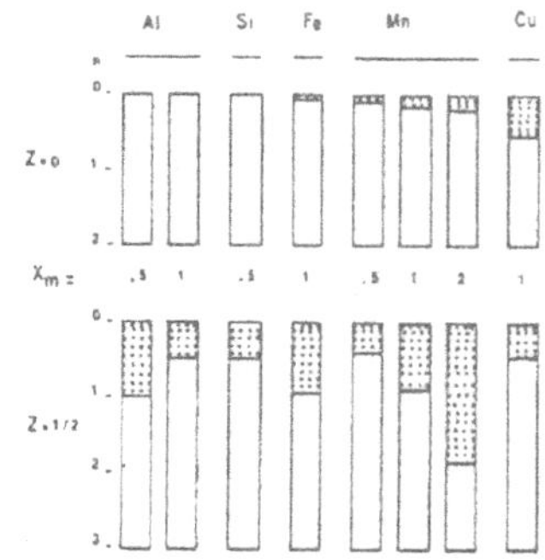

Figure 17. The distribution of the various substitutional element in the $LaNi_{5-x}M_x$ solid solutions on the nickel 3g and 2c sites in the hexagonal $CaCu_5$ structure.

The heats of formation of $LaNi_5$ [63], $LaNi_4Al$ [63], $LaNi_4Mn$ [68], $LaNi_4Fe$ [68], and $LaNi_{5-x}Cu_x$, with x=1-5 [68-70], have been measured by calorimetric methods. Their values have also been calculated with the Miedema model [54] by using a linear interpolation of the values estimated for $LaNi_5$ and LaM_5. A comparison of the experimental and calculated values, see Table 4, reveals good agreement except for $LaNi_4Al$. Because the stability of these compounds is representative of the cohesive energy resulting from both electronic and geometric factors, it is not surprising, that the most stable alloys have the smallest unit cell volume, again with the exception of $LaNi_4Al$.

It has been shown that the heat of formation of the $AB_{5-x}C_x$ compounds is better understood if one takes into account the mixing effect due to the substitution of B by C [69]. The mixing effect can be estimated by adding to the heat of formation of the alloy, an additional term, ΔH_l, given by

Table 4. Comparison of the experimental and calculated 298K heats of formation of some AB_5 compounds arranged in order of increasing unit cell volume [68,70].

Compound	ΔH_f, kJ/mol experimental	ΔH_f, kJ/mol, calc. from Miedema model[a]	Unit cell volume, $Å^3$
$LaNi_5$	−158.9	−165.9	86.9
$LaNi_4Co$	−147.3	−148.1	87.4
$LaNi_4Cu$	−143.1	−142.1	87.9
$LaNi_4Fe$	−129.7	−130.8	88.8
$LaNi_3Cu_2$	−128.3	−125.4	89.4
$LaNi_4Al$	−246.8	−169.7	89.6
$LaNi_4Mn$	−133.9	−131.7	91.5
$LaNi_2Cu_3$	−107.5	−108.7	91.25
$LaNiCu_4$	−93.3	−91.96	92.95
$LaCu_5$	−75.7	−75.2	95.77

[a] $\Delta H_f = f(c)\{-P_e(\Delta\Phi^*)^2 + Q(\Delta n\omega)^2\}$

$$\Delta H_l = \Delta H_f(AB_{5-x}C_x) - [(5-x)/5]\Delta H_f(AB_5) - (x/5)\Delta H_f(AC_5).$$

By assuming that all the contributions to the energy of the intermetallic compounds result from pair interactions between A, B, and C neighboring atoms, and by knowing the distribution of the substitutional atoms in the structure, ΔH_l may be calculated. When B and C are both transition metals, ΔH_l is small and its calculation by either the Pasturel [71] or the Miedema [70] methods yields good agreement with the measured heat of formation. However for $LaNi_4Al$, only the ΔH_f value of −259.4kJ/mol, calculated by Pasturel, gives good agreement with the experimental value, see Table 4. Apparently in this case ΔH_l is especially important.

5. 2. 2. *Hydride Compounds*. Since the work of Van Mal et al. [72], who demonstrated the change in the hydrogen absorption properties in $LaNi_5$ resulting from the partial replacement of nickel by cobalt, the systematic investigation of the pseudobinary $LaNi_5$ compounds has become particularly interesting. Some of these studies have included $LaNi_4M$, where M is Cr, Fe, Co, Cu, Pd, and Ag, and $La_{0.8}R_{0.2}Ni_5$, where R is Y, Zr, Nd, Gd, Er, and Th, [73], $LaNi_{5-x}M_x$, where M is Al [63, 74-79], Cu [76, 80-83], Mn [64, 84], Ga, In, Sn [85], Si, Ge, B [61], Fe [76,86], and Pt [78], and $La_{1-x}R_xNi_5$, where R is Ce [19], Misch metal, Ca [76], Yb [87], Pr, Nd, and Sm [88].

In the $LaNi_{5-x}M_xH_y$ hydrides, the hydrogen capacity, y, may be strongly affected by the type and extent of the metal, M, substitution. When aluminum is substituted for nickel, y decreases from 6.2 in $LaNi_{4.9}Al_{0.1}$ to 4.8 in $LaNi_4Al$ [63]. In contrast, when manganese is substituted for nickel, y is nearly constant at 6 over the range of x values between 0 and 2. Further, the hydrogen content is less affected by the substitution of lanthanum. A nearly constant y value of approximately 6.2 at 40°C is found for the series of $La_{0.8}R_{0.2}Ni_5$ compounds, where R is Y, Nd, Gd, and Er [73]. In these substituted hydrides, the unit cell volume expansion is smaller than in $LaNi_5$ and is 20% in $LaNi_4MnH_6$ and 14.4% in $LaNi_4AlH_{4.8}$. These expansions correspond to 3.3 and 3.0Å^3 per hydrogen atom as compared with the 3.8Å^3 per hydrogen atom observed for $LaNi_5H_{6.5}$ [65]. The formation of these hydrides leads to internal lattice microstrains which do not relax during the hydrogen desorption process. This effect is clearly demonstrated by the broadening observed in the Bragg diffraction peaks of these hydrides. In $LaNi_5$, these strains are important and highly anisotropic, but they rapidly become less important and nearly isotropic as the extent of nickel substitution increases in the $LaNi_{5-x}M_x$ solid solutions [66].

5. 2. 3. *Thermodynamics of the Hydride Compounds*. Because $LaNi_5$ forms only one concentrated β-phase hydride, its pressure-composition-temperature isotherms have only three regions, the α- and β-phases, and the $\alpha + \beta$ plateau. The two phase region shows a hysteresis between absorption and desorption, and the thermodynamic values given for this region usually refer to the desorption process. It has been shown that P_e is very sensitive to the sample stoichiometry and increases exponentially with the nickel content in $LaNi_x$. This increase is given by the expression, $RT\ln P_e = \alpha x + \beta$, where α is 6.23kJ/mol and β is −27.9kJ/mol at 40°C [72]. Partial replacement of lanthanum in $LaNi_5$ generally leads to an increase in P_e whereas a partial replacement of nickel leads to a decrease in P_e, a decrease which strongly depends on the type and extent of substitution, as is shown in Figure 18. Manganese and aluminum substituted compounds form very stable hydrides at 40°C with P_e values of 5×10^{-3}atm for $LaNi_4MnH_6$ and 1.6 $\times$ 10^{-2}atm for $LaNi_4AlH_{4.8}$. These values should be compared with the 3.7atm value for $LaNi_5H_{6.4}$. Exceptions occur with platinum and palladium substitutions, for which an

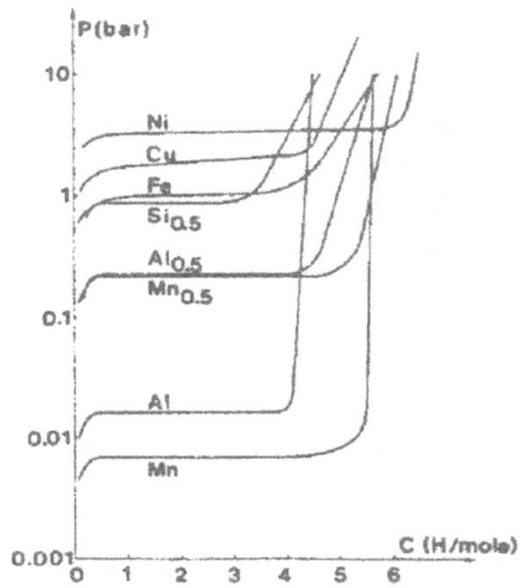

Figure 18. The 40°C hydrogen desorption isotherms for some $LaNi_{5-x}M_x$ solid solutions.

increase in P_e is observed [73]. Figure 19 represents schematically the pressure ranges observed for the substituted compounds.

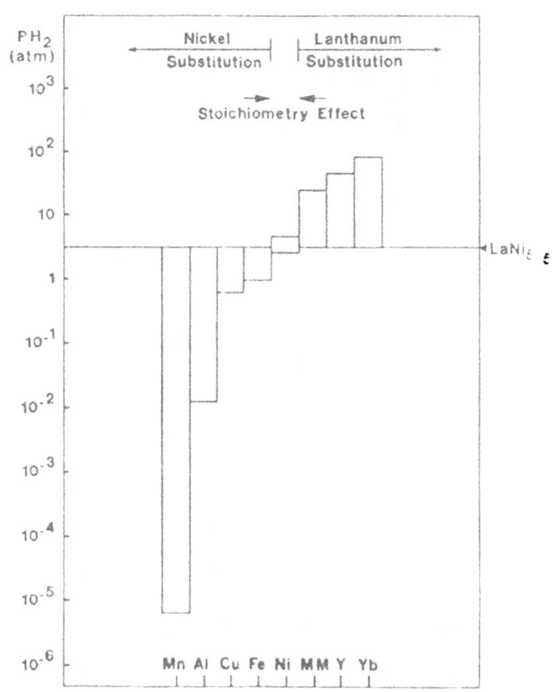

Figure 19. The influence of substitution on the range of equilibrium pressure in several intermetallic hydrides.

In the two phase region of the pressure-composition-temperature isotherms of $LaNi_5$ hydride, the standard enthalpy of formation of −30.9kJ/mol and the standard entropy of formation of −108.7J/K mol have been obtained [89] from the Van't Hoff plot. A more accurate value of the standard enthalpy of formation of −31.83±0.09kJ/mol has been obtained from differential heat flow calorimetry measurements [90]. The variation in these thermodynamic quantities, induced by the substitution of lanthanum and nickel, has only been studied in the plateau region of the pressure-composition-temperature isotherms. The standard enthalpy of the hydrogenation reaction typically decreases when nickel is partially replaced by aluminum. For instance, the standard enthalpy of hydrogenation of $LaNi_4Al$ is −47.6kJ/mol which is substantially lower than the −31.7kJ/mol value for $LaNi_5$. In contrast, when lanthanum is partially substituted [88], the standard enthalpy of hydrogenation increases. For instance, the value for $CeNi_5$ is −17kJ/mol. Further, the entropy change upon hydrogenation has been found to be nearly independent of metal substitution. This results because the entropy associated with hydrogenation is dominated by the entropy of hydrogen gas, 130J/K mol at room temperature, an entropy which is lost upon the absorption of hydrogen by the alloy [89].

5. 2. 4. *Hydride Structures*. The structure of β-$LaNi_5$ deuteride has been described in Section 4. 2. In contrast with the numerous studies devoted to the structure of $LaNi_5$ hydrides, the distribution of deuterium atoms in the deuterides of intermetallic compounds has been much less extensively studied. For $LaNi_{4.5}Al_{0.5}D_{4.5}$ the *P6/mmm* space group provides the best refinement [91] with two deuterium sites. The *P321* structural model with two deuterium sites, proposed earlier [92, 93] for $LaNi_5D_6$, has been used for the structural refinement of $LaNi_4AlD_{4.1}$. Gurewitz et al. [67] have proposed that $LaNi_4CoD_4$ has an orthorhombic *Cmmm* structure with two different deuterium sites. There have also been systematic structural investigations of $LaNi_{4.5}Mn_{0.5}D_{6.6}$, $LaNi_4MnD_{5.9}$, $LaNi_3Mn_2D_{5.95}$, $LaNi_4AlD_{4.8}$ [65], $LaNi_{4.5}Al_{0.5}D_{5.4}$, $LaNi_4CuD_{5.1}$, and $LaNi_{4.5}Si_{0.5}D_{4.3}$ [66]. Because the diffraction patterns show regular diffraction lineshapes [65], all of these structures could be refined by using the Rietveld lineprofile analysis technique. For all of these deuterides, the best refinements were obtained with the same five site deuterium model and the *P6/mmm* space group shown in Figure 20. However, the number of significantly occupied sites varied from five in $LaNi_4MnD_{5.9}$ to two in $LaNi_4AlD_{4.8}$, as is indicated in Table 5. When trying to refine the two site model with the *P31m* space group, satisfactory R factors could not be obtained. For example, for $LaNi_4MnD_{5.9}$, the structure refined in this space group led to an R factor of 13.5 percent, a value which is markedly higher than the 6.7 percent value obtained with the *P6/mmm* space group. Even in the most favorable case of $LaNi_4AlD_{4.8}$, in which only two sites are occupied by deuterium, the R-factor was 8.2 percent for the *P31m* space group as compared with 7.1 percent for the *P6/mmm* space group.

5. 2. 5. *Stability*. The models established by several authors to correlate the stability and the hydrogen content of the hydrides with the properties of the parent compounds are based on geometric, thermodynamic, or electronic considerations. None of these models are strictly quantitative, but they do give an indication of the variations in the hydride properties and do permit some predictions. Herein we discuss the validity of these models for the pseudobinary $LaNi_5$ compounds.

Lundin et al. [95] have established that a linear relationship exists between the free energy of formation, ΔG, of the hydrides and the radius calculated for an AB_3 tetrahedral interstice. The hydride stability increases with increasing interstice size.

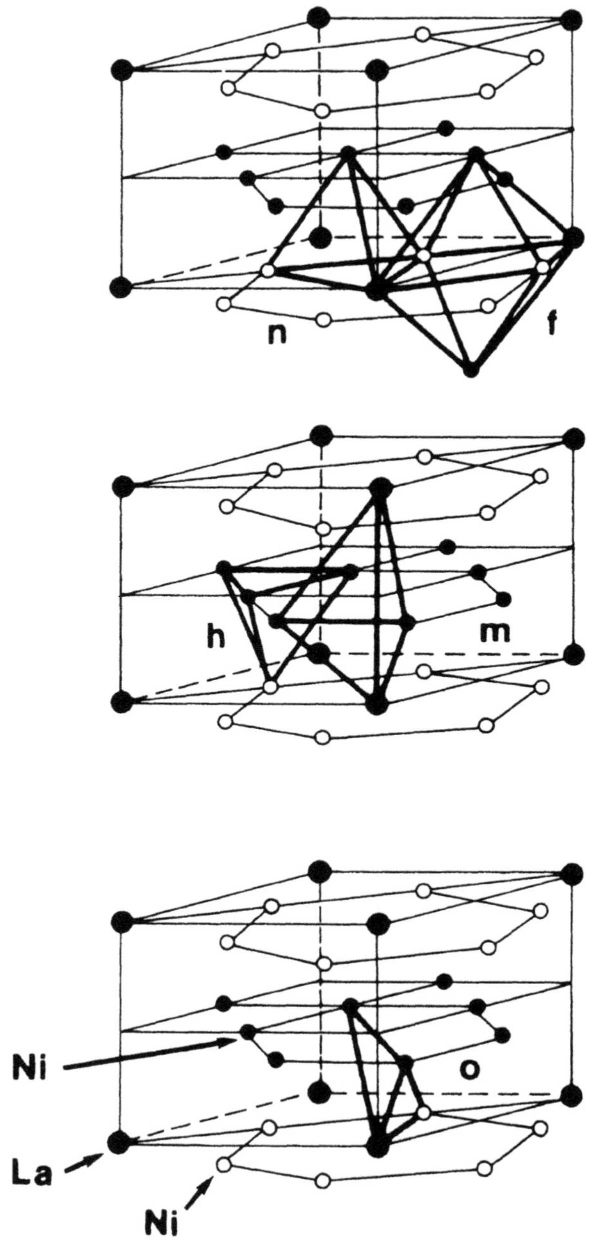

Figure 20. The hexagonal *P6/mmm* structure of $LaNi_5$ showing the five sites occupied by hydrogen. The f, h, m, n, and o symbols correspond to the Wyckoff site notation.

Table 5. The number and types of deuterium sites occupied in $LaNi_{5-x}M_x$ compounds.

Compound	Deuterium content	Number of sites	Type of occupied sites				
$LaNi_5$	6.5	5	6m	12n	4h	12o	3f
$LaNi_4Mn$	5.9	5	6m	12n	4h	12o	3f
$LaNi_{4.5}Mn_{0.5}$	6.6	4	6m	12n	4h	12o	
$LaNi_{4.5}Al_{0.5}$	5.4	4	6m	12n	4h	12o	
$LaNi_4Cu$	5.1	3	6m	12n	4h		
$LaNi_{4.5}Si_{0.5}$	4.3	3	6m	12n	4h		
$LaNi_4Al$	4.8	2	6m	12n			

Because later structural studies of pseudobinary $LaNi_5$ compounds have shown that the hydrogen atoms are distributed over several interstitial sites, this correlation is not justified. However, it does introduce a more general relationship, which has been proposed by several authors. At a given temperature, the equilibrium pressure of the hydride decreases exponentially as the unit cell volume of the intermetallic parent compound increases [74, 96, 97]. This exponential behavior is shown for several intermetallic compounds in Figure 21. This correlation is equivalent to the one proposed by Mendelsohn et al. [98], who have shown that the magnitude of the Gibbs free energy increases linearly with the unit cell volume. These rules are valid for most of the pseudobinary AB_5 compounds, but deviations from these general rules have been observed for silicon, platinum, iron, and copper substitutions [97, 99]. In contrast, the stability of the ternary hydrides has been related to the heat of formation of their parent compounds by Van Mal et al. [75], who found that the more stable is the AB_5 intermetallic compound, the less stable is its hydride. In this work, the authors have

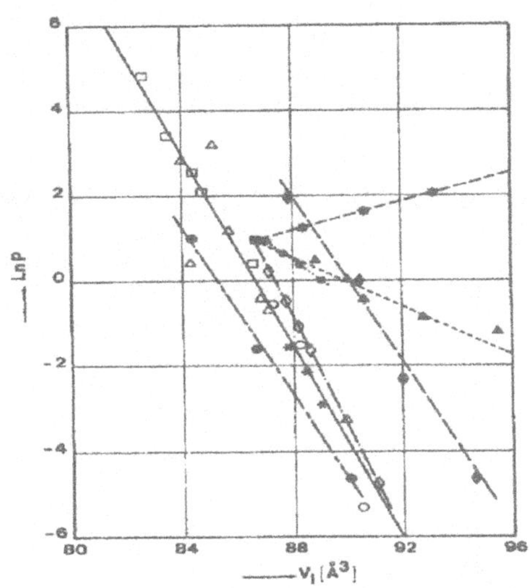

Figure 21. The plateau pressure versus unit cell volume of $LaNi_5$ hydrides for various partial substitutions of nickel with Al, Mn, Fe, Cu, and Pt.

estimated the heat of formation of the intermetallic compounds from the Miedema cellular model [71], a model which is only valid for binary compounds of transition metals.

Diaz et al. [63] and Pasturel et al. [68-70] have attempted to check the validity of the Miedema "rule of reversed stability" when nickel in $LaNi_5$ is partially replaced by Al, Mn, Fe, and Cu. Experimental values of the enthalpies of formation of the intermetallic compounds, and their respective hydrides, are given in Table 6. The results indicate that the Miedema model does not apply to these pseudobinary compounds. Indeed, although the intermetallic compounds of $LaNi_4M$, where M is Mn, Fe, Ni, and Cu, exhibit very similar enthalpies of formation, their enthalpies of hydrogenation are quite different, especially for manganese. Furthermore the very stable $LaNi_4Al$ also forms a very stable hydride. Pasturel et al. [68-70] have developed a model to calculate the enthalpy of hydrogenation of these pseudobinary compounds. Starting from the Miedema relationship,

$$\Delta H_{reaction} = \Delta H_f(AH_m) + \Delta H_f(BH_m) - \Delta H_f(AB_n),$$

they make the following assumptions. First, for 2m = 6, all the A-B metal bonds in AB_nH_6 are destroyed, whereas, for 2m < 6, some of the A-B metal bonds coexist with A-H and B-H bonds. Second, when B is partially substituted by C, the heat of formation of the intermetallic compounds includes a mixing enthalpy term, ΔH_l. Thus, the enthalpy of the hydrogenation reaction for the $AB_{5-x}C_x$ compounds is given by

$$\Delta H_{reaction} = (1-m/3)\Delta H_f(AB_{5-x}C_x) - \Delta H_f(AB_5) + (m/3)\Delta H_f(AH_3)$$
$$+ [m(5-x)/15]\Delta H_f(B_5H_3) + (mx/15)\Delta H_f(C_5H_3).$$

The enthalpies of formation of the binary hydrides, AH_3, B_5H_3, and C_5H_3 are either experimental values [100] or values estimated by Bouten and Miedema [101]. The $\Delta H_{reaction}$ values, calculated from these values and the above expression, are given in Table 6. The calculated values are in agreement with the experimental values obtained from pressure-composition-temperature isotherms, except for the $LaNi_2Cu_3$ and $LaNiCu_4$ hydrides for which the experimental pressures cannot be accurately determined by this method [82].

Table 6. The 298K enthalpy of formation and the experimental and calculated enthalpies of hydrogenation for several LaM_5 compounds.

Compound	$\Delta H_f(LaM_5)$, kJ/mol	$\Delta H_{react}(LaM_5H_x)$ expt., kJ/mol	$\Delta H_{react}(LaM_5H_x)$ calc., kJ/mol
$LaNi_5$	−158.9	−31.7	−29.7
$LaNi_4Cu$	−143.1	−33.9	−34.3
$LaNi_4Fe$	−129.7	−34.3	−37.2
$LaNi_3Cu_2$	−128.3	−37.6	−36.4
$LaNi_4Al$	−246.8	−47.6	−49.7
$LaNi_4Mn$	−133.9	−48.5	−46.0
$LaNi_2Cu_3$	−107.5	−30.9	−38.8
$LaNiCu_4$	−93.3	−30.5	−43.5
$LaCu_5$	−75.7	−44.3	−47.2

With reference to the model previously established for the binary hydrides by Switendick [102,103], Gelatt et al. [104] and Griessen and Driessen [105] have proposed relationships between the stability of the hydrides and the electronic properties of the intermetallic parent compounds. By using photoemission studies [106] on $LaNi_{5-x}Cu_x$, and by assuming that the d-band structure of $LaNi_5$ is very similar to that of elemental nickel, Malik et al. [107] and Weaver et al. [108] have found that the stability of the $LaNi_5$-type hydrides depends upon the energy of the d-band. The higher the energy of this band, the more stable is the hydride. Bucur and Lupu [109] have reported a quantitative model for $LaNi_{5-x}M_x$, where M is Cr, Fe, Co, Cu, Pd, Ag, and Pt. They found that the difference between the free energy of formation of the hydrides of these compounds and the hydride of $LaNi_5$ depends linearly on the average energy of the lowest band of the transition metal, M, and the concentration, x. In contrast, Takeshita et al. [110] have concluded from their low temperature heat capacity measurements that there is no simple relationship between the stability of the hydrides and the density of states at the Fermi level of the intermetallic compounds. However, they have suggested that compressibility may play a role in the hydriding characteristics of the RM_5 compounds.

5. 2. 6. *Hydrogen Content.* To date no band structure calculations have been undertaken for the $LaNi_5$ hydrides. However, Wallace and Pourarian [106] by refering to the results found for the simple monohydrides [102, 103, 111], have suggested that, for the substituted $LaNi_5$-type compounds, the amount of hydrogen which can be absorbed is controlled by the number of metal-hydrogen bonding states. The reduction in hydrogen absorption capacity is associated with the reduction in the number of d-states near the Fermi surface.

Gschneider et al. [97] have pointed out the important role of the 3d electrons in determining the hydrogen absorption capacity. From an analysis of the electronic specific heats of some $ANi_{5-x}B_x$ alloys, where A is Y, La, and Th and B is Al and Cu, they have suggested that the amount of hydrogen absorbed in excess of one-half hydrogen per metal atom depends upon the number of unpaired nickel 3d electrons.

Westlake [12] has developed a method based on the geometric considerations described in Section 2.3. Our structural study has shown that the reduction in the amount of hydrogen absorbed by some pseudobinary $LaNi_{5-x}M_x$ compounds, where M is Si, Al, Mn, and Cu, is associated with the reduction in the number of occupied interstitial sites, see Table 5. Independently of the substituted metal, M, hydrogen atoms exhibit a strong preference for the larger 6m and 12n sites. With the exception of the silicon substituted compound, there are always approximately 4.4 to 5 hydrogen atoms on these two sites. In contrast, the filling of the 3f, 4h, and 12o sites is strongly dependent on the substitution. These results can be only partially understood on the basis of the Westlake geometric model [12]. Although the 6m and 12n sites always obey the minimum site size criterion, such is not the case for the other three sites. In addition, the 4h and 12o sites are in close contact with at least two metallic atoms in the $z=1/2$ plane where most of the substitution occurs. However, because these substitutions are random, they may induced local effects which we cannot model when predicting the hydrogen site occupation. Further, these geometric factors and the non-filling of the 4h and 12o sites, may be due to the number of valence electrons of the substitutional M atoms, as has been suggested by Gschneidner et al. [97]. Obviously the geometric and electronic criteria cannot be separated.

In conclusion, substitutions into $LaNi_5$ are a convenient way to change its equilibrium hydrogen pressure. The geometric criterion allows us to predict rather easily, on the basis of the intermetallic unit cell volume, the stability of the hydride. It is the heat of formation of the intermetallic compound, which reflects their cohesive energy. Thus, the intermetallic compounds with the largest unit cell volume are less stable, but lead to more stable hydrides. The exception when aluminum is substituted for nickel indicates the important contribution of electronic factors, factors which require further investigations. The limitation of the Westlake geometric model [12] and the results of our structural studies also point to the need for a more detailed investigation of the electronic properties of these intermetallic compounds.

References

[1] W. M. Mueller, J. P. Blackledge, and G. G. Libowitz, eds., *Metal Hydrides*, Academic Press, New York (1968).

[2] *Fifth International Symposium on the Properties and Applications of Metal Hydrides*, Maubuisson, France, 1986, A. Percheron Guégan and M. Gupta, eds., J. Less-Common Metals **129-131** (1987), with references to earlier conferences in this series.

[3] *Hydrogen in Metals,* Proceedings of the International Symposium, Belfast, 1985, F. A. Lewis and E. Wicke, eds., Z. Phys. Chemie N. F. **143-147** (1986), with references to earlier conferences in this series.

[4] *Metal-Hydrogen Systems. Fundamentals and Applications,* Uppsala, 1992, D. Noréus, S. Rundquist, and E. Wicke, eds., Z. Phys. Chemie N. F. **179-183** (1993-1994), with references to earlier conferences in this series.

[5] G. Alefeld and J. Völkl, eds., *Hydrogen in Metals I and II,* Topics Appl. Phys., Vol. 28 and 29, Springer, Heidelberg (1978).

[6] L. Schlapbach, ed., *Hydrogen in Intermetallic Compounds I and II,* Topics Appl. Phys., Vol. 63 and 67, Springer, Heidelberg (1988 and 1992).

[7] Y. Fukai, *The Metal Hydrogen System,* Springer Series in Materials Science, Vol. 21, Springer, Heidelberg (1992).

[8] G. Sandrock, S. Suda, and L. Schlapbach, in *Hydrogen in Intermetallic Compounds II,* L. Schlapbach, ed., Topics Appl. Phys., Vol. 67, Springer, Heidelberg (1992) p. 197.

[9] A. Percheron Guégan and J. M. Welter, in *Hydrogen in Intermetallic Compounds I,* L. Schlapbach, ed., Topics Appl. Phys., Vol. 63, Springer, Heidelberg (1988) p. 11.

[10] T. B. Flanagan and W. A. Oates, in *Hydrogen in Intermetallic Compounds I,* L. Schlapbach, ed., Topics Appl. Phys., Vol. 63, Springer, Heidelberg (1988) p. 49.

[11] A. C. Switendick, Z. Phys. Chem. N. F. **117**, 89 (1979).

[12] D. G. Westlake, J. Less-Common Metals **91**, 1 (1983).

[13] T. Graham, Phil. Trans. Roy. Soc. (London) **156**, 415 (1866).

[14] F. A. Lewis, *The Palladium Hydrogen System*, Academic Press, London, (1967).

[15] T. B. Flanagan, Engelhard Ind. Tech. Bull. **7**, 9 (1966).

[16] H. Frieske and E. Wicke, Ber. Bunsenges. Physik. Chem. **80**, 22 (1976).

[17] J. E. Worsham, M. K. Wilkinson, and C. G. Shull, J. Phys. Chem. Solids **3**, 303 (1957).

[18] G. G. Libowitz, H. F. Hayes, and T. R. P. Gibb, J. Phys. Chem. **62**, 76 (1958).

[19] J. H. N. Van Vucht, F. A. Kuijpers, and H. C. A. M. Bruning, Philips Res. Reports **25**, 133 (1970).

[20] J. J. Reilly and R. H. Wiswall, Inorg. Chem. **13**, 218 (1974).

[21] J. J. Reilly and R. H. Wiswall, Inorg. Chem. **7**, 2254 (1968).

[22] D. Shaltiel, I. Jacob, and D. Davidov, J. Less-Common Metals **53,** 117 (1977).

[23] Y. Ishido, N. Nishimiya, and A. Suzuki, Denki Kagaru **45,** 52 (1977) and Energy Developments in Japan **1**, 207 (1978).
[24] J. J. Murray, M. L. Post, and J. B. Taylor, J. Less-Common Metals **80,** 210 (1981).
[25] H. Wenzl and E. Lebsanft, J. Phys. F: Met. Phys. **10**, 2147 (1980).
[26] J. R. Johnson, J. Less-Common Metals **73,** 345 (1980).
[27] J. J. Reilly, in *Hydrides for Energy Storage,* A. F. Andresen and A. J. Maeland, eds., Pergamon, Oxford (1978) p. 301.
[28] K. Yvon and P. Fischer, in *Hydrogen in Intermetallic Compounds I,* L. Schlapbach, ed., Topics Appl. Phys., Vol. 63, Springer, Heidelberg (1988) p. 87.
[29] H. M. Rietveld, J. Appl. Crystallogr. **2**, 65 (1969).
[30] D. Noréus, L. G. Olsson, and P. E. Werner, J. Phys. F: Met. Phys. **13**, 715 (1983).
[31] A. Percheron Guégan, C. Lartigue, J. C. Achard, P. Germi, and F. Tasset, J. Less-Common Metals **74,** 1 (1980).
[32] C. Lartigue, A. Percheron Guégan, J. C. Achard, and J. L. Soubeyroux, J. Less-Common Metals **113,** 127 (1985).
[33] P. Thompson, J. J. Reilly, L. M. Corliss, J. M. Hastings, and R. Hempelmann, J. Phys. F: Met. Phys. **16** , 679 (1986).
[34] P. Thompson, J. J. Reilly, F. Reidinger, J. M. Hastings, and L. M. Corliss, J. Phys. F: Met. Phys. **9** , L61 (1979).
[35] W. Schäfer, G. Will, and T. Schober, Mat. Res. Bull. **15**, 627 (1980).
[36] P. Fischer, J. Schefer, K. Yvon, L. Schlapbach, and T. Riesterer, J. Less-Common Metals **129,** 39 (1986).
[37] D. Noréus, and P. E. Werner, J. Less-Common Metals **97,** 215 (1984).
[38] J. L. Soubeyroux, D. Fruchart, A. Mikou, M. Pezat, and B. Darriet, Mat. Res. Bull. **19**, 119 (1984).
[39] P. Zolliker, K. Yvon, J. D. Jorgensen, and F. Rotella, Inorg. Chem. **25**, 3590 (1986).
[40] J. J. Didisheim, K. Yvon, D. Shaltiel, and P. Fischer, Solid State Comm. **31**, 47 (1979).
[41] J. J. Didisheim, K. Yvon, D. Shaltiel, P. Fischer, P. Bujard, and E. Walker, Solid State Comm. **32**, 1087 (1979).
[42] K. Yvon, J. Less-Common Metals **130,** 53 (1984).
[43] E. Wicke and H. Brodowky, in *Hydrogen in Metals II,* G. Alefeld and J. Völkl, eds., Topics Appl. Phys., Vol. 29, Springer, Heidelberg (1978) p. 73.
[44] E. Wicke and G. H. Nernst, Ber. Bunsenges. Phys. Chem. **68**, 224 (1964).
[45] H. Brodowsky and E. Poeschel, Z. Physik. Chem. N. F. **144**, 143 (1965).
[46] H. Brodowsky and H. Husemann, Ber. Bunsenges. Phys. Chem. **70**, 626 (1966).
[47] D. Artman, J. F. Lynch, and T. B. Flanagan, J. Less-Common Metals **45,** 215 (1976).
[48] R. C. Phutela and O. J. Kleppa, J. Chem. Phys. **76**, 1525 (1982).
[49] T. B. Flanagan, G. Gross, and J. D. Clewley, in the*Proceedings Second International Congress on Hydrogen in Metals,* Pergamon, Oxford (1977) p. 1C3.
[50] R. Feenstra, D. G. de Groot, R. Griessen, J. P. Burger, and A. Menovski, J. Less-Common Metals **130,** 375 (1987).
[51] R. Burch and R. G. Buss, J. Chem. Soc., Faraday Trans. I **71**, 922 (1975).
[52] J. D. Clewley, J. F. Lynch, and T. B. Flanagan, J. Chem. Soc., Faraday Trans. I **73**, 494 (1977).
[53] H. Brodowky, Ber. Bunsenges. Phys. Chem. **76**, 740 (1972).
[54] H. Husemann and H. Brodowki, Z. Naturforsch. **23a**, 1693 (1968).
[55] Y. Sakamoto, T. B. Flanagan, and T. Kuji, Z. Phys. Chem. **143**, 61 (1985).

[56] M. LaPrade, K. D. Allard, J. F. Lynch, and T. B. Flanagan, J. Chem. Soc., Faraday Trans. I **70**, 1615 (1974).
[57] K. D. Allard, A. Maeland, J. W. Simons, and T. B. Flanagan, J. Phys. Chem. **72**, 136 (1968).
[58] K. D. Allard, J. F. Lynch, and T. B. Flanagan, Z. Phys. Chem. **93**, 15 (1974).
[59] H. Wenzl, Intern. Metals Rev. **27**, 140 (1982).
[60] A. Percheron Guégan, C. Lartigue, and J. C. Achard, J. Less-Common Metals **109**, 287 (1985).
[61] M. H. Mendelsohn and D. M. Gruen, in *Rare Earths in Modern Science and Technology*, G. J. MacCarthy, and J. J. Rhyne, eds., Plenum, New York (1980) p. 593.
[62] J. Lamloumi, A. Percheron Guégan, J. C. Achard, G. Jehanno, and D. Givord, J. Phys. **45**, 71 (1984).
[63] H. Diaz, A. Percheron Guégan, J. C. Achard, C. Chatillon, and J. C. Mathieu, Int. J. Hydrogen Energy **4**, 445 (1979).
[64] C. Lartigue, A. Percheron Guégan, J. C. Achard, and F. Tasset, J. Less-Common Metals **75**, 23 (1980).
[65] A. Percheron Guégan, C. Lartigue, J. C. Achard, P. Germi, and F. Tasset, J. Less-Common Metals **74**, 1 (1980).
[66] J. C. Achard, A. J. Dianoux, C. Lartigue, A. Percheron Guégan, and F. Tasset, in *Rare Earths in Modern Science and Technology*, Vol. 3, G. J. MacCarthy, J. J. Rhyne, and H. B. Silver, eds., Plenum, New York (1982) p. 481.
[67] E. Gurewitz, H. Pinto, M. P. Dariel, and H. Shabed, J. Phys. F: Met. Phys. **13**, 545 (1983).
[68] A. Pasturel, C. Chatillon-Collinet, A. Percheron Guégan, and J. C. Achard, J. Less-Common Metals **84**, 73 (1982).
[69] A. Pasturel, Ph.D. Thesis, University of Grenoble , 1983.
[70] A. Pasturel, F. Liautaud, C. Colinet, A. Percheron Guégan, and J. C. Achard, J. Less-Common Metals **96**, 93 (1984).
[71] A. R. Miedema, J. Less-Common Metals **32**, 117 (1973).
[72] H. H. Van Mal, K. H. J. Buschow, and F. A. Kuijpers, J. Less-Common Metals **32**, 289 (1973).
[73] H. H. Van Mal, K. H. J. Buschow, and A. R. Miedema, J. Less-Common Metals **35**, 65 (1974).
[74] J. C. Achard, A. Percheron Guégan, H. Diaz, and F. Briaucourt, *Proceedings of the Second International Conference on Hydrogen in Metals,* Paris, June 1977, p. 1E2.
[75] M. H. Mendelsohn, D. M. Gruen, and A. E. Dwight, Nature **269**, 45 (1977).
[76] G. D. Sandrock, *Proceedings of the Second World Hydrogen Energy Conference,* Zurich, August 1978.
[77] G. D. Sandrock, *Proceedings Twelveth Intersociety Energy Conversion Engineering Conference of the American Nuclear Society*, Vol. 1 (1977) p. 951.
[78] Y. Chung, T. Takeshita, O. D. McMasters, and K. A. Gschneidner Jr., J. Less-Common Metals **74**, 217 (1980).
[79] T. Takeshita, S. K. Malik, and W. E. Wallace, J. Solid State Chem. **23**, 271 (1978).
[80] J. Shinar, I. Jacob, D. Davidov, and D. Shaltiel, in the *Proceedings of the International Symposium on Hydrides for Energy Storage,* Pergamon, Oxford (1978) p. 337.
[81] D. Shaltiel, J. Less-Common Met. **62**, 407 (1978).

[82] J. Shinar, D. Shaltiel, D. Davidov, and A. Gravyevsky, J. Less-Common Met. **60**, 209 (1978).
[83] T. Takeshita, G. Dublon, O. D. McMasters, and K. A. Gschneidner Jr., in *Rare Earths in Modern Science and Technology*, Vol. 3, G. J. MacCarthy, J. J. Rhyne, and H. B. Silver, eds., Plenum, New York (1982) p. 487.
[84] C. E. Lundin and F. E. Lynch, Int. Rep. Denver Research Institute, 1977.
[85] M. H. Mendelsohn, D. M. Gruen, and A. E. Dwight, Mater. Res. Bull. **13**, 1221 (1978).
[86] J. Lamloumi, C. Lartigue, A. Percheron Guégan, and J. C. Achard, in *Rare Earths in Modern Science and Technology*, G. J. MacCarthy, J. J. Rhyne, and H. B. Silver, eds., Plenum, New York (1980) p. 563.
[87] J. C. Achard, A. Percheron Guégan, J. Sarradin, and G. Bronoel, in *Proceedings International Symposium on Hydrides for Energy Storage,* Pergamon, Oxford (1978) p. 485.
[88] H. Uchida, M. Tada, and Y. Huang, J. Less-Common Metals **88**, 81 (1982).
[89] H. H. Van Mal, Ph. D. Thesis, Technical University Delft; Philips Res. Rep. Suppl. (1976) p. 1.
[90] J. J. Muray, M. L. Post, and J. B. Taylor, J. Less-Common Metals **80**, 201 (1981).
[91] C. Crowder and W. J. James, J. Appl. Phys. **53**, 2637 (1982).
[92] V. A. Yartis, V. V. Burnasheva, S. E. Tsirkunova, E. N. Kozlov, and K. N. Semenenko, Kristallografiya **27**, 242 (1982).
[93] V. A. Yartis, V. V. Burnasheva, K. N. Semenenko, N. V. Fadeeva, and S. P. Solovev, Int. J. Hydrogen Energy **7**, 957 (1982).
[94] C. Lartigue, Ph. D. Thesis, University of Paris VI, 1984.
[95] C. E. Lundin, F. E. Lynch, and C. B. Magee, J. Less-Common Metals **56**, 19 (1977).
[96] G. Busch, L. Schlapbach, and A. Seiler, in *Hydrides for Energy Storage*, A. F. Andresen and A. J. Macland, eds., Pergamon, Oxford (1978) p. 293.
[97] K. A. Gschneidner Jr., T. Takeshita, Y. Chung, and O. D. McMasters, J. Phys. F: Met. Phys. **12**, L1 (1982).
[98] M. H. Mendelsohn, D. M. Gruen, and A. E. Dwight, J. Less-Common Metals **63**, 193 (1979).
[99] J. Lamloumi, Ph. D. Thesis, University of Paris VI, 1982.
[100] G. G. Libowitz, *The Solid State Chemistry of Binary Metal Hydrides*, W. A. Benjamin, New York (1965).
[101] P. C. Bouten and A. R. Miedema, J. Less-Common Metals **71**, 147 (1980).
[102] A. C. Switendick, Adv. Chem. **167**, 265 (1978) and references therein.
[103] A. C. Switendick, in *Hydrogen in Metals*, G. Alefed and J. Völkl, eds., Springer-Verlag, Berlin (1978) p. 101.
[104] C. D. Gelatt, H. Ehrenreich, and J. A. Weiss, Phys. Rev. B **17**, 1940 (1978).
[105] R. Griessen and A. Driessen, Phys. Rev. B **30**, 4372 (1984).
[106] W. E. Wallace and F. Pourarian, J. Phys. Chem. **86**, 4958 (1982).
[107] S. K. Malik, F. J. Arlinghaus, and W. E. Wallace, Phys. Rev. B **25**, 6488 (1982).
[108] J. H. Weaver, A. Franciosi, D. J. Peterman, I. Takeshita, and K. A. Gschneider Jr., J. Less-Common Metals **86**, 195 (1982).
[109] R. V. Bucur and D. Lupu, J. Less-Common Metals **90**, 203 (1983).
[110] T. Takeshita, K. A. Gschneidner Jr., D. K. Thome, and O. D. McMasters, Phys. Rev. B **21**, 5636 (1980).
[111] M. Gupta, J. Less-Common Metals **88**, 221 (1982).

Chapter 6

PROGRESS IN METAL - HYDRIDE TECHNOLOGY

P. DANTZER
CNRS - URA 446 - ISMA
Université Paris-Sud, Bât 415
91405 Orsay-Cedex, France

1. Introduction

In the seventies and eighties, public interest was aroused in the possibilities of diversifying energy resources, driven by feared shortages of petroleum supplies and by the continuous expansion of the world energy needs. The present decade is focusing attention upon the serious problem of atmospheric pollution: the 'greenhouse' effect that has been correlated with the sharp increase in the production of chlorofluorocarbon (CFC) molecules that are very stable and are more active than carbon dioxide in destroying the ozone layer.

Metal-hydrides offer alternative energy possibilities. The fact that hydrogen plays a prominent part as a non polluting fuel in internal combustion engines, as well as a non polluting working fluid in chemical heat pump systems, lends strong arguments to supporting research in clean energy systems based on metal hydrides. In fact, as stated by Reilly and Sandrock [1], 'The spectrum of possible applications for rechargeable metal hydrides is limited only by the imagination of the inventor'.

The development of these applications and the interest in metal-hydrides really started in 1970 with the discovery of $LaNi_5$ [2] at Philips, and FeTi [3] at Brookhaven National Laboratory. Although the first intermetallic compound, ZrNi, and intermetallic hydride $ZrNiH_3$, were already prepared in 1958 by Libowitz [4], it took some time before it was realised how many possibilities could be opened up by controlling the absorption properties of metallic hydrides by varying the composition of the different metallic components of the alloys.

Several national research programs were launched around 1975 in the US, Japan, and Germany, in the aftermath of the first petroleum crisis. These programs were mainly oriented toward solving hydrogen storage problems. It should be said that current commercial applications are limited for economic reasons, because the commercially available storage reservoirs are small units used mainly for laboratory purposes. However, progress has been made in materials production to the point where it is now possible to obtain, on an industrial scale, a large spectrum of metal alloys of sophisticated composition. Beginning in 1980, the energetic aspect associated with energy transfer and storage was proposed in the US [5] and in Japan [6] with the innovation of new heat management processes, the hydride chemical heat pumps. The largest contribution to this particular field came from the Japanese scientists, but the proposed machines are still at the prototype level. In 1990 the first hydride batteries reached the marketing stage.

The commercial and economic aspects will not be discussed here, but will remain as a shadow for metal-hydrides technology. In general, the situation is as follows. With the

F. Grandjean et al. (eds.), Interstitial Intermetallic Alloys, 107–150.

exception of isotope separation, where expense is not a limiting factor for development, most other hydride applications are in competition with well established and cheaper solutions, at least for the present time. In face of this economic fact, it should be clear, now that metal-hydrides have been proven viable, the effort to optimise the various known processes should be continued to increase the competitiveness of the proposed systems.

The purpose of this chapter is to evaluate the current status of research and development in the field of applications, and to draw attention to investigations which must be encouraged to improve our understanding of hydrogen behaviour in metal-hydrogen systems with respect to non-equilibrium processes occuring during hydriding and dehydriding reactions. Collections of the latest major works on applied research are available in books [7-10] and in the proceedings of international symposia [11-14]. We should also mention the work of experts who have been continously reporting, for more than 20 years, on a future hydrogen-based economy, including hydrides [15-18]. National programs related to hydride applications may be found in several works [19-21]. Throughout this chapter, the term 'metal-hydrogen', 'intermetallic compound', and 'intermetallic hydrides' will be abbreviated as 'MH', 'IMC', and 'IMH' respectively.

2. Current Problems in Hydride Applications

The two basic properties which make metal hydrides attractive are their high and reversible storage capacity per mole of compound, and the high energy stored per unit volume. The mass of hydrogen that can be stored by unit volume of hydride is almost twice what can be stored in liquid form, and the energy contained may run as high as one MJ per litre of hydride. These figures fully justify the current applied research on hydrides as fuel or for energy transfer. The appeal of the former stems from the fact that the automobile industry is the world's foremost civilian industry. The latter subject arouses greater interest and support in energy dependent countries and those heavily engaged in environmental politics.

Hydride devices have two operating modes, those operating in 'closed systems', where the same hydrogen is cycled over and over, such as heat pumps and detectors, and those operating in 'open systems', where the storage unit is fed with new hydrogen once the existing charge is spent, or is fed continuously with hydrogen. Open systems are more sensitive to impurities in the hydrogen gas than the closed systems. This implies that the stability of the IMC in the presence of chemical impurities should be well established, as it may be a detrimental or a quality factor, depending on the type of open system application, storage reservoir or purification device.

The main drawback to hydrides is their low hydrogen weight percentage, with the exception of magnesium hydride, it is never more than 2%. Of course this weight limitation is of less importance in hydrogen purification processes, as well as in actuators or miniature detectors or, to an even lesser extent, in stationary hydrogen storage. The first step in hydride technology is to find an alloy with appropriate properties offering optimum performance in order for it to produce a marketable unit. The ability to maintain absorption properties over the long term will be of paramount importance. Consider applications in cars with an average lifetime of 200000 km. A hydride reservoir fuelling a car for 200 km will have to be filled only 1000 times with H_2 gas. For a chemical heat pump, in continuous operation in 10 minutes cycles, over an expected 10 year lifetime, the number of required cycles will be 60000. For batteries, the number of charge / discharge cycles should be of the order of a few hundred.

One thing is certain, of the many alloys proposed, none is universal. Moreover, because hydrides always operate under drastic non-equilibrium conditions, the 'equilibrium' pressure, composition, temperature values (P,C,T) are relevant only in the preliminary selection phase of studies determining the comparative properties of the IMH systems. In an ideal world all information influencing the long term retention of the absorption properties of a given system would be well established before attempting to develop a hydride device. Unfortunately this is not always possible if we consider the hundreds of alloys available. Finally, as each application exhibits its own quite definite technical and economical constraints, the alloys selected should provide the best overlap between constraints and property changes during the hydriding and dehydriding cycles, for extensive use. This suggests that a clear inventory is needed and that the factors influencing the absorption properties must be understood.

3. Relevant Properties

The parameters used in selecting an IMC have been subjected to many investigations and are classified according to their metallurgical aspects, including elaboration, structural characterisation and activation, their thermochemical properties, including thermodynamic parameters, their chemical stability with regard to impurities and cycling, their dynamic behaviour, i.e. mass and heat transport properties, and the technological problems of handling the materials.

The metallurgical problems have recently been discussed by Percheron-Guegan and Welter [22], who described the preparative techniques of the most representative families of IMCs. Manchester and Khatamian [23] have recently discussed the activation mechanisms.

3.1. THERMO-CHEMICAL REACTIVITY

From a technological viewpoint, the thermochemical reactivity concerns the IMC in its fully activated state, which usually means a mechanically disrupted finely divided powder, with grain sizes of the order of a few microns. This is the starting material for the investigations devoted to applied purposes.

3.1.1. *Thermodynamics.* P,C,T measurements by volumetric devices presented in the form of isotherms should, in principle, suffice to generate all the necessary thermodynamic data for activated materials, although there are some difficulties comparing the thermochemistry of IMH systems with that of pure MH systems [24a-b]. The fundamental thermodynamics of IMH was discussed by Flanagan and Oates [25], who introduced para-equilibrium (PE) and complete equilibrium (CE) notation for the low metastable state and high temperature state of the IMC [26-27]. In the following the systems are considered as behaving in the PE configuration, that is as pseudo-binary hydrogen systems.

The most important properties are related to the presence of the 'plateau pressure' which represents the coexistence of two condensed phases with the gaseous phase. The phase transformation for the absorption reaction is then written as,

$$(x_2 - x_1)^{-1} MH_{x_1} + \frac{1}{2} H_2 \rightarrow (x_2 - x_1)^{-1} MH_{x_2}, \quad (1)$$

and the following well known relations are deduced,

$$\Delta\mu_H^{abs} = \frac{1}{2} RT \ln\left(p_{H_2}^{abs}\right), \quad (2)$$

and

$$\Delta H_H^{\alpha\rightarrow\beta} = \frac{-R\partial \ln\left(p_{H_2}^{abs}\right)}{\partial\left(1/T\right)} \quad (3)$$

Similar equations hold for the desorption reaction. Further detailed informations on the thermodynamic properties of these materials can be found in the chapter written by Flanagan in this book.

Equation (3) relates the amount of heat evolved during the formation of a hydride compound to the plateau pressure. In a first step we can take $\Delta H^{\alpha\rightarrow\beta}$ to be a good approximation of the heat of formation, ΔH^f, which determines the stability of the hydride. Predicting the heat of formation is a very difficult task. Semi-empirical models have been reviewed by Griessen and Riesterer [28], who also made a compilation of the heats of formation of various IMHs and MHs. However the values for the heats of solution and heats of formation of the metal-hydrogen systems have not all been updated, in particular for the V, Nb, Ta, Pd, Ti, La, Y, and Th-H_2 systems [29a-h].

Remembering the van't Hoff relation, written as,

$$\frac{1}{2} RT \ln\left(p_{H_2}^{abs}\right) = \frac{\Delta H^f}{RT} - \frac{\Delta S^f}{R}, \quad (4)$$

where ΔS^f is the entropy of formation, we then have a useful way to estimate the success of a given IMH in an application. Equation (4) assumes the reaction is reversible, at equilibrium, and so should be restricted to use in a first principle energy calculation. Using this relation to describe the dynamics of processes, overextends its usefulness and leads to errors, because irreversibilities have to be introduced when modelling hydride cycles (second principle induced effects).

Hysteresis and sloping plateau effects are largely responsible for the difficulties in determining accurate and reproducible data, which, as a consequence, currently limits the validity of predictions of long term thermodynamic behaviour for the IMH in use. Hysteresis complicates the experimental observations, because several parameters control the path followed during thermodynamic characterisation. Different reported values of the ratio p^{abs}/p^{des} may also be due to artifacts, such as a deviation from the stoichiometry in the IMC. This effect has been shown by Buschow and van Mal [30] for the $LaNi_{5\pm\delta}$ system. Park et al [31a] have pointed out that the size of the aliquot of hydrogen used in determining isotherms, influences the reported thermodynamic data. Non-reproducibility in the thermodynamics properties results from experimental observations having undefined imposed parameters, internal or external to the system, leading to various time dependent thermodynamic states. Recently, Dantzer et al [31b] have identified and quantified the parameters acting on the $LaNi_5$-H_2 system. Controlling the mass flow rate of hydrogen makes it possible to maintain quasi-isothermal conditions, < 0.1°C, with a fine control of the driving forces during the phase transformation. Then, reproducibility has been reached and a memory effect has been pointed out. One other point deserves more attention in that it is directly related to the determination of the appropriate state equations for hydride systems, that being the transition from the quasi-static state to the dynamic state, with continuous absorption of H_2. Work in this domain was initiated by Goodell et al [32], Josephy and Ron [33a-b], and Groll et al [34]. The latest contributions to our understanding of hysteresis are due to Flanagan et al [35], Mc. Kinnon [36], Quian et al [37], and Shilov et al [38].

Different reasons have been offered to explain the sloping plateau effect such as inhomogeneity within the compound introduced during its preparation and degradation imposed by a highly stressed material correlated to the extensive volume variations during cycling. X-ray analysis of cycled compounds showed broadening of lines [39], whereas analysis of the line profiles showed the presence of residual strains after dehydriding, for specific symmetry axes of the compounds.

Because the problems involved in thermodynamics have now been relatively well investigated, one can expect more rigorous determination of thermodynamic data in the future. Noting that, for the $LaNi_5$-H_2 system, the first calorimetric studies above room temperature were reported in 1989 [40], and a first tentative phase diagram was proposed in the same year by Shilov et al [41], more accurate thermodynamic data are certainly necessary above room temperature to contribute to a better knowledge and more efficient use of IMHs.

3.1.2. *Chemical Contamination.* The large specific area after activation (0.2-$5 m^2/g$) corresponds to a highly reactive powder that acts as a getter attracting all contaminant molecules present in the hydrogen gas. Keeping in mind the technological aspect, the important points to be investigated here are the surface modifications leading to surface segregation, the deterioration of the catalytic effects, and the slowing down of absorption kinetics.

The time evolution of metal-H_2 impurity interactions has been studied extensively and their effects quantified by Sandrock and coworkers [42a-d] for the support of AB_5 and AB prototype alloys. The alloy impurity effects were classified as poisoning, retarding, reactive, and innocuous. These studies were complemented with an analysis of the alloys capability of recovering their absorption properties. A model was developed correlating the cyclic loss of H_2 storage capacity to a damage function and to a contaminant concentration dependence. This work has been corroborated by other studies in this area [43a-e].

Interest in metal-H_2 surface interactions grew as batteries were developed and with the idea of hydrogen storage through the support of thin film metal-hydrides. The latter field was initiated in 1983 by Wenzl [44], who prepared a thin film of NbV solid solution alloy. The greatest advantage to this approach resides in its overall heat transfer capability, as heat transfers much more easily in this configuration. Jain [45] recently reviewed the hydrogen storage in thin film metal-hydride. Thin film deposition has also been used to produce a highly reactive surface, thereby removing the barrier to hydrogen dissociative absorption, such as occurs with Pd-coated Mg samples [46]. This achieves a catalytic and protective effect. A $LaNi_5$ thin film has been used by Shirai et al [47a-b] to inject protons into an amorphous WO_3 thin film which changes its colour with the absorption of hydrogen. By using this phenomena, different metal films can be inserted between $LaNi_5$ and WO_3 films for investigations of their permeability. This technique was recently applied to study the penetration of hydrogen through an amorphous V_2O_5 film deposited on WO_3 and covered with $LaNi_5$. Thin film samples of FeTi, TiNi, TiPd, $TiMn_2$, and $LaNi_5$ were prepared, layer by layer, by controlled metal evaporation [48] while the effects of oxygen precoverage on the reduction of the hydrogen absorption rate was investigated. The results show trends similar to those found for metal films and metal / titanium sandwich films.

Even if contamination is unavoidable, there is no longer any major problem in holding the impurities down to the lowest level, thereby increasing considerably the lifetime and high performances of the IMCs, at least for solid-gas applications. The degradation encountered in metal hydride electrodes will be discussed in the relevant section.

3.1.3. *Stability and Cycling.* Disproportionation may occur in cycling, and will result in phase separation and the formation of more stable hydrides, thus reducing the H_2 storage capacity of the material. Cycling experiments are time consuming and obviously require great care to avoid coupling with other phenomena that will accelerate the disproportionation process, such as segregation induced by gaseous impurities. Different methods of ageing / cycling are currently being reported, such as temperature or pressure induced cycling, or cycling using two coupled reactors, as in a heat pump procedure [53a]. A static technique was recently proposed by Sandrock et al [49] for studying disproportionation; the IMC is loaded at high temperature while the hydrogen gas pressure remains above the dissociation plateau pressure of the hydride. To date, no general explaination has been given to a better understanding of the disproportionation process, although the results of the experiments seem to depend upon the alloys used and upon the method adopted. The safest way of determining the long-term stability of an IMH is still by the experiment.

The first report on temperature cycling concerned the $EuRh_2$-H_2 system, in 1978 [50], and this was followed by similar studies on $(La_{0.9}Eu_{0.1})Ni_{4.6}Mn_{0.4}$ [51a-c]. Degradation of this compound was reported as the temperature was increased from room temperature up to 300°C. After 1500 cycles, the hydrogen absorption capacity was only 26% of the initial value. It was confirmed that disproportionation was an intrinsic process and was not induced by any oxidation of the sample, as that would have been detected by Mössbauer spectral evidence of Eu_2O_3. As cycling continues, the metal atom diffusion processes are enhanced by the large lattice strains induced by the volume changes during hydriding. Moreover, the heat generated during absorption can lead to local temperature gradients, which in turn, promote diffusion and rearrangement of the metal sublattice [50].

Buschow [52] developed a model to describe the tendency of an IMC to disproportionate. Although the model is based on an oversimplified description, it does show that decomposition does not depend on the stability of the ternary hydride, nor on the stability of the corresponding IMC, but suggests that it is correlated with a metal atom diffusion activation energy in ternary hydrides. Following this reasoning, low temperature cycling and good heat transfer capability of the reactors are needed in order to prevent degradation. These conditions were satisfied in ageing experiments performed with a dual bed configuration for the $LaNi_5$ and $LaNi_{4.7}Al_{0.3}$ compounds between 25 and 80°C, experiments which avoided any over pressure on the samples [53a-b]. It was shown that the compounds retained their hydrogen absorption capacity up to 1250 cycles, at which point the experiments were stopped.

The same compounds studied by Lee and coworkers [54a-b], by thermal cycling, between 30 and 185°C, at a cycling frequency of 10 minutes, showed reduced capacity of the order of 50 and 12% after 2500 cycles for $LaNi_5$, and $LaNi_{4.7}Al_{0.3}$, respectively. Similar results were obtained by Gamo et al [55], who reported degradation of the $LaNi_5$ during pressure cycling, where fresh hydrogen gas was introduced in each cycle. The stability of multicomponent mischmetal-nickel compounds were also investigated by cycling with impurities [56-57].

Long term cycling ability experiments have also been carried out on TiFe. Reilly subjected TiFe to 13000 absorption cycles without noticeable deterioration [58], but thermally induced cycling performed by Lee et al [59a-b] between 30 and 200°C produces a drastic reduction in storage capacity. The cycling stability of $TiFe_{0.8}Ni_{0.2}H_x$ was tested between 20 and 185°C by Ron and coworkers [60a-b] up to 65000 cycles where a full cycle lasted for about 8 minutess. Periodically X-ray, reaction rate, and PCT measurements were taken. After 65000 cycles, the system exhibits a slight increase in absorption pressure, 0.1 atm at H/M=0.3 and 55°C, and a reduction of about 16% in the hydrogen uptake.

Nonetheless, degradation in cycling does not seem to be an insurmountable problem, although some IMC's are more sensitive than others. However, because it is suggested that degradation is accelerated by thermally activated processes, such as metallic diffusion, the use of IMC's should currently be limited in use to low temperature domains, 100-150°C. Unless additional metallic elements contribute to reducing the decomposition tendency, this will be an improvement to the development of multicomponent alloys for use in commercial devices.

3.2 TRANSPORT PROPERTIES

3.2.1. *Kinetics Survey.* It has been argued that cycling rates will not be limited by the chemical kinetics of the reactions, but rather, by the capacity to add or remove heat from the hydride bed. This argument has often been used to support the assumption that quasi-thermodynamic equilibrium is reached instantaneously, greatly simplifying the modelling of hydride reactor dynamics. Though this is true for most IMC's, this assumption should be reconsidered is some cases, especially for low temperature uses of hydrides. Note that, for any hydride storage unit, once we know the kinetic law and the temperature dependence, we can determine the theoretical limits for the exchange rate of the device.

It is beyond the scope of this chapter to review such a broad field as heterogeneous IMH kinetics in detail. However, this does deserve comment in the context of hydride device optimisation. IMH kinetics have been extensively studied, and since 1983, considerable effort has been made to improve the experimental observations, chiefly to satisfy quasi-isothermal conditions for which only intrinsic kinetic parameters are involved. The excursion of several decades in rate constant for the $LaNi_5$-H_2 system, as compiled by Goodell et al [61], has now been reduced to one decade, but there is still poor agreement in the reaction mechanisms and in the overall kinetics laws suggested.

Here again, the difficulties result from the presence of activated powder of undefined grain shape, and from the presence of hydrides, which are unstable at room temperature. So there is some difficulty in associating the kinetics experiments with metallographic studies to identify the mechanisms controlling the reactions, as was done by Mintz and Bloch, who studied the hydriding reactions mechanisms of rare-earth hydrogen systems for a given sample geometry [62].

The fact that the rate equation has to account for more than one process seriously complicates the interpretation of kinetics experiments, to the point that modelling is the most serious tool of progress. Goodell and Rudman [63] proposed extending the nucleation and growth formulation to solve the kinetics of $LaNi_5$. The model was developed on the basis of three intrinsic processes operating in series, which were finally reduced to a surface and a bulk process, leading to the identification of four parameters, two rate constants and two activation energy parameters. It is interesting to note the similarity of this model with the unreacted core model [64] widely used in chemical reaction engineering, a model which involves three consecutive steps, surface limitation, penetration, and diffusion through the reacted layer to the surface of the unreacted core, and transformation reaction.

Others have analysed the experimental data by a simpler model yet, where the various processes involved are described by an overall empirical law of the form

$$\text{Rate} = k.f(p).f(x), \tag{5}$$

where k is the rate constant, which should be only temperature dependent to reflect the mobility of the reaction through an Arrhenius term, f(p) is the thermodynamic driving force through a pressure dependency function, and f(x) is the function expressing the dependency of the transformed product. Although such an expression is apparently simpler to solve numerically, and certainly easier for modelling hydride reactors, the recent literature does not provide any such unambiguous law. So we will make no attempt here to discuss the different mechanisms proposed, but will limit discussion to the proposed laws published for AB_5 and AB_2 compounds. The literature concerning earlier studies will be found in the various papers listed.

The $LaNi_5$-H_2 system was investigated by Josephy and Ron [65], Koh et al [66], and Han and Lee [67] who also studied the $LaNi_{4.7}Al_{0.3}$-H_2 system, as did Wang and Suda [68a-c]. All authors claimed to work with high-performance reactors that maintained isothermal conditions, while the experiments were performed under isobaric conditions for references 65-67 and isochoric and variable pressure conditions for references 68a-d and over a temperature range extending from room temperature to 90°C.

Josephy and Ron [65] found that the decomposition of the hydride satisfied a first-order type of reaction with a linear pressure dependence. The interesting point is that they were able to identify an intrinsic rate constant from which an activation energy of 29 $kJmol^{-1}$ could be determined from the temperature dependence. The experiments of Koh et al [66] are very similar, but their data did not satisfy a first-order type of plot, while a logarithmic dependence of pressure is suggested and activation energies of 27 and 37 $kJmol^{-1}$ are obtained for the absorption and desorption processes, respectively. These values cannot be compared with the previous ones, because the rate constant includes a pressure dependence. Although the thermal ballast technique was used to insure good heat transport in the bed, the data indicate temperature variations for both processes, variations which definitively influence the interpretation of the experiments [69a-b]. Han and Lee [67] do not proposed overall rates, but analyse the various possible mechanisms in depth.

Suda et al derived empirical equations where f(p) and f(x) in equation (5) are transformed into an analytical function $f(x^a, p^b)$ where a and b are assimilated to a reaction order with respect to hydrogen concentration and hydrogen pressure. This technique allowed them to discern different kinetic behaviours in the single and two phases region. For the two phase region, they derived an intrinsic rate constant with activation energies of

36.8 and 54 $kJmol^{-1}$ for the absorption and desorption processes respectively. Recently, Wang and Suda [70] reported a new experimental technique using a highly heat-sensitive reactor, with controlled cooling by injection of liquid carbon dioxide. The results for the hydriding reaction in the two-phase region are reasonably close to the earlier results, where the difference amounts to 6% for the rate constant and for the activation energy (39.2 $kJmol^{-1}$). The only common feature for all the reported works is that the absorption process is faster than the desorption.

Bernauer et al [71] investigated the reaction kinetics of the multicomponent system $Ti_{0.98}Zr_{0.02}V_{0.43}Fe_{0.09}Cr_{0.05}Mn_{1.5}$ at -90 to -100°C and found that the data could be explained by first-order kinetics with two different rate constants for low and high hydrogen concentrations.

Iron, cobalt and copper have been partially substituted for nickel in the $LaNi_5$ system to study the effect this has on the reaction rates [72a-b]. The effect of surface contamination on the hydriding behaviour of $LaNi_5$ has also been thoroughly analysed by Uchida and coworkers [73a-b]. These authors were able to correlate the changes in hydriding rates with the alterations of the highly reactive surface due to contamination, and they made a critical analysis of the different rate equations which are currently used.

Reilly [74a-b] proposed to use metal hydride slurries, hydride particles suspended in n-undecane, to study the isothermal kinetics of fast chemical processes. $LaNi_5$-H_2 and $Pd_{0.85}Ni_{0.15}$-H_2 systems were investigated. It was shown that for both systems the kinetics were found to be in agreement with a phase boundary controlled model. The rate limiting process is the phase transformation that takes place at the interface between the unreacted hydride and the hydrogen-saturated layer of metal product.

In view of experimental progress over recent years, we look forward to coming studies that will provide the appropriate intrinsic kinetic parameters which, when combined with the other physical properties, will make it possible to model the dynamics of hydride processes with greater precision.

3.2.2. *Thermal Properties and Confinement.* The low thermal conductivity of powdered hydride beds, (packed beds or porous media, as it is termed in thermal physics) was recognised early on as a key parameter in reactor design. There is no question that the greater the thermal conductivity of the sample, the shorter the cycling rates that will be tolerated, thereby increasing the performance of the reactors.

Steady state and transient methods have been used to determinate the thermal properties of packet hydride beds, and the thermal conductivity has been correlated with hydrogen concentration and pressure. Suda et al [75a-b] pointed out that in non-steady state experiments, the uncertainties in the thermal conductivity are closely related to the determination of the reaction rate use in solving the heat equations. The effective (overall) thermal conductivity, K_e, of powder hydride beds will rarely exceed 1-2 $Wm^{-1}K^{-1}$, while all measurements reported in a plot of K_e versus the log p_{H_2} lie on an 'S' shaped curve. This reflects gaseous transport properties, where different heat flow regimes exist in the powder, depending on the mean free path of the gaseous molecules relative to the size of the pores of the beds.

Effective thermal conductivity is essentially interpreted using theoretical models, which should be able to predict this property. In view of the numerous parameters affecting K_e, such as the individual thermal conductivity's of the solid and gas phases, the bed porosity, and the size distribution and shape of the particles, each of which are considered to affect the functional dependence of K_e differently, care should be taken in selecting the model. Heat flow analyses have, to date, essentially been conducted using models formulated in the sixties, such as those developed by Kunii-Smith [76] and Yagi-Kunii [77a-b] (referred to as K.S. and Y.K. in the following). A recent survey of the thermal conductivity of packed beds was published by Tsotsas and Martin [78].

Suda et al selected the K.S. model to identify the K_e of the reference material, but retained an empirical polynomial form for the expression of K_e for the hydride beds. A similar representation was maintained in succeeding reports, where a three-dimensional structure [79] and then an inclusion of a copper wire matrix [80] were used to enhance heat transmission. Suissa et al. [81] based their interpretation on the Y.K. model and came to the conclusion that K_e remains within a very narrow interval even if the thermal conductivity of the solid changes over a very broad range. Finally the Batchelor model (1977) used by Kempf [82] seem to be inappropriate as the value it produces for K_e is too low.

Detailed heat flow analysis is not entirely without interest as it does give some idea of order of magnitude of K_e, but the various techniques for improving thermal conductivity should now be approached in a more rigorous way than by trial and error. The different heat transfer mechanisms have to be analysed more deeply, as proposed by Pons et al [83a-c], who developed a new technique for measuring the effective thermal conductivity and the wall heat transfer coefficient. The experiments are simulated by a numerical model of the heat transfer in the whole reactor, with validated assumptions. The simulation results agree with the experiments, whereas the analysis shows that the temperature gradients induce composition gradients within the hydride bed, making the local temperatures, composition or reactions rates very different from the averaged values.

Various techniques have been tested to improve the thermal characteristics of the beds. The methods use different ways of binding a metallic element to the powder to form a composite mixture, such as aluminium foam added to the powder by Groll et al [84], preparation of composite compacts (PMH) by Ron et al [85], Tuscher et al [86] and Wang et al [87], a cold pressing and sintering technique developed by Töpler et al [88], and micro encapsulation within a cooper layer a few microns thick by Ishikawa et al [89a-c]. So the problems of confinement of small particles in materials can be eliminated as long as cracks do not form in the pressurised hydride pellets. A gain of a factor of 10 to 50 in K_e is reported for the PMH of Ron, while the technique of Töpler provides a material with K_e of the order of 7 to 9 $Wm^{-1}K^{-1}$, a value that is also the same order of magnitude as that reported in reference 80.

In their careful and extensive study, Ron and coworkers [90a-c], showed that while K_e can be greatly increased by compacting the IMC with a metallic matrix, unfortunately the permeability of the bed decreases drastically above a critical amount of the binding metal. This leads to mass transfer limitations which, in turn, affect the heat rate output of the reactor. They were able to describe K_e for the composite material with an empirical law and then concluded that K_e and the permeability factor would have to be optimised to obtain high hydrogen and thermal yields. The conclusion should rather have been that the coupling between heat and gaseous mass transfers are affected by numerous parameters, such as different heat flow mechanisms, bed porosity and permeability, and particle size and shape, all which should be taken into account in analysing and optimising the performance of packed beds.

Other methods have been employed to increase the heat transfer capacity of the IMH, such as the obvious one of modifying the reactor design. Different reactor configurations have been proposed by Lewis [91], and Aoki et al [93a-b]. A metallurgical approach was attempted by Ogawa et al [94] who increased the alloy's resistance to disintegration by unidirectional solidification on the $LaNi_5$-Ni eutectic alloy, with the idea of achieving a 'fibrous' composite material. Uchida et al [95] mixed the IMC powder with silicone rubber, making a sheet that could be given various shapes for battery applications. Finally, metal-hydride slurries have been proposed by Johnson and Reilly [96] to overcome the main difficulties with powder beds. Thermal conductivity is increased, bed expansion eliminated and reactor designs simplified. There are other disadvantages due to the solvent vapour pressure, solvent degradation, agitation of the bed, and larger reactor size. The $LaNi_5$-H_2-n-undecane system has retained greater attention.

3.2.3. *Dynamic properties.* From the properties discussed so far, we may now advance the basic criteria for further development of hydride technology. Considering the dynamic behaviour means that the real performance of a hydride has to be evaluated in terms of the mass of hydrogen stored or transferred, and power produced or required. The discussion now shifts to the present situation in modelling of hydride reactors.

It is well agreed that metal hydride reaction beds have to progress in two directions, increasing the available hydrogen content or the hydrogen mass really transferred, and increasing the thermal conductivity of the beds. The former goal may be reached by searching for new alloys or improving existing ones, while the latter calls for more experimental investigations and a better understanding of the thermal transport phenomena as explained above. The best way to clarify the state of research and development of hydride reactors consists of adopting an engineering viewpoint using two hypothetical ideal coupled reactors (R_1, R_2), which are associated with heat sources (HS_1, HS_2) and heat exchangers (HE_1, HE_2), as shown in Fig.1. The equations governing the overall process of hydrogen transfer are written below for the three levels.

Heat Source 1
$$M_{s1} Cp_{s1} \frac{dT_{i1}}{dt} = \dot{m}_1 Cp_{s1}\left(T_{o1} - T_{i1}\right) + W_1 \tag{6}$$

Heat Source 2
$$M_{s2} Cp_{s2} \frac{dT_{i2}}{dt} = \dot{m}_2 Cp_{s2}\left(T_{o2} - T_{i2}\right) + W_2 \tag{7}$$

Heat Exchanger 1
$$\dot{m}_1 Cp_{s1}\left(T_{o1} - T_{i1}\right) = \dot{m}_1 Cp_{s1}\left(T_{i1} - T_{r1}\right) \eta_1 \tag{8}$$

Heat Exchanger 2
$$\dot{m}_2 Cp_{s2}\left(T_{o2} - T_{i2}\right) = \dot{m}_2 Cp_{s2}\left(T_{i2} - T_{r2}\right) \eta_2 \tag{9}$$

Reactor 1
$$\left[\sum mCp\right]_1 \frac{dT_{r1}}{dt} = \left|\Delta H_1\right| \frac{dMH_1}{dt} + \dot{m}_1 Cp_{s1}\left(T_{i1} - T_{r1}\right) \eta_1 \tag{10}$$

Reactor 2
$$\left[\sum mCp\right]_2 \frac{dT_{r2}}{dt} = \left|\Delta H_2\right| \frac{dMH_2}{dt} + \dot{m}_2 Cp_{s2}\left(T_{i2} - T_{r2}\right) \eta_2 \tag{11}$$

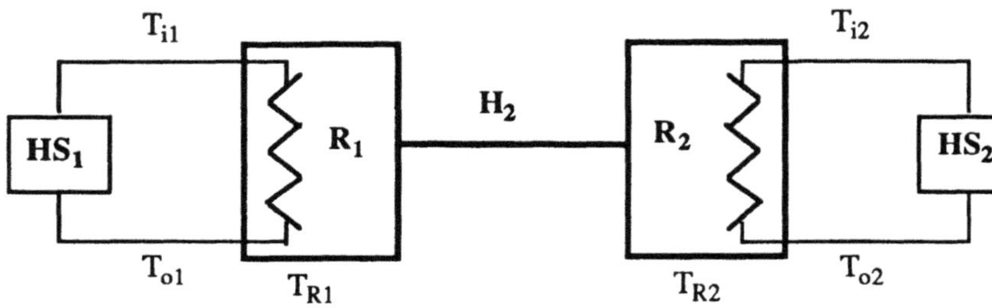

minimum physical parameters needed :

Heat Source, HS_i :	M_{si} ,	mass of fluid i
	$\dot{m}_{si}$	mass flow rate of fluid i
	Cp_{si}	specific heat of fluid i
	W_i	heating, cooling rate of the source i
Heat Exchanger, HE_i :	T_{ij} , T_{oj}	inlet, outlet temp. of fluid j
	η_j	efficiency of the exchager j
Reactors, R_i	T_{ri}	temperature of hydride
	$\left[\sum mCp\right]_i$	total heat capacity of reactor
	MH_i	mass of hydride i
	$\lvert\Delta H_i\rvert$	heat of formation of hydride i

Fig. 1. Representation of ideal coupled hydrides reactors.

The energy balance equations (6-9) assume the operational technical parameters are known, such as the power of the heat source mass flow rate of the fluid exchanger, efficiency of the heat exchanger, while equations (10-11) represent the thermal balance for the chemical reactions evolved in each reactor. The first term of the second member corresponds to the chemical heat source, part of which is transferred through the heat exchager and part of which is accumulated. The energy equations are complemented by the mass balance equations, which include the state equation for each IMH and the corresponding kinetics laws. This set of ordinary differential equations describes the full aggregate system and the solution is found numerically. Depending on the accuracy with which the physico-chemical and thermal properties of the materials are expressed in the model, the numerical methods can be used for elementary dimensioning or simulation or, in the best of cases, for process optimisation.

The last point has not been reached to date. The proposed models, for coupled or single reactors, can be used for dimensioning and for reasonably good numerical simulations, but no more, because the set of equations are based on oversimplified assumptions. For example the hydrides are assumed to remain homogeneous with no spatial distribution of temperature, pressure or hydrogen concentrations, with all resistance to heat transfer within the hydride bed reduced into one overall heat transfer coefficient. The other weakness of the models is related to the extended use of approximated empirical state equations and kinetic laws.

At this point, it is important to point out that much research is needed to achieve thermodynamic optimisation of dynamic hydride systems. This work undoubtedly will have a positive counterpart in the future development of reactor design and the development and optimisation of adaptive hydride systems.

4. Applications

4.1. HYDROGEN STORAGE

A common feature of all the various means of hydrogen production is the need for storage. This need can be satisfied by 'conventional' methods, such as pressurised gas and liquid storage, or by 'recent' technologies, such as metal hydrides systems and cryoadsorber storage, where hydrogen is stored at 70K in specially manufactured activated carbon [97].

The notion of size has to be introduced when speaking of storage systems, because, depending upon what use is considered for the hydrogen, different constraints have to be obeyed. For example, large stationary storage units should, in principle, be available at in the production site, and have a storage capability of the order of 2000-3000 m^3 S.T.P.. Smaller stationary storage units will take over for the local distribution of hydrogen, while mobile storage systems will be reserved for transport and distribution to different sites, or for hydrogen fuelling vehicles.

Table 1 - Storage Properties of Various Materials

Material	H_2 Weight Density wt%	H_2 Volume Density gdm^{-3}	Energy $MJkg^{-1}$	Density $MJdm^{-3}$
MgH_2	7	101	9.9	14
$FeTiH_{1.95}$	1.75	96	2.5	13.5
$LaNi_5H_{6.7}$	1.37	89	2	12.7
Liq. H_2 (20K)	100	70	141	10
Gas H_2 (100atm)	100	7	14	1
Cryoadsorber (77K)	3.8 --> 5.2	15-->30		

Hydride storage capacity is evaluated in terms of the mass of hydrogen stored either per kilogram of hydride, or per liter. This is summarised in Table 1 for the most familiar IMHs, and is compared with other storage methods. Compared with 'conventional' storage, metal hydrides offer obvious advantages of compactness in reactor design, moderate pressure and safe uses, and energy savings relative to the energy costs of liquefaction. Other parameters also favour hydride storage, such as the range of storage unit sizes possible, and the quality of the gas.

Since the first studies on metal hydride storage around 1973, storage has been proven viable, but, unfortunately, its economical feasibility has not been achieved. This is particularly true for large storage units, for which development has practically ceased. The high cost is blamed on the cost of the alloys, but also on the investments in heat exchangers to remove heat from or supply it to the reservoirs. Smaller storage units (2-50 m^3 S.T.P) are now commercially available, and whether or not they will come into extensive use depends on whether or not applications can be identified where the advantages mentioned above will raise hydride storage over cryogenic and compressed gas methods of storage. After a short survey of earlier work on hydrogen storage, the commercial units will be discussed.

4.1.1. *Large, medium Sized Stationary Storage.* The first research into hydrogen storage began in 1973 at Brookhaven National Laboratory (BNL), where the possibility of storing electric energy produced in off-peak hours by producing hydrogen from an electrolyser followed by hydrogen storage was investigated [98]. A demonstration prototype was built by Public Service Electric and Gas Co. of NewJersey, with the cooperation of BNL. The system underwent 60 cycles without difficulty. The hydride storage received hydrogen from an electrolyser via a compressor. Then hydrogen was released from the reservoir to a fuel cell to produce electricity. The storage unit contained 400 kg of FeTi and an effective charge of 6.4 kg of hydrogen (72 m^3 S.T.P). Heat was removed by water circulation through an internal heat exchanger operating between 15 and 45°C. It is clear that this earlier prototype did not produce the optimum performance but it did have the merit of showing that when combined with other hydrogen related technologies, large scale hydride applications were possible.

This idea was soon taken up by Suzuki et al [99], who loaded a storage reservoir with $MmNi_{4.5}Mn_{0.5}$, with a storage capacity of 16 m^3 S.T.P. In a demonstration plant at Kogakuin University, the hydride storage units insure purification and compression up to 7 MPa [6] with a capacity of 240 m^3 S.T.P. Ono et al [100] reported a semi-commercial scale reservoir based on $LaMmNi_{4.8}Al_{0.2}$ and a hydrogen capacity of 175 m^3 S.T.P. Though other storage units have been developed since, they will be mentioned in other types of applications, such as commercial size hydride heat-pump reservoirs.

The largest storage system for which information is now available was developed with the support of the European Economic Community by HWT, a division of Daimler-Benz and Mannesmann [101-102]. It has a capacity of 2000 m^3 S.T.P stored in 10000 Kg of the AB_2 Laves phase compound, $Ti_{0.98}Zr_{0.02}V_{0.49}Fe_{0.09}Cr_{0.05}Mn_{1.5}H_3$. The hydrogen is loaded and unloaded within one hour, provided the appropriate energy is supplied for cooling and heating during the absorption and desorption processes. The plant has an assembly of 32 reservoirs containing the active material. Each storage unit is an expansion of smaller reservoirs, which are described below. This storage facility was constructed mainly to investigate the generation of ultra pure hydrogen on an industrial scale, but it also provides the opportunity to examine the full potential of large scale hydride storage, as well as other possibilities, such as compression, and recovering hydrogen from industrial processes.

4.1.2. *Small Stationary Storage*. These units can be purchased from companies such as Ergenics [103] and HWT. Their main applications are in laboratories, where a strong argument in their favour is in their purifying effect. As shown in technical notes [104a-b] a hydride storage reservoir fed with industrial gas will release hydrogen with a total contamination of the order of 1 ppm. The storage characteristics of the proposed reservoirs are listed in Table 2.

Of course, manufacture and characterisation of the IMCs is associated with hydride technology. For low temperature storage, <100°C, and moderate pressure, < 7MPa, Ergenics [105] produces a family of AB_5 and AB, HY-STOR alloys, and HWT has developed a series of AB_2 multicomponent compounds. The basic research on the synthesis, structural characterisation, and thermo-physico-chemical properties of the latter compounds are summarised by Bernauer et al [71].

The heat transfer media has an influence on the hydrogenation rate, so the storage units are proposed with still-air, forced-air, or water regulation for operation. The heat exchanger is external to the reactor containing the alloys, and consists of an external jacket with circulating water in the Ergenic models, and a longitudinally finned core for the HWT units [106]. The difficult problem of pulverisation and expansion during hydriding has been solved by including a specially designed micro filter and confining the alloy in Al capsules which, if well adjusted, act as fins that increase the heat flow from the hydride beds to the external walls and within the hydride (HWT). These hydride devices represent the state of the art in the technology of hydride storage. No restriction can be made on the quality of the hydrogen gas supplied and the variety of alloys makes it possible to adapt the storage units to each particular requirement, as proposed by Ergenics.

We will conclude this section with a question. What laboratory or company could afford to invest 3000DM for $1m^3$ storage reservoir from HWT, where $20m^3$ of compressed gas cost roughly 300DM?

Table 2. Commercial H_2 Storage Reservoirs

	Type	Storage Capacity m^3	Total Weight kg	Working Pressure 10^6 Pa	Flow Rate dm^3/min	Working Temp. °C
Ergenics	ST-1	0.04	0.5	< 2.5		depends
	ST-45	1.275	24			on alloy
	ST-90	2.500	36		0 - 85	selected
HWT	KL 114-1	1	12	< 4	0 - 20	20 - 40
	KL 114-2	2	22	< 4		or
	KL 114-3	3	33	< 4		20 - 100
	KL 114-5	5	51	< 4		

4.1.3. *Mobile Transportation*. Magnesium-based hydrides were proposed to be used for vehicular hydrogen fuel storage in 1969 by Hofman et al [107]. The exhaust heat of the engine produced the energy required to decompose the hydrides. Unfortunately, the stored hydrogen could not be fully discharged at a high enough rate because the heat had to be supplied at a high temperature (300°C) and the amount of energy needed was too high to be supplied by the rate of heat generated by the engine. As a consequence, the hydrogen rate was insufficient to follow the engine demand. This problem was partly solved in 1973 with the discovery of the less stable $FeTiH_x$ hydride. Although $FeTiH_x$ presented a lower specific mass and lower energy density than the Mg alloy, it does have the advantage of using low grade heat, low temperature, for discharging hydrogen, while high-grade heat, high temperature, could be restricted to the Mg alloys. A compromise had to be made when the dual bed configuration was employed. The perspective of using metal-hydride for fuel storage generated great interest up to the beginning of the 1980s. During that period, some gasoline vehicles were converted to hydrides, and efforts in that field were largely sponsored by publics funds. The leader and pioneers were Billings Corporation, now Hydrogen Energy Corporation of Kansas City, and Daimler-Benz. A list of different hydride vehicles is given in Table 3.

In spite of their weight penalty, hydride vehicles compare favourably both with the other alternative power sources mentioned at the beginning of the section, and with electric propulsion. The most serious competitor is still liquid hydrogen storage, which has been the subject of much continuing research in Germany and Japan over the past ten years. Eventhough boil off losses and safety problems have not yet been solved, progress has already been made in engine operation [108].

Table 3. Hydride Vehicles

Type of vehicles converted to hydrides :
Billings Corp. (BC), now Hydrogen Energy Corp. of Kansas City: Buick Century, Peugeot sedan, AMC Jeep, Winnebago buses, pick up truck, tractor.
Hydrogen Consultants Inc. (DRI): Pick-up truck, caterpillar, forklift.
Characteristics: see references 10, 18, 110

Mercedes - Benz : (test fleet)	MB 280 TE sedan H_2 -gasoline	MB 310 van H_2
Engine	2.8 dm^3	2.3 dm^3
Total weight, kg	2350	3500
Payload, kg	400	700
Range, km	150	120
H_2 storage	280 kg, Ti(VMn) 11 dm^3 gasoline	560 kg, Ti(VMn) 22 dm^3 gasoline
hydride containers	2	4
Toyota - Japan Steel	forklift (for sale since 1990)	

During the same period the feasibility of hydrides for automobiles had to be proved. This included the development of new alloys, tests on long-term behaviour, stability with regards to gaseous contaminants, container development, optimisation and safety, and real road tests.

The problem of high mass flow rate of hydrogen was not fully solved with the FeTi alloy which also has the disadvantage of having two different plateau pressures, which increases the difficulties in controlling the hydrogen gas mass flow rate. Daimler-Benz, after trying this possibility, investigated the AB_2 compounds mentioned above, which were used on their second generation of vehicles. The tests were conducted using five hydrogen powered vans and five sedans with mixed gasoline-hydrogen reservoirs. The results of the test fleet were reported by Töpler and Feucht [109] after the vehicles had travelled a total of over 250000 km . The report includes an investigation of various parameters of importance for technical purposes, such as the stability over extensive cycling, sensitivity to gaseous impurities, and their effects on the kinetics of hydrogen exchange. As for most IMC's, the most harmful contaminants are O_2, CO, CO_2, H_2O. They reduce the capacity and cycling stability, whereas N_2 and CH_4 affect only the kinetics of absorption. The other technical problems associated with refuelling the cars gave satisfactory results over a total of 3900 cycles. In general the tanks were charged to a level 80% in 10 minutes.

No data concerning the environmental impact of hydrogen have been furnished, although it would have been of importance to compare this with gasoline cars equipped with catalyzers. Hydride vehicles should contribute only to NO_x pollutants, due to the combustion of hydrogen with the air mixture, while hydrocarbons would essentially be reduced to traces, owing to lubricant coatings, and that is assuming that hydrogen is not produced from coal. In the present economic situation, experimental hydride vehicles remain at this stage of development.

It is easily understandable that further development of hydride vehicles is still a long term affair that can be supported only by public funds, and weighted by environmental problems. As already mentioned by others, if there is a future for hydride vehicles, it will be for fleet vehicles, like bus transportation, taxis, and delivery vans. Note that, in some cases, where ballast is required to improve traction or stability, such as for farm tractors, forklift, caterpillar engines, in which the weight 'penalty' of the hydride reservoir transforms into a weight 'advantage' [110-111]. For these systems the transition from prototype to commercial unit, only depends on the industrial choice.

4.2. CHEMICAL SELECTIVITY

4.2.1. *Getters.* Since gettering was first established by Langmuir, getter materials are widely used in the area of vacuum-sealed electronic devices, in vacuum pumping, as well as in other practical applications, such as keeping vacuum insulation in dewars or in thermostated bottles. Getters are classified in two groups, the first comprises evaporable getters such as Ti evaporated film and the second, non-evaporable or bulk getters, which have the advantages of compactness, and of being manufacturable in almost any desired shape. Research and development on bulk getters has received greater attention in the last ten years because of the potential applications in thermonuclear fusion reactors where fuel quality and impurity controls are absolutely necessary. The development of these materials benefited directly from the possibilities offered by IMC reactivity with hydrogen and other chemical elements.

Different commercial materials are available from SAES [112], Ergenics, and HWT. Depending on temperature and pressure they are capable of sorbing irreversibly reactive species like O_2, N_2, CO, and CO_2. Other molecules, such as H_2O, NH_3, and CH_4 can be decomposed at given temperatures where the basic elements, O, N, and C are then absorbed. As H_2 and its isotopes, D_2 and T_2, are pumped reversibly at low temperature, getters act as a storage material from which the isotopes can be released at higher temperatures at a later time. The characteristics of getters, their selectivity and recommended temperatures, can all be found in technical notes provided by the companies mentioned above.

SAES markets multicomponent alloys based on Zr metals, under the trade names ST 101, Zr-Al alloys, STI 98, Zr-Fe alloys, and ST 707, Zr-Fe-V alloys. Mendelsohn and Gruen [113] investigated alloys of $Zr(V_{1-x}Fe_x)_2$ composition and demonstrated their usefulness as bulk getters. The products proposed by Energics are referenced under the trade names HY-Stor 140 and 105 for H_2 and H_2O gettering, while HY-Stor 402 covers a broad range of active species such as H_2, N_2, O_2, H_2O, CO, and CO_2. These series of alloys are developed with Zr as base materials. Gettering with rare earth based alloys is proposed with HY-Stor 501.

The properties of the multicomponent materials developed so far for fusion applications are not fully satisfactory if we consider the overall process, of storing, supplying and recovering tritium gas. The absorption pressure is too high at room temperature; the alloys should be able to satisfy an absorption pressure of 10^{-7} Pa at room temperature and one atmosphere at 500°C [114].

The problem of gettering hydrogen isotopes has been investigated by Cann et al [115], for preventing hydride cracking in Zr-2.5% Nb pressure tubes in nuclear power reactors. During operation of the reactor, the tubes absorb deuterium from the heavy water coolant, through corrosion. Yttrium, selected as getter, was encapsulated in Zr alloy and fixed to the tubes. The effect of yttrium hydride formation on gettering rates and hydrogen isotope concentration in the Zr alloy near the Zr alloy / yttrium interface were determined at 586 K for different hydrogen fluxes. The conditions required to prevent delayed hydride cracking were thereby established.

4.2.2. *Purification.*

The boundary between purification and separation may be set by convention according to the levels of impurities to be removed. 'Purification' will be used if the impurities do not exceed 0.1%, otherwise 'separation' is a more appropriate term to define the process.

The commercial multicomponent alloys mentioned above for gettering processes are the materials used for gas purifiers. The commercial units proposed today offer wide possibilities in many areas ranging from the laboratory to industrial uses. They are well suited to the manufacture of semiconductor components, optical fibers, sensors, etc. Several hydride purifiers are marketed by Ergenics [116] and HWT [117], from simple reactor / canisters to the fully controlled device. These systems are able to produce an output with a level of impurities as low as one part per billion. Such performance of course requires that the parts of the system attached to the output gas must be consistent with the handling of ultra high purity gases. Improvements in electropolished surfaces, high quality welding, high quality bellow valves, purifier elements assembled in a dust-free rooms make ultra high purification systems very competitive with conventional purification procedures.

The models proposed by Ergenics remove H_2 and its isotopes at room temperature, while other reactive impurities are removed by applying the appropriate temperatures and gas flow rates to the purifier. The temperature ranges from 250°C for O_2 to 450-500°C for CO and CH_4. Hydrogen purification is achieved while the gettering alloys operate in a hydrided state. HWT has developed a flexible system which can be used with industrial gases, with a relatively high level of contamination. The principle of the units is represented schematically in Figure 2. Depending on the quality of the gas input (output), the gas is introduced (withdrawn) at different levels of the purification unit. Then the lowest quality

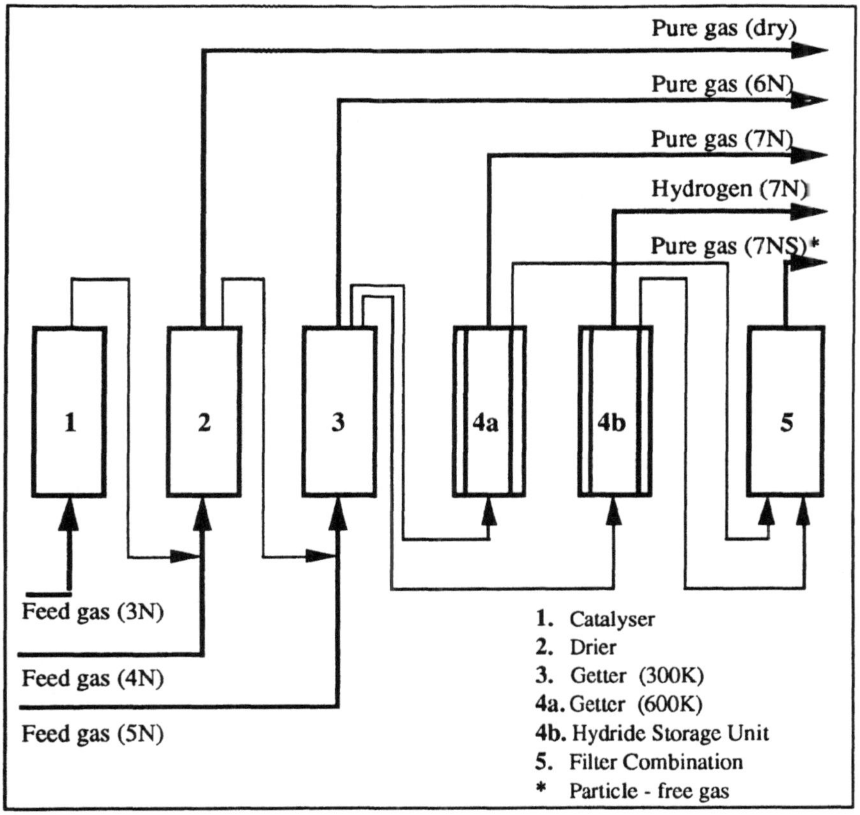

Fig. 2. HWT purification concept.

input gas is pre-treated by conventional procedures, including catalysts (1) and dryers (2), before starting the purification with the getter alloy (3), which eliminates all reacting elements at room temperature. At that point, the residual of O_2, CO, and H_2O are already below 0.1 ppm, where a 6N quality is achieved. The high temperature treatment starts at steps 4a, 4b. So for ultra high purity hydrogen production, metal hydrides with high absorption pressures are needed that can trap all the other reactive constituents, while for inert gas production, a stable hydride with low absorption pressure is needed to trap H_2 and the other impurities. The systems operate on the trough flow principle. Finally the dust-free gas is produced by passing it through specially designed filters. The unit can process gas continuously at flow rates of 15 m^3 S.T.P per minute.

A small hydrogen purification process has been developed at the University of Unicamp, Brasil [118], which insures high quality hydrogen distribution over their laboratories.

4.2.3. *Separation.* The removal of one species of gas from a mixture containing constituents other than hydrogen is of prime importance in terms of economic production. This problem concerns the production of helium from a mixture composed of He-H_2, and the production of H_2 from the bleed gases of an ammonia plant. Another well known case is tritium recovered from a mixture of hydrogen isotopes. These two points will be discussed separately.

4.2..3.a. Industrial.

The first work on H_2 separation based on intermetallic compounds was presented by Reilly and Wiswall [119]. This preliminary laboratory scale research was developed with $LaNi_5$ and FeTi family compounds. Later, a pilot plant based on the selective absorption properties of metal hydrides was developed by Air Product and Chemical Inc. and Ergenics. It was designed for the recovery of H_2 from the purge gas stream of an ammonia synthesis loop [120]. Note that this mixture is favourable to separation because it does not contain large amounts of O_2 and CO that are highly damaging to IMC's. Flow trough reactors are loaded with pelleted $LaNi_5$. The pellets are bonded by sintering with a Ni powder, which acts as a thermal ballast and reduces the heat effects of the hydriding / dehydriding reactions. The incorporation of Ni therefore enhanced the heat capacity and resulted in controlled temperature swings. Ideally, the reactor can behave adiabatically, which reduces the heat flow problems. The process was tested with a mixture of $60\%H_2$-$3\%NH_3$-$37\%N_2$ and pilot tests were run continuously for 6 months. During that time hydrogen recoveries range from 90 to 93%, with a purity of 99%. The advantage of such treatment is that it tolerates ammonia, whereas other processes, as cryogenic or zeolite methods require that the feed gas be free of ammonia.

The idea of hydrogen recovery from the bleed gases of an ammonia plant based on IMC was taken up by Wang et al, who also proposed to recover helium. Unfortunately, the authors reported experiments carried out with several IMCs of the AB_5 family for which the recovery of H_2 was tested only with Ar-H_2 mixtures of different compositions [121].

An industrial separation approach with a metal hydride slurry was reported by Zwart et al [122]. The program resulted in a collaboration between the University of Twente and DSM.RES.BV. of the Netherlands. The IMC selected was $MmNi_{4.5}Al_{0.5}$. The method was tested with different processes to insure continuous recovery of hydrogen. Evaluations were carried out for two gases in an ammonia plant, synthesis gas (75 vol% H_2) and gas from cryogenic recovery system (92 vol% H_2). The configurations investigated represent simple slurry circulation (absorber / desorber), use of a striper, (the slurry is stripped of dissolved inert in counter current with part of the product gas), and increase of slurry concentration; this later step requires heat supplied to maintain temperature imposed at the inlet of the absorber. When compared with other methods, such as pressure swing absorption and membrane separation, a metal hydride slurry offers the advantage that high purity levels of 99.99 vol % H_2 can be achieved with lower purge-stream losses resulting in higher chemical efficiencies.

Recently, Albrecht et al [123] described an experimental program in the framework of tritium technology research at the Next European Torus (NET). The objective was to investigate the separation of gas impurities from a helium-hydrogen or helium-tritium gas mixture for possible applications of commercial gas purifiers. The gases treated had the composition 85-95% Q_2 (Q=H, D, T), 5% He and up to 10% impurities, such as CO, CO_2, N_2, NQ_3, and Q_2O. The ultimate aim was to recover chemically bound hydrogen isotopes from hydrocarbons, ammonia, and water. The separation / purification works as follows, the first step involves separating a high percentage of molecular hydrogen isotopes. This uses metal getters, including a Pd/Ag diffuser, where NQ_3 is decomposed catalytically at the surface of the membrane. The second step decomposes CQ_4 and Q_2O and removes the N, C, and O elements through the getter beds. In the last step the gases are passed through a Pd/Ag diffuser to separate out the remaining Q_2 from He. The facilities were operated to test the performance of several IMC hydride getters, and to obtain the major process parameters.

4.2.3.b. Isotopic.

The first important application of metal-hydrogen systems was realised for the nuclear industry, due to the potential of hydrides for use as moderators and shielding in nuclear reactors. In 1968, Huffine [124] described the hydride manufacturing techniques that had been developed for these purposes. The discovery of IMH had opened up new opportunities for this industry where, the future fusion reactors will require a careful and safe way of handling tritium. The quantities of tritium that will have to be handled in fusion reactors will be of the order of several kilograms.

All the advantage so far claimed for hydride applications were being stressed even more for tritium confinement, where safety must be paramount. Thus purification and separation of isotope mixtures and storage of high purity tritium could all be achieved with a suitable metal or IMC, avoiding at the same time all the problems inherent in large pressure vessels, pumps, compressors, contamination with the walls of the vessel, and tritium decay, which produces a mixture of T_2-^3He. The latter problem can be solved by bulk storage with temperature scans of the tritide reservoir followed by bleeding-off of the residual ^3He gas. A review of storage material for tritium has recently been written by Lässer and Schober [125]. In the following the recent progress in the separation of hydrogen isotopes through the use of bulk materials will be presented after a short survey of the isotopic properties. The notation protium (H) is used in the present section, while Q represents H, D or T.

The isotopic dependence of a metal or of an alloy is evidenced in various properties, such as its thermodynamic properties, diffusion coefficients, and kinetics. Thermodynamics shows that, for the same temperature and composition, the solubility limits and the pressure vary with the isotope and with the host material investigated. Note that the observed effects may be in opposite directions as observed with the well known f.c.c. Pd and b.c.c. Nb, V, and Ta metals. For example, in the Pd-Q_2 system, as the mass of the isotope increases, the solvus line between the α and $\alpha+\beta$ regions is translated to higher concentration, while for the Nb-Q_2 system it is shifted to lower concentration. The desorption pressure within the phase region increases in the order H, D, T for Pd, whereas for Nb the heavier isotopes have lower equilibrium pressures. Concerning the diffusion coefficients of hydrogen isotopes, it is well established that tritium diffuses faster than deuterium and protium in Pd above 100K, while tritium diffuses slower than protium in Nb. The kinetics will obviously be influenced by the bulk diffusion processes, but also by other parameters such as surface exchange rates of the different isotopes for which particle sizes, surface area and catalytic properties of the material are important factors governing the processes.

In view of the complex behaviour of hydrogen isotopes reported for the most extensively investigated metals, it should be no surprised that isotopic dependence in IMCs does not make the determination of the effects any simpler. An example is $CaNi_5Q_2$ (Q=H or D) system that, at room temperature, can form three distinct hydrides phases [126]. Deuteride phases of the same compositions are reported, and different isotopic pressure effects for each phase are observed, a desorption isotherm may show inverse ($P_{H_2} > P_{D_2}$), null or normal effects ($P_{H2} < P_{D2}$). Recently Wei [127] reported opposite equilibrium isotope effects for the solid solution and the hydride phase of $LaNi_5$.

The difficulties in determining accurate thermodynamic data for IMHs apply as well for the isotopic effect determinations. If standard procedures are not established, it will be almost impossible to compare results obtained from different groups. At present the IMCs investigated do not give full satisfaction for tritium storage, separation, or long-term behaviour of the tritide. To satisfy this particular demand, new materials must still be looked for, and efforts are still necessary to contribute to a better understanding of the complex isotope surface exchange phenomena which result in mixtures of HD, HT, and DT [128a-c]. Westinghouse has investigated new materials for separation. According to Fischer [129], $NdCo_3$ is a promising candidate for the separation of hydrogen isotopes.

Among the techniques proposed, chromatographic processes have been undergoing tests for hydride isotope separation for many years. Gluckhauf and Kitti [130], in 1957, studied the separation of a H_2-D_2 mixture through a column of Pd. In 1981, Altridge [131] obtained a patent for a chromatographic procedure based on an AB_5 compound. Several different chromatographic techniques have been brought to bear on the problem of isotopic separation, including conventional and displacement chromatography, pressure and temperature swing methods using multi-stage reservoirs, mainly for gas-solid separation, and parametric pumping for gas-liquid-solid separation [127].

Displacement chromatography is the most common technique and is widely used in the laboratory. The Joint European Torus (JET) has selected it for their separation requiring a maximum daily throughout of 5 moles of tritium and 15 moles of deuterium with a minor presence of protium. For large quantities of protium in the mixture, the separation is insured by a cryodistillation system. The column contains Pd on a porous alumina substrate operating at 20°C and 1 atm. in the β phase of PdQ. The efficiency of the separation is evaluated by the separation factor,

$$\alpha_H^Q = \frac{\left(Q/H\right)^G}{\left(Q/H\right)^S} \qquad (13)$$

where Q and H represent the concentration of the isotopes in the gas, G, and solid, S, phases, respectively. The column operates in four steps. The column is fed at room temperature first with He, then with the gas mixture, and finally with the protium eluant, and is then regenerated at high temperature. Details of the procedure are found in reference 132.

The pressure and temperature swing absorption techniques were extensively studied by Hill and co-workers [133a-b], with a vanadium metal separator. They also contributed greatly to modelling the processes. At the operating temperatures selected, tritium was preferentially absorbed by the solid, while in the kinetic effect tritium was absorbed and released more slowly than protium in an absorption / desorption cycle. The two effects were then opposed and were used to define different operating conditions. For example, for short cycles, two minutes, the kinetic isotope effect controlled the separation and the process produced a high-pressure product enriched in T_2 and a low-pressure product depleted in T_2, with the system operating at 373K. For long cycles, four hours, the system was operated at lower temperature and the process was controlled by the equilibrium isotope effect. The systems investigated required multistage separation, where the stages are assembled in series to form a continuous cuntercurrent cascade. The enriched stream from each stage flows in one direction and the depleted stream flows in the opposite direction. This was modelled to determine the number of stages required to accomplish a specified separation.

The possibility of using a ZrV_2 bed as a means of recovering hydrogen isotopes from an inert gas has been investigated by Mitsuishi et al. [134]. The column was studied experimentally from room temperature up to 300°C for absorption, and regeneration at temperatures between 400-800°C. Parametric analysis of the concentration of the species during the flow, the so called 'breakthrough curves', showed that they were greatly affected by heat generation during absorption by the mass transfer coefficient of the fluid (solid) phase and by the interface area per unit volume of the bed. Recently, the flow of hydrogen isotopes through a chromatographic Pd column coupled to a reflux tank was computed by Nichols [135]. This process operates semicontinuously by thermal cycling while the desorbed gas is collected in the reflux tank. It is then recycled to the cold column for another purification. This problem is analogous to a reflux stream in a distillation column.

Isotope separation using metal-hydrides or IMH is certainly feasible on an industrial scale. As far as safety is concerned for the handling of tritium, the search for new materials is still an open field. More investigations are also needed to understand the dynamics of the exchange reactions taking place in order to maintain efficient separation.

4.3. HYDRIDES AND THERMODYNAMIC DEVICES

The potential of hydrides now extends to those applications where a secondary effect, such as compression, heat generation, or heat pumping is important. The hydride reservoirs are the basic tools, and the hydrogen gas acts as the working medium. System performance is evaluated in terms of thermodynamic efficiency which has to take into account the many parameters influencing the process.

The technology described for hydrogen storage and purification is still appropriate for validating new concepts. But higher performance reactors will be needed if hydride engines are to be developed on a larger scale, because, here again, they are at a heavy disadvantage in the face of widespread and well established devices. Thus, while improvements are still being made in compressor and hydride heat management devices, efforts should be also be brought to bear on hydride systems for specific domains where only hydrides have the required performances, as this will bolster interest in this technology. A good example is the production of water vapour from low grade energy, as well as the refrigeration of frozen food. Advantages, such as compactness and high energy density, should be kept in mind when hydride heat pumps are compared with other heat pump devices. For compression purposes, hydride compressors are noiseless, clean, and vibration free systems.

4.3.1. *Thermochemical Hydride Compressors*

4.3.1.a. Principle.

The principle of the thermochemical hydride compressor is shown in Figure 3. The temperature dependence of the plateau pressure, Equation 4, is used to compress the hydrogen between the low temperature, T_l, low pressure plateau, P_l, and the high temperature, T_h, high pressure plateau, P_h. In operation, the reactor is repeatedly heated and cooled between T_l and T_h, resulting in a cycle that follows the path ABCD. Except for the

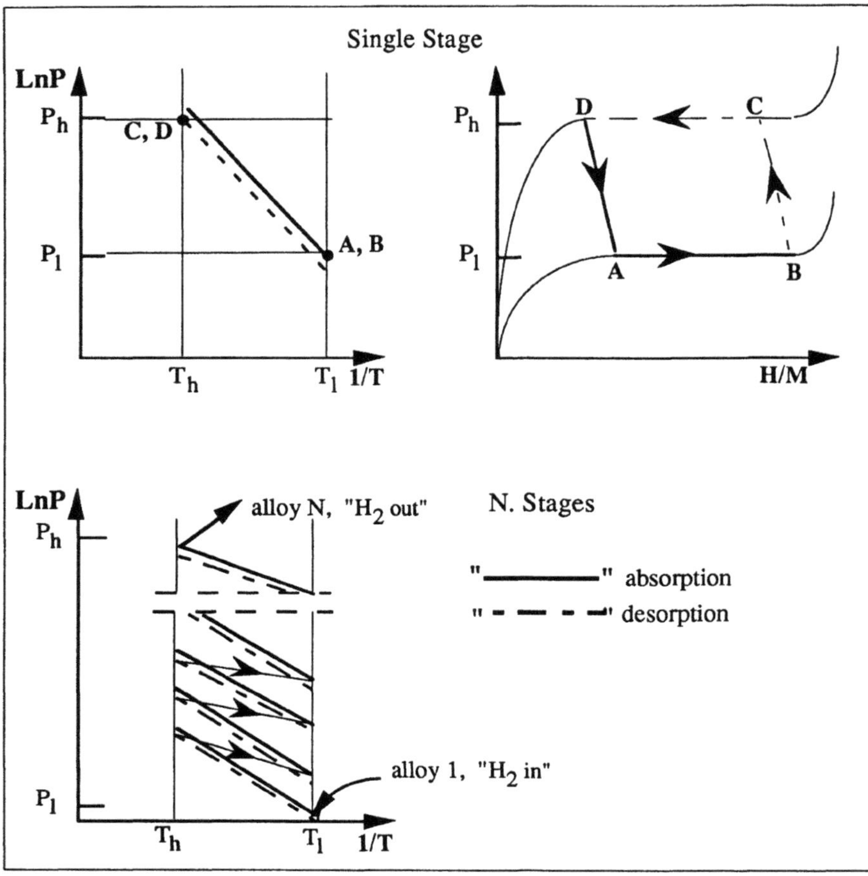

Fig. 3 . Principle of the hydride compressor (see text for explainations)

control valves, such a compressor has no moving parts and is free of vibrations. The compression ratio depends on the alloy selected, on the temperature range, and on the overall system dynamics. A 10 to 1 compression ratio is of the order of what can be expected from present hydride compressors. The thermal energy needed to do the work of compression can be produced from waste heat or solar energy.

The maximum efficiency, the Carnot efficiency, of the thermal compressor is given by the relation,

$$\eta_c = \frac{W}{Q_h} = \frac{T_h - T_l}{T_h} \tag{13}$$

Of course this theoretical limit will not be reached and the efficiency of the compressor is closely dependent on the design of the heat exchanger. Solovey [136] published an inventory of the limiting factors in thermal hydride compressors, and quantified these factors to arrive at an evaluation of the true efficiencies that can been expected from such systems.

4.3.1.b. Development.

Around 1971, Reilly et al [137] described one of the first applications of hydrides to compressors. The compressor used vanadium hydride and was built to drive periodically a mercury column acting as a pump for other gases. The first IMC based hydride compressor using this principle was built at Philips [138]. It comprises three containers filled with $LaNi_5$. Hydrogen was compressed from 4 to 45 atm. with a thermal load at about 160°C and a heat sink at 15°C. The total cycle time reported was about 10 minutes. Another experimental hydride compressor was built by Bertin Co. [139] that operated between 20 and 80°C. This compressor was tested under different operating conditions, and it was shown, as expected, that the useful power did not increase when the cycle time was reduced as expected. The mass of hydride required for one mechanical Watt was of the order of 40g, ten times more than expected. As for early prototypes at Philips, this one demonstrated the feasibility of hydride compressors and proved that the efficiencies could be greatly increased if thermal inertia is minimised. The solution to this problem remains in the design of new heat exchangers.

A great improvement in the overall design of hydride compressors was produced by Golben [140] from Ergenics, who developed the first commercial systems. The device is a four stage metal hydride / hydrogen compressor that uses hot water at 75°C as its energy source. Four stages filled with different alloys from the $LaNi_5$ family are assembled in series in two different beds. The hydrogen transfer is insured automatically by unidirectional valves. The author claims an expected compressor life of 20 years, assuming near continuous operation, if the device is set at a 2 minutes cycle time. Based on this work, two models are commercially proposed by Ergenics, a single-stage electrically powered unit that is able to compress H_2 between 0.5 and 3.5 MPa at an average rate of 3 STL/min, and a four-stage hot water (75°C) heated unit that compresses H_2 from 0.4 to 4.1 MPa at a rate of 40 STL/min for a weight of 64 kg [141].

4.3.1.c. Uses of Hydride Compressor.

Hydride compressors can be coupled to low-pressure hydrogen sources, such as hydrogen electrolysers. They can be used both to recover the boil off losses from liquid hydrogen and to repressurize a hydrogen tank if necessary. One of the promising possibilities that was first demonstrated by Philips, concerned the coupling of their prototype compressor to a Joule-Thomson system with the high pressure hydrogen gas precooled to 78K with liquid nitrogen. Though the refrigeration capacity was reduced by heat exchanger losses, the prototype demonstrated a cold production of about 0.6-0.8 W at 26K. The low reported efficiency of 9.5% resulted from different parameters that were not optimised. According to the authors, 20% of Carnot efficiency, may be achieved by more careful design. Following this idea, Kumano et al [142] worked on the hydrogen liquification problem, developing a new closed hydrogen liquifying system, where the hydride compressor is a three stage compressor, based on the same principle as the one developed by Golben [140].

The system was able to reach a minimum temperature of 16.5K, while tests show that a cold production of 1.2 W at 20K is possible in continuous operation. Further development of the unit has been announced by the authors. Sorption refrigerators based on this concept have aroused interest at Aerojet Electrosystems Co., Azusa, CA, USA and are in the process of development for aerospace applications [143].

One very useful application of hydride compressors was proposed by Golben [144] to solve the problem of water pumping to meet the requirements, defined by the World Bank, for small-scale water pumps. Thermal energy is supplied from flat-plate solar collectors, while the water being pumped is used as a heat sink. The compressed hydrogen applies pressure to a pair of piston cylinders that operate a reciprocating water pump cylinder. The hydride compressor - water pump was sized to meet the 150 W required, but in addition gives an extra 70 W of electrical power to run the operating valves. This system is now commercially available.

4.3.2. *Heat Management: Hydride Chemical Engines.* An important field for the industrial use of hydrides has arisen over the last ten years is that of hydride chemical engines, HCE, which can be used for several possibilities, such as heat pumps for cooling, heating, or both, and heat transformer systems. The pioneers in this particular area are Van Mal [138] in Delft and Gruen et al [145] at Argonne, who developed a machine for cooling purposes, the famous HYCSOS program [146]. This exploratory work was followed by extensive investigations in Japan and to a lesser extent by other groups. Efforts have continued in this direction in Germany and Israel.

Overall, the factors limiting the development of chemical engines are both the cost, and the limited number of materials proposed. The last decade has witnessed many attempts to find a new chemical system, new thermodynamic cycles, and also optimum conditions for a specific use of an engine. However, the absorption-chemical heat pumps that are commercially available, such as lithium bromide / water, activated carbon / methanol, and zeolite / water systems, all operate at restricted temperatures, which ranges from 0-5°C to 100-120°C. This range could theoretically be greatly extended with the use of metal hydrides. Three other major parameters have helped spawn interest in these new engines, the excellent energy density and the concomitant compactness, the high chemical reaction rate and the resulting low cycle time and high power production, and finally the lack of moving parts in a closed system which is only driven by heat.

These HCE are being developed for home, heating and cooling purposes, but they could also satisfy a wide range of applications in industry, where they could make profitable use of thermal effluents. Chemical industries require heat at 150-400°C, but agrofood industries require heat of no more than 150°C. Refrigeration for frozen foods is still a growing market, with the important problem raised by the imposed decrease in chlorofluorocarbon production. For the particular case of food transportation by truck or rail, the machines should be able to produce 'deep cold' during all phases of distribution and storage, which means compact refrigeration systems must be found.

4.3.2.a. Thermodynamic Principles.

The principle of an HCE in its most general sense is to transfer energy from one temperature level to another. Hydride machines work cyclically with energy being transferred via the hydrogen gas from the high pressure level, P_1, to the low pressure level, P_2. Depending upon the direction of the hydrogen gas transfer and the temperature levels at which the transfers occur, the type of engine is different and will satisfy different demands, such as cooling, heating, or both, or heat upgrading. The different possibilities are illustrated schematically in Figure 4 with the help of van't Hoff plots.

Any hydride cycle requires four steps then, which are the following for the heat pump mode: hydrogen at high pressure and high temperature is first transferred from the saturated hydride, AH_x, at T_h, to the hydrogen-free alloy, B, at medium temperature, T_m. Next alloy A is cooled from T_h to T_m, hydride BH_x from T_m to T_l, and then hydrogen at low pressure and low temperature is transferred from hydride BH_x, at T_l, to alloy A, at T_m. Finally the total system is heated to return to the initial state.

Disregarding the dynamic aspect of hydride engines for the moment and following Dantzer and Orgaz [147a-c], if one assumes that the mass of hydrogen transferred at the pressures P_l and P_h occurs with no driving force, $\Delta P=0$ between the reactors, and that the hydriding / dehydriding reactions are ideal then we obtain relations which will allow us to develop simple thermodynamic models that will be of help in selecting the materials.

Figure 4 shows that, for a fixed temperature T_m, the hydride cycle is ideally defined. The T_h, T_m, T_l with the corresponding pressures P_h, P_l provide the optimum alloy pair for the optimum operating conditions. In such a case there is no other hydride pair offering higher efficiency than the single solution, which is in accordance with our description, while the ideal efficiencies are,

$$COP_c = \frac{\Delta H_B}{\Delta H_A}, \qquad COA_c = \frac{\Delta H_A + \Delta H_B}{\Delta H_A}, \quad \text{and} \quad CHT_c = \frac{\Delta H_A}{\Delta H_A + \Delta H_B} \qquad (14)$$

It can easily be shown that these ideal efficiencies correspond to the Carnot efficiencies of the engines represented in Figure 4, for cooling, the coefficient of performance, COP_c, for heating, the coefficient of heat amplification, COA_c, and heat upgrading, the coefficient of heat transformation, CHT_c, ΔH_i being the heat of formation of hydride i.

4.3.2.b. Alloys Selection and Expected Temperature Ranges.

The future of HCEs depends very much on what targets are selected. It is hopeless to try to compete with a zeolite-water heat pump that operates in the temperature range of 5-80°C, for which the cost of zeolite, as well as of the technology, are already much lower [148]. Of course the earlier hydride prototypes had to prove the viability of hydride systems, but this is now well established for single stage systems.

The problem of matching optimum hydride properties with a use in a specific engine is important, because there is no an 'universal' pair of hydrides which satisfy all the many applications. As a consequence, in view of the large number of compounds being investigated, modelling was needed to define tentatively the possibilities offered by the hydride heat pumps and heat transformers, to determine their best mode of operation, i.e., single-stage or multi-stages, and finally to provide the optimised hydrides producing the maximum efficiency for the required temperatures. These problems were analysed by Dantzer and Meunier [149a] who succeeded in classifying the performances of pairs of hydrides according to the temperature ranges required to satisfy European Union guidelines.

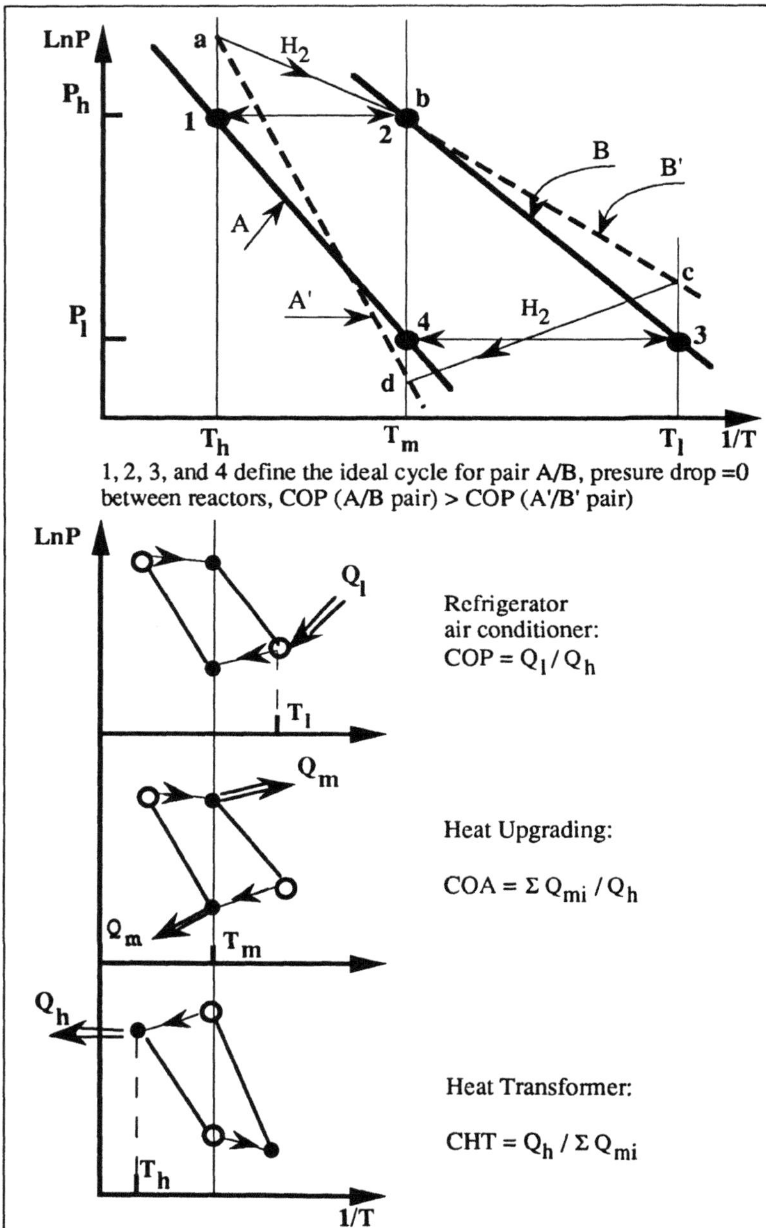

Fig. 4 : Principle of the hydride chemical heat pump (top). Different modes of functionning (bottom), the Qi indicate the thermal energy of interest (pumped or produced) for the selected mode

The method proposed does not suffer from any restriction concerning the number of alloys submitted to the selection, but is limited only by the lack of accurate data concerning the physical and chemical properties of certain alloys. Details of the method, as well as the proposed targets for selected pairs of alloys, are given in reference [149a]. There are two levels of simulation. First, a coarse screening is needed because of the larger number of combinations possible. Here an oversimplified description of the thermodynamic cycle is made. This elementary approach may be used to estimate roughly the performance that can be expected from the HCE. This was the approach also adopted by Balakunar et al [149b-c], who reported a comparative study of nine preselected pairs of alloys for heat pump and heat transformer applications. Then a second screening is made in which the chemical constraints are now introduced into the model. These parameters are deduced from a precise and thorough thermochemical analysis of the $LaNi_5$ / $LaNi_{4.77}Al_{0.3}$ pair [147a]. The final selection insures that the pairs produce machines that will operate within an assigned temperature range δT_m, which, according to the uniqueness of the solution for optimum operation determines the boundaries T_h and T_l. The alloy selection is then reduced to a problem of comparing set temperature thresholds and ordering proposed pairs according to the computed efficiencies. This is easily solved by microcomputers. This is at present the most appropriate simulation / selection based on the results of quasi-static experiments. It is clear that the next step will have to be of an experimental test of the proposed solutions.

Results show that cogeneration of hot water (70°C) and chilled water (2°C), cogeneration of steam (120°C) and chilled water (2°C), of hot water (70°C) and refrigeration (-10°C) are quite improbable with single stage cycles. A temperature increase of 70°C for the cold side of the heat pump is the upper bound that can be hoped for with acceptable efficiencies. Two targets are possible with hydride heat pumps. One is cooling in a range running from air conditioning to refrigeration, down to -30°C. The advantage of HCEs compared with other systems is that they operate at very low evaporating temperatures which is not the case for solid adsorption heat pumps. But kinetics and heat transfer may then become limiting effects. If these problems are overcome, refrigeration at -30°C may be a promising field of application. Another application is as a heat transformer, vapour water production which would require some progress in the materials for use as a low grade heat source. The high temperature of 300-400°C may be reached with magnesium alloy hydrides. But we will disregard these possibilities for the present, knowing that work is going on in several directions concerning the synthesis of the alloys. Their potential for high temperature use is presented in Section 4.4.

4.3.2.c. Prototype Development

The simplicity of the hydride chemical heat pump is only in the concept. Unfortunately the real processes that go on between two coupled hydride reactors involve a relatively large number of parameters which, up to now, have delayed the production of high performance systems. In spite of the lingering gap between predicted and real performance, there is constant progress in the prototypes, which have now reached the size of large, potentially commercial units. Recent investigations and improvements can be seen in patent applications filled in the last few years [150a-e].

HCEs have been the subject of recent articles. Suda summarises the activities and achievements in Japan and describes the major contribution made by his group [6]. Examples of prototypes operating in Japan are given in Table 4. The prototypes studied include all the possibilities of hydride heat pumps from the laboratory scale up to

Table 4 - Examples of metal hydride heat pumps installed in Japan

Output (kcal/h)	Cycle, Th, Tm, Tl (°C)	Alloy (kg)	COP
9600 cooling + hot water supply 15000 heating + hot water supply	Heat Pump / Refrigeration: 92 - 110, 30, 15	R.E. base: 90	0.9 cool. 1.4 heat.
2000 chilled water generation	Refrigeration: 164, 20, 13	R.E. base: 40	
1500 cooling	Refrigeration: 140, 40, 10	R.E. base: 38	0.4
4000 chilled water generation	Double effect / Refrigeration: 150, 45, 10	R.E. base: 48	1.7
3000 heat source	Heat transformer: 90, 70, 15	R.E. base: 46	0.38
1000 refrigeration	Refrigeration:	R.E. base: 18 Ti base: 17	
6000 cooling + heating	Heat Pump / Refrigeration: 280, 144, 5	Ti base: 180	0.8 cool. 1.7 heat.
13000 cooling + heating	Heat Pump / Refrigeration: 72, 50, 2 - 5	Ca base: 800	
150000 cooling + heating	Compression type: heat source Temperature during heating : 60 during cooling : 30	Ca base: 4000	6 - 7
350000 cooling + heating*	Heat Pump: 80-85,65, city water Cooling : 80-85, 25, 5	Ca base: 3600	
66000 steam generation	Double effect, Temperature raise: 150, 80, city water	R.E. base: 2600	
1000 heat source	Heat transformer: 100, 80, 20	R.E. base: 12**	
***	Refrigeration: 140 -170, 45, 5	R.E. base: 10000	
18000 cooling 42000 heating	Heat Pump / Refrigeration:	Ti base / Zr base: 600	0.45 cool. 1.1 heat.

* Single cylinder heat exchanger

** Copper clad hydrogen absorbing alloy utilized

*** Mainly intended to store heat (35000 cal) from nighttime power commercial size units. Ron and Josephy [151] have quantified the parameters for hydride chemical heat pumps feasibility, and described the main parameters affecting the specific thermal power output. The net power output at the useful load per unit weight of hydride is the relevant parameter that has to be determined if we want to compare the performance between prototypes. Though predictions are currently feasible, real evaluation is not a trivial problem and requires a properly instrumented system.

Reports on the operational characteristics of hydride chemical heat pumps are found in a limited number of publications. Nagel et al [152] investigated the performances of a hydride chemical heat pump in a refrigeration cycle. They obtained a cooling output of 1.28kW, or 120 W/kg of desorbing alloy, for an optimum cycle time of 13 minutes. Elsewhere Argabright [153] reported on a joint project involving the South California. Gas Co., Solar Turbine Inc., and BNL, for the design of high performance heat exchangers for hydride heat pumps. The relatively low cooling power was attributed to the unoptimized designs of certain parts of the system, which increased the heat capacity and affected the cooling performance of the heat pump. Ron et al [154a-c] designed an hydride chemical heat pump for air conditioner use in buses. The high temperature source was the waste heat of the exhaust gases of the motor. Laboratory tests of this system showed that the lowest possible temperature was in the range -2°C < T < 7°C, with a temperature increase of 27°C. The cooling power is the highest reported up to date at 250W/kg.

Yanoma et al [155] described the design and the operation of a commercial size heat transformer with heat upgrading from 70 to 90°C. Although the advantage of producing hot water at 90°C from a heat source at 70°C does not seem to open a wide field of application, the prototype is of technical interest for the future. Note that, once the system has been designed and assemblied, the only experimental factors that can be varied are the temperature levels of the heat sources, the mass flow rate of the fluid through the heat exchangers, the fluid itself, and the cycling time. Given these parameters, the experimental studies will determine the best operating conditions for the machine, conditions which do not necessarily correspond to the optimum performance of the alloys. In this particular case the engine produced 174 kW output, or 49 W/kg of hydride at a cycle time of 10 minutes.

A two-stage laboratory scale hydride heat transformer lab model was developed by Werner and Groll [156a-b] to study the dynamic behaviour of the reaction beds. Suda et al [157] published informations related to multistage prototypes.

4.3.2.d. Dynamics of Hydride Beds, Experiments, Models and Problems.

From the literature, the total cycle time that has been determined that is best for system operation is of the order of 10 to 15 min. The estimation of useful power, will depend on the output load temperature, on the alloy selected, and on the cycle time. The results of the simulation of Dantzer and Meunier [149a] show that 50 to 150 W/kg could be reached for refrigeration from -30 to 0°C on a 12 minutes cycle time. This can surely be done with today's technology. Because hydride heat pumps will certainly have to operate with shorter cycle times to increase power and competitiveness with respect to other proposed systems, this will require accurate control of the reactions dynamics and the operating parameters. The present stage of development of hydride reactors, and the associated problems, were analysed in Section 3.2.3, so the following discussion is limited to the presentation of the work already produced in this field.

Tusher et al [158] reported the dynamic characteristics of single and dual hydride beds for a cylindrical configuration of the reactors filled with a compact porous metal hydride.

The experiments, conducted in 1983, demonstrated how difficult it was to achieve isothermal conditions. The authors report that the dynamic isotherms are deeply influenced by the ability to transfer heat between the hydride bed and the heat exchanger. This type of observation has been corroborated by others. Thus Nagel et al [159], by changing the operational conditions for several operational modes, showed that the amount of hydrogen transferred is correlated with the real dynamic behaviour of each hydride. A detailed analysis of the data is found in reference 160. The problem of dynamic pressure was also investigated by Josephy et al [33], Groll et al [34].

In 1986, Bjurstrom investigated in detail the dynamic behaviour of coupled hydride reactors with the support of experimental and numerical studies [161a-c]. The experimental work was done in Suda's group, and led to the conclusion that the hydriding kinetics, and the corresponding dehydriding kinetics in the second reactor are both controlled by heat transfer. A mathematical model was constructed using averaged parameters to describe the heat transfer and reaction processes. The model is based on five equations, two per reactor, including the kinetic and energy balance equations and a mass balance equation. Satisfactory agreement was reported between experiment and the simulated process in spite of the oversimplified assumptions. This is the most advanced model to date, and it may be of very great use in predicting the dynamic behaviour of coupled hydride beds if the heat transfer and kinetics parameters are known. Although it is oversimplified for purposes of optimisation, the model could be used in the first step of the design of new reactors. This type of modelling lead Gambini [162] to make a parametric analysis of an hydride chemical heat pump, an analysis which confirmed the previous results, whereas Da Wen Sun and Deng [163] tried a purely numerical analysis of the dynamics of hydride beds.

4.4. HIGH TEMPERATURES APPLICATIONS

Magnesium hydride is the prototype material for high temperature applications, but to date its low dehydriding kinetics has restricted its use to purely academic applications. However, since Bogdanovic and co-workers [164] developed new methods of preparation, the magnesium hydride properties have been greatly improved. They succeeded in developing a nickel-doped, catalytically activated, non-pyrophiric magnesium hydride with negligible hysteresis, in which the reaction kinetics are much faster. Following the Bogdanovic method, Raisi et al [165] also obtained doped magnesium hydride with comparable behaviour. Heat storage applications have been considered for this magnesium hydride prepared in this new way where it is integrated into a small-scale solar thermal power station [166]. The development of this prototype is being pursued in a joint project by the Max Planck Institute, Mühleim, Bomin Solar Gmbh Co, Lörrach and IKE, Stuttgart. Detailed information on the solar power station has been given by Groll [167]. A specially designed solar collector is connected to Stirling engines to generate electricity. The use of magnesium hydride for heat storage allow the engine to maintain constant power output during cloudy weather and after sunset. The heat is transferred between the different parts of the system, the solar absorber, Stirling engine, and thermal energy storage reactors, by heat pipes. Solar energy produced during daylight hours is transferred to the Stirling engine and to the high temperature source, 200-300°C, of magesium hydride, which decomposes to a secondary, low temperature, hydride. During this step, useful heat is produced at 40 to 80°C. The reverse operation produces refrigeration for the low temperature alloy and heat for the Stirling engine via absorption of hydrogen on magnesium, if necessary. The system has been designed to satisfy a maximum electricity power production of about 1 kW. The amount of nickel doped magnesium will be of the order of 20 kg. Successfull tests of the prototype led to a second generation engine with a maximum electrical power output of 4 kV.

4.5 BATTERIES and FUELS CELLS

4.5.1 *Batteries*. Research into favour of the practical uses of hydride storage materials for electrochemical applications began about 30 years ago. Around 1967, Lewis [168] proposed a Pd-H electrode, and the Battelle Institute, in Geneva, began studying the use of TiNi electrodes for hydrogen storage purposes [169].

In 1970, the need for power sources for space applications led to the development of metal hydrogen batteries to supply energy to satellites. These batteries, which are a combination of a regular storage battery and a fuel cell battery, offered the advantages of high reliability and stability. The disadvantage, due to the presence of relatively large hydrogen gas pressure of 30 bars could be avoided by storing the hydrogen in an intermetallic compound, as proposed and tested by Earl and Dunlop [170]. This was demonstrated by Justi et al [171a-b] by using TiNi electrodes, and later was tested with $LaNi_5$. In 1975, a US patent was filed by Will [172] for a battery in which $LaNi_5$ was the negative electrode and NiOH the positive electrode. A French patent was obtained by Percheron et al [173] for electrode materials based on $LaNi_5$ substituted compounds, and electrochemical use of such materials. This patent was extended in 1977 to new compounds and in 1978 lead to a US patent [174a-b]. The search for new compounds up to 1978, are discussed in reference [175a-c]. In 1984, Willems [176a] reported on the work performed at Philips on metal-hydride electrodes developed with $LaNi_5$ related compounds.

For technical reasons, and because of physical and chemical problems [177], the development of hydride batteries has been delayed; but it seems that environmental considerations, or the possible shortage of Cd, due to the continued and extensive demand of small power supplies, for electronic equipment, computers, or for consumer purposes, are now pressing industrial concerns to accelerate their development. According to Sakai et al [178] the nickel-metal hydride cell was ready for the market in 1991.

nickel electrode: $$Ni(OH)_2 + OH^- \underset{\leftarrow discharge}{\overset{charge \rightarrow}{\rightleftharpoons}} NiOOH + H_2O + e^-$$

hydride electrode: $$LaNi_5 + xH_2O + xe^- \underset{\leftarrow discharge}{\overset{charge \rightarrow}{\rightleftharpoons}} LaNi_5H_x + xOH^-$$

overall reaction : $$LaNi_5 + xNi(OH)_2 \underset{\leftarrow discharge}{\overset{charge \rightarrow}{\rightleftharpoons}} xNiOOH + LaNi_5H_x$$

The concept of sealed rechargeable nickel / Metal hydride batteries can be found in the chapter written by Notten in this book and in reference 176b. The Ni-MH battery consists of a negative hydride electrode and a positive nickel hydroxide electrode, in an alkali electrolyte. Taking the prototype $LaNi_5$ as an electrode material, the reactions that take place to produce current are written above.

According to Bronël et al [175], the measured effective capacity for $LaNi_5H_6$ is of the order of 320 mAh/g, whereas Willems [176] calculated a storage capacity of 370 mAh/g,. The corresponding value for Cd is only 270 mAh/g. This high energy density, and the possibility of high rates of charging and discharging, were soon recognised, together with the low operating temperature of the electrodes. The major difficulties concerned the poor long term behaviour, due to surface damage of the alloy electrode, leading to a drastic drop in capacity after a few cycles. To overcome these problems, a great deal of research has gone into developing appropriate multicomponent alloys to test the large scale process of pulverising and coating the material [179-180], and to study the effect of binding materials in shaping the alloy electrodes [181a-c]. Severall groups [175], [176], [182], and [183] reported the substitution effects for the AB_5 types compounds. The electrochemical properties of Zr-Ni alloys have been studied by Sawa et al [184], who also studied the effects of oxidation on this material. Gamo et al [185] investigated the effects of partial substitution of V and Mo with Ni in ZrV_2, $ZrMo_2$ for electrodes applications. Amorphous Ni-Zr, taken as-cast from a meltspinning machine, was used by Ryan et al [186] in an alkaline solution, with no catalytic agent added.

Efforts have also been directed towards the search for new electrolytes. A solid protonic conductor was used by Poinsignon et al [187] with $TiNiH_x$ as electrode. A novel composite electrolyte configuration has been studied by Liaw and Huggins [188]. Kumar et al [189] showed the usefulness of measuring electrode potential as a function of cathodic charging for the study of the hydrogen absorption characteristics of an IMC electrode, while Jordy et al [190] proposed an evaluation of the metal hydride electrode potential based, on analysis of the isotherm curves.

The degradation of hydride electrode storage capacity with the number of charge / discharge cycles has been extensively studied by various groups. Sakai et al [182] analysed the cycling lives of AB_5 based alloys electrodes, and concluded that $LaNi_{2.5}Co_{2.5}$ offered the highest durability. On the basis of experimental studies, they gave the following qualitative interpretation. The pulverisation rate of the grains may be one of the dominant factors influencing the capacity decay, because pulverisation increases the area of the surface, which then oxidises continuously. Boonstra et al. [191a-c] at Philips concentrated their studies on an $LaNi_5$-Cu pellet electrode in which the degradation processes, the effect of the electrolyte, and the effect of the pretreatment of powder were successively quantified. The overall mechanism proposed to explain the reduced decay closely follows the model proposed by Schlapbach [192] for gas-solid degradation in the presence of water vapour and contaminant oxygen, when a newly cleaned, catalytically active, coated Ni surface was regenerated after formation of lanthanum hydroxide and oxide. The authors showed that the oxidised layers at the surface of the particles are formed by reaction with the electrolyte and that encapsulation of $LaNi_5$ particles by copper diminished the oxide growth without affecting the storage capacity of the electrodes. Chirkov et al [193] developed a model to compute the charge and discharge curves which characterise the electrochemical process. The authors solved the coupled equations which govern the mass transport at the metal / electrolyte interface and the hydrogen diffusion through the bulk and established the limiting amount of charge that could be extracted at a given grain.

4.5.2. *H_2 Fuel Cells*. Development of hydrogen fuel cell is driven by space exploration, with manned-missions. The highly efficient conversion of chemical energy to electricity, in practice, of the order of 60%, and the water production make the H_2 fuel cell very attractive for such use. A by product of this research should occur for vehicular applications. But, for extensive applications, the cells must overcome their weight handicap and they must be capable of air operation. According to Staschevski [194], no system exhibits a weight to power ratio of less than 10kg/kW. Current problems and studies of fuel cells are found in reference 195.

The H_2 fuel cell can be considered as an energy converter working with an invariant electrode / electrolyte system, fuelled by O_2 and H_2 stored in reservoirs. Ergenics [196a-c] developed an ion exchange membrane fuel cell combined with hydride storage vessels. Although the system was developed for a space station shuttle bus, Ergenics extended their studies to the design of high energy density, portable power equipment. Four models are now proposed with from 200W, 12V d.c. to 1000W, 24V d.c. power, while the cell reactants are stored as compressed gases, cryogenic liquids or, for H_2 as a rechargeable metal hydride.

4.6. SENSORS, DETECTORS

Welter [197] proposed the use of metal hydrides as active materials for sensors and regulators. A first example is given a thermostatic expansion valve, in which the temperature sensing element is filled with a metal hydride. The hydrogen pressure variation on temperature is used to drive a membrane or bellows. When compared with conventional sensors filled with a volatile organic liquid, the metal hydride shows an advantage in avoiding recondensation phenomena in the cooler section. The second case is an electrothermal resistor that is used as a level and flow monitor for liquids.

It is appropriate to mention in this section the most successful application of a metal-H_2 system, that is as a fire detector which is currently used in most airplanes throughout the world [198]. The sensor is based on the Ti-H_2 system enclosed in a long stainless steel tube which controls a switch. Any heat generation along the tube will cause a pressure increase that will activate the switch.

5. CONCLUSIONS

All the applications which have been proposed for metal-hydrides have proved to be viable. The limitations are economic. But considering that 15 to 20 years is a very short time for developing a new technology, and in view of the progress already accomplished, we may reasonably expect that the use of hydrides at the consumer level will expand in many areas. The best examples are provided by the fire detector, the commercial hydride reservoirs, purification systems and the hydride batteries.

References

1. J. J. Reilly, G. Sandrock: Scientific Americain **242**, 118 (1980).

2. J. Van Vucht, F. A. Kuijpers, H. Bruning: Philips Res. Reports **25**, 133 (1970).
3. J. J. Reilly, R. H. Wiswall: Inorg. Chem. **13**, 218 (1974) .
4. G. G. Libowitz, H. F. Hayes, T. Gibb: J. Phys. Chem. **62**, 76 (1958).
5. D. M. Gruen, F. Schreiner, I. Sheft: Int. J. Hydrogen Energy **3**, 303 (1978).
6. S. Suda: Int. J. Hydrogen Energy **12**, 323 (1987).
7. W. M. Mueller, J. P. Blackledge, G. G. Libowitz: *Metal Hydrides* (Academic Press, New York, 1968).
8. G. Alefeld, J. Völkl, eds: *Hydrogen in Metals II*, Application - Oriented Properties, Topics Appl. Phys. Vol 29 (Springer, Berlin 1978).
9. R. Barnes, ed.: *Hydrogen Storage Materials*, Materials Science Forum, Vol 31 (Trans Tech Publications, Switzerland, 1988).
10. G. Sandrock, S. Suda, L. Schlapbach: "Applications" in *Hydrogen in Intermetallic Compounds II*, ed. by L. Schlapbach, Topics Appl. Phys. Vol 67 (Springer, Berlin, 92).
11. *Hydrogen in Metals*, Proc. of the Int. Symp., Belfast, Ireland, 1985, eds. F. A. Lewis, E. Wicke: Z. Phys. Chemie N.F. **143-147** (1986) with references to earlier conferences in this series.
12. *Int. Symp. on the Properties and Applications of Metal Hydrides V*, Maubuisson, France, 1986, eds. A. Percheron-Guegan, M. Gupta: J. Less-Common Met. **129-131** (1987), with references to earlier conferences in this series.
13a. *Metal-Hydrogen Systems : Fundamentals and Applications*, Proc. of the first Int. Symp. combining "Hydrogen in Metals" and "Metal Hydride", Stuttgart, Germany (1988) eds. R. Kirchheim, E. Fromm, E. Wicke, Z. Phys. Chemie N.F. **163-164** (1989).
13b *Metal-Hydrogen Systems : Fundamentals and Applications*, Proc. of the 1990 Int. Symp. Banff, Alberta, Canada, ed. F. D; Manchester: J. Less-Common Met. **172-174** (1990),
13c *Metal-Hydrogen Systems : Fundamentals and Applications*, Proc. of the 1992 Int. Symp. Uppsala, Sweden, eds. D. Noréus, S. Rundqvist, E. Wicke: Z. für Phys. Chem. **179-183** (1992).
14. *Hydrogen Energy Progress VII*, Proc. of the 7th World Hydrogen Energy Conf., Moscow, USSR, 1988, eds. T. N. Veziroglu, A. N. Protsenko, Vol I,II,III, (Pergamon Press, New York, 1988).
15. C. J. Winter: Int. J. Hydrogen Energy **12**, 521 (1987).
16. H. Quadflieg: Int. J. Hydrogen Energy **13**, 363 (1988).
17. J. O'M. Bockris, J. C. Wass: "About the real economics of massive hydrogen production at 2010 AD, in *Hydrogen Energy Progress VII*, **1**, 101 (Pergamon Press, New York, 1988).
18. M. A. DeLuchi: Int. J. Hydrogen Energy **14**, 81 (1989).
19. O. Bernauer: "Development of hydrogen-hydride technology in the FRG" in *Hydrogen Energy Progress VII*, **1**, 181 (Pergamon Press, New York, 1988).
20. A. N. Podgorny: Int. J. Hydrogen Energy **14**, 599 (1989).
21. H. Tagawa, T. Otha: "Japan hydrogen energy and technology program" in *Hydrogen Energy Progress VII*, **1**, 153 (Pergamon Press, New York, 1988).
22. A. Percheron-Guegan, J. M. Welter: "Preparation of Intermetallics and Hydrides" in *Hydrogen in Intermetallic Compounds I*, ed. by L. Schlapbach, Topics Appl. Phys. Vol 63 (Springer, Berlin, 1988) pp. 1-48.

23. F. D. Manchester, D. Khatamian: "Mechanisms for Activation of Intermetallic Hydrogen Absorbers" in ref. 9, pp. 261-296.

24a H. Uchida, K. Terao, Y. C. Huang: Z. Phys. Chemie N.F. **164,** 1275 (1989).

24b H. Uchida, A. Hisano, K. Terao, N. Sato, A. Nagashima: J. Less-Common Met. **172-174**, 1018 (1991).

25. T. B. Flanagan, W. A. Oates: "Thermodynamic of Intermetallic Compound Hydrogen Systems" in *Hydrogen in Intermetallic Compounds I*, ed L. Schlapbach, Topics Appl. Phys. Vol 63 (Springer, Berlin, 1988) pp. 49-85.

26. W. A. Oates, T. B. Flanagan: Mat. Res. Bull. **19**, 1397 (1984).

27. T. B. Flanagan,W. A. Oates: J. Less-Common Met. **100,** 299 (1984).

28. R. Griessen, T. Riesterer: "Heat of Formation Models" in *Hydrogen in Intermetallic Compounds I*, ed. by L. Schlapbach, Topics Appl. Phys.Vol 63 (Springer, Berlin, 1988) pp. 219-284.

29a O. J. Kleppa, P. Dantzer, M. E. Melnichak: J. Chem. Phys. **61**, 4048 (1974).

29b P. Dantzer, O. J. Kleppa: J. of Solid State Chem. **24**, 1 (1978).

29c G. Boureau, O. J. Kleppa, P.Dantzer: J.Chem. Phys. **64,** 5241 (1976).

29d G. Boureau, O. J. Kleppa: J. Chem. Phys. **65**, 3915 (1976).

29e P. Dantzer: J. Phys. Chem. of Solid **44,** 913 (1983).

29f P. Dantzer, O. J. Kleppa: J. of Solid State Chem. **35**, 34 (1980).

29g P. Dantzer, O. J. Kleppa: J. Chem. Phys. **73**, 5259 (1980).

29h C. Picard, O. J. Kleppa: High Temperature Science **12**, 89 (1980).

30. K. H. J. Buschow, H. H. van Mal: J. Less-Common Met. **29**, 203 (1972).

31a. C. N. Park, T. B. Flanagan: Ber. Bunsenges. Phys. Chem. **89**, 1300 (1985).

31b P. Dantzer, M. Pons, A. Guillot: Z. für Phys. Chem. **183**, 205 (1994)

32. P. D. Goodell, G. Sandrock, E. L. Huston: J.Less-Common Met. **73**, 135 (1980).

33a Y. Josephy, M. Ron: Z. Phys. Chemie N.F. **147**, 233 (1986).

33b M. Ron, Y. Josephy: J. Less-Common Met. **131**, 51 (1987).

34. M. Groll, W. Supper, R. Werner: Z. Phys. Chemie N.F. **164**, 1485 (1989).

35. T. B. Flanagan, J. D. Clewley, T. Kuji, C. N. Park, D. H. Everett: J. Chem. Soc. Faraday Trans1 **82**, 2589 (1986).

36. W. R. Mc. Kinnon: J Less-Common Met. **91**, 293 (1983).

37. S. Quian, D. O. Northwood: Int. J. Hydrogen Energy **13**, 25 (1988).

38. A. L. Shilov, N. T. Kuznetsov: J. Less-Common Met. **152**, 275 (1989).

39. A. Percheron-Guegan, C. Lartique, J. C. Achard, P. Germi, F. Tasset: J. Less-Common Met. **74**, 1 (1980).

40. P. Dantzer, E. Orgaz, V. K. Sinha: Z. Phys. Chemie N.F. **163**, 141 (1989).

41. A. L. Shilov, M. E. Kost, N. T. Kuznetsov: J. Less-Common Met. **144**, 23 (1988).

42a G. Sandrock, P. D. Goodell: J. Less-Common Met. **73,** 161 (1980).

42b P. D. Goodell: J Less-Common Met. **89**, 45 (1983).

42c F. G. Eisenberg, P. D. Goodell: J. Less-Common Met. **89**, 55 (1983).

42d G. Sandrock, P. D. Goodell: J. Less-Common Met. **104,** 159 (1984).

43a L. Schlapbach, J. C. Achard, A. Percheron-Guegan: Proc. 3ième Congrès Int. Hydrogène et Matériaux, 7-11 juin 1982, A6.

43b R. Suzuki, J. Ohno, H. Gondho: J. Less-Common Met. **104,** 199 (1984).

43c P. Selvam, B. Viswanathan, C. S. Swamy, V. Srinivasan: Z. Phys. Chemie N.F. **164**, 1199 (1989).

43d J. I. Han, J. Y. Lee: J. Less-Common Met. **152**, 319 (1989).

43e J. I. Han, J. Y. Lee: J. Less-Common Met. **152**, 329 (1989).
44. H. Wenzl, K. H. Klatt, P. Meuffels, K. Papathanassopoulos: J. Less Common Met. **89**, 489 (1983).
45. I. P. Jain, Y. K. Vijay, L. K. Malhotra, K. S. Uppadhyay: Int. J. Hydrogen Energy **13**, 15 (1988).
46. A. Krozer, B. Kasemo: J. Less-Common Met. **160**, 323 (1990).
47a H. Shirai, H. Sakaguchi, G. Adachi: J. Less-Common Met. **159**, L17 (1990).
47b H. Sakaguchi, H. Seri, G. Adachi: J. Phys. Chem. **94**, 5313 (1990).
48 H. G. Wultz, E. Fromm: J. Less-Common Met. **118**, 315 (1986).
49. G. Sandrock, P. D. Goodell, E. L. Huston, P. M. Golben: Z. Phys. Chemie N.F. **164**, 1285 (1989).
50. R. L. Cohen, K. W. West, R. H. J. Buschow: Solid State Communications **25**, 293 (1978).
51a R. L. Cohen, K. W. West, J. H. Wernick: J. Less-Common Met. **70**, 229 (1980).
51b R. L. Cohen, K. W. West, J. H. Wernick: J. Less-Common Met. **73**, 273 (1980).
51c R. L. Cohen, K. W. West: J. Less-Common Met. **95**, 17 (1983).
52 K. H. J. Buschow: Mat. Res. Bul. **19**, 935 (1984).
53a P. Dantzer: J. Less-Common Met. **131**, 349 (1987).
53b J. Bonnet, P. Dantzer, B. Dexpert, J. N. Esteva, R. Karnatak: J. Less Common Met. **130**, 491 (1987).
54a J. M. Park, J. Y. Lee: Mat. Res. Bull. **22**, 455 (1987).
54b J. I. Han, J. Y. Lee: Int. J. Hydrogen Energy **13**, 577 (1988).
55. T. Gamo, Y. Moriwaki, N. Yanagihara, T. Iwaki: J. Less-Common Met. **89**, 495 (1983).
56. Y. G. Kim, J. Y. Lee: J. Less-Common Met. **144**, 331 (1988).
57. Q. D. Wang, O. J. Wu, N. Qiu: Z. Phys. Chemie N.F. **164**, 1305 (1989).
58. J. J. Reilly: *Proc. Int.Symp on Hydrides for Energy Storage,* Geilo, 1977, (Pergamon, New York, 1978), p. 301.
59a H. J. Ahn, S. M. Lee, J. Y. Lee: J. Less-Common Met. **142**, 253 (1988).
59b J. Y. Lee, S. S. Park: Z. Phys. Chemie N.F. **164**, 1337 (1989).
60a M. Ron, Y. Josephy: Z. Phys. Chemie N.F. **164**, 1343 (1989).
60b E. Bershadsky, Y. Josephy, M. Ron: J. Less-Common Met. **172 - 174**, 1036 (1991).
61. P. D. Goodell, R. S. Rudman: J. Less-Common Met. **89**, 117 (1983).
62. M. H. Mintz, J. Bloch: Prog. Solid State Chem. **16**, 163 (1985).
63. P. S. Rudman: J. Less-Common Met. **89**, 93 (1983).
64. S. Yagi, D. Kunii: Chem. Eng. Japan: **19**, 500 (1955).
65. Y. Josephy, M. Ron: J. Less-Common Met. **147**, 227 (1989).
66. J. T. Koh, A. J. Goudy, P. Huang, G. Zhou. **153**, 89 (1989).
67. J. I. Han, J. Y. Lee: Int. J. Hydrogen Energy **14**, 181 (1989).
68a X. L. Wang, S. Suda: Z. Phys. Chemie N.F. **164**, 1235 (1989).
68b X. L. Wang, S. Suda: J. Less-Common Met. **159**, 83 (1990).
68c X. L. Wang, S. Suda: J. Less-Common Met. **159**, 109 (1990).
69a. P. Dantzer, E. Orgaz: J Less-Common Met. **147**, 27 (1989).
69b C. Bayane, E. Sciora, N. Gérard, M. Bouchdoug: Thermochimica Acta **224**, 193 (1993)
70. X. L. Wang, S. Suda: J. Less-Common Met. **172-174**, 969 (1991).

71. O. Bernauer, J. Töpler, D. Noreus, R. Hempelman, D. Richter: Int. J. Hydrogen Energy **14**, 187 (1989).
72a A. Zarynow, A. J. Goudy, R. B. Schweibenz, K. R. Clay: J. Less-Common Met. **172 - 174**, 1009 (1991).
72b K. R. Clay, A. J. Goudy, R. B. Schweibenz, A. Zarynow: J.Less-Common Met. **166**, 153 (1990).
73a H. Uchida, M. Ozawa: Z. Phys.Chemie N.F. **147**, 77 (1986).
73b Y. Ohtani, S. Hashimoto, H. Uchida: J. Less-Common Met. **172-174**, 841 (1991).
74a J. J. Reilly, Y. Josephy, J. R. Johnson: Z. Phys. Chemie N.F. **164**, 1241 (1989).
74b Z. Gavra, J. R. Johnson, J. J. Reilly: J. Less-Common Met. **172-174**, 107 (1991).
75a S. Suda, N. Kobayashi, R. Yoshida, Y. Ishido, S. Ono: J. Less-Common Met. **74**, 127 (1980).
75b S. Suda, Y. Komazaki, E. Morishita, N. Takemoto: J. Less-Common Met. **89**, 325 (1983).
76. D. Kunii, J. M. Smith : A.I.Ch.E.J. **6**, 71 (1960).
77a S. Yagi, D. Kunii : A.I.Ch.E.J. **3**, 373 (1957).
77b S. Yagi, D. Kunii : A.I.Ch.E.J. **6**, 97 (1960).
78. E. Tsotsas, H. Martin: Chem. Eng. Process **22**, 19 (1987).
79. S. Suda, Y. Komazaki, N. Robayashi: J. Less-Common Met. **89**, 317 (1983).
80. M. Nagel, Y. Komazaki, S. Suda: J. Less-Common Met. **120**, 35 (1986).
81. S. Suissa, I. Jacob, Z. Hadari: J. Less-Common Met. **104**, 287 (1984).
82. A. Kempf, W. R. B. Martin: Int. J. Hydrogen Energy **11**, 107 (1986).
83a M. Pons, P. Dantzer: J. Less-Common Met. **172-174**, 1147 (1991).
83b M. Pons, P. Dantzer: Z. für Phys. Chem. **183**, 213 (1994).
83c M. Pons, P. Dantzer: Z. für Phys. Chem. **183**, 225 (1994).
84. W. Supper, M. Groll, U. Mayer: J.Less-Common Met. **104**, 279 (1984).
85. M. Ron, D. M. Gruen, M. H. Mendelsohn, I. Sheft: J. Less-Common Met. **74**, 445 (1980).
86. E. Tuscher, P. Weizierl, O. J. Eder: Int. J. Hydrogen Energy **8**, 199 (1983).
87. Q. D. Wang, J. Wu, H. Gao: Z. Phys. Chemie N.F. **164**, 1367 (1989).
88. J. Töpler, O. Bernauer, H. Buchner: J. Less-Common Met. **74**, 385 (1980).
89a H. Ishikawa, K. Oguro, A. Kato, H. Suzuki, E. Ishii: J. Less-Common Met. **107**, 105 (1985).
89b H. Ishikawa, K. Oguro, A. Kato, H.Suzuki, E. Ishii: J. Less-Common Met. **120**, 123 (1986).
89c H. Ishikawa, K. Oguro, A. Katho, H. Suzuki, E. Ishii, T. Okada, S. Sakamoto: Z. Phys. Chemie N.F. **164**, 1409 (1989).
90a Y. Josephy, Y. Eisenberg, S. Perez, A. Bendavid, M. Ron: J. Less-Common Met. **104**, 297 (1984).
90b E. Bershadsky, Y. Josephy, M. Ron: Z. Phys. Chemie N.F. **164**, 1373 (1989).
90c E. Bershadsky, Y. Josephy, M. Ron: J. Less-Common Met. **153**, 65 (1989).
91 G. Anevi, L. Jansson, D. Lewis: J. Less-Common Met. **104**, 341 (1984).
92 N. P. Kherani, W. T. Shmayda, A. G. Heics: Z. Phys. Chemie N.F. **164**, 1421 (1989).
93a H. Aoki, H. Mitsui: "Investigation of annular Metal hydride Reaction beds: Characteristics of reaction and heat transfer" Reports from Toyota Central Research & Development labs. Inc. Japan (1988).

93b H. Aoki and H. Mitsui: "Investigation of tubular heat exchangers with internal longitudinal fins for metals hydride reactor beds" Reports from Toyota Central Research & Development labs. Inc. Japan (1988).
94 . T. Ogawa, R. Ohnishi, T. Misawa: J. Less-Common Met. **138**, 143 (1988).
95. H. Uchida, T. Ebisawa, K. Terao, N. Hosoda, Y. C. Huang: J. Less-Common Met. **131**, 365 (1987).
96. J. R. Johnson, J. J. Reilly: Z. Phys. Chemie N.F. **147**, 263 (1986).
97. J. S. Noh, R. K. Agarwal, J. A. Schwarz: Int. J. Hydrogen Energy **12**, 693 (1987).
98. R. Wiswall: in ref. 8 pp 230-233.
99. H. Suzuki, Y. Osumi, A. Kato, K. Oguro, M. Nakane: J. Less-Common Met **89**, 545 (1983).
100. S. Ono, Y. Ishido, E. Akiba, K. Jindo, Y. Sawada, I. Kitagawa, T. Kakutani : Proc. 5th World Hydrogen Energy Conf. V 3, 1291-1301 (1984).
101. HWT: Wiesenstraβe 36, d-4330, Mülheim an der Ruhr.
102. O. Bernauer, C. Halene: J. Less-Common Met. **131**, 213 (1987).
103. ERGENICS: 247 Margaret King Av., Ringwood, N.J. 07456, USA.
104a ERGENICS : catalogue for hydrogen storage unit ST-1, ST-45, ST-90, US patent 4,396,114.
104b HWT: Catalogue N° HUT 8602 E, laboratory hydride storage units.
105. E. L. Huston, G. D. Sandrock: J. Less-Common Met. **74**, 435 (1980).
106. O. Bernauer: Int. J. Hydrogen Energy **13**, 181 (1988).
107. K. C. Hofman, W. E. Winsche, R. H. Wiswall, J. J. Reilly, T. V. Sheehan, C. H. Waide: "Metal Hydrides as a source of fuel for Vehicular Propulsion", SAE technical paper series N°690232, 1969.
108. S. Furuhama: Int. J. Hydrogen Energy **14**, 907 (1989).
109. J. Töpler, K. Feucht: Z. Phys. Chemie N.F. **164**, 1451 (1989).
110. F. E. Lynch : J. Less-Common Met. **172-174**, 943 (1991).
111. D. Davidson, M. Fairlie, A. E. Stuart: Int. J. Hydrogen Energy **11**, 39 (1986).
112. SAES : Milano, Italy.
113. M. H. Mendelsohn, D. M. Gruen: J. Less-Common Met. **74**, 449 (1989).
114. K. Watanabe, M. Matsuyama, K. Ashida, H. Miyake: J. Vac. Sci. Technol. A **7**, 2725 (1989).
115. C. D. Cann, E. E. Sexton, A. A. Bahurmuz, A. J. White, G. A. Ledoux: J. Less-Common Met. **172-174**, 1297 (1991).
116. Ergenics: Catalogue HY-Pure, Gas Purifiers.
117. HWT : Catalogue N° HUT 8604 E, gas purification systems.
118. E. Peres Da Silva: "Industrial prototypes of hydrogen compressor based on metallic hydrides technology", in Proc. of the 8th World Hydrogen Energy Conf., 1990, ed. T. N. Veziroglu.
119. J. J. Reilly, R. H. Wiswall : "Hydrogen Storage and Purification Systems", Rep. BNL 17136 (Brookhaven Nat Lab., Upton, N-Y, 1972).
120. J. J. Sheridan, F. G. Eisenberg, E. J. Greskovich, G. D. Sandrock, E. L. Huston: J. Less-Common Met. **89**, 447 (1983).
121. Q. D. Wang, J. Wu, C. P. Chen, Y. Zhou: J. Less-Common Met. **131**, 321 (1987).
122. R. L. Swart, J. T. Tinge, W. Meindersma: Z. Phys. Chemie N.F. **164**, 1435 (1989).
123. H. Albrecht, U. Kuhnes, W. Asel: J. Less-Common Met. **172-174**, 1157 (1991).
124. C. L. Huffine: "Fabrication of Hydrides" in ref. 7, pp. 675-747.

125. R. Lässer, T. Schober: "Some aspects of tritium storage in hydrogen storage materials", in ref. 9, pp. 40-75.
126. G. Sandrock, J. J. Murray, M. L. Post, J. B. Taylor: Mat. Res. Bull. **17**, 887 (1982).
127. Z. W. Wei: Int. J. Hydrogen Energy **12**, 337 (1987).
128a. B. M. Andreev, G. H. Sicking: Ber. Bunsenges. Phys. Chem. **91**, 177 (1987).
128b B. Jungblut, G. Sicking: Z. Phys. Chemie N.F. **164**, 1177 (1989).
128c M. Karas, E. P. Magomedbekov, G. H. Sicking: J. Less-Common Met. **159**, 307 (1990).
129. I. A. Fischer: J. Less-Common Met. **172-174**, 1320 (1991).
130. E. Glueckhauf, G. P. Kitt : "Chromatographic Separation of Hydrogen Isotopes", in Vapor phase Chromatography, (Butterworths, London, 1957).
131. T. Altridge : "Chromatographic Hydrogen Isotope Separation", U.S. Patent 4, 276,060 june 1981.
132. F. Botter, J. Gowman, J. L. Hemmerich, B. Hircq, R. Lässer, D. Leger, S. Tistchenko, M. Tschudin: Fusion Technology **14**, 562 (1988).
133a Y. W. Wong, F. B. Hill, Y. N. I. Chan : Separ. Sci. Technol. **15**, 423 (1980).
133b F. B. Hill, V. Grzetic : J. Less-Common Met. **89**, 399 (1983).
134 N. Mitsuishi, S. Fukada, H. Tokuda, T. Nawata, Y. Takai: in ref. 14, Vol 1, pp 647-671.
135 G. S. Nickols: J. Less-Common Met. **172-174**, 1338 (1991).
136. V. V. Solovey: "Metal hydride thermal power installations" in ref 14, Vol 1, pp 1391-1399.
137. J. J. Reilly, A. Holtz, R. H. Wiswall: Rev. Sci. Ins. **42**, 1485 (1971).
138. Van Mal : Thesis, technological University, Delft, May 1976.
139. M. Blondeau, M. Bonneton, M. Jannot: Revue générale de thermique, France, N°257, 411 (1983).
140. P. M. Golben: "Multi-Stage 8ydride-Hydrogen Compressor" Proc. 18th Intersociety Energy Conversion Engineering Conf. 1983, pp. 1746-1753.
141. Ergenics catalogue: Model 1.5-15-40 : Hydride/Hydrogen, Four Stage Compressor, Model HC-6, Hydride/Hydrogen Electric Compressor..
142. T. Kumano, B. Tada, Y. Tsuchida, Y. Kuraoka, T. Ishige, H. Baba: Z. Phys. Chemie N.F. **164**, 1509 (1989).
143. L. Wade, K. Karperos: "Compactness and Reliability key interests in Sorption Refrigerators" in Research and Development, oct. 1987, pp. 45-47.
144 P. M. Golben: "Solar Energy, Hydrogen Sponge, Keys to Water Pump Operation, in Design News, Fluid Power ed. F. Yeapple, Nov.23 (1987).
145. D. M. Gruen, M. H. Mendelsohn, I. Sheft: Solar Energy **21**, 153 (1978).
146. R. Gorman, P. S. Moritz:"Metal Hydride Solar Heat Pump and Power System (HYCSOS)", AIAA/ASERC Conf. on Solar Energy: Technology Status, Phoenix, Ariz, Nov 1978, pp 1-6.
147a P. Dantzer, E. Orgaz: J. Chem. Phys **85**, 2961 (1986).
147b P. Dantzer, E. Orgaz: Int. J. Hydrogen Energy **11**, 2961 1986.
147c E. Orgaz, P. Dantzer: J. Less-Common Met. **131**, 385 1987.
148 N. Douss, F. Meunier, L. M. Sun: Ind. Eng. Chem. Res. **17**, 310 (1988).
149a P. Dantzer, F. Meunier: "What Materials to use in Hydride Chemical Heat Pumps" in ref 9, pp. 1-17.

149b M. Balakumar, S. Srinivasa Murthy, M. V. Krishna Murthy: Heat Recovery Systems & CHP: 7, 221, 1987.

149c S. Srinivasa Murthy, M. V. Krishna Murphy, M. V. C.Sastri: "Two-stage metal hydride heat transformer: a thermodynamic study", in ref. 14, Vol II, pp 1253-1265.

150a Nomura, Hideo, Iguchi, Kazuyuki: "Heat pump type air conditioning unit utilizing a hydrogen storage alloy", Daikin Industries Ltd, Osaka, Japan Patent 63-29163(A), 6 feb. 1988.

150b Iwasaki, Masahide, Nakajima, Masao, Inokuma, Yoshihiko, Kawai, Shigemasa: "High temperature steam generating heat pump", Sekisui Chemical Co. ltd Osaka, Nihon Kagaku Gijutsu Co. ltd. Osaka, Japan Patent 63-34461(A), 15 feb 1988.

15Oc Ebato, Kazuo, Tamura, Keiji, Yoshida, Hiroshi, Yasunaga, Tomohiro: "Heat exchanger unit for hydrogen storage alloy" Nippon Yakin Kogyo Co ltd. Tokyo, Japan Patent 63-259, 300. 26 Oct. 1988.

15Od Kusakabe, Hiroshi: "Air Conditioner", Sanyo Electric Co. Ltd Moriguchi, Japan Patent 63-231,155. 27 Sept. 1988.

15Oe Mochizuki, Kaorou Sato, Tatsuo: "Air Conditioner" Toshiba Co.Kawasaki, Japan Patent 63-210,566. 1 sept. 1988.

151. M. Ron, Y. Josephy: Z. Phys. Chemie N.F. **164**, 1475 (1989).

152. N. Nagel, Y. Komazaki, M. Uchida, S. Suda, Y. Matsubara: J. Less Common Met. **104**, 307 (1984).

153. T. A. Argabright: "Heat/Mass Flow Enhancement Design for a Metal Hydride Assembly", Final Report, 1985, available from Nat. Technical Information Service, U.S.Dept. of Commerce, Springfield, VA. 22161.

154a M. Ron: J. Less-Common Met. **104**, 259 (1984).

154b M. Ron, Y. Eisenberg, Y. Josephy, M. Gutman, U. Navon: "Gas-Solid Reaction Heat Exchanger for Vehicle Engine Exhaust Vaste Heat Recovery" SAE technical paper series N°860588, Feb 24-28, 1986.

154c M. Ron and Y. Josephy: Z. Phys. Chemie N.F. **147**, 241 (1986).

155. A. Yanoma, M. Yoneto, T. Nitta, T. Okuda: "Design and operation of the commercial size chemical heat pump system using metal hydrides" in Proc. of the joint Conf. ASME-JSNE, Honolulu, March 1987, pp. 431-437.

156a R. Werner, N. Groll: "Design aspects of metal hydride heat transformers". in Proc. Int. Congress on High Performances Heat Pumps, ed. B. Spinner, (Perpignan, France, Sep. 1988) pp. 320-329.

156b R. Werner, M. Groll: J. Less-Common Met. **172-174**, 1122 (1991).

157. S. Suda, Y. Komazaki: J. Less-Common Met. **172-174**, 1130 (1991).

158. E. Tuscher, P. Weizierl, O. J. Eder: J. Less-Common Met. **95**, 171 (1983).

159. M. Nagel, Y. Komazaki, S. Suda: J. Less-Common Met. **120**, 35 (1986).

160. M. Nagel, Y. Komazaki, Y. Matsubara, S. Suda: J. Less-Common Met. **123**, 47 (1986).

161a H. Bjurström, Y. Komazaki, S. Suda: J. Less-Common Met. **131**, 225 (1986).

161b H. Bjurström, D. Lewis, S. Suda: "Simulation of periodic heat pumps based on metal hydrides", in Proc. of Workshop, Absorption Heat Pumps, held in London 12-14 April 88, eds. P. Zegers, J. Miriam C.E.C. pp. 110-120.

161c H. Bjurström, S. Suda: Int. J. Hydrogen Energy **14**, 19 (1989).

162. M. Gambini: Int. J. Hydrogen Energy **14**, 821, (1989).

163a D. W. Sun, S. J. Deng: J. Less-Common Met. **141**, 37 (1988).
163b D. W. Sun, S. J. Deng: J. Less-Common Met. **155**, 271 (1989).
164. B. Bogdanovic, B. Spliethoff: Int. J. Hydrogen Energy **12**, 863 (1987).
165. Ali T. Raissi, D. K. Slattery, M. J. Axelrod, R. Zida: "Synthesis and Characterisation of Nickel doped Magnesium Hydride" in Proc of the 8th World Hydrogen Energy Conf., 1990, ed. T. N. Veziroglu.
166. B. Bogdanovic, B. Spliethoff, A. Ritter: Z. Phys Chemie N.F. **164**, 1497 (1989).
167. M. Wierse, R. Werner, M. Groll: J. Less-Common Met. **172-174**, 1111 (1991).
168. F. A. Lewis: *The Palladium Hydrogen System*, (Academic Press, London, 1967).
169. M. A. Gutjahr, H. Buchner, K. D. Beccu, H. Säufferer: Power Sources N°4, 79 (1973), ed. D. H. Collins.
170 M. Earl, J. Dunlop: COMSAT Tech. Rev. 3, 437 (1973).
171a E. W. Justi, H. H. Ewe, A. W. Kalberlak, N. M. Saridakis, M. H. Schaeffer: Energy Conversion **10**, 183 (1970).
171b H. Ewe, E. W. Justi, K. Stephan: Energy Conversion **13**, 109 (1973).
172. F. G. Will: U.S. Patent N° 3874928 (1975).
173. A. Percheron-Guegan, J. C. Achard, J. Loriers, M. Bonnemay, G. Bronoel, J. Sarradin, L. Schlapbach: "Nickel base alloys and their electrochemical applications", French patent N° 7516160 (May 1975).
174a A. Percheron et al: extension of the french Patent 7516160, N°7706138, (march 1977).
174b A. Percheron-Guegan, J. C. Achard, J. Sarradin, G. Bronoel: "Electrode Materials based on Lanthanum and Nickel and Electrochemical uses of such materials" US patent 4107405, August 1978.
175a G. Bronoel, J. Sarradin, M. Bonnemay, A. Percheron-Guegan, J. C. Achard, L. Schlapbach: Int. J. Hydrogen Energy **1**, 251 (1976).
175b A. Percheron-Guegan, F. Briaucourt, H. Diaz, J. C. Achard, J. Sarradin, G. Bronoel: "Substitutions in $LaNi_5$ compound-comparative study of related hydrides in electrolytic and solid gas reactions" in Proc. of the 12th Rare Earth Res. Conf., ed. Lundin, Vai Colorado, July 1976, pp. 300-308.
175c G. Bronoel, J. Sarradin, A. Percheron-Guegan, J. C. Achard: Mat. Res. Bull. **13**, 1265 (1978).
176a. J. J. G.Willems: Philips J. of Research **39**, Suppl. N°1 (1984).
176b P. H. L. Notten, R. E. F. Einerhand: Adv. Mat. **3**, 343 (1991).
177. H. Tamura, C. Iwakura, T. Kitamura: J. Less-Common Met. **89**, 567 (1983).
178. T. Sakai, T. Hazama, H. Miyamura, N. Kuriyama, A. Kato, H. Ishikawa: J. Less-Common Met. **172-174**, 1175 (1991).
179. T. Sakai, A. Yuasa, H. Ishikawa, H. Myamura, N. Kuriyama: J. Less-Common Met. **172-174**, 1194 (1991).
180. D. E. Hall, J. M. Sarver, D. O. Gothard: Int. J. Hydrogen Energy **13**, 547 (1988).
181a H. Uchida, T. Ebisawa, Y. C. Huang: "A New Treatment for the Pulverization of Hydrogen Storage Alloys", Int. Symp. on Hydrogen Produced for Renewable Energy, Oct. 1985, Florida, USA.
181b Y. Matsumara, L. Sugiara, H. Uchida: Z. Phys. Chemie N.F. **164**, 1545 (1989).
181c T. Sakai, A. Takagi, N. Kuriyama, H. Miyamura: J. Less-Common Met. **172-174**, 1185 (1991).

182. T. Sakai, K. Oguro, H. Miyamura, N. Kuriyama, A. Kato, H. Ishikawa, C. Iwakura: J. Less-Common Met. **161**, 193 (1990).
183. Y. Q. Lei, Z. P. Li, Y. M. Wu, J. Wu, Q. D. Wang: "Multicomponent Mischmetal-Nickel Rechargeable Hydride Electrodes" in Proc. Int. Renewable Energy Conf. (IREC), Honolulu, Sept. 1988, pp. 391-399.
184a H. Sawa, K. Ohseki, M. Otha, H. Nakano, S. Wakao: Z. Phys. Chemie N.F. **164**, 1521 (1989).
184b H. Sawa, M. Otha, H. Nakano, S. Wakao: Z. Phys. Chemie N.F. **164**, 1527, (1989).
185. Y. Moriwaki, T. Gamo, H. Seri, T. Iwaki: J. Less-Common Met. **172-174**, 1211 (1991).
186. D. N. Ryan, F. Dumais, B. Patel, J. Kycia, J. O. Ström-Olsel: J. Less-Common Met. **172-174**, 1246 (1991).
187. C. Poinsignon, M. Forestier, M. Anne, D. Fruchart, S. Miraglia, A. Rouault, J. Pannetier: Z.Phys.Chemie N.F. **164**, 1515 (1989).
188. B. Y. Liaw, R. A. Huggins: Z. Phys. Chemie N.F. **164**, 1533 (1989).
189. J. Kumar, S. Saxena: "Electrochemical Investigation of hydrogen absorption in Titanium based intermetallics": in ref. 14, Vol pp. 629-646.
190. C. Jordy, A. Percheron-Guegan, J. Bouet, P. Sanchez, J. Leonardi: J. Less-Common Met. **172-174**, 1236 (1991).
191a A. N. Boonstra, G. J. M.Lippits, T. N. M. Bernards: J. Less-Common Met. **155**, 119 (1989).
191b A. H. Boonstra, T. N. M. Bernards: J. Less-Common Met. **161**, 245 (1990).
191c A. H. Boonstra, T. N. M. Bernards: J. Less-Common Met. **161**, 355 (1990).
192. L. Schlapbach: J. Phys. F, Metal Phys. **10**, 2477 (1980).
193. V. N. Zhuravleva, A. G. Pshenichnikov, Y. U. G. Chirkov, K. V. Shnepelev: "Mass transfer in the hydrogen-absorbing metals grain", Hydrogen Energy Progress VII, 445, VI, (1988).
194. D. Staschevski : Int. J. Hydrogen Energy **11**, 279 (1986).
195. K. Kordesh, K. Holz, P. Kalal, M. Reindel, H. Steininger: Int. J. Hydrogen Energy **13**, 475 (1988).
196a M. J. Rosso, O. J. Adlhart, J. A. Marmolejo: "A fuel cell energy storage for space station extravehicular activity" 18th Intersociety Conf. on environmental Systems, San Francisco, SAE technical paper series N°881105 (1988).
196. O. J. Adlhart, M. J. Rosso, and J. A. Marmolejo: "A fuel Cell Energy Storage System Concept for the space Station Freedom extravehicular mobility unit", Inter. Congress and Exposition, Detroit, SAE technical paper series N°891582 (1989).
196c O. J. Adlhart, M. J. Rosso, and J. A. Marmolejo: "Design and Performance of an air-cooled ion exchange membrane fuel cell", 33th Int. Power Sources Symposium, 13-16 june 1988.
197. J. M. Welter: J. Less-Common Met. **104**, 251 (1984).
198. D. E. Warren, K. A. Faughnan, R. A. Fellows, J. W. Godden, B. M. Seck: J. Less-Common Met. **104**, 375 (1984).

Chapter 7

Rechargeable nickel-metalhydride batteries: a successful new concept

P.H.L. Notten
Philips Research Laboratories
Prof. Holstlaan 4 - 5656 AA Eindhoven
The Netherlands

1. Introduction

Soon after the discovery in the late sixties that intermetallic compounds, such as $SmCo_5$ and $LaNi_5$, were able to absorb and also desorb large amounts of hydrogen gas [1,2], it was realized that electrodes made of these materials could serve as a new electrochemical storage medium [3,4]. In the last few years the hydride-forming electrode has indeed proven to be a serious alternative for the cadmium electrode, which is nowadays widely employed in rechargeable nickel-cadmium (NiCd) batteries [5-7]. Particularly, the 30-50 % higher energy storage capacity and the non-toxic properties of the chemical elements from which these hydride-forming materials are composed are of benefit compared to the cadmium electrode. Application of sealed rechargeable batteries in cordless versions of consumer electronics such as shavers, video cameras and portable telephones will become even more important in the near future than already is the case to date. Detailed knowledge about the electrochemical performance of various battery systems and of nickel-metalhydride (NiMH) batteries, in particular, is therefore indispensable.

In order to obtain more insight into the functioning of sealed rechargeable NiMH batteries, the basic principles underlying their operation and the resulting electrochemical characteristics will be outlined in Section 2 of this chapter. Besides the basic electrochemical reactions, which will be described in Sect. 2.1, special emphasis is put on the possible side-reactions in Sect. 2.2. These side-reactions are of essential importance for the proper functioning of such a sealed battery, especially under both overcharging and overdischarging conditions. Their influence on the battery internal gas pressure and temperature will be discussed. The factors, which are responsible for the well-known phenomenon of battery self-discharge under open-circuit conditions will be covered in

F. Grandjean et al. (eds.), Interstitial Intermetallic Alloys, 151–195.

Sect. 2.3. The various parasitic electrochemical reactions, contributing to the battery self-discharge will be considered.

Section 3 is devoted to the (electro)chemical characteristics and performance of the Ni electrode. In particular the solid-state charge transfer mechanism, characteristic for this electrode, will be discussed. In addition, the various electroactive nickeloxide modifications involved in the energy storage process will be considered.

Several types of metalhydride-forming (MH) compounds can, in principle, be employed as electrode materials in rechargeable NiMH batteries, such as, for example, TiFe (AB), $ZrNi_2$ (AB_2) and $LaNi_5$ (AB_5). The developments of AB_5-type hydride-forming compounds have contributed to a large extent to the realization of this new battery system. AB_5-type compounds have proven to be one of the most promising storage materials in the above-mentioned application and these materials are nowadays mainly employed in the commercially available, small-sized, NiMH batteries. We will therefore focus in Section 4 on the materials research of this class of hydride-forming compounds. The similarity between hydride formation initiated via the gas phase and via an electrochemical charge transfer will be discussed in Sect.4.1. Special attention will be given in Sect. 4.2 to the improvement of the electrochemical long-term cycling stability of $LaNi_5$ in alkaline solution. The kinetics of the charge transfer reaction is another important parameter, determining the charge/discharge capability of the MH electrode. Various methods to improve the electrocatalytic activity of the compounds will be discussed in Sect. 4.3. The underlying fundamental processes, determining both the long-term cycling stability and the charge/discharge kinetics will be dealt with in detail.

2. The concept of sealed rechargeable NiMH batteries

2.1. BASIC REACTIONS

The concept of a NiMH battery very closely resembles that of a NiCd battery. A schematic representation of a NiMH battery containing an AB_5-type hydride-forming electrode is shown in Figure 1. The electrodes are electrically insulated from each other by a separator. Both separator and electrodes are impregnated with an alkaline solution which provides for the ionic conductivity between the two electrodes. The overall electrochemical reactions, occurring at both electrodes during charging (ch) and discharging (d) can, in their most simplified form, be represented by

$$Ni(OH)_2 + OH^- \underset{d}{\overset{ch}{\rightleftarrows}} NiOOH + H_2O + e^- \qquad (1)$$

$$AB_5 + xH_2O + xe^- \underset{d}{\overset{ch}{\rightleftarrows}} AB_5H_x + xOH^- \tag{2}$$

During charging divalent Ni^{II} is oxidized in the trivalent Ni^{III} state and water is reduced to hydrogen atoms, which are, subsequently, absorbed by the hydride-forming compound. The reverse reactions take place during discharging. The net effect of this reaction sequence is that hydroxyl ions in the electrolyte are transported from one electrode to the other and hence that no electrolyte consumption takes place during current flow. The basic reactions are also indicated in Fig. 1. A more detailed description of the physico-chemical aspects of both the Ni electrode and the hydride-forming electrode will be described in Sections 3 and 4, respectively.

In general exponential relationships between the partial anodic/cathodic currents and the applied electrode potential are observed under kinetically-controlled conditions, as is depicted schematically in Fig. 2 (dashed curves). The potential scale is given with respect to a Hg/HgO (6 N KOH) reference electrode. The equilibrium potential of the Ni-electrode under standard conditions is far more positive (E^0_{Ni} = +439 mV) than that of the MH-electrode [8], which is found to be dependent on the plateau pressure of the hydride-forming material used (E^0_{MH} ranges between -930 and -860 mV), as we will discuss in Sect. 4.1. This implies that the theoretical open-circuit potential of a NiMH

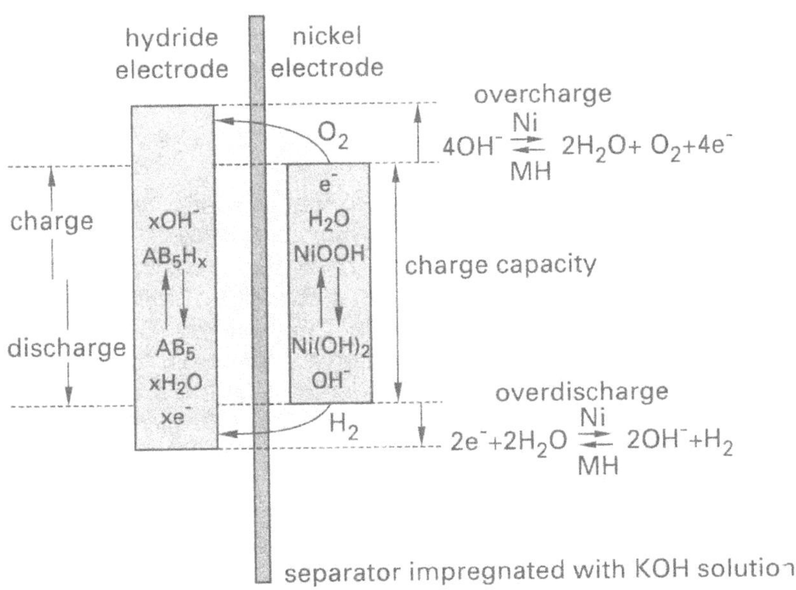

Fig. 1. Schematic representation of the concept of a sealed rechargeable NiMH battery.

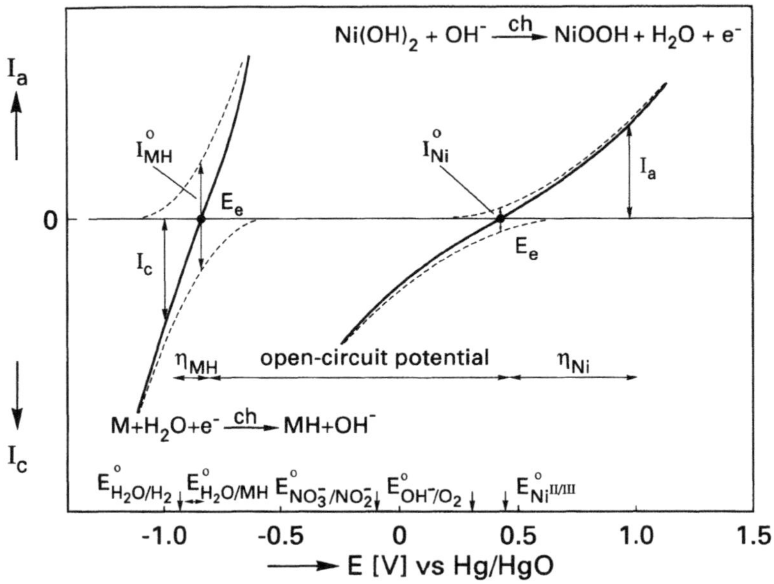

Fig. 2. Schematic representation of the current-potential curves for a Ni and a MH electrode (solid lines), assuming kinetically controlled charge transfer reactions. The partial anodic and cathodic reactions are indicated as dashed lines. The exchange currents (I^0) are defined at the equilibrium potentials (E_e). Potentials are given with respect to a Hg/HgO reference electrode. Besides the standard redox potentials (E^0) of the main electrode reactions and of some side-reactions are also indicated.

battery is approximately 1.3 V, very similar to that of a NiCd battery. This makes these two different battery systems indeed very compatible.

During galvanostatic charging of the battery with a constant current an overpotential (η) will be established at both electrodes. The magnitude of each overpotential component (η_{Ni} and η_{MH} in Fig. 2) is determined by the kinetics of the charge transfer reactions. An electrochemical measure for the kinetics of a charge transfer reaction is generally considered to be the exchange current I^0, which is defined at the equilibrium potential, E_e, where the partial anodic current equals the partial cathodic current (see Fig. 2). In case of the Ni-electrode I^0 is reported to be relatively low ($I^0_{Ni} = 10^{-7}$ A/cm^2 [9,10]), which implies that at a given constant anodic current, I_a, the overpotential at the Ni-electrode is relatively high (see Fig. 2). In contrast, the kinetics of the MH-electrode is reported to be strongly dependent on the materials composition [11,12]. Assuming a highly electrocatalytic hydride-forming compound, this implies that the current-potential curves, characteristic for the MH-electrode are very steep in comparison to those for the Ni-electrode, resulting in a much smaller value for

η_{MH} at the same cathodic current I_c, as is schematically shown in Fig. 2. It is evident that the battery voltage under current flow is a summation of the open-circuit potential and the various overpotential components. This includes the ohmic potential drop (η_{IRe}) caused by the electrical resistance of the electrolyte (R_e). The reverse processes occur during discharging, resulting in a cell voltage lower than 1.3 V. Clearly, since the potential of both electrodes may change considerably, the absolute values of these potentials cannot be directly deduced from the cell voltage. The use of a reference electrode is therefore inevitable in order to interpret the current-potential dependencies in an appropriate way.

2.2. SIDE REACTIONS

2.2.1. Overcharging/Overdischarging. To ensure the well-functioning of sealed rechargeable NiMH batteries under a wide variety of conditions, the battery is designed in such a way that the Ni electrode is the capacity-determining electrode. Such a configuration forces side-reactions to occur at the Ni electrode both during overcharging and overdischarging, as we will show below.

During overcharging OH^- ions are oxidized and oxygen evolution starts at the Ni electrode, according to

$$4OH^- \overset{Ni}{\rightleftarrows} O_2 + 2H_2O + 4e^- \tag{3}$$

As a result, the partial oxygen pressure inside the sealed cell starts to rise. Advantageously, oxygen can be transported to the metalhydride electrode, where it can be reduced at the MH/electrolyte interface in hydroxyl ions at the expense of the hydride-formation reaction (2):

$$O_2 + 2H_2O + 4e^- \overset{MH}{\rightleftarrows} 4OH^- \tag{4}$$

This implies that the partial oxygen pressure inside the battery can, in principle, be very low, assuming that this recombination mechanism is functioning properly. In the steady-state, the amount of oxygen evolved at the Ni electrode is equal to the amount of oxygen recombining at the metalhydride electrode. This implies that all electrical energy supplied to the battery during overcharging is completely converted into heat.

The formation of heat (W) inside a battery has been represented by

$$W = i\left\{\frac{-T\Delta S}{nF} + \sum |\eta| + iR_e\right\} \tag{5}$$

where i is the current flowing through the battery, T the temperature, n the number of electrons involved in the overall charge transfer reaction (summation of Eqs. (1) and (2)) and F is the Faraday constant [13]. The factors which contribute to the evolved heat during current flow can be easily recognized in Eq. (5):

(i) The entropy change (ΔS) brought about by the electrochemical reactions.

(ii) The summation term is composed of the various overpotential components and has to include the various electrochemical reactions.

(iii) The internal battery resistance, whose contribution may be significant, especially when high currents are applied, as the heat evolution due to this effect is proportional to the square of the current.

As long as the basic electrochemical reactions (Eqs. (1) and (2)) proceed inside the battery, both overpotential components are relatively small. This implies that the heat contribution, resulting from the electrode reactions is limited. However, this situation changes drastically as soon as the oxygen recombination cycle at the MH electrode starts. Since the MH electrode potential is at least 1 V more negative with respect to the standard redox potential of the OH^-/O_2 couple (see Fig. 2), this implies that the established overpotential for the oxygen recombination reaction is extremely high (>1.2 V). Considering Eq. (5), it is therefore indeed to be expected that the heat evolved inside a battery will sharply increase as soon as the oxygen recombination cycle starts. Although the recombination cycle moderates a considerable pressure rise inside the NiMH battery, it is essential to avoid prolonged overcharging in order to prevent a considerable temperature rise, which may negatively affect other electrode properties. In conclusion we may say that, dependent on the kinetics of the oxygen recombination reaction, i.e. dependent on the competition between reaction (2) and (4), the gas pressure and/or temperature of the battery will rise during overcharging.

Protection against overdischarging is also of considerable importance, especially when NiMH batteries, which inevitably reveal small differences in storage capacities, are used in series. Under these circumstances water is forced to be reduced at the Ni electrode (see Fig. 1), according to

$$2H_2O + 2e^- \overset{Ni}{\rightleftarrows} 2OH^- + H_2 , \qquad (6)$$

which also results in a pressure build-up inside the battery when no precautions are taken. As the chemical affinity of the metalhydride electrode towards hydrogen gas is excellent, it is evident that this gas can be again converted into water at the MH electrode during overdischarging:

$$H_2 + 2OH^- \overset{MH}{\rightleftarrows} 2H_2O + 2e^- \qquad (7)$$

Whether conversion of molecular hydrogen occurs directly at the MH electrode or atomic hydrogen is oxidized indirectly after chemical adsorption and/or absorption has taken place, is not clear. It is, however, clear that, in both cases, high demands are put on the physical properties of the electrode/electrolyte interface. In conclusion, a hydrogen recombination cycle controls the pressure rise under overdischarging conditions also. As hydrogen evolution and hydrogen oxydation at the separate electrodes take place in the same potential region (see Fig. 2) it is obvious that the battery voltage is very close to zero volt under these conditions. This strongly contrasts to NiCd batteries, for which a large potential reversal is generally observed during prolonged overdischarging.

2.2.2. Self-discharge reactions. It is well-known that charged NiMH batteries, similar to NiCd batteries, to a certain extent loose their stored charge under open-circuit conditions . The self-discharge rates are strongly dependent on external conditions, such as, for example, the temperature of the batteries. Typical self-discharge rates at room temperature are of the order of 1% of the nominal storage capacity per day. Various mechanisms contribute to the overall self-discharge rate. These mechanisms are all electrochemical in nature. The mechanisms operative in NiMH batteries occur mainly via the gas phase and can be divided into processes initiated by the Ni electrode or by the MH electrode. The most important mechanisms contributing to the overall self-discharge rate will be treated below:

(i) Considering the redox potentials of the Ni electrode and that of the competing oxygen evolution reaction (Eq. (3)), it is obvious that trivalent Ni^{III} is thermodynamically unstable in an aqueous environment (see redox potentials in Fig. 2). As a consequence, NiOOH will be reduced by hydroxyl ions at the open-circuit potential, according to

$$NiOOH + H_2O + e^- \rightarrow Ni(OH)_2 + OH^- \quad (8)$$

$$4OH^- \xrightarrow{Ni} O_2 + 2H_2O + 4e^- \quad (9)$$

The electrons released by the OH^- ions are transfered to the Ni electrode at the electrode/electrolyte interface. Although the Ni^{III} species are thus in principle unstable, electrical charge can, however, be stored in the Ni electrode. This is due to the fact that the kinetics of the oxygen evolution reaction are, fortunately, relatively poor, so that it takes quite a while before capacity loss due to battery self-discharge becomes appreciable. Subsequently, the produced oxygen gas can be transported to the MH electrode, where it can be converted again into OH^- ions at the expense of charge stored in the MH electrode, i.e.

$$O_2 + 2H_2O + 4e^- \xrightarrow{MH} 4OH^- \quad (10)$$

$$MH + OH^- \rightarrow M + H_2O + e^- \quad (11)$$

The ultimate result is that stored charge in both the Ni and MH electrode is released through a gas-phase shunt, in this case oxygen gas.

(ii) A different type of gas-phase shunt is initiated by the MH electrode and is caused by the presence of hydrogen gas inside the battery. As the storage capacity of the MH electrode is considerably larger than that of the Ni electrode (see Fig. 1) and the MH electrode contains a certain amount of precharge (in the form of hydride), a minimum partial hydrogen pressure is inevitably established inside the NiMH battery, according to the chemical equilibrium

$$MH \rightleftarrows M + \tfrac{1}{2}H_2\uparrow \qquad (12)$$

The minimum H_2 pressure is dependent on the condition of the battery but will often be determined by the hydrogen plateau pressure. As a result, H_2 is in contact with the Ni electrode. Since the standard redox potential of the OH^-/H_2 redox couple is much more negative than that of the Ni^{II}/Ni^{III} couple, hydrogen can be oxidized at the Ni electrode, whereas the Ni electrode is simultaneously reduced, according to

$$H_2 + 2OH^- \xrightarrow{Ni} 2H_2O + 2e^- \qquad (13)$$

$$NiOOH + H_2O + e^- \rightarrow Ni(OH)_2 + OH^- \qquad (14)$$

This electrochemical process occurs under open-circuit conditions at the Ni electrode and will be strongly influenced by the partial hydrogen pressure inside the battery. It has indeed been reported that the self-discharge rate at the Ni electrode is proportional to the partial hydrogen pressure [14]. For this reason it is of importance that the hydrogen pressure inside the battery is kept as low as possible, i.e. to employ hydride-forming compounds which are characterized by a relatively low hydrogen plateau pressure. Again, according to Eqs. (12) and (14), the chemical energy stored in both the MH and Ni electrode is wasted by a gas-phase shunt and can no longer be employed for useful energy supply.

(iii) The third self-discharge mechanism is related to the fabrication process of the nickeloxide electrode. These solid-state electrodes are generally prepared by electrolytic reduction of an acidic salt electrolyte, often $Ni(NO_3)_2$ [15,16]. During this process NO_3^- ions are reduced to NH_4^+ ions. This results in a significant increase in pH near the electrode/electrolyte interface. The solubility product of $Ni(OH)_2$ will be exceeded and, as a result, $Ni(OH)_2$ will subsequently precipitate on the substrate. A consequence of this process is that, despite the fact that the as-prepared electrodes are thoroughly washed, a certain amount of nitrate ions are inevitably incorporated into the Ni electrodes, which can be leached out during the battery cycle-life. These NO_3^- ions, dissolved in the liquid phase, form the basis of this third self-discharge mechanism. These ionic species can be reduced to lower oxidation states [16]. It is generally assumed that a so-called nitrate/nitrite shuttle is operative in alkaline rechargeable batteries [7]. The standard redox potential of the nitrate/nitrite redox couple [16] is much more positive than that of the MH electrode ($E^0_{NO2/NO3}$=-91 mV vs

Hg/HgO, see also Fig. 2). This implies that NO_3^- ions delivered by the Ni electrode can be reduced at the MH electrode under open-circuit conditions, according to

$$NO_3^- + H_2O + 2e^- \xrightarrow{MH} NO_2^- + 2OH^- \quad (15)$$

$$MH + OH^- \rightarrow M + H_2O + e^- \quad (16)$$

The produced nitrite ions can diffuse to the Ni electrode. As the electrode potential of the Ni electrode is more positive than the redox couple of the nitrate/nitrite couple NO_2^- can be converted to nitrate again while NiOOH is simultaneously reduced:

$$NO_2^- + 2OH^- \xrightarrow{Ni} NO_3^- + H_2O + 2e^- \quad (17)$$

$$NiOOH + H_2O + e^- \rightarrow Ni(OH)_2 + OH^- \quad (18)$$

This reaction sequence can proceed continuously, as the electroactive nitrate and nitrite species are continuously produced at both electrodes. The final result is again that charge stored in both the MH and Ni electrode is consumed and no longer is available for useful energy supply.

3. The nickeloxide electrode

Numerous studies have been devoted to the Ni-electrode over the last few decades. Excellent reviews on this topic have recently been written [15,17]. These reviews include the various preparation methods and manufacturing processes of active electrode materials. In contrast to what is suggested by Eq. (1), the charge and discharge reactions of a Ni-electrode are much more complex. This is further accentuated by the fact that different electroactive modifications of both divalent and trivalent Ni have to be considered. The existance of these modifications has first been extensively discussed by Bode et al. [18], and has, since then, been the subject of many investigations [19-23].

Fig. 3 summarizes the various Ni species, most relevant for energy storage applications. Starting with metallic Ni the highly hydrated α-modification of divalent $Ni(OH)_2$ can be electrochemically prepared at relatively negative potentials [20], in accordance with thermodynamic considerations [8]. Upon cycling and/or ageing at more positive potentials this modification is dehydrated into the β form, which can be further oxidized to trivalent β-NiOOH. During overcharging, β-NiOOH can be irreversibly converted into the hydrated γ-modification. γ-NiOOH can be electrochemically reduced into α-$Ni(OH)_2$.

The two β-type modifications have been identified as the active materials in battery electrodes. Both compounds have a well-defined crystalline structure of the so-called brucite C6-type. Two unit cells of this hexagonal structure type are

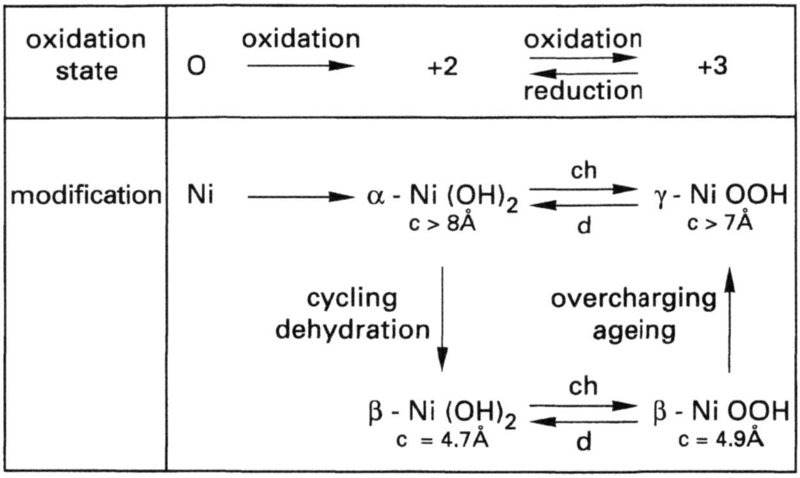

Fig. 3. Reaction scheme proposed by Bode et al. [18] for the Ni electrode reactions in alkaline solutions.

shown for $Ni(OH)_2$ in Fig. 4a. Characteristic is the intercalated structure in which stacking layers of only Ni atoms and of O and H atoms can clearly be recognized. NiOOH can be considered as the H-deficient form of $Ni(OH)_2$ (see Fig. 4). It has been pointed out that the conductivity of both well-crystallized β-modifica-

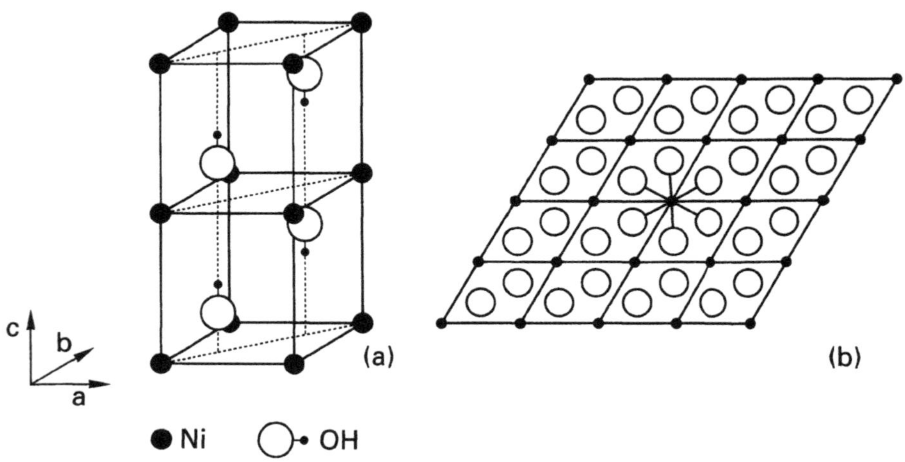

Fig. 4. Hexagonal crystal structure of β-$Ni(OH)_2$ in a three-dimensional view (a) and in projection on the basal plane (b) of several unit-cells. The hydrogen atoms are omitted in (b).

tions is relatively poor and that these compounds have semiconducting properties [15], as is not uncommon for many metaloxides [24]. However, by both infrared spectroscopic and gravimetric studies it has been revealed that these compounds in their precipitated form as commonly used in electrodes, contain small amounts of H_2O, through which the conductivity is somewhat improved.

Both hydrated forms, i.e. α-$Ni(OH)_2$ and γ-NiOOH, are proposed to have a layered brucite structure, similar to that shown in Fig. 4, except for the additional intercalation layers of water between the Ni layers. This results in an expansion of the crystal lattice in a very specific way; in contrast to the a-axis of the hexagonal structure, which remains more or less constant for all modifications (a≈3 Å), the c-axis is considerably increased, as indicated in Fig. 3. The presence of these water layers enables ionic transport within the crystal lattice to occur. This phenomenon, which has been proven to take place by gravimetric measurement, in particular by utilizing the quartz crystal microbalance [25,26], would be responsible for a further improvement of the conductivity of these hydrated modifications.

The charge transfer reaction between the β-modifications of a nickeloxide battery electrode is generally considered to occur via a solid state transition mechanism. This mechanism is schematically outlined in Fig. 5a. During charging both protons and electrons are liberated within the β-$Ni(OH)_2$ solid:

$$Ni(OH)_2 \underset{d}{\overset{ch}{\rightleftarrows}} NiOOH + H^+ + e^- \qquad (19)$$

The electrons are transported to the current collector at the back of the electrode. On the other hand, protons are transported through the solid by means of diffusion to the solid/solution interface, where they react with OH^- ions to give water:

$$H^+ + OH^- \underset{d}{\overset{ch}{\rightleftarrows}} H_2O \qquad (20)$$

Since the electrolyte consists of a strong alkaline solution, supply of OH^- is generally not considered to be a rate-limiting factor. Despite the fact that the electronic conductivity of β-$Ni(OH)_2$ is found to be poor, solid state diffusion of (hydrated) cations is reported to be the rate-limiting step in the charge transfer reaction [27,28]. It should, however, be noted that not only hydrated protons but other hydrated cations present in the electrolyte, such as Li^+, can also be incorporated into the solid and may thus be involved in the solid state diffusion process. Incorporation of "foreign ions" has indeed been indentified by sophisticated gravimetric measurements [25,26].

The morphological implications of the above discussed mechanism during charging are shown in Fig. 5b. Starting with a completely discharged electrode

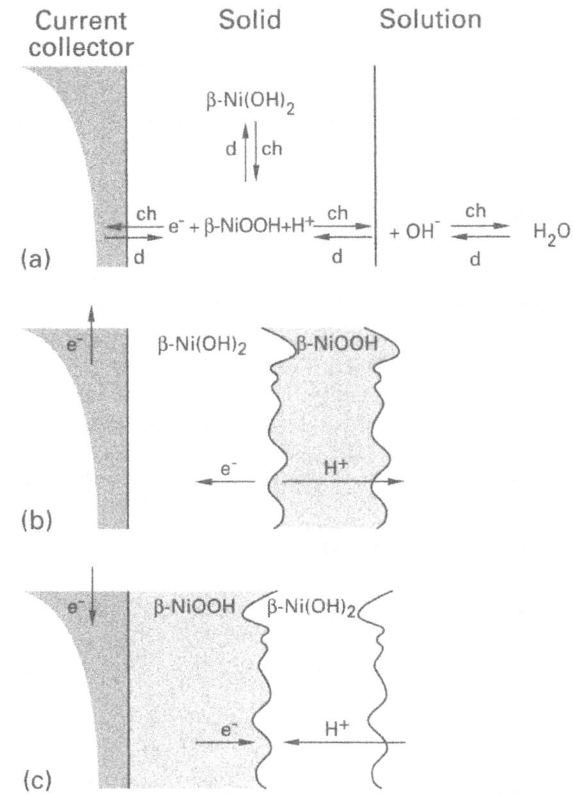

Fig. 5. Schematized solid-state transition mechanism for the $Ni(OH)_2$/NiOOH charge transfer reaction (a). Morphological implications of such mechanism during both charging (b) and discharging (c).

it seems likely that in the initial stages of the oxidation reaction, liberation of electrons and protons from the electroactive $Ni(OH)_2$ starts near the solid/solution interface. This results in the formation of a β-NiOOH surface layer. As the charging process proceeds, the thickness of the surface layer will grow and, consequently, the diffusion layer thickness through which cations must be transported. This will hamper the overall reaction (see Fig. 5b). During discharging the reverse processes take place. The cations must now be transported through a β-$Ni(OH)_2$ layer, which is initially formed at the electrode surface, as illustrated in Fig. 5c. The NiOOH/$Ni(OH)_2$ interface will now move inwards the electrode during the course of the discharge process. It has been analyzed that diffusion through the oxidized solid is somewhat faster in comparison to the diffusion rate within the reduced solid [27]. This would imply that charging can proceed somewhat faster than discharging.

It is clear from the above considerations that cycling between the reduced and oxidized state induces the electrode to swell and shrink. This continuous process puts some serious demands upon the Ni electrode. It has, for example, to be avoided that parts of the electrode become electrically insulated, through which the storage capacity will seemingly decline. In addition, short-circuiting of the Ni and MH (or Cd) electrodes within the battery is, among other things, also strongly related to this swelling and shrinking process. So, a mechanical stable electrode is essential. In order to improve the mechanical stability, several chemical additives, such as Co and Cd-species, have shown to be very effective [15,29]. These additives not only improve the mechanical stability but also the electrochemical electrode characteristics to some extent [30,31]. It should, however, be noted that one of the main advantages of NiMH batteries over NiCd batteries is the environmental aspect. In this respect it is obvious that the Ni electrode employed in NiMH cells may not contain heavy metal species either. To design a properly functioning, Cd-free, nickel electrode was one of the important research challenges in the last few years.

It is evident that the largest influence on the electrochemical characteristics is caused by the electroactive species itself. It has been shown that the potentials at which oxidation and reduction of the various Ni-modifications occur differ significantly [19,20]. This sometimes results in the formation of different discharge plateaus.

Another important point which has already been addressed in the previous section is the decomposition reaction of H_2O, as represented by Eq. (3). Fig. 2 reveals that the standard redox potential of the oxygen evolution reaction ($E°_{OH-/O2}$ = +0.3 V vs Hg/HgO) is considerably more negative than that of the competing Ni reaction (1). This, in fact, implies that the water decomposition reaction would be more favourable than the Ni^{II} oxidation reaction and, in addition, that trivalent Ni^{III} is thermodynamically unstable in this aqueous environment. Fortunately, the kinetics of the oxygen evolution reaction are found to be relatively poor, which enables the Ni redox reaction to occur at a dominant rate under moderate conditions. However, a competition between the two reactions has always to be taken into account. Obviously, this becomes of special importance when the amount of $Ni(OH)_2$ in the Ni-electrode becomes smaller, as is the case at the end of the charging process.

In conclusion, we may state that characterization of the Ni electrode by determination of the mixed electrode potential is extremely difficult due to the thermodynamical instability of this electrode and due to the presence of various electroactive nickeloxide modifications. This is manifested by the poorly defined, unstable, open-circuit electrode potentials generally observed for the Ni electrode.

4. Hydride-forming electrode materials

4.1 INTRODUCTION

The similarity between the hydride formation/decomposition reaction via the gas phase and via the electrochemical charge transfer reaction is very close, especially as far as the materials bulk properties are concerned. This is schematically represented in Fig. 6. In the case of a gas phase reaction hydrogen gas is brought into contact with a hydride-forming compound and the gaseous molecules dissociate at the solid/solution interface (a). Atomic hydrogen is strongly adsorbed at the metal surface, as the metal-hydrogen interaction is energetically very favourable [32]. The adsorbed hydrogen atoms can, subsequently, be converted into the absorbed state by jumping into the interstitial sites beneath the first atomic layer. Further transportation into the bulk of the solid occurs via diffusion. The occurrence of a possible phase transformation is omitted in this symplified representation.

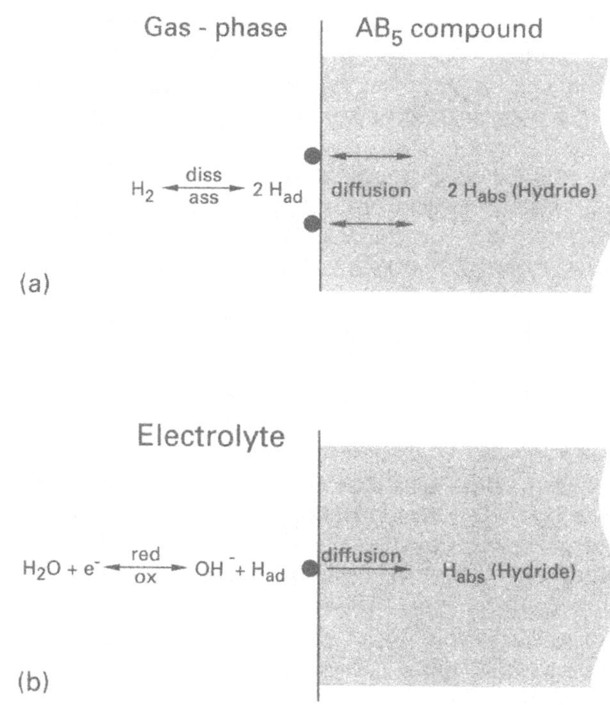

Fig. 6. Schematic representation of the hydride formation/decomposition process occurring via the gas phase (a) and electrochemical charge transfer reaction (b).

The electrochemical analogy is shown in Fig. 6b. The overall electrochemical reaction has already been represented by Eq. (2). Here we will go into some detail and consider the various partial steps involved in the electrochemical hydride formation/decomposition reaction. When we restrict ourselves, for the moment, to the reduction process, the following steps can be successively distinguished:

(i) The supply of reactants by means of (convective) diffusion from the bulk (b) of the electrolyte to the solid/solution interface (s)

$$H_2O_{(b)} \rightleftarrows H_2O_{(s)} \tag{21}$$

(ii) The charge transfer reaction occurring at the interface, which can be represented in alkaline solution by

$$H_2O_{(s)} + e^- \underset{k_a}{\overset{k_c}{\rightleftarrows}} H_{(ad)} + OH_{(s)}^- \tag{22}$$

where k_c and k_a are the reaction rate constants for the reduction and oxidation reaction, respectively. As a result of the reduction reaction, atomic hydrogen $H_{(ad)}$ is again chemically adsorbed on the surface. The formation of this energetically favourable M-H bond is very important with respect to the kinetics of the charge transfer reaction, as we will discuss in more detail in Sect. 4.3.

(iii) The removal of the electrochemically formed reaction products from the interface by means of diffusion. This includes the absorption of the adsorbed hydrogen by the AB_5 compound ($H_{(abs)}$) through which a hydride is formed

$$H_{(ad)} \underset{k_2}{\overset{k_1}{\rightleftarrows}} H_{(abs)} \tag{23}$$

and transport of OH^- ions into the bulk of the electrolyte

$$OH_{(s)}^- \rightleftarrows OH_{(b)}^- \tag{24}$$

(iv) Depending on the materials composition and on the hydrogen concentration in the solid, either an α phase or a β hydride phase in equilibrium with the α phase is formed (not considered in Fig. 6b).

$$H_{(abs(\alpha))} \rightleftarrows H_{(abs(\beta))} \tag{25}$$

(v) Finally, recombination of two $H_{(ad)}$ atoms has to be taken into account. This leads to the formation of H_2 which is released from the electrode surface as a gas

$$2H_{(ad)} \rightarrow H_2 \tag{26}$$

Whether this side-reaction (Eq. (26)) takes place after the absorption process (Eq. (23)) is completed or both reactions occur simultaneously, depends on the thermodynamic properties of the hydride-forming intermetallic compound [33].

Evidently, the reverse reaction sequence takes place during the oxidation (discharging) reaction with the exception of Eq. (26). From the above consideration it is clear that it makes no difference for the solid state kinetics whether hydride formation is initiated via the gas phase or electrochemically. Due to the different surface reactions, the overall kinetics of both hydride-formation processes may, however, be quite different.

The hydrogen absorption and desorption properties of intermetallic compounds are generally characterised by pressure-composition isotherms. It has long been recognized that there is a direct relationship between the equilibrium pressure for absorption and desorption of hydrogen gas (P_{H2}) and the equilibrium potential (E_e) of a metalhydride electrode [33]. This relationship has been expressed by

$$E_e = -\frac{RT}{nF}\ln P_{H_2} \tag{27}$$

where R is the gas constant, n the number of electrons involved in the hydrogen evolution reaction (Eqs. (22) + (26)) and E_e is measured against a reference electrode [33]. A comparison of the electrochemically determined hydrogen pressure and that determined via the gas phase is shown in Fig. 7. The results of a pressure-composition isotherm for an AB_5-type compound, measured in the gas phase with a conventional Sievert's-type equipment are represented by the filled circles in curve (a). The open circles (curve (b)) refer to the hydrogen pressures as calculated, according to Eq. (27), from the measured equilibrium potentials at different states-of-charge. These results show that the similarity in the plateau region is indeed very close. The differences appearing at higher hydrogen contents are related to the fact that, in the case of electrochemical charging (curve (b)), the hydrogen evolution reaction (Eq. (26)) competes with the hydride-formation reaction (Eq. (23)). Measuring the equilibrium potential is nowadays a frequently used method to determine the hydrogen plateau pressure of hydride electrodes [34,35]. It should, however, be noted that due to the logarithmic dependence given by Eq. (27) the accuracy of this electrochemical method is not as good as direct pressure measurements.

The plateau pressure of the hydride-forming material employed in a rechargeable NiMH battery may not be too high for different reasons: *(i)* the gas pressure inside a battery should be relatively low for obvious safety reasons *(ii)* to reduce the battery self-discharge rate *(iii)* competition between the hydrogen evolution reaction and the hydride-formation reaction becomes more severe when high plateau materials are employed. On the other hand, an important electrochemical requirement to be met for a NiMH battery is a high

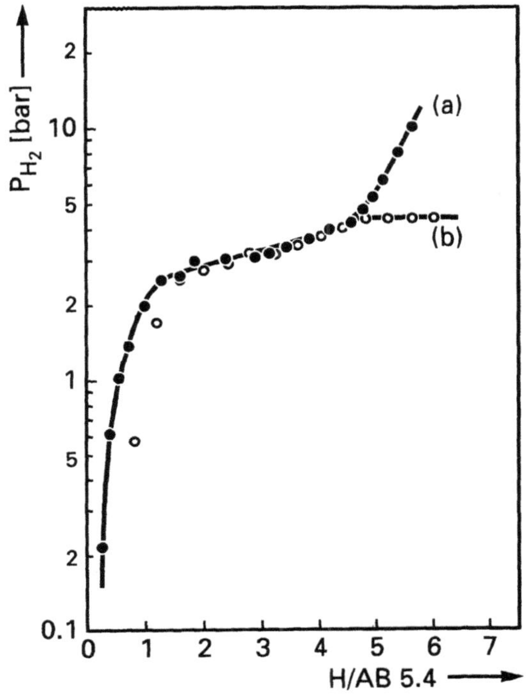

Fig. 7. Hydrogen absorption isotherms for $LaNi_{4.4}Cu$ (an $AB_{5.4}$ compound) as obtained from the gas phase (filled circles in curve (a)) and calculated from equilibrium potential measurements for the electrochemical case (open circles in curve (b)) at 20 °C.

cell voltage, i.e. employing a MH electrode with a relatively negative open-circuit potential, as already pointed out in relation to Fig. 2. Inspection of Eq. (27) reveals that such requirement is fulfilled when the plateau pressure is relatively high. Considering these various prerequisites, it is desirable to employ MH electrode materials with a hydrogen plateau pressure in the range of 0.1 to 1 bar.

4.2 ELECTROCHEMICAL CYCLING STABILITY

4.2.1. Oxidation sensitivity. Another important parameter for a hydride-forming electrode is its storage capacity and the capability to retain this capacity upon frequent electrochemical cycling. The best way to investigate the properties of MH electrodes separately, independent from those of the Ni electrode, is to test these electrodes in a so-called half-cell configuration. As already discussed in

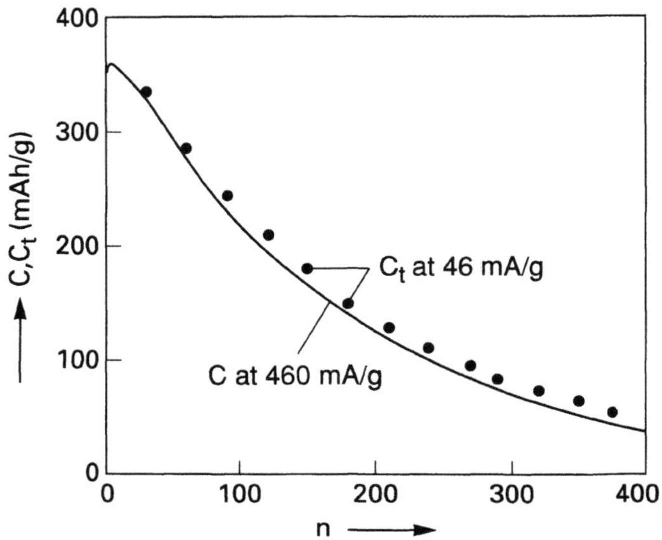

Fig. 8. Cycle-life plot of a MH electrode prepared by mixing $LaNi_5$ powder with copper powder and pressing this mixture into a pellet [33]. Cycling regime: charging (1 hour) and discharging with 460 mA/g (solid line). Discharging was ended when the electrode potential reached a value of 900 mV vs Hg/HgO. Additionally discharging was performed with a current of 46 mA/g every 30 cycles to obtain the deep-discharge capacities.

Sect. 2.1 the potential of the MH electrode is monitored against a reference electrode in such a configuration.

The cycling behaviour of an electrode made of the parent compound $LaNi_5$ is, unfortunately, somewhat disappointing. Fig. 8 shows that the storage capacity (C) obtained with a high discharge current (solid line) is initially very high, approximately 360 mAh/g. This value is very similar to the storage capacity measured in the gas phase and indeed corresponds to about 6 H per formula $LaNi_5$. However, the capacity decreases very rapidly upon repeatedly charging and discharging. Almost 90% of the initial storage capacity ($C_t(0)$) has become inactive after 400 charge and discharge cycles, resulting in a stability factor, S(400), of only 12% [33]. Herewith S(400) is defined as

$$S(400) = \frac{C_t(400)}{C_t(0)} \tag{28}$$

where the $C_t(n)$ values refer to the so-called deep-discharge capacities obtained every 30 cycles after additionally discharging the electrode with a ten-fold lower current and n represents the cycle number. This degradation process has been attributed to oxidation of $LaNi_5$. This can be recognized in Fig. 9, which reveals

Fig. 9. Scanning electron micrographs of $LaNi_5$ electrodes dismounted at different stages of the electrode cycle-life (from Ref. [33]). The cycle number is indicated. One bar represents 1 μm.

the electrode morphology after electrochemical cycling at different stages of the cycling process. Compared to the starting material (Fig. 9a), the size of the powder particles of the hydrogen-cycled electrode (Fig. 9b) is clearly diminished. It has been shown that this particle size reduction is induced by the hydrogen absorption process, more particularly, by the α-to-β phase transformation occurring inside the solid particles [33,36]. This phase transformation is accompanied by a considerable volume expansion of the crystal lattice. In the case of $LaNi_5$ the overall volume expansion amounts to approximately 25%

[36]. It is obvious that the large amounts of stress/strain induced within the solid, cause the particles to crack. The particle size reduction is not only restricted to electrochemical cycling but occurs during gas phase cycling also. The final average particle size for $LaNi_5$ powder has been determined to be about 3 μm [37,33]. After prolonged cycling (Fig. 9c) the still-electroactive powder particles are completely covered with oxidation products. After 350 cycles most of the original intermetallic compound has become inactive due to oxidation, resulting in a very low electrochemical storage capacity (see Fig. 8).

In recent work we have derived an equation which can account for oxidation of AB_5-type compounds in alkaline media [38]. Assuming oxidation to occur by chemical attack of water molecules [33] via an irreversible first order reaction, this process can, in general terms, be represented by

$$AB_x + 3H_2O \rightarrow A(OH)_3 + xB + 3/2H_2 \tag{29}$$

where AB_x is a general notation, including not only intermetallic compounds of the $LaNi_5$ family with stoichiometric composition but also compounds having a non-stoichiometric composition. The presence of the proposed solid-state reaction products after prolonged electrochemical cycling has been confirmed by XRD [33]. When we consider the AB_x powders to consist of uniform spherical particles with an initial radius r_o, which oxidize at a constant reaction rate determined by the reaction rate constant k_{ox}, the oxide layer thickness will gradually grow in time and, as a consequence, the amount of the underlying, still active, hydride-forming material will simultaneously decrease. The kinetics of the oxidation reaction can then be represented by

$$-\frac{dM_{ABx}}{dt} = k_{ox} a_{H_2O} \tag{30}$$

where dAB_x/dt is the molar oxidation rate per unit area, k_{ox} the oxidation rate constant and a_{H2O} the activity of water. For simplicity, the reaction order of water is assumed to be unity. Obviously, the decrease of the absolute amount of active material, expressed by the active particle volume (V), is proportional to the continuously changing surface area of the active material ($4\pi r_t^2$) and the reaction rate constant, i.e.

$$-\frac{dV}{dt} = \frac{k_{ox} a_{H_2O} 4\pi r_t^2 M_{AB_x}}{\rho_{AB_x}} \tag{31}$$

in which M_{ABx} and ρ_{ABx} are the molecular weight and the density of the AB_x compounds, respectively. As the volume of a spherical particle (V) is related to its radius ($V = 4/3\pi r_t^3$), r_t can be eliminated in Eq. (31) and after rearrangement we obtain

$$-V^{-2/3}.dV = \frac{k_{ox} a_{H_2O} M_{AB_x} 4^{1/3} \pi^{1/3} 3^{2/3}}{\rho_{AB_x}} . dt \tag{32}$$

Integrating this equation with the boundary conditions

$V = V_o$ at $t = 0$ and
$V = V_t$ at $t = t$

leads to

$$V_t^{1/3} = V_o^{1/3} - \frac{k_{ox} a_{H_2O} M_{AB_x} 4^{1/3} \pi^{1/3}}{\rho_{AB_x} 3^{1/3}} . t \tag{33}$$

or

$$V_t^{1/3} = V_o^{1/3} - \frac{V_o^{1/3} k_{ox} a_{H_2O} M_{AB_x}}{r_o \rho_{AB_x}} . t \tag{34}$$

when the initial particle radius r_o is introduced. The storage capacity at any moment in the electrode cycle-life is proportional to the total amount of uncorroded active material relative to the initial amount and the initial storage capacity:

$$C_t = \frac{V_t}{V_o} . C_t(0) \tag{35}$$

When we further define the specific surface area of the starting material (A_o) as

$$A_o = \frac{3}{r_o \ \rho_{AB_x}} \tag{36}$$

[4] and subsequently eliminate V_t and r_o in Eq. (34) by means of Eqs. (35) and (36) an expression for C_t is eventually obtained [38]:

$$C_t^{1/3} = C_t(0)^{1/3} - \frac{C_t(0)^{1/3} A_o k_{ox} a_{H_2O} M_{AB_x}}{3} . t \tag{37}$$

As the oxidation of these materials is a relatively slow process we may assume the oxidation reaction to be surface-controlled and thus that diffusion limitation of water does not occur. This implies that a_{H2O} in Eq. (37) can be regarded as a constant. The same holds for both $C_t(0)$ and M_{ABx} which, for a given compound, are material constants. In previous work an empirical exponential relationship was employed to characterize the cycle-life of a MH electrode [33,38]. It can, however, easily be shown that the mathematically derived Eq. (37) is, under certain conditions, very similar to the empirical relationship used before [38].

It has been shown qualitatively by Boonstra et al. [39] that the cycle life of $LaNi_5$ electrodes is strongly influenced by the particle size and thus, according to Eq. (36), to the specific surface area of the powder. A more quantitative example is shown in Fig. 10 for an electrode composed of an $AB_{5.4}$ compound, using different particle size fractions. The measured values for $C_t(0)^{1/3}$ are plotted versus time. Excellent linearities are found in all cases. Since both $C_t(0)$ and k_{ox} can be considered material constants in these experiments, the different slopes of the lines must be attributed to differences in A_o. It is important to note that the particles of the powders used in this experiment are all smaller than the material intrinsic particle size as accomplished by the hydridization process.

In conclusion we can say that despite the large initial storage capacity and the high electrocatalytic activity of $LaNi_5$, this compound is unsuitable for application in rechargeable NiMH batteries due to its poor long-term electrochemical cycling stability. In order to improve the cycling stability it is obvious from the above considerations (Eq. (37)) that two ways can be followed: either the oxidation rate constant, k_{ox}, or the specific surface area, A_o, of the MH powder must be lowered. This will be the central point in the following two subsections.

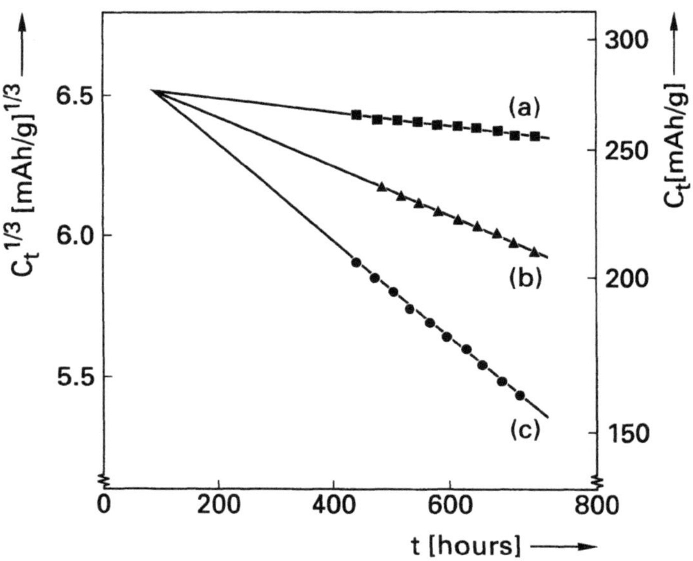

Fig. 10. Cycle-life plots (discharge current 35 mA/g), plotted according to Eq. (37), for various $LaNi_{4.4}Cu$ ($AB_{5.4}$) electrodes made of different partical size fractions. Particle size diameters: 36-71 μm (curve (a)); 5-20 μm (curve (b)); <5 μm (curve (c)) (from Ref. [38]).

4.2.2. Multicomponent compounds. In the mid-eighties Willems and Van Beek succeeded in stabilizing $LaNi_5$ and accomplished a breakthrough in the development of NiMH batteries [33,40]. They showed that the volume expansion of hydride-forming compounds, resulting from hydrogen absorption, could be substantially reduced by partly substituting the La and Ni atoms in the $LaNi_5$ crystal structure by other elements [33,40]. This volume expansion was identified to be the main cause for material degradation during electrochemical cycling. Fig. 11a reveals the two atomic positions in the AB_5-type crystal lattice. Multicomponent hydride-forming compounds were prepared by replacing part of the La atoms by other lanthanide atoms and Ni atoms by transition metals within the AB_5 stoichiometry.

The as-produced multicomponent compounds were found to be much more stable during electrochemical cycling than $LaNi_5$, as indicated by the increased values for S(400) in Fig. 12. In particular, introduction of Co atoms on the B-type atom positions in the crystal lattice was found to be very beneficial and improved S(400) from 12% for $LaNi_5$ to up to 60% for $LaNi_2Co_3$ (curve (a)). Addition of either Al or Si further improved the long-term cycling stability to values close to 100% (see Fig. 12, curve (b)). Apart from the cycling stability other requirements must be met by the MH electrode material. A pentary AB_5

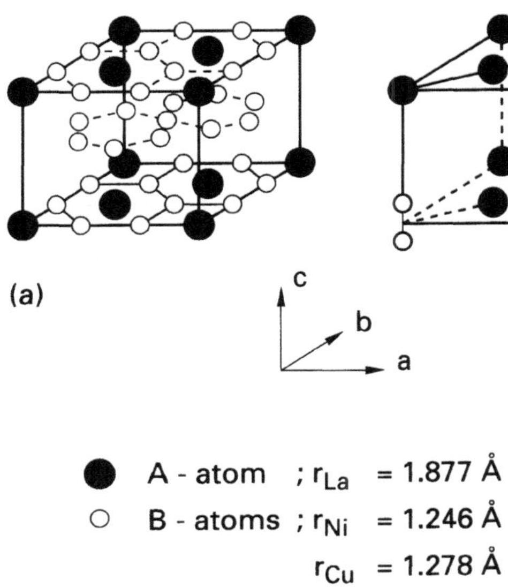

Fig. 11. Crystal lattice of stoichiometric AB_5-type compounds (a) revealing the lanthanide atoms (●) and transition metals (○). Only the A-type positions are considered in structure (b) where some La atoms are replaced by pairs of B-type atoms oriented along the c-axis, resulting in non-stoichiometric AB_x compounds (see also section 4.2.3.).

compound, for example, with composition

$$La_{0.8}Nd_{0.2}Ni_{2.4}Co_{2.5}Si_{0.1} \tag{38}$$

was found to combine an excellent long-term cycling stability and a sufficiently low hydrogen plateau pressure [33,40]. This makes this compound very attractive for application in rechargeable NiMH batteries.

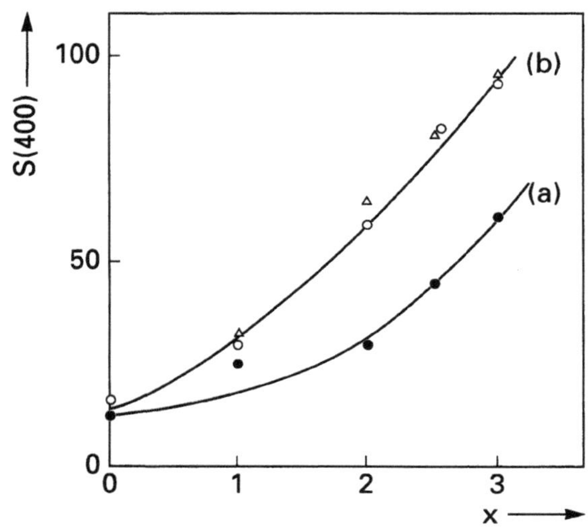

Fig. 12. Dependence of the long-term electrochemical cycling stability (S(400)) on the degree of Ni substitution by Co atoms in stoichiometric AB_5 compounds with composition $LaNi_{5-x}Co_x$ (curve (a)), $LaNi_{5-x}Co_xSi_{0.1}$ and $LaNi_{5-x}Co_xAl_{0.1}$ (triangles and open circles, respectively in curve (b)), where the Co content is given by x (data from Ref. [33]).

The hydrogen absorption/desorption isotherm for an activated powder with the above-mentioned composition is shown in Fig. 13. A clear hydrogen plateau region is found at a pressure of about 0.2 bar. Only a very small hysteresis is found. These results are in agreement with the results reported before [33,41]. In recent work we described that *in situ* XRD measurements are a useful tool to investigate the crystallographic changes inside the MH particles [36]. *In situ* XRD patterns were measured at this activated powder along the absorption/desorption isotherm under steady-state conditions after pressure stabilization had taken place.

The most informative part of the XRD results obtained at the various hydrogen concentrations, indicated in the absorption isotherm of Fig. 13a, are shown in Fig.14. In the initial stages of the absorption process only one set of XRD (α phase) reflections are found. The same holds for the β phase region at

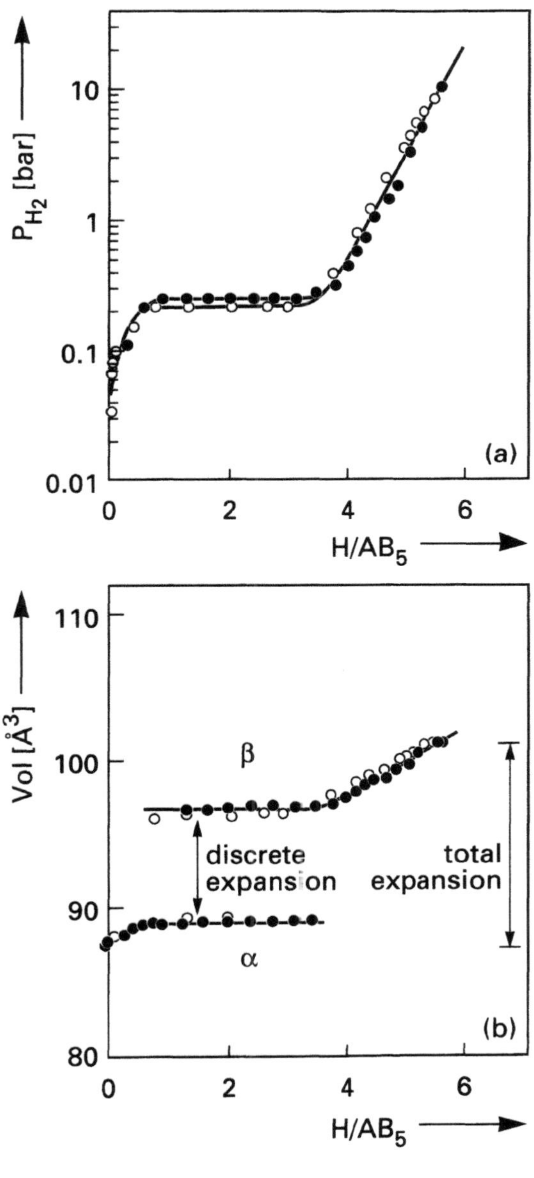

Fig. 13. Simultaneously measured hydrogen absorption/desorption isotherm (a) and volumetric unit-cell dimension (b) as obtained from *in situ* XRD at $La_{0.8}Nd_{0.2}Ni_{2.4}Co_{2.5}Si_{0.1}$ (sample annealed for 1 week at 1200 °C) under steady-state pressure conditions at 20°C. Absorption and desorption results are indicated by filled and open circles, respectively.

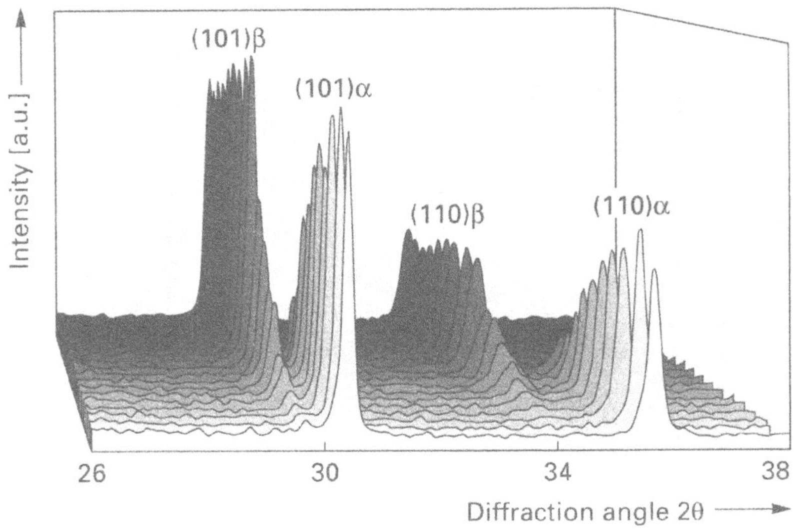

Fig. 14. Most informative part of the *in situ* XRD plots for $La_{0.8}Nd_{0.2}Ni_{2.4}Co_{2.5}Si_{0.1}$ (annealed for one week at 1200°C) powder at different stages of the hydrogenation process. The successive scans are obtained under steady-state conditions and correspond to the absorption isotherm of Fig. 13.

higher hydrogen contents. In the plateau region two sets of XRD reflections can be recognized, which indicates that the α and β phase coexist in this concentration range. In addition, it is clear that the intensity of the XRD reflections changes completely, indicating that the amount of both individual phases changes drastically along the hydrogen plateau, as expected. The XRD patterns allow the calculation of the unit-cell dimensions. The as-obtained crystallographic data are shown in Fig.13b and indeed reveal that both the α and β phase coexist in the plateau region. The total volumetric expansion (at 10 bar) amounts to 15.7%, whereas the discrete expansion, which has been defined as the normalized difference in lattice expansion between the β and α phase in the two-phase region (see Fig. 13b), is still appreciable and of the order of 9% [42].

The cycle-life plot for this optimized material is shown in Fig. 15. The electrochemical storage capacity is again very similar to that measured in the gas phase. Furthermore, Fig. 15 indeed reveals that the cycling stability has improved considerably compared to that of $LaNi_5$. A value for S(400) of 86% is calculated for this "standard alloy". It was, however, found that the rate of the charge transfer reaction for these compounds is retarded substantially, as can be recognized by the much lower discharge efficiencies (difference between curves (a) and (b) in Fig. 15; compare also with the results shown for $LaNi_5$ in Fig. 8).

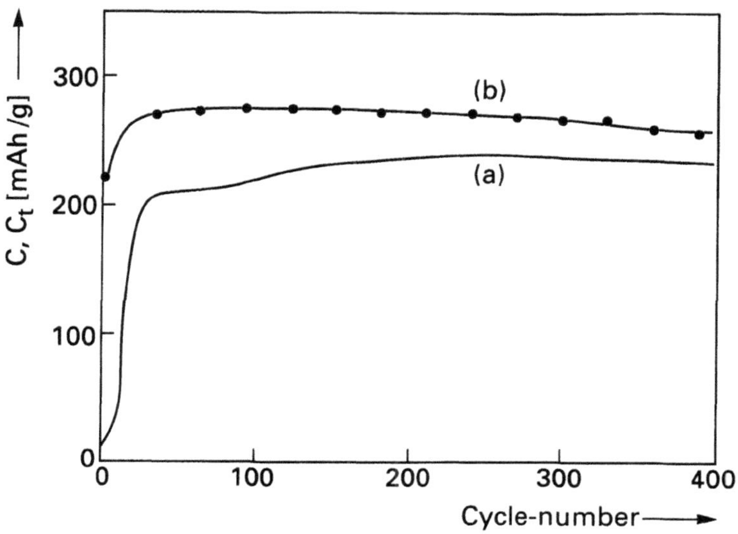

Fig. 15. Cycle-life plot of a $La_{0.8}Nd_{0.2}Ni_{2.4}Co_{2.5}Si_{0.1}$ electrode charged and discharged with 350 mA/g (curve (a)). Additionally discharging with 35 mA/g was performed every 30 cycles, resulting in the total storage capacity (C_t) dependence of curve (b) (from Ref. [33])

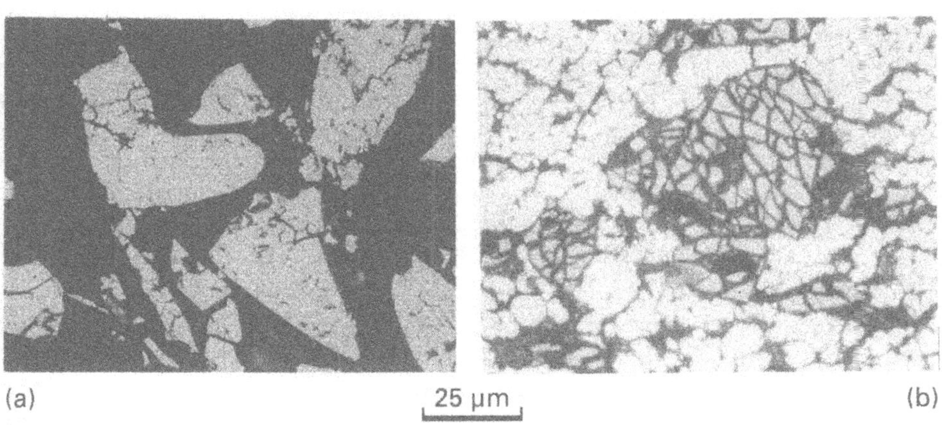

Fig. 16. Morphology inside hydride-forming particles after 5 gas phase cycles (a) and 428 electrochemical cycles (b). Alloy composition is $La_{0.8}Nd_{0.2}Ni_{2.4}Co_{2.5}Si_{0.1}$

Fig. 16 shows the morphology of the powder particles after both gas phase cycling (a) and electrochemical cycling (b). These photographs clearly reveal that the original unhydrided particles, which were of the order of a few tens of a μm, are cracked to a large extent. It has been shown that the discrete lattice expansion is responsible for the particle size reduction [42]. The final particle size is in the range of 1-5 μm and does not depend on whether the hydride is formed via the gas phase or electrochemically. The final particle size of this multicomponent compound is very similar to that reported for $LaNi_5$. These results are also in agreement with reported BET measurements, which is a well-known gas adsorption method for surface area determination [33]. Furthermore, Fig. 16b shows that in the case of electrochemical cycling the size-reduced MH particles are all covered by a porous oxide layer. Since the specific surface area of this compound has not changed significantly with respect to that of $LaNi_5$, these results imply, according to Eq. (37), that the intrinsic material property, as represented by the oxidation rate constant, k_{ox}, has improved substantially, simply by atom substitution within the AB_5 stoichiometry. Since the discovery of these stable compounds, substitution within the AB_5 stoichiometry has proven to offer a wide variety of prospects to battery manufacturers and research institutes to further optimize the various important electrode characteristics [5,6,12,45,46].

4.2.3. Non-stoichiometric compounds. In recent work we have shown that the oxidation sensitivity can also be significantly improved by utilizing non-stoichiometric (AB_x) compounds. These AB_x compounds can be easily prepared from the melt either by rapid solidification [47] or by annealing the solidified compounds in the homogeneity regions of their phase diagram [48]. Characteristic for these compounds is that part of the A-type atoms in the crystal lattice are replaced by dumbbell pairs of B-type atoms [49]. This typical substitution mechanism is schematically shown in Fig. 11b. Assuming the dumbbell pairs to be oriented along the c-axis, it can be easily understood that these materials are characterized by a rather anisotropically-deformed crystal lattice, i.e. considering the atomic dimensions given in Fig. 11b, that these compounds show a significantly reduced a-axis and a strongly elongated c-axis. The number of La replacements by pairs of B-type atoms was found to be strongly influenced by the chemical composition of the alloy [38,48]. For example, only alloys with an overall composition of up to $AB_{5.4}$ can be produced with the binary $LaNi_x$ system [50], whereas the Cu-containing ternary $La(Ni/Cu)_x$ system allows one to prepare compounds with overall compositions in excess of $AB_{6.0}$ [38].

Fig. 17 shows the cycle-life plots of electrodes made of various ternary non-stoichiometric compounds. All AB_x alloys are single-phase and have the same Cu content. The alloy composition can be represented by $LaNi_{x-1}Cu_{1.0}$, where x is varied in the range of 5.0 to 6.0. It is clear that the long-term cycling stability is drastically increased with increasing B-content, i.e. with a larger number of

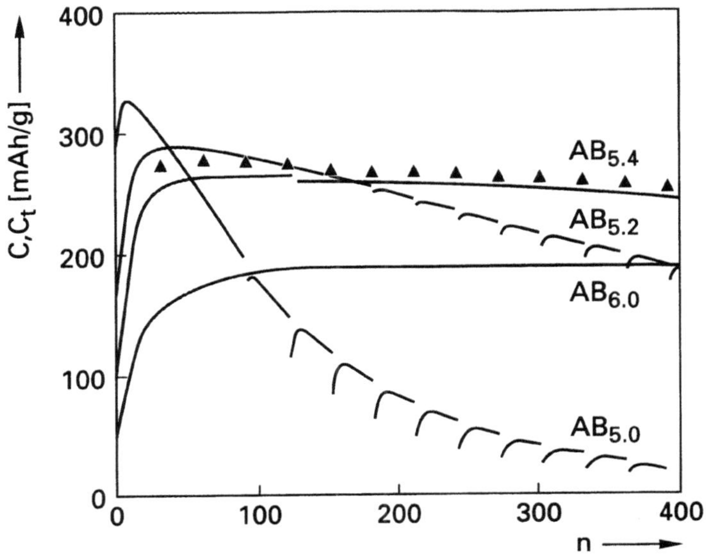

Fig. 17. Storage capacity (C), resulting from discharging (350 mAh/g), as a function of the number of complete charge/discharge cycles for single-phase non-stoichiometric AB_x compounds with constant Cu content ($LaNi_{x-1}Cu$). Total storage capacities (C_t) as obtained from deep-discharging (35 mA/g) are indicated by triangles for $LaNi_{4.4}Cu$ only (from Ref. [38]).

introduced dumbbell pairs. Above a non-stoichiometric composition of $AB_{5.4}$ the capacity decay is very small. An advantage of this substitutional mechanism is that the electrocatalytic activity remains extremely high, resulting in high discharge efficiencies even under extreme conditions (e.g. low temperatures) [48,42]. As the cycling stability of the stoichiometric $LaNi_{4.0}Cu$ is also very poor, of the same order of magnitude as that of $LaNi_5$ (compare with Fig. 8), it is evident that the improved stability must indeed be attributed to the deviation from the AB_5 stoichiometry. It should be noted that the long-term cycling stability of the binary $LaNi_{5.4}$ compound is also very similar to that of $LaNi_5$, illustrating the special role of Cu in these non-stoichiometric alloys.

The electrode morphology after more than 750 electrochemical charge/discharge cycles have been investigated by SEM. Strikingly, large differences between the electrodes are found, which clearly correlate with the electrode cycling stability. Two extreme examples are shown in Fig. 18. The oxidation-sensitive $LaNi_{4.0}Cu$ electrode (Fig. 18a) reveals that many cracks are traversing all the original alloy particles and that the surface of these newly formed surfaces are all covered with a porous oxide layer, as has been confirmed by Electron Probe Microanalyses [38]. These results are indeed very similar to those found for $LaNi_5$ and $LaNi_{5.4}$ (see e.g. Fig. 9). On the other hand, the more stable non-stoichiometric compounds show that a pronounced particle size reduction, resulting from electrochemical cycling is absent (Fig. 18b). In terms of Eq. (37) these results imply that the improved long-term cycling stability of the stable non-stoichiometric compounds must be attributed to a considerably lowered specific surface area (A_o). Although the hydride-forming particles are covered by an oxidic surface layer also, it is clear that the impact of a lower surface area is that the absolute amount of oxides is substantially reduced, resulting in a much lower storage capacity loss during electrochemical cycling (Fig. 17). This contrasts the effect of the conventional substitution mechanism described in the previous section, which resulted in multicomponent compounds. The origin of the long-term cycling stability of multicomponent compounds was there argued to result from an improvement of k_{ox}.

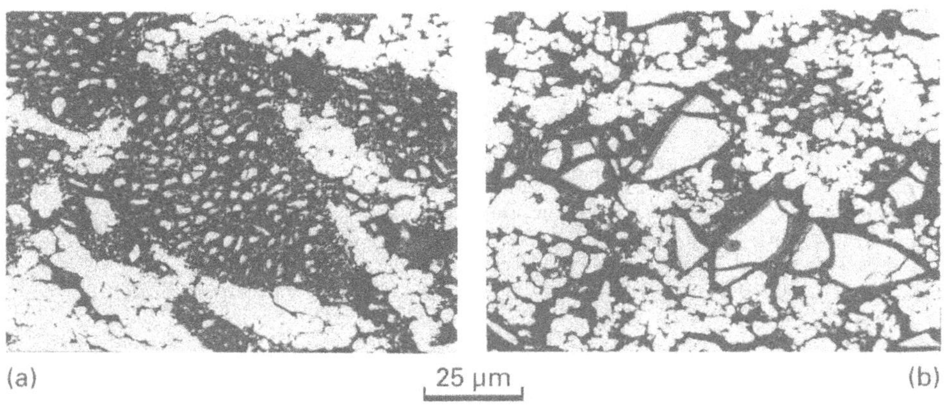

Fig. 18. SEM photographs of polished $LaNi_{4.0}Cu$ (a) and $LaNi_{4.4}Cu$ (b) electrodes, which have been electrochemically cycled for more than 750 complete charge/discharge cycles.

The nature of the reduced particle size reduction has been investigated by *in situ* XRD [42] and can best be visualized on the basis of results obtained with $LaNi_{4.0}Cu$ and $LaNi_{4.4}Cu$ powders in the gas phase. Large differences in hydrogen absorption behaviour appear between the two materials. In the case of $LaNi_4Cu$ (Fig. 19a) a clear phase transition is observed throughout the absorption/desorption process. The α and β reflections are largely separated from each other, implying significant dimensional differences between the two crystallographic phases. On the other hand, the electrochemically stable $LaNi_{4.4}Cu$ does not exhibit such a pronounced phase separation (Fig. 19b). Instead an almost continuous shift of the XRD reflections towards smaller angles is found. In only one of the XRD scans (indicated by the arrows) the α phase is found to be in equilibrium with the β phase. It should, however, be noted that an appreciable phase separation is absent in this case as the

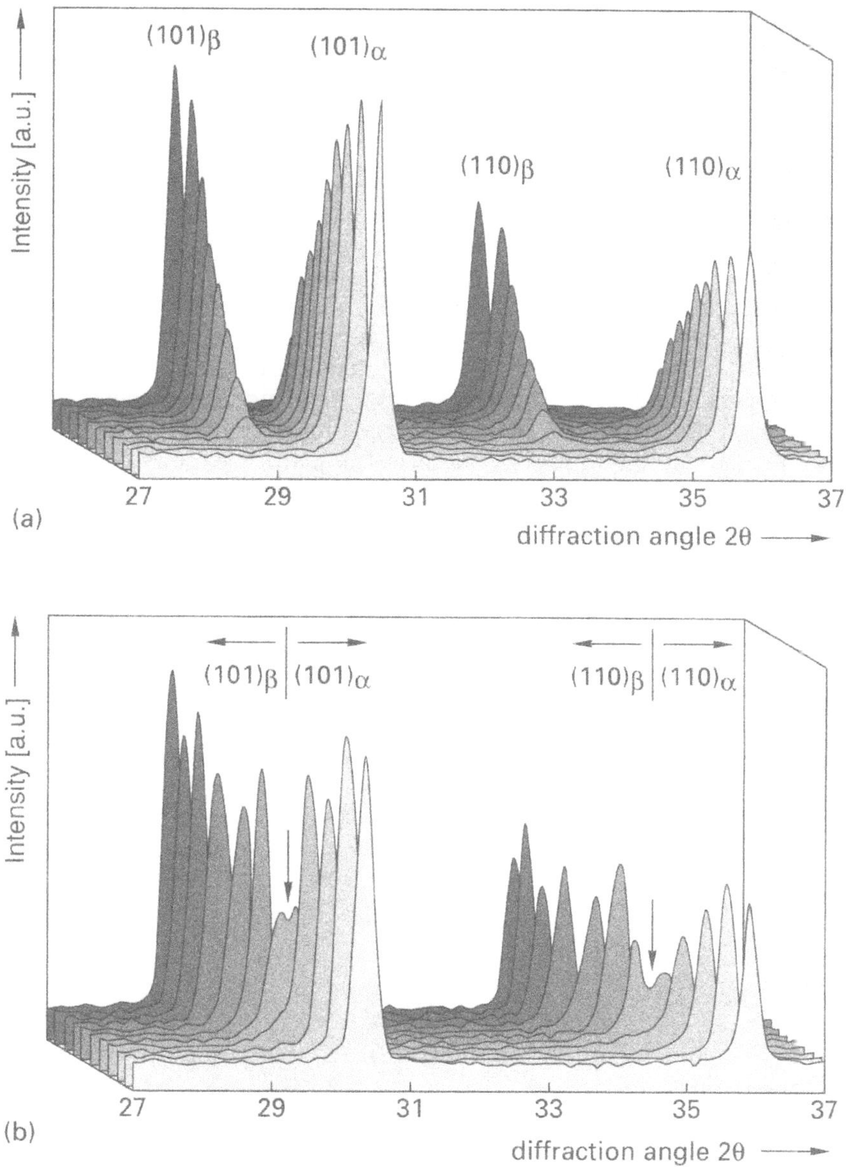

Fig. 19. *In situ* XRD plots obtained under steady-state conditions for stoichiometric $LaNi_{4.0}Cu$ (a) and non-stoichiometric $LaNi_{4.4}Cu$ (b) at different stages of the hydrogenation process (from Ref. [42]).

characteristic reflections of both phases are found to be very close to one another. The dimensional implications of these phases are therefore expected to be much lower than for the corresponding stoichiometric material (compare with Fig. 19a).

On the basis of the *in situ* XRD plots the crystallographic unit-cell dimensions can be calculated as a function of the hydrogen content. These results are shown in Fig. 20. It is clear that the discrete lattice expansion, as defined in Fig. 13, is largest for the stoichiometric compound and strongly decreases with increasing degree of non-stoichiometry, whereas the total lattice expansion is appreciable in all cases. Simultaneously, the extent of the plateau region decreases to finally disappear for the $AB_{5.4}$ alloy. These results become even more clear when they are represented in a phase diagram, which is shown in Fig. 21. A broad miscibility gap appears for the stoichiometric alloy but the width of the gap decreases rapidly with increasing degree of non-stoichiometry to become very small for the $AB_{5.4}$ compound. Above the as-denoted critical composition (C_c) no two-phase region can be identified, as is, for example, the case for the $AB_{6.0}$ compound. Consequently, no discrete lattice expansion is found at all for this compound [42].

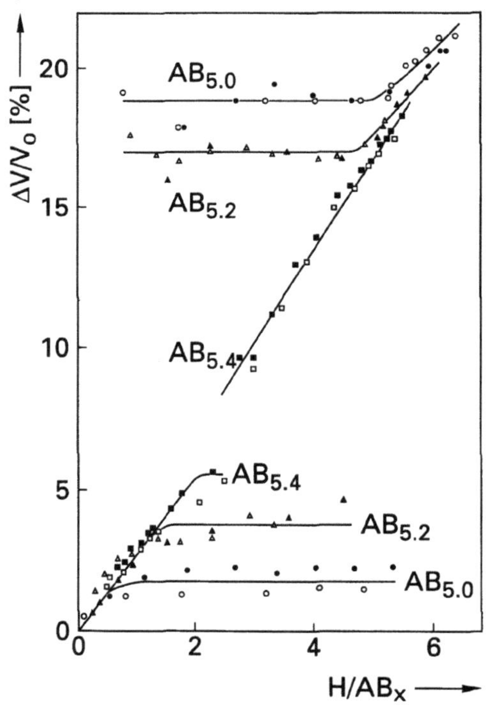

Fig. 20. Normalized volumetric crystal lattice expansion as a function of the hydrogen content in various non-stoichiometric ternary compounds as obtained from *in situ* XRD The lower set of curves correspond to the α phase and the higher set to the β phase Absorption and desorption results are represented by filled and open symbols, respectively (from Ref [42])

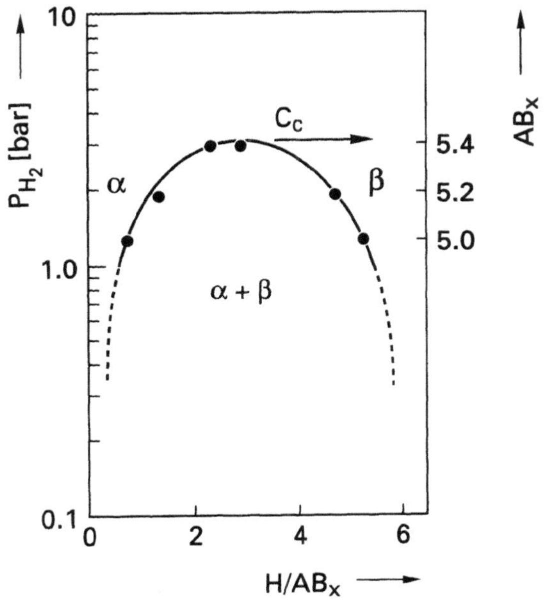

Fig. 21. Phase diagram for the ternary AB_x-hydrogen system at 20°C as derived from in situ XRD measurements and absorption/desorption isotherms. The two-phase region disappears at the as-denoted critical composition (C_c).

The above results show that the differences in mechanical stability between the various compounds must be related to whether or not the hydrogenation/decomposition process is accompanied by a large volumetric expansion of the phases involved. This is schematically illustrated in Fig. 22. It seems likely that, due to the forbidden hydrogen concentration range, the transition region between the *α* and *β* phase is extremely small for materials revealing a large discrete lattice expansion (Fig. 22a). This induces the formation of defect structures within the bulk of the powder particles, as was recently shown for $LaNi_5$ [36]. The capability to accommodate the lattice expansion of these defect structures is, however, not unlimited in these brittle materials and the excess of stress/strain will inevitably be released by the formation of cracks within the powder particles, as has experimentally been found (Fig. 18a). It should be emphasized that during the dynamic stages of the hyrogenation/decomposition reaction the transition region is continuously moving until a steady-state situation is reached. On the other hand, when we consider solids outside the miscibility gap (see Fig. 21), the induced stress/strain level is expected to be much lower, as any hydrogen concentration range within the bulk of these

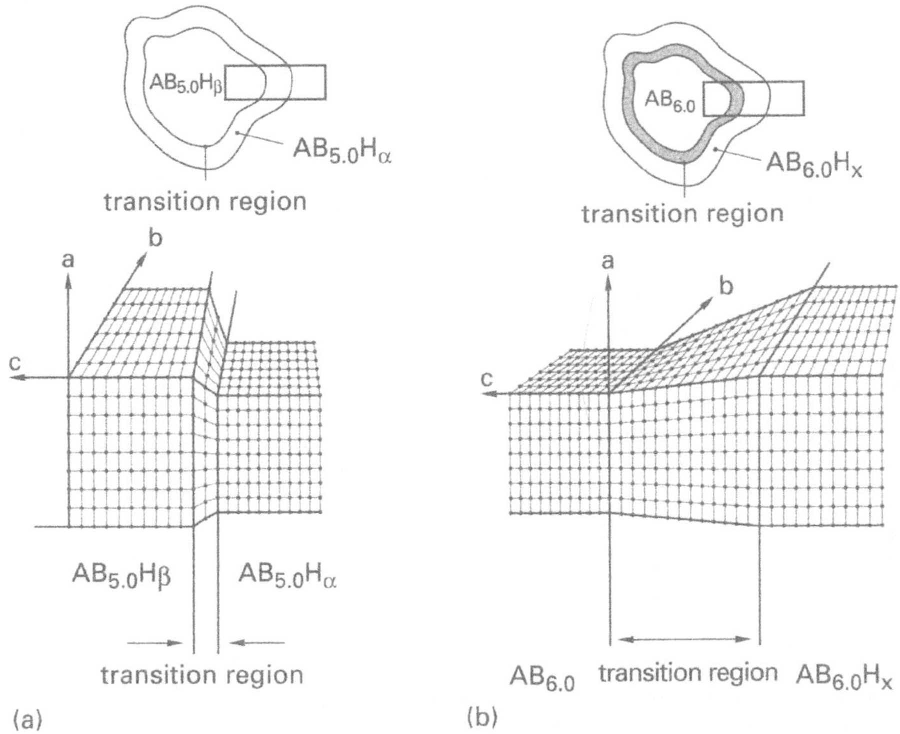

Fig. 22. Crystallographic implications of the hydride formation/decomposition mechanism for a material, which is characterized by a α-β phase transition (a) and a compound lacking such clear phase transition (b). The examples refer to the AB_5 and AB_6 compound, respectively.

compounds is allowed (Fig. 22b). It is obvious that materials lacking a clear α/β phase transition and thus lacking extreme stress/strain levels, are much more resistant to particle size reduction (Fig. 18b), and consequently, according to Eq. (37), show a much improved long-term cycling stability as was illustrated in Fig. 17.

4.3. KINETICS

4.3.1. Theoretical considerations. In Sect. 2.1. we already discussed in a rather qualitative way that an exponential relationship between the current and the MH electrode potential is to be expected (see dashed curves in Fig. 2). In this section we will go into more quantitative detail. In order to reduce the complexity of characterizing the kinetics of the hydride formation reaction we do not consider a phase transformation. Furthermore, the hydrogen evolution reaction is assumed not to take place, which can easily be accomplished experimentally under certain conditions. The reactions represented by Eqs. (25) and (26) will therefore be omitted in the following mathematical derivation.

As the electrode potential at the point of zero charge does not contribute to the reaction rate of the electrochemical charge transfer reaction, the rate (r) of both the reduction and oxidation reactions, as represented by Eq. (22), are given by

$$r_a = A_a \theta^x a_{OH^-}^y \cdot \exp\left\{-\frac{E_a}{RT}\right\} \tag{39}$$

and

$$r_c = A_c a_{H_2O}^z \cdot \exp\left\{-\frac{E_c}{RT}\right\} \tag{40}$$

in which A_a and A_c are preexponential factors: the surface coverage θ is a measure for the amount of chemically adsorbed hydrogen; a_{OH^-} and a_{H_2O} are the activities of the indicated electroactive species at the solid/solution interface; x, y and z are the reaction orders of $H_{(ad)}$, OH^- and H_2O, respectively, and E_a and E_c the activation energies for both partial reactions. The rate constants k_a and k_c are given by

$$k_a = A_a \cdot \exp\left\{-\frac{E_a}{RT}\right\} \tag{41}$$

and

$$k_c = A_c \cdot \exp\left\{-\frac{E_c}{RT}\right\} \tag{42}$$

respectively. Applying a potential (E) to the electrode influences both activation energies according to

$$E_a^* = E_a - \alpha nFE \tag{43}$$

and

$$E_c^* = E_c + (1-\alpha)nFE \tag{44}$$

and hence the rates of both reactions. Herewith α is a proportionality constant called the charge-transfer coefficient which is, in general, a constant having a value between 0 and 1. Combining Eqs. (39) and (40) with Eqs. (43) and (44) leads to expressions

$$r_a = A_a \theta^x a_{OH^-}^y \,.\, \exp\left\{\frac{-E_a + \alpha nFE}{RT}\right\} \tag{45}$$

and

$$r_c = A_c a_{H_2O}^z \,.\, \exp\left\{\frac{-E_c - (1-\alpha)nFE}{RT}\right\} \tag{46}$$

Rearrangement of Eqs. (45) and (46) and conversion of the reaction rates into current densities yields

$$i_a = nF\theta^x a_{OH^-}^y \exp\left\{\frac{\alpha nFE}{RT}\right\} * A_a \exp\left\{-\frac{E_a}{RT}\right\} \tag{47}$$

and

$$i_c = nFa_{H_2O}^z \exp\left\{\frac{-(1-\alpha)nFE}{RT}\right\} * A_c \exp\left\{-\frac{E_c}{RT}\right\} \tag{48}$$

As in the case of the electrode storage capacity, it is more relevant for energy storage applications to refer to currents normalized per units of weight instead of current densities. The specific surface area A_o is therefore introduced in Eqs. (47) and (48). If we further introduce the reaction rate constants k_a and k_c, given by Eqs. (41) and (42), and consider n=1 for the electrochemical charge transfer reaction (22), a general description for the kinetics of the partial oxidation and partial reduction reaction, as represented by the dashed curves in Fig. 2, is obtained.

$$I_a = FA_o k_a \theta^x a_{OH^-}^y \exp\left\{\frac{\alpha FE}{RT}\right\} \tag{49}$$

and

$$I_c = FA_o k_c (1-\theta)\, a_{H_2O}^z \exp\left\{\frac{-(1-\alpha)FE}{RT}\right\} \tag{50}$$

The values for the partial currents, I_a and I_c in Eqs. (49) and (50), are then expressed per units of weight (mA/g).

The exchange current I_o is defined at the equilibrium potential where I_a is equal to I_c (see Fig. 2). From Eqs. (49) and (50) it follows that under this condition I_o can be represented by

$$I_o = FA_o k_a \theta^x a_{OH^-}^y \exp\left\{\frac{\alpha F E_e}{RT}\right\} \tag{51}$$

$$= FA_o k_c (1-\theta)\, a_{H_2O}^z \exp\left\{\frac{-(1-\alpha) F E_e}{RT}\right\} \tag{52}$$

An expression for E_e can be obtained from Eqs. (51) and (52)

$$E_e = \frac{RT}{F} \ln\left\{\frac{k_c (1-\theta)\, a_{H_2O}^z}{k_a \theta^x a_{OH^-}^y}\right\} \tag{53}$$

After rearrangement we obtain

$$E_e = \frac{RT}{F} \ln\left\{\frac{k_c (1-\theta)}{k_a}\right\} + \frac{RT}{F} \ln\left\{\frac{a_{H_2O}^z}{\theta^x a_{OH^-}^y}\right\} \tag{54}$$

which is easily recognized as the well-known Nernst equation, in which the first term on the right-hand side represents the standard redox potential (E^o) and the second term takes into account the concentration dependence of the electrode potential.

When an overpotential (η) is applied to the hydride-forming electrode, for which holds that

$$\eta = E - E_e \tag{55}$$

Eqs.(49) and (50) can be converted into

$$I_a = FA_o k_a \theta^x a_{OH^-}^y \exp\left\{\frac{\alpha F E_e}{RT}\right\} * \exp\left\{\frac{\alpha F \eta}{RT}\right\} \tag{56}$$

and

$$I_c = FA_o k_c (1-\theta)\, a_{H_2O}^z \exp\left\{\frac{-(1-\alpha) F E_e}{RT}\right\} * \exp\left\{\frac{-(1-\alpha) F \eta}{RT}\right\} \tag{57}$$

The summation of the partial anodic and the partial cathodic current-potential curves (Eq. (56) + Eq. (57)) yields a mathematical expression for the charge transfer reaction, represented by the solid line in Fig. 2. When, in addition, the exchange current as represented by Eqs. (51) and (52), are introduced we finally obtain the well-known Butler-Volmer equation [44]

$$I = I_o\left[\exp\left\{\frac{\alpha F\eta}{RT}\right\} - \exp\left\{\frac{-(1-\alpha)F\eta}{RT}\right\}\right] \tag{58}$$

The kinetics can best be characterized by the exchange current. Eliminating E_e in Eq. (51), using Eq. (53), leads to a general expression for I_o

$$I_o = FA_o k_a^{(1-\alpha)} k_c^{\alpha} \theta^{x(1-\alpha)} (1-\theta)^{\alpha} a_{OH^-}^{y(1-\alpha)} a_{H_2O}^{\alpha z} \tag{59}$$

Since a_{OH^-} and a_{H_2O} can be regarded as constants in strong alkaline solutions, I_o is dependent on the specific surface area, the rate constants for reaction (22) and the surface coverage. For an appropriate comparison of I_o values of different materials a particular value of the surface coverage is required. Since the surface coverage is related to the amount of absorbed hydrogen, via Eq. (23), this can be accomplished by choosing a constant hydrogen level in the bulk of the hydride-forming AB_5 compound.

The exchange current is obviously a powerful parameter as a measure for the kinetics of the electrochemical hydrogen reaction. Although this parameter is defined in the equilibrium state it cannot be determined directly under these conditions. I_o can, however, simply be deduced from non-equilibrium measurements [44]. Provided that mass transport of the electroactive species in both the solid and electrolyte is not a limiting factor, I_o can be determined by measuring the dependence of the current on the overpotential. When η is changed within a small range both exponential terms in Eq. (58) can be expanded as two series. The quadratic and higher order terms can be neglected when $\eta < RT/\alpha F$ and we obtain a simplified version of the Butler-Volmer equation (low field approximation)

$$I = I_o \frac{F}{RT}\eta \tag{60}$$

Since F/RT is a constant for a given temperature, a linear relationship between the current and the overpotential is to be expected. I_o can then be calculated from the slope of such a line. This has indeed been confirmed experimentally [11].

4.3.2. Substitution. It is well-known that the rate constants for the electrochemical hydrogen reaction are strongly dependent on the metal at which the reaction occurs [43]. The strength of the metal-hydrogen bond is largely responsible for this behaviour as is shown in the classical Volcano plot of Fig. 23. A maximum is found for the elements Pd and Pt. These metals have a 170 times higher value for I_o than Ni and Co, which are the most abundant constituents of the stoichiometric multicomponent compound with composition given by Eq. (38). Thus, replacing small amounts of, for example, Co by Pd or Pt should

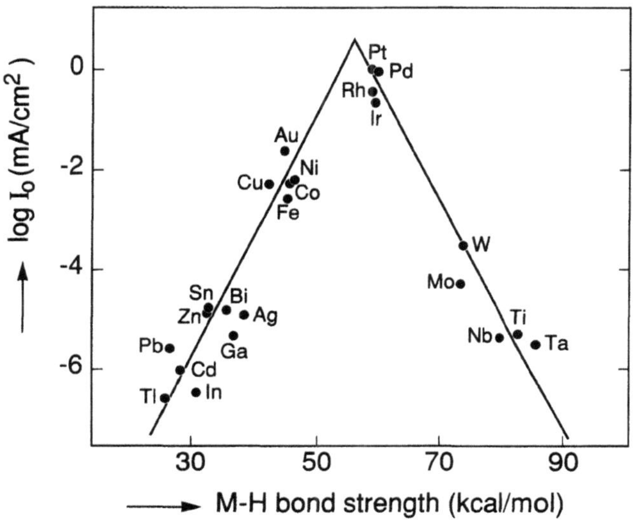

Fig. 23. Dependence of the exchange current density (i_o) of the electrochemical hydrogen reaction on the metal-hydrogen (M-H) bond strength (from Ref. [43]).

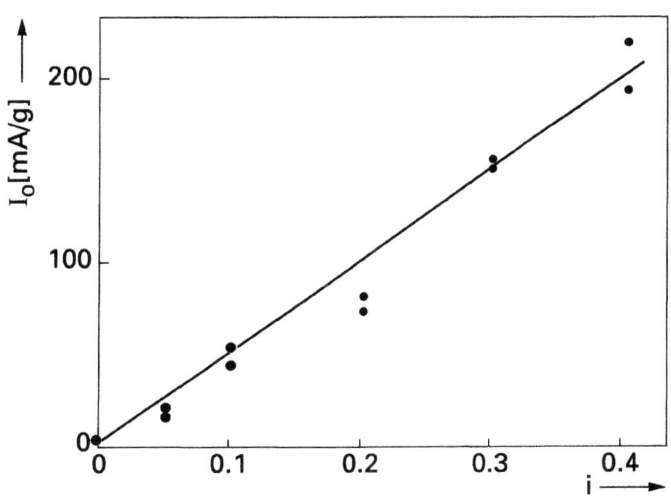

Fig. 24. Exchange current (I_o) measured at freshly prepared electrodes as a function of the Pd content (i) in single-phase alloys with composition $La_{0.8}Nd_{0.2}Ni_{2.5}Co_{2.4}Pd_iSi_{0.1}$ (from Ref. [12]).

significantly improve the electrocatalytic activity of the single-phase standard alloy.

Fig. 24 shows the results of the I_o values measured at freshly prepared electrodes as a function of the Pd content sustituted in the AB_5 standard alloy. The electrocatalytic activity is indeed strongly increased with increasing Pd content, as expected on the basis of Fig. 23. A linear dependence is found. At the highest Pd content a value for I_o was measured which was about 60 times higher than that found for the standard alloy. A disadvantage of this method is, of course, that these precious metals are substituted throughout the solid whereas they only are needed at the solid/solution interface in order to improve the kinetics of the hydrogen reaction.

4.3.3. double-phase compounds. An alternative approach to improve the kinetics of the charge transfer reaction is to make use of catalysts. Various investigations have shown that when the appropriate combinations of transition metals are alloyed, highly electroactive compounds are obtained which exceed

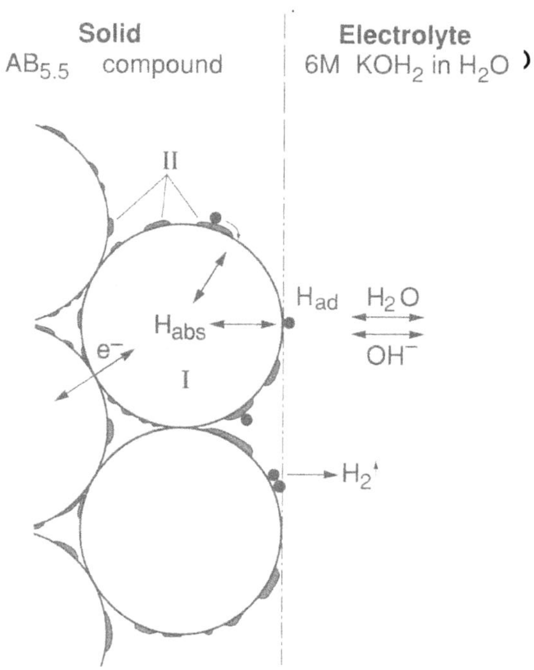

Fig. 25. Schematic representation of a double-phase $AB_{5.5}$ powder electrode in contact with an electrolyte. The hydride-forming bulk phase (I) and the highly electrocatalytic phase (II), decorating the powder surface, provides for a fast charge transfer reaction.

the electrocatalytic activity of the individual elements [51,52]. Alloying with relatively inactive elements, like for example Mo with Co or Ni (see Fig. 23), results in the formation of highly symmetrical and electroactive compounds with composition $MoCo_3$ and $MoNi_3$. To improve the overall electrocatalytic activity of hydride-forming compounds use has been made of these catalysts, which results in the formation of so-called double-phase compounds [11,12]. The desired material morphology of the proposed double-phase compounds is shown in Fig. 25. The bulk phase of these powders still consists of a hydride-forming compound, in the present case an AB_5-type compound, and a second catalytic phase, decorating the exterior of the powder particles, is responsible for the enhanced charge transfer reaction.

Considering the phase diagram of the La-Ni system it has been demonstrated that the desired double-phase structure can be easily formed from the melt during the solidification process when an excess of B-type elements are incorporated in the melt [11]. In the initial stages of the solidification process crystallites are grown with the AB_5-type composition, whereas at the end of the solidification process the B-enriched second-phase precipitates are formed at the grain bounderies. Obviously, with regard to the hydrogen-storage capacity it is desirable that the ratio between the amount of hydride-forming AB_5 compound and that of the secreted precipitates is as large as possible. Use has therefore been made of melts with an overall composition slighly over the stoichiometric composition, for instance with a melt composition of $AB_{5.5}$. Two examples of as-produced solids, in which an excess amount of Ni was added to the melt, are shown in the cross-sections of Fig. 26. The $AB_{5.5}$ melt was

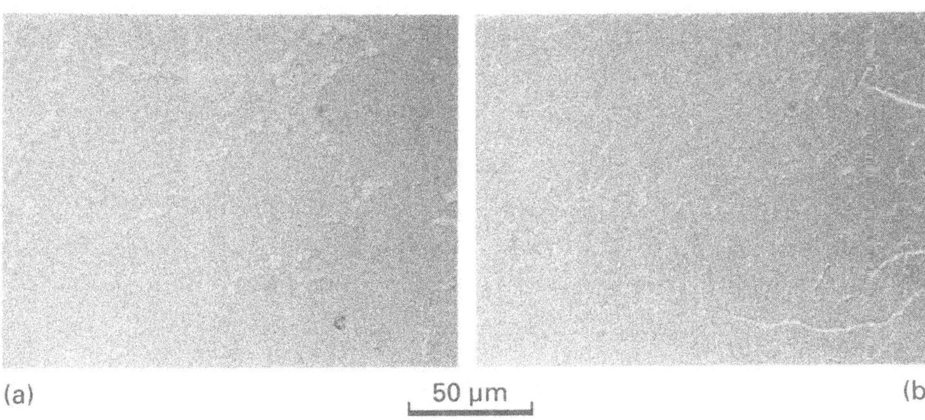

Fig. 26. Optical photomicrographs revealing the surface morphology of two double-phase $AB_{5.5}$ compounds formed during the solidification process from the melt. The size of the formed crystallites is dependent on the solidification rate: (a) refers to a low rate and (b) to a high solidification rate. Melt composition is $La_{0.8}Nd_{0.2}Ni_{3.0}Co_{2.4}Si_{0.1}$ (from Ref. [11]).

solidified at different cooling rates. The double-phase structure can be recognized in both cases. In accordance with metallurgical considerations the size of both the hydride-forming crystallites and the second-phase precipitates is found to be strongly dependent on the solidification rate. In the case of a slow solidification rate the crystallites are in the range of 50-100 μm (Fig. 26a), whereas much smaller crystallites, in the range of 10-30 μm (Fig. 26b), are formed when the melt is cooled down more rapidly. The chemical composition of both phases has been analysed by EPMA and was found to be in agreement with the desired composition [11].

The charge and discharge kinetics of the various investigated double-phase hydride-forming compounds was indeed found to be significantly improved with respect to those of the corresponding single phase alloys [11,47]. The kinetics of these compounds has been quantified, according to Eq. (60), by measuring the exchange current [11]. Fig. 27 shows the exchange currents of several double-phase compounds to be considerably higher than that of the corresponding standard alloy. The beneficial influence of the higher solidification rate (compare triangles in Fig. 27) can be attributed to the higher specific surface

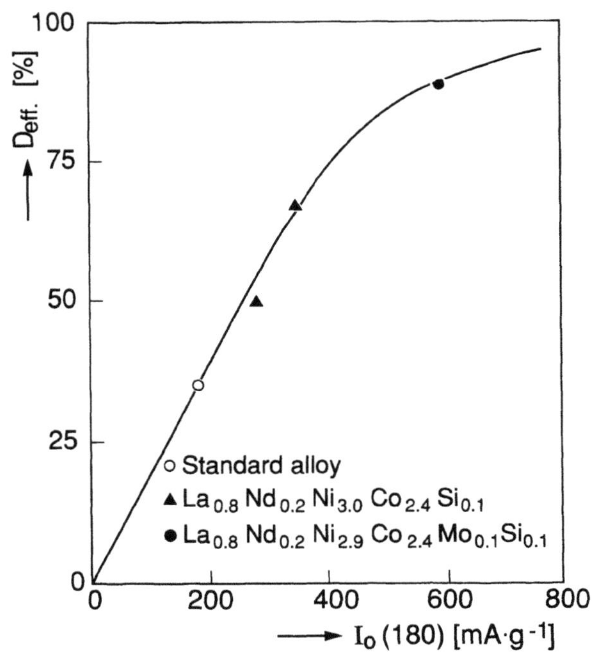

Fig. 27. The discharge efficiencies for different $AB_{5.0}$ and $AB_{5.5}$-type electrode materials at 0 °C using a high discharge current of 840 mA/g as a function of the measured exchange current, I_o. Triangles refer to two different solidification rates (from Ref. [11]).

area of the second phase precipitates. In line with the higher values for I_o are the higher discharge efficiencies measured under rather extreme conditions at 0 °C and using a very high discharge current. Compared to the low efficiency of 34 % reported for the standard alloy, a large improvement of up to 90 % is achieved with the most electroactive Mo-based alloy (filled circle in Fig. 27).

5. Concluding remarks

In this work we mainly described the research results on AB_5-type intermetallic compounds in which the lanthanides are in a pure form. For practical battery applications, however, it is essential to employ mischmetal-based lanthanides in order to reduce the cost of NiMH batteries. It has over the last years been shown by battery manufacturers that the main principles as outlined in this chapter are also valid for mischmetal-based alloys. In addition, it should be noted that the principles to improve the electrochemical long-term cycling stability and the charge/discharge kinetics are not only restricted to the class of AB_5-type hydride-forming compounds but can very likely be expanded to other classes of hydride-forming compounds as well.

Finally, it should be emphasized that the performance of NiMH batteries is not only determined by the MH electrode but that the properties of the Ni electrode has always to be taken into consideration. In order to obtain environmentally friendly NiMH batteries it was an essential step to remove the cadmium species from the Ni electrode. It is, however, well-known that additives, such as Cd-species, have a strong influence on both the mechanical and electrochemical properties of the Ni electrode and hence on the overall properties of NiMH batteries.

References

[1] H. Zijlstra, F.F. Westendorp, *Solid State Commun.*, **7** (1969) 857.

[2] J.H.N. van Vucht, F.A. Kuijpers, H.C.A.M. Bruning, *Philips Res. Rep.*, **25** (1970) 133.

[3] P.A. Boter, *United States Patent* 4,004,943 (1977).

[4] H.J.H. van Deutekom, *United States Patent* 4,214,043 (1980), *Ibid,* 4,312,928 (1982).

[5] H.Ogawa, M.Ikoma and I. Matsumoto, Proc. 16th Int. Power Sources Symp., Bournmouth, Edts. T. Keily and B.W. Baxter, *Power Sources,* **12** (1989) 393.

[6] M. Nogami, K. Moriwaki and N. Furukawa, *3rd Int. Rechargeable Battery Sem.*, Deerfield Beach, Florida (1990).

[7] U. Köhler and Ch. Klaus, in *"Dechema Monochraphien"*, **128** (1993) 213.

[8] M. Pourbaix, *"Atlas d'Équilibres Électrochemiques à 25 °C"*, Gauthiers-Villars, Paris (1963).
[9] R.M. LaFollette and D.N. Bennion, *J. Electrochem. Soc.*, **137** (1990) 3693.
[10] A.J. Bard, *"Encyclopedia of Electrochemistry of Elements"*, Vol. 1, Marcel Dekker, New York, p. 206 (19).
[11] P.H.L. Notten and P. Hokkeling, *J. Electrochem. Soc.*, **138** (1991) 1877.
[12] P.H.L. Notten and R.E.F. Einerhand, *Adv. Mat.*, **3** (1991) 343.
[13] C.H. Hamann and W. Vielstich, *"Elektrochemie II"*, Verlag Chemie, Weinheim (1981).
[14] Y.J. Kim, A. Visintin, S. Srinivasan and A.J. Appleby, *J. Electrochem. Soc.*, **139** (1992) 351.
[15] J. McBreen, *"Modern aspects of Electrochemistry"*, **21** (1990) 29.
[16] A.J. Bard, R. Parsons and J. Jordan (Edts), *"Standard redox potentials in aqueous solutions"*, Marcel Dekker, New York (1985).
[17] J. Halpert, *Proc. Electrochem. Soc.*, **90-4** (1990) 3.
[18] H. Bode, K. Dehmelt and J. Witte, *Electrochim. Acta*, **11** (1966) 1079.
[19] Y.J. Kim, *J. Cor. Sci. Korea*, **18** (1989) 132.
[20] F. Hahn, B. Beden, M.J. Croissant and C. Lamy, *Electrochim. Acta*, **31** (1986) 335.
[21] F. Hahn, D. Floner, B. Beden and C. Lamy, *Electrochim. Acta*, **32** (1987) 1631.
[22] R.D. Armstrong and Hong Wang, *Electrochim. Acta*, **36** (1991) 759.
[23] D. Guay, G. Tourillon, E. Dartyge, A. Fontaine, J. Mc. Breen, K.I. Pandya and W.E.O'Grady, *J. Electroanal. Chem.*, **305** (1991) 83.
[24] S.R. Morrison, *"Electrochemistry at semiconductor and oxidized metal electrodes"*, Plenum Press, New York (1980).
[25] S.I. Cordoba-Torresi, C. Gabrielli, A. Hugot-Le Goffard, R. Torresi, *J. Electrochem. Soc.*, **138** (1991) 1548.
[26] P. Bernard, C. Gabrielli, M. Keddam, H. Takenouti, J. Leonardi and P. Blanchard, *Electrochim. Acta*, **36** (1991) 743.
[27] D.M. Mc. Arthur, *J. Electrochem. Soc.*, **117** (1970) 729.
[28] G.W.D. Briggs and P.R. Snodin, *Electrochim. Acta*, **27** (1982) 565.
[29] W. Randszus and H.A. Kiehne (Edts), *"Gasdichte Nickel-Cadmium Akkumulatoren"*, Varta Batterie AG, Hannover (1978).
[30] A.H. Zimmerman and R. Seaver, *J. Electrochem. Soc.*, **137** (1990) 2662.
[31] B.B. Ezlov and O.G. Malandin, *J. Electrochem. Soc.*, **138** (1991) 885.
[32] L. Schlapbach (Edt), *"Topics in applied physics: Hydrogen in intermetallic compounds I and II"*, volume **63** and **67**, Springer Verlag, New York (1988).
[33] J.J.G. Willems, *Philips J. Res. Suppl.*, **39** (1984) 1.
[34] H. Yayama, K. Hirakawa and A. Tomokiyo, *Mem. Fac. Eng. Kyushu Univ.*, **45** (1985) 25.
[35] T. Sakai, H. Miyamura, N. Kuriyama, A. Kato, K. Oguro and H. Ishikawa, *J. Electrochem. Soc.*, **137** (1990) 795.

[36] P.H.L. Notten, J.L.C. Daams, A.E.M. De Veirman and A.A. Daams, *J. Alloys Comp.*, in press (1994).
[37] H.H. van Mal, *Philips Res. Rep. Suppl.*, **No. 1** (1976) 1.
[38] P.H.L. Notten, R.E.F. Einerhand and J.L.C. Daams, *J. Alloys Comp.*, in press (1994).
[39] A.H. Boonstra, T.N.M. Bernards and G.J.M. Lippits, *J. Less-Common Met.*, **159** (1990) 327.
[40] J.R.G.C.M. van Beek, J.J.G. Willems and H.C. Donkersloot, Proc. 14 Int. Power Sources Symp., Brighton, L.C. Pearce (Edt), *Power Sources*, **10** (1985) 317.
[41] T.B. Flanagan, W. Luo and J.D. Clewley, *Electrochem. Soc. Proc.*, **92-5** (1992).
[42] P.H.L. Notten, J.L.C. Daams and R.E.F. Einerhand, *J. Alloys Comp.*, in press (1994).
[43] S. Trassati, *J. Electroanal. Chem.*, **39** (1972) 163.
[44] K.J. Vetter, *"Electrochemical kinetics"*, Academic Press, Ltd., London (1967).
[45] T. Sakai, A. Takagi, T. Hazama, H. Miyamura, N. Kuriyama, H. Ishikawa and C. Iwakura, *"Batteries for Utility Energy Storage"*, Proc. 3 Int. Conf., Kobe (1991) 499.
[46] C. Iwakura and M. Matsuoka, *Progr. Batt. & Batt. Mater.*, **10** (1991) 81.
[47] P.H.L. Notten, J.L.C. Daams and R.E.F. Einerhand, *Ber. Bunsenges. Phys. Chem.*, **96** (1992) 656.
[48] P.H.L. Notten, R.E.F. Einerhand and J.L.C. Daams, *Z. Phys. Chem.*, **183** (1994) 267.
[49] W. Coene, P.H.L. Notten, F. Hakkens, R.E.F. Einerhand and J.L.C. Daams, *Phyl. Mag. A*, **65** (1992) 1485.
[50] K.H.J. Buschow and H.H. van Mal, *J. Less-Common Met.*, **29** (1972) 203.
[51] M.M. Jakšić, *Electrochim. Acta*, **29** (1984) 1539.
[52] M.M. Jakšić, *Int. J. Hydrogen Energy*, **11** (1986) 519.

Chapter 8

Introduction to lattice dynamics

H.R.Schober
Institut für Festkörperforschung
Forschungszentrum Jülich
D-52425 Jülich
Germany

1. Introduction

The atoms in a solid will at any temperature oscillate around their equilibrium configuration, at zero temperature due to their zero point motion. This motion of the atoms is the major heat bath in solids and is responsible for important properties such as heat capacity, thermal conductivity, lattice expansion, displacive phase transitions etc. The theory of these lattice vibrations is often referred to as lattice dynamics. In its present form it is now eighty years old, originating in papers by Born and von Karman in 1912. Lattice dynamics is closely related to the thermodynamics of solids and elasticity theory which have both been studied much before that date.

The present chapter is intended to give a short introduction into the field and to give a reminder of some important formulae. More detailed accounts can be found in many textbooks, see e.g [1; 2; 3; 4; 5; 6; 7]. Additionally vibrational properties of specific classes of substances are often discussed in books and reviews on these substances.

Lattice dynamics is infamous for the multitude of indices: cartesian coordinates, atom numbers, unit cell numbers, sublattice numbers etc. In different areas different notations are useful. An ideal crystal is best described in terms of cell and sublattice numbers whereas such a numbering scheme makes no sense for a glass. Nevertheless many basic formulae are valid for both. In order not to have to repeat the equations we have adopted, where appropriate, the more general scheme of indexing. In order to avoid confusion different letters are used for the different indices:$\alpha, \beta, \ldots$: cartesian index, ℓ: atom index, $\mathbf{m}, \mathbf{n}, \ldots$: unit cell number, $\mu, \nu, \ldots$: sublattice number and L: combined atomic and cartesian index. As far as possible we have indicated summations explicitly and have resorted to a matrix notation only where otherwise this would have complicated the formulae beyond all reason.

In the following section first the basic equations of lattice dynamics are introduced, starting from Newton's equation of motion of the atoms. These are used to derive the harmonic approximation and the notion of phonons and how these can

F. Grandjean et al. (eds.), Interstitial Intermetallic Alloys, 197–223.

be derived from phenomenological models for ideal lattices. The third section gives a survey of the properties of the ideal harmonic crystal. In harmonic approximation the contributions of the single phonons are often additive and can therefore be expressed by a spectrum function. The scattering of particles by the lattice, on the other hand, depends on the "structures" of the phonons and can be described by correlation and Green's functions. These are convenient tools in the study of effects of anharmonicty and defects such as interstitial atoms. We do not treat anharmonicity but restrict ourselves to a few remarks on the the quasiharmonic approximation, which only accounts for the shift of phonon energies but not for the finite lifetimes. Finally the concepts of localized and resonant modes are introduced. The dynamics of defects in the lattice can qualitatively differ from the one of the host. Light strongly coupled atoms will tend to vibrate with high frequencies, giving rise to localized vibration modes. Heavy and weakly coupled impurities will tend to low frequency vibrations, renonance modes. Both kinds of modes can also occur simultaneously, given an appropriate geometry.

2. Elements of Lattice Dynamics

2.1. EQUATIONS OF MOTION

Lattice dynamics is conventionally based on the adiabatic or Born - Oppenheimer approximation which eliminates the electrons from the equations of motion of the nuclei. The basic idea behind this approximation is that typically the velocities of the electrons are much higher than the ones of the nuclei, $\approx 10^6 m/sec$ and $\approx 10^3 m/sec$, respectively. One can therefore assume that, on the timescale of the motions of the nuclei, the electronic configuration adjusts itself instantaneously to the configuration of the nuclei. The electrons remain always in their respective groundstate configurations. The energy of the electronic ground state changes with the positions of the nuclei and thus acts, in addition to the direct interaction, as an effective potential between the nuclei.

After elimination of the electronic degrees of freedom the dynamics of the atoms in the solid is determined by a general many-body potential Φ depending on the instantaneous atomic positions. For the simple case of a pair interaction $U(R)$ between the atoms one has

$$\Phi = \sum_{\ell_1 \ell_2} U\left(\left|\mathbf{R}^{\ell_1} - \mathbf{R}^{\ell_2}\right|\right), \tag{1}$$

where $\mathbf{R}^\ell$ denotes the instantaneous position of atom ℓ. Obviously no exact solution of the resulting equations of motion is possible. In lattice dynamics one is interested in vibrations of the atoms around their equilibrium positions. The amplitudes of these vibrations will in general be small compared to the interatomic distances. We expand, therefore, the potential energy, Φ, in powers of these displacements :

$$\mathbf{R}^\ell = \mathbf{X}^\ell + \mathbf{u}^\ell \tag{2}$$

where $\mathbf{X}^l$ denotes the equilibrium position and $\mathbf{u}^l$ the displacement. The Taylor expansion of Φ can then formally be written as

$$\Phi = \Phi_0 + \frac{1}{2} \sum_{\substack{\ell_1 \ell_2 \\ \alpha_1 \alpha_2}} \Phi^{\ell_1 \ell_2}_{\alpha_1 \alpha_2} u^{\ell_1}_{\alpha_1} u^{\ell_2}_{\alpha_2} + \sum_{n=3}^{\infty} \frac{1}{n!} \sum_{\substack{\ell_1 \ldots \ell_n \\ \alpha_1 \ldots \alpha_n}} \Phi^{\ell_1 \ldots \ell_n}_{\alpha_1 \ldots \alpha_n} u^{\ell_1}_{\alpha_1} \ldots u^{\ell_n}_{\alpha_n} \tag{3}$$

Here the constant term Φ_0 is the energy of the static lattice which has no influence on the dynamics and will be omitted further on. The second term constitutes the harmonic approximation and the remaining terms the anharmonic corrections. There is no first order term in Eq.(3) since we expanded the potential energy around an equilibrium configuration. The equations of motion for the atoms follow as

$$M^{\ell_1} \ddot{u}^{\ell_1}_{\alpha_1} = - \sum_{\ell_2 \alpha_2} \Phi^{\ell_1 \ell_2}_{\alpha_1 \alpha_2} u^{\ell_2}_{\alpha_2} - \sum_{n=3}^{\infty} \frac{1}{(n-1)!} \sum_{\substack{\ell_2 \ldots \ell_n \\ \alpha_2 \ldots \alpha_n}} \Phi^{\ell_1 \ldots \ell_n}_{\alpha_1 \ldots \alpha_n} u^{\ell_2}_{\alpha_2} \ldots u^{\ell_n}_{\alpha_n} \tag{4}$$

where M^l denotes the atomic mass of atom ℓ. The coefficients $\Phi^{\ell_1 \ell_2}_{\alpha_1 \alpha_2}, \ldots, \Phi^{\ell_1 \ldots \ell_n}_{\alpha_1 \ldots \alpha_n}$ are called second, third,.. order force constants or coupling constants. They are defined as derivatives of the many-body interatomic potential with respect to atomic displacements from the equilibrium position

$$\Phi^{\ell_1 \ldots \ell_n}_{\alpha_1 \ldots \alpha_n} = \frac{\partial^n \Phi}{\partial u^{\ell_1}_{\alpha_1} \ldots \partial u^{\ell_n}_{\alpha_n}} \tag{5}$$

In lowest order (harmonic approximation) the second order (harmonic) force constant $\Phi^{\ell_1 \ell_2}_{\alpha_1 \alpha_2}$ is the force on atom ℓ_1 in direction α_1 due to a unit displacement of atom ℓ_2 in direction α_2 with all other atoms in their equilibrium position.

To avoid the large number of indices one often uses a matrix notation. In this notation Eq.(4) reads

$$\mathbf{M}\ddot{\mathbf{u}} = -\boldsymbol{\Phi}\mathbf{u} - \frac{1}{2}\boldsymbol{\Phi}^{(3)}\mathbf{u}\mathbf{u} + \ldots \tag{4a}$$

The force constants have to obey a number of symmetry requirements. First it follows immediately from their definition that the force constants are invariant against permutations of the index pairs (ℓ_j, α_j). Second the invariance of the potential energy Φ and its derivatives against translations ($\mathbf{u}^\ell \to \mathbf{u}^\ell + \mathbf{t}$) of the solid as a whole requires

$$\sum_{\ell_1 \alpha_1} \Phi^{\ell_1 \ell_2 \ldots}_{\alpha_1 \alpha_2 \ldots} = 0 \tag{6}$$

and additional conditions are imposed by rotational invariance.

Third in addition to these relations due to rigid displacements of the solid as a whole a given form of the interaction potential can introduce additional relations between the elements of the force constants.

The conditions discussed so far are valid for any collection of atoms. For crystals additional restrictions are imposed by the structural symmetries. In an infinite ideal crystal the equilibrium positions of the atoms can be given as

$$\mathbf{X}^{\mathbf{m}}_{\mu} = m_1 \mathbf{a}_1 + m_2 \mathbf{a}_2 + m_3 \mathbf{a}_3 + \boldsymbol{\kappa}^{\mu} \tag{7}$$

where $\mathbf{m}$ denotes the triple of integers m_i, the $\mathbf{a}_i$ are primitive translation vectors of the lattice which define the Bravais lattice and $\boldsymbol{\kappa}^\mu$ is the position of atom μ with respect to the cell origin. In primitive lattices the cell origin is normally placed at the site of the atom ($\kappa = 0$) and the index μ is dropped. The primitive vectors in Eq.(7) do not necessarily make the lattice symmetry obvious. It is therefore often more convenient to use instead of the translation vectors $\mathbf{a}$ a more symmetry adapted set, e.g. in an fcc lattice the cell vectors are $a/2(1,1,0)$, $a/2(1,0,1)$ and $a/2(0,1,1)$ but atomic sites are normally given by $X_\alpha^{\mathbf{m}} = (a/2)m_\alpha$ where a is the lattice constant and the m_α are integers with $\sum_\alpha m_\alpha =$ even. Eq.(7) can now be used to replace the arbitrary atom indices ℓ_i in the coupling parameters by the more systematic enumeration $\genfrac{}{}{0pt}{}{\mathbf{m}}{\mu}$. In these new indices the coupling parameters exhibit the invariance of the lattice against symmetry operations from its space group. These symmetry operations not only leave the potential energy and its derivatives invariant but also the undistorted lattice. Under these symmetry operations, therefore, not only the numerical value of the potential energy but also the form of the expression for it, Eq.(3), is invariant. Most important is the invariance of the lattice against translation by a basis vector $\mathbf{a}_i$. This invariance implies that the coupling parameters of the ideal lattice do not depend on the absolute positions of the primitive cells but only on their distance vector. For the harmonic force constants one has

$$\Phi_{\alpha\beta}^{\genfrac{}{}{0pt}{}{\mathbf{m}\,\mathbf{n}}{\mu\,\nu}} = \Phi_{\alpha\ \beta}^{\genfrac{}{}{0pt}{}{\mathbf{0}\ \mathbf{n}-\mathbf{m}}{\mu\quad \nu}} = \Phi_{\alpha\ \beta}^{\genfrac{}{}{0pt}{}{\mathbf{n}-\mathbf{m}}{\mu\quad \nu}} = \Phi_{\alpha\beta}^{\genfrac{}{}{0pt}{}{\mathbf{h}}{\mu\,\nu}} \tag{8}$$

where we have used the standard notation of one uppermost vector index $\mathbf{h} = \mathbf{n} - \mathbf{m}$. The number of independent nonzero elements of the force constants is further reduced by point group symmetry operations and more complicated symmetry elements such as screw axes and glide planes. Under such a symmetry operation a lattice point $\genfrac{}{}{0pt}{}{\mathbf{m}}{\mu}$ is transformed into an equivalent one $\genfrac{}{}{0pt}{}{\mathbf{m}'}{\mu'}$. As an example let us consider the nearest neighbour coupling in a simple cubic lattice. The six equivalent positions are given by $\pm a(1,0,0)$, $\pm a(0,1,0)$ and $\pm a(0,0,1)$ and the coupling matrix between nearest neighbours has only two independent elements

$$\boldsymbol{\Phi}^{(1,0,0)} = -\begin{pmatrix} \alpha & 0 & 0 \\ 0 & \beta & 0 \\ 0 & 0 & \beta \end{pmatrix}. \tag{9}$$

The offdiagonal elements vanish due to symmetry against the inversions $y \to -y$ and $z \to -z$ while the yy- and zz- elements are identical because of the fourfold symmetry around the x-axis. The coupling to the neighbors at a distance $a(3,2,1)$ has in general the maximum number of 6 independent elements since this distance vector cannot be transformed into itself by a symmetry element. The couplings to equivalent neighbours are related by symmetry operations.

The symmetry conditions of the coupling parameters are important technically because they reduce the calculational effort. In any model starting from a reasonable form of the potential they are automatically fulfilled. If, however, the coupling constants are used as fit-parameters, as is commonly done, the symmetries have to be built in to safeguard against unphysical results.

2.2. HARMONIC APPROXIMATION

Taking only the harmonic terms the equation of motion Eq.(4) can be solved exactly for any assembly of atoms. For simplicity we combine the indices $\begin{pmatrix} l \\ \alpha \end{pmatrix}$ of Eq.(2) and $\begin{pmatrix} \mathbf{m} \\ \mu \\ \alpha \end{pmatrix}$ of Eq.(8) to a single index L. The equation of motion can then be written in symmetrized form as

$$\sqrt{M_L}\ddot{u}_L = -\sum_{L'} D_{LL'}\sqrt{M_{L'}}u_{L'} \tag{10}$$

where for N atoms the index L takes $3N$ values, $M_L = M_l$ and

$$D_{LL'} = \Phi_{LL'}/\sqrt{M_L M_{L'}}. \tag{11}$$

The $3N \times 3N$ matrix $\mathbf{D}$ is called *dynamical matrix*. It is real and symmetric. It should be noted that for spatially periodic systems the term dynamic matrix is also used for the spatial Fourier transform of $D_{LL'}$, see below.

For the displacements $u_L(t)$ we make an ansatz

$$\sqrt{M_L}\; u_L(t) = a\, e_L\, e^{i\omega t} \tag{12}$$

where a stands for an arbitrary amplitude. Eq.(10) is thus reduced to an eigenvalue problem for the matrix $\mathbf{D}$:

$$\omega^2 e_L = \sum_{L}^{\prime} D_{LL'}\, e_{L'} \tag{13}$$

There are $3N$ solutions with eigenvalues, ω_j^2, and eigenvectors $\mathbf{e}^j$ to this problem. Since the dynamical matrix is real and symmetric the eigenvalues are real. The harmonic system is stable if all eigenvalues are positive. An eigenmode with frequency ω_j and eigenvector $\mathbf{e}^j$ is called the jth normal mode of the system. The eigenvectors can be chosen to be real and orthogonal to each other

$$\sum_L e_L^j\, e_L^{j'} = \delta_{jj'} \tag{14}$$

and satisfy the closure condition

$$\sum_j e_L^j\, e_{L'}^j = \delta_{LL'} \tag{15}$$

The $3N$ amplitudes, a_j, of the normal modes, Eq. (12), obey independent harmonic oscillator equations

$$\ddot{a}_j = -\omega_j^2 a_j. \tag{16}$$

The mode j "moves" in the harmonic oscillator potential $\frac{1}{2}\omega_j^2 a_j^2$. The transformation to eigenvectors (modes) transforms thus the real space potential Φ to effective potentials for the modes. These two notions of potentials should not be confused.

The actual atomic amplitudes in the normal modes are gained from Eq.(12) where the weighting with $\sqrt{M}$ expresses the fact that, assuming similar force constants, the amplitudes of heavy atoms are smaller than the ones of light atoms.

It is sometimes convenient to use complex eigenvectors instead of real ones, e.g., in a translationally invariant lattice the normal modes are plane waves which are mostly characterized by eigenvectors of the form $\exp\{i\mathbf{kR}\}$. The reality of the atomic displacements then requires the conditions

$$(\mathbf{e}^j)^* = (\mathbf{e}^{-j}) \ , \ a_j^* = a_{-j} \ , \ \omega_j = \omega_{-j} \tag{17}$$

The energy of the harmonic system is

$$\begin{aligned} E &= \sum_L p_L^2/2M_L + \frac{1}{2}\sum_{LL'} \Phi_{LL'}\, u_L\, u_{L'} = \\ &= \frac{1}{2}\sum_L M_L\, \dot{u}_L\, \dot{u}_L + \frac{1}{2}\sum_{LL'} \Phi_{LL'}\, u_L\, u_{L'} = \\ &= \frac{1}{2}\sum_j \left[\dot{a}_j^*\, \dot{a}_j + \omega_j^2\, a_j^*\, a_j\right] \end{aligned} \tag{18}$$

The derivation given above also holds in quantum mecahanics. The possible values of the energy then become quantized

$$E_{n_1,n_2\ldots n_{3N}} = \sum_{j=1}^{3N} \hbar\,\omega_j\,(n_j + \frac{1}{2}) \tag{19}$$

with n_j the occupation number of vibration (phonon) j. The term $\frac{1}{2}$ stands for the zero point energy. Eq.(19) reflects again the independence of the normal modes, the phonons, in the harmonic approximation.

The diagonalisation of the dynamical matrix can be done at present for systems of a few thousand atoms numerically. For larger systems without atomic order it is not feasible. For an infinite periodic system, such as an ideal crystal, translational symmetry can be used to blockdiagonalize the dynamical matrix by a Fourier transformation with respect to the cell vectors $\mathbf{X}^{\mathbf{m}}$, introduced in Eq.(7). Due to periodicity the eigenvectors have the form

$$e_L = e_{\alpha}^{\overset{\mathbf{m}}{\mu}} = \frac{1}{\sqrt{N_c}}\, e_{\alpha}^{\mu}(\mathbf{q}, j)\, e^{i\mathbf{q}\boldsymbol{\kappa}^{\mu}}\, e^{i\mathbf{q}\mathbf{X}^{m}} \tag{20}$$

where N_c is the number of cells and the vectors $\mathbf{q}$ are restricted to one unit cell of the reciprocal lattice. It is usual to use a symmetric unit cell, the first Brillouin zone of the reciprocal lattice. $\mathbf{q}$ is then the wavevector of the vibration. $\lambda = 2\pi/q$ is the wavelength of the vibrational wave in the lattice which travels in direction $\mathbf{q}/q$. $\mathbf{e}(\mathbf{q}, j)$ is the polarization vector; it determines the direction of the atomic vibrations. For lattices with more than one atom in the unit cell $\mathbf{e}(\mathbf{q}, j)$ determines together with the phase factor $\exp(i\mathbf{q}\boldsymbol{\kappa}^{\mu})$ the relative displacements of the atoms in the unit cell. The polarization index j takes $3s$ values with s the number of atoms in the unit cell, see below. The polarization vectors depend in general on $\mathbf{q}$.

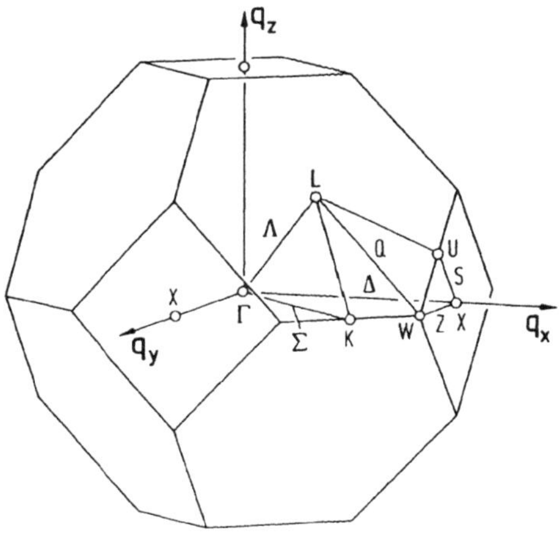

Fig. 1: First Brillouin zone of the fcc lattice. The symmetry points are in units of $2\pi/a$; Γ : (0,0,0); X : (0,1,0); $L : (\frac{1}{2}, \frac{1}{2}, \frac{1}{2})$, $W : (\frac{1}{2},1,0)$, $K : (\frac{3}{4}, \frac{3}{4}, 0)$, $U : (\frac{1}{4}, 1, \frac{1}{4})$.

The choice of the first Brillouin zone as the unit cell of the reciprocal lattice has the advantage that it is invariant under the spacegroup of the crystal. Fig. 1 shows the Brillouin zone of an fcc crystal with the standard notation of symmetry points and directions. At least for the simpler structures the polarization vectors in the main symmetry directions are determined purely by symmetry, independent of q. Similarly also degeneracies in the normal modes can be determined. These symmetry arguments are very important for the determination of the phonons from experiments. The unit vectors, $\mathbf{b}_i$, of the reciprocal lattice are derived from the ones of the direct lattice by

$$\mathbf{a}_i \mathbf{b}_j = 2\pi \, \delta_{ij} . \tag{21}$$

As the vectors $\mathbf{a}$ span the real lattice so the vectors $\mathbf{b}$ span the reciprocal lattice by

$$\boldsymbol{\tau}^{\mathbf{m}} = m_1 \mathbf{b}_1 + m_2 \mathbf{b}_2 + m_3 \mathbf{b}_3 . \tag{22}$$

The eigenvectors Eq.(20) are not changed if one adds a reciprocal lattice vector $\boldsymbol{\tau}$ to $\mathbf{q}$.

In the previous derivations we have always assumed a finite number of atoms. For an infinite periodic lattice the allowed values of $\mathbf{q}$ become dense, $\mathbf{q}$ is a continuous variable and summations over $\mathbf{q}$ have to be replaced by integrations

$$\sum_{\mathbf{q}} \to \frac{V_c}{(2\pi)^3} \int d^3\mathbf{q} \tag{23}$$

with V_c the cell volume.

Inserting the eigenfunction Eq.(20) into the equation of motion Eq.(10) the problem of calculating the phonons is reduced to an eigenvalue problem of dimension $3s$ for each $\mathbf{q}$, where s is the number of atoms in the unit cell

$$\omega_j^2(\mathbf{q}) \, e_\alpha^\mu(\mathbf{q}, j) = \sum_{\nu,\beta} D_{\alpha\beta}^{\mu\nu}(\mathbf{q}) \, e_\beta^\nu(\mathbf{q}, j) \tag{24}$$

where j denotes the $3s$ possible polarizations and $\mathbf{D}$ is a new dynamical matrix

$$D^{\mu\nu}_{\alpha\beta}(\mathbf{q}) = \frac{1}{\sqrt{M^\mu M^\nu}} \sum_{\mathbf{h}} \Phi^{\mathbf{h}}{}^{\mu\nu}_{\alpha\beta}\, e^{i\mathbf{q}\mathbf{X}^{\mathbf{h}}} \tag{25}$$

The dynamical matrix, $\mathbf{D}$, is hermitian and has due to Eq.(17) the properties

$$D^{\mu\nu}_{\alpha\beta}(\mathbf{q}) = (D^{\nu\mu}_{\beta\alpha}(\mathbf{q}))^* = [D^{\mu\nu}_{\alpha\beta}(-\mathbf{q})] \tag{26}$$

and hence

$$e^{\mu}_{\alpha}(\mathbf{q}, j) = (e^{\mu}_{\alpha}(-\mathbf{q}, j))^* \qquad \text{and} \qquad \omega^2_j(\mathbf{q}) = \omega^2_j(-\mathbf{q}) \tag{27}$$

If two $\mathbf{q}$-values in the Brillouin zone can be mapped onto each other by a symmetry operation of the lattice the phonon frequencies for both $\mathbf{q}$-values are identical with the polarization vectors transformed correspondingly. Thus for a cubic lattice all phonons can be calculated from the $\mathbf{q}$–values of $1/48^{th}$ of the Brillouin zone.

In the long wavelength limit, $\mathbf{q} \to 0$, three of the $3s$ phonon branches become sound waves, known from continuum theory. In lattice dynamics these phonon branches are called acoustic phonons. In these acoustic phonons the atoms in the unit cell vibrate in phase with each other. The polarization vectors of these branches depend in the limit $\mathbf{q} \to 0$ only on $\mathbf{q}/q$ and the frequencies are

$$\lim_{q\to 0} \omega_j(\mathbf{q}) = c_j(\mathbf{q}/q) \cdot q \tag{28}$$

where c_j is the (phase) sound velocity which depends in a lattice on direction and polarization. In the limit $\mathbf{q} = 0$ the acoustic phonons become rigid translations of the lattice as a whole. The frequencies of the other $3s - 3$ phonon branches go to a finite limit for $\mathbf{q} \to 0$ and the atoms of the unit cell vibrate against each other. If the atoms carry a charge as in ionic crystals such a vibration causes a macroscopic oscillation of the electric dipole moment which can interact with electromagnetic radiation. These modes are therefore called optical modes.

In Figs. 2 – 4 the phonon dispersion curves, i.e. the dependence of the frequency on the value of q, in the main symmetry directions are shown for four typical metals. Such curves have normally a characteristic shape depending on the lattice structure, e.g. fcc, and on the type of material, e.g. metallic, ionic or molecular crystal. The absolute magnitude of the frequencies depends on the specific material, e.g. it scales with $\sqrt{M}$ according to Eq.(12). The dispersion curves of two different materials with the same electronic structure will be particularly similar and one can define a homology relation between the frequencies ω_1 and ω_2 of the two materials $\omega_1/\omega_2 = \sqrt{M_2 a_2^2}/\sqrt{M_1 a_1^2}$ where a stands for the lattice constant. Such a homology relation is obeyed e.g. between Na and K with an average deviation of only 3%. We will now discuss some of the basic properties of the phonon dispersion curves using the three examples shown.

The phonon dispersion of Cu, Fig. 2, is representative for an fcc metal with short range forces. The fit to the experimental values shown is by a sixth neighbour interaction model with 12 parameters but a nearest neighbour model would also give a reasonable fit. Since Cu has a Bravais lattice there are three phonon branches

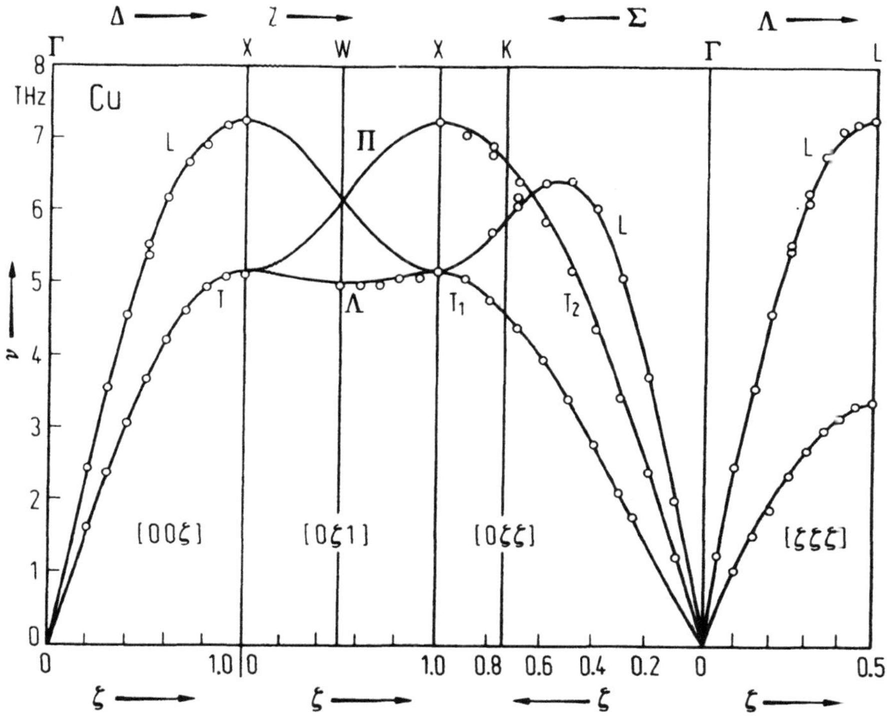

Fig. 2: Phonon dispersion curves of Cu at T = 49K plotted against the reduced wavevector $\zeta = a\dot{\mathbf{q}}/2\pi$. The continuous lines represent a sixth neighbor axially symmetric Born–v.Karman fit to the experimental points, after [8].

(polarizations) for each $\mathbf{q}$–value. A number of properties follows directly from the symmetries of the dynamical matrix which are a consequence of the symmetries of the Brillouin zone. In the main symmetry directions (Δ, Σ, Λ Fig.1) the polarization vectors are completely determined by symmetry and one can distinguish between longitudinal (L), $\mathbf{e}\|\mathbf{q}$, and transversal (T) modes, $\mathbf{e} \perp \mathbf{q}$. The latter ones are degenerate in Δ and Λ directions. This distinction of longitudinal and transversal modes is not possible for general directions but, nevertheless, often holds approximately. At the X and L points the dispersion curves have horizontal shapes since these points are centres of inversion symmetry between the first Brillouin zone and the ones adjoining ($\omega(\mathbf{q}) = \omega(\mathbf{q} + \boldsymbol{\tau})$). The K–point lies on the boundary of the first Brillouin zone but is no such centre. Continuing in Σ direction one reaches the X–point of a neighbouring zone which in turn explains the degeneracy of the $[0, 1, 1]L$ and $[0, 1, 1]T_1$ phonons which are identical to the $[0, 0, 1]T$ phonon. This seeming contradiction between longitudinal and transversal character stems from the continuation of the Σ branch into a second Brillouin zone where one has $\mathbf{e}\|(\mathbf{q} + \boldsymbol{\tau})$ which in general does not imply $\mathbf{e}\|\mathbf{q}$.

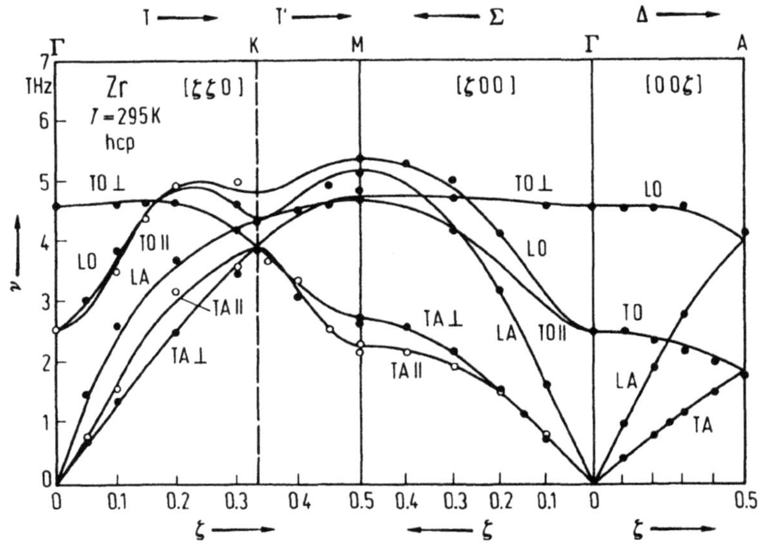

Fig. 3: Phonon dispersion curves of hcp Zr at $T = 295$K plotted against the reduced wavevector ζ, $\mathbf{q} = (4\pi/a\sqrt{3},\ 4\pi/a\sqrt{3},\ 2\pi/c)$. The continuous lines represent a sixth neighbor axially Born–v.Karman fit to the experimental points, after [9].

As an example for the phonon dispersion of a metal with hcp structure we show in Fig. 3 the phonon dispersion of Zr. There are two atoms in the Bravais cell and, therefore, three optic and three acoustic branches. The acoustic branches have a structure similar to the one of Cu due to the fcc unit cell. Compared to the sine–like shape of the acoustic branches the optic ones are much flatter. In a number of places optic and acoustic branches cross each other. Depending on the symmetries the branches can intersect or hybridize with each other, e.g. at the K-point. Symmetry implies again various degeneracies.

An example for a more complicated structure is shown in Fig. 4 where the dispersion curves of the Laves phase of $CeAl_2$ are depicted. There are six atoms in the unit cell and correspondingly eighteen Phonon branches. Due to the large mass difference between Ce and Al there is a gap between the optic modes. This complicated dispersion can be described quite well by a ten parameter model.

Phonon dispersion curves have been measured for most elemental crystals and for many compounds. Compilations can be found for elemental metals [11; 12] for alloys [13] and for insulators [14]. The harmonic theory discussed so far does not allow for a variation of the frequencies with temperature and therefore does not provide for the lattice expansion. Measured phonon dispersion curves at some given temperature have to be understood in the framework of the quasiharmonic approximation discussed further down, where one does the expansion of the energy around the equilibrium configuration at this temperature. This way the phonon

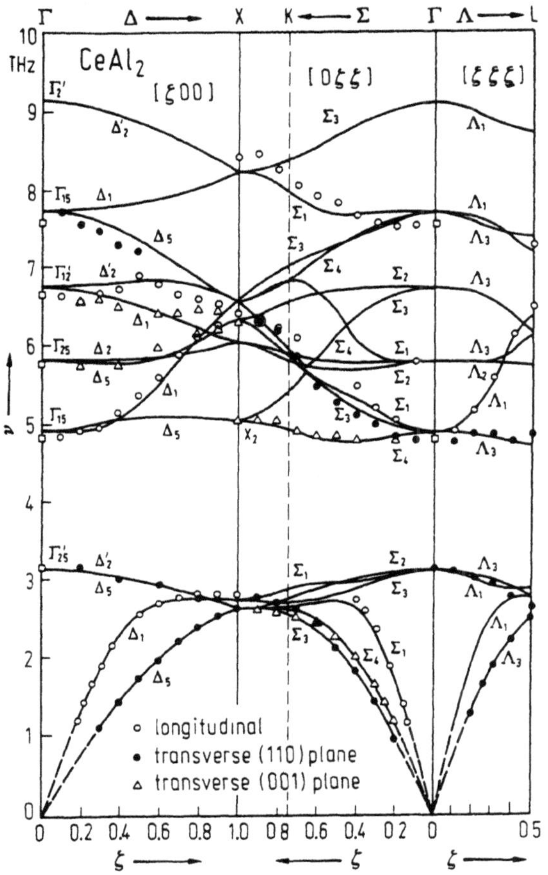

Fig. 4: Phonon dispersion curves of Laves phase $CeAl_2$ T = 296K plotted against the reduced wavevector. The continuous lines represent a ten parameter axially symmetric Born–v.Karman fit to the experimental points, after [10].

frequencies become temperature dependent and so strictly speaking do the empirical models fitted to them. These effects become important near phase transitions.

2.3. PHENOMENOLOGICAL MODELS OF LATTICE DYNAMICS

In applications one is normally not interested in a specific phonon but wants to know the influence of all phonons or of a group of phonons on some property of the lattice, e.g. specific heat, heat conductivity, thermal atomic displacements, electron scattering etc. One needs, therefore, a fast means to calculate the dynamical matrix. First principle calculations are at present much too time consuming and one has to resort to phenomenological models. The parameters of these models are fitted to experimentally determined phonon frequencies, elastic constants and electric polarizabilities. The models serve to extrapolate from these properties to the dispersion curves in the entire Brillouin zone. This extrapolation will be the more reliable the more information on the nature of the material is built into it. Consequently there is an ever increasing multitude of models ranging from simple

Born–von Karman models to nearly first principle calculations. The compilation of 1981 [11] gives already some sixty models for the phonon dispersion of Cu and the number keeps increasing. In the following, two main classes of models will be briefly discussed, Born–von Karman model variants and nearly free electron models.

General Born-von Karman (B.v.K.) model (Tensor Force Constant Model)

In principle the Born-Oppenheimer approximation allows the dynamic matrix of any system to be expressed by force constants which only have to obey the symmetry requirements discussed before. In practice this can be done only for very short range interactions in lattices of high symmetry or when the analytic form of the interaction is known. To describe a general nearest neighbour interaction in an fcc lattice 3 independent force constants are needed. An extension to twice the range (4th neighbours) needs already 12 parameters. If one has two nonequivalent atoms in the unit cell (e.g. hcp) these numbers are about doubled. On the other hand the shape of the dispersion curves often necessitates fits by long range interactions, giving high Fourier coefficients. In order to reduce the number of parameters additional restrictions are introduced, e.g. axial symmetry, which can often be justified from first principle calculations. Fits with models of different range show that in general only the first few force constants are reasonably stable against the cutoff. Force constants to the more distant neighbours fluctuate in sign and magnitude and are purely fit parameters without any physical meaning. The parameters are mostly fitted to phonons in symmetry directions and this can cause inaccuracies for the off symmetry phonons. B.v.K. fits only to the phonons alone reproduce the elastic constants often very badly, therefore these are often included in the fit as constraints. In q–space localized anomalies, e.g. Kohn anomalies, of course cannot be reproduced by short range real space models. B.v.K. models cannot be used to extrapolate from one structure to another without additional assumptions on the underlying interaction. A determination of the real physical couplings involves not only an exact knowledge of the phonon frequencies but of the polarizations of off–symmetry phonons. Particularly when there is more than one atom in the unit cell one can obtain equally good fits to the phonon frequencies with different models resulting in qualitatively different polarization vectors, i.e. different amplitude patterns, see e.g. [15] for hcp Zn.

Axially Symmetric Born-von Karman Model

In the axially symmetric model, used to obtain the fit in Fig. 2, one assumes that between a pair of atoms there is only one radial (longitudinal) and one transversal force constant, f_r and f_t respectively. Such a model can be thought of as derived from a central pair potential $U(R)$ and the two force constants are related to the derivatives of U:

$$f_r = \frac{\partial^2 U(R)}{\partial R^2} \qquad \text{and} \qquad f_t = \frac{1}{R}\,\frac{\partial U(R)}{\partial R} \tag{29}$$

The coupling matrix then takes the form

$$\Phi_{\alpha\beta}^{mn} = -(f_r - f_t)\,\frac{X_\alpha^{mn}\,X_\beta^{mn}}{(X^{mn})^2} - f_t\,\delta_{\alpha\beta} \qquad (30)$$
$$= -K\,\frac{X_\alpha^{mn}\,X_\beta^{mn}}{(X^{mn})^2} - C\,\delta_{\alpha\beta}$$

In case the interaction is only via a pair potential, the transversal force constants between different neighbours are restricted by the equilibrium conditions of the lattice. $f_t(R_\alpha^{\mathbf{m}} - R_\alpha^{\mathbf{n}})$ is the force exerted by atom $\mathbf{m}$ on atom $\mathbf{n}$. In equilibrium the sum of all the forces has to vanish for each atom. For infinite lattices with inversion symmetry this condition is automatically satisfied. Additional relations between the transversal force constants result from the condition that, in equilibrium, the lattice energy has to be minimal with respect to homogeneous deformations, i.e. changes of the unit vectors $\mathbf{a}_i$ (Eq.(7)) and hence $\frac{\partial E}{\partial a_{i\alpha}} = 0$. For the transversal force constants this means

$$\sum_{\mathbf{h},\mu,\nu} f_t^{\mathbf{h}}\,(X_\alpha^{\mathbf{h}} + \kappa_\alpha^\mu - \kappa_\alpha^\nu)\,(X_\beta^{\mathbf{h}} + \kappa_\beta^\mu - \kappa_\beta^\nu) = 0 \qquad (31)$$

This condition implies the Cauchy relation for the elastic constants. It is violated if many body forces, or volume forces are present.

For hexagonal crystals a **modified axially symmetric model** is often employed where one uses in Eq.(30) different parameters for directions in the basal plane and perpendicular to it.

Nearly Free Electron Models

For simple metals the electronic structure can be calculated by perturbation theory starting from plane waves for the conduction electrons and eliminating the tightly bound core electrons by a pseudopotential. These theories provide a well founded framework which is treated in many textbooks and articles, e.g. [16]. First principle pseudopotentials are nonlocal electron–ion potentials. In actual calculations one assumes mostly a specific form, local or nonlocal, with fitted parameters. For a local pseudopotential the dynamical matrix takes a particularly simple form which makes these models popular for the description of phonons. In a Bravais lattice the dynamical matrix can be written as

$$D_{\alpha\beta}(\mathbf{q}) = \frac{V_c}{M}\sum_{\boldsymbol{\tau}}(\boldsymbol{\tau}+\mathbf{q})_\alpha\,\left[F^c(|\boldsymbol{\tau}+\mathbf{q}|) + F^{bs}(|\boldsymbol{\tau}+\mathbf{q}|)\right]\,(\boldsymbol{\tau}+\mathbf{q})_\beta$$
$$-\frac{V_c}{M}\sum_{\boldsymbol{\tau}\neq 0}\tau_\alpha\,\left[F^c(\tau) + F^{bs}(\tau)\right]\,\tau_\beta \qquad (32)$$

Here the Coulomb term F^c (Ewald-term) stems from the direct Coulomb interaction of the ions with a compensating uniform background of electrons

$$F^c(q) = \frac{Z^2e^2}{\epsilon_o V_c^2}\,\frac{1}{q^2} \qquad (33)$$

whereas the bandstructure term represents the indirect interaction of the ions via the response of the conduction electrons to the ionic movement in the adiabatic approximation

$$F^{bs}(q) = \frac{\epsilon_o q^2}{e^2} \mid w_o(q) \mid^2 \frac{\epsilon_o - \epsilon(q)}{\epsilon(q)} \tag{34}$$

with $w_o(q)$ the Fourier transform of the local pseudopotential and $\epsilon(q)$ the electric permittivity of an homogeneous electron gas of the average density of the conduction electrons. Through the permittivity the dynamical matrix depends explicitly on the cell volume of the crystal. In this description the constant term in the energy expansion Eq.(3) depends explicitly on the volume. Due to these volume dependencies these models do not have the correct limit for all long wavelength phonons. This deficiency can be corrected by including higher order terms in the pseudopotential. The model can also be formulated in real space. One then gets a long range oscillating pair potential. The equilibrium condition again involves explicit volume dependencies and one has therefore a long range axially symmetric model without the restriction of the Cauchy relations. With these models the phonon dispersion of simple metals can be fitted with very few (in some cases only two) parameters. This, however, does not guarantee their physical significance. It has been found that a good fit in second order as in Eq.(32) can be destroyed by higher order terms. The same parameters need not be appropriate for different structures. To simulate the repulsion of the ionic cores the expression in Eq.(32) is often augmented by nearest neighbour B.v.K. parameters. Such models can then also be applied to fit the phonons in transition metals. There is a large number of models using simplified expressions for the form factor (Eq.(34)) plus some short range B.v.K. models, e.g. [17]. These models can be quite successful as far as the quality of the fit is concerned. One sacrifices, however, physical rigour for the sake of computational simplicity. Even further simplified approaches omit all terms with $\boldsymbol{\tau} \neq 0$ and thus violate the periodicity condition of the phonons.

3. Properties of the Ideal Harmonic Crystal

3.1. FREQUENCY SPECTRUM AND RELATED PROPERTIES

In the harmonic approximation the normal modes, i.e. the phonons, are independent. The total wavefunction is the product of the single mode wavefunctions and the total energy and the thermodynamic functions are sums of the single mode contributions. These sums can be expressed as averages over the frequency spectrum, $g(\omega)$, defined such that $g(\omega)d\omega$ is the fraction of eigenfrequencies in the interval $[\omega, \omega + d\omega]$:

$$g(\omega) = \frac{1}{3N} \sum_{\mathbf{q},j} \delta(\omega - \omega_j(\mathbf{q})) = \frac{V_c}{(2\pi)^3} \frac{1}{3s} \sum_{j=1}^{3s} \int d^3\mathbf{q}\delta(\omega - \omega_j(\mathbf{q})) \tag{35}$$

and

$$\int_0^\infty d\omega\, g(\omega) = 1. \tag{36}$$

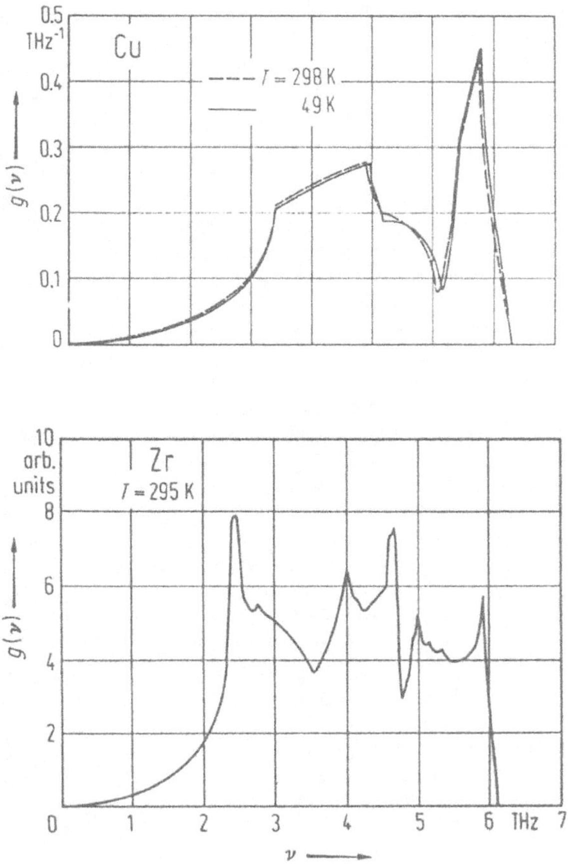

Fig. 5: Frequency spectra of Cu (top) and Zr (bottom). The spectra were calculated from Born-v. Karman fits [11].

Fig. 5 shows the frequency spectra of Cu and Zr, corresponding to the dispersion curves shown in figs. 2 and 3, respectively. At low frequencies the spectrum increases as ω^2 which is a direct consequence of the linear dispersion of the long wavelength acoustic modes Eq.(28). The spectrum of Cu shows, typical for fcc–metals, two maxima: one related to the short wavelength transverse phonons the other one to the longitudinal phonons. It should be remembered that the number of phonons for a given magnitude of q increases as q^2. Reflecting the more complicated structure of the lattice the phonon spectrum of Zr is much more structured. A numerical calculation of the frequency spectrum requires an integration over the whole Brillouin zone or over its irreducible part, if symmetry is used.

The thermodynamic functions can be derived from the partition function

$$Z = tr\{e^{-\beta H}\} \tag{37}$$

where tr stands for trace, H is the Hamiltonian and $\beta = 1/k_bT$ with k_b the Boltzmann constant and T the absolute temperature. In the harmonic approximation the trace can be evaluated for each mode separately resulting in products of sums over occupation numbers and finally

$$Z = \prod_{q,j} \frac{\exp[-\frac{1}{2}\beta\hbar\omega_j(\mathbf{q})]}{1-\exp[-\beta\hbar\omega_j(\mathbf{q})]}. \tag{38}$$

The Helmholtz free energy per atom (N = number of atoms)

$$\begin{aligned} F &= -\frac{1}{N}k_b T \ln Z = \frac{1}{N}k_b T \sum_{\mathbf{q}j} \ln[2\sinh(\hbar\omega_j(\mathbf{q})/2k_b T)] \\ &= 3k_b T \int d\omega \ln[2\sinh(\hbar\omega_j(\mathbf{q})/2k_b T)]\, g(\omega) \end{aligned} \tag{39}$$

From this the internal energy E, the specific heat at constant volume C_v and the vibrational entropy S^H of the crystal, all per atom, can be calculated by standard thermodynamic relations

$$\begin{aligned} E &= F - T\left(\frac{\partial F}{\partial T}\right)_V \\ &= 3\frac{\hbar}{2}\int d\omega \coth(\hbar\omega/2kT)\,\omega\, g(\omega) \end{aligned} \tag{40}$$

$$\begin{aligned} C_v &= \left(\frac{\partial E}{\partial T}\right)_V \\ &= 3k_b \int d\omega \left[\left(\frac{\hbar\omega}{2k_b T}\right)^2 / \sinh^2(\hbar\omega/2k_b T)\right] g(\omega) \end{aligned} \tag{41}$$

$$\begin{aligned} S^H &= -\left(\frac{\partial F}{\partial T}\right)_V = \\ &= 3k_b \int d\omega \left\{\frac{\hbar\omega}{2k_b T}\coth(\hbar\omega/2k_b T) - \right. \\ &\qquad \left. - \ln[2\sinh(\hbar\omega/2k_b T)]\right\}\, g(\omega) \end{aligned} \tag{42}$$

These expressions are strictly limited to the harmonic approximation. If the phonon frequencies depend on the temperature as in the quasiharmonic approximation (see next section) the spectrum $g(\omega)$ becomes temperature dependent $g(\omega, T)$ and Eqs.(41–43) have to be augmented by additional terms involving $\partial g(\omega, T)/\partial T$ accounting for the temperature dependence of the spectrum. Since there is no lattice expansion in the harmonic approximation C_p and C_v are equal.

Fig. 6 shows as a typical example the vibrational part of the specific heat of Cu. Up to about 200 K the harmonic theory works very well, at higher temperatures anharmonic corrections become important. The high and low temperature limits of the thermodynamic functions are given in table I. The specific heat and entropy obey a T^3–law at low temperatures. In metals an additional term $C_e \sim T$ is due to the specific heat of the electronic system. For high temperatures the harmonic specific heat becomes constant $\approx 3k_b$ per atom. The increase in the actually observed

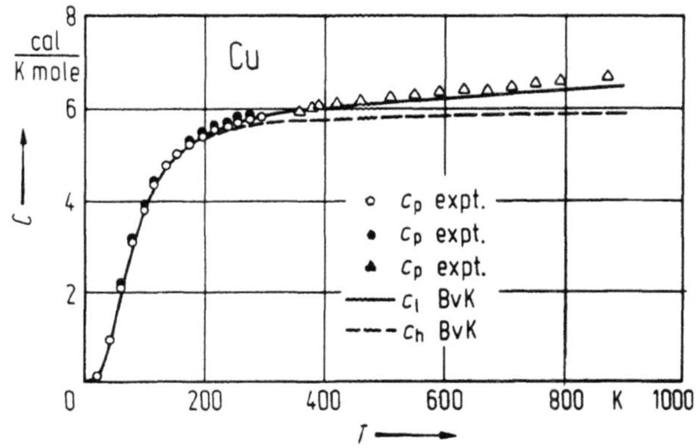

Fig. 6: Lattice specific heat in Cu (dashed line = harmonic and full line = quasi-harmonic approximation; symbols, caloric experiments by various authors), after [18].

TABLE I
Low and high temperature limits of the thermodynamic functions/atom in the harmonic approximation. The constant Θ is the (zero temperature caloric) Debye temperature.

	$T \to 0$	$T \to \infty$
F	$3\left\{\int \frac{1}{2}\hbar\omega\, g(\omega)d\omega - \frac{\pi^4}{15}\left(\frac{T}{\Theta}\right)^3 k_bT\right\}$	$3\,k_bT \int \ln(\beta\hbar\omega)g(\omega)d\omega$
E	$3\left\{\int \frac{1}{2}\hbar\omega\, g(\omega)d\omega + \frac{\pi^4}{5}\left(\frac{T}{\Theta}\right)^3 k_bT\right\}$	$3\,k_bT$
C_v	$3\frac{4\pi^4}{5}\, k_b\left(\frac{T}{\Theta}\right)^3$	$3\,k_b$
S^H	$3\frac{4\pi^4}{15}\, k_b\left(\frac{T}{\Theta}\right)^3$	$3\,k_b - 3\,k_b \int \ln(\beta\hbar\omega)\, g(\omega)d\omega$

specific heat is due to the softening of the phonons with temperature and can be explained within the quasiharmonic approximation.

The low temperature behaviour can be approximated very well by a simplified form of the spectrum, the Debye spectrum. At low temperatures the thermodynamics of the crystal will be dominated by the low frequency phonons. The spectrum for low frequencies is always proportional to ω^2 stemming from the linear dispersion for small $\mathbf{q}$. In the Debye approximation one assumes a purely quadratic behaviour of $g(\omega)$ up to a cutoff given by the normalization condition Eq.(36).

$$g_D(\omega) = \begin{cases} \frac{3}{\omega_D^3}\,\omega^2 & \omega \le \omega_D \\ 0 & \omega > \omega_D \end{cases} \tag{43}$$

The Debye frequency ω_D is often expressed in terms of a Debye temperature by

$$k_b\,\Theta_D = \hbar\,\omega_D \tag{44}$$

In essence the Debye approximation is a low temperature approximation and the only parameter in $g_D(\omega)$, the Debye frequency should be determined from the low frequency part of the true spectrum $g(\omega)$ or equivalently from the sound velocity. It is, however, popular to use the Debye spectrum as a simple reference system to describe some physical property of the lattice, most often the specific heat. The parameter ω_D is fixed by the condition that the Debye spectrum should give the same value of the specific heat as the true spectrum. Since the specific heat is temperature dependent one obtains thus a temperature dependent Debye frequency $\omega_D(T)$ and Debye temperature $\Theta_D(T)$. At low temperatures $\omega_D(T)$ will be a true representation of the acutal $g(\omega)$ whereas with increasing temperature it becomes a somewhat complicated weight of the spectrum which quite strongly depends on the fitted quantity, e.g. C_v in case of the usual caloric Debye temperature.

The expression for the harmonic specific heat can be rewritten for the Debye spectrum as

$$C_v(T) = 9Nk_b \left(\frac{T}{\Theta_D(T)}\right)^3 \int_0^{\Theta_D(T)/T} dx \, \frac{x^4 \, e^x}{(e^x - 1)^2} \tag{45}$$

Equating this result with the specific heat, calculated from the true spectrum or measured experimentally, fixes the caloric Debye temperature $\Theta_D(T)$. Fig. 7 shows the comparison of the Debye temperature of Cu determined experimentally to the ones calculated from three different Born–von Karmann models. There is a variation of about 10 % of the Debye temperature with temperature. Spectra containing parts at very high frequency would give stronger variations.

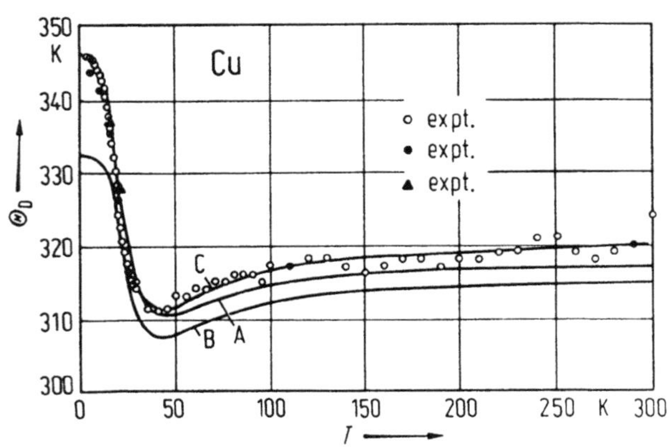

Fig. 7: Comparison of the caloric Debye temperature $\Theta_D(T)$ of Cu calculated from the phonon spectra with caloric experiments. Curves A, B and C are calculated from spectra at 49K, 298K and 80K, respectively. (after [19]).

Another approximation is gained if one calculates the vibrations of the single atoms with the rest of the lattice fixed to their equilibrium positions. For each atom one obtains thus three frequencies, called Einstein frequencies. If one takes the average of all these atomic Einstein frequencies one gets the Einstein frequency of the lattice, ω_E. In a cubic Bravais lattice the atomic Einsteinfrequencies are all identical and equal to ω_E. In general they are the eigenvalues of the "self interaction" part of the real space dynamical matrix, $D_{\alpha\beta}^{\ell_1\ell_2}$, for $\ell_1 = \ell_2$. The average Einsteinfrequency ω_E is given by the trace of the total dynamical matrix and can due to the invariance of the trace under unitary transformation be expressed by the spectrum:

$$\omega_E^2 = \frac{1}{3N}\, tr\,\mathbf{D} = \int \omega^2\, g(\omega)\, d\omega \tag{46}$$

In the Einstein approximation the continuum of phonon frequencies is replaced by the Einstein frequency

$$g_E(\omega) = \delta(\omega - \omega_E) \tag{47}$$

This very simple model can be used to estimate the lattice properties for temperatures $T \geq \Theta_E = \frac{\hbar\omega_E}{k_b}$ but it breaks down for $T \to 0$.

Often one is not interested in the average spectrum of the crystal but in the partial spectrum, say of the vibrations of atom ℓ in direction α. This can be defined in terms of the eigenvectors Eq.(13) of the dynamical matrix as

$$g_\alpha^\ell(\omega) = \sum_j (e^j)_\alpha^\ell\, (e^j)_\alpha^\ell\, \delta(\omega - \omega_j) \tag{48}$$

also called local spectrum of atom ℓ in direction α. Each mode is weighted by the projection of its amplitude on atom ℓ in direction α. In the case where all $g_\alpha^\ell(\omega)$ are equal this definition is identical to the one of the average spectrum Eq.(35). These local spectra can be used to calculate the temperature dependent mean square displacement of an atom

$$< (u_\alpha^\ell)^2 > = \int \frac{\hbar}{2M^\ell}\, \frac{1}{\omega}\, \coth\, \frac{1}{2}\, \beta\hbar\omega\; g_\alpha^\ell(\omega)\, d\omega \tag{49}$$

This is the simplest form of the equal time displacement correlation functions discussed in the next subsection. Expanding Eq.(45) one obtains for low temperatures

$$< (u_\alpha^\ell)^2 > = 2\,\frac{\hbar}{M^\ell} \int \frac{1}{\omega}\, g_\alpha^\ell(\omega)\, d\omega + O(T^2) \tag{50}$$

and for high temperatures

$$< (u_\alpha^\ell)^2 > = \frac{k_b T}{M^\ell} \int \frac{1}{\omega^2}\, g_\alpha^\ell(\omega)\, d\omega + O(\frac{1}{T}) \tag{51}$$

At low temperatures the mean square displacement has a constant value determined by the zero point motion. For high temperatures it increases linearly with temperature, with its magnitude given by the ω^{-2} moment of the local spectrum. We will see that this is the static displacement Green's function. In Fig. 8 the mean square displacement for Cu in the harmonic approximation is depicted. Also shown is the deviation at higher temperatures due to the softening of the phonons.

The atomic mean square displacement determines the Debye–Waller factor observed in scattering experiments.

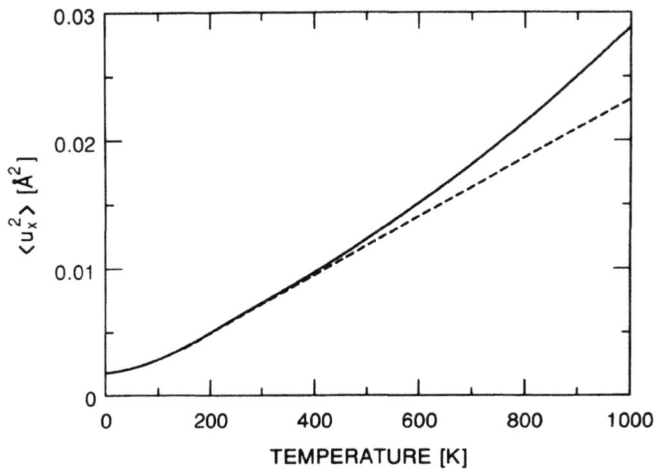

Fig. 8: Atomic mean square displacement in Cu (dashed line: harmonic values, solid line: quasiharmonic values) [7].

3.2. GREEN'S FUNCTIONS

Whereas it is sufficient to know the frequency spectrum to calculate the thermodynamic functions, more information is needed to calculate the response of the lattice to some perturbation, e.g. the scattering of neutrons or X-rays. This lattice response can be expressed most easily in terms of the two time Green's function or equivalently in terms of two time correlation functions. Detailed derivations are given in references [20] and [21] which we largely follow here. Under the influence of an outside perturbation which we write as a force $\mathbf{F}^{\ell}$ acting on the atoms the equation of motion (Eq.(4)) in the harmonic approximation is given by

$$M^{\ell_1} \ddot{u}^{\ell_1}_{\alpha_1} + \sum_{\ell_2 \alpha_2} \Phi^{\ell_1 \ell_2}_{\alpha_1 \alpha_2} \, u^{\ell_2}_{\alpha} = F^{\ell_1}_{\alpha_1}(t) \tag{52}$$

or in matrix notation

$$\mathbf{M}\, \ddot{\mathbf{u}} + \boldsymbol{\Phi}\, \mathbf{u} = \mathbf{F}(t). \tag{50a}$$

A partial solution of this equation of motion can be written in terms of the Green's function $G^{\ell_1 \ell_2}_{\alpha_1 \alpha_2}(t)$

$$u^{\ell_1}_{\alpha_1}(t) = \sum_{\ell_2 \alpha_2} \int_{-\infty}^{\infty} dt' \, G^{\ell_1 \ell_2}_{\alpha_1 \alpha_2}(t - t') \; F^{\ell_2}_{\alpha_2}(t') \tag{53}$$

where the retarded Green's function is the response to a δ–force

$$M^{\ell_1} \, \ddot{G}^{\ell_1 \ell_2}_{\alpha_1 \alpha_2}(t) + \sum_{\ell_3 \alpha_3} \Phi^{\ell_1 \ell_3}_{\alpha_1 \alpha_3} \; G^{\ell_3 \ell_2}_{\alpha_3 \alpha_2}(t) = \delta_{\alpha_1 \alpha_2} \, \delta_{\ell_1 \ell_2} \delta(t) \tag{54}$$

or

$$\mathbf{M}\; \ddot{\mathbf{G}}(t) + \boldsymbol{\Phi}\; \mathbf{G}(t) = \mathbf{1}\; \delta(t) \tag{45a}$$

with the retardation condition $\mathbf{G}(t) = 0$ for $t < 0$. This condition ensures that the displacement at a time t does not depend on forces later than t. Taking the Fourier transform with respect to time (Eq.(12)) $\mathbf{G}(\omega)$ can be written formally as

$$\mathbf{G}(\omega) = \frac{1}{\boldsymbol{\Phi} - \mathbf{M}(\omega + i\eta)^2} \qquad \eta \to 0 \tag{55}$$

The infinitesimal quantity η guarantees the retardation. We have chosen the sign of $\mathbf{G}$ so that the static Green's function $\mathbf{G}(\omega = 0)$ is positive. In quantum theory often the opposite sign is used in the definition of $\mathbf{G}$.

The Green's function $\mathbf{G}(\omega)$ is the inverse of $\boldsymbol{\Phi} - \mathbf{M}\omega^2$. The inversion can be done in real space for finite systems only. It can always be expressed in terms of the eigenvalues and eigenvectors of the dynamical matrix Eq.(13).

$$G^{\ell_1\ell_2}_{\alpha_1\alpha_2}(\omega) = \frac{1}{\sqrt{M^{\ell_1}M^{\ell_2}}} \sum_j \frac{(e^j)^{\ell_1}_{\alpha_1}\,(e^j)^{\ell_2}_{\alpha_2}}{\omega_j^2 - (\omega + i\eta)^2} \tag{56}$$

Using the standard relation

$$\frac{1}{x - i\eta} = \frac{x^2}{x^2 + \eta^2} + i\,\frac{\eta}{x^2 + \eta^2} = P(\frac{1}{x}) + i\pi\,\delta(x)$$

the Green's function can be split into their real and imaginary parts which are connected by the Kramers–Kronig relation

$$\mathrm{Re}\{\mathbf{G}(\omega)\} = \frac{1}{\pi} \int_0^\infty d\omega'\, \frac{1}{\omega'^2 - \omega^2}\, \mathrm{Im}\{\mathbf{G}(\omega')\} \tag{57}$$

which reduces the task of calculating $\mathbf{G}$ to calculating the imaginary part

$$\mathrm{Im}\{G^{\ell_1\ell_2}_{\alpha_1\alpha_2}(\omega)\} = \mathrm{sign}(\omega)\, \frac{1}{\sqrt{M^{\ell_1}M^{\ell_2}}}\, \pi \sum_j (e^j)^{\ell_1}_{\alpha_1}\, (e^j)^{\ell_2}_{\alpha_2}\, \delta(\omega^2 - \omega_j^2) \tag{58}$$

The diagonal elements are, apart from factors, the local spectra Eq.(48). The physical meaning of the real and imaginary parts of $\mathbf{G}$ can be understood by evaluating Eq.(52) for a point force. The real part describes the displacements in phase with the force and the imaginary part describes displacements lagging behind in phase by $\pi/2$. These latter ones are connected with the energy dissipation into the lattice. The larger the spectral density and hence $\mathrm{Im}\mathbf{G}(\omega)$ for a given frequency, the more energy can dissipate into the lattice.

4. Beyond the infinite harmonic crystal

4.1. ANHARMONICITY

The harmonic approximation discussed in the previous section gives a satisfactory description of the vibrational properties of most solids at low temperatures when the vibrational amplitudes are small. The neglect of the anharmonic terms in Eq.(3) has, however, a number of consequences, notably there is no thermal expansion whence the phonon frequencies, force constants and elastic constants are also independent of temperature. Additionally the specific heat becomes constant at high temperatures and the constant volume and constant pressure specific heats are equal, $C_p^H = C_V^H$. As eigenstates of the harmonic Hamiltonian the phonons have, in harmonic approximation, infinite lifetimes, zero linewidths, infinite free paths and hence the thermal conductivity of a perfect harmonic crystal is infinite. To correct the above shortcomings, anharmonicity has to be taken account of. This can be done on different levels. Effects of the temperature variation on the phonon frequencies can be treated adequately relatively easily by the *quasiharmonic approximation* where the lifetime of the phonons is still infinite. In this approximation the lattice constants are taken as temperature dependent and the expansion in Eq.(3) is then done from the temperature dependent equilibrium positions

$$\mathbf{X}_{\mu}^{\mathbf{m}} = m_1\mathbf{a}_1(T) + m_2\mathbf{a}_2(T) + m_3\mathbf{a}_3(T) + \boldsymbol{\kappa}^{\mu}(T) \tag{59}$$

This results in temperature dependent quasiharmonic force constants, whence the derived properties become temperature dependent. Born–v.Karman parameters derived from experiment have to be understood in this approximation.

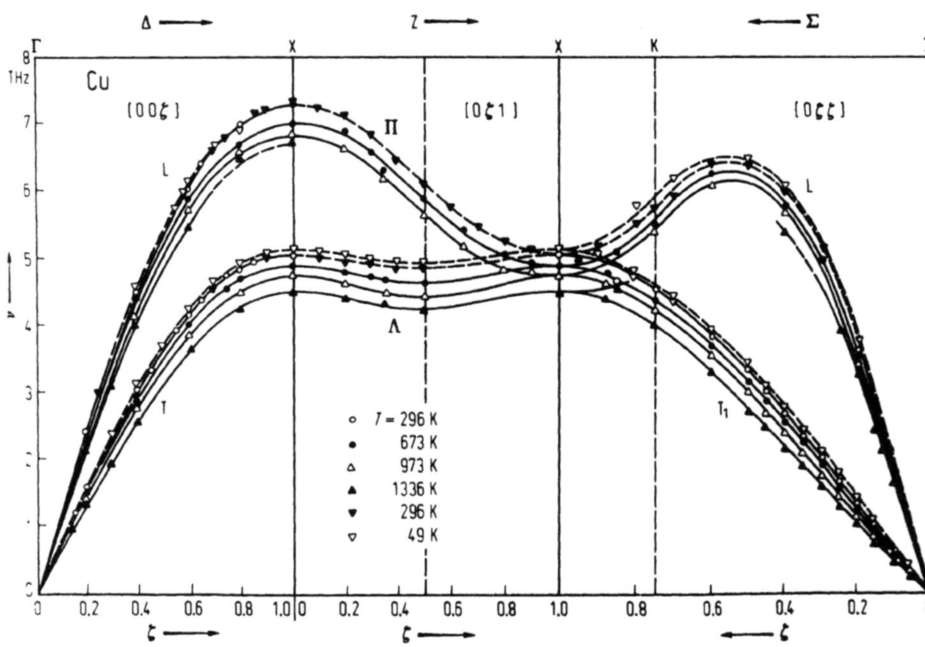

Fig. 9: Temperature dependence of the phonon dispersion in Cu. The lines are guides to the eye. (after [22])

As an example Fig. 9 shows the dispersion curves of Cu measured at temperatures ranging from 49K to 1336K, i.e. 20K below the melting point. The longitudinal frequencies soften over the whole temperature range by about 10% and the transverse ones by about 15%. This smooth variation is also reflected in the shifts of the corresponding frequency spectra, Fig. 10.

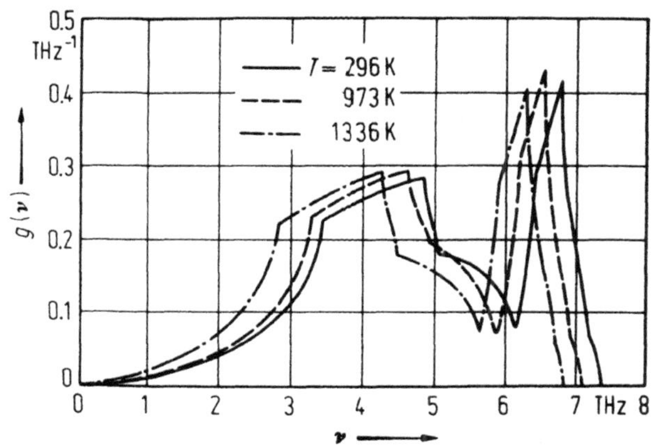

Fig. 10 Frequency spectra of Cu at different temperatures [11].

Even in simple structures the behaviour is not always so simple. For example one finds in the high temperature phase of Zr (bcc β-Zr) two groups of phonons. The majority of the phonons soften with increasing temperature as expected for a normal metal. The phonons in certain regions of the Brillouin zone show, however, the opposite behaviour: they harden considerably.The frequency of some phonons increases by about 50% in a temperature range of 700K. This inverse temperature behaviour is related to the hcp-bcc martensitic phase transition of Zr at 1138K and is an example of the connection between phase transitions and phonon anharmonicity [23]

To obtain the finite lifetimes of the phonons one has to include properly the interaction between the phonons (anharmonic terms) or with other excitations such as electrons. The lowest order anharmonic terms involve three phonons and describe processes such as the decay of the phonon $(\mathbf{q}_1, j_1)$ into two phonons $(\mathbf{q}_2, j_2)$ and $(\mathbf{q}_3, j_3)$. Energy conservation requires

$$\hbar\omega(\mathbf{q}_1, j_1) - \hbar\omega(\mathbf{q}_2, j_2) - \hbar\omega(\mathbf{q}_3, j_3) = 0 \tag{60}$$

and quasi-momentum conservation

$$\hbar\mathbf{q}_1 - \hbar\mathbf{q}_2 - \hbar\mathbf{q}_3 = \hbar\boldsymbol{\tau} \tag{61}$$

where $\boldsymbol{\tau}$ is a reciprocal lattice vector. If $\boldsymbol{\tau} = 0$ one speaks of a normal (N-) process while processes with $\boldsymbol{\tau} \neq 0$ are referred to as “Umklapp” (U-) processes. The evaluation of anharmonic effects is rather complicated and goes far beyond the scope of this lecture.

4.2. IMPERFECT CRYSTALS

The perfect periodic lattice treated in the previous sections is an idealisation of the real solid. In real crystals periodicity is destroyed by a number of imperfections: point defects (substitutional impurities, vacancies , interstitial atoms and small defect clusters) or extended defects such as dislocations, stacking faults, grain boundaries and surfaces. The local dynamics of these impurities can differ considerably from that of the host material. Often one finds local vibrational modes at the impurity: resonant and localized modes of point defects or surface waves. Even a small concentration of defects can strongly affect the dynamics of a crystal. In the following we will illustrate these using the most simple defect, the substitutional isotope effect, and a self interstitial in an fcc lattice. More details will be given in the lecture on hydrogen dynamics.

The vibrational amplitudes of any atom in the lattice can in harmonic approximation be described by the local spectrum Eq.(48). These local spectra can deviate strongly from the average vibrations. As an example let us take a substitutional impurity in an otherwise perfect lattice.

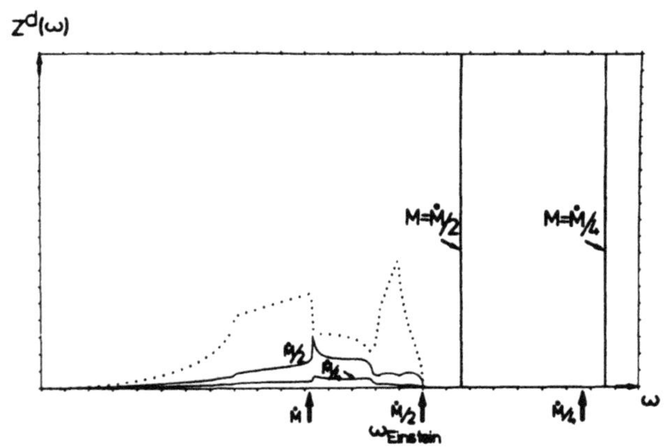

Fig. 11: Local frequency spectrum of light isotopes in an fcc lattice with atomic mass M (solid lines). The localiced modes are indicated by vertical lines, labeled with the defect mass. The dotted line shows the ideal spectrum and the arrows indicate the Einstein frequencies for the different masses [21].

We have seen earlier that the frequencies of an ideal lattice scale with the mass as $\omega \propto 1/\sqrt{M}$. The vibrations of the impurity will follow this tendency. For heavy impurities the local spectrum will be softer and for light ones harder. For sufficiently small masses the impurity will vibrate not only with frequencies found in the ideal host lattice but also with frequencies above the maximum host frequency. Such a vibration is called localized. Since there are no host phonons of similar frequency a localized vibration is localized in space at the impurity and has, in harmonic approximation, a sharp frequency. Fig. 11. shows the local spectrum of a light impurity in an fcc lattice.

Similar effects are found for very heavy impurities when the number of host phonons at the "natural" frequency of the impurity is small. One finds then a strong peak in the local spectrum which becomes narrower with increasing mass see Fig. 12. The width of the peak is a measure of the dissipation of the vibrational energy if the impurity is excited with this resonant frequency. Such a resonant vibration is spatially localized, alas much more weakly than a proper localized vibration. Resonant vibrations are often referred to as quasi-localized.

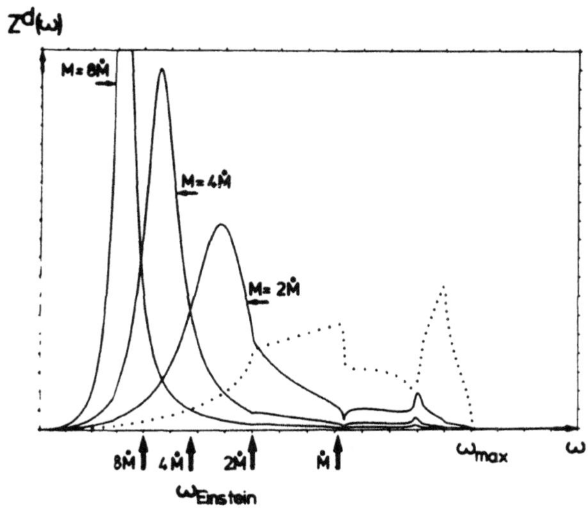

Fig. 12: Local frequency spectrum of heavy isotopes in an fcc lattice with atomic mass M (solid lines). The dotted line shows the ideal spectrum and the arrows indicate the Einstein frequencies for the different masses [21].

Interstitial atoms show a much richer behaviour which depends strongly on their mass,their coupling to the host lattice and the local geometry. A very strong coupling to the host lattice will give rise to localized vibrations wheras a very weak coupling leads to resonant vibrations. For interstitial atoms both cases can occur simultaneously. Let us consider the example of the self-interstitial atom in an fcc metal, Fig.13.

For our purposes it can be envisaged as a substitutional Cu_2 molecule where the two atoms are strongly coupled causing a localized bond stretching vibration (A_{1g}). At the same time the restoring forces against libration (E_g) or translation (A_{2u}) of the molecule as a whole are weak causing resonant vibrations. This example shows the importance of the tensor character of the coupling. The simultaneous occurrence of localized and resonant modes is typical for "large" defects which compress the lattice locally and thus cause a tendency towards instability in some directions This example shows that not only weakly coupled defects can have resonant vibrations but that a defect which would be normally considered as strongly coupled to the host lattice can have low frequency resonant modes due to the local atomic arrangement. The defect configuration can even become unstable and is then often split into configurations of lower symmetry. The configuration of the self interstitial atom can be envisaged as the result of an instability of the octahedral configuration which would have higher symmetry.

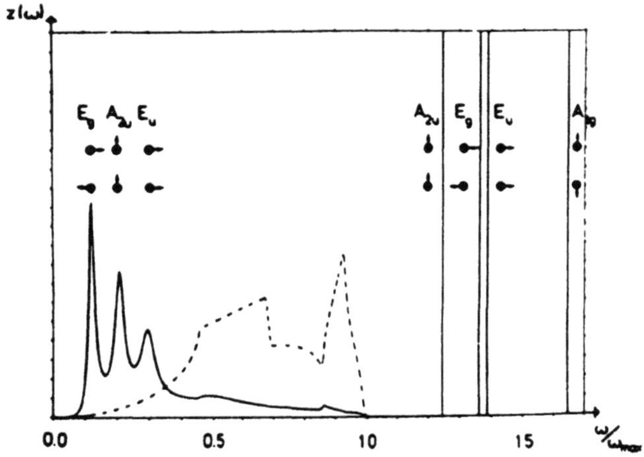

Fig. 13
Local frequency spectrum of a self-interstitial in the fcc Cu lattice (solid line). The localiced modes are indicated by vertical lines. The modes are labelled according to their symmetry, which is shown schematically. The broken line indicates the ideal spectrum [21].

References

[1] Böttger, H. (1983), Principles of the Theory of Lattice Dynamics, Berlin: Akademie-Verlag.
[2] Born, M., Huang, K. (1954), Dynamical Theory of Crystal Lattices, Oxford: Clarendon Press.
[3] Brüesch, P. (1982, 1986, 1987) Phonons: Theory and Experiments I, II, III, Berlin: Springer Verlag.
[4] Horton, G.K., Maradudin, A.A. (eds.) (1974-91), Dynamical Properties of Solids, Vols. 1-6, Amsterdam: North Holland.
[5] Maradudin, A.A., Montroll, E.W., Weiss, G.H., Ipatova, I.P. (1971), Theory of Lattice Dynamics in the Harmonic Approximation, Solid State Physics, Supplement 3, New York: Academic Press.
[6] Srivastava, G.P. (1990),The Physics of Phonons, Bristol (England): Adam Hilger.
[7] Schober, H.R. and Petry, W. (1993), in: Materials Science and Technology, Vol. 1, (V. Gerold ed.), Weinheim: Chemie Verlag.
[8] Nicklow, R.M., Gilat, G., Smith, H.G., Raubenheimer, L.J., Wilkinson, M.K. (1967), Phys. Rev. **164**, 922.
[9] Stassis, C., Zarestky, J., Arch, D., McMasters, O.D., and Harmon, B.N. (19787), Phys. Rev. **B18**, 2632.
[10] Reichardt, W. and Nücker, N. unpublished.
[11] Schober, H.R., Dederichs, P.H. (1981), in: Landolt-Börnstein, Vol. III/13a, (K.-H. Hellwege and J.L. Olsen eds.), Berlin:
[12] Kress, W. (1987), Phonon Dispersion Curves up to 1985, Physics Data, Karlsruhe: Fachinformationszentrum Karlsruhe. Springer Verlag.
[13] Kress, W. (1983), in: Landolt-Börnstein, Vol. III/13b, (K.-H. Hellwege and J.L. Olsen eds.), Berlin: Springer Verlag.
[14] Bilz, H., Kress, W. (1979), Phonon Dispersion Relations in Insulators, Berlin: Springer Verlag.
[15] Chesser, N.J., Axe, J.D. (1974), Phys. Rev. **B9**, 4060.
[16] Cohen, M.L., Heine, V. (1970), Solid State Physics (F. Seitz, D. Turnbull, H. Ehrenreich eds.) **24**, 38.
[17] Krebs, K. (1964), Phys. Lett. **10**, 12.
[18] Miiller, A.P., Brockhouse, B.N. (1971), Can. J. Phys. **49**, 704.
[19] Nilsson, G., Rolandson, S. (1973), Phys. Rev. **B7**, 2393.
[20] Leibfried, G., Breuer, N. (1978), Point Defects in Metals, Springer Tracts in Modern Physics

81, Berlin: Springer.
[21] Dederichs, P.H., Zeller, R. (1980), Point Defects in Metals II, Springer Tracts in Modern Physics **87**, Berlin: Springer.
[22] Larose, A., Brockhouse, B.N. (1976), Can. J. Phys. **54**, 1990.
[23] A. Heiming, W. Petry, J. Trampenau, M. Alba, C. Herzig, H.R. Schober and G. Vogl (1991), Phys. Rev. B **43**, 10948.

Chapter 9

AN INTRODUCTION TO NEUTRON SCATTERING

W. B. YELON
University of Missouri
Research Reactor
Columbia, MO 65211

A. Introduction

It has been stated that the neutron is the most versatile probe of condensed matter systems. While some may dispute this, it is clear that the unique properties of the neutron make it an extremely important, and in many instances, unique probe of interstitial alloy systems. Neutron scattering can not only examine the structure of the alloys, including unambiguous determination of the location and occupancy of the interstitial species, but can measure the diffusion of the interstitial species and determine the shape and depth of the potential in which the interstitial atom is found. Furthermore, neutron scattering can follow, in detail, the changes in magnetic ordering and coupling which accompany the insertion of interstitial species into magnetic phases. Since the neutron is a neutral, weakly interacting probe, it allows investigation of the bulk versus near surface properties and the easy study of alloys in low and high temperature environments (mK-2000°C) and in high pressure environments of several kbar. While some of the information determined from neutron measurements may be extracted from other methods, including x-ray scattering, Mössbauer spectroscopy, NMR and other techniques, none provide the breadth and microscopic detail of neutron scattering. In order to appreciate the vast range of opportunity afforded by neutron scattering, it is useful to review the fundamentals of neutron scattering and then to consider, in a second chapter, specific examples of neutron studies of interstitial alloy systems. Comprehensive texts on neutron scattering have been written by Bacon [1], Windsor [2] and others. In addition, a three volume set in "Experimental Methods in Physics" [3] is dedicated to the subject, and includes numerous reviews of methods and results, covering many of the subjects discussed here. An earlier book also covers many of these topics [4].

B. Neutron Fundamentals: Cross Sections

With the exception of magnetic scattering, the interaction of the neutron with matter is determined by neutron-nuclear cross sections. Although there are some systematic trends in the magnitude of the scattering cross sections versus atomic number, this dependence is far weaker than for x-rays, for which the scattering is proportional to the number of electrons. It is also characterized by wide variations about the systematic trend line In addition, the

F. Grandjean et al. (eds.), Interstitial Intermetallic Alloys, 225–248.

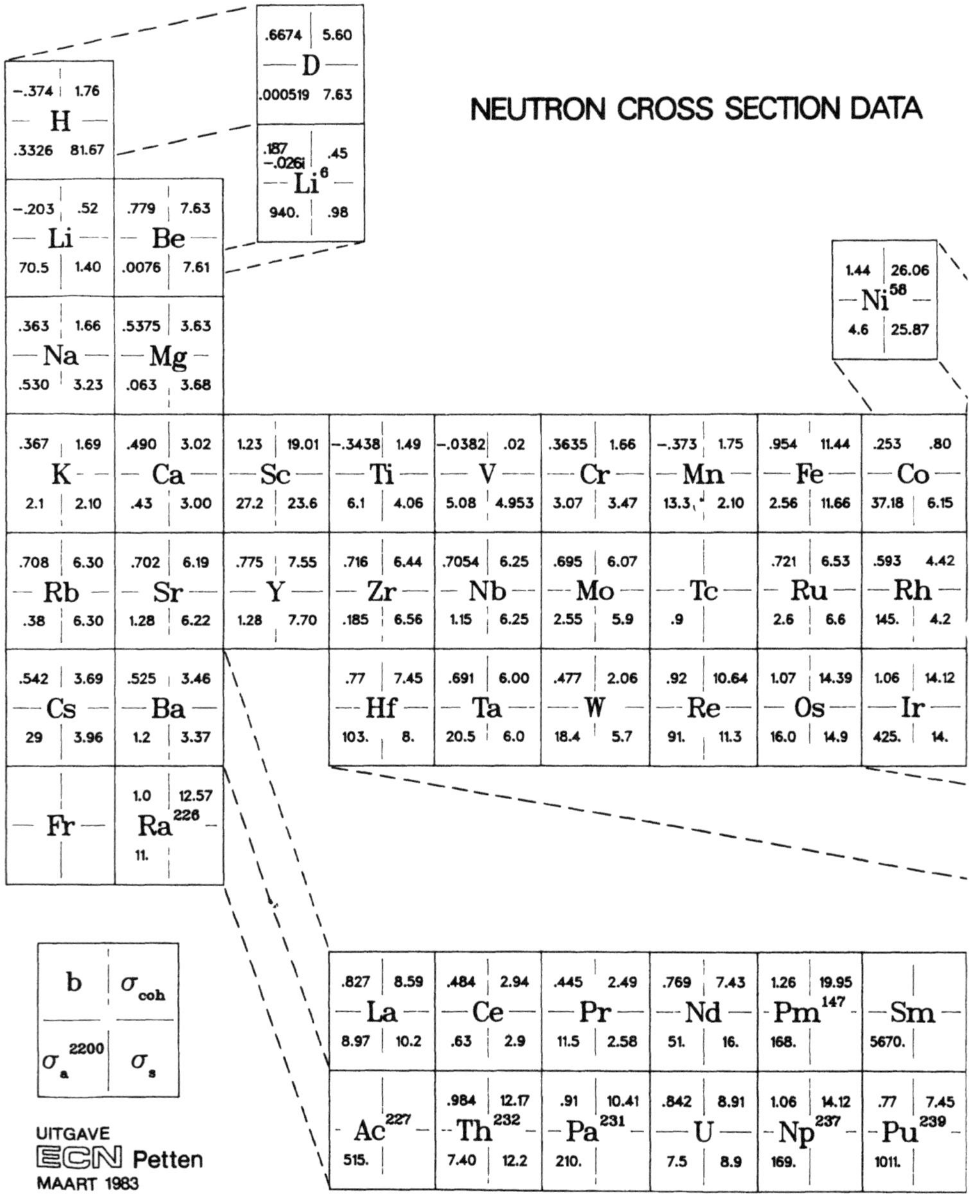

Figure 1. Periodic chart of the elements showing neutron cross section data. **b** is the scattering cross section and σ_a is the absorption cross section for neutrons with velocity

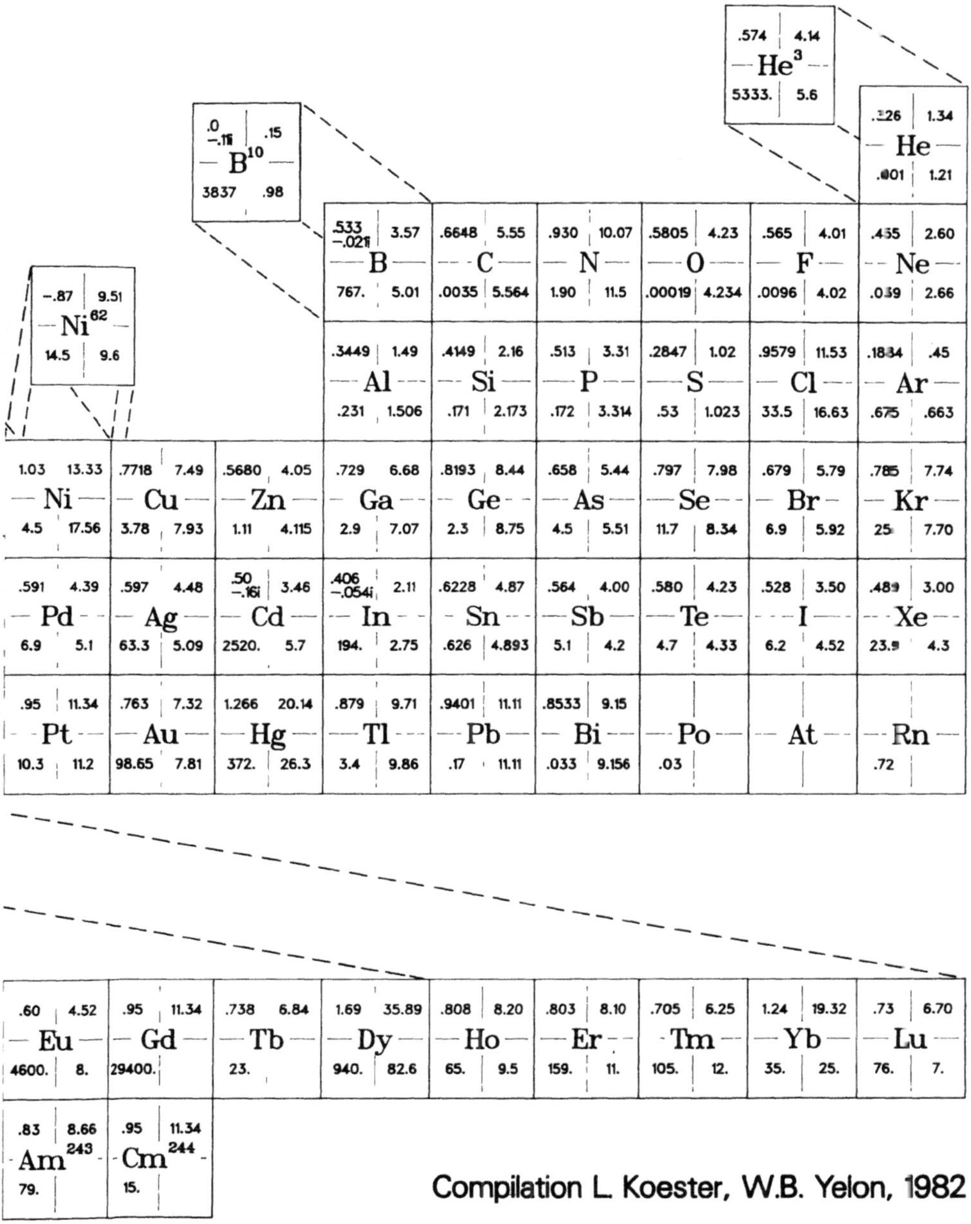

scattering length (10^{-12}cm), σ_{coh} is the coherent scattering cross section, σ_s is the total 2200m/s. Cross sections are x 10^{-24}cm.

scattering cross section consists of two components, incoherent and coherent, and a third cross section, the absorption, is needed to fully characterize these interactions. Magnetic scattering, through the interaction of the neutron magnetic moment with the unpaired spins of the scattering atom (both electrons and nuclei) leads to coherent magnetic scattering, for ordered systems, and to incoherent scattering for nuclei with unpaired spins, such as H and V, and to paramagnetic scattering in materials with disordered magnetic moments, as above the Curie point in ferromagnets. These interactions will be considered in turn, in order to appreciate the possibilities they create.

B.1. COHERENT CROSS SECTIONS

The periodic table of the elements is given in Figure 1 [5]. Shown for each element and a few selected isotopes are four quantities. The coherent cross section, σ (in barns), is given in the upper right hand corner. This quantity determines the strength of Bragg scattering, phonon and other coherent, nonmagnetic processes. In the upper left corner is found the scattering length **b** (10^{-12} cm), which is simply related to σ by:

$$\sigma = 4\pi b^2 \ . \tag{1}$$

This quantity plays the same role as the electron number, Z, in x-ray scattering. It is more convenient to examine this quantity than the cross section, since not all scattering lengths are positive. The sign difference reflects a 180^0 phase shift. In fact, those few atoms with negative scattering lengths have zero phase shift on scattering, while those with positive lengths have 180^0 shifts. Since the vast majority have this shift, convention has assigned them a positive sign. Unlike the x-ray form factor, the nuclear scattering does not decrease with angle, since the scattering is effectively point-like and does not cause destructive interference of the neutron plane wave across the dimensions of the scatterer (Figure 2). As can be seen from Figure 1, the scattering lengths show significant variation from atom to atom, and even within isotopes of a given atom, especially hydrogen and deuterium, for which the magnitudes differ by nearly a factor of two and the signs are different. Thus, in most alloy systems, when the scattering between a hydrogen site and the metal site(s) are in phase, in the conventional crystallographic sense, the total scattering amplitude will be reduced, while a deuterium atom at the same site will lead to an increase in the scattering power. Since the scattering lengths of hydrogen and deuterium have the same order of magnitude as those of the metal atoms, this may lead to quite large changes in Bragg intensities. Furthermore, the scattering lengths of nearby elements may be quite different (e.g. V, Cr, Mn, Fe and Co), leading to contrast in ordered alloy systems where little or no contrast may be seen in an x-ray experiment. Finally, it should be noted that, with the exception of V and a few heavy elements, the total variation in scattering length is only about a factor of five. Thus, in general, all atoms contribute significantly to the Bragg intensities, making the observation of light elements, including interstitial species, relatively straightforward with neutrons. This is quite different from the x-ray case, in which the scattered intensities are dominated by the heavy elements, and the observation of light atoms may be difficult, if not impossible. The coherent cross section leads directly to the observed Bragg intensities and to excitation spectra arising from the correlated motions of particles (phonons).

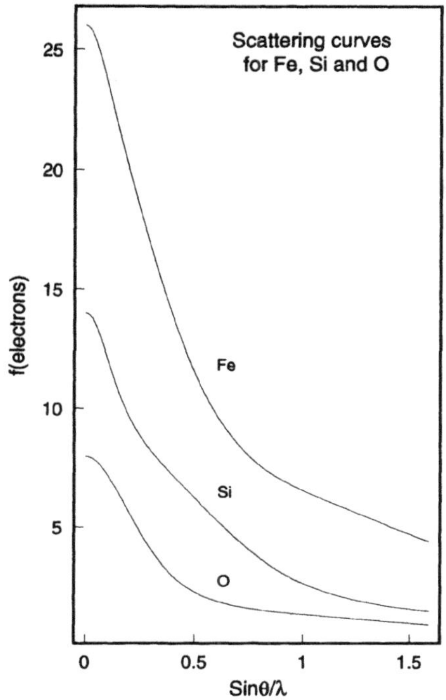

Figure 2. Atomic (x-ray) scattering powers for a few elements showing the strong angular dependence. By comparison, neutron scattering curves are horizontal lines.

B.2. INCOHERENT CROSS SECTION

The values given in the lower right hand corner of the element symbols in Figure 1 are the total scattering cross sections for the respective elements. In many cases these values are quite close to the coherent cross sections, but there are notable differences in others. The difference is the incoherent cross section, which arises from two different causes. These are both similar to the so-called alloy effect seen for x-rays. In that effect, two atoms, with different scattering factors, occupying the same site, give coherent scattering proportional to the weighted average of their cross sections, while an additional background is seen, arising from the difference in cross section. In the neutron case, even a single element (e.g. Ni) may cause such effects due to the differences in coherent cross sections of the isotopes of that element. This effect is avoidable only through the use of separated isotopes of those elements, but is usually not of great concern in scattering experiments, and may be used for measurement of incoherent (single particle) effects. A more important cause of the difference in cross sections is scattering of neutrons by nuclei in a non-zero spin state, through the formation of compound intermediate states. For example, scattering of a neutron by a hydrogen atom is quite different if the neutron spin is parallel or anti-parallel to that of the hydrogen nucleus. If both the nuclei and the neutron beams were fully polarized, then the scattering would be from only one spin state and would be fully coherent. In the usual experiment, however, the coherent cross section is the weighted average of the scattering lengths for these two spin states. In a case such as hydrogen, where the two scattering lengths have opposite signs, the coherent cross section is small compared to the incoherent cross section arising from the difference.

The coherent and incoherent cross sections in Figure 1 are for the "bound atom": assumed to be fixed at the origin. The neutron may, in fact, induce recoil in the scattering nucleus, and in the gaseous state the scattering cross section will be modified to that of the free atom:

$$\sigma_{free} = \left(\frac{A}{A+1}\right)^2 \sigma_{bound} \,, \tag{2}$$

where A is the nuclear mass number. In most solids this distinction is not significant, and the bound atom cross section is used for all but hydrogen atoms. The cross section for hydrogen is usually found to be intermediate to the two limits given above; typically 40 barns. This large incoherent cross section is particularly important for studying the free particle motion of hydrogen atoms in alloy systems. It is the probe which allows direct measurement of the potential in which the hydrogen is trapped, and of the diffusion and reorientation directions and rates, as H moves between interstitial sites.

B.3. ABSORPTION CROSS SECTION

The final lower left cross sections given in Figure 1 are for true absorption, leading to the production of a nucleus with one additional neutron. The values given correspond to absorption of neutrons with a velocity v of 2200 m/s or a wavelength of 1.8 Å. Except for special cases in which there is a resonance close to thermal neutron energies, this cross section will have a 1/v dependence and can be calculated for the specific experiment from the tabulated value. The most important feature to note is that, with the exception of a relatively few elements including B, Cd, Gd, Sm and Eu, the absorption is quite small and is often less than the scattering cross section. Thus, the neutron is a highly penetrating probe, able to enter many mm or even cm into most materials of interest. This has several important consequences. First, the surface state of a sample is far less important than for x-rays or other probes, such as electron diffraction, and neutron scattering is generally going to average over a large sample volume. Secondly, it is relatively simple to construct sample environments out of normal engineering materials, such as Al alloys, for high and low temperature and other experimental conditions. In addition, by careful control of incident and exit beam slits, it is possible to "look" at an internal volume of a bulk material, as is done in neutron stress scanning experiments.

In considering the attenuation of the neutron beam, either in the sample or in its environment, it is necessary to consider all the cross sections, since the beam will be attenuated both through scattering and through the true absorption. This is especially important in hydrogenous media, for which the incoherent cross section may dominate the absorption.

B.4. MAGNETIC SCATTERING

The neutron has a magnetic moment and will interact with the magnetic moments of the atoms and the nuclear spins in a material. For simplicity, only the coherent magnetic diffraction will be considered here, although in a disordered magnetic material the scattering is incoherent and strongly peaked in the forward direction. In ordered materials, (e.g. ferro or antiferromagnets), the differential magnetic cross section is, following the notation used in Bacon,

$$d\sigma_m = q^2 S^2 \left(\frac{e^2\gamma}{mc^2}\right)^2 f^2 \ , \tag{3}$$

where **q** is the magnetic interaction vector defined by

$$\mathbf{q} = \varepsilon(\varepsilon \cdot \mathbf{K}) - \mathbf{K} \ , \tag{4}$$

where **K** is a unit vector in the atomic magnetic moment direction and ε is a unit vector perpendicular to the reflecting planes (the scattering vector). S is the spin of the scattering atom, γ is the neutron magnetic moment and f is the magnetic form factor. This last term is similar to the well known x-ray form factor and arises from the spatial extent of the unpaired electrons responsible for the magnetic moment of the scattering atom. It, thus, has a similar shape, but tends to fall off more rapidly than the x-ray form factor since not all electrons, (and especially not the innermost electrons) contribute to the magnetic scattering. The term described in equation 4 leads to a strong dependence of the magnetic scattering on the scattering plane, even in polycrystalline material. For example, in uniaxial magnets, Bragg peaks corresponding to planes perpendicular to the unique axis will have maximum magnetic contributions, while planes along the unique axis will have no contribution. In keeping with the treatment of the nuclear scattering it is more useful to write the magnetic scattering length, p, as

$$p = \frac{e^2\gamma}{mc^2} gJf = \frac{e^2\gamma}{mc^2} \mu f \ , \tag{5}$$

where μ is the magnetic moment of the atom, and g is the Lande splitting factor. In this way we can write the total coherent cross section as

$$d\sigma = b^2 + 2bp(\boldsymbol{\lambda} \cdot \mathbf{q}) + p^2 q^2 \ , \tag{6}$$

where λ is a unit vector in the direction of the neutron magnetic moment. For unpolarized neutrons, the middle term of the expression will average to zero and the nuclear and magnetic contributions add independently.

C. Wavelength-Energy Relations

It is well known that characteristic x-rays have wavelengths comparable to the interatomic distances in solids, and are, therefore, useful probes of the crystallographic structure.

Unfortunately, the energies of the same x-rays (several keV) are much larger than the characteristic energies in these solids (phonons, magnons, etc.) and other forms of electromagnetic radiation (light, Raman or infrared spectroscopies) must be employed to probe those excitations. These alternate sources are, consequently, quite restricted in the momenta probed, because of the mismatch in wavelength and lattice dimension. The neutron is unique in being matched, simultaneously, to both the energy and wavelength scales. Thermal neutrons, produced by moderating fission or spallation neutrons, have wavelengths varying from about 0.8 to 5 Å and energies between about 130 and 3 meV. The velocities of the neutrons range from 5000 to 800 m/s. By special modification of the neutron moderators, this range may be extended even further. For reference, the relations between these quantities are given:

$$\lambda = \frac{3955.4}{v} = \frac{2\pi}{k} ,$$

$$E = 2.0717\, k^2 = \frac{81.787}{\lambda^2} = 5.2276 \times 10^{-6}\, v^2 ,$$

where λ is in Å, E in meV, k in $Å^{-1}$ and v in m/s. As a consequence, the neutron is not only a probe of the structure, but is readily used to probe the motion of the atoms, and the magnetic spins within the solid, (and sometimes within the same sample). Since a change in neutron energy, during an interaction with a solid specimen, leads to a change both in velocity and wavelength, a wide variety of instruments can be used to investigate such processes. In addition, through clever design of instruments, the dynamic range of scattering is truly extraordinary: energy transfers of less than 1 µeV up to 1 eV have been studied. Some of the principles of these instruments will be discussed later. Application of these techniques to interstitial systems is particularly important, as they allow elucidation of diffusion rates, atom potentials, and changes in bonding of the host lattice with insertion.

D. Neutron Production and Scattering Instrumentation

There are two basic methods of neutron production, nuclear reactors and spallation sources [6]. Both produce high energy neutrons which must be slowed to thermal energies before they can be used for scattering experiments, but differences in their modes of production lead to significantly different methods for scattering research. These differences and the consequent instrumentation are discussed below.

D.1. STEADY STATE SOURCES

Neutron production at flux densities high enough to be useful for scattering research is possible only in compact core reactors in which the central flux is above $10^{13}/cm^2s$. Consequently, special designs are employed to produce power densities of 1/3 MW/liter or above. The neutrons produced by the fission reaction have energies between 2 and 6 MeV and these must be slowed (moderated) to thermal energies to be used for scattering. Typically, light water (H_2O), heavy water (D_2O) and graphite can be used as moderators. Beam tubes will be located in the moderator-reflector as close as possible to the peak flux point, and the beams transported to the experimental hall (Figure 3). Special moderators, cooled to cryogenic temperature (LH_2, LD_2, etc) or heated to near 2000°C are used to shift the spectra to lower or higher energies, respectively. The flux distribution from such a reactor is well described by a Maxwellian spectrum of the form (Figure 4)

$$\frac{dN}{d\lambda} = \frac{2}{\lambda}\left(\frac{E}{kT}\right)^2 e^{-E/kT}\, d\lambda = \frac{2}{\lambda}\left(\frac{h^2}{2m\lambda^2 kT}\right)^2 e^{-(h^2/2m\lambda^2 kT)}\, d\lambda \quad . \tag{7}$$

Scattering instruments will typically select, from this spectrum, a "monochromatic" slice ($\Delta\lambda/\lambda = 10^{-2}$) through Bragg diffraction from a monocrystal. This will then fall on the sample whereupon the scattered neutrons will be measured in appropriate ways.

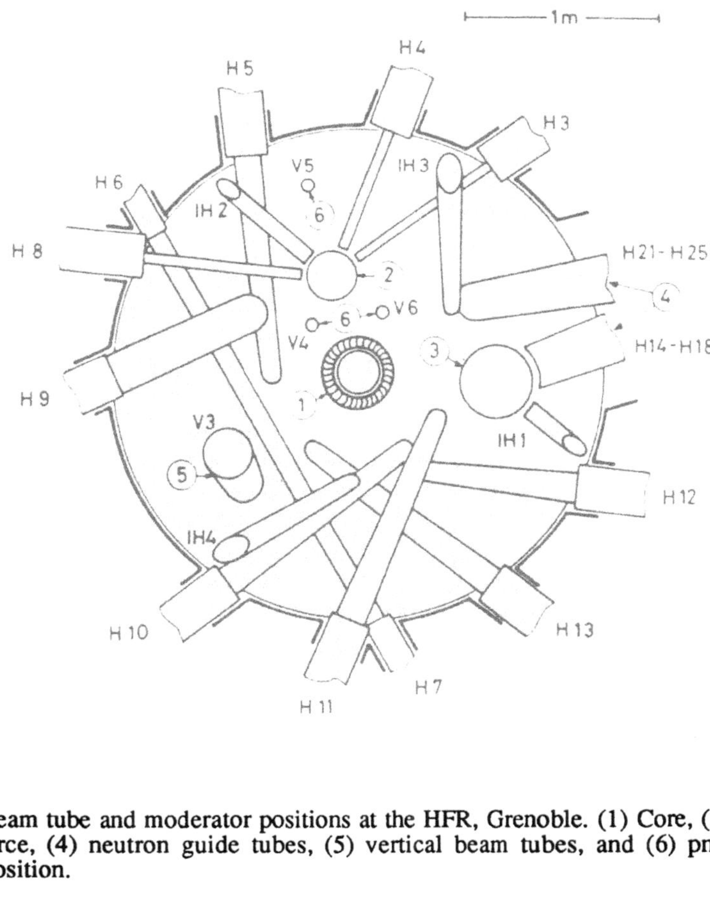

Figure 3. Beam tube and moderator positions at the HFR, Grenoble. (1) Core, (2) hot source, (3) cold source, (4) neutron guide tubes, (5) vertical beam tubes, and (6) pneumatic tube irradiation position.

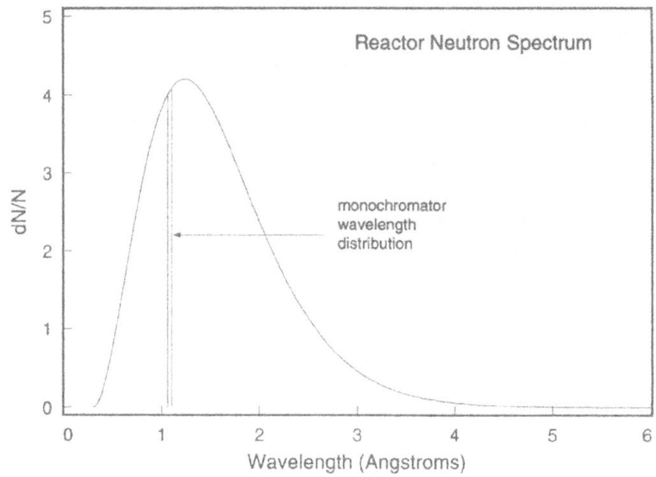

Figure 4. Maxwellian flux distribution for a thermal neutron source showing a typical monochromator "slice".

D.2. PULSED SOURCES

An alternative method of neutron production uses spallation, in which high energy protons, typically 800 MeV, are accelerated onto a heavy metal target. Copious quantities of neutrons are "boiled" off the target, at somewhat higher energies than fission neutrons, and subsequently moderated using similar media as in fission reactors. Thus, the spectra are peaked at similar energies. There are two significant differences between these sources and reactors. First, the protons are not delivered onto the spallation target in a continuous beam; rather, they come in pulses with a repetition rate of the order of 30 Hz. Secondly, in order to preserve the good time structure for the pulses of thermalized neutrons, the moderators are made thin, thereby allowing a high flux of epithermal neutron (1/E tail) to escape the moderator and enter the beam tubes. These differences lead to a radically different style of research methodology; experiments use the neutron flight time, rather than wavelength properties, as an energy analyzer, and often can use the entire beam on the sample, rather than the monochromatic slice used with the reactor. This compensates effectively for the lower time averaged neutron flux delivered by the pulsed sources, compared to reactors. In addition, experiments can be carried out at higher incident neutron energy than at reactors, creating special new opportunities.

Although the two types of sources are quite different, there are certain experiments, such as powder diffraction which can be carried out equally well at either, while there are others that are uniquely suited to one or the other. This complementarity has helped to create the broadest range of possibilities for the experimentalist.

D.3. SCATTERING INSTRUMENTATION

Broadly speaking there are two classes of scattering instruments: elastic scattering (diffraction) and inelastic scattering instruments, including quasielastic. The former type of instrument accepts all of the intensity from the sample, within a given angle or time channel, without attempting to distinguish between those neutrons which have been elastically scattered and those which may have gained or lost energy. In the latter class, the energy of the scattered neutrons is measured, either through Bragg scattering or through the measurement of the flight time from sample to detector. These two types of instruments provide different information about the sample. The former measures the static properties (structure) including the particle distribution on the scale of 1-1000 nm and the atomic level (crystal and amorphous structures). The latter probes the dynamic properties of the sample: movements of the lattice (phonons), of the magnetic moments (magnons), and of individual particles within the system (diffusion, tunneling, localized vibrations). All of these are important to the study of interstitial phases and a few of the principles of the techniques are outlined below.

D.3.a. *Elastic Scattering*. A typical powder diffractometer at a steady state reactor is illustrated in Figure 5. A monochromatic beam is selected from the polychromatic neutron spectra with a bent Si crystal, and scattered onto the sample. In order to achieve high data collection rates, the scattered neutrons are measured with a position sensitive detector, which accepts a 20° span and is moved in steps of the same size to cover the entire scattering diagram. Alternative designs may replace the position sensitive detector by a large number of Soller slit collimators in front of individual detectors. For pulsed neutron sources similar data is collected using a detector(s) at fixed angle and measuring the arrival time of the scattered neutrons (Figure 6). In either case, the sample may be found inside a cryostat, furnace or pressure cell. The last option is particularly interesting at the pulsed sources, since fixed windows may be used for the incident and exit beams. A relatively straightforward adaptation of this type of instrument is also used for the study of liquids and amorphous materials. For single crystal diffraction, from small unit cells, only one reciprocal lattice point will lie on the sphere of reflection at a given time and therefore only a single detector is needed to measure the scattered intensity at a steady state source. An Eulerian cradle will be provided to orient the specimen in the necessary directions to probe all lattice planes. At the pulsed source this becomes a "Laue" instrument, with order separation accomplished in time-of-flight.

A second class of elastic scattering instrument is the so-called small angle scattering (which should properly be called small-q scattering). In this technique (Figure 7) a highly collimated, but poorly monochromated beam, falls on a sample and the scattering is registered in an area detector centered on the direct beam. The observed scattering will generally be at angles below the first Bragg peak for the sample. The data are analyzed to determine the size and shape of microscopic fluctuations in the sample, such as precipitates and voids in metals, dislocation clusters, etc. At the pulsed sources, this experiment will be carried out at a fixed angle and the different scattering vectors separated in time.

A final type of elastic scattering instrument is concerned with the scattering between the Bragg peaks: diffuse scattering instruments. These represent an adaptation of the single

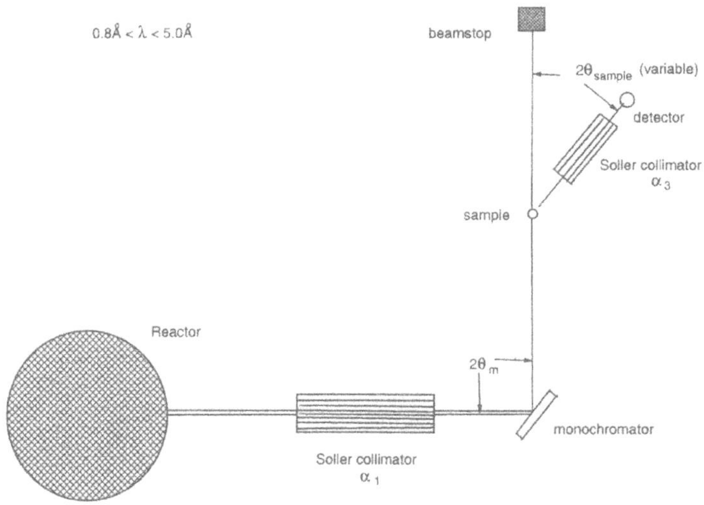

Figure 5a. Schematic of a typical steady state reactor powder diffractometer.

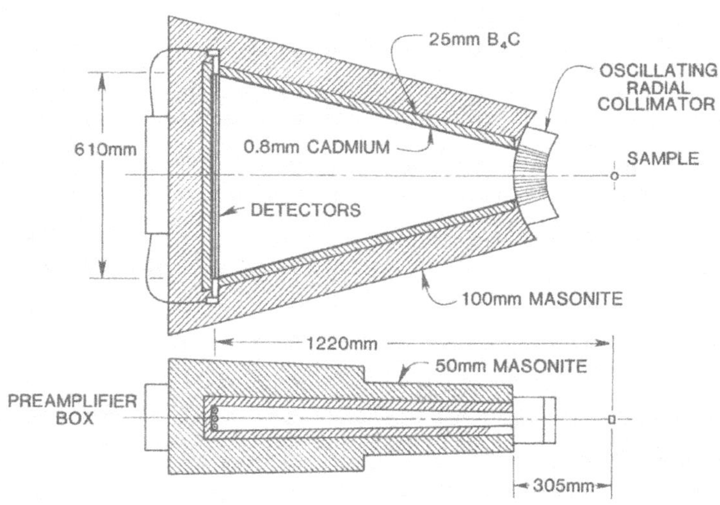

Figure 5b. Schematic of a position sensitive detector used to enhance the data acquisition rate for a powder diffractometer.

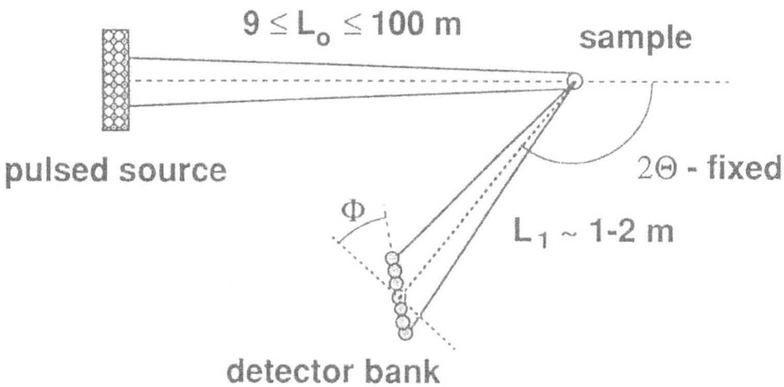

Figure 6. Schematic of a neutron powder diffractometer using the time-of-flight method. The detector bank uses "time focussing" to increase the data rates.

crystal instrument in which large area detectors are used to examine the scattering arising from lattice defects, short range order and other effects.

D.3.b. *Inelastic and Quasielastic Scattering*. The determination of the energy change of the neutron upon scattering takes a variety of forms, and while there is strong overlap between pulsed source and steady state instruments for elastic scattering, there is a greater degree of separation for inelastic scattering. The most well known inelastic instrument is the triple axis spectrometer, used at reactors (Figure 8). The neutron beam is prepared as for diffraction, but the scattered neutrons are interrogated by a second monochromator crystal (analyzer) to determine their final wavelength. The elastically scattered part will, of course, have the incident wavelength, but, as the analyzer is scanned, additional peaks, representing peaks in $\sigma(\mathbf{q},\omega)$, may be found due to phonons or magnons created or annihilated during the neutron interaction. The triple axis spectrometer is best suited for the study of excitations with modest to strong dispersion, i.e. with energy varying with the wavevector **q**. Because the triple axis measures only one point in $(\mathbf{q},\omega)$ space at a time, and the cross sections for inelastic scattering are typically 3 or more orders of magnitude smaller than for elastic scattering, the experiments with triple axis require large, single crystal specimens and long measuring times. One exception to this is the use of the triple axis to measure phonon densities of state, $\sigma(q,\omega)$, for which polycrystalline samples may be employed. Inelastic scattering on the pulsed source requires additional preparation of the beam. In one arrangement, the incident beam is chopped such that only one energy neutron strikes the sample, at a known time. The final energy can then be determined by the arrival time at a detector (Figure 9). The experiment can gain intensity by measuring at many angles simultaneously, although this gain factor may be smaller than at first sight, since most of the angles will contain information only over a

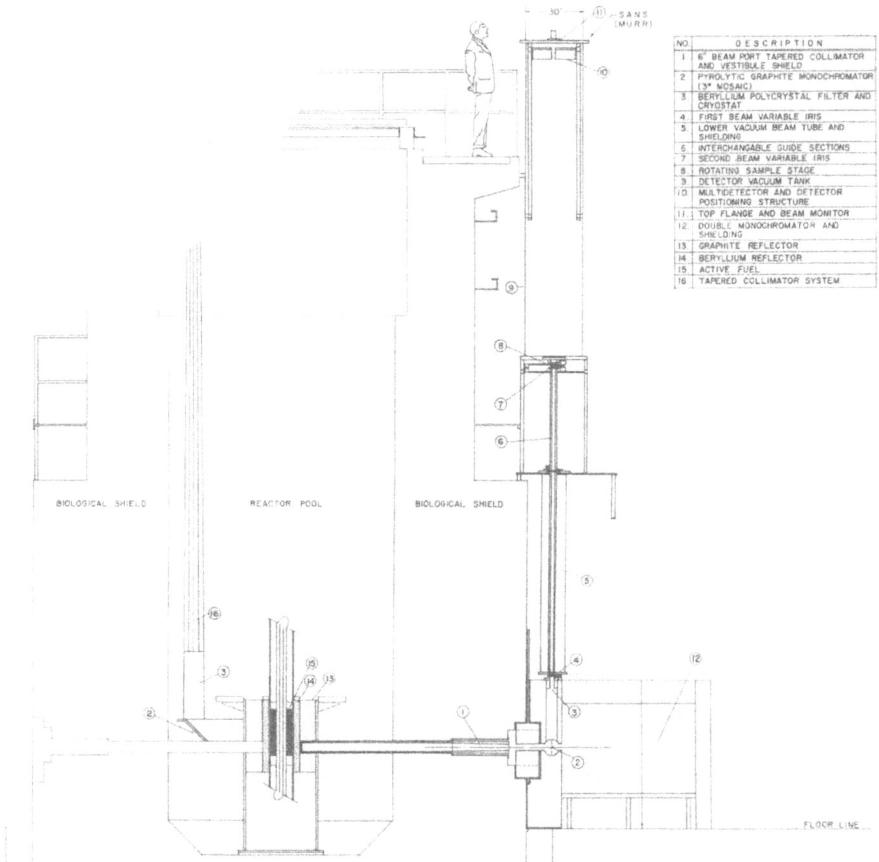

Figure 7. Schematic of a small angle neutron scattering spectrometer using a graphite monochromator for wavelength selection.

very limited time, due to the "emptiness" of the scattering space. In comparison, the triple axis can be set to measure only in regions of interest. An alternative geometry for the pulsed source consists of using the full beam on sample, but employs crystal analyzers to fix the energy of the final neutrons (Figure 10). The energy transfer is then determined from the

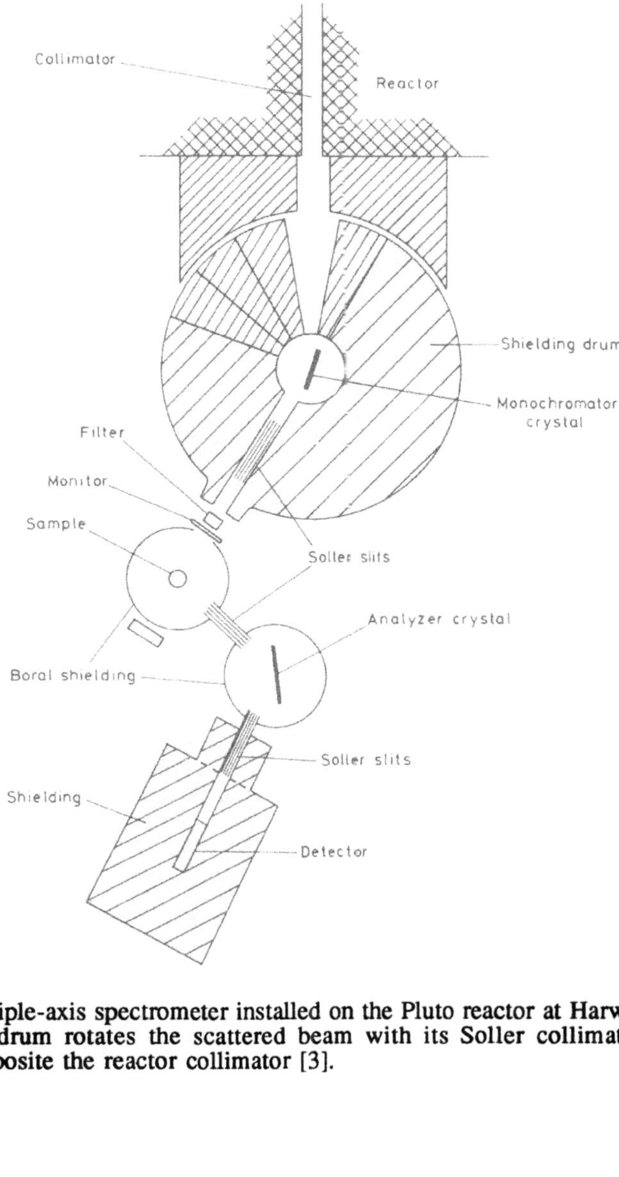

Figure 8. The triple-axis spectrometer installed on the Pluto reactor at Harwell. The massive monochromator drum rotates the scattered beam with its Soller collimator. Sectors rise automatically opposite the reactor collimator [3].

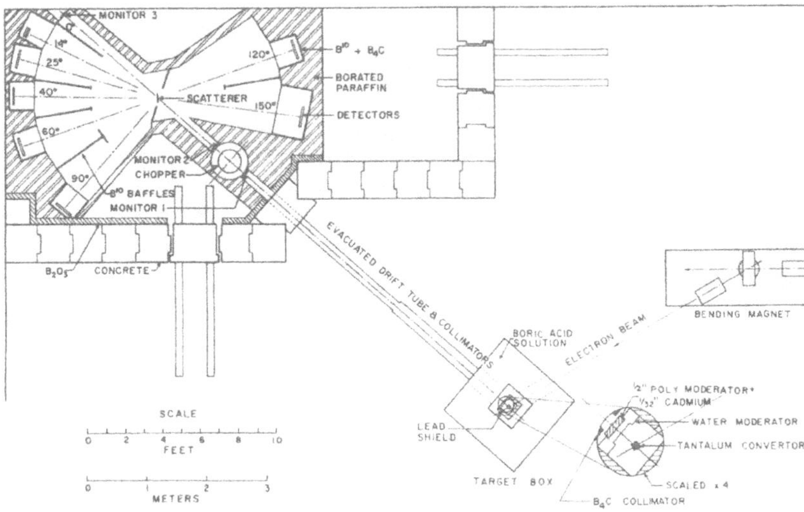

Figure 9. The inelastic rotor spectrometer constructed on the electron linac at the Rensselaer Polytechnic Institute [2].

arrival time. As in the chopper spectrometers, many angles can be measured simultaneously. There are two respects in which the pulsed source instruments are superior to the triple axis. First, the 1/E tail of the neutron spectrum allows measurement to much higher energy transfers. Second, when the excitations are relatively dispersionless, as with most optical or vibrational modes, data from many angles may be combined to significantly improve the data rates. Of course, it is possible, on a steady state source, to use a chopper, in connection with a monochromatic beam, and then to employ a large detector bank in much the same manner.

Two special variations of the triple axis spectrometer are employed to study quasielastic scattering and very low energy excitations such as tunneling modes. The first of these is backscattering (Figure 11), in which the incident, polychromatic beam is diffracted at $2\theta \approx 180°$, resulting in extreme monochromaticity ($\Delta\lambda/\lambda = 10^{-4}$). The scattered beam is analyzed using a similar geometry, although many crystals may be employed. The energy is scanned either by Doppler shifting the monochromator crystal, or by thermally varying its lattice spacing. In this way, resolution of about 1 μeV can be achieved. Another technique, known as spin echo uses the precession of the neutrons in a magnetic field to determine the energy change. In this case, both the incident and scattered neutron must pass through precession fields, and thus only one angle is measured at a time. This technique does not require the

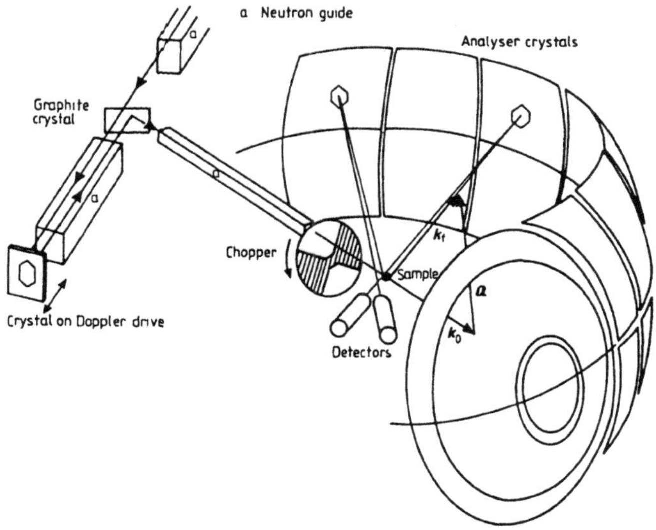

Figure 10. Schematic of the backscattering spectrometer IN10 at the Institute Laue Langevin. The incident beam is backscattered by the monochromator mounted on a Doppler drive, deflected to the sample, scattered and then backscattered again by the analyzers to the detectors [9].

same degree of monochromation as does backscattering and thus the higher flux on the sample compensates for the smaller detection surface.

D.3.c. *Magnetic Polarization*. The discussion (except for spin echo), thus far, is independent of whether the beam is magnetically polarized or not. Use of polarized neutrons creates yet additional experimental capabilities. If the monochromator is magnetically aligned, then there are different cross sections for neutrons with spin parallel or antiparallel to the polarization direction. The respective scattering lengths are

$$b_+ = b + p$$

$$b_- = b - p \quad . \tag{8}$$

If the monochromator material is chosen such that b = p for a particular reflection, then only neutrons with parallel spins will be scattered and a fully polarized beam (with 1/2 the intensity of an unpolarized beam) will be produced. This beam can then be used in either elastic or inelastic scattering instruments. If one uses a magnetically aligned sample, and measures the elastic scattering ratio of intensities when the neutron is polarized parallel or antiparallel to the

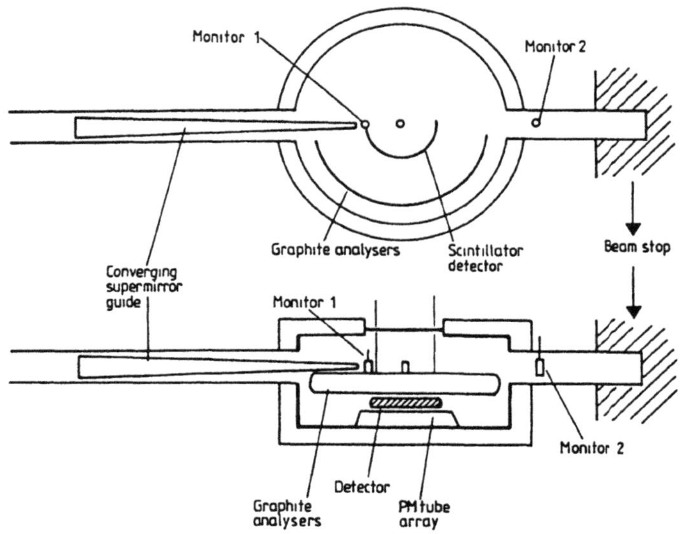

Figure 11. Schematic of the High Resolution Inelastic Spectrometer at the Spallation Neutron Source of the Rutherford Appleton Laboratory [9].

sample alignment, then it is possible to separate the magnetic and nuclear scattering from each Bragg peak. In addition, if one analyzes the spin state of scattered neutrons for a nonmagnetic sample, one can separate the spin-flip scattering from the non-spin flip scattering. Since the former can arise only from the spin incoherence, that contribution, which may be quite large for hydrogen-containing samples, may be reduced. This technique has not been widely used, but could be of importance in cases where the signal to noise ratio suffers seriously from the large incoherent background.

The power of polarized neutron scattering is particularly great for inelastic scattering experiments. Both lattice vibrations (phonons) and spin waves (magnons) can give observable inelastic intensity with unpolarized neutrons. The former excitations do not cause a flip of the neutron spin, while the latter do. Thus, analysis of the spin of the scattered neutrons permits unambiguous separation of these two excitations.

The applicability of polarized neutrons, particularly combined with polarization analysis, is significantly wider than can be discussed here and reference should be made to the bibliography [1,3]. However, in practice, there has been less use made of polarized neutrons than might be expected because of technical problems. Chief among these is the problem of beam polarization. The best monochromator for polarizing short wavelength neutrons is a single crystal of CoFe. Due to the relatively large absorption of Co, which limits the reflectivity width, and the loss of 1/2 inherent in any polarization process, the intensity of beams from this type of crystal is much lower than for non-polarized beams. With polarization analysis this loss is squared. For longer wavelengths, Heusler alloy crystals may be used, but production of good quality, large crystals has been problematical. A solution to this problem may be emerging with the development of neutron supermirrors: artificial lattices (e.g. Fe-Si) which reflect one spin state and transmit the other. These will, eventually, allow for intensities comparable to conventional instruments. For pulsed neutron sources the problem of polarization is more difficult still, since one needs to polarize the entire spectrum. For that, polarizing filters are under development, but still far from implementation on a routine basis.

E. Applications of Neutron Scattering

Before discussing specific cases of the use of neutron scattering in the study of interstitial alloys, it is useful to describe the principles of the different techniques, describing the type of information which can be extracted from various experiments and the methods used to arrive at those results. After describing the methods, a second chapter will describe the results of a wide range of neutron scattering experiments on interstitial alloys.

E.1. STRUCTURE ANALYSIS

If the magnetic scattering is ignored, for the present, then structure analysis with neutrons is virtually identical to x-ray diffraction; the Bragg peaks are measured and some form of least squares analysis is carried out to match the computed structure factors to the observed quantities. When single crystals are available, this is a relatively simple matter, as each structure factor is individually measured. If light elements are present, or if two or more species with similar Z are present, then neutron scattering will have a distinct advantage over the x-ray case in locating the light elements or distinguishing between nearby elements. Unfortunately, many, if not most of the systems of interest may not be found in single crystal form, and only powder diffraction data may be obtainable. This is particularly true for pseudobinary alloys in which more than one element may occupy a single crystallographic site, or in interstitial compounds, in which insertion of the interstitial species causes a strong lattice expansion and therefore causes the sample to become a powder, even if a single crystal is used as the starting sample. If the powder diagram is sufficiently simple, then structure factors may be obtained by integrating the intensities under the various peaks, but this is the exception rather than the rule. In recent years a new technique, known as the "Rietveld method" has emerged to enable the analysis of even highly complex structures from powder diffraction data [7]. The basis of the method is the description of the diagram not in individual structure factors, but, instead, by the angular (or time, for pulsed neutron research,) dependence of the intensity, written as

$$I(\theta) = \sum_n c_n F_n^2 \chi(\theta - \theta_n) \quad \text{and} \quad F_n = \sum_j b_j e^{i\mathbf{k}_n \cdot \mathbf{r}_j} e^{-2W} + \text{magnetic terms} \ , \tag{9}$$

where the summation is over all of the Bragg peaks which contribute intensity at the angle in question. F_n is the structure factor of the nth reflection and χ gives the fraction of the peak intensity contributed at the angle in question, which in general is different from the peak position by the angle $\theta-\theta_n$. The summation on j is over the atoms in the unit cell, k_n is the reciprocal lattice vector for the nth peak, r_j are the coordinates of the jth atom and W is the Debye-Waller term. For steady state neutron diffraction, the peaks are found to be nearly Gaussian and, thus, this quantity is easily calculated. The structure factors are calculated from an initial model structure, which may then be parameterized in terms of atomic coordinates, temperature factors, and site occupancies. These parameters will then be refined in a least squares procedure. In addition to the structural parameters, which determine the structure factors, the peak positions will depend upon the lattice parameters and instrument constants (wavelength, zero point), while the peak widths will depend on other instrument parameters (collimation, monochromator angle) and upon strain and particle size effects in the sample. All of these may be included in the least squares problem as parameters and refined accordingly. With modern, high resolution diffractometers, diagrams with many hundred reflections, and of the order of 200 parameters may be refined. An example of such a refinement is given in Figure 12. Over 1100 independent reflections contribute to the diagram, and 45 position parameters were refined along with the magnetic, site occupancy, lattice and other parameters. For pulsed source instruments, the peak shape is substantially more complex and the diagram must be normalized for the spectral dependence of the incident intensity, but these problems have been well resolved, and incorporated into the

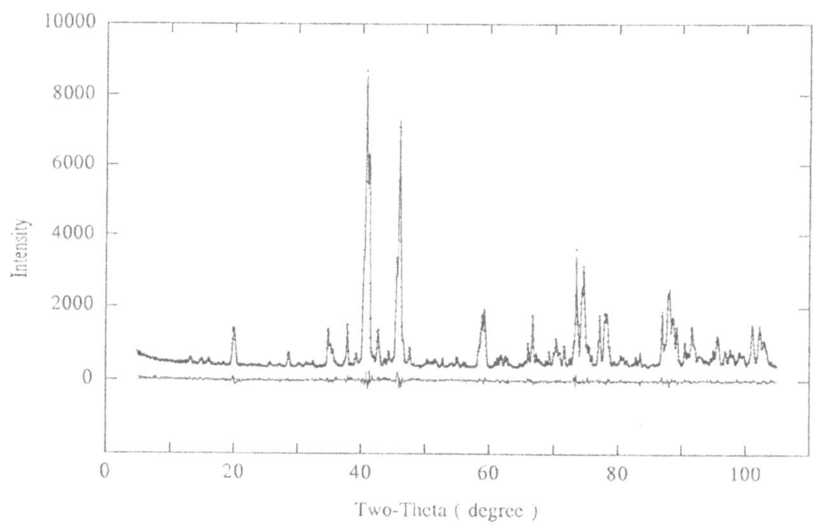

Figure 12. Neutron Rietveld refinement of $Nd_3Fe_{29-x}Ti_x$. The data are points on the upper curve, the solid line is the calculated scattering pattern and the lower curve is the difference between observed and calculated, plotted on the same scale.

various codes available for refinement. In addition, the codes have been expanded to permit the simultaneous refinement of more than one phase which may be present in a sample. This method has become a mainstay of research in inorganic systems because of the simplicity of sample preparation and data collection. As the refinement codes become more sophisticated and user friendly, the utilization of the method will be limited only by the unavailability of beam time at the various reactors and pulsed neutron sources. For this reason, much effort has gone into development of larger detector arrays and improved neutron optics to deliver optimized beams on sample. At the present time, at the University of Missouri Research Reactor, data is typically collected in 24 hours on 1 gm specimens, although, for simple problems, less time or sample would suffice. Upgrade of the diffractometer, presently in the planning stage, could reduce these requirements by as much as a factor of 10, thereby allowing almost unlimited access.

E.2. MAGNETIC STRUCTURES

If single crystals are available, the most powerful method for magnetic structure analysis is the polarized beam method, in which the magnetic and nuclear contributions are uniquely determined. Since this is not often the case, in the systems of interest here, one is forced to use a powder method. As shown in the previous section, the magnetic and nuclear intensities add independently and thus the Rietveld method is easily applied. In fact, in some versions of the Rietveld codes, the magnetic intensities are treated as a separate phase, in which the atom coordinates are constrained to those of the nuclear phase refinement of the same component. If the structure is ferromagnetic, then the magnetic intensities will be found at the same positions as the nuclear Bragg peaks. If the structure is antiferromagnetic, with a unit cell larger than the crystallographic cell, then new reflections will be found which are purely magnetic in character. One restriction on the use of powders for magnetic structure refinement is the result of polycrystalline averaging over the different equivalent reflections. It can be shown that no information can be extracted about the magnetic moment direction, for cubic structures, while for uniaxial magnets, no information is available about the direction of the magnetic moment component in the plane perpendicular to the unique direction [8]. Only for orthorhombic or lower symmetry, can powder diffraction be used to uniquely determine the moment direction. However, it is possible to magnetically align the powder, provided the aligning field does not cause the moment to rotate from the zero field direction, and then to determine the unique direction from the diffraction measurement. These measurements can provide information about the magnitude of the moment on each site, as well as information about non-collinear structures.

E.3. SMALL ANGLE SCATTERING

Small angle scattering examines fluctuations in the scattering density of materials on a length scale of 1-1000 nm. These may arise from many origins, including porosity, precipitate formation, incomplete diffusion of interstitial species, dislocations and grain boundaries in metals, and from variations in chemical composition in biological and polymer systems, among other causes. X-ray small angle scattering is of little use in interstitial alloys because the additional scattering contributed by the interstitial atoms is usually negligible. In contrast, the relatively large scattering power of H, D, C, or N, and especially the negative scattering length of H, may lead to large changes in scattering density. The scattering is typically featureless, (Figure 13) falling smoothly with increasing angle. The slope of the curve, in

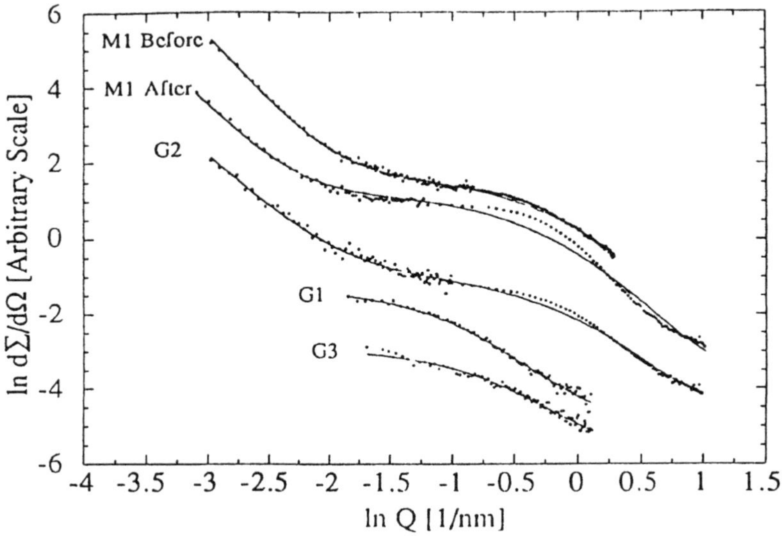

Figure 13. Small angle neutron scattering (SANS) data for "red" Si. The broad humps are interference peaks.

different regions, gives the information about size and shape of the density fluctuations, but generally, a model is required for interpretation of the data. In addition, comparison of "blank" samples with the sample of interest is essential for eliminating extraneous effects.

E.4. QUASIELASTIC SCATTERING—DIFFUSION

Coherent lattice excitations and magnons are usually found at energies of a few meV and above. Quasielastic scattering and certain tunneling excitations can be found close to zero energy. The neutron will probe, through the incoherent cross section, the diffusive motion of individual particles. A recent text by Bee [9] is a comprehensive treatment of the subject. The neutron momentum will be changed, to first order, only by the component of the momentum along the neutron wavevector, and thus the energy shift will not be by a discrete amount, but rather will be in a band around zero energy transfer, since the neutron can either gain or lose energy, depending upon the direction of the particle motion. If the diffusion is not random, but rather has a directional dependence, as from interstitial site to site, then the scattering will also show a **q** dependence, permitting a determination of the jump direction [10,11]. A review through 1986 of the theory and results of studies of hydrogen in metals can be found in a chapter by Springer and Richter [12]. A few key results are discussed here.

If one assumes a simple Bravais lattice, in which the residence time, τ, of the interstitial atom on a site is long compared to the jump time, and that there are n adjacent sites to which the atom may jump, then the resulting incoherent scattering is found to be a Lorentzian:

$$S_i(Q,\omega) = \frac{f(Q)/\pi}{\omega^2 + f^2Q} , \tag{10}$$

and the half width at half maximum is

$$\Gamma = f(Q) = \frac{1}{n\tau} \sum_{k=1}^{n} [1 - e^{-Q \cdot d_k}] , \tag{11}$$

where $\mathbf{d}_k$ is a jump vector. This expression will be modified if there are inequivalent sites to which the interstitial atom can jump. The term in f(Q) contains the directional dependence of the jumps. This expression is only valid in the low concentration limit. As the concentration, c, of interstitial species increases, jumps will be "blocked." If the concentration is not too high, a mean field approach is valid, and $1/\tau$ in equation 11 is replaced by $(1\text{-}c)/\tau$. In the long wavelength limit ($Q \rightarrow 0$), the linewidth always decreases as DQ^2, with a coefficient identical to the single-particle diffusion constant. Of the interstitial atoms of interest, only H has a large enough incoherent cross section to permit easy measurement of the incoherent scattering. However, the coherent scattering also has a contribution due to the correlated diffusive motion of the atoms in the solid. The coherent scattering function becomes:

$$S_c(Q,\omega) = C(1 - C)\frac{f_c(Q)/\pi}{\omega^2 + f^2(Q)} + \text{Bragg terms} . \tag{12}$$

For small Q, the quasielastic width is found to be:

$$\Gamma_c = \frac{d^2}{n\tau} Q^2 = D_{ch} Q^2 . \tag{13}$$

The collective diffusion constant, D_{ch} can be related to the single particle diffusion constant D. Since the coherent cross sections of D and N are significantly larger than the incoherent cross sections, studies of alloys with these species have generally probed the coherent scattering and made corrections for the incoherent part.

In addition to the quasielastic contribution, which is used to measure diffusive motion, very low energy excitations corresponding to tunneling modes may be observed. These can be associated with the motion of a single atom between two potential minima of slightly different depths, or of a group of atoms, such as a tetrahedra, flipping between two different potentials, perhaps related to the orientation of the tetrahedra in adjacent cells.

E.5. INELASTIC SCATTERING

The interaction of the metal host lattice and the interstitial species (especially H or D) is generally quite strong. As a result, the vibrational excitations of the interstitial atoms will be at relatively high energy, and the lattice vibrations of the host lattice will be modified by the interactions. Both of these effects can be studied with neutrons. The lowest lying vibrational mode of H in metals is typically above 100 meV. With reactor hot sources and spallation neutron sources, it is possible to measure not only the fundamental modes but also higher harmonics. The deviation of the harmonic energies, from multiples of the fundamental, is a direct reflection of the anharmonicity of the potential. This potential has been found, in certain cases, to be bell-shaped, and the harmonic energies are lowered, while in others the potential approaches a square well and the harmonic energies are raised. Different sites also show different potentials, and the vibrational frequencies can be used to distinguish between possible occupancy models. In addition, the large mass difference between H and its isotopes leads to strong renormalization of the vibrational energies. The anharmonic effects enter into this, and comparison of H and D modes in the same system give results consistent with results from the H harmonic measurements. An important conclusion derived from these studies is that the rapid diffusion of H in most metals cannot be a thermally activated process, because of the deep potential. Instead, it is concluded that the jump diffusion must be a phonon assisted process, in which vibrations of the lattice help to carry the H from site to site.

The phonons of the lattice will also be affected in several ways. At high concentrations, the interstitial atom (especially H) will alter the band structure and consequently the near-neighbor metal-metal interactions. In addition, the interstitial species generally will cause a lattice expansion, reducing the metal-metal interactions, even at relatively low concentration. This latter effect will lead, principally, to a general softening of the phonon modes. Similar effects must be found for the magnetic interactions, but these have not been studied extensively. The observation of increased Curie temperatures, in many rare-earth Fe interstitial compounds, points to an enhanced exchange interaction as a result of the lattice expansion, but band effects may also play a significant role. The general unavailability of single crystals of the interstitial compounds of interest, has, unfortunately, prevented exploration of these issues. It is possible to measure the initial slope of the magnon dispersion curves and the magnon gap, even in polycrystalline samples, and this may be worth a serious effort.

References

1. G.E. Bacon, *Neutron Diffraction*, Clarendon Press, Oxford (1975).
2. C.G. Windsor, *Pulsed Neutron Scattering*, Taylor and Francis, London, (1981).
3. K. Sköld and D.L. Price, *Methods of Experimental Physics*, Vol. **23**, Parts A-C, Academic Press, Orlando, (1986-1987).
4. G. Kostroz, *Treatise on Materials Science and Technology*, Vol. **15**, Academic Press, New York, (1979).
5. L. Koester and W.B. Yelon, *Compilation 1982*, ECN. V. F. Sears, in *Methods of Experimental Physics*, eds. K. Sköld and D. L. Price, Vol. **23**, Part A, Academic Press, Orlando (1986) pp. 521-550. L. Koester, H. Rauch, E. Seymann, Compilation 1992, *Neutron News*, Gordon and Breach, Switzerland.
6. J.M. Carpenter and W.B. Yelon, in *Methods of Experimental Physics*, eds. K. Sköld and D.L. Price, Vol. **23**, Part A, Academic Press, Orlando (1986) pp. 99-196.
7. H.M. Rietveld, J. Appl. Cryst. **2**, 65 (1969).
8. G. Shirane, Acta Cryst. **12**, 282-5 (1959).
9. M. Bee, *Quasielastic Neutron Scattering*, Adam Hilger, Bristol and Philadelphia, (1988).
10. C.T. Chudley and R.J. Elliot, Proc. Phys. Soc. London **77**, 353 (1961).
11. K. Sköld and G. Nelin, J. Phys. Chem. Solids **28**, 2369 (1967).
12. T. Springer and D. Richter, in *Methods of Experimental Physics*, eds. K. Sköld and D.L. Price, Vol. **23**, Part B, Academic Press, Orlando (1987) pp. 131-182.

Chapter 10

Statics and dynamics of hydrogen in metals

H.R.Schober
Institut für Festkörperforschung
Forschungszentrum Jülich
D-52425 Jülich
Germany

ABSTRACT. The interaction of interstitial hydrogen with its metallic host lattice is described by the standard Greens' function method. Experimentally the hydrogen site occupancy is either determined by scattering off the H or by its influence with the lattice. Some typical experiments are discussed. The vibrational dynamics of H is dominated by its localized vibrations. It will be shown that besides these also the vibrations with lattice frequencies (band modes) are important for the understanding of H-dynamics. The influence of quantum effects is shown.

1. Introduction

In this chapter we will discuss the statics and the vibrational dynamics of hydrogen. A more extensive treatment of this subject can be found in the book by Fukai [1]. Due to the small mass of hydrogen a split into static and dynamic properties is not really possible. Nevertheless we will take the classical route of first discussing the statics and then the dynamics.

Interstitial hydrogen represents a small defect. The neighbours are displaced typically by 0.1Å. This allows one to treat the host lattice to a good approximation by harmonic lattice dynamics, discussed in a previous chapter. This leads to a description in terms of Green's functions and Kanzaki forces, see e.g. [2; 3]. One thus gains a direct connection to macroscopic quantities. We will only give some basic formulae. A detailed derivation would go far beyond the scope of this chapter.

Over the years a great variety of methods have been employed to determine the site occupancy of the H atoms. There are methods where one looks directly at the H atoms, e.g. channeling and neutron scattering. Other methods probe the effects of the hydrogen on the lattice, e.g. volume expansion, x-ray scattering and mechanical relaxation.

The usual tool to study the dynamics is again Green's function methods [4; 3]. The local properties of the H are then described by a defect Green's function which in turn is given by the one of the ideal lattice and the additional couplings due to the insertion of the H into the host lattice. Isolated hydrogen atoms show pronounced localized vibrations whose frequencies are much higher than the maximal

F. Grandjean et al. (eds.), Interstitial Intermetallic Alloys, 249–266.

ones of their hosts. These vibrations can, to a good approximation, be considered as vibrations of the H in a rigid cage given by the surrounding metal atoms.

Even though these localized vibrations are the most conspicuous ones in all experiments, the vibrations of the hydrogen together with the host atoms are, due to their much lower frequencies, essential for a quantitative understanding of quantities such as the mean square displacement. They determine the temperature variation.

With increasing concentration the localized modes merge to optic phonons. From the dispersion of these phonon branches one can deduce information on the strength of the interaction between the single H atoms.

Due to the small mass of H, quantum effects become important and have to be included in many cases. These effects will be briefly discussed in a final section. Their importance will become also evident in the subsequent chapter on hydrogen diffusion.

2. Structure and Elastic Constants

2.1. THEORY

Before calculating the dynamics of a hydrogen in a metal it is necessary to calculate the structure surrounding it in the crystal. Ideally this should be done by a full electronic calculation. This meets, at present, with considerable difficulties. One is that hydrogen cannot be described by the usual pseudo-potential and its $1/R$ interaction with the conduction electrons makes it a strong perturbation of the host lattice. Furthermore the small mass of the hydrogen forces one to take the hydrogen motion into account, as we will see later on. One, therefore, takes recourse to empirical interactions which are fitted to experimentally measured quantities, such as lattice expansion, effects on elastic constants and hydrogen vibrations. It is therefore necessary to connect those quantities with the properties of the interaction. This is usually done in the framework of Green's function techniques where an ideal harmonic crystal is taken as the reference system.

Low concentration

First we consider low hydrogen concentrations when the problem can be reduced to the one for a single atom and macroscopic properties are found by random superposition of the single atom effects. Let us assume a hydrogen atom localized at some site. The total potential energy of the crystal with this defect at its origin, Fig. 1, consists of the potential energy of the ideal host lattice, U, and the potential energy due to the defect, U^{def}. The latter need not only contain a direct interaction between the defect and the host but also defect induced changes in the host-host interaction. Later on we will see that in some cases it is necessary to include also the vibrational energy of the hydrogen in U^{def}. For the sake of simplicity we assume a Bravais lattice for the host. The lattice vector of atom $\mathbf{m}$ is given as

$$\mathbf{R}^{\mathbf{m}} = \mathbf{X}^{\mathbf{m}} + \mathbf{s}^{\mathbf{m}}. \tag{1}$$

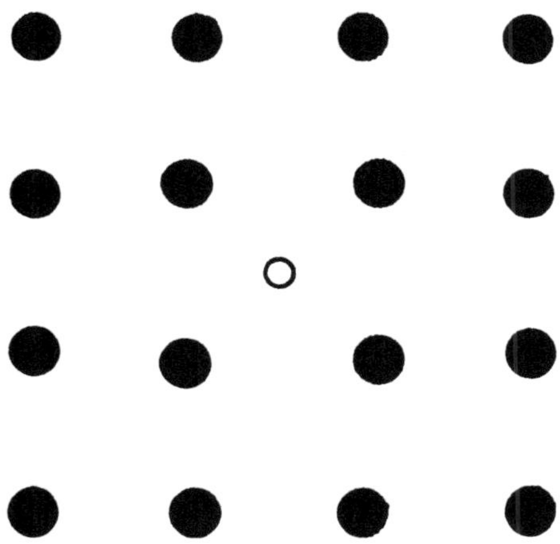

Fig. 1: Schematic view of interstitial H (∘) in a lattice of metal atoms (•).

here $\mathbf{X}^{\mathbf{m}}$ denotes the ideal lattice position of atom $\mathbf{m}$ and $\mathbf{s}^{\mathbf{m}}$ is here the static displacement as opposed to the dynamic displacement $\mathbf{u}^{\mathbf{m}}$. The ideal host potential is expanded into its harmonic part plus the anharmonic corrections $U^{anh}(\{\mathbf{R}\})$

$$U^{total} = \Phi_0(\{\mathbf{R}\}) + \frac{1}{2} \sum_{\substack{\mathbf{mn} \\ \alpha\beta}} s_\alpha^{\mathbf{m}} \Phi_{\alpha\beta}^{\mathbf{mn}} s_\beta^{\mathbf{n}} + U^{anh}(\{\mathbf{R}\}) + \sum_{\mathbf{m}} U^{def}(\mathbf{R}^{\mathbf{m}}). \tag{2}$$

The lattice equilibrium condition at T=0, neglecting quantum corrections, is obtained by differentiation with respect to the atomic positions

$$-\frac{\partial}{\partial R_\alpha^{\mathbf{m}}} U^{def}(\mathbf{R}^{\mathbf{m}}) - \frac{\partial}{\partial R_\alpha^{\mathbf{m}}} U^{anh}(\mathbf{R}^{\mathbf{m}}) \equiv K_\alpha^{\mathbf{m}} = \sum_{\mathbf{n}\beta} \Phi_{\alpha\beta}^{\mathbf{mn}} s_\beta^{\mathbf{n}} \tag{3}$$

K_α^m is a generalized Kanzaki force. It is defined as that force which causes in the ideal harmonic crystal the same displacements, as the real force does in the real crystal. For small static displacements ,s, (for "small" defects) where the anharmonicity is unimportant, the Kanzaki force becomes equal to the real force, taken at the relaxed lattice positions, as it was introduced originally [5]. The changes of the host–host interaction induced by the hydrogen are included in the Kanzaki force. Also effects due to the hydrogen vibration can be included.

Eq. (3) can be solved by the static Green's function, $\mathring{G}_{\alpha\beta}^{\mathbf{mn}}$, of the ideal lattice, i.e. the $\omega = 0$ limit of the dynamic Green's function introduced in the chapter on lattice dynamics.

$$s_\alpha^{\mathbf{m}} = \sum_{\mathbf{n}\beta} \mathring{G}_{\alpha\beta}^{\mathbf{mn}} K_\beta^{\mathbf{n}} \tag{4}$$

For large distances from the defect the lattice Green's function converges towards its continuum counterpart $\mathring{G}_{\alpha\beta}(\mathbf{R})$ which is a function of the elastic constants. For a description of the long range displacement field, the force pattern may be expanded into multi-poles. The zero moment (the sum over all Kanzaki forces) vanishes due to translational invariance and the first moment, the dipole contribution, is the leading term

$$s_\alpha(\mathbf{X^m}) \simeq \sum\nolimits_{\mathbf{n}\beta} \mathring{G}_{\alpha\beta}(\mathbf{X^m} - \mathbf{X^n}) K_\beta^{\mathbf{n}} \simeq \sum_{\beta\gamma} \frac{\partial}{\partial X_\gamma^{\mathbf{m}}} \mathring{G}_{\alpha\beta}(\mathbf{X^m}) P_{\alpha\gamma} \tag{5}$$

with the (force) dipole tensor

$$P_{\alpha\beta} = \sum_{\mathbf{n}} X_\beta^{\mathbf{n}} K_\alpha^{\mathbf{n}} \tag{6}$$

The symmetry of the dipole tensor $P_{\alpha\beta}$ is determined by the point symmetry of the defect. In a bcc lattice both octahedral and tetrahedral sites have tetragonal symmetry. Hence

$$P = \begin{pmatrix} A & 0 & 0 \\ 0 & A & 0 \\ 0 & 0 & B \end{pmatrix} \tag{7}$$

The octahedral site in an fcc lattice has cubic symmetry i.e. $A = B$.

Since the Green's function falls off as $1/R$, the displacements decay slowly as $1/R^2$. They are, in general, anisotropic due to the anisotropy of the host lattice, through $\mathring{G}_{\alpha\beta}(\mathbf{X})$ and additionally for many types of interstitial atoms due to the anisotropy of $P_{\alpha\beta}$. In the latter case the difference of the long range displacement fields of differently oriented defects will also be $\propto 1/R^2$. The Kanzaki forces are localized at the defect, their range given by the defect force and the lattice anharmonicity. Dipole tensors $P_{\alpha\beta}$ have been measured for many hydrogen interstitials and are used as a main parameter in fitting interactions.

So far we have tacitly assumed an infinite crystal. In a finite crystal there is an additional lattice expansion, ΔV^{im}, by the so called image forces, i.e. due to the missing restoring forces at the surface. Independent of the crystalline shape the defect relaxation volume is given by [2].

$$\Delta V = \sum_\alpha P_{\alpha\alpha}/(3\kappa) = \Delta V^\infty + \Delta V^{im} \tag{8}$$

with κ the bulk modulus and ΔV^∞ the defect relaxation volume in an infinite crystal. The various relaxation volumes are related by the Eshelby factor γ ($\Delta V^\infty = \gamma \Delta V$) which is e.g. for Cu $\gamma = 0.69$. The term ΔV^∞ gives the volume expansion localized around the hydrogen atom, whereas the image contribution corresponds to a more uniform lattice expansion.

The introduction of hydrogen changes the elastic constants of the host. First the (diaelastic) change of the elastic constants, ΔC, due to a random distribution of noninteracting hydrogen atoms is related to the changed harmonic couplings in the lattice. It is obtained from the energy change under an applied external strain $v_{\alpha\beta}$. Inserting Eq. (4) into Eq. (1) and expanding in $v_{\alpha\beta}$ we obtain

$$V_c \Delta C_{\alpha\beta\gamma\delta} = -c \frac{\partial P_{\alpha\beta}}{\partial v_{\gamma\delta}} - c \Delta V^{im} \kappa \frac{\partial C_{\alpha\beta\gamma\delta}}{\partial p} \tag{9}$$

with V_c the cell volume, c the defect concentration and p pressure. Here the first term represents the defects polarizability and the second one the contribution of the image volume expansion. Interstitial defects are often highly polarizable under strain. This is related to resonant defect vibrations.

In analogy with electrostatics, there can be, in addition to this diaelastic polarizability, also a paraelastic one due to the reorientation of anisotropic defects in a external strain field [6].

$$V_c \Delta C^{para}_{\alpha\beta\gamma\delta} = -\frac{c}{kT}\{\langle P_{\alpha\beta} P_{\gamma\delta}\rangle - \langle P_{\alpha\beta}\rangle\langle P_{\gamma\delta}\rangle\} \tag{10}$$

where $\langle\rangle$ indicates averaging over initially equivalent orientations. This effect is, contrary to the diaelastic one, strongly temperature dependent. It can only be observed when the defect can reorient either by thermally activated hopping or by tunneling. The resulting temperature and frequency dependence is utilized in ultrasonic attenuation experiments.

Finite concentrations

With increasing concentration of hydrogen atoms the interaction between them gains importance and will eventually lead to ordered phases. The different hydrogen atoms will interact directly ("chemical interaction") as well as indirectly via their displacement fields ("elastic interaction"). Both interactions have to be included to derive phase diagrams.

The elastic interaction is the work done by the forces exerted by one atom (A) against the displacements of the other atom (B). In lowest order it can be expressed with the help of Eq. (3) as

$$E^{(AB)} = \frac{1}{2}\sum_{\mathbf{m}\alpha}\left[K_\alpha^{\mathbf{m}(A)} s_\alpha^{\mathbf{m}(B)} + K_\alpha^{\mathbf{m}(B)} s_\alpha^{\mathbf{m}(A)}\right] \tag{11}$$

and using Green's function, Eq. (5), this becomes in symmetrized form,

$$E^{(AB)} = \sum_{\mathbf{mn}\alpha\beta} K_\alpha^{\mathbf{m}(A)} \mathring{G}_{\alpha\beta}^{\mathbf{mn}} K_\beta^{\mathbf{n}(B)}. \tag{12}$$

For large distances this can be expressed using Eq. (5) by the elastic (continuum) Green's function and the dipole tensors of the two atoms. The interaction energy falls for large distances off as $\propto 1/R^3$. In general it will vary in sign and magnitude for the different directions.

2.2. EXPERIMENT

Information on the structure of interstitial hydrogen in metals can be obtained from a large number of experiments. Here we will discuss briefly only a few important techniques and their results. Fig. 2 shows the octahedral and tetrahedral sites for fcc, bcc and hcp lattices.

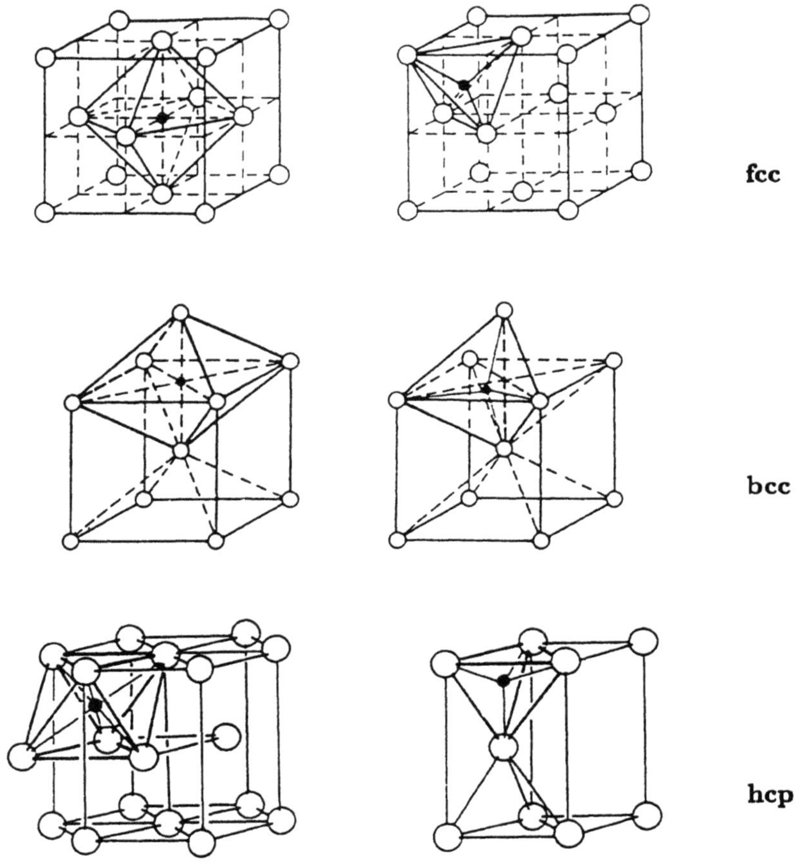

Fig. 2. Octahedral (left) and tetrahedral (right) sites in fcc, bcc and hcp structures [7] (∘: host, •: hydrogen)

Volume of hydrogen in metals

A large amount of data on the volume expansion, ΔV, Eq. (8), due to the uptake of hydrogen has been accumulated [1]. Typically the values lie in the range 2.8 ± 0.2 Å^3. Exceptions are hydrogen atoms in lanthanoid metals which cause a much larger lattice expansion and hydrogen atoms on the octahedral sites in bcc and hcp lattices where one finds $\Delta V = 2.1 \pm 0.2$Å^3. Deviations from these typical values at very high concentration might be due to a partial occupancy of substitutional sites.

The isotope dependence of ΔV is usually inside the error bars.

X-ray diffraction

X-rays are scattered by the electrons. The hydrogen atoms themselves are, therefore, hardly detectable in such experiments. In X-ray scattering experiments the change of the host lattice due to the presence of the hydrogen is studied.

The intensity of the X-rays of wave-vector $\mathbf{Q}$ scattered by the hydrogen atoms (H) and by the displaced lattice atoms ($\mathbf{m}$) is

$$I_X(\mathbf{Q}) \propto \left| \sum_H f_H(\mathbf{Q}) \exp i\mathbf{Q}\mathbf{R}^H + \sum_{\mathbf{m}} f_{\mathbf{m}}(\mathbf{Q}) \exp i\mathbf{Q} \cdot (\mathbf{X}^{\mathbf{m}} + \mathbf{s}^{\mathbf{m}}) \right|^2 \tag{13}$$

where $f_H(\mathbf{Q})$ and $f_{\mathbf{m}}(\mathbf{Q})$ are the scattering amplitudes of the hydrogen and the metal atoms, respectively. Since $f_H(\mathbf{Q}) \ll f_{\mathbf{m}}(\mathbf{Q})$ the effect arises mainly from the metal atom displacement. The uniform part of the strain, i.e. the change in average lattice constant causes a shift of the Bragg peaks. The nonuniform displacements effect diffuse scattering close to the Bragg peaks (Huang scattering) and also between the peaks.

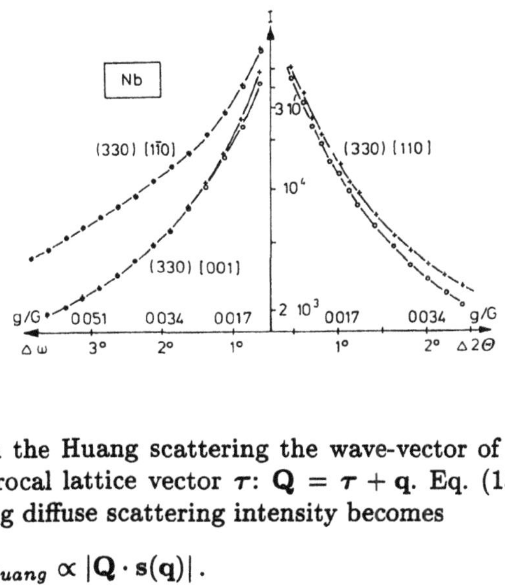

Fig. 3: Diffuse scattering intensities of H in Nb near the (330) reflection in [110], [1$\bar{1}$0] and [100] directions [8]. (o pure Nb, + $NbH_{0.023}$)

In the Huang scattering the wave-vector of the scattered X-rays is close to a reciprocal lattice vector $\boldsymbol{\tau}$: $\mathbf{Q} = \boldsymbol{\tau} + \mathbf{q}$. Eq. (13) can then be expanded and the Huang diffuse scattering intensity becomes

$$I_{Huang} \propto |\mathbf{Q} \cdot \mathbf{s}(\mathbf{q})| \,. \tag{14}$$

The Huang scattering originates thus from the small $\mathbf{q}$ components of the Fourier transform of the displacements $\mathbf{s}^{\mathbf{m}}$ which correspond to the long-range displacement field, Eq. (5). Thus the Huang scattering is well suited to determine the dipole tensor Eq. (6). An example for such an experiment is shown in Fig. 3 for H in Nb. The introduction of H changes the intensity in [110] direction but not in [001] and [1$\bar{1}$0] directions. Analyzing Eq. (14) one derives from the missing contribution in [001] direction that P is diagonal and from the [1$\bar{1}$0] direction $(A - B)^2 = 0$ in Eq. (7). This means the long range distortion field of H in Nb is cubic, despite the strong deviation of the tetrahedral site in a bcc lattice from an ideal tetrahedron. The dipole tensor, Eq. (6), cannot be described by central forces from the H to its nearest Nb neighbours only. A common description is by central forces up to next nearest neighbours which "accidentally" produce a cubic dipole tensor. Another possibility would be non-central forces which could result from changes in the interaction between the Nb atoms surrounding the H. This isotropy of the dipole tensor has attracted much attention and caused speculations on hydrogen delocalization. Recent ab initio calculations support a static occupation of the tetrahedral site and give only a small deviation from a cubic P [10].

The remaining component of P is determined from the absolute scattering intensity in the [110] direction and $P_{\alpha\beta} = \delta_{\alpha\beta}(3.37 \pm 0.1)\text{eV}$. This is in excellent agreement with the volume expansion measurements.

Neutron diffraction

Neutrons are scattered by the nuclei, hence unlike X-rays neutrons are scattered strongly by the hydrogen. Many structures have been determined by this technique. The sensitivity of the neutrons to the hydrogen allow even to measure the density distribution of the H in the lattice. The observed intensities reflect the wave-functions of the H. This method has been used e.g. to study changes in site occupations with temperature [9]. More details on neutron scattering are given in a separate chapter of this volume.

Channeling

When a beam of ions with typical energies of MeV is incident on a single crystal most ions are scattered by the near surface layer and eventually stopped or backscattered. If the incidence direction coincides with a main symmetry direction of the lattice the penetration depth is strongly increased, the ions see open channels between the rows of atoms. The fast ions moving in the channels are focused in the channels by the potentials of the atoms forming the channels. The effective potential seen by the moving ions should have a minimum at the center of the channel and the incident beam will get concentrated there, see Fig.4.

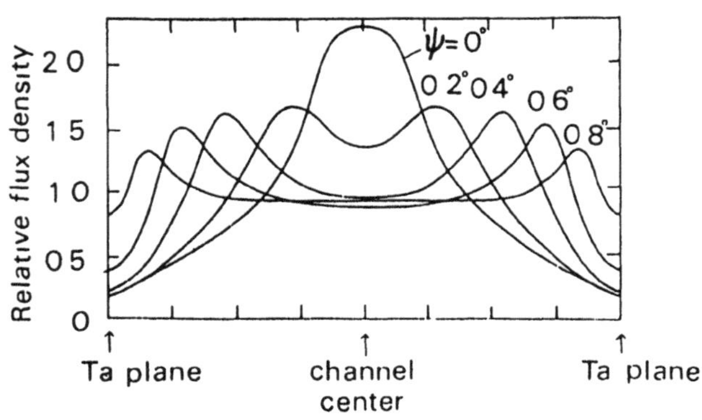

Fig. 4: Relative flux density of ^3He (750 keV) in a (100) planar channel in Ta [11].

This results in a reduced backscattering. Interstitial atoms can now sit in the channels and block them, thus increasing again the backscattering yield. If one choses suitable nuclear reactions of the incident ions with H one can utilize the larger penetration in the channel to determine the H sites. For the site location of protons one uses e.g. the H(^{3}He,p)α reaction. The yield increases then in the channels blocked by H. For the tetrahedral and octahedral sites in the simple lattices the reaction yield profiles can qualitatively be understood from the projections of the sites along the channels. Fig. 5 shows the experimental data for D in Pd. The

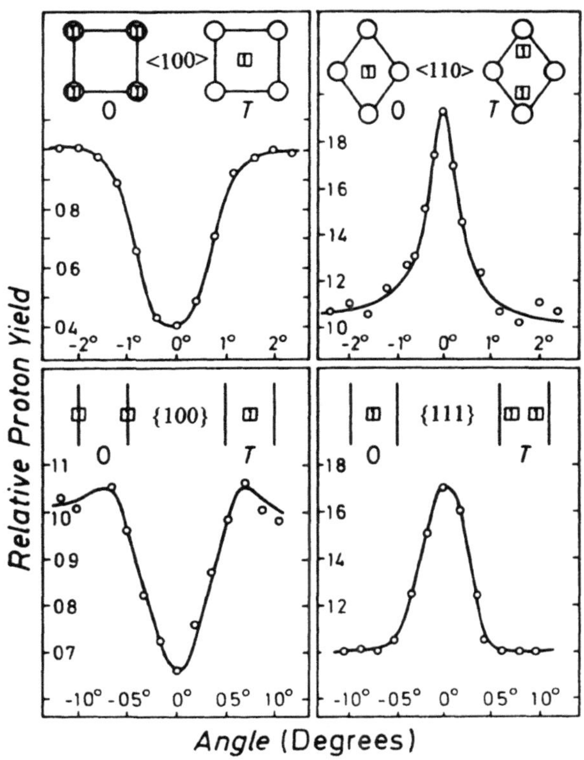

Fig. 5: D in Pd yield profiles of the D(^{3}He,p) ^{4}He reaction from angular scans across the axes and planes indicated in the figure (750 keV $^3He^+$, $PdD_{0.007}$, 295K; D: □, Pd: ○)[12].

results clearly indicate octahedral site occupancy. The yield is maximal in those channels where the D is seen and minimal when it is blocked by lattice atoms. The accuracy of this method allows also to measure the mean square displacements of the D.

Mechanical relaxation

Mechanical relaxation experiments are used mainly to determine mobilities. Here one utilizes the coupling of the long-range displacement fields of the H to external strain fields. If the H has an anisotropic dipole tensor, Eq. (6), the degeneracy of the energies of the different orientations can be lifted by applying an external strain. If the temperature is sufficiently high to allow reorientations, this leads to an anelastic relaxation called *Snoek effect.* This effect is connected with the paraelastic elastic constant, Eq. (10). The absence of the Snoek effect is evidence of a cubic dipole tensor. This can be due to site symmetry as in PdH or to other reasons as in NbH. If a dilatation gradient is applied to a sample the hydrogen will diffuse to the dilated end. The resulting concentration gradient causes an additional anelastic strain the *Gorski effect.* The effect is proportional to $(\mathrm{tr}P)^2$. However this quantity is easier gained by lattice expansion measurements.

Elastic constants

The introduction of hydrogen leads to marked changes of the elastic constants Eq. (9). E.g. for Nb the c_{44} shear modulus hardens whereas the c' shear modulus softens due to hydrogen uptake. The compressibility increases [13].

3. Vibrational dynamics

3.1. SINGLE DEFECT

The tool to calculate the vibrations of an isolated hydrogen is again Green's function theory. If one wants to avoid it and has sufficient computer power, brute force diagonalization of a sufficiently large lattice can be a substitute. To treat an interstitial by Green's function technique one defines, for the lattice plus defect, a total Green's function G which can be calculated from the ideal Green's function $\mathring{G}$ and the local coupling changes $\Delta\Phi$. G is defined in partitioned matrix notation [3] as

$$\left[\begin{pmatrix} M^H & 0 \\ 0 & M^L \end{pmatrix}\omega^2 - \begin{pmatrix} \Delta\Phi^{HH} & \Delta\Phi^{HL} \\ \Delta\Phi^{LH} & \Phi^{LL}+\Delta\Phi^{LL} \end{pmatrix}\right]\begin{pmatrix} G^{HH} & G^{HL} \\ G^{LH} & G^{LL} \end{pmatrix} = 1 \tag{15}$$

where the superscripts H and L stand for the hydrogen and host lattice subspaces, respectively. M is the ionic mass and Φ the ideal lattice coupling. G^{HH}, G^{HL} denote the response of the H to a force acting on the H itself or the host lattice, respectively. They can be calculated by simple substitution using $\Phi^{LL}\mathring{G}^{LL} = 1$.

$$G_{ii}^{HH} = \left[M^H\omega^2 - \Delta\Phi^{LL} - \Delta\Phi^{LH}\left(1-\mathring{G}^{LL}(\omega)\Delta\Phi^{LL}\right)^{-1}\mathring{G}^{LL}(\omega)\Delta\Phi^{LH}\right]^{-1} \tag{16}$$

The local spectrum, $Z_i(\omega)$, of the hydrogen vibrations in direction i is given in terms of this Green's function as:

$$Z_i^H(\nu) = 2\pi Z_i^H(\omega) = 4M^H\omega\ ImG_{ii}^{HH}(\omega). \tag{17}$$

In typical materials the metal mass is about two orders of magnitude larger than the hydrogen mass. The hydrogen is a light impurity. Additionally experiment has shown the hydrogen – lattice coupling to be strong. The vibrational spectrum Z^H, therefore, consist of two parts: localized vibrations, far above the maximal host frequency, and vibrations with lattice frequencies (band modes). The localized vibrations can be qualitatively described as vibrations of the H in the fixed cage of the surrounding metal atoms while the band modes are deformed lattice modes with participation of the H.

Localized modes

The localized vibrations of hydrogen have been measured for a number of materials [14; 1]. Most measurements are by incoherent inelastic neutron scattering utilizing the large incoherent scattering cross section of H.

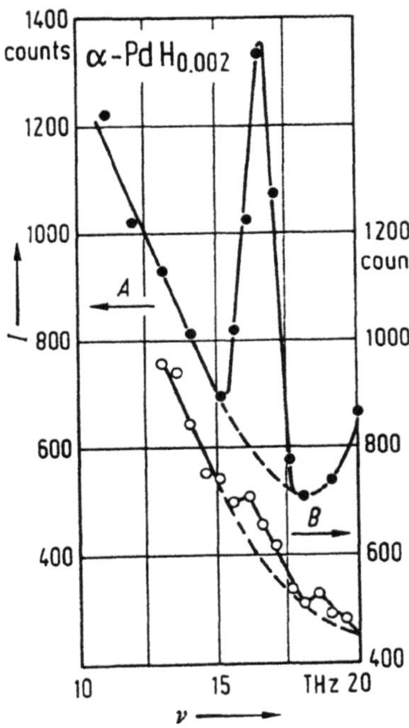

Fig. 6: Inelastic neutron scattering intensity of the localized hydrogen vibration in $PdH_{0.002}$, after [15]. Curves A and B represent different experimental setups.

As an example Fig. 6 shows the localized vibration of H in Pd. Hydrogen occupies octahedral sites in the fcc lattice of Pd. Due to the cubic symmetry of this site there is only one, triple degenerate localized mode. Its frequency of $\nu = 16\,\mathrm{THz}$ is well above the maximal frequency of the Pd lattice , $\nu_{max} = 7\,\mathrm{THz}$.

As a typical representative of H in a bcc metal Fig.7 shows the localized vibration in NbH. There are two localized modes The ratio of their intensities is about 1:2. The higher mode is degenerate. This is compatible with the occupancy of a tetrahedral site in bcc. If one assumes a purely longitudinal coupling between the H and its nearest neighbour Nb atoms and neglects the lattice distortion one obtains a ratio of $\sqrt{2}$ between the two frequencies. The ratio of the experimental values is only slightly larger, 1.6. For the case of octahedral site occupancy in bcc lattices the lower of the two localized modes would be degenerate. The frequencies of the two modes, $\nu^1_{loc} = 27\,\mathrm{THz}$ and $\nu^{2,3}_{loc} = 41\,\mathrm{THz}$, are much higher than in PdH. Surprising is the large observed width of the modes. Harmonic theory predicts a zero width for localized modes in clear contradiction to the experiment. The origin of the large

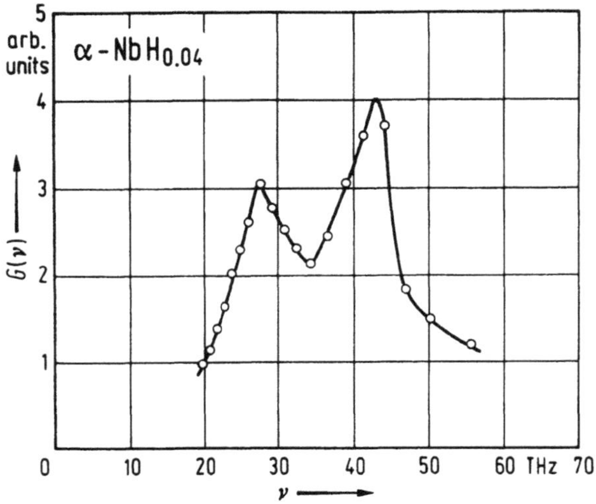

Fig. 7: Vibrational density of stat α-NbH$_{0.04}$, after [16].

broadening is not quite understood. It could be related to a strong variation of the frequency with lattice distortions.

Qualitatively the localized vibrations in hcp transition metals are similar to the ones in the bcc metals. In hcp metals at low concentration H occupies tetrahedral sites. In LuD$_{0.19}$, for the localized vibration in the basal plane, a frequency of $\nu_{loc}^{a} =$ 24 THz and for the one along the hexagonal axis $\nu_{loc}^{c} = 17$ THz have been reported [17]. The splitting between the two modes is larger than anticipated from the c/a ratio of the lattice and the undistorted geometry. Including lattice relaxations the splitting is reproduced. There is a tendency of the hydrogen to form pairs which causes a further splitting of the lower of the two localized modes.

Band modes

Only about the fraction M^H/M^L of the total spectrum is contained in the band modes. In lowest approximation one would assume the H to have the same vibrational amplitudes as the host. The geometry of the site and the coupling can, however, enhance the H vibrations. Such an effect has been observed in NbH [18].

Fig. 8 shows calculated spectra of the hydrogen vibrations in transition metals. For comparison the host spectrum is also shown. Since the amplitudes are inversely proportional to the atomic masses, see "Introduction to lattice dynamics", the spectra were normalized by the respective masses. Shown are two models. Model II only reproduces the localized hydrogen modes and is a lower limit estimate for the band modes. Model I was additionally fitted to the experimental values of the dipole tensor and the elastic constant change.

Despite their small contribution to the spectrum, the band modes cannot be neglected for the total vibrational amplitude. Due to their lower frequencies they determine the mean square displacement, up to room temperature. The mean square displacement, i.e. the average thermal amplitude, consists of contributions from the

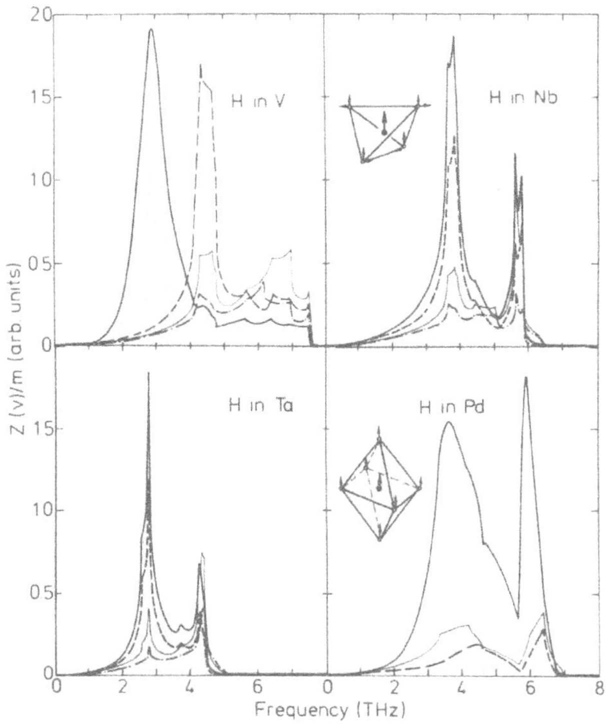

Fig. 8: Band spectra of hydrogen in transition metals. —— $Z_z^H(\nu)$ model I, - - - $Z_z^H(\nu)$ model II, · — ·— $Z_x^H(\nu)$ model II, —— host spectrum. (Inserts: schematic representation of z-vibration in bcc and fcc lattices) [19].

localized and band modes: $\langle u^2\rangle^H = \langle u^2\rangle^H_{loc} + \langle u^2\rangle^H_{band}$.

Fig. 9 shows the mean square displacements corresponding to the spectra of the previous figure. For the bcc metals the values are averaged over the three possible orientations of the tetrahedral sites. The calculated values agree well with experiment. In the bcc materials the contribution of the localized modes is nearly temperature independent ($T < 700$K) whereas the band contribution already behaves classically at fairly low temperatures: $\langle u^2\rangle^H_{band} \sim kT\langle\omega^{-2}\rangle^H_{band}$. This indicates the importance of these band modes for the temperature dependence of the migration process.

In Pd, due to the lower localized mode frequency, also the contribution $\langle u^2\rangle^H_{loc}$ is strongly temperature dependent and can be treated classically above $T \approx 200$K.

The mean square displacement can be used to estimate an activation energy for a lattice activated diffusion. The values deviate approximately by 10% from the experimental ones. In the bcc materials where a change in apparent activation energy is observed one obtains, in this way, the low temperature value and not, as one might assume, the high temperature one.

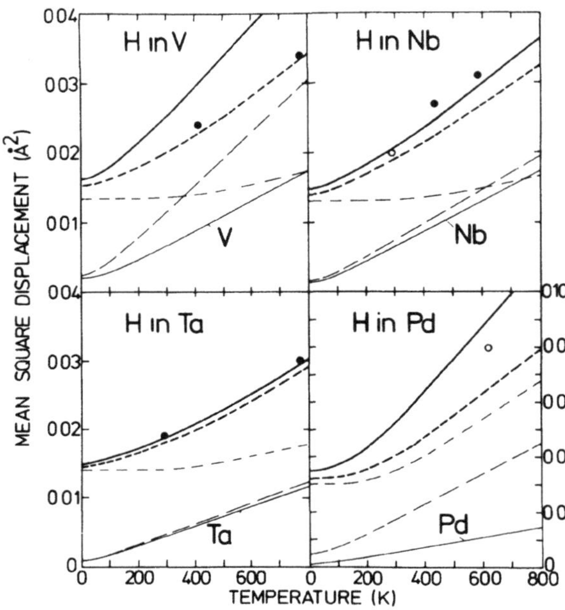

Fig. 9: Mean direction averaged square displacements [19]. —— $\langle \bar{u}_\alpha^2 \rangle^H$ model I, - - - $\langle \bar{u}_\alpha^2 \rangle^H$ model II, – – $\langle \bar{u}_\alpha^2 \rangle^H_{band}$ model I, - - - $\langle \bar{u}_\alpha^2 \rangle^H_{loc}$ model I, —— $\langle \bar{u}_\alpha^2 \rangle^{host}$; exp. values: • [22; 23], ∘ [24].

3.2. FINITE CONCENTRATION

The effects of finite concentrations of defects on the phonon distribution can again be calculated by Green's function techniques. We will restrict ourselves here to a few experimental examples. With rising H concentration the localized vibrations at the single sites merge into optic modes. As examples Figs. 10 and 11 show the dispersion curves of PdD and NbD, respectively. Such measurements are done by coherent inelastic neutron scattering. For these measurements D is more favourable than H. The optic mode frequencies should be multiplied by the root of the mass ratio, $\sqrt{M_D/M_H} = \sqrt{2}$ to obtain the corresponding hydrogen frequencies.

The optic modes of NbD are very flat and show nearly no dispersion. This is a direct consequence of the strong localization connected with the high frequency and indicates that the direct interaction between the H is weak. The optic modes of PdD are only little above the maximal Pd frequency and show a pronounced dispersion. This reflects the weaker localization of the lower frequency localized modes and a higher interaction between the D atoms.

The "resonant-like" band modes will eventually merge to flat phonon modes. However, interactions for frequencies in the band of the host lattice frequencies are much stronger than for localized modes. This is a direct consequence of the

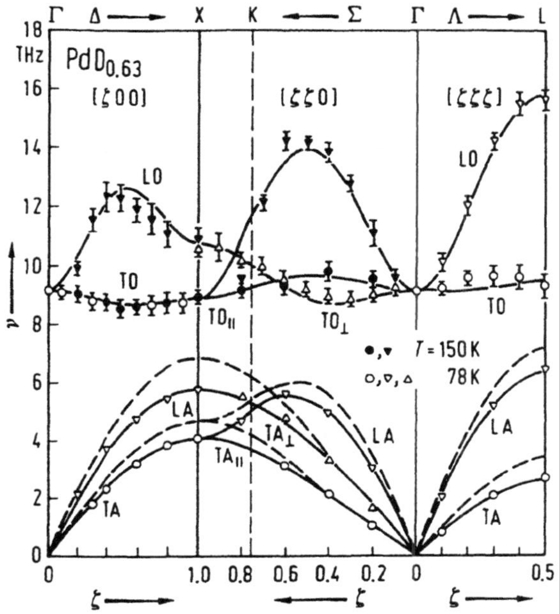

Fig. 10: Phonon dispersion curves of $PdD_{0.63}$ at 150 and 78K. The symbols are the experimental data at the two temperatures. The values agree within error bars. The full line represents a Born - v.Karman fit. The dashed line shows the dispersion of pure Pd at 120K [20].

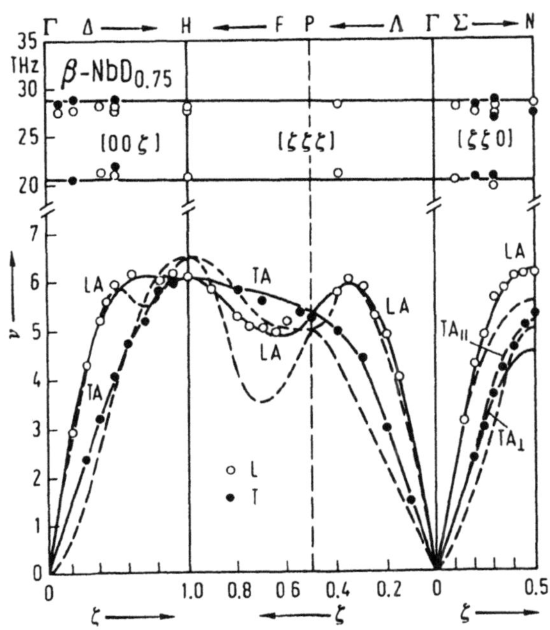

Fig. 11: Phonon dispersion curves of $NbD_{0.75}$ in the ordered β-phase at room temperature. The symbols are the experimental data. The full line represents a Born - v.Karman fit. The dashed line shows the dispersion of pure Nb [21].

large number of extended phonons available to transmit interactions with those frequencies. Resonant modes are not properly localized and the hydrogen band modes are not even proper resonant modes. They only show some "resonant-like"

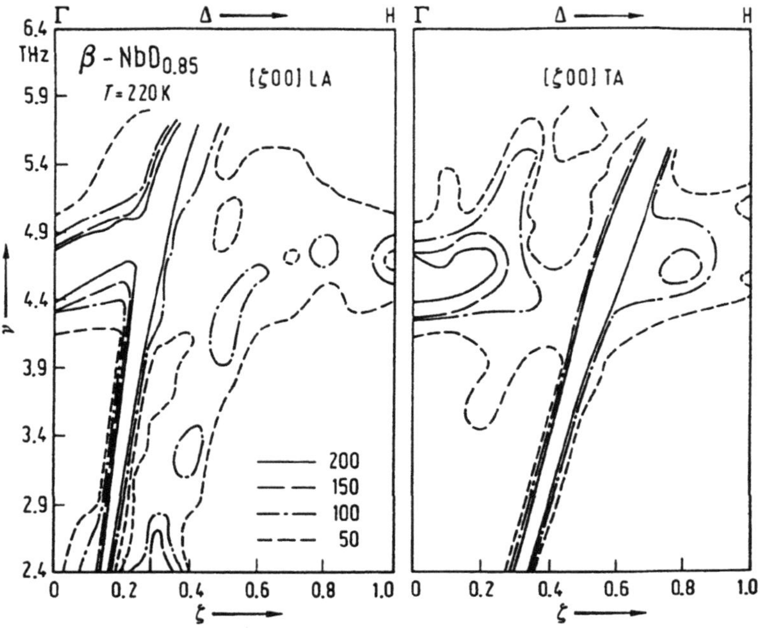

Fig. 12: Equal intensity contours for longitudinal and transverse branches in the [ζ00] direction in β-$NbD_{0.85}$ at 220K [25]

enhancement of some amplitudes. The disorder in the site occupations will cause broad intensity distributions in neutron scattering experiments which could, in principle, be used to get information on local ordering, see Fig. 12.

4. Quantum effects

So far we have considered the hydrogen basically as a mass point at a given lattice site which exerts forces on its neighbours. This is the usual description of interstitials appropriate for heavy atoms at $T \neq 0$. Due to its small mass, hydrogen is at the limit of validity of such a classical description and quantum effects are important in many cases. Experimentally they manifest themselves e.g. in quantum diffusion and tunneling which will be discussed in a separate chapter.

The relative importance of quantum effects can be assessed by comparing typical "classical" energies such as the lattice relaxation energies with the quantum mechanical zero point vibrational energy. For typical "large" interstitial atoms, such as a self interstitial in Cu the relaxation energy (potential energy) is of order of eV. The zero point vibration energy, corresponding to an Einstein vibration of $\approx$ 4THz corresponding to $\approx$ 16meV is $E_{zero} \approx 3 \times 8$meV, much less than the relaxation energy. Further, the configurations are only influenced by energy changes (Eq. 3) which reduces a possible influence of zero point motions even more. A purely classical treatment is, therefore, adequate

The situation is more complicated for interstitial hydrogen. Hydrogen is a "small" interstitial and the potential energy relaxation is only of the order of a few tenths of an eV. This number has to be compared to the zero point vibrational energy. In the case of PdH we have $E_{zero} \approx 0.1$eV, still less than the potential energy relaxation. In NbH we find from the localized vibrations $E_{zero} \approx 0.2$eV which is already very close to the potential energy relaxation. Quantum effects have to be accounted for.

As long as the hydrogen can be considered to be localized at some site the zero point energy can be included as an extra term in the defect potential energy term U^{def} of Eq. (2). The equilibrium condition will then be shifted out of the minimum of the potential energy to reduce E_{zero}. In NbH the relaxation energy is then estimated as about 2/3 potential energy and 1/3 zero point energy. The lattice relaxation reduces the localized mode frequencies by about 50% from their values in the unrelaxed lattice [26]. In PdH the effects are small.

The relaxation energy is the gain in energy due to the lattice relaxation. In the context of hydrogen it is referred to as "self trapping energy". One envisages the hydrogen placed in the unrelaxed lattice, where all sites are equivalent. By causing a lattice relaxation, the hydrogen lowers the energy at its site which is then no longer equivalent to the other sites. Translational invariance is destroyed. If one considers the sites plus the deformation fields one regains translational invariance "polaron effect".

If one now excites a local vibration, an energy $h\nu_{loc}$ is added. In NbH this is about 100meV which is of the same order as the activation energy for a classical hopping over a barrier to a nearest neighbour site. A well localized excited state is no longer stable.

Full quantum mechanical calculations of the hydrogen wave-function is rather complicated. The general consensus is, however, that at least in the groundstate hydrogen is localized at a given site. For a more extensive discussion see e.g.[1]

At higher temperatures the fast motion of the H between different sites can also destroy the picture of a fully relaxed lattice. The hydrogen at one site still "remembers" its previous position. This can lead to a (partially) coherent motion between subsets of sites. This can also happen in a classical picture when subsequent jumps are highly correlated with each other. Such a motion was suggested as a reason for the observed isotropy of the dipole tensor (Eq. 7) in the bcc transition metals [27].

References

[1] Fukai, Y. (1993), *The Metal-Hydrogen System*, Berlin: Springer Verlag.

[2] Leibfried, G., Breuer, N. (1978), *Point Defects in Metals*, Springer Tracts in Modern Physics **81**, Berlin: Springer.

[3] Dederichs, P.H., Zeller, R. (1980), *Point Defects in Metals II*, Springer Tracts in Modern Physics **87**, Berlin: Springer.

[4] Maradudin, A.A., Montroll, E.W., Weiss, G.H., Ipatova, I.P. (1971), *Theory of Lattice Dynamics in the Harmonic Approximation*, Solid State Physics, Supplement 3, New York: Academic Press.

[5] Tewary, V.K. (1973), Adv. Phys. **22**, 758.

[6] Robrock, K.-H (1990), *Mechanical Relaxation*, Springer Tracts in Modern Physics 118, Berlin: Springer Verlag.

[7] Carstanjen, H.D. (1989), Z. Phys. Chem. NF **165**, 141.

[8] Metzger, H., Peisl, H. and Wanagl, J. (1976), J. Phys. F: Metal Phys. **6**, 2195.

[9] Hirabayashi, M. and Asano, H. (1981) in: *Metal Hydrides* (G. Bambakidis ed.), p. 53, New York:PLenum.
[10] Elsässer, C., Fähnle, M., Schimmele, L., Chan, C.T. and Ho, K.M. (1994), Phys. Rev. B to be published.
[11] Carstanjen, H.D. (1980), Phys. Stat. Sol. (a) **59**, 11.
[12] Carstanjen, H.D., Dünstl, J., Löbl, G. and Sizmann, R. (1978), Phys. Stat. Sol. (a) **15**, 529.
[13] Magerl, A., Berre, B. and Alefeld, G. (1976), Phys. Stat. Sol. (a) **36**, 161.
[14] Kress, W. (1983), in: Landolt-Börnstein, Vol. III/13b, (K.-H. Hellwege and J.L. Olsen eds.), Berlin: Springer Verlag.
[15] Drexel, W., Murani, A., Tocchetti, D., Kley, W., Sosnowska, I. and Ross, D.K. (1976), J.Phys. Chem. Sol. **37**, 1135.
[16] Verdan, G., Rubin, R. and Kley, W. (1968) in: *Neutron Inelastic Scattering* Vol. I, p. 223, Vienna:IAEA.
[17] Blaschko, O., Krexner, G. Pintschovius, L., Vajda, P. and Daou, J.N. (1988), Phys. Rev. B **38**, 9612.
[18] Lottner, V., Schober, H.R. and Fitzgerald, W.J. (1979), Phys. Rev. Lett. **42**, 1162.
[19] Schober, H.R. and Lottner, V. (1979), Z. Phys. Chem. NF **114**, 203.
[20] Glinka, C.J., Rowe, J.M., Rush, J.J., Rahman, S.K., Sinha, S.K. and Flotow, H.E. (1978), Phys. Rev. B **17**, 488.
[21] Lottner, V., Kollmar, A., Springer, T., Kress, W. and Bilz, H. (1978) in: *Lattice Dynamics 1977* (M. Balkanski ed.), p. 247, Paris: Flammarion.
[22] Lottner, V. (1979), Z. Phys. B **32**, 157.
[23] Gissler, W., Jay, B., Rubin, R. and Vinhas, L.A. (1973), Phys. Lett. **43a**, 279.
[24] Rowe, J.M., Rush, J.J,, de Graaf, L.A. and Ferguson, C.A. (1972), Phys. Rev. Lett. **29**, 1250.
[25] Shapiro, S.M., Richter, D., Noda, Y. and Birnbaum, H. (1981), Phys. Rev. **23**, 1594.
[26] Schober, H.R. and Stoneham, A.M. (1991), J. Less Comm. Met. **172**, 538.
[27] Dosch, H.,Schmid, F., Wiethoff, P. and Peisl,J. (1992), Phys. Rev. B **46**, 55.

Chapter 11

FUNDAMENTALS OF NUCLEAR MAGNETIC RESONANCE

P C Riedi and J S Lord
J F Allen Research Laboratories
Department of Physics and Astronomy
University of St Andrews
North Haugh
St Andrews
Fife
KY16 9SS

1. Introduction

Nuclear Magnetic Resonance is now such an enormous subject that only a simplified account of some selected topics can be given here. A semi-classical vector model of the nuclear magnetism will be sufficient for our purposes but a complete account of NMR requires the density matrix of quantum mechanics. The standard texts [1,2] provide the detailed derivation of the basic equations and more specialised topics are discussed in [3-11]. The emphasis of this chapter is on the NMR of metals, although the early sections are quite general. The high resolution NMR of liquids, a topic of enormous importance in chemistry and biology, is only mentioned in passing.

The object of NMR is to manipulate the equilibrium nuclear magnetisation of the material, which may be many orders of magnitude smaller than the magnetisation due to the electrons, by radio frequency (RF) fields and then to use the information gained from the response of the system to understand the physics of the material. This information may be of static properties e.g the electron density sensed by the nucleus or of the spectrum of fluctuating magnetic fields in the material. There will not be space in this article to describe much of the underlying solid state physics that can be studied by NMR, this is discussed in other chapters. Instead the emphasis is on the basic features of NMR, section 2, followed by a brief account of experimental techniques in section 3.

The usually small, but enormously important, difference between the applied magnetic field and that measured at the nucleus is covered in section 4 along with a discussion of NMR relaxation in metals. In ordered magnetic materials a static magnetic field is induced at the nucleus, even when no external field is applied to the material, and the difference from conventional NMR in this case is discussed in section 5.

The examples of experimental work used in this chapter have mostly come from the St Andrews laboratory, as a matter of convenience, but many other equally good NMR spectra, taken elsewhere, could, of course, have been used in their place.

F. Grandjean et al. (eds.), Interstitial Intermetallic Alloys, 267–292.

2. Principles of Magnetic Resonance

2.1 MOTION OF AN ISOLATED MAGNETIC MOMENT

In the semi-classical description of NMR, the nucleus is treated as a magnetic moment $\boldsymbol{\mu}_\mathbf{n}$ which precesses about a magnetic field $\mathbf{B_o}$ that is conventionally taken to be in the $\hat{\mathbf{z}}$ direction ie $B_0\hat{\mathbf{z}}$. Quantum mechanics shows that $\boldsymbol{\mu}_\mathbf{n}$ is related to the angular momentum of the nucleus $\mathbf{I}$ by the equation,

$$\boldsymbol{\mu}_\mathbf{n} = \gamma_n \mathbf{I} \tag{2.1}$$

where γ_n, called the gyromagnetic ratio, varies widely for different nuclei, table 1.

The torque on the magnetic moment in a uniform field $\mathbf{B_o}$ is given by the rate of change of angular momentum,

$$\begin{aligned} \frac{d\mathbf{I}}{dt} &= \boldsymbol{\mu}_\mathbf{n} \times \mathbf{B_o} \\ \therefore \frac{d\boldsymbol{\mu}_\mathbf{n}}{dt} &= \boldsymbol{\mu}_\mathbf{n} \times \gamma_n \mathbf{B_o} \end{aligned} \tag{2.2}$$

The vector product shows that the direction of $\boldsymbol{\mu}_\mathbf{n}$ changes at right angles to $\boldsymbol{\mu}_\mathbf{n}$ and $\mathbf{B_o}$ and therefore, considering one end of $\boldsymbol{\mu}_\mathbf{n}$ to be fixed in space, the vector $\boldsymbol{\mu}_\mathbf{n}$ generates a cone with axis along $\mathbf{B_o}$.

The most convenient mathematical treatment of equation (2.2) is to transform from the static laboratory frame (x, y, z) to a rotating reference frame (x', y', z') where z' is coincident with z. The field seen by the moment in the frame rotating with angular frequency ω' is [1-4],

$$\mathbf{B_{rot}} = \left(B_o - \frac{\omega'}{\gamma_n} \right) \hat{\mathbf{z}}'. \tag{2.3}$$

When

$$\omega' = \omega_o = \gamma_n \; B_o, \tag{2.4}$$

the moment sees no effective field and is fixed along one direction in the rotating frame.

Transforming to the laboratory frame the angular frequency of the precession of the moment about $\mathbf{B_o}$ is therefore,

$$\omega_\mathbf{o} = -\gamma_n \mathbf{B_o}, \tag{2.5}$$

and the frequency for resonance is,

$$\nu_o = \left|\frac{\omega_o}{2\pi}\right| = \left|\frac{\gamma_n B_o}{2\pi}\right|. \qquad (2.6)$$

Equation (2.3) contains the essence of NMR, the frequency for resonance is proportional to the field at the nucleus, but we shall see that the effective field at the nucleus is not exactly equal to the external field B_o, although the difference for most materials is less than one per cent, section 4.

2.2 NUCLEAR MAGNETISM AT THERMAL EQUILIBRIUM

Nuclear magnetic moments are much smaller than the magnetic moment of the electron, they are of the order of the nuclear magneton,

$$\mu_N = \frac{e\hbar}{2m_p} \approx 5.05 \times 10^{-27}\,\mathrm{JT}^{-1}, \qquad (2.7)$$

where m_p is the mass of the proton, so the nuclear magnetism in thermal equilibrium in a large external field B_o can be calculated from the high temperature limit of elementary statistical mechanics.

The quantization of angular momentum restricts I in equation (2.1) to integer and half integer values. The energy of the nucleus in the field B_o is given by,

$$\varepsilon = -\gamma_n \hbar B_o m$$

where m goes in integer steps from -I to I.

Table 1. Nuclear spin, abundance and approximate gyromagnetic ratio for selected nuclei.

Nucleus	2I	$\gamma_n/2\pi$ (MHz/T)	Abundance %
^1H	1	43	100
^2D	2	6.5	1.6×10^{-2}
^{17}O	5	5.8	4×10^{-2}
^{55}Mn	5	10.6	100
^{59}Co	7	10.1	100
^{63}Cu, ^{65}Cu	3	11.3, 12.1	69,31
^{89}Y	1	2.1	100
^{147}Sm, ^{149}Sm	7	1.5, 1.2	15,14

Consider a nucleus with spin 1/2 for simplicity, there are only two levels and the nuclear magnetisation depends upon the difference in the probability of occupation of the two levels. This is given by the Boltzmann factor and hence,

$$M = n\frac{\gamma_n \hbar}{2}\left(\frac{1-e^{-\beta\Delta}}{1+e^{-\beta\Delta}}\right)$$

where $\beta=1/k_BT$, $\Delta = \gamma_n \hbar B_o$ and n is the number of nuclei m^{-3}. The high temperature limit, $\beta\Delta \ll 1$, can always be taken in conventional NMR so,

$$M = \frac{n\gamma_n^2\hbar^2}{4k_B}\left(\frac{B_o}{T}\right)\hat{z}. \tag{2.8}$$

In general, for spin I, the result is

$$M = \frac{n\gamma_n^2\hbar^2 I(I+1)}{3kT} B_o \tag{2.9}$$

and the static susceptibility is defined as,

$$\chi_o = \frac{M}{H} = \frac{\mu_o M}{B_o}. \tag{2.10}$$

The nuclear magnetisation is so small, at 300K in a field of 10T the difference in population of the two levels for protons is only ~3ppm, that it can usually only be detected against the background electronic magnetisation by using NMR.

Notice that **M** is along the external field $\mathbf{B_o}$. In the x-y plane the random phases of the precessing nuclei average to zero. Nuclear magnetic resonance may be viewed as a technique for establishing and detecting a coherently precessing nuclear magnetisation in the x-y plane.

2.3 LONGITUDINAL (SPIN-LATTICE) RELAXATION

In section 2.2 an expression for the equilibrium nuclear magnetisation in a field $\mathbf{B_o}$ was derived. It is clear however that if the sample was suddenly placed in the field $\mathbf{B_o}$ the magnetisation could not be established instantly, in fact it is commonly found to come to equilibrium as,

$$M(t) = M(\infty)\left(1 - e^{-t/T_1}\right). \tag{2.11}$$

The longitudinal, or spin lattice, relaxation time, T_1, can vary from μs to hours, depending upon the material and the temperature, and has a strong influence on the design of a NMR experiment. Two other, transverse, relaxation times will be considered in section 2.5.

2.4 LINEAR AND CIRCULAR POLARISATION OF RF FIELDS

The motion of a magnetic moment in a steady field $\mathbf{B_o}$ was seen in section 2.1 to be a precession on the surface of a cone with axis parallel to $\mathbf{B_o}$. The RF field produced by an alternating current in a solenoid is however linearly polarised along the axis of the solenoid. The linearly polarised field can be represented as two counter rotating fields of the same angular frequency but half the amplitude,

$$2B_1\cos(\omega_{RF}t) = B_1\left[e^{i\omega_{RF}t} + e^{-i\omega_{RF}t}\right] \qquad (2.12)$$

In NMR the RF field is set at right angles to the static field i.e. in the x-y plane. On transforming to the rotating reference frame described in section 2.1 we see that, for

$$\omega_o = \omega_{RF} = \omega',$$

one sense of rotation of the RF field is fixed in the x'-y' plane, conventionally along x', and the other rotates at $2\omega_o$. Only one sense of rotation can therefore interact strongly with the nuclear moments and the effective field in the rotating frame,

$$\mathbf{B_{rot}} = \left[\left(B_o - \frac{\omega'}{\gamma_n}\right)\hat{\mathbf{z}}' + B_1\ \hat{\mathbf{x}}'\right] \qquad (2.13)$$

is simply $B_1\ \hat{\mathbf{x}}'$ at exact resonance.

We shall see in the next section that in pulsed NMR the resonance at ω_o may be excited when ω_{RF} is close to, but not necessarily equal to, ω_o. In this case the rotating reference frame is usually considered to rotate at ω_{RF}, leading to a small component of the field $(B_o - \omega_{RF}/\gamma_n)$ along $\hat{\mathbf{z}}'$.

2.5 FREE INDUCTION DECAY

The application of the RF field given by equation (2.12), with $\omega_{RF}=\omega_o$, to a magnetic system in thermal equilibrium in a static field $B_o\hat{\mathbf{z}}$, will cause the moments to precess about the effective field in the rotating frame, which at exact resonance was seen in equation 2.13 to be simply $B_1\hat{\mathbf{x}}'$. After a time t_w the spins will have rotated through an angle,

$$\theta = \gamma_n B_1 t_w. \qquad (2.14)$$

If the pulse is turned off when,

$$\gamma_n B_1 t_w = \frac{\pi}{2}, \qquad (2.15)$$

a "$\pi/2$ pulse", the moments will lie along y' in the rotating reference frame, figure 1. In the laboratory frame there is now a nuclear moment M in the x-y plane precessing at ω_o. The changing magnetic flux produced by the rotating M will induce a voltage proportional to $\omega_o M$ in a pick up coil, figure 2.

In liquids a finite magnetisation may exist for a time of order T_1 in the x-y plane, which may be seconds, but in solids this Free Induction Decay (FID) is typically lost in the receiver noise after 100μs and may be unobservable if the decay is so fast that the receiver has not recovered from the effects of the RF pulse. The simplest form of the transverse relaxation is

$$A(t) = A(0)e^{-t/T_2} \tag{2.16}$$

where T_2, the transverse relaxation time, is a measure of how fast the spins de-phase.

The recovery of the longitudinal magnetisation may be measured using the FID. One possible pulse sequence is, $\pi/2$-t-$\pi/2$-FID. The magnitude of the FID at a fixed time after the second pulse is proportional to,

$$M_z(t) = M_z(\infty)\left(1 - e^{-t/T_1}\right) \tag{2.17}$$

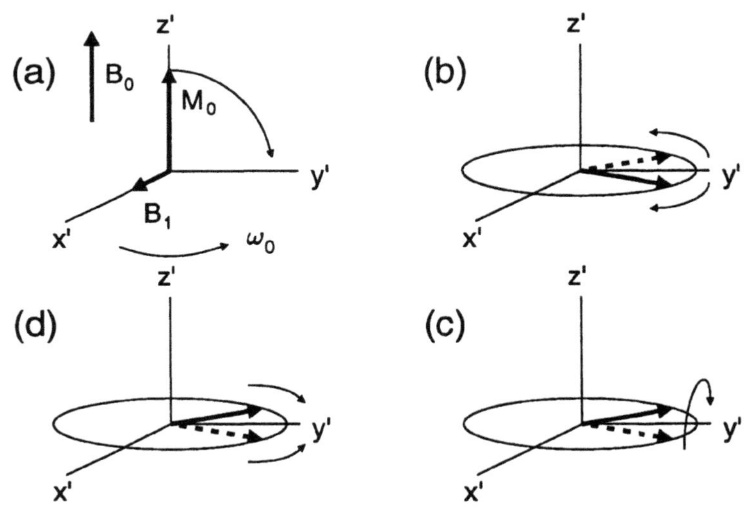

Figure 1(a) A $\pi/2$ pulse along x' at t=0 takes M_Z to y' in the rotating frame. (b) Free Induction Decay. (c) A π pulse after time t. (d) An inhomogeneously broadened line produces a spin echo after a time 2t.

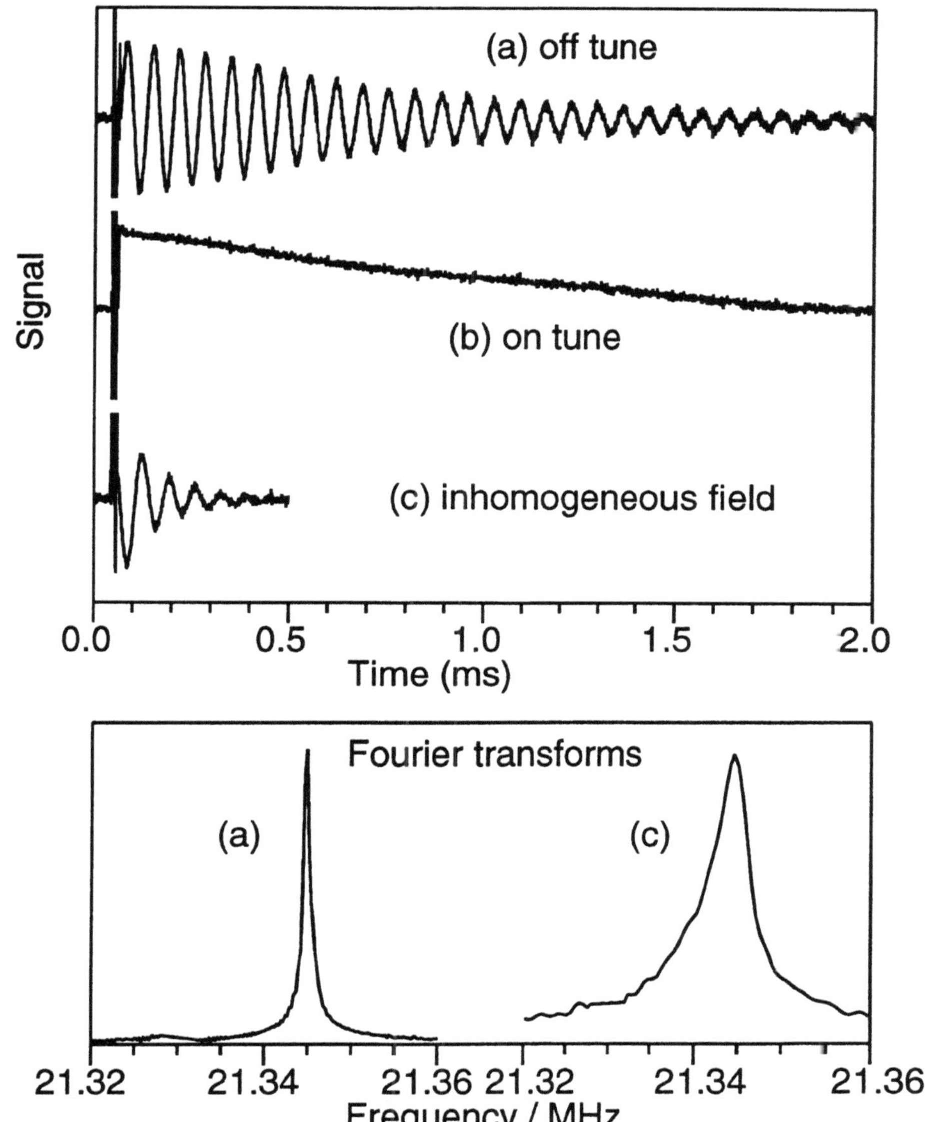

Figure 2. Free Induction Decay (FID) of ^{1}H in glycerine measured using a coherent pulsed spectrometer, section 3.2.

The repetition time of the pulse sequence must be greater than $5T_1$ to allow M_z to return to its full value.

An alternative pulse sequence is, π-t-π/2-FID, after which,

$$M_z(t) = M(\infty)\left(1 - 2e^{-t/T_1}\right) \qquad (2.18)$$

There is an important distinction to be made between the decay of transverse magnetisation due to spin-spin interactions and that due to a distribution of static fields in the sample. The static fields may be due to poor magnet homogeneity or to intrinsic field gradients due to, e.g. at low temperature, a paramagnetic atom. In this case the transverse relaxation is written as T_2^* and, for $T_2^* < T_2 < T_1$, the nuclear moments are still in the x-y plane after the FID has decayed as,

$$A(t) = A(0)e^{-t/T_2^*}, \qquad (2.19)$$

although their vector sum is zero.

2.6 SPIN ECHOES

One of the most important discoveries in NMR was made by Hahn when he noticed that the macroscopic transverse nuclear magnetisation could be recovered for nuclei in a distribution of static fields by applying a second RF pulse. Although an echo can be obtained for any second pulse that is not a multiple of 2π in equation 2.15, the simplest picture is obtained after the sequence (π/2-t-π), figure 1. At time 2t the nuclear moments that have not been subject to a T_2 relaxation process come back into phase and the magnitude, of the spin echo, is given by

$$A(2t) = A(0)e^{-2t/T_2}. \qquad (2.20)$$

The true transverse relaxation time can therefore be recovered by measuring A(2t) as a function of time. The width of the echo is of order T_2^* since it is equivalent to two FID back to back, figure 3.

The large RF pulse needed to produce a π/2 pulse with a sufficiently short t_w, section 2.7, saturates the receiver and the NMR signal cannot be observed for a "dead time" typically of order 10μs. The great experimental advantage of the spin echo measurement over the FID occurs when $T_2^* << T_2$, as in strongly magnetic materials for example. It may then be impossible to observe the FID, because the signal is far below the receiver noise level after the dead time, but the spin echo can still be observed.

There are many more complicated pulse sequences, see [2-5], at the end of which either the FID or spin echo is observed in order to measure T_1 and T_2 under particular circumstances, and the diffusion constant may be measured directly if a field gradient is applied across the sample. This topic will be covered in a later lecture on the NMR of mobile hydrogen in metals.

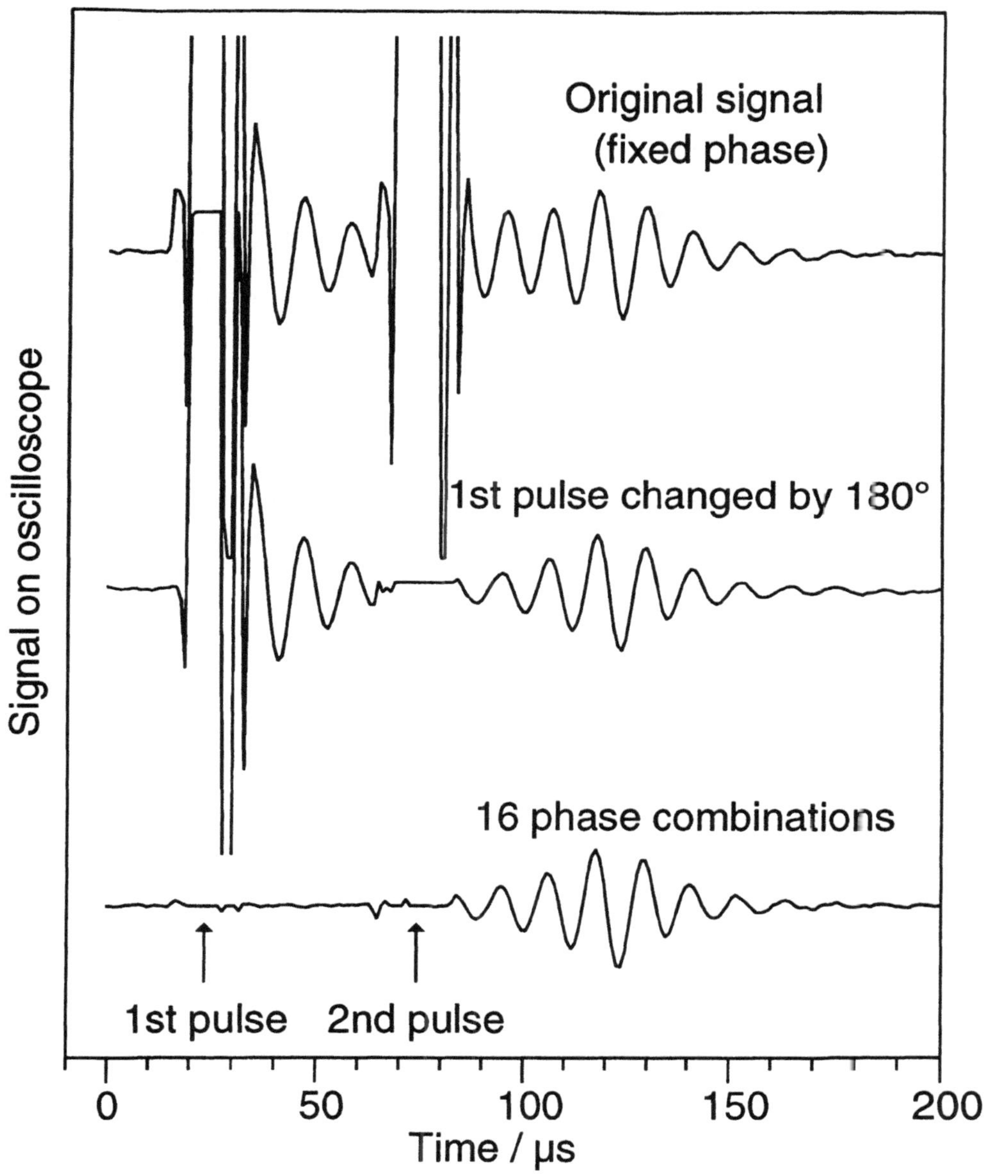

Figure 3. Phase cycling of spin echo experiment. The coherent spectrometer is not exactly at resonance, see section 3.2 and figure 6.

2.7 CONTINUOUS WAVE AND PULSED NMR

In principle, any resonance phenomena can be observed by stimulating it with either a weak continuous wave field or a sufficiently short pulse. The complete characterisation of the resonance requires a knowledge of the phase and magnitude of the response relative to that of the stimulus. This phase information is usually provided by writing the frequency response as a complex susceptibility $(\chi' - i\chi'')$ where χ'' is the in phase and χ' the quadrature response to the stimulus.

In early NMR measurements the sample was held in a weak continuous wave RF field $2B_1 \cos\omega_{RF}t$, with ω_{RF} chosen so that $B_o = \omega_{RF} / \gamma_n$ was at a convenient value, and the resonance displayed by sweeping the field over a range centred on B_o. The effect of χ'' is to change the power absorbed by the tuned circuit in which the sample is placed and χ' changes the tuning (dispersion) of the circuit as indicated in figure 4. (In practice a small low frequency modulation field was added to B_o and, after phase sensitive detection, the derivative of the line shapes shown in figure 4 recorded.)

The shape of χ' and χ'' shown in figure 4 was first calculated by Bloch for the NMR of liquids. They are known as Lorentzian lines and have the form,

$$\chi' = (\omega_o - \omega)T_2\, \chi'' \qquad (2.21)$$

$$\chi'' = \frac{\chi_o \omega_o T_2}{2} \, \frac{1}{1 + (\omega - \omega_o)^2 T_2^2} \qquad (2.22)$$

where χ_o is the static susceptibility, section 2.2.

Note that the full width at half maximum (FWHM) of χ'' is equal to $\Delta\omega = 2 / T_2$ or, equivalently, $\Delta\nu = 1 / \pi T_2$. In solids the lines are usually neither Lorentzian nor Gaussian but do have the general shape shown in figure 4.

A tremendous advance in NMR technique occured when it was realized [1-5] that the FID following a sufficiently short pulse was related to the CW response by a Fourier transform. An exponential decay of the FID for example, $\exp(-t/T_2)$, leads to a Lorentzian line shape while a decay of the form, $\exp(-t^2)$, leads to a Gaussian line. Ideally, the pulse should be a delta function ie effectively of zero width, but in practice the requirement is that $t_w << T_2$ or T_2^* so that relaxation is negligible during the pulse, figure 2.

The great advantage of pulsed NMR in high resolution NMR is that there may be many sharp lines in a frequency range $\omega_o \pm \delta\omega$ for a given B_o and, provided $1/t_w$ is much greater than $\delta\omega$, all the spins can be rotated by a single pulse with $\omega_{RF} \approx \omega_o$. A complex Fourier transform of the FID then provides the complete spectrum. It is also possible to Fourier transform the spin echo.

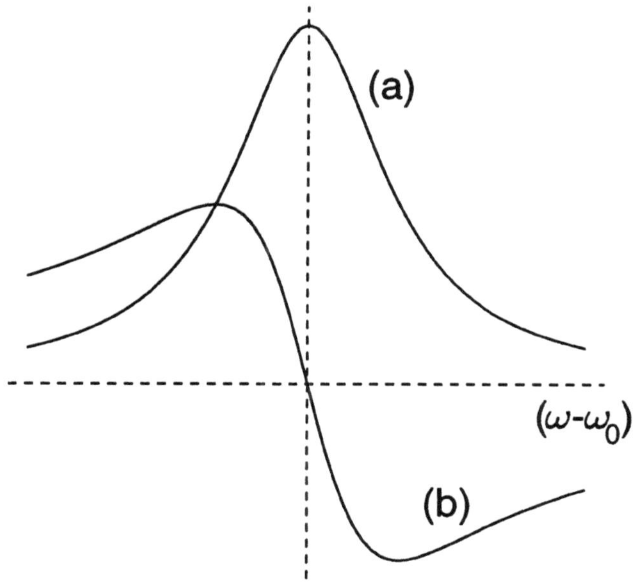

Figure 4(a) The in phase (absorption) and (b) the quadrature (dispersion) components of the Lorentzian line shape, equations 2.21 and 22.

When the NMR lines are very broad, as in single domain ferromagnets, it is not possible to provide enough RF pulse power, many kilowatts, to make $t_W << T_2^*$ for a $\pi/2$ pulse, but a histogram of the distribution of effective fields seen by the nuclei may be obtained by integrating the echo at a series of frequencies across the spectrum. In order to allow for the phase shifts in the circuitry at different frequencies it is necessary to integrate the echo at (unknown) phase ϕ and then exactly at $\phi+\pi/2$ and to plot the phase insensitive quantity S as a function of frequency, where

$$S = \left(S_\phi^2 + S_{\phi+\pi/2}^2\right)^{1/2} \qquad (2.23)$$

2.8 ELECTRIC FIELD GRADIENTS (EFG)

A nucleus with spin 1/2 may be considered to be a spherical charge distribution and will therefore not be influenced by the EFG. Nuclei with spin greater than 1/2 may be treated as spheroidal charge distributions with quadrupole moment Q. The effect of the EFG on a nucleus of spin greater than 1/2 in a site of less than cubic symmetry is to tend to split the spectrum into 2I lines, although in a polycrystalline sample in a magnetic field a continuous distribution will be observed. The height of the peaks in the spectrum is therefore reduced compared to that calculated by assuming all the nuclei have the same frequency for resonance and restricts the use of e.g ^{2}D (I=1) NMR in metal hydrides.

The two limiting cases of the effect of the EFG will be discussed. In the next two sections the EFG will be considered as a perturbation on the NMR of single crystals and poly-crystalline materials and then the spectrum when there is no external magnetic field (NQR) discussed.

2.8.1 *EFG in Single Crystals* The standard notation for the study of the effects of the EFG on NMR spectra is to write the EFG in a coordinate system (X,Y,Z) such that cross products of the form

$$\frac{\partial^2 V}{\partial X \partial Y} = V_{XY} = 0.$$

The EFG along the principal axes are related by Laplace's equation

$$V_{XX} + V_{YY} + V_{ZZ} = 0, \tag{2.24}$$

and the convention is to choose the axes such that,

$$|V_{ZZ}| \geq |V_{YY}| \geq |V_{ZZ}|.$$

In view of equation (2.24) the EFG may be defined in terms of two parameters,

$$eq = V_{ZZ} \tag{2.25}$$

$$\eta = \frac{V_{YY} - V_{XX}}{V_{ZZ}} \tag{2.26}$$

The parameter η is called the axial asymmetry and is restricted to the range $0 < \eta < 1$. Only the case of η=0, axial symmetry, will be discussed in detail because of the algebraic complexity of the general case. The required general equations have been derived and are important, since the value of η may make it possible to identify the particular crystal sites at which the nuclei are located [6].

When the nuclear interaction with the magnetic field is much greater than that with the EFG the nuclear energy levels may be derived by first order perturbation theory,

$$E_m = \gamma_n B_o m + \frac{h\nu_Q}{4}\left(3\cos^2\theta - 1\right)\left[m^2 - \frac{I(I+1)}{3}\right] \tag{2.27}$$

where,

$$\nu_Q = \frac{3e^2 qQ}{2hI(2I-1)} \tag{2.28}$$

and θ is the angle between $B_o\hat{z}$ and the Z axis of the EFG. NB Equation (2.28) is the most frequently used definition of ν_Q but e^2qQ/h is sometimes used.

The NMR spectra for a single crystal is therefore found to consist of 2I lines at frequencies shifted from $\gamma_n B_o$ by

$$\nu_m^{(1)} = -\nu_Q\left(m - \frac{1}{2}\right)\frac{1}{2}\left(3\cos^2\theta - 1\right) \qquad (2.29)$$

There is an important difference between the spectra for half integer and integer spin. Taking I=3/2 for example, there will be an unshifted central line and two lines at $\pm\ \nu_Q\left(3\cos^2\theta - 1\right)/2$. The central line is therefore insensitive to small random variations due to strains or impurities in the crystal, although in second order perturbation theory its frequency does depend upon the ratio $\nu_Q^2|\gamma_n B_o$. The spectra for integer spin, eg I=1` as for ^2D, are far weaker in practice than those of half integer spin because the central line $\gamma_n B_o$ is not present and the lines at $\pm\nu_Q\left(3\cos^2\theta - 1\right)/4$ are both sensitive to defects.

2.8.2 *Powder spectra* The NMR spectrum for a polycrystalline material is obtained by averaging over all angles between the random distribution of EFG and the magnetic field. A continuous spread of frequencies is then obtained with 2I features. In the case of I=1 the strong features are at $\pm\nu_Q/4$ from $\gamma_n B_o$ and for I=3/2 the central line is unshifted in first order perturbation theory and is flanked by features displaced from it by $\pm\nu_Q/2$. In the presence of strong line broadening by the EFG only the central $\pm1/2$ line for half integer spin may be resolved and the other features predicted by the theory may be either below the noise level or else impossible to study in detail.

2.8.3 *Nuclear Quadropole Resonance* The nuclear energy levels are split by the EFG, for nuclei with spin greater than one half, even in the absence of an external magnetic field. The NQR signals then have to be sought by sweeping the spectrometer frequency, using a spectrometer of the type discussed in section 3.2 for example.

The nuclear energy levels are given, for η=0, by

$$E_m = \frac{e^2qQ}{4hI(2I-1)}\left[3m^2 - I(I+1)\right] \qquad (2.30)$$

$$= \frac{\nu_Q}{6}\left[3m^2 - I(I+1)\right] \qquad (2.31)$$

Using the usual selection rule, $|\Delta m| = 1$, it will be seen that a single line spectrum is predicted for both I=1 and I=3/2,

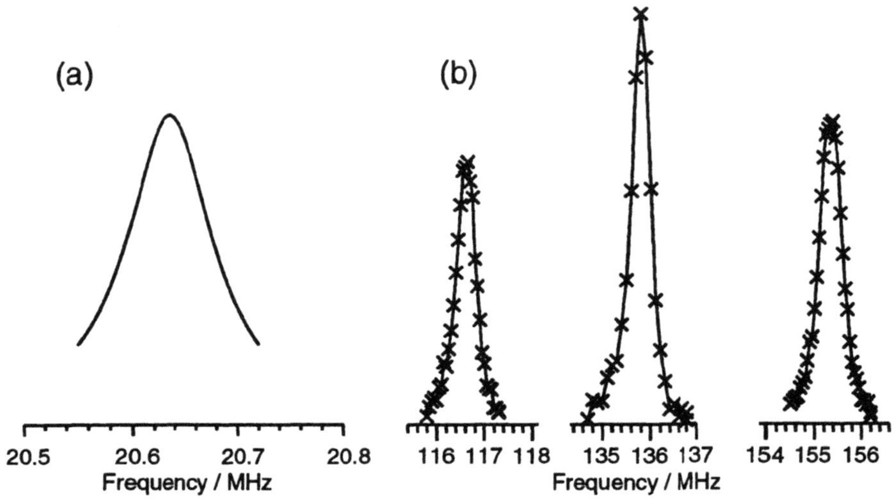

Figure 5. The ^{63}Cu resonance in CuO: (a) NQR at room temperature in the paramagnetic state, (b) NMR at 77K in the antiferromagnetic state.

$$\nu(I=1) = \frac{\nu_Q}{2} \tag{2.32}$$

$$\nu\left(I=\frac{3}{2}\right) = \nu_Q \tag{2.33}$$

where ν_Q is defined in equation 2.28. Notice that there is no need to use single crystals because the axis of quantization is simply the Z axis in each crystallite of a polycrystalline material.

As an example of the above theory we consider the NMR and NQR of CuO, figure 5. At room temperature the material is paramagnetic so only the single NQR line of ^{63}Cu (I=3/2) is observed but in the ordered antiferromagnetic state at 77K an effective magnetic field of some 12T is induced at the Cu nucleus, see section 5, and a three line spectrum is observed.

3. Experimental

3.1 MAGNITUDE OF NMR SIGNALS

Provided that the NMR linewidth is sufficiently narrow for all the nuclear spins to be rotated by a RF pulse of width t_w, the FID signal at a time t from the centre of the pulse will be given by an expression of the form,

$$S(t) = \text{const.}\,\omega_o M f(t) \tag{3.1}$$

where the constant depends upon the circuit characteristics, ω_o is the NMR frequency, M is the equilibrium nuclear magnetization, equation 2.9, and f(t) is the decay function for the FID, e.g exp $(-t/T_2)$ or exp $(-t/T_2^*)$ for a Lorentzian line, section 2.7.

Substituting from equation 2.9,

$$S(t) = \text{const.}\frac{\omega_o n\gamma_n^2 I(I+1)B_o}{T}\,f(t) \tag{3.2}$$

$$= \text{const.}\,\frac{n\gamma_n^3 I(I+1)B_o^2}{T}\,f(t) \tag{3.3}$$

$$= \text{const.}\,\frac{n\gamma_n I(I+1)\omega_o^2}{T}\,f(t) \tag{3.4}$$

Equations (3.3) and (3.4) allow the magnitude of the NMR signal to be scaled from one nucleus to another at constant field and constant frequency respectively, provided the spectrum is not split by an EFG.

The NMR receiver is saturated, "dead", for typically, 3-10μs after the RF pulse so the FID decay may be unobservable, if T_2 or T_2^* is very short. However, when the decay is due to a distribution of static fields, $T_2^* < T_2$, the signal may still be detectable via the spin echo, section 2.6. It will be seen from equation 3.2-4 that the signal is proportional to the reciprocal of the absolute temperature so studies of nuclei with low concentration or small γ_n may only be practical at low temperature.

When the width, $\delta\omega$, of a NMR line is so great that the $\pi/2$ pulse condition, equation 2.14,

$$\frac{\pi}{2} = \gamma_n B_1 t_w \tag{3.5}$$

cannot be satisfied for $\delta\omega t_w << 1$, using the maximum RF power available, only a packet of spins near ω_{RF} contribute to the FID or spin echo. It is still possible to reconstruct the NMR spectrum by integrating the spin echo, equation 2.23, at either a series of values of ω_{RF} for a given B_o or alternatively sweeping B_o across the spectrum at fixed ω_{RF}.

In practice NMR signals range from mV for ^1H in water at room temperature to fractions of μV for nuclei of low isotopic abundance with small γ_n or materials with wide NMR lines. In the latter case extensive signal averaging is required to extract the signal from thermal noise and spurious background signals in the receiver following the strong RF pulse.

3.2 THE NMR SPECTROMETER

Conventional pulsed NMR is usually carried out using a rather complicated commercial machine but the essential features of a spectrometer are shown in figure 6. This spectrometer was designed [12] to work over the frequency range 10-1,000 MHz with particular application to the NMR of ordered magnetic materials, section 5.

A stable frequency is provided by a synthesized oscillator whose output is divided into two parts. One part provides a reference signal for the phase sensitive detection of the FID or spin echo, the other goes to a RF switch. The pulses from the switch go to a power amplifier and then to the sample which is in either a tuned circuit, for maximum sensitivity, or a wide band delay line structure.

The diodes (D1) protect the receiver by shorting the large RF pulses from the transmitter to ground. The diodes (D2) block noise from the transmitter but pass the RF pulses with little attenuation. The FID or echo is amplified, phase sensitive detected in a double balanced mixer (DBM) and amplified in a video amplifier. The signal is captured in a digital transient recorder or signal averager and passed to a computer. The lineshape for a narrow line is obtained from the Fourier transform of the FID, figure 2, or spin echo. A broad distribution is plotted point by point as described in section 2.7 and discussed further in section 5.

The spin echo may be separated from the FID and the spurious signals following the RF pulses by cyling the phases of the first and second pulses in steps of $\pi/2$. A π change of phase of the first pulse for example inverts the spin echo, figure 3, and after 16 phase combinations only the echo is visible.

4. Resonance and Relaxation

4.1 EFFECTIVE FIELD AT THE NUCLEUS

The magnetic field at the nucleus, B_{eff}, is not exactly equal to the applied field, B_o, as has been assumed so far, although the difference is usually less than 1% except in ordered magnetic materials, section 5. The difference arises because the electronic magnetization set up by B_o induces an additional field at the nucleus. We shall present only a simplified account of B_{eff}, for example the chemical shift defined below is a tensor, in general, and refer the reader to the references for details of the theory [1-8].

In diamagnetic insulators the *chemical shift*, σ, is defined by the equation,

$$B_{eff}=B_o(1-\sigma), \qquad (4.1)$$

where σ is independent of temperature and typically in the range 10^{-4}–10^{-5}. The NMR lines of liquids in a good magnet may be measured to ≈ 1 part in 10^9, ie many orders

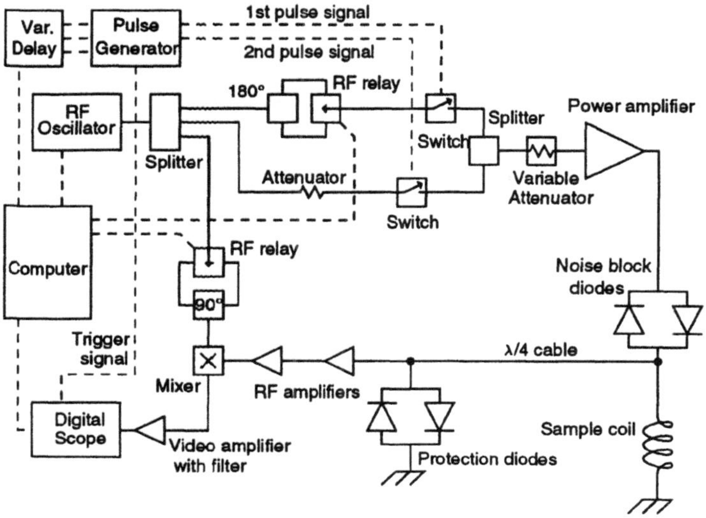

Figure 6 A 10-1,000 MHz spectrometer, see[12].

of magnitude less than σ, and the NMR spectrum from a molecular group such as CH_2 easily distinguished from that of eg CH_3 or OH in C_2H_5OH.

In simple metals the most important contribution to B_{eff} comes from the Fermi contact interaction between the nuclei and the Pauli spin paramagnetism of the conduction electrons. The Hamiltonian has the form

$$\mathcal{H} = A\ \mathbf{I.S}, \tag{4.2}$$

where A is a constant, I the nuclear spin and S the electron spin. The *Knight shift* is defined by

$$B_{eff} = B_o(1+K), \tag{4.3}$$

where K is independent of temperature for simple metals and is $\simeq 10^{-3}$ for metals such as Na or Al [6].

In rare earth and transition metals and compounds there will be a number of contributions to the Knight shift, due to electronic orbital and spin interactions with the nuclei, and the Knight shift is then written,

$$K = \sum_i K_i = \sum_i a_i \chi_i \tag{4.4}$$

$$\chi = \sum_i \chi_i \tag{4.5}$$

where χ is the measured susceptibility. In metals where correlated electron effects are important, eg heavy fermion, intermediate valence, condensed Kondo compounds and high T_c superconductors, K and χ may be highly anisotropic, figure 7. Frequently, only one term in equation (4.4) or (4.5) is temperature dependent and then K is a linear function of χ, figure 8. The a_i are related to the contributions B_i to the hyperfine field (HFF) at the nucleus by,

$$a_1 = \frac{dK_1}{d\chi_i} = \frac{B_1}{N_A \mu_o \mu_B} \tag{4.6}$$

where N_A is Avogadro's number and χ_i a molar susceptibility.

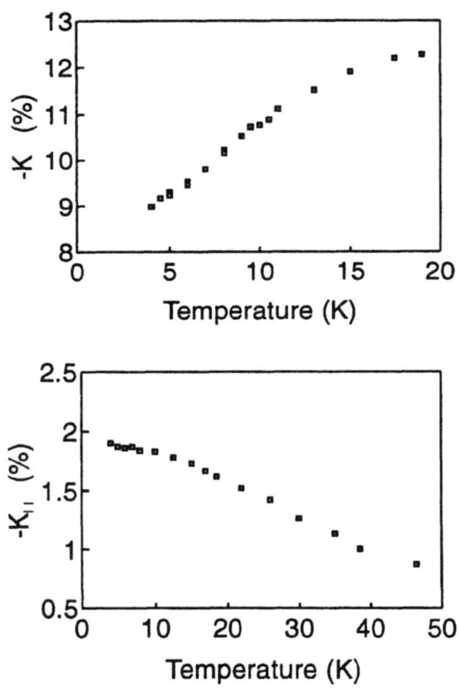

Figure 7. The temperature dependence of the ^{195}Pt NMR shift for UPt_3 for $B_o \perp$ and // to the a axis.

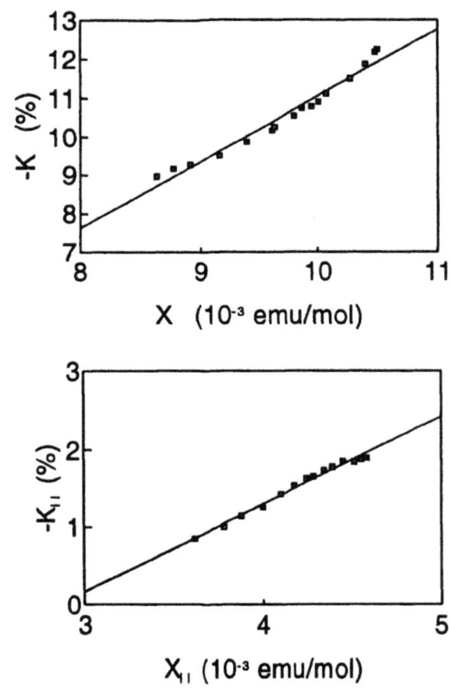

Figure 8. The Knight shift as a function of the dc susceptibility for UPt_3. After [13].

4.2 NUCLEAR RELAXATION

The relaxation times T_1 and T_2, of spin 1/2 nuclei provide information on the fluctuations of the magnetic fields in the material. Fluctuating electric field gradients are frequently the most important mechanism for the relaxation of nuclei of spin greater than 1/2 but will not be considered here.

In the rotating reference frame, fluctuations of the transverse fields along x' and y' will contribute to T_1, while fluctuations along y' or z' contribute to T_2. Therefore in the laboratory frame it is fluctuations at ω_o that are important for T_1 but fluctuations at both ω_o and zero frequency contribute to T_2.

In simple metals the contact interaction, equation 4.2, is the most important mechanism for T_1, leading, for free conduction electrons, to the relationship with the Knight shift,

$$K^2T_1T = \frac{\mu_B^2}{\pi\hbar\gamma_n^2 k_B} = \zeta_o, \tag{4.7}$$

first derived by Korringa. The experimental value of K^2T_1T / ζ_o is typically about 2 for metals such as Cu or Al [6] due to the effect of electron-electron interactions.

In the more complicated magnetic metals with correlated electrons discussed at the end of section 4.1, the experimental value of,

$$\frac{\zeta}{\zeta_o} = \frac{K^2T_1T}{\zeta_o} \tag{4.8}$$

is anisotropic and a function of temperature, figure 9.

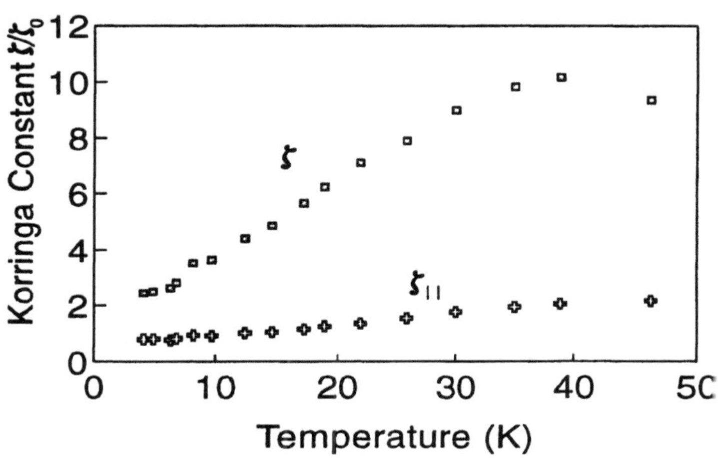

Figure 9. Temperature dependence of ζ / ζ_o, as defined in equation 4.8, for UPt_3, after [13].

5. NMR Of Ordered Magnetic Materials

5.1 THE EFFECTIVE FIELD AT THE NUCLEUS

In an ordered magnetic material B_{eff} may be huge (-33T in Fe, 305T in Tb) even when B_o is zero. The effective field may be broken down into many distinct contributions [8-11] but for the systems we wish to discuss in detail we may group these for convenience as

$$\mathbf{B_{eff}} = \mathbf{B_s} + \mathbf{B_{or}} + \mathbf{B_{dp}} + \mathbf{B_t} \tag{5.1}$$

where B_s is the hyperfine field (HFF) due to the spin moment at the site, B_{or} is due to the orbital moment, B_{dp} is the classical dipole sum about the site and B_t is the transferred contribution from other sites in the crystal. The important feature of equation (5.1) is that B_s produces a negative contribution to the effective field, ie antiparallel to the magnetisation and B_{or} a positive contribution and, in general, all the terms are anisotropic.

In rare earth ions which are not in a S state B_{or} is the dominant term but for iron based compounds $|B_s| >> |B_{or}|$ while for ^{59}Co NMR in intermetallic compounds $|B_s| \sim |B_{or}|$. A non-magnetic impurity will have an appreciable B_{eff} due to the last two terms in equation (5.1) eg -5.5T for Al in Fe and one of the strengths of NMR is that it can be carried out for almost every element in the periodic table provided the element has sufficient solubility in the system of interest (typically ~0.01-0.1%).

5.2 THE ENHANCEMENT EFFECT

The integrated intensity of a NMR line is a constant under a given set of experimental conditions, so the greater the linewidth the smaller the peak height of the signal. A simple calculation of the dipole field of a proton at a lattice spacing of a few angstroms correctly shows that the NMR linewidth in a non-magnetic solid will be ~10^{-4}T. In a material containing electronic moments (which are three orders of magnitude greater than nuclear moments) it might be expected that the linewidth would be so great that the peak signal would not be detectable against thermal noise. It was found, however [14], that while the linewidth of ^{59}Co in fcc Co was indeed ~0.1T the signal was as strong as in electron spin resonance. The anomalously strong signal arises because the radio frequency field drives the ordered electronic moments of the ferromagnet as well as the nuclear moments and NMR is observed as a modulation of the motion of the electronic magnetisation.

In a spherical particle consisting of a single domain the magnetisation will lie along the easy (z) direction, figure 10, and the HFF is also along z. A linear RF field along x of amplitude $2B_1$ will tip the magnetisation through a small angle $2B_1/B_A$ where B_A is the anisotropy field. Since B_{eff} is tied to the direction of the magnetisation, a linear field of magnitude $2B_1(B_{eff} + B_A)/B_A$ in the x direction is induced at the nucleus.

This field is normal to B_{eff} and will induce nucleus transitions so the field is said to be enhanced by the factor,

$$\eta_d \sim B_{eff} / B_A \tag{5.2}$$

which for fcc Co is ~100.

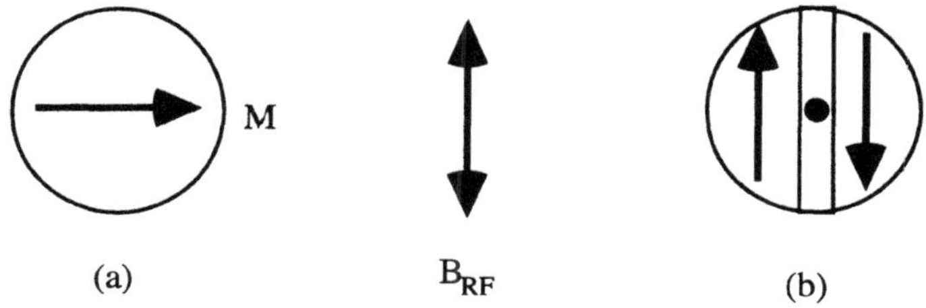

Figure 10. The geometry for a maximum NMR signal from (a) a domain, (b) a domain wall.

If an external magnetic field is applied along the easy direction

$$\eta_d \sim B_{eff} / (B_o + B_A). \tag{5.3}$$

Unless the sample is in the form of very small particles (~100Å) it will normally divide into domains as considered below. The domain structure may be removed by a sufficiently large external field but from equation (5.3) this reduces the enhancement considerably.

The ^{59}Co signals from fcc Co observed in [14] were far stronger than would be expected from an enhancement of ~100 and it was shown that in a multidomain particle the nuclei in the domain walls could experience a greater enhancement of the RF field than those in domains.

When a linear RF field is applied in the z direction to a sample containing a 180° wall in the z-y plane the wall may move in the x direction, leading to an enhancement that varies throughout the wall, given by [8],

$$\eta_w \sim \frac{VB_{eff}}{2\mu_0 MA} \chi_e \left(\frac{d\Psi}{dx}\right) \tag{5.4}$$

where $\Psi(x)$ is the angle of the local magnetisation within the wall relative to the domain magnetisation for a particle of volume V, average wall susceptibility χ_e and wall area A. In fcc Co the maximum value of η_w, at the centre of the wall, is $\sim 10^4$.

It was emphasised above that the wall only "may" move because wall motion is restricted by strain and defects, particularly in materials of high anisotropy, and in addition the wall may be thought of as a damped simple harmonic oscillator with its own natural resonant frequency (ν_w) so the motion will depend upon the ratio of ν_w to the NMR frequency.

In deriving equation (5.4) it was assumed that a planar wall moved rigidly under the influence of the RF field but Stearns has suggested [15] that in powders the wall edge may be pinned and the motion is then that of a drum head. Significant differences between the enhancement effect in a good single crystal of YIG and a polycrystal have in fact been reported which suggest that the wall motion is planar only in the single crystal [16].

The most obvious consequence of equation (5.4) is that in domain wall enhanced NMR there can be no unique RF field. The value of η_w will be a maximum at the wall centre and decreases as the wall blends into the domain. It appears that in many multidomain materials the signals come from nuclei at the wall centre and wall edges rather than from the bulk of the domain.

A more serious problem for domain wall enhanced NMR arises when the magnetisation is strongly anisotropic, since the anisotropy of the HFF may then lead to a different NMR frequency and a different linewidth for wall centre and wall edge or domain signals. As will be seen in figure 11 the NMR spectrum of a single crystal of hcp Co is a function of the RF power for pulses of fixed length because the domain wall centre and edge signals have different frequencies and enhancements.

The interpretation of the domain wall enhanced NMR spectra of strongly anisotropic materials with large unit cells can therefore be extremely difficult. A complicated spectrum may be due to different Co moments at each site or to a range of NMR lines from different parts of the domain wall.

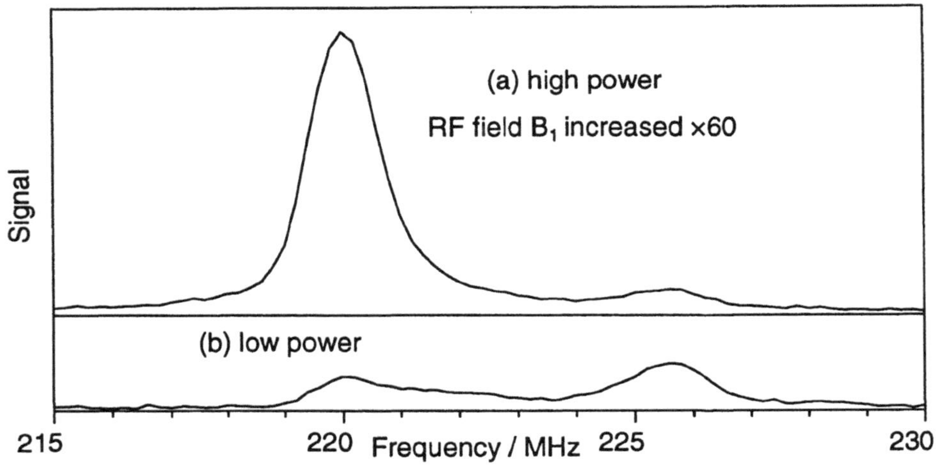

Figure 11 The ^{59}Co NMR of a single crystal of hcp Co at 4.2K. The domain wall centre signal near 226 MHz requires less power, ie has higher enhancement, than the domain wall edge signal near 220 MHz.

5.3 NMR AT HIGH PRESSURE

The HFF in a ferromagnet is often assumed to be simply proportional to the magnetisation. At constant volume and at low temperature this conclusion is probably correct, but the volume dependence of the hyperfine field may be quite different from that of the magnetisation and therefore provides a good test of the accuracy of first principles computer calculations of the ground state. The agreement between computer calculations [17] and experiment [18] for nickel, which showed that while the magnetisation decreases under pressure, the magnitude of the HFF increases, was particularly striking at the time.

A more subtle example is the intermetallic compound YFe_2 which was once generally considered to have no moment on the Y site. The pressure dependence of the HFF of ^{57}Fe is qualitatively similar to that of the magnetisation, but has the opposite sign to Y [19]. Computer calculations [20] showed that there is a moment around the Y site $\cong -0.4\mu_B$ (i.e. antiparallel to the Fe moment of $1.6\mu_B$) whose magnitude decreases under pressure and leads to an increase in the magnitude of the HFF at the Y site as the pressure is increased.

5.4 SOME APPLICATIONS OF NMR OF ORDERED MAGNETIC MATERIALS

A *magnetic phase transition* may be monitored by the differences in the relaxation times and the NMR spectrum on either side of the transition. In figure 5 for example the room temperature ^{63}Cu NQR spectrum is a single line because CuO is in the paramagnetic state but at 77K, in the antiferromagnetic state, a hyperfine field of some 12T splits the spectrum into three lines.

The presence of *specific defects* may be detected if they produce extra lines in the NMR spectrum. The cubic Mn(I) site of Mn_4N for example [21] has a single line spectrum near 134 MHz at 4.2K, figure 12, but lines at 115 (101) MHz are produced if the Mn is next to 1(2) N vacancies. The vacany breaks the cubic symmetry of the

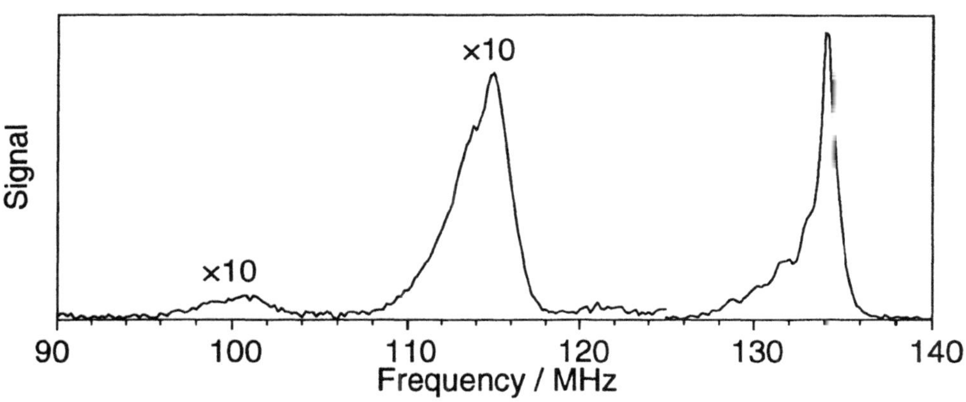

Figure 12. The ^{55}Mn NMR of the Mn(I) site of Mn_4N at 4.2K. The lines at 100 and 115 MHz are the result of N vacancies.

Mn site but the EFG is not great enough to split the line into 2I+1 resolvable components. The EFG can however be measured because it produces, figure 13, a modulation of the spin echo decay [22].

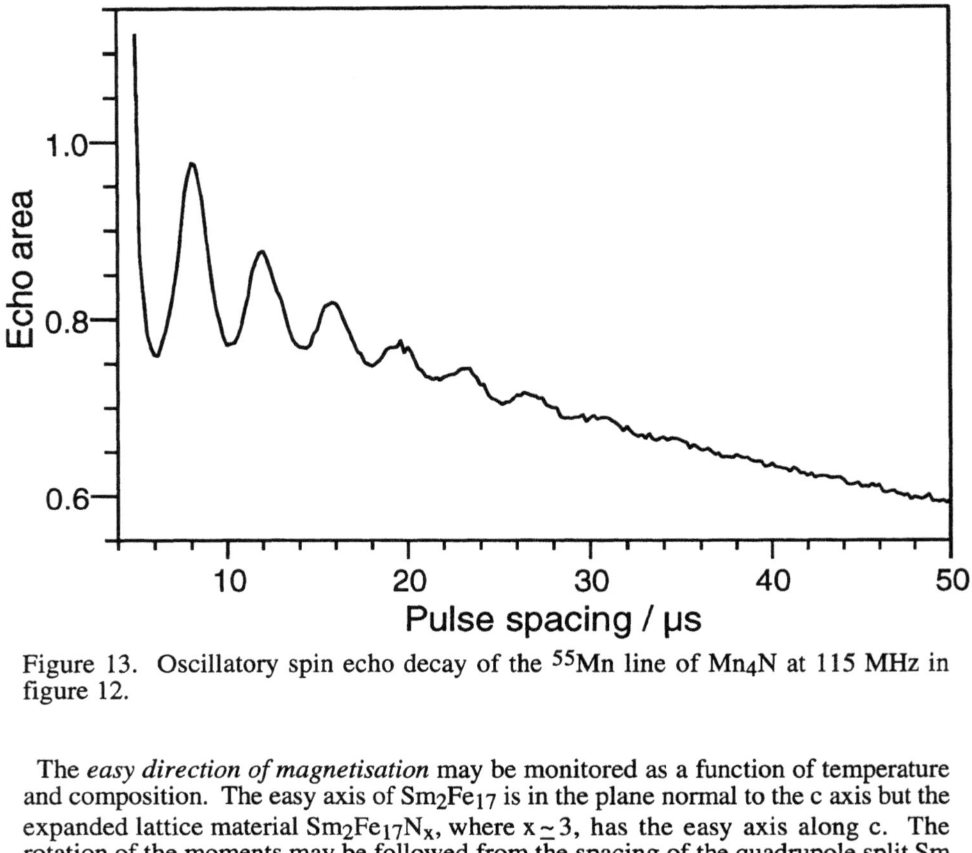

Figure 13. Oscillatory spin echo decay of the ^{55}Mn line of Mn_4N at 115 MHz in figure 12.

The *easy direction of magnetisation* may be monitored as a function of temperature and composition. The easy axis of Sm_2Fe_{17} is in the plane normal to the c axis but the expanded lattice material $Sm_2Fe_{17}N_x$, where $x \simeq 3$, has the easy axis along c. The rotation of the moments may be followed from the spacing of the quadrupole split Sm NMR spectrum, figure 14.

The *interface region* of a multilayer produces a characteristic spectrum depending upon how close it is to ideal. At a planar boundary between (111) fcc Co and Cu for example, Co atoms will have 9 Co nearest neighbours (n.n) and 3 Cu n.n. The ^{59}Co spectrum would then consist of a line characteristic of fcc Co, 217.4 MHz at 4.2K, and a line near 170 MHz for the interface. It will be seen, figure 15, that the spectrum of a good MBE Co/Cu multilayer approximates to the ideal but there is evidence for other configurations. The frequency of the main line is also shifted downwards from the bulk value because the Co is under extensive strain in the plane due to the lattice constant of Cu being greater than that of Co.

Chapter 12

Diffusion of hydrogen in metals

H.R.Schober
Institut für Festkörperforschung
Forschungszentrum Jülich
D-52425 Jülich
Germany

ABSTRACT. Hydrogen diffuses in metals much faster than any other atom. Two different (tracer and chemical) diffusion coefficients are distinguished. Dependent on the host lattice and site occupation diffusion is by nearly classical jumps or by a tunneling mechanism (quantum diffusion). Trapping of hydrogen by octahedral interstitials in the bcc metal opens the possibility to study tunneling mechanism in details. Besides phonon assisted tunneling nonadiabatic electron effects are observed. For an understanding of the diffusion mechanisms a number of different energy scales are relevant. Standard small polaron theory is introduced and its extensions to higher and lower energies are discussed. For larger hydrogen concentrations the diffusion coefficient becomes strongly concentration dependent.

1. Introduction

This chapter deals with the diffusion of hydrogen in metals which has been studied extensively both experimentally and theoretically. Hydrogen diffusion shows many unique features. Due to the small mass hydrogen diffusion is strongly affected and in some materials and temperature ranges even controlled by quantum effects. The large mass ratio between the hydrogen isotopes leads to large easily observable isotope effects in its diffusion. Hydrogen is highly mobile, even at low temperatures. This makes it accessible to studies over large temperature ranges with a large variety of experimental techniques.

The diffusional behaviour is of great technological importance in many fields, e.g. hydrogen gas as energy carrier, hydrogen storage and purification, fusion reactor technology etc.

There are several reviews on hydrogen diffusion [1; 2; 3] in general and on the theory [4; 5; 6]. This chapter can only give a short survey.

After introducing the chemical and tracer diffusion coefficients the experimental results for different materials are discussed. Depending on the site occupied the diffusion is found to be mainly classical, with small quantum corrections, or to be dominated by quantum effects.

Hydrogen dynamics is governed by a large number of energies: self trapping energy, nearest neighbour site energy, coincidence energy, classical diffusion activation

F. Grandjean et al. (eds.), Interstitial Intermetallic Alloys, 293–316.

energy and vibrational energy. The relative magnitudes of these energies determine which mechanism dominates diffusion.

The standard theory of quantum diffusion (small polaron theory) [7] uses a mechanism of phonon activated tunneling. An analytic evaluation is only possible with some very stringent approximations. For higher temperatures these approximations no longer hold. A numerical simulation has shown quantum effects to persist even at room temperature [8].

At low temperatures the quantum diffusion rate is no longer determined by the phonons alone but electronic excitations have to be included explicitly [9].

A short final section will show the dependence of the diffusion coefficient on hydrogen concentration.

2. Experimental survey

The diffusion of hydrogen can be considered to take place on a regular lattice of interstitial sites. This makes it simpler than the diffusion of the host atoms themselves which involves additionally vacancies. The diffusion can be characterized by two different diffusion coefficients: the chemical diffusion D and the tracer (or self) diffusion coefficient D_t. The chemical diffusion coefficient describes according to Fick's law the hydrogen flux in a concentration gradient:

$$\mathbf{j}(\mathbf{r},t) = -D\nabla\rho(\mathbf{r},t) \tag{1}$$

where $\rho(\mathbf{r},t)$ is the hydrogen density. Using the continuity equation

$$\dot{\rho}(\mathbf{r},t) + \nabla\mathbf{j}(\mathbf{r},t) = 0 \tag{2}$$

one obtains the diffusion equation

$$\dot{\rho}(\mathbf{r},t) - D\nabla^2\rho(\mathbf{r},t) = 0. \tag{3}$$

The chemical diffusion coefficient is observed in experiments where one monitors the decay of a non-equilibrium hydrogen distribution. Typical experimental data consist of measurements of hydrogen permeation, desorption and absorption, resistance relaxation, volumetric and Gorski effect measurements.

The tracer diffusion coefficient D_t on the other hand describes the diffusive behaviour of a given hydrogen interstitial. The probability $p(\mathbf{r},t)$ to find it at time t at position $\mathbf{r}$ is governed by

$$\dot{p}(\mathbf{r},t) = D_t\nabla^2 p(\mathbf{r},t). \tag{4}$$

The tracer diffusion coefficient is measured directly in incoherent neutron scattering and in field gradient methods in NMR.

The two diffusion constants agree only in the limit of low concentrations. In general the ratio between them can be written as

$$\frac{D}{D_t} = \frac{1}{H_R}\frac{\rho}{kT}\frac{\partial\mu}{\partial\rho} \tag{5}$$

where H_R is the Haven's ratio. It is always smaller than unity. It accounts for the interaction between neighbouring hydrogen interstitials. With rising concentration

H_R decreases to values typically between 0.8 and 0.5. The quantity $(\rho\partial\mu/\partial\rho)/(kT)$ is called the thermodynamic factor with μ the chemical potential.

The diffusion of H is an example of a jump diffusion process. It is characterized by the mean residence time τ of the H at a given site or, correspondingly by a jump rate $1/\tau$ to neighbouring sites. This jump rate is calculated in the theories of hydrogen diffusion. Under the simplified assumption that successive jumps are uncorrelated the jump rate is related to the diffusion constant by

$$D = \frac{4d^2}{6\tau} \tag{6}$$

where $2d$ is the jump length.

In classical approximation the jump rate for an interstitial atom can be written as

$$\tau^{-1} = \nu e^{-(G_s - G_0)/kt} \tag{7}$$

where ν is an attempt frequency of order of the Debye frequency of the lattice and G_s and G_0 are the free enthalpies in the saddle point and in the equilibrium site, respectively. In both sites the unstable mode across the migration barrier is thought to be omitted. Writing the free enthalpy as $G = E - ST$, with S the vibrational entropy, one gets

$$\tau^{-1} = e^{-(E_s - E_0)/kT} e^{(S_s - S_0)/k} \tag{8}$$

Here $E_s - E_0 = E_a$ can be interpreted as an activation energy. For the entropy we use the high temperature expansion, see chapter lattice dynamics, and combine the attempt frequency with the vibrations in equilibrium and obtain the Vineyard expression [10].

$$\tau^{-1} = \frac{\prod \nu_0}{\prod {}' \nu_s} e^{-E_a/kT}. \tag{9}$$

Here the $'$ denotes that the product is over the stable modes only. Typically the prefactor of the exponential term has again the magnitude of the Debye frequency. For heavy interstitials the activation energy is given by the change in potential energy and independent of isotopic mass. In the case of hydrogen the vibrational energy has to be included resulting in a weak isotope dependence. In a logarithmic plot versus the inverse temperature (Arrhenius plot) Eq. (9) gives a straight line. Slight curvatures can be explained in the quasi harmonic approximation by temperature dependencies of the vibrational frequencies affecting the energies and entropies [11]. Large changes of slope indicate a change of diffusion mechanism.

In the derivation of Eq. (9) it was assumed that all degrees of freedom, bar the unstable one, are thermalized, both in equilibrium and at the saddle point. Further E_a is assumed to be much larger than $h\nu$. The latter condition holds approximately for PdH but is clearly violated for NbH where diffusion has to be described quantum mechanically. We will see later that an Arrhenius behaviour can also result from quantum diffusion.

2.1. EXPERIMENTAL RESULTS

Hydrogen diffuses in metals faster than any other atoms in solids. Fig. 1 compares the diffusion of H in V with that of C and with self-diffusion. The tracer diffusion coefficient of H is orders of magnitudes larger than the one for the other atoms. It is comparable to the typical value of atomic diffusion in liquids: $D = 2.6 \times 10^{-5}\text{cm}^2\text{s}^{-1}$ for molecular diffusion in water at room temperature [2]. The activation energy is very small, e.g. $E_a \approx 45$meV for H in V.

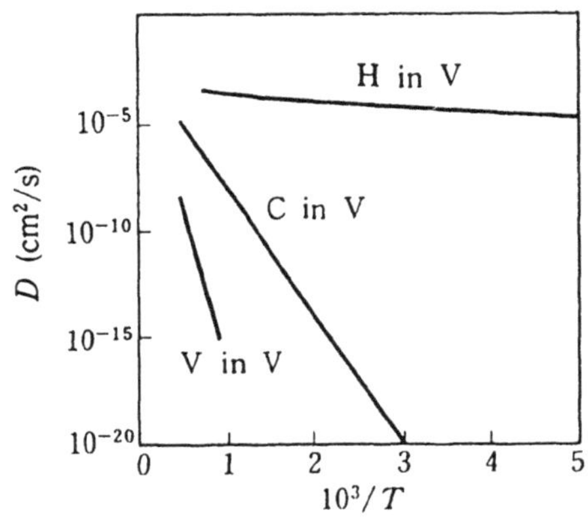

Fig. 1: Diffusion coefficients of H and C in V and self-diffusion of V [2].

Hydrogen diffusion is measured by a great variety of techniques, both macroscopic and microscopic techniques. Non-equilibrium distributions of H are used e.g. in permeation, resistivity relaxation and Snoek effect measurements. Data on diffusion in equilibrium distributions are gained by Snoek effect experiments, NMR, quasi-elastic neutron scattering and Mössbauer effect. Some of the latter techniques provide direct information not only on jump rates but also on diffusion paths.

The diffusion rate of H depends strongly on the host lattice and is related to the site occupancy. In the following we give typical examples for the different classes of material. The values refer to the limit of low concentrations where chemical and tracer diffusion are expected to be equal.

Diffusion in fcc metals

In those materials where the site occupancy has been established, e.g. for Pd [12], octahedral occupancy has been found and is assumed to be the general case. Diffusion is then on the lattice of octahedral sites which again is an fcc lattice, see Fig. 2. The distance between nearest neighbour sites ranges between 2.5 and 2.9 Å.

The most thoroughly studied fcc metal is Pd. One observes up to temperatures of 1220 K an Arrhenius behaviour, $D = D_0 \cdot \exp(E_a/kT)$, with an activation energy of 0.23 eV and a prefactor $D_0 = 2.9 \cdot 10^{-3}\text{cm}^2\text{s}^{-1}$. For deuterium (D) the data are

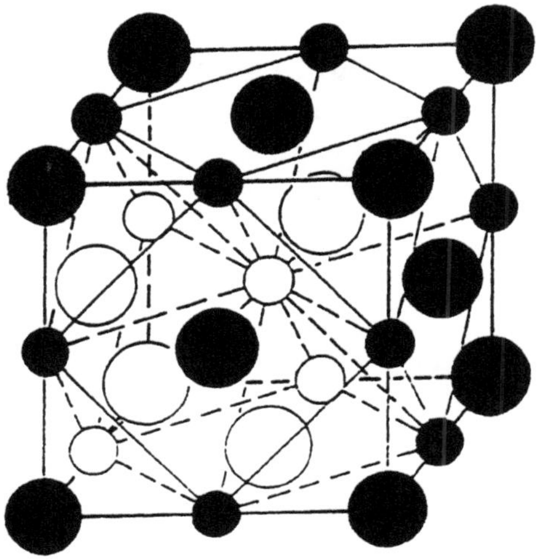

Fig. 2: Fcc metal lattice (●) and lattice of octahedral sites (•) [8].

reduced to $E_a = 0.206\text{meV}$ and $D_0 = 1.7 \cdot 10^6 - 3\text{cm}^2\text{s}^{-1}$ and for tritium (T) to $E_a = 0.185\text{meV}$ and $D_0 = 7.3 \cdot 10^{-4}\text{cm}^2\text{s}^{-1}$. There is some evidence of a lower activation energy betow $T = 150\text{K}$. The measurements were done on dilute PdFe alloys and, therefore, trapping effects might be involved [16].

The observed activation energies in Cu and Ni are higher, i.e. $E_a \approx 0.4\text{meV}$. The isotope effects are of similar magnitude and are comparable to the changes in vibrational energy between the isotopes. In all cases the activation energy is much larger than the energy of the localized hydrogen vibration. The diffusion should, therefore, be classical with some quantum effects.

Diffusion in bcc metals

At least in the low concentration limit (α-phase) hydrogen is situated on tetrahedral sites in all cases where site occupancy has been determined. The lattice of tetrahedral sites in bcc is much denser than the corresponding lattice of octahedral sites in fcc. There are six tetrahedral sites per metal atom. The distance between sites is about 1.0 to 1.2Å, about half the distance of the sites in fcc lattices. Already in classical diffusion a shorter jump distance means a higher jump rate for a given attempt frequency. This effect is enhanced for H due to the overlap of the H wave-functions at neighbouring sites which allows for tunneling as we will discuss later.

The typical best investigated examples are VH, NbH and TaH. Fig. 5 shows a compilation of Gorsky effect results for the diffusion coefficients of hydrogen, deuterium and tritium in these materials (note the different ordinate scales). These materials have a high solubility for H which reduces the probability of trapping by impurities per H atom.

The diffusion coefficients are very large. V exhibits together with fe the highest reported diffusion coefficients for hydrogen. This may be attributed to their small

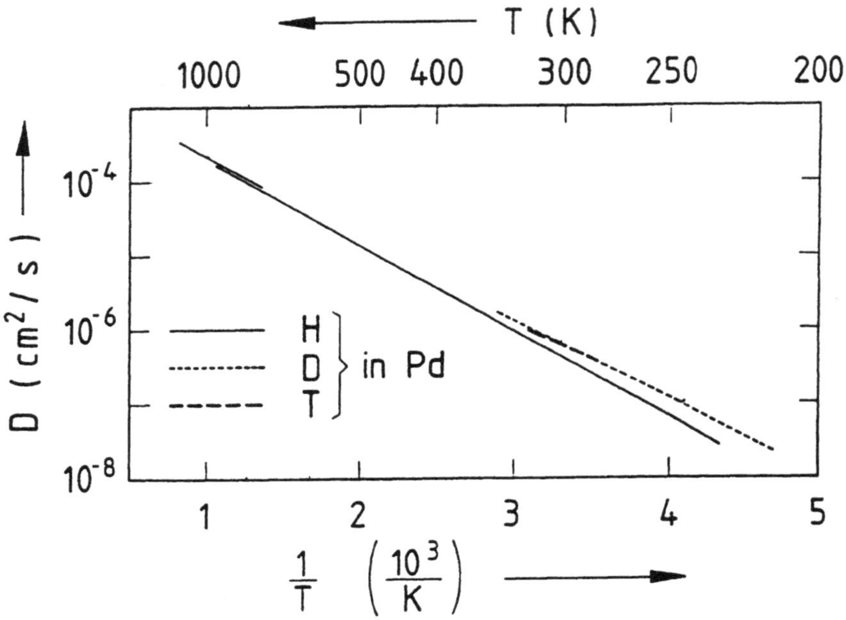

Fig. 3: Arrhenius plot results for the diffusion coefficient of H, D and T in Pd [3]. Experimental values from H: [1; 13; 14], D:[13], T: [15].

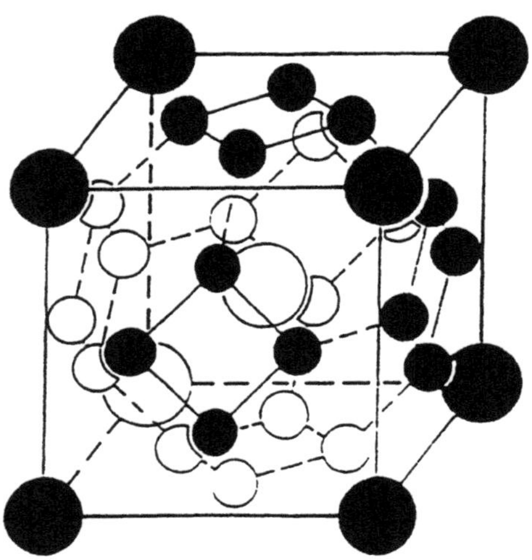

Fig. 4: Bcc metal lattice (●) and lattice of tetrahedral sites (•) [3].

lattice constant. Particularly at low temperatures one sees a very large isotope effect. In some cases a pronounced change of slope, i.e. a change in activation energy is observed. The activation energy changes from values of $E_a \approx 40-60$meV at lower

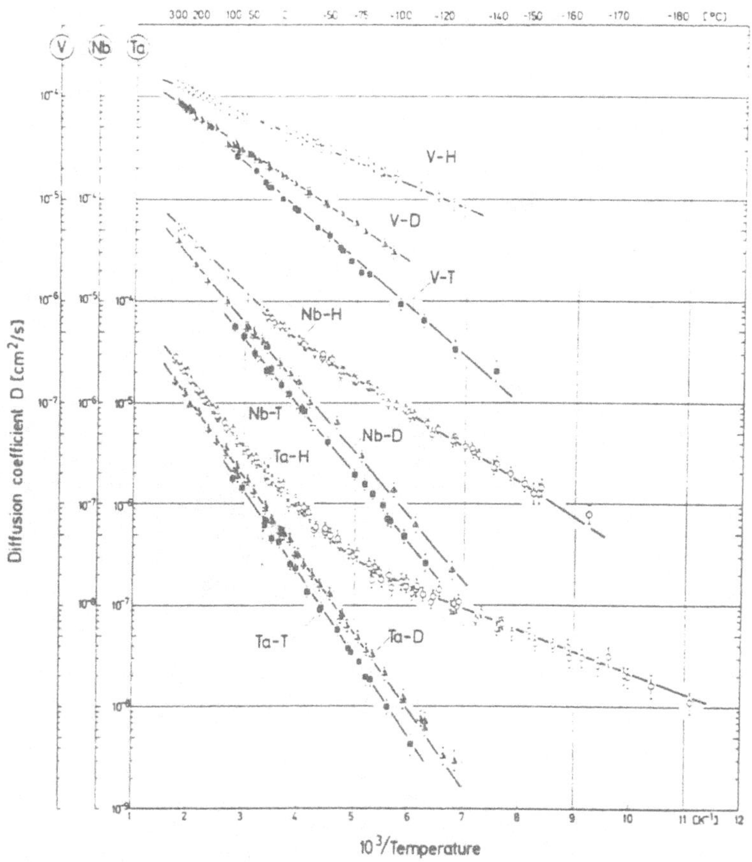

Fig. 5 Arrhenius plot of the diffusion coefficients of hydrogen, deuterium and tritium in V, Nb and Ta [17] (note the different ordinate scales).

temperatures to values of 100 – 160meV at higher temperatures. It is interesting to note that the high temperature activation energies have values similar to the localized vibration energies and the low temperature values are considerably less than the energy needed to excite a localized vibration. It will be shown later that diffusion for these materials is dominated by quantum mechanical effects ("quantum diffusion"). From NMR experiments lower values of the activation energy at low temperatures have been derived [18].

Diffusion in hcp metals

In low concentrations (α-phase) hydrogen occupies tetrahedral sites in these materials. The diffusion path is also via octahedral sites, Fig. 6. A consequence of the

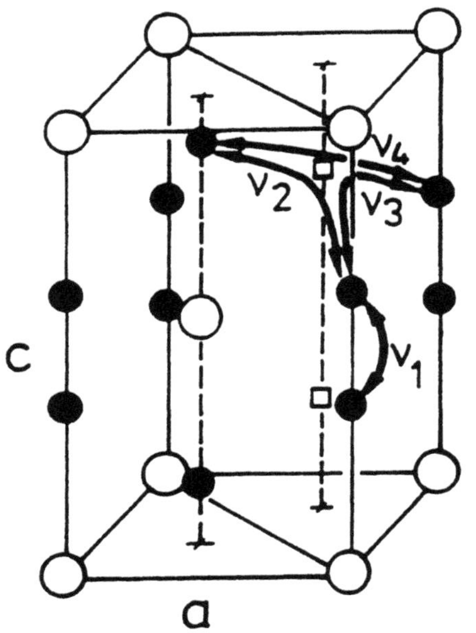

Fig. 6: Primitive unit cell of the hcp lattice (metal atoms: open circles, tetrahedral sites: full circles, octahedral sites: open squares). The arrows indicate different diffusion paths. [19]

hexagonal symmetry is that the diffusion tensor has two independent components, D_c for the diffusion in the direction of the hexagonal axis and D_a for the diffusion in the basal plane.

As an example Fig. 7 shows the diffusion constant of H in Lu. The anisotropy is very small. The activation energies $E_a \approx 0.58$eV are relatively large. Diffusion is limited by the jump rate through the octahedral site. The jump rate between the two neighbouring tetrahedral sites (ν_1 in Fig. 7) is at least a factor of ten larger. This experimental result is plausible considering the much shorter jump distance between the two neighbouring tetrahedral sites. H in hcp metals thus exhibits fast, local jumps between the pairs of tetrahedral sites and additionally much slower jumps needed for long range diffusion. Diffusion will be dominated by mainly classical dynamics, whereas the local motion will be highly non-classical.

Diffusion in alloys

Fig. 8 shows a compilation of results for the diffusion coefficients of hydrogen in some binary alloys. The data refer to alloys with CsCl, A-15 and (disordered) fcc structure in the low H concentration range. The values for the three alloys with CsCl structure vary strongly. The activation energies are $E_a \approx 0.49$eV in FeTi and $E_a \approx 0.48$eV in NiTi but only $E_a \approx 0.035$eV in PdCu. It is known that H occupies octahedral sites in FeTi and the same holds probably for NiTi. The low activation energy in PdCu most likely results from a different site occupation. The similarity to the values observed in the bcc metals suggests tetrahedral site occupation.

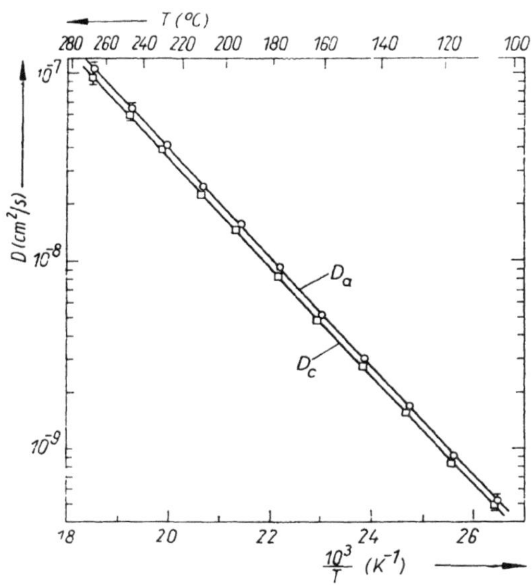

Fig. 7: Arrhenius plot of the two diffusion coefficients D_c and D_a for diffusion of H in single-crystalline Lu [19].

Trapped hydrogen

Like all interstitials hydrogen can be easily trapped by impurities. A trapped hydrogen can only diffuse if the binding energy is overcome. This can be an explanation for discrepancies in the measured diffusion constants particularly at low temperatures where often only very small concentrations of hydrogen are soluble. If the H concentration becomes comparable to the concentration of other impurities and defects one no longer measures the pure diffusion but a combination of diffusion and de-trapping.

An interesting situation occurs if the trapped H can move between some equivalent sites without de-trapping. This local dynamics can be studied with methods which are otherwise ruled out by long range diffusion. If the local geometry does not differ strongly, the information on the local jump process will give valuable insights also into the diffusional jump process.

A much studied example is hydrogen in Nb trapped by O, N or C impurities. In this case, the O and the N, and presumably also the C are located at octahedral site and the trapped H occupies tetrahedral sites. Fig. 9 shows two unit cells of Nb with an O interstitial at an octahedral site. The letters denote equivalent tetrahedral sites.

There are in total 16 equivalent sites (e) which are arranged in 8 pairs of nearest neighbour tetrahedral sites with a distance of about 1.17Å. This has the consequence that jump processes will occur on two time-scales a fast one for jumps (or tunneling) between the paired sites and a much slower one for jumps to different pairs of e-

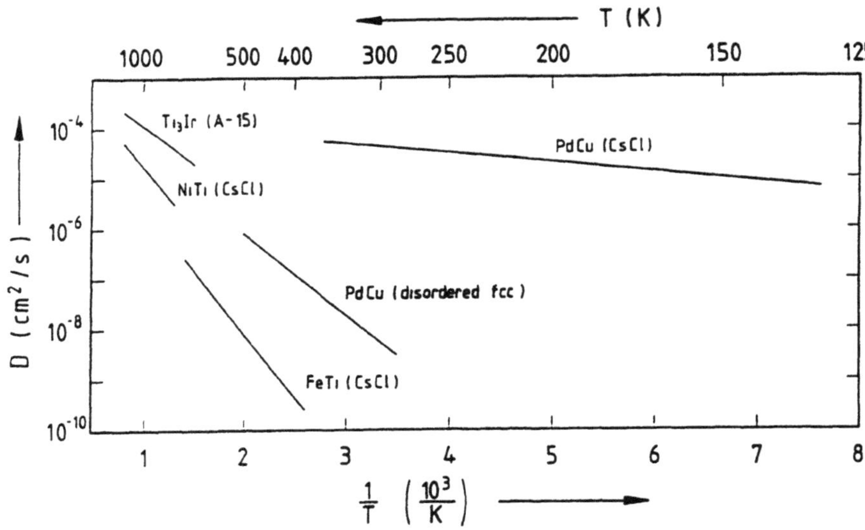

Fig. 8: Arrhenius plot of the diffusion coefficients in some binary alloys. The figure shows also the structure of the alloys [3]. Experimental values: PdCu [1], FeTi [20], NiTi [21], Ti_3Ir [22].

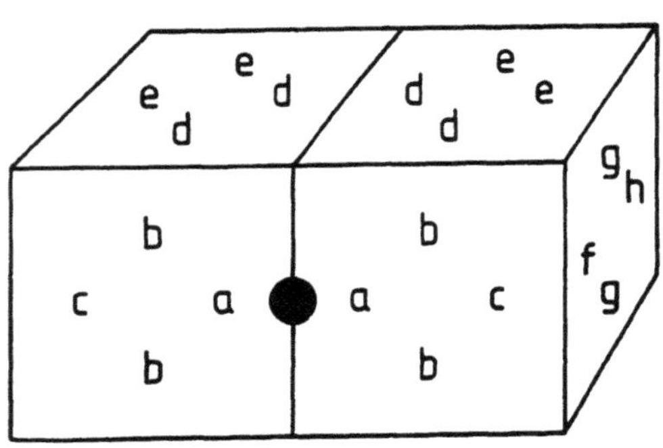

Fig. 9: Two bcc unit cells with an O, N or C atom (full circle) at an octahedral site. The letters $a - h$ indicate the tetrahedral sites on the visible surfaces of the unit cells. The H is likely to occupy e-sites [3].

sites. Experiments have shown that below $\approx$ 10K H tunnels coherently between the two sites of a pair and above $\approx$ 10K the motion becomes diffusive like incoherent tunneling. Different to the quantum diffusion at higher temperatures, see Fig. 5, the jump rate at these low temperatures is not only determined by phonon effects but also by a non-adiabatic influence of the electrons. This coupling to the electrons causes an increase of the jump rates towards low temperatures, Fig. 10.

In the limit $T \to 0$ one can test for the influence of the conduction electrons on the tunneling using superconductivity. In the superconducting state the influence

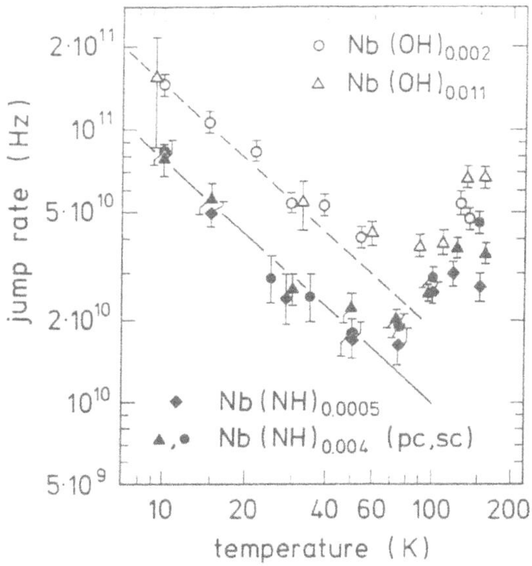

Fig. 10: Jump rate ν of trapped H. Open and full symbols refer to H trapped by O and N, respectively. The lines are theoretical predictions for a dominant non-adiabatic interaction with conduction electrons. [23]

of the conduction electrons is suppressed by formation of Cooper pairs. Using a magnetic field it is possible to switch between the superconducting and the normal state. For H trapped in NbO one finds a tunnel splitting of $J(0) = 0.226$meV in the superconducting and $J(0) = 0.206$meV in the normal conducting state [24]. The relaxation rate, on the other hand, is larger in the normal conducting state than in the superconducting one.

3. Theory of hydrogen diffusion in metals

In the previous section we have seen that the diffusion of hydrogen in metals at temperatures above $T \approx 50$K is, depending on the site occupation, either by a nearly classical jump mechanism or by a phonon assisted tunnelling process. In the first case there will be only a small isotope effect on the activation energy which can be caused by the change in vibrational energy. For a light interstitial the change in the localized vibration energy along the diffusion path has to be added to the usual saddle point energy determined from the potential energy. In the quantum diffusion case the jump rate is strongly determined by the tunnel splitting which depends exponentially on the isotopic mass. Hence the isotope effect on diffusion is expected to be much larger.

In this section we will give an introduction to the theory of quantum diffusion. First we will treat the technologically more important regime of moderate and high temperatures where the diffusion is dominated by the phonons. The interesting non-adiabatic coupling to the electrons will be treated only very briefly.

3.1. ENERGIES

Standard theories of diffusion and tunneling of hydrogen in metals start from two interrelated ideas. The first is that the hydrogen is, on a relevant time-scale, localized at interstitial sites of the host lattice and that tunneling is from one such site to the next. The second idea is that of self-trapping, meaning that an interstitial hydrogen will distort the host lattice and thus lower the total energy. To get an idea of the various energies involved let us take here and in the following Nb:H as an example. The hydrogen occupies tetrahedral sites and the four nearest neighbour atoms are displaced by about 0.1 Å. From model calculations one estimates the difference between the total energies of the undistorted and the relaxed lattices, respectively, to $E_{\mathrm{ST}} \approx 450$ meV. This so called self-trapping energy is the sum of the change in potential energy (300 meV) and the change in zero-point vibrational energy. Since the displacements of the Nb atoms are not too large, a harmonic model for the Nb-Nb interaction should be reasonably accurate. The potential energy part (elastic energy) of the energy gain by self-trapping at site i can then be related to the force $\mathbf{F}^{(i),m}$ exerted on the Nb atom m due to the presence of the H. This can be a direct force from H on Nb but could also include some induced Nb-Nb forces. The energy gain by this relaxation, the elastic self-trapping energy, is

$$E_{\mathrm{ST}}^{\mathrm{elastic}} = \frac{1}{2} \sum_{m,n} \mathbf{F}^{(i),m} \mathbf{G}^{mn} \mathbf{F}^{(i),n} \tag{10}$$

where $\mathbf{G}$ is the harmonic lattice Green's function, the inverse of the matrix of coupling constants which can be calculated from the measured phonon dispersion. The forces have to be calculated self-consistently by minimizing the total energy including its vibrational part. The static displacements $\mathbf{s}^{(i),m}$ of the Nb atoms due to the hydrogen are given by

$$\mathbf{s}^{(i),m} = \mathbf{G}^{mn} \mathbf{F}^{(i),n} \tag{11}$$

The dynamics of the hydrogen on its site is dominated by the localized vibration modes with frequencies of $\hbar\omega_0 = 100$ meV and 170 meV, respectively, where the latter mode is double degenerate. The frequencies of the localized vibrations are strongly affected by lattice distortions, they would be nearly 50% higher in the unrelaxed lattice. The localized modes account for nearly 99% of the spectral intensity of the hydrogen. Additionally, H participates in the host lattice vibrations (band modes). These modes are at much lower frequencies, the relevant energy scale is given by the Debye frequency of Nb, $\hbar\omega_D = 22$ meV. The amplitude of H is strongly enhanced for some of these vibrations and shows a resonant like behaviour. These latter vibrations determine, because of their lower frequencies, the temperature dependence of the thermal mean square displacements of the H below 1000 K which is a strong indication that they will also be important for the hydrogen mobility. The resonant like vibrations of H result from the near neighbour geometry of the tetrahedral sites and consequently couple to short wavelength phonons. The coupling of the hydrogen to the long wavelength phonons is much weaker, the vibrational amplitude of the hydrogen equals that of the host atoms for these modes. It is important to distinguish between the three different groups of vibrations: The frequencies of the localized modes determine the order of magnitude of the tunneling matrix element between sites, the weakly coupled low frequency phonons give the temperature dependence in the standard small polaron theory and the excitation of resonant and localized vibrations causes the breakdown of this description towards higher energies.

In addition to these intra-site energies a number of inter-site energies are relevant for diffusion. First, if the hydrogen is moved from its site i to a nearest neighbour site f without allowing the lattice to relax, the potential energy will be raised by

$$E_{\mathrm{NN}}^{\mathrm{elastic}} = \sum_{m,n} (\mathbf{F}^{(i),m} - \mathbf{F}^{(f),m})\ \mathbf{G}^{mn}\ \mathbf{F}^{(f),n}. \tag{12}$$

In model calculations one finds for this term values of typically 100 to 150 meV. The change of the localized modes will add an additional contribution.

This asymmetry between sites prevents tunneling between them. To allow tunneling between sites i and f either the hydrogen has to tunnel together with its deformation field (dressed tunneling, small polaron) or the lattice has to be deformed such that sites i and f become equivalent. There is of course an infinite number of such deformations. The minimal energy needed is called coincidence energy. It can be estimated by hypothetically putting half the hydrogen into each of the two sites. The elastic energy is then determined by the average of the two forces $\mathbf{F}^{(i)}$ and $\mathbf{F}^{(f)}$. Relative to the self-trapped state the coincidence energy in then

$$E_c^{\mathrm{elastic}} = \tfrac{1}{4} \sum_{m,n} (\mathbf{F}^{(i),m} - \mathbf{F}^{(f),m}) \mathbf{G}^{mn} \mathbf{F}^{(i),n} = \tfrac{1}{4} E_{\mathrm{NN}}^{\mathrm{elastic}} \tag{13}$$

To this elastic energy one has to add again a contribution due to the change in zero point motion. Models give values for the total coincidence energy E_c between 35 and 50 meV of which approximately 10 meV is the vibrational part. This coincidence energy is a measure of the strength of the phonon dressing effects.

Finally one can define a classical activation energy $E_M^{\mathrm{classical}}$. This is the energy needed to move the hydrogen by an adiabatic deformation of the lattice from one site to the next. The saddle point configuration is then the lowest coincidence configuration for which the distance between the two sites vanishes. The model calculations give values between 70 and 130 meV as classical activation energy, of which again typically 10 meV is the vibrational part. This energy is of the same order or even smaller than the energies of the localized vibrations and the usual condition $E_M^{\mathrm{classical}} \gg \hbar\omega_0$ does not hold for the hydrogen vibrations but only for the lattice vibrations. Hence, deviations from a classical hopping behaviour have to be expected in the Nb:H system even at high temperatures. Fig. 11 depicts schematically the potential energy curves for the hydrogen atom for fixed lattice configurations as discussed above.

A basic quantity characterizing quantum effects is the tunneling matrix element J in the coincidence configuration. Assuming a sine like potential between the two sites, the tunneling matrix element can be evaluated using the eigenvalues of the Mathieu equation

$$J = \hbar\Delta = \hbar\omega \frac{4}{\pi\sqrt{\pi}} \sqrt{\sigma} \exp{(8/\pi^2)\sigma} \tag{14}$$

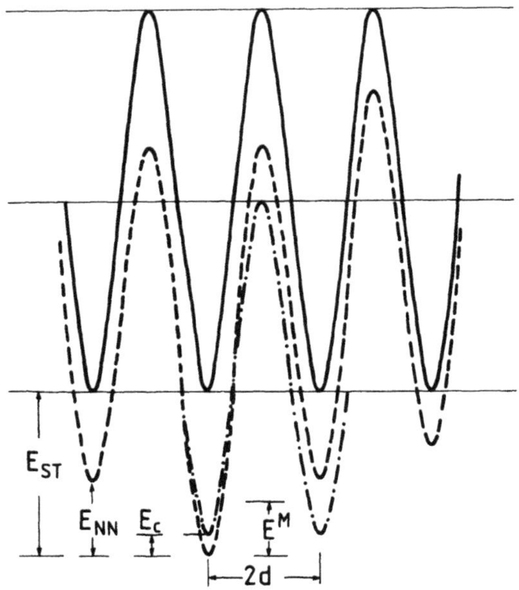

Fig. 11: Schematic potential energy curves for a hydrogen interstitial ($2d$ = distance between neighbouring interstitial sites). Solid line: metal host lattice unrelaxed, broken line: host lattice relaxed (self trapped configuration), dot-dashed line: coincidence configuration, $E_{\rm ST}$ = self trapping energy, $E_{\rm NN}$ = nearest neighbour site energy, E_c = coincidence energy, $E_{\rm M}$ = classical migration energy.

with

$$\sigma = 4m\omega d^2/\hbar \tag{15}$$

where Δ is the tunneling frequency, m is the mass of the tunneling particle, ω the effective vibrational frequency and $2d$ is the distance between the two equilibrium sites. σ scales with $\sqrt{m}$ and due to the exponential dependence the tunneling matrix element shows a very strong isotope effect. Kehr [25] has evaluated Eq. (14) for Nb ($2d$ =1.17 Å) and gets the estimates J_H = 1.5 meV, J_D = 0.14 meV and J_μ = 85 meV for hydrogen, deuterium and μ^+, respectively. These values are large enough to make tunneling observable and for diffusion to be dominated by quantum effects at least at low temperatures. The observed tunneling matrix element is of course reduced from this bare (naked) one by the effects of phonon and electron dressing which are the main subject of this section.

In a typical fcc metal such as Pd the situation is different. Due to the larger distance between the octahedral sites occupied by the hydrogen ($2d$ = 2.7 Å) the tunneling matrix element is much smaller. Kehr's estimates are $J_H = 4 \times 10^{-6}$ meV, $J_D = 2 \times 10^{-9}$ meV and J_μ = 1.3 meV, respectively. Quantum effects will only be observable for the light μ^+. The localized mode frequency is much lower than in Nb. For hydrogen one finds $\hbar\omega_0$ = 64 meV which is much smaller than the observed activation energy for diffusion $E_{\rm M}$ = 230 meV. Diffusion will be classical apart from small effects due to the vibrational zero point energy.

In hcp metals the situation is more complex. The hydrogen is thought to occupy tetrahedral interstitial sites. The diffusion involves a jump between nearest-neighbour tetrahedral sites and another jump to octahedral sites. For the first jump one has a similar situation as in the bcc metals and quantum effects will be important. The diffusion itself is limited by the longer second jump which is mainly classical, similar to the fcc case, with a relatively large activation energy of $E_{\rm M} =$ 570 meV for Y.

3.2. QUANTUM DIFFUSION

As a first approximation the dynamics of interstitials in metals can be treated within the Born–Oppenheimer approximation which assumes that the electron distribution adjusts instantaneously to the ionic configuration, i.e. the electrons always occupy the momentary ground-state and there is a smooth transition between these states. The only elementary excitations are then the phonons whose energies can be calculated well in this approximation. In a metal there are of course additional electronic excitations with arbitrarily low energies and these will affect the dynamics at very low temperatures where the phonons are frozen out due to the ω^2–dependence of their spectrum.

In the Born–Oppenheimer approximation the dynamics of the hydrogen metal system is determined by the kinetic energy and an effective inter-ionic potential. The Hamiltonian can be written as

$$H = T_H + T_M + V_M + V_{HM} \ . \tag{16}$$

Here T_H and T_M are the kinetic energies of the hydrogen and the metal ions, respectively, V_M is the potential energy of the ideal metal host lattice and V_{HM} is the hydrogen-metal interaction.

The solution of the tunneling problem involves the calculation of the matrix element

$$\begin{aligned} M_{n_i n_f} &= \left\langle \Psi^{(i)}_{n_i}(\mathbf{R}^H, \mathbf{s}) \left| H - \tfrac{1}{2}(H^{(i)} + H^{(f)}) \right| \Psi^{(f)}_{n_f}(\mathbf{R}^H, \mathbf{s}) \right\rangle \\ &= \left\langle \Psi^{(i)}_{n_i}(\mathbf{R}^H, \mathbf{s}) \left| V_T(\mathbf{R}^H, \mathbf{s}) \right| \Psi^{(f)}_{n_f}(\mathbf{R}^H, \mathbf{s}) \right\rangle . \end{aligned} \tag{17}$$

Here the wavwfunctions $\Psi^{(i)}_{n_i}$ and $\Psi^{(f)}_{n_f}$ are eigenfunctions of the Hamiltonians $H^{(i)}$ and $H^{(f)}$, respectively, where the potential energy in Eq. (16) was modified such as to produce stable potential energy wells for the hydrogen in sites (i) and (f),respectively. The indices n_i and n_f denote the vibrational states. The kinetic energy terms cancel in this expression and the transition operator V_T is the change in the potential energies only. To calculate the matrix elements one has to integrate the potential energy surface over all coordinates.

At sufficiently high temperatures ($T \geq 50$ K) hydrogen tunneling is dominated by phonon effects and can therefore be described by the Hamiltonian (13). The effects due to non-adiabatic electronic excitations are small corrections.

The transition probability can then be found by the use of the Fermi golden rule expression for quantal transition rates:

$$\Gamma = \ll w^{if} \gg_{n_i} \tag{18}$$

with the partial rates

$$w^{if} \;=\; \frac{2\pi}{\hbar}\left|\left\langle \Psi^{(i)}_{n_i}\,(\mathbf{R}^H,\mathbf{s})\left|V_T(\mathbf{R}^H,\mathbf{s})\right|\Psi^{(f)}_{n_f}(\mathbf{R}^H,\mathbf{s})\right\rangle\right|^2\;\delta(E^{(i)}-E^{(f)}) \tag{19}$$

which follows directly from (14). Here $E^{(i)}$ and $E^{(f)}$ are the total energies of the initial and final states, respectively and $\ll\;\;\gg_{n_i}$ denotes thermal averaging over initial states n_i and summation over final states n_f.

The validity of this expression is restricted to the purely incoherent regime. To study the transition from coherent to incoherent tunneling different methods have to be employed. Also correlations between single jumps which become important at high temperatures are not included.

Since in the region of incoherent tunneling, the hopping rate Γ is the same whether the interstitial moves in a double well potential or in a periodic crystal potential, the rate also determines the quantum diffusion coefficient D. For a crystal with cubic symmetry one has

$$D = \frac{z}{6}(2d)^2\Gamma \tag{20}$$

where z is the number of neighbouring interstitial sites and $2d$ the distance between two sites.

The standard solution of Eq.(18) in the framework of small polaron theory [7] involves four additional approximations, all of them valid only at low temperatures. First the *adiabatic approximation* for the hydrogen motion, as in the Born–Oppenheimer approximation for the electrons, it is assumed that the hydrogen wave-function follows the metal ions instantaneously. At any given time the hydrogen tunneling is determined by the frozen lattice configuration at that time. The total wave-function is then approximated by

$$\Psi^{(i)}_n\,(\mathbf{R}^H,\mathbf{s}) \;=\; \Phi^{(i)}_n\,(\mathbf{R}^H;\mathbf{s})\;\chi^{(i)}_n(\mathbf{s}) \tag{21}$$

where $\Phi^{(i)}_n$ represents the hydrogen wave-function which depends parametrically on the lattice configuration. Second, since the adiabatic approximation is only valid at low temperatures, the hydrogen wave-function is restricted to the ground state $\Phi^{(i)}$ (*truncation approximation*).

The hydrogen wavefunction $\Phi^{(i)}$ satisfies

$$\left(T_H + V^{(i)}_{HM}(\mathbf{R}^H,s)\right)\Phi^{(i)}(\mathbf{R}^H;s) = \epsilon^{(i)}\Phi^{(i)}(\mathbf{R}^H;s) \tag{22}$$

and the lattice wavefunctions $\chi^{(i)}_n$ satisfy

$$\left(T_M + V + M + \epsilon^{(i)}\right)\chi^{(i)}_n(s) = E_n\chi^{(i)}_n(s). \tag{23}$$

With this wave-function we can define a "naked" tunnel splitting

$$J(\mathbf{s}) \;=\; 2\left\langle\Phi^{(i)}(\mathbf{R}^H;\mathbf{s})\left|V_T(\mathbf{R}^H,\mathbf{s})\right|\Phi^{(f)}(\mathbf{R}^H;\mathbf{s})\right\rangle \tag{24}$$

where the integration is over the hydrogen coordinates only. This tunnel splitting depends still on the lattice configuration. At low temperatures when only long wavelength phonons are excited this dependence is weak and the tunnel splitting is taken as a constant (*Condon approximation*).

The matrix element in Eq.(17) is obtained by averaging $J(\mathbf{s})$ over the lattice wave-functions $\chi^{(i)}_{n_i}(\mathbf{s})$ and $\chi^{(f)}_{n_f}(\mathbf{s})$

$$M_{n_i n_f} = \tfrac{1}{2} \left\langle \chi^{(i)}_{n_i}(\mathbf{s}) \left| J(\mathbf{s}) \right| \chi^{(f)}_{n_f}(\mathbf{s}) \right\rangle \tag{25}$$

or simpler in the Condon approximation

$$M_{n_i n_f} = \tfrac{1}{2}\, J \left\langle \chi^{(i)}_{n_i}(\mathbf{s}) \mid \chi^{(f)}_{n_f}(\mathbf{s}) \right\rangle . \tag{26}$$

The naked tunneling term is dressed by the overlap of the initial and final state lattice wave-functions, the polaron term.

As a next step, in accordance with the harmonic Ansatz for the inter-metallic coupling, we expand $\epsilon^{(i)}(\mathbf{s})$ in powers of the displacements

$$\epsilon^{(i)}(\mathbf{s}) = \epsilon^{(i)} - \sum_{m\alpha} F^{(i)m}_{\alpha} s^m_{\alpha} + \frac{1}{2} \sum_{\substack{mn\\ \alpha\beta}} s^m_{\alpha} V^{(i)mn}_{\alpha\beta} s^n_{\beta} + \ldots \tag{27}$$

The first term is a constant "chemical" binding energy of the hydrogen, the second is the relaxation energy in lowest order and the third term is a change of the vibrational coupling parameters to lowest order. Taking the displacements $\mathbf{s}^m$ from the coincidence configuration instead of the ideal lattice position, the expansion (22) converges much more rapidly. The force term is reduced to

$$\tilde{\mathbf{F}}^{(i)m} = \mathbf{F}^{(i)m} - \mathbf{F}^{(c)m} \approx \frac{1}{2}\left(\mathbf{F}^{(i)m} - \mathbf{F}^{(f)m}\right) \tag{28}$$

and similarly the force constant change

$$\tilde{\mathbf{V}}^{(i)mn} = \mathbf{V}^{(i)mn} - \mathbf{V}^{(c)mn} \approx \frac{1}{2}\left(\mathbf{V}^{(i)mn} - \mathbf{V}^{(f)mn}\right). \tag{29}$$

The bulk of the force constant change is now absorbed in $\mathbf{V}^{(c)mn}$ which is no longer translationally invariant. Expressing $\mathbf{s}^n$ by phonon creation and annihilation operators, one sees that Eq.(25) is an expansion in powers of phonon operators. Normally, only the first order term is taken (*linear coupling approximation*). Two phonon terms are important in the low temperature limit of nonmetallic crystals [6] whereas in metals the electronic terms, not yet included, prevail in this limit.

At low temperatures one is interested in the coupling to long wavelength phonons which can still be described by plane waves which are slightly shifted between the initial and final states. Expressing the long-range displacements by the respective dipole tensors and using phonon creation and annihilation operators (b^+ and b) the linear term becomes in the long wavelength limit

$$\sum_{n\alpha} \tilde{F}^{(i)n}_{\alpha} s^n_{\alpha} = \left(\frac{\hbar}{2NM}\right)^{1/2} \frac{1}{2} \sum_{\mathbf{q}\lambda} \left[i\left(P^{(i)}_{\alpha\beta} - P^{(f)}_{\alpha\beta}\right) \frac{e(\mathbf{q},\lambda)_{\alpha}}{\sqrt{\omega_{\mathbf{q}\lambda}}}\, q_{\beta} \cos(2\mathbf{q}) \right] + \left[\left(P^{(i)}_{\alpha\beta} + P^{(f)}_{\alpha\beta}\right) \frac{e(\mathbf{q},\lambda)_{\alpha}}{\sqrt{\omega_{\mathbf{q}\lambda}}}\, q_{\beta} \sin(2\mathbf{q}) \right] (b_{\mathbf{q}\lambda} + b^{+}_{-\mathbf{q}\lambda}). \tag{30}$$

To lowest order in $\mathbf{q}$ this gives in view of $\omega_{\mathbf{q}\lambda} \sim q$

$$\sum_{n,\alpha} \tilde{F}_\alpha^n \, s_\alpha^n \;=\; \sum_{\mathbf{q}\lambda} c_{\mathbf{q}\lambda}(b_{\mathbf{q}\lambda} + b^+_{-\mathbf{q}\lambda}) \tag{31}$$

where

$$c_{\mathbf{q}\lambda} \;=\; u_\lambda \, \omega_{\mathbf{q}\lambda}^{s-\frac{1}{2}}. \tag{32}$$

Here $s = 1$ applies to tunneling between sites with different orientation and therefore changing the dipole tensor $\mathbf{P}$, e.g. tetrahedral sites in bcc lattices, and $s = 2$ applies if the site symmetry does not change, e.g. octahedral sites in fcc lattices. It should be noted, however, that in Nb and Ta a nearly isotropic dipole tensor is found so that the $s = 2$ term dominates despite the site symmetry.

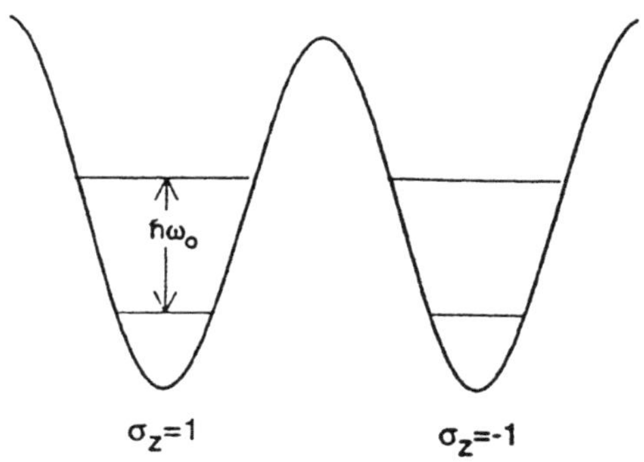

Fig. 12 Double well potential.

Under the above approximations, which are reasonable at low temperatures, the hydrogen metal system can be described by a much simpler Hamiltonian. The hydrogen can be considered to move in a double well potential of a form as depicted in Fig. 12. In view of the large single well excitation energy $\hbar\omega_0$, the hydrogen localized in either well can occupy only the vibrational ground state. This may be described in terms of a pseudo–spin by assigning eigenvalues of a Pauli spin matrix to these states, say, $\sigma_z = -1$ to the initial state (i) and $\sigma_z = 1$ to the final state (f). Of course, the two states are coupled by a tunneling matrix element J which is supposed to include dressing effects by short wave-length phonons not treated properly by Eq.(29). The simplified Hamiltonian is called *spin boson Hamiltonian.*

3.3. EVALUATION OF DIFFUSION RATES

Employing the adiabatic, Condon and linear coupling approximations, described above, one gets the simple expression

$$\Gamma \;=\; \frac{\pi}{2\hbar}\, J_0^2 \;\ll |< \chi^{(i)}(\mathbf{s})|\chi^{(f)}(\mathbf{s}) >|^2 \;\; \delta(E^{(i)} - E^{(f)}) \;\gg_{n}, \tag{33}$$

where J_0 is the tunnel splitting of the naked hydrogen in the fixed potential of the host ions and the temperature dependence is given by the averaged overlap of the total vibrational wave-functions of the host ions with the hydrogen in its initial and final position, respectively. Even in this approximation the evaluation of the transition rate is non-trivial since any vibrational mode in the initial configuration will have a non-vanishing overlap with many modes in the final configuration. Only in the long-wavelength limit will the modes be only weakly affected by the local defect geometry and it is reasonable to approximate them by identical normal modes with only the mean positions of the atoms changed. The total overlap integrals, Eq.(24), factorizes then into products of the single mode overlaps which can be calculated. The transition rate simplifies to [7]

$$\Gamma = \frac{J_0^2}{4\hbar^2} \int_{-\infty}^{+\infty} dt \exp\left\{ -\sum_q S_q \left[\coth(\frac{\hbar\omega_q}{2k_BT})[1-\cos(\omega_q t)] + i\sin(\omega_q t) \right] \right\}. \quad (34)$$

Here we have used the index q to denote the phonon **q**-vector as well as the polarisation. Each mode contributes with a weight given by its Huang-Rhys factor, S_q which is given in terms of projections, d_q, onto the modes q of the difference of the total displacements in the initial and final configurations

$$S_q = \frac{1}{2}\frac{m\omega_q}{\hbar} \mid d_q^{(i)} - d_q^{(f)} \mid^2 \quad (35)$$

with m the host atom mass. These Huang - Rhys factors can be evaluated from the forces $\mathbf{F}^{(i)}$ and $\mathbf{F}^{(f)}$ in Eq. (27). The leading terms of the Huang - Rhys factors are $\propto q$ in the isotropic case and $\propto q^{-1}$ in the anisotropic case.

The above result, Eq. (34), for the transition probability contains a divergent part due to diagonal transitions, i.e. transitions without phonon excitation. These transitions are not dealt with properly in this model. In fact the main effect of the interaction with the conduction electrons is a suppression of the diagonal transitions. They may, therefore, be subtracted giving Γ after some transformations as follows

$$\Gamma = \frac{J_0^2}{4\hbar^2} e^{-S(T)} \int_{-\infty}^{\infty} dt \left[\exp\left\{ \sum_q S_q \frac{\cos(\omega_q t)}{\sinh(\hbar\omega_q/2k_BT)} \right\} - 1 \right]. \quad (36)$$

The average Huang-Rhys factor $S(T)$ describes the average phonon dressing of the tunneling matrix element and is given by

$$S(T) = \sum_q S_q \coth\left(\frac{\hbar\omega_q}{2k_BT}\right). \quad (37)$$

For low temperatures $S(T)$ is of the form

$$S(T) = S(0) + W(T) \quad (38)$$

where the temperature dependent part $W(T)$ is proportional to T^{2s}, with the exponent s introduced in Eq. (32). The zero temperature dressing factor may be absorbed into a renormalized tunneling frequency

$$\Delta = \frac{J_0}{\hbar} e^{-S(0)/2} \quad (39)$$

and the transition rate then takes the form

$$\Gamma = \frac{\Delta^2}{4} e^{-W(T)} \int_{-\infty}^{+\infty} dt \left[e^{-\Xi(T,t)} - 1 \right] \tag{40}$$

in which

$$\Xi(T,t) = -\sum_q S_q \frac{\cos(\omega_q t)}{\sinh(\hbar\omega_q/2k_B T)}. \tag{41}$$

For low temperatures the exponential functions can be expanded. The lowest order contribution is then by two phonon processes giving a transition rate $\propto T^7$ in the isotropic case [7]. For non-symmetric transitions, i.e. for anisotropic defects in an external field, one can get also contributions from one phonon processes giving transition rates $\propto T$ [26]. However, these low temperature results have to be addressed with caution, since the electronic influence may already be substantial in this range of temperatures. The high temperature limit is treated in the Debye approximation of the phonon spectrum and Γ takes the simple form [7].

$$\Gamma \ = \ (\pi/16\hbar^2 E_c k_B T)^{1/2} \ J^2 \ \exp\left(-E_c/k_B T\right) \tag{42}$$

This describes a near Arrhenius behaviour with the coincidence energy E_c Eq.(13) as activation energy. Even by fitting the values of J and E_c this result cannot reproduce the measured diffusion constants, in particular the observed change in slope of the measured diffusion constant in Nb and Ta.

To remedy this shortcoming, the more serious approximations made, in particular the Condon approximation, have to be avoided. In the "occurrence probability" method [27] the adiabatic approximation is partially retained but one averages over possible coincidence configurations. Also transitions between differently excited localized H states are included. Since different coincidence configurations have different tunneling probabilities for the "naked" H, an additional temperature dependence is found which can explain the change in slope of the Arrhenius-plot of diffusion of H in Nb. Gillan [28] uses quantum molecular dynamics to identify the symmetric coincidence configurations. A direct calculation of the transition rate is not yet possible and the influence of asymmetric coincidence configurations is difficult to estimate. Both methods allow a comparison with classical diffusion.

A more rigorous approach is the embedded cluster method [7]. There one makes use of the spatial localization of the resonant and localized modes. We do not separate the hydrogen and host degrees of freedom but treat a cluster of atoms (typically 6-21 atoms) explicitly and use the above approximations only for the embedding of the cluster into the rest of the host crystal. The limitations of this method are mainly due to computer capacity. The number of degrees of freedom one can treat explicitly is limited owing to the number of integrations involved in calculating a single transition element w^{if}, Eq.(22). On the temperature side one is limited by the rapid increase in terms w^{if} contributing to the total transition rate, Eq.(15).

Fig. 13 shows the result of such a calculation for Nb:H. The harmonic approximation for the wave-functions already gives a qualitative description. To get a quantitative picture anharmonic corrections have to be taken.

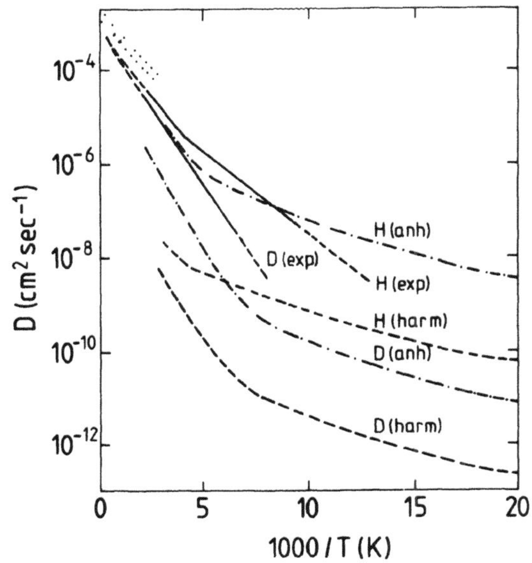

Fig. 13 Diffusion coefficients for H and D in Nb. Solid line: experimental results [17], broken line: calculation with harmonic wavefunctions, dash-dotted line: calculation including anharmonic corrections, dotted line Vineyard approximation for high temperatures [29].

According to this model the diffusion of H at room temperature is still influenced strongly by quantum mechanics. The calculation is not sufficiently accurate to see whether and how the classical limit is reached at higher temperatures. For comparison Fig. 13 shows also the results of a calculation in the Vineyard approximation [10] for classical diffusion. The shift in the prefactor between H and D is about what one would expect from the mass difference. The activation energy is 97 meV and 96 meV for H and D, respectively. Here 90 meV is the potential energy part. The discrepancy between our quantum results and these classical ones is not sufficient to draw any definite conclusion.

3.4. NON-ADIABATIC ELECTRONIC EFFECTS

At low temperatures, typically $T < 50\text{K}$, the number of phonons which can contribute to the transition becomes very small due to the $Z(\nu) \propto \nu^2$ dependence of the phonon density of states. This also causes the divergence of the diagonal term in Eq.(34). The electrons, on the other hand, have in metals a finite density of states at the Fermi surface which in turn provides a finite density of states for excitations with energy $\rightarrow 0$. This opens additional transition channels involving electronic excitations. Formally the electronic excitations can be treated like the phonons in the long wavelength limit. In addition to the Huang-Rhys factors in the integrand in Eq.(34) there will then be a corresponding electronic term. In the electron dominated low temperature regime the transition rate becomes

$$\Gamma = \frac{\Gamma(K)}{\Gamma(1-K)} \frac{\Delta}{2} \left(\frac{2\pi k_B T}{\hbar \Delta} \right)^{2K-1} \tag{43}$$

where Δ is a renormalized tunnel frequency and K is the Kondo parameter which characterizes the coupling of the tunnel system to the electronic excitations. With this expression the low temperature jump rates (Fig. 10) can be reproduced.

4. Finite concentrations

In lowest order approximation one expects, with increasing concentration of hydrogen, a decrease of the diffusion coefficient due to site blocking. A hydrogen atom blocks its site and possibly also the neighbouring sites to all other atoms and thus reduces the jump probability. The diffusion constant becomes

$$D = p(x) D_0 \exp - E_a/kT, \tag{44}$$

where p(x) is the blocking factor. For hydrogen diffusion in PdH_x this blocking factor was found to be simply $p(x) = 1 - x$, indicating that each H atom only blocks its own site [30]. For $x = 1$ all octahedral sites in the fcc lattice are occupied and no diffusion by single jumps between such sites is possible any longer.

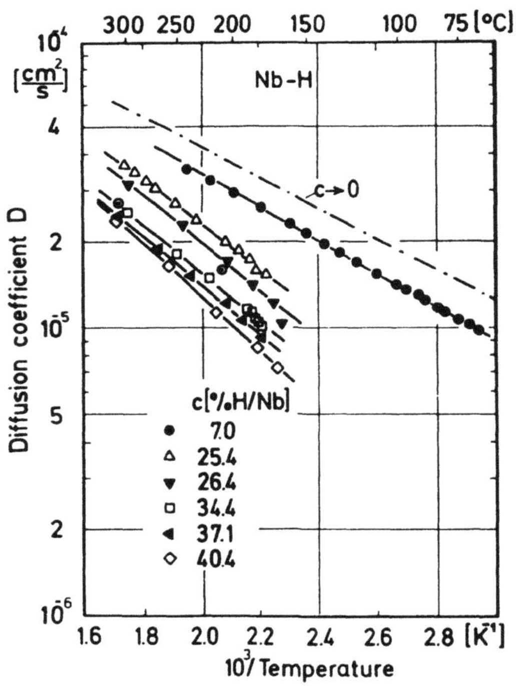

Fig. 14 Tracer diffusion coefficient of H in Nb for different concentrations [32].

In other systems a number of effects complicate the behaviour. Most important an increase in concentration can lead to a phase transition thus altering the structure. Partial ordering and hydrogen – hydrogen interactions will also alter the mobility. A single hydrogen atom will block more than one site, particularly when possible sites are close to each other as the terahedral sites in the bcc lattice. The hydrogen atoms interact with each other directly ("chemical interaction") as well as indirectly via the host lattice (elastic interaction). This can lead to correlated motions of hydrogen atoms. This was observed in experiments on NbH_xD_y [31]. Keeping $x + y = 0.60$ constant and increasing the fraction of deuterium atoms, one observes a slowing down of the diffusion of H. The slower diffusion of D affects the diffusion of H. Since the static interaction should not be strongly isotope dependent, this is a dynamic effect.

The various effects can in general cause a dependence of the prefactor as well as the activation energy on concentration. As an example Fig. 14 shows measurements of the tracer diffusion coefficient of hydrogen in Nb at concentrations of up to 40% [32]. One observes a clear decrease of the diffusion coefficient with increasing concentration. The apparent activation energy increases.

The general case can show a rather complicated behaviour. The diffusion coefficients can even depend on the shape of the sample [32]

ACKNOWLEDGEMENT

The author is grateful to Prof. Wipf for allowing him to use his review prior to publication.

References

[1] Völkl, J. and Alefald, G. (1978) in: *Hydrogen in Metals I* (G. Alefeld and J. Völkl eds.), p. 321, Berlin:Springer Verlag.
[2] Fukai, Y. (1993), *The Metal-Hydrogen System*, Berlin: Springer Verlag.
[3] Wipf, H. to be published in: *Hydrogen in Metals III* (H. Wipf ed.), Berlin: Springer Verlag.
[4] Grabert, H. and Schober, H.R. to be published in:*Hydrogen in Metals III* (H. Wipf ed.), Berlin: Springer Verlag.
[5] Manchester, F.D. (ed.) (1991) *Metal Hydrogen Systems*, J. Less. Comm. Metals bf 172-174.
[6] Kagan, Yu. and Prokof'ev, N.V. (1992) in: *Quantum Tunneling in Condensed Media* (Yu. Kagan and A.J. Legget eds.), p. 37, Amsterdam: North-Holland.
[7] Flynn, C.P. and Stoneham, A.M. (1970), Phys. Rev. B **1**, 3966.
[8] Schober, H.R. and Stoneham, A.M. (1988), Phys. Rev. Lett. **60**, 2307.
[9] Kondo, J. (1984), Physica B **125**, 279 and **126**,377.
[10] Vineyard, G.H. (1957), J. Phys. Chem. Sol. **3**, 121.
[11] Schober, H.R., Petry, W. and Trampenau, J. (1992), J. Phys: Condens. Matter **4**, 9321.
[12] Wicke, E., Brodowsky, H. and Züchner, H. (1978) in: *Hydrogen in Metals II* (G. Alefeld and J. Völkl eds.), p. 73, Berlin:Springer Verlag.
[13] Völkl, J., Wollenweber, G., Klatt, K.-H. and Alefeld, G. (1971), Z. Naturforsch. **26a**, 922.
[14] Katsuta, H., Farraro, R.J. and McLellan, R.B. (1979), Acta Met. **27**, 1111.
[15] Sicking, G., Glugla, M. and Huber, B. (1983), Ber. Bunsenges. **87**, 418.
[16] Higelin, G., Kronmüller, H. and Lässer, R. (1984), Phys. Rev. Lett. **53**. 2117.

[17] Qi, Zh., Völkl, J., Lässer, R. and Wenzl, H. (1983), J. Phys. F: Met. Phys. **13**, 2053.
[18] Messer, R., Blessing, A., Dais, S., Höpfel, D., Majer, G., Schmidt, C., Seeger, A. and Zag, W. (1986), Z. Phys. Chem. NF Suppl.-H **2**,61.
[19] Völkl,J., Wipf, H., Beaudry, B.J. and Gschneidner, K.A. (1987), Phys. Stat. Sol. (b) **144**, 315.
[20] Arnold, G. and Welter, J.-M. (1983), Metall. Tans. A **14**, 1573.
[21] Schmidt, R., Schlereth, M., Wipf, H., Assmus, W. and Müllner, M. (1989), 2473.
[22] Beisenherz, D., Guthardt, D. and Wipf, H. (1991), J. Less-Common Met. **172-174**, 693.
[23] Steinbinder, D., Wipf, H., Dianoux, A.-J., Magerl, A., Neumaier, K., Richter, D. and Hempelmann, R. (1991), Europhys. Lett **16**, 211.
[24] Wipf, H., Steinbinder, D., Neumaier, K., Gutsmiedl, P., Magerl, A. and Dianoux, A.-J. (1987), Europhys. Lett. **4**, 1379.
[25] Kehr, K. (1978) in: *Hydrogen in Metals I* (G. Alefeld and J. Völkl eds.), p. 197, Berlin:Springer Verlag.
[26] Teichler, H. and Seeger, A. (1981), Phys. Lett. A **82**, 91.
[27] Klamt, A. and Teichler, H. (1986), Phys. Stat. Sol. (b) **134**, 533.
[28] Gillan, M.J. (1987), Phys. Rev. Lett. **58**, 563.
[29] Schober, H.R. and Stoneham, A.M. (1991), J. Less. Comm. Met. **172-174**, 538.
[30] Cotts, R.M. (1978) in: *Hydrogen in Metals I* (G. Alefeld and J. Völkl eds.), p. 227, Berlin:Springer Verlag.
[31] Fukai, Y., Kubo, K. and Kazama, S. (1979), Z. Phys. Chem. **115**, 181.
[32] Bauer, H.C., Völkl, J., Tretkowski, J. and Alefeld, G. (1978), Z. Phys. B **29**, 17.

Chapter 13

AN INTRODUCTION TO THE STUDY OF HYDROGEN MOTION IN METALS BY NUCLEAR MAGNETIC RESONANCE

P C Riedi and J S Lord
J F Allen Laboratories
Department of Physics and Astronomy
University of St Andrews
North Haugh
St Andrews
Fife
KY16 9SS

1. Introduction

At room temperature the diffusion constant of hydrogen in many metals is comparable to that for ionic diffusion in water. The extensive theory of the NMR of liquids therefore provides a natural starting point for the discussion of hydrogen motion in metals but it should be recognized that there will be points of distinction between the two processes. For example, in a metal MH_x the hydrogen moves between specific interstitial sites and therefore the motion of a given particle may be restricted as x increases because a neighbouring site is already occupied.

The spin of the proton is 1/2 so the NMR properties of ^1H in MH_x are determined by the motion of the hydrogen and the magnetic dipole fields due to both metal and hydrogen atoms. The relaxation of the M nucleus may also be determined by the fluctuating fields set up by the hydrogen motion. The NMR of MD_2 is however more complicated because, with a spin of 1, the ^2D NMR is also sensitive to electric field gradients. We will therefore concentrate on ^1H NMR but briefly consider the problem of ^2D NMR in Section 7.

In the next section we discuss the three quantities of interest for the diffusion of ^1H in MH_x: the diffusion constant (D), the mean square step length $\overline{\ell^2}$ and the mean residence time at a site (τ_D). The relaxation process for ^1H NMR in MH_x depends upon the existence of fluctuating magnetic fields due to the particle motion so in section 3 an introduction is given to the essential features of stationary random processes. The original theory of NMR relaxation in liquids due to Bloembergen, Purcell and Pound (BPP) is sketched in Section 4 and then the modifications required for MH_x considered. The NMR linewidth of ^1H is also sensitive to hydrogen motion and this effect is considered in Section 5.

The linewidth and relaxation times of the ^1H NMR are functions of τ_D and a model of the hydrogen motion is required before D can be calculated. However D can be measured directly by performing pulsed NMR in a magnetic field gradient as shown in Section 6. The features of ^2D NMR that distinguish it from ^1H NMR, and make it less generally useful, are considered in Section 7 and the NMR of M Nuclei in Section 8. Finally, ^1H NMR in paramagnetic metals is considered in Section 9.

F. Grandjean et al. (eds.), Interstitial Intermetallic Alloys, 317–332.

The whole topic of hydrogen in metals has recently been reviewed by Fukai [1] and another review will appear shortly [2]. The application of NMR to MH_x and MD_x has been reviewed by Cotts [3,4]. We will not therefore present proofs of e.g. the equations relating T_1 to the fluctuation spectrum or always refer to original publications. In particular, it should be noted that no attempt has been made to make a complete survey of NMR in MH_x and the experimental results shown are meant to be illustrative rather than the first or the best of their type.

As always in physics the fullest information about a system is obtained by combining many techniques to test the theoretical basis of the subject and, apart from NMR, valuable information on hydrogen motion in metals has been obtained from e.g. quasi-elastic neutron scattering, internal friction, acoustic attenuation, and muSR. The papers presented at the International Symposium on Metal-Hydrogen Systems provide a useful review of the subject [5].

2. Diffusion

For time intervals long compared to the mean residence time (τ_D) at a site, the simplest expression for the diffusion constant (D) is given by a random walk in three dimensions,

$$D = \frac{\overline{\ell^2}}{6\tau_D} \tag{2.1}$$

for a mean square step size $\overline{\ell^2}$, see e.g. [1,2].

The diffusion constant defined above is sometimes called the intrinsic diffusion constant because it refers to the motion of a labelled particle in a uniform material rather than to the flow down a concentration gradient.

The motion of a particle on a lattice is never completely random, as implied in equation (2.1) and two corrections may have to be made for ^1H motion in MH_x. Firstly, the right hand side of equation (2.1) must be multipled by the tracer correlation factor, f_t, because an atom has a greater probability of reversing its last jump then of jumping to other sites. The maximum value of f_t (unity) occurs in the limit that x goes to nought. Secondly, the mean residence time will increase if an attempted jump is frustrated because the intended site is already occupied. The value of τ_D appropriate in the limit that x goes to nought is increased by a site blocking factor that, for single site blocking, is linear in x and also depends upon the site geometry. In the case of motion between octahedral interstitial sites in a fcc metal lattice for example, τ_D is increased, and therefore D decreased, by a factor (1-x), see section 6 and figure 7.

The temperature dependence of τ_D and D in MHx is often found to follow the classical Arrhenious expression for thermally activated hopping,

$$\tau_D = \tau_\infty e^{E_a/kT} \tag{2.2}$$

$$D = D_o e^{-E_a/kT} \tag{2.3}$$

where τ_∞ and D_o are constants and E_a is the activation energy. The agreement with the classical theory is however rather superficial. The value of E_a may depend upon the range of temperature over which the data is fitted and the value of D does not scale with the mass of ^1H and ^2D as expected classically. The motion is therefore determined by quantum mechanical tunneling between interstitial sites rather than classical jumps over an energy barrier [1,2].

The values of $D_o(E_a)$ for ^1H are very different for bcc and fcc metals, see [1] for a review. The bcc values are typically $(1\text{-}5)\times10^{-4}cm^2s^{-1}$ (0.05-0.15eV) and the fcc, $(2\text{-}10)\times10^{-3}cm^2s^{-1}$ (0.20-0.40eV).

3. Stationary Random Processes

The behaviour of a randomly fluctuating phenomena, such as thermal noise in an electrical circuit, is well defined statistically but not predictable in detail. A stationary random process is one for which the statistical properties depend upon time differences but not upon the absolute time. Specifically, we consider stationary phenomena for which positive and negative fluctuations are equally probable and therefore the average mean value of the variable V(t) is zero, ie

$$\begin{aligned} < V(0) > &= < V(t) >= 0 \\ < V(0)^2 > &= < V(t) >^2 \neq 0 \end{aligned}$$

The autocorrelation function is defined as

$$G(t) =< V(t')V(t' + t) > \tag{3.1}$$

From our definition of a stationary random process we know that G(t) is independent of t' and that,

$$\begin{aligned} &G(0) =< V(t'^2) > \neq 0 \\ &G(t) \to 0 \quad (t >> \tau_c) \end{aligned}$$

where τ_c is some characteristic (correlation) time for the system.

The simplest form for G(t),

$$G(t) = G(0)e^{-t/\tau_c} \tag{3.2}$$

was used in the original (BPP) theory of dipolar relaxation discussed in Section 4.

Equations 3.1 and 3.2 define the statistical properties of the random process in the time domain. The Fourier transform of equation 3.1 is called the power spectrum or spectral density function. In electrical engineering texts the power spectrum is written $P(\omega)$ but in BPP theory it is usually written as $J(\omega)$ where,

$$J(\omega) = \int_{-\infty}^{\infty} G(t)\ e^{-i\omega t} dt \tag{3.3}$$

and, from equation 3.2,

$$J(\omega) = G(0) \frac{2\tau_c}{1+\omega^2\tau_c^2} \tag{3.4}$$

The two important features of $J(\omega)$ are, (1) it is independent of frequency "white" for $\omega^2\tau_c^2 << 1$, figure 1, and (2), the total noise power integrated over all frequencies, i.e. the area under the $J(\omega)$ curve, is independent of τ_c since,

$$\begin{aligned} \int_{-\infty}^{\infty} J(\omega)d\omega &= 2G(0) \int_{-\infty}^{\infty} \frac{\tau_c}{1+\omega^2\tau_c^2}\ d\omega \\ &= 2\pi G(0) \end{aligned}$$

We shall see in section 4 that the important fluctuations for the longitudinal NMR relaxation time, T_1, are those with frequency components near the NMR angular frequency ω_o and therefore the minimum value of T_1 from fluctuating fields will occur when $\omega_o\tau_c \approx 1$. A fuller account of random processes, with particular application to NMR, will be found in [6-8].

4. Dipolar Relaxation

4.1 BLOEMBERGEN, PURCELL AND POUND (BPP) THEORY

The motion of the ^1H through MH_x will lead to it sensing fluctuating dipole fields from other protons and also from the M nuclei. In a strongly paramagnetic or ferromagnetic material there will be much bigger fluctuating fields from the electronic M moments but we delay consideration of this feature until Section 9. The relationship between the fluctuating fields, with correlation time τ_c, and the mean dwell time of the protons(τ_D) depends upon the origin of the fields. Since all the protons are in motion the correlation time for ^1H-^1H interactions is given by $\tau_D/2$ while the correlation time for ^1H-M interactions is simply τ_D.

The original BPP theory [9] was derived for nuclear relaxation in a liquid and is clearly inappropriate in detail for solids where the diffusion process is limited to a particular set or sub-set of interstitial sites. The modifications required for diffusion in

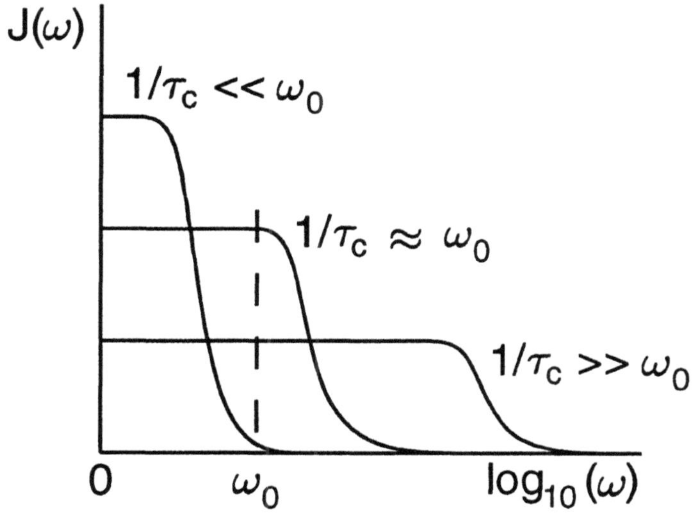

Figure 1. The power spectrum (spectral density function) as a function of log |ω| for different correlation times, τ_c.

a solid will be considered in section 4.2 but the original BPP form is adequate for the measurement of the activation energy (E_a).

The BPP equations for longitudinal relaxation (T_1), transverse relaxation (T_2) and spin lattice relaxation in the rotating frame ($T_{1\rho}$) may be written in terms of the spectral density function. In many cases, eg PdH_x, the ^{1}H-M interaction is negligible and only H-H interactions need to be considered.

The BBP equations are then:

$$(T_1)^{-1} = 4C\left[J^{(1)}(\omega_o) + J^{(2)}(2\omega_o)\right] \tag{4.1}$$

$$(T_2)^{-1} = C\left[J^{(0)}(o) + 10J^{(1)}(\omega_o) + J^{(2)}(2\omega_o)\right] \tag{4.2}$$

$$(T_{1\rho})^{-1} = C\left[J^{(0)}(2\omega_1) + 10J^{(1)}(\omega_o) + J^{(2)}(2\omega_o)\right] \tag{4.3}$$

$$C = 3\gamma_n^4\hbar^2 I(I+1)/8$$

where γ_n is the gyromagnetic ration, I is 1/2, ω_o is the NMR frequency and $\omega_1 = \gamma_n B_1$, where $2B_1$ is the applied RF field, see [7] for details. The definition and measurement of $T_{1\rho}$ is discussed in section 4.3.

In order to evaluate $J^{(p)}(\omega)$ it is necessary to evaluate the autocorrelation function for the dipole interactions. As suggested in section 3, the simplest form for the autocorrelation function is given by equation 3.2 and then,

$$J^{(p)}(\omega) = G^{(p)}(o)\ \frac{2\tau_c}{1+\omega^2\tau_c^2} \tag{4.4}$$

On taking an average over all crystal orientations

$$G^{(o)}(o):G^{(1)}(o):G^{(2)}(o) = 6:1:4 \tag{4.5}$$

and $G^{(p)}(o)$ involves a dipole-dipole sum ($1/r^6$) over the interstitial sites occupied by hydrogen [7].

The important feature of equations 4.1 - 4.5 is that $(T_1)^{-1}$ has a maximum for $\omega_o\tau_D \approx 1$ and $(T_{1\rho})^{-1}$ a maximum for $\omega_1\tau_D \approx 1$. Since the field for resonance is $\approx 1T$, much greater than the RF field, $\approx 10^{-2} - 10^{-3}T$, a combination of T_1 and $T_{1\rho}$ measurements allows a wide range of $\tau_D(10^{-5} - 10^{-8}s)$ to be measured. This is important because, particularly in fcc metals, the form of τ_D commonly found,

$$\tau_D = \tau_\infty e^{E_a/kT} = \tau_\infty e^{T_a/T} \tag{4.6}$$

with $E_a \approx 0.38eV$, $T_a \approx 4400K$, ensures a change of τ_D of the order of 2500 between say, 250 and 450K.

4.2 RELAXATION IN METALS

Relaxation via nuclear dipole-dipole interactions is only one of a number of possible relaxation mechanisms for ^1H in a metal. In general we may write for the observed relaxation rate,

$$T_1^{-1} = (T_{1e})^{-1} + (T_{1d})^{-1} + (T_{1m})^{-1} \tag{4.7}$$

where $(T_{1e})^{-1}$ is the relaxation rate due to the contact interaction with conduction electrons, $(T_{1d})^{-1}$ is given by equation (4.1), and $(T_{1m})^{-1}$ is the relaxation rate due to electronic moments. This last term may be due to paramagnetic impurities in MH_x, where M is Sc for example, and is then only important over a limited range of temperature, or may be dominant due to eg Pr^{3+} in PrH_x, section 9. The contact interaction has the form $T_{1e}T$= constant and can be subtracted from the measured relaxation rate if measurements are made over a sufficiently wide range of temperature.

The temperature and frequency dependence of T_{1d} is in quite good agreement with BPP near the minimum in T_{1d}, figure 2. Taking $\omega_o\tau_d$ as a constant at the minima for different ω_o, the value of E_a in equation (4.6) may be found, figure 3. The value of E_a is found to be consistent with the more sophisticated calculations described below but the value of τ_∞ is about a factor of 2 too high.

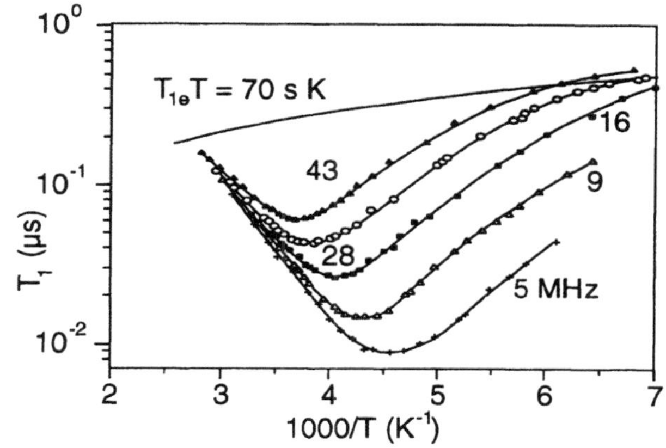

Figure 2. T_1 as a function of 1/T for ^1H in $PdH_{0.76}$ at several frequencies. The electron contribution to T_1 is shown as T_{1e}. After [10].

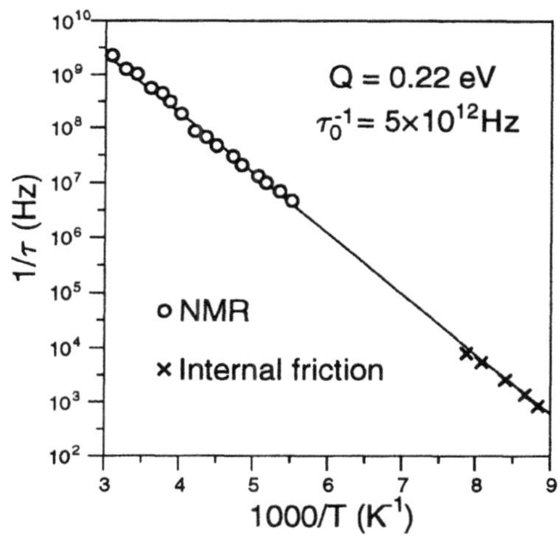

Figure 3. The inverse dwell time as a function of 1/T for $PdH_{0.76}$. NMR (o), internal friction (Δ). After [10].

A very general analysis, using dielectric theory, of experimental relaxation data has shown [11] the weakness of BPP theory but is in excellent agreement with a new expression for $J^{(p)}(\omega)$ at low frequency given by Sholl [12]. Sholl finds that $J^{(p)}(\omega)$ may be written, for $\omega\tau_D << 1$,

$$J^{(p)}(\omega) = J^{(p)}(o) - A_p\ \omega^{1/2} \tag{4.8}$$

rather than,

$$J^{(p)}(\omega) = J^{(p)}(o) - A\omega^2$$

as in BPP theory, and hence

$$T_1^{-1}(\omega) = T_1^{-1}(o) - A'\omega^{1/2} \tag{4.9}$$

The form of equation (4.9) is in excellent agreement with experimental values of T_1 for $\omega\tau_D << 1$, figure 4, and the diffusion constant could also be deduced from the value of A'.

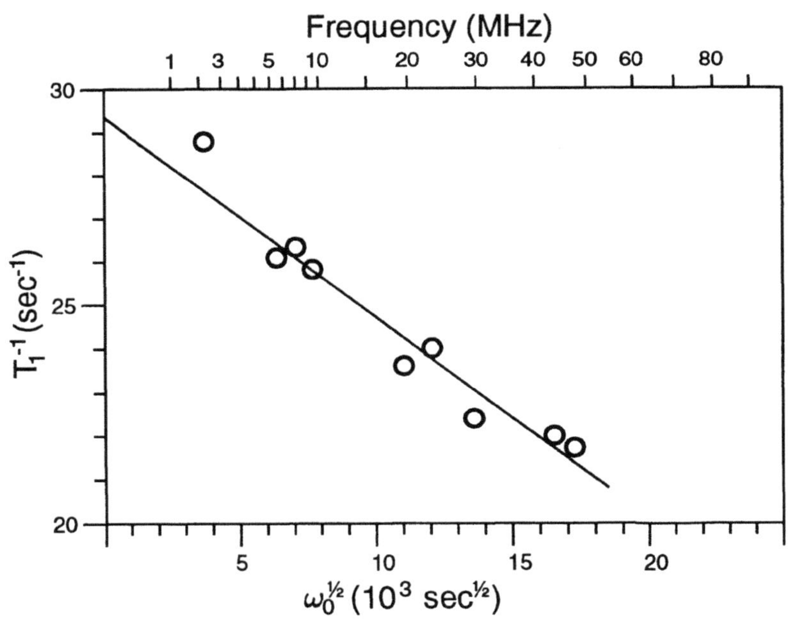

Figure 4. Frequency dependence of $1/T_1$ for $TiH_{1.63}$ at 725K for $\omega_o\tau_D < 1$. After [13].

4.3 EXPERIMENTAL

The pulse sequences for a simple measurement of T_1 have been discussed in an earlier chapter. Either a $\pi/2-t-\pi/2$ sequence or a $\pi-t-\pi/2$ sequence followed by a measurement of the Free Induction Decay (FID) has been used successfully. In measurements on host nuclei with quadrupole moments the recovery of the FID is however often found to be non-exponential after either of these pulse sequences. An exponential decay may be recovered if the resonance is saturated by a rapid sequence of n $\pi/2$ pulses rather than the initial single pulse.

The value of $T_{1\rho}$, the spin lattice relaxation time in the rotating frame, is measured using the pulse sequence shown in figure 5. A $\pi/2$ RF pulse of magnitude $2B_1$ is applied at exact resonance. Taking the rotating component of the RF field, B_1, to be along x' in the rotating frame, the initial nuclear magnetization, $M_i\hat{\mathbf{z}}$, will now be along y'. The phase of the RF is now changed as quickly as possible by $\pi/2$ so that B_1 is parallel to M_i. The nuclear magnetization is said to be spin-locked to the RF field B_1, and in the rotating frame at exact resonance this is the only field seen by the nuclei.

The nuclear magnetization at equilibrium is proportional to the magnetic field and therefore begins to relax exponentially in the rotating frame towards $M_f = M_i B_1 / B_o$, where $B_o\hat{\mathbf{z}}$ is the static field for resonance, with time constant $T_{1\rho}$. The value of $T_{1\rho}$ is found by measuring the FID as a function of the spin-lock time.

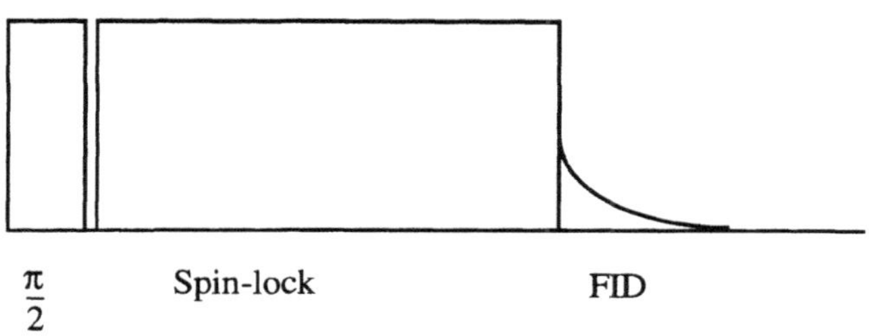

Figure 5. The RF pulse sequence for the measurement of $T_{1\rho}$. A $\pi/2$ pulse along x' is followed by a spin-lock pulse along y' in the rotating reference frame.

5. Linewidths

The most widely used technique for the evaluation of τ_D and its temperature dependence has been the measurement of T_1 and $T_{1\rho}$ but the motion of the hydrogen in MH_x also has consequences for the ^1H linewidth [7]. At low temperature, where τ_D is long, the random distribution of nuclear dipole moments leads to a rigid lattice linewidth of $\approx 10^{-3}$T. As the temperature is increased the rapid motion of the ^1H

leads to motional narrowing until finally the linewidth is determined by the homogeneity of the magnetic field across the sample. The simple expression [7],

$$\Delta\omega = \frac{\tau_D}{2}\left\langle \Delta\omega_d^2 \right\rangle \qquad (5.1)$$

where $\Delta\omega$ is the measured linewidth and $\left\langle \Delta\omega_d^2 \right\rangle$ is the dipolar second moment of the line, is sufficient to obtain a value of τ_D from measurements of the strongly narrowed line, figure 6. The range of τ_D that can be measured by this technique is only about 30 compared to 10^3 from T_1 and $T_{1\rho}$ measurements.

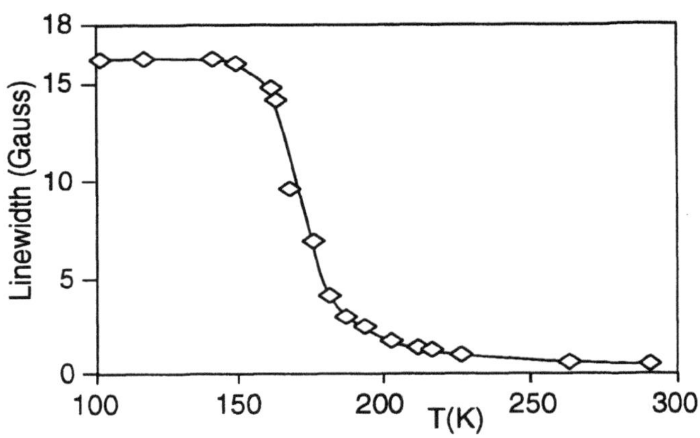

Figure 6. Linewidth of ^1H NMR of $NbH_{0.709}$ as a function of temperature. After [14].

6. Diffusion Measurements

In order to calculate the diffusion constant (D) from relaxation measurements it is necessary to evaluate τ_D and the mean square step size of the particle motion and then to use equation 2.1. The resulting value for D is therefore model dependent. It was appreciated by Hahn, in the first paper on spin echoes, that the $\pi/2$ - π pulse sequence would not re-focus spins that had moved into a region with a different field. The value of D could therefore be measured from the dependence of the magnitude of the echo on the field gradient (G) in the same direction as the original uniform field, which is of the form,

$$M(2t) = M(o)\exp\left[-(2t/T_2) - \left(2\gamma_n^2 G^2 D t^3/3\right)\right] \qquad (5.1)$$

However, there are experimental problems when D is small, and therefore G has to be large, if G is present when the echo is observed. A superior technique is to turn the field gradient on between the $\pi/2$ and π pulse and between the π pulse and the echo, see Cotts [3] and [4] for a discussion of this complicated technique.

The values of D measured by the pulsed field gradient method are found to follow the Arrhenius law, equation 2.3, with the same value of E_a found for τ_D from relaxation measurements, figure 7.

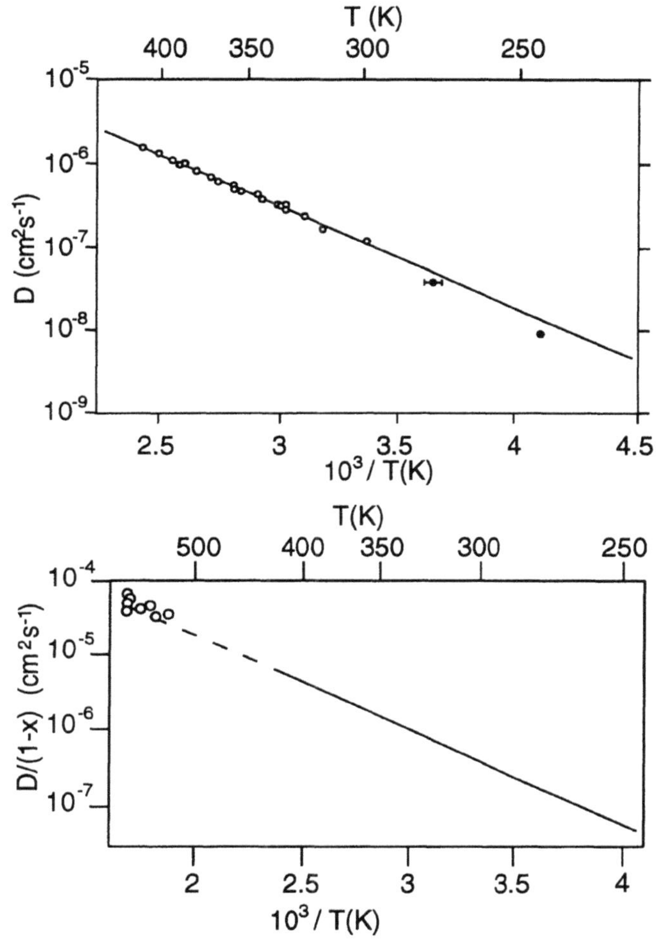

Figure 7 (a) D as a function of 1/T for $PdH_{0.7}$ from Pulsed Field Gradient (o) and T1 minimum (•) measurements.

7 (b) D for α'-PdH_x, corrected for site blocking by (1-x), as a function of 1/T. The full line is from NMR, points from neutron scattering. After [15].

A combination of pulse field gradient measurements and T_1 measurements, to give τ_D, enables the mean square step size of the motion to be calculated, equation 2.1, and compared with possible paths in the metal. The hydrogen ions in $\gamma - TiH_x$ for example randomly occupy the tetrahedral intersitial sites of the fcc Ti lattice. The tetrahedral sites form a simple cubic lattice. The two most likely jump paths are to nearest or third nearest neighbour tetrahedral sites. As will be seen from figure 8 the computer calculation of D is sufficiently accurate to make clear that the ^1H jumps to the nearest neighbour sites.

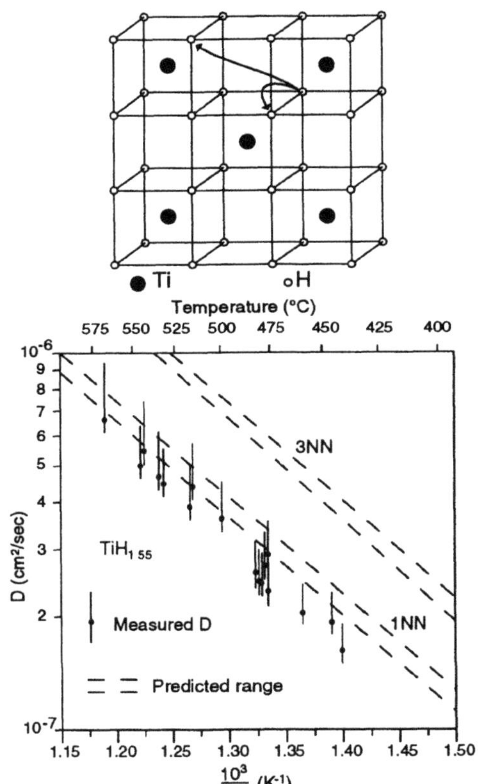

Figure 8a Crystal structure of $\gamma - TiH_x$. The hydrogen atoms randomly occupy the tetrahedral interstitial sites. The most probable jumps are to nearest and third nearest neighbour tetrahedral sites as shown.

8b Measured and calculated values of D for $\gamma - TiH_x$ for the two paths shown in (a). After [16].

7. NMR of ^2D

The properties of ^2D and ^1H are summarized in table 1. The dipole-dipole interaction, proportional to γ_n^2, is some 50 times weaker for ^2D than for ^1H and the ^2D linewidth is usually determined by the interaction between its quadrupole moment and the electric field gradients in the material. The linewidth will still be motionally narrowed, for sufficiently fast motion, if the ^2D samples all the interstitial sites of a particular type and therefore an unnarrowed line may provide evidence for restricted motion through a sub-set of sites.

The NMR spectrum obtained, for spin greater than 1/2,from a polycrystalline material in a large magnetic field is broadened and shifted by the electric quadrupole interactions if the nuclei are immobile. The predicted shapes of the spectra for different values of the spin have been calculated [17] and can be used to determine the strength and symmetry of the nuclear interaction with the electric field gradients.

Table 1. The properties of the ^1H and ^2D isotopes of hydrogen.

	^2D	^1H
Natural abundance (%)	1.56×10^{-2}	99.98
Spin	1	$\frac{1}{2}$
$(\gamma_n/2\pi)$ (MHz/T)	6.54	42.6
Quadrupole moment ($\times 10^{-24} cm^2$)	2.8×10^{-3}	0

The maximum intensity of the ^2D spectra will be much smaller than that of ^1H at the same concentration, because of the smaller value of γ_n and the increased width of the spectra due to the quadrupole interaction, and therefore there have been relatively few experiments on MD_x. A review of the data has been given by Cotts [3].

8. Relaxation of Metal NMR

When the NMR relaxation of the metal nucleus is sensitive to ^1H motion it is possible to test the consistency of the NMR theory by measuring the relaxation rate for each type of nucleus. A good example is the relaxation of ^{45}Sc (I=7/2) and ^1H in ScH_x [18]. In this case the most important relaxation mechanism for T_1 of ^1H is the fluctuations of the dipole field of the ^{45}Sc due to the ^1H motion. The T_1 of the ^{45}Sc is due to a coupling between the electric quadropole moment of the ^{45}Sc and fluctuating electric field gradients due to the ^1H motion.

The ^{45}Sc T_1 data was partitioned,

$$T_1^{-1} = \left(T_{1Q}\right)^{-1} + \left(T_{1e}\right)_{Sc}^{-1}$$

and for the ^1H, as usual,

$$T_1^{-1} = \left(T_{1d}\right)^{-1} + \left(T_{1e}\right)_{H}^{-1}.$$

The conduction electron relaxation rates in ScH_x are a function of x, because $T_{ie}T$ is proportional to the density of states at the Fermi level, $N(\varepsilon_F)$, and $N(\varepsilon_F)$ was found to decrease as x increased. The value of T_{1d} for ^1H was however independent of x showing that it was the interaction with the ^{45}Sc that determined the relaxation rate.

The values of the activation energy for τ_D, Ea of equation 2.2, obtained from the analysis of the ^1H data, 0.54 eV, was identical to that found from the ^{45}Sc NMR. There was, as usual, less good agreement for the value of τ_∞. The ^{45}Sc value for small x was about 1×10^{-14}s, while the ^1H value was $(0.5\text{-}2.0) \times 10^{-14}$s. The overall agreement between the ^{45}Sc and ^1H NMR analysis and the inelastic neutron scattering data was considered to be very satisfactory.

9. Relaxation by electronic moments

The two processes considered so far for the ^1H relaxation rate in MH_x, $\left(T_{1d}\right)^{-1}$ and $\left(T_{1e}\right)^{-1}$, are negligible in comparison with the contribution $\left(T_{1m}\right)^{-1}$ from the dipolar coupling between ^1H and the electronic moment of a rare earth ion. The ^1H NMR is therefore an excellent probe of the electronic magnetization. As an example we consider $PrH_{2.0}$ and $PrH_{2.5}$ [19]. In both cases the Pr can be considered as Pr^{3+} but the 9-fold degenerate ground state is split by a cubic crystalline electric field (CEF) in $PrH_{2.0}$, due to hydrogen on tetragonal interstitial sites, while in $PrH_{2.5}$ the CEF is orthorhombic, because half the tetrahedral sites are also occupied. The correlation time of the ^1H relaxation of both $PrH_{2.0}$ and $PrH_{2.5}$ is due to the fluctuation moment of the

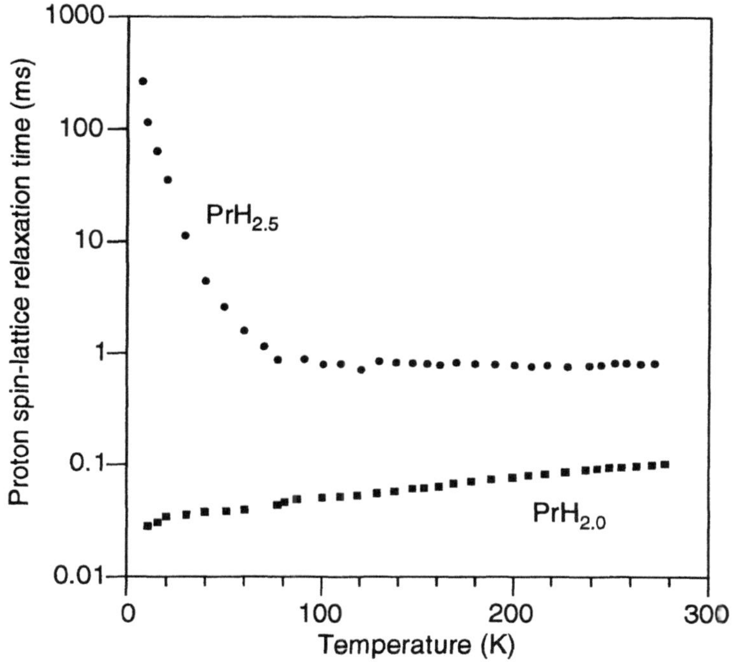

Figure 9. Temperature dependence of the proton spin-lattice relaxation time of $PrH_{2.0}$ and $PrH_{2.5}$. After [19].

Pr^{3+} ion but the different CEF lead to a completely different behaviour of T_{1m} as a function of temperature, figure 9. The ground state of $PrH_{2.0}$ is magnetic, a triplet state, and T_{1m} decreases rapidly as the temperature is lowered, while the ground state of $PrH_{2.5}$ leads to the opposite behaviour. In both cases a fit to the NMR measurements gave values of the lower CEF levels in good agreement with neutron data.

Conclusion

The NMR of MH_x has been shown to provide reliable measurements of the diffusion constant and mean residence time for ^1H and ^2D in metals. It is also sometimes possible to establish the particular type of interstitial sites that the ^1H or ^2D move between.

The NMR results must be combined with other techniques, such as internal friction muSR and neutron scattering, in order to characterise the hydrogen motion over the widest range of temperature.

REFERENCES

[1] Y. Fukai, *The Metal-Hydrogen System*, Springer-Verlag, Berlin (1993).
[2] H. Grabert, H. R. Schober, and H. Wipf, in *Hydrogen in Metals III*, H. Wipf, ed., Springer, Heidelberg (to be published).
[3] R. M. Cotts, in *Hydrogen in Metals I, Topics in Applied Physics 28*, G. Alefield and J. Völkl, eds., Springer-Verlag, Berlin (1978) p. 227-265.
[4] R. M. Cotts, in *Electronic Structure and Properties of Hydrogen in Metals*, P. Jena and C. B. Satterthwaite, eds., Plenum, New York (1983) p. 451-464.
[5] International Symposium on Metal-Hydrogen Systems, J. Less-Common Metals **172-174**, 1-465 (1991).
[6] A. G. Marshall and F. R. Verdun, *Fourier Transforms in NMR, Optical and Mass Spectrometry*, Elsevier, Amsterdam (1990).
[7] A. Abragam, *Principles of Nuclear Magnetism*, Oxford University Press, Oxford (1961, 1983).
[8] C. P. Slichter, *Principles of Magnetic Resonance*, 3rd ed., Springer-Verlag, Berlin (1989).
[9] N. Bloembergen, E. M. Purcell, and R. V. Pound, Phys. Rev., **73**, 679 (1948).
[10] G. K. Schoep, N. K. Poulis, and R. R. Arons, Physica **75**, 297 (1974).
[11] T. K. Halstead, K. Metcalfe, and T. C. Jones, J. Magn. Res., **47**, 292 (1982).
[12] C. A. Sholl, J. Physics C: Solid State Phys., **14**, 447 (1981).
[13] N. Salibi and R. M. Cotts, Phys. Rev., **B27**, 2625 (1983).
[14] B. Stalinski and O. J. Zogal, Bull. Acad. Polon. Sci. **13**, 397 (1965).
[15] P. P. Davis, E. F. W. Seymour, D. Zamir, W. D. Williams, and R. M. Cotts, J. Less-Common Metals **49**, 159 (1976).
[16] L. D. Bustard, R. M. Cotts, and E. F. W. Seymoour, Phys. Rev., **B22**, 12 (1980).
[17] G. C. Carter, L. H. Bennett, and D. J. Kahan, *Metallic Shifts in NMR*, Pergamon Press, Oxford (1973).
[18] J. W. Han, C. T. Chang, D. R. Torgeson, E. F. W. Seymour, and R. G. Barnes, Phys. Rev., **B36**, 615 (1987).
[19] M. Belhoul, R. J. Schoenberger, D. R. Torgeson, and R. G. Barnes, J. Less-Common Metals **172-174**, 366 (1991).

Chapter 14

DIFFUSION OF HYDROGEN IN AMORPHOUS AND NANOCRYSTALLINE ALLOYS

M. HIRSCHER
Max-Planck-Institut für Metallforschung
Institut für Physik
Heisenbergstr. 1
D-70569 Stuttgart
Fed. Rep. of Germany

ABSTRACT: Experimental results on short-range and long-range hydrogen diffusion are presented for nanocrystalline $Fe_{90}Zr_{10}$, produced by crystallization of melt-spun amorphous ribbons, and compared with those for amorphous alloys. Both the amorphous and nanocrystalline alloys show a strong concentration dependence on the hydrogen diffusion kinetics, which can be explained by a potential with a distribution of site, as well as saddle-point enthalpies, according to the Kirchheim model. In the nanocrystalline alloys these topologically disordered regions are assigned to the interfaces. The average activation enthalpy of hydrogen diffusion in the interfaces is about 0.1 eV lower than in amorphous $Fe_{91}Zr_9$, indicating lower potential barriers in the interfaces of the nanocrystalline material than in the amorphous state.

1. Introduction

The diffusion of hydrogen (H) in metals and alloys is strongly influenced by the microstructure of the material. The nature of this microstructure can be phenomenologically described by the potential which acts on the proton along its diffusion path. Fig. 1 gives a schematic overview of the potentials for different structures without including typical defects, such as interstitials, vacancies, impurity atoms, dislocations, grain or phase boundaries. In the case of a crystalline pure metal the potential is characterized by a unique activation enthalpy (Fig. 1a). For a crystalline binary alloy with chemical long-range order, two site and two saddle-point enthalpies already exist (Fig. 1b). The variety of site and saddle-point enthalpies drastically increases for chemically disordered, but topologically ordered, i.e. crystalline, alloys leading to a series of activation enthalpies for the local H jump processes (Fig. 1c). For example, in a fcc binary alloy ten energetically different configurations of octahedral interstitial sites are possible [1, 2]. In a chemically as well as topologically disordered system, such as amorphous alloys, a statistical distribution of site and

F. Grandjean et al. (eds.), Interstitial Intermetallic Alloys, 333–347.

saddle-point enthalpies is expected giving rise to a broad continuous spectrum of activation enthalpies (Fig. 1d).

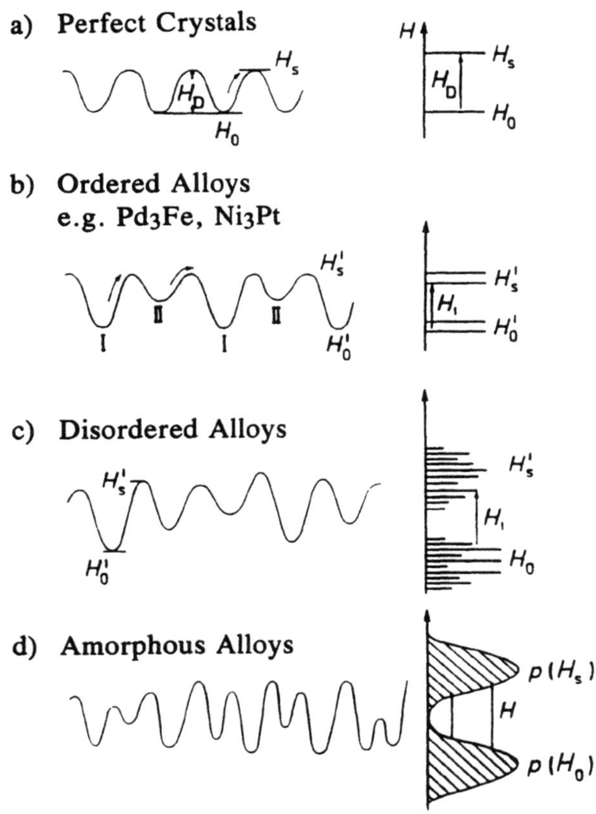

Fig. 1: Schematic representation of potential curves and of distributions of site and saddle-point enthalpies for different microstructures. a) perfect monoatomic crystals, b) binary ordered alloys, c) chemically disordered alloys, d) amorphous alloys.

While searching for new materials with better magnetic or mechanical properties, a large effort has been made into the development of nanostructured or nanocrystalline materials, since their grain size is comparable or even smaller than characteristic physical interaction ranges, e.g. the ferromagnetic exchange length. These materials possess an average grain size of less than 100 nm, in most cases even only 10 to 20 nm. The volume fraction of distorted atoms belonging to the interfacial regions is around 30 % assuming a boundary thickness of about 1 nm. To a first order approximation, these nanocrystalline alloys can be considered as a heterogeneous system containing both crystalline regions within the grains and distorted amorphous-like regions within the interfaces.

The H atom may be used as a probe to study the microstructure on an atomistic level due to its high mobility, even at low temperatures [3]. For ferromagnetic metals and alloys the highly sensitive technique of magnetic after-effect (MAE) measurements allows even single jumps of hydrogen atoms between neighbouring interstitial sites, to be detected [4]. For studies of H diffusion at low temperatures this method has successfully been applied to many systems and types of defect structures [2, 5, 6, 7]. Recent investigations on heterogeneous alloys show that the topological as well as the chemical microstructure, have a strong influence on the H diffusion kinetics [8, 9].

Thermal desorption spectrometry (TDS) measurements are an excellent method to determine the long-range H diffusion kinetics. Recently, this technique has even been applied to detect the H degassing of thin amorphous GdFe films, and to characterize the chemical surrounding of the interstitial sites occupied by H atoms [10]. In bulk samples a comparison of the TDS results, with the short-range diffusion kinetics measured by the MAE, yields additional information about the microstructures participating in the diffusion path [8, 9].

In this paper the important characteristics of hydrogen diffusion in amorphous alloys are briefly reviewed and explained for new measurements of H in amorphous NiFeB alloys [11]. Special emphasis is put on the behaviour of hydrogen in nanocrystalline alloys. Measurements of the concentration dependence of the short-range and long-range H diffusion in nanocrystalline $Fe_{90}Zr_{10}$ [12] are compared with the behaviour of H-charged amorphous alloys.

2. Experimental Techniques

2.1. MAGNETIC AFTER-EFFECT MEASUREMENTS

The magnetic after-effect (MAE) is defined as the time-dependent change of the reluctivity, $r(t)$, after demagnetization, where $r(t)$ is the reciprocal of the initial susceptibility, $\chi_a(t)$ [13, 14]. The measurements were performed in a fully automated a.c. measuring device [15]. The experimental results are represented as isochronal relaxation curves of the reluctivity amplitude defined as:

$$\frac{\Delta r(t_1,t_2,T)}{r(t_1,T)} = \frac{r(t_2,T) - r(t_1,T)}{r(t_1,T)} \quad . \tag{1}$$

In the present measurements $t_1 = 0.5\,\mathrm{s}$ and t_2 is varied from 1.5 to 179.5 s.

In the case of hydrogen, the MAE is based on the interaction energy of the hydrogen atoms with the spontaneous magnetization within the domain walls [4]. In the case of the so-called orientation MAE, the local surrounding of the H atom is anisotropic, due to defects, chemical or topological disorder. The interaction energy between this anisotropic complex and the spontaneous magnetization may be decreased by a reorientation of the anisotropy axis of the complex, due to the H atoms hopping to energetically lower neighbouring configurations. In general, these jumps are thermally activated and can be described by the Arrhenius law

$$\tau(T,Q) = \tau_o \exp[Q/kT] \quad , \tag{2}$$

where τ_o is the pre-exponential factor, Q the activation enthalpy and k the Boltzmann constant. In the case of a Debye process, the time dependence of the reluctivity is given by

$$r(t,T) = r(0) + [r(\infty,T) - r(0,T)]G(t,\tau) \quad , \tag{3}$$

where the relaxation function is represented by

$$G(t,\tau) = 1 - \exp[-t/\tau] \quad . \tag{4}$$

For a distribution of activation enthalpies, $G(t,\tau)$ depends on the distribution function, $p(\tau)$, of relaxation times

$$G(t,\tau) = \int_0^\infty p(\tau)(1 - \exp[-t/\tau])\mathrm{d}\tau \quad ; \quad \int_0^\infty p(\tau)\mathrm{d}\tau = 1 \quad . \tag{5}$$

Numerically the MAE spectrum is analysed by a superposition of i box-type distributions [16] in the logarithm of the relaxation time τ, which according to Eq. (2) is equivalent to i box-type distributions in the activation enthalpy between $Q \leq Q_i \leq Q + \Delta Q$ [5, 17]. The relaxation function is then given by

$$G(t,\tau) = 1 + \sum_i p_i \frac{kT}{\Delta Q} \left[\mathrm{Ei}(-t/\tau_i) - \mathrm{Ei}(-t/\tau_{i+1}) \right] \quad , \tag{6}$$

where Ei denotes the exponential integral [18], all boxes posses the same width, ΔQ, and their weights, p_i, are normalized to one, $\sum_i p_i = 1$.

2.2. THERMAL DESORPTION SPECTROMETRY

The thermal desorption spectrometry (TDS) of hydrogen was studied via a manometric method. The specimens were heated at a constant rate of typically 5 K/min from 300 up to 600 K in a quartz recipient under high vacuum (10^{-3} Pa). The pressure rise was detected with a very sensitive pressure gauge (Baratron MKS). Owing to a cooling trap kept at 25 K, which was placed in front of the gauge in order to freeze out additional gases, such as N_2, O_2, or H_2O, this rise can be attributed to the partial pressure of hydrogen. The system was calibrated by connecting a reference volume filled with hydrogen, and its resolution was better than 0.002 μg H_2.

3. Phenomenological Background

In an amorphous, i.e. chemically and topologically disordered, alloy the H atom diffuses with a potential which has a statistical distribution of site and saddle-point enthalpies. For thermodynamic equilibrium the H atoms occupy the lowest site potentials (Fig. 2). Since each interstitial-like site can only be occupied by one H atom, the occupation probability obeys Fermi-Dirac statistics according to the Kirchheim model [19]. Therefore, charging with H results in a successive filling of the interstitial sites. The occupation probability of the i-th interstitial site with a free enthalpy G_i is then given by

$$f_i = \left(1 + \exp[(G_i - E_F)/kT]\right)^{-1} \quad , \tag{7}$$

where E_F denotes the Fermi level of the H atoms, which corresponds to the chemical potential μ. A realistic description for the spectrum of site enthalpies G_i may be a

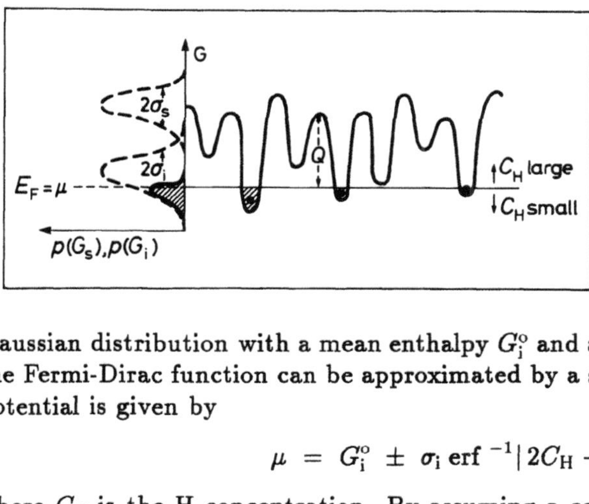

Fig. 2: Schematic potential model with a Gaussian distribution of interstitial site and saddle-point enthalpies, $p(G_i)$ and $p(G_s)$, with 1/e width σ_i and σ_s, respectively. At low temperatures the sites are occupied by H atoms up to the Fermi level, i.e. chemical potential, $E_F = \mu$.

Gaussian distribution with a mean enthalpy G_i^o and a width σ_i. At low temperatures the Fermi-Dirac function can be approximated by a step function, then the chemical potential is given by

$$\mu = G_i^o \pm \sigma_i \, \mathrm{erf}^{-1} | 2C_H - 1 | \quad , \tag{8}$$

where C_H is the H concentration. By assuming a constant saddle-point energy, G_s, the mean activation enthalpy of diffusion is found to be

$$\overline{Q} = Q_o \pm \sigma_i \, \mathrm{erf}^{-1} | 2C_H - 1 | \quad , \tag{9}$$

where $Q_o = G_s - G_i^o$. The concentration dependence of the pre-exponential factor, D_o, is much weaker, i.e., $\ln D_o \propto \left(\mathrm{erf}^{-1} | 2C_H - 1 |\right)^2$ [19]. By neglecting this C_H dependence of D_o, the maximum temperature of the diffusion-controlled degassing rate is given by

$$T_{max} = T_{max}^o \pm \sigma_T \, \mathrm{erf}^{-1} | 2C_H - 1 | \quad , \tag{10}$$

where T_{max}^o and σ_T are constants similar to Q_o and σ_i in Eq. (9).

4. Amorphous Alloys

4.1. SPECIMEN PREPARATION

The most commonly and technologically applied method to produce amorphous alloys is the melt-spinning technique, in which a jet of melt is cast onto a rotating copper wheel reaching typical cooling rates of 10^5 to 10^6 K/s [20]. Under these conditions many systems forming amorphous alloys, with a chemically and topologically disordered structure, have been found during the last 15 years. The systems can be classified into two groups, the transition-metal—metalloid glasses, including small atoms of B, P, Si etc. as glass forming elements, and binary systems, such as rare-earth—transition-metal, FeZr, AlCa, etc., which typically possess a low eutectic melting temperature in their phase diagram. Since almost all systems of both classes have been investigated and in general show similar features for H diffusion [7, 19, 21, 22, 23, 24], only results of H in the NiFeB system will be presented as an example.

The amorphous alloys of composition $(Ni_{75}Fe_{25})_{82}B_{18}$ were produced by the melt-spinning technique. The amorphicity of the ribbons was confirmed by transmission-electron-microscopy studies (Fig. 3) and measurements of the magnetic coercive force yielding values between $H_c = 58$ to 92 mOe, which are characteristic for amorphous soft-magnetic alloys possessing structurally neither long-range order nor any large anisotropy.

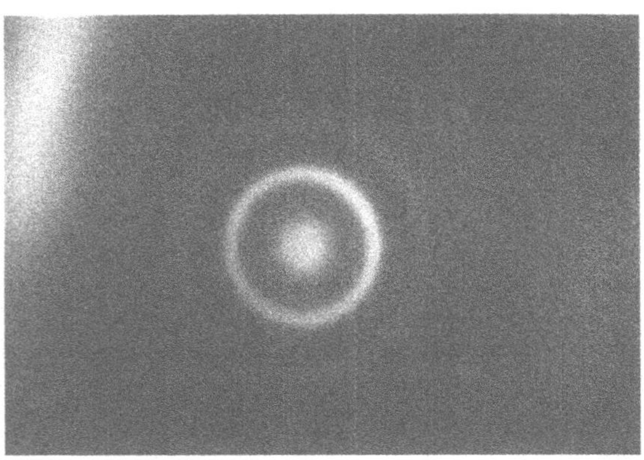

Fig. 3: Electron-diffraction pattern of amorphous $(Ni_{75}Fe_{25})_{82}B_{18}$ obtained in the transmission electron microscope.

The specimens were electrolytically hydrogen charged in 1 n H_2SO_4 with current densities of between 50 μA/cm^2 to 5 mA/cm^2 for times from 30 to 60 min. The total H concentrations were determined to be between 0.2 and 2.85 at.%, as measured by TDS experiments up to a temperature of 600 K.

4.2. SHORT-RANGE H DIFFUSION

After H charging, the $(Ni_{75}Fe_{25})_{82}B_{18}$ alloys show a broad relaxation maximum around 110 K in the MAE spectrum ($C_H = 2.8$ at.%). With increasing annealing temperature (T_A), i.e. decreasing H concentration, the maximum shifts to higher temperatures, e.g. to 170 K after $T_A = 350$ K, which corresponds to $C_H = 0.3$ at.% (Fig. 4). A numerical evaluation using the time law of the orientation MAE and a spectrum of relaxation times, yields mean activation enthalpies between 0.4 eV and 0.32 eV depending on the H content. The pre-exponential factor is in the range of 10^{-12} s and the average 1/e width of the enthalpy distribution about 0.2 eV, both within the experimental uncertainty independent of H concentration.

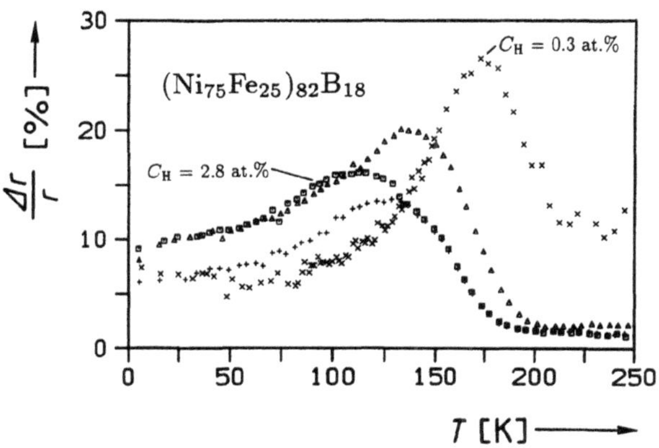

Fig. 4: MAE spectra of H-charged amorphous $(Ni_{75}Fe_{25})_{82}B_{18}$ for different H concentrations, C_H, between 0.3 and 2.8 at.% ($t_1 = 0.5$ s, $t_2 = 179.5$ s).

This characteristic concentration dependence for the short-range hydrogen diffusion was observed in many amorphous alloys, and was attributed to a successive filling of the distribution of site potentials with hydrogen atoms, according to the Fermi-Dirac statistics (Kirchheim model). In this model, only hydrogen atoms occupying interstitial sites near the Fermi level find neighbouring, energetically equivalent, empty sites into which they can jump. The interstitial sites far below the Fermi level are completely occupied and reorientation processes of hydrogen atoms over such sites are blocked. As reflected in the magnitude of the pre-exponential factor of about 10^{-12} s, only short-range diffusion processes, over typically one or two atomic jump distances, are induced, giving rise to the orientation MAE [4]. Owing to the existence of a certain short-range order in amorphous alloys, two energetically nearly equivalent neighbouring interstitial sites will always be present. By the interaction of the anisotropy axis of the interstitial complex, with the direction of the spontaneous magnetization, a splitting of energetically degenerate sites occurs leading to reorientation jumps of hydrogen atoms [4]. Since the magnetic interaction energy is very small compared to the thermal energy and to the site potentials, jumps are only possible if only one of the potential valleys is occupied, i.e. for sites near the

Fermi level. Therefore, the broad activation enthalpy spectra found for fixed hydrogen concentrations are caused by a distribution of saddle-point enthalpies with a $1/e$ width of about $\sigma_s = (0.20 \pm 0.05)\,\mathrm{eV}$.

In the Kirchheim model, with increasing hydrogen concentration the interstitial sites are successively filled starting with the lowest enthalpy sites. A very detailed analysis of MAE measurements on $Fe_{80}B_{20}$, $Fe_{40}Ni_{40}P_{14}B_6$, and $Fe_{91}Zr_9$ in the framework of the Kirchheim model is given by Hofmann and Kronmüller [7]. The concentration dependence of the mean activation enthalpy can be directly related to the width of the distribution of site enthalpies by Eq. (9). The value $\sigma_i = (0.15 \pm 0.05)\,\mathrm{eV}$ obtained for $(Ni_{75}Fe_{25})_{82}B_{18}$, is in good agreement with typical values found in other amorphous systems by MAE measurements [7]. Corresponding results were found by other experimental techniques, e.g. internal friction measurements [22, 23, 24], on a variety of H-charged amorphous alloys.

4.3. LONG-RANGE DIFFUSION

For H-charged $(Ni_{75}Fe_{25})_{82}B_{18}$, heated at a constant rate of $\dot{T} = 5\,\mathrm{K/min}$ in high vacuum, the degassing rate is shown in Fig. 5 for two different initial H concentrations. The maximum temperature shifts to lower temperatures for higher initial H contents, i.e. $T_{max} = 407\,\mathrm{K}$ for $C_H = 0.22\,\mathrm{at.\%}$ and $T_{max} = 343\,\mathrm{K}$ for $C_H = 2.85\,\mathrm{at.\%}$. As shown in the insert of Fig. 5, the high temperature shoulder of the degassing rate coincides for both samples within the experimental accuracy.

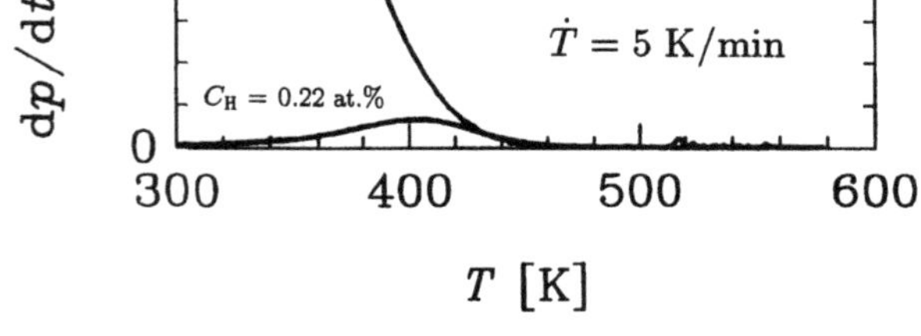

Fig. 5: H degassing rate, dp/dt, as a function of the temperature, T, for amorphous $(Ni_{75}Fe_{25})_{82}B_{18}$ with two different initial H concentrations, C_H ($\dot{T} = 5\,\mathrm{K/min}$).

In the Kirchheim model the thermal desorption experiment is equivalent to a successive emptying of the interstitial sites by H atoms. First the shallow sites and then the more deeper potentials occupied by H atoms become vacant. Therefore the desorption curve of a fully charged specimen represents all possible site enthalpies, i.e. on the low temperature side the shallow ones contribute to the degassing rate and on the high temperature side the deep sites. A specimen charged with a lower H content starts to degas at higher temperatures since only the deeper sites are occupied by H atoms. If the heating rate is not faster than the diffusion to the surface, the degassing curves of the samples, H-charged to low and high concentrations, should coincide within the high temperature region, which was found to be the case (see insert in Fig. 5). This characteristic concentration dependence of the H degassing, can only be attributed to a distribution of site enthalpies. For several amorphous systems similar distributions were found in TDS experiments [7], and diffusivity measurements applying an electrochemical method [19, 21].

5. Nanocrystalline Alloys

5.1. SPECIMEN PREPARATION

Until recently measurements of hydrogen diffusion in grain and phase boundaries were impossible, owing to the small volume fraction of boundaries in conventionally

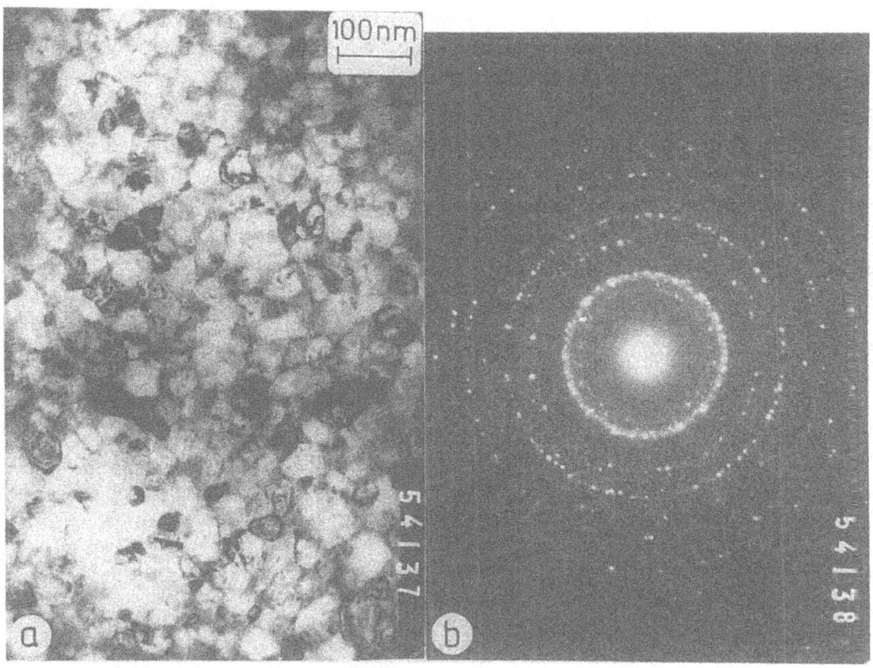

Fig. 6: Transmission-electron-microscopy bright-field image (a) and electron-diffraction pattern (b) of nanocrystalline $Fe_{90}Zr_{10}$.

processed crystalline metals and alloys. In the last few years nanocrystalline materials with a large volume fraction of grain and phase boundaries have become available. The first processing method was developed by Gleiter et al. [25]. This gas-phase technique produces small crystallites by solidification of metal vapor and a following compression to pellets. In another procedure melt-spun amorphous ribbons are crystallized by a heat treatment to prepare thermally very stable, nanocrystalline alloys [26]. This new method is already technologically applied to produce soft-magnetic materials [27].

Following the second approach, amorphous ribbons of the composition $Fe_{90}Zr_{10}$ were prepared by the melt-spinning technique in vacuum. These ribbons were annealed at 870 K for 48 h in high vacuum to achieve 100 % crystallized specimens with an average grain size of 15 nm as determined by transmission electron microscopy (Fig. 6).

The specimens were electrolytically charged with hydrogen in an aqueous solution of 1 n H_2SO_4 to total H concentrations of between 0.05 to 2.6 at.%, as detected by TDS measurements.

5.2. SHORT-RANGE DIFFUSION

After hydrogen charging, nanocrystalline $Fe_{90}Zr_{10}$ exhibits a broad MAE relaxation maximum in the temperature range from 60 K to 200 K, e.g. $C_H = 1.1$ at.% (Fig. 7). With increasing H concentration the relaxation maximum shifts from about 190 K down to 105 K (Fig. 8). A numerical evaluation using the time law of the orientation after-effect and a spectrum of relaxation times yields mean activation enthalpies between 0.4 eV and 0.2 eV depending on the H content (Fig. 9). The pre-exponential factor lies in the range of 10^{-12} s and the 1/e width of the spectra is about 0.15 eV independently of H concentration.

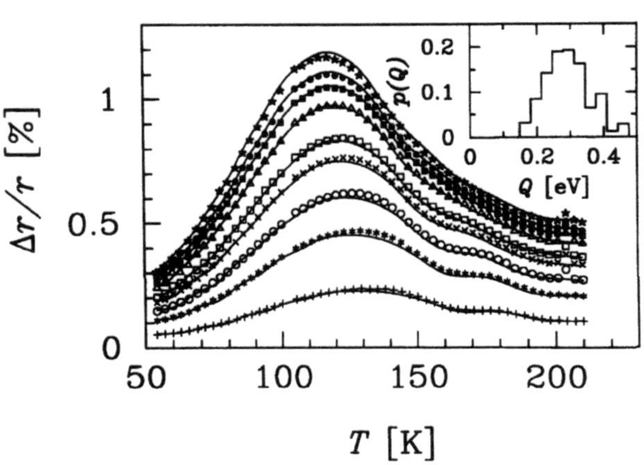

Fig. 7: MAE spectrum of H-charged ($C_H = 1.1$ at.%) nanocrystalline $Fe_{90}Zr_{10}$. Experimental data are represented by symbols for $t_1 = 0.5$ s and $t_2 = 1.5$ (lowest curve), 4.5, 9.5, 19.5, 29.5, 59.5, 89.5, 119.5, 179.5 s (highest curve). Solid lines correspond to a numerical fit yielding to the distribution of activation enthalpies, $p(Q)$, shown in the upper right corner.

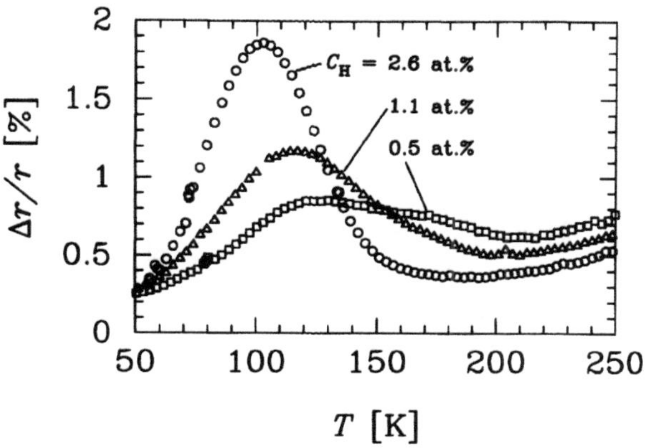

Fig. 8: MAE spectra of H-charged nanocrystalline $Fe_{90}Zr_{10}$ for different H concentrations, C_H ($t_1 = 0.5$ s, $t_2 = 179.5$ s).

In contrast to conventionally produced crystalline alloys, the MAE spectrum of hydrogen in nanocrystalline $Fe_{90}Zr_{10}$ can only be described by a broad spectrum of relaxation times. Such spectra are a common feature for relaxation processes in topologically disordered systems, such as amorphous alloys. Therefore the Kirchheim model can be applied in the interpretation of the experimental data. Firstly, the magnitude of the pre-exponential factor of about 10^{-12} s is typical for reorientation processes over distances between one or two interstitial sites. Again, only hydrogen atoms occupying interstitial sites near the Fermi level find neighbouring, energetically equivalent, empty sites into which they can jump. Sites far below the Fermi level are completely occupied and reorientation processes are blocked. Accordingly, the activation enthalpy spectra found for fixed H concentrations are due to a distribution of saddle-point enthalpies with a 1/e width of about $\sigma_s = (0.15 \pm 0.04)$ eV.

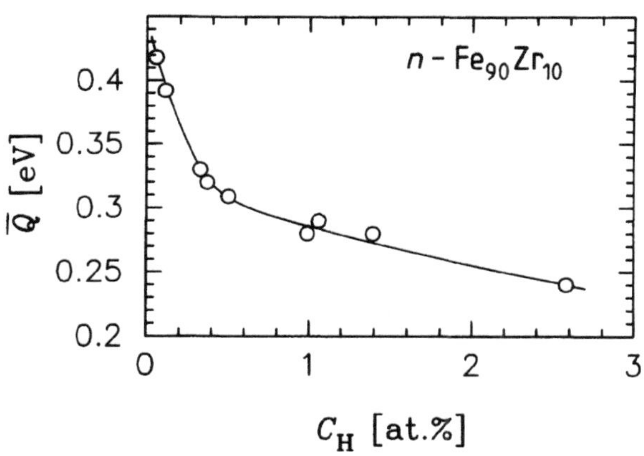

Fig. 9: Mean activation enthalpy of local reorientations, $\overline{Q}$, as a function of the H concentration, C_H.

The strong concentration dependence of the mean activation enthalpy (Fig. 9) can also be analysed according to the Kirchheim model by Eq. (9) as shown in Fig. 10.

The slope of the line corresponds to a 1/e width $\sigma_i = (0.18 \pm 0.04)\,\mathrm{eV}$ of the interstitial site distribution. This value is in good agreement with measurements of the H diffusivity in nanocrystalline Pd pellets, yielding a Gaussian distribution of site enthalpies for the grain boundary regions with a 1/e width of $\sigma_i \approx 0.21\,\mathrm{eV}$ [28]. A comparison of nanocrystalline $Fe_{90}Zr_{10}$ with measurements on amorphous $Fe_{91}Zr_9$

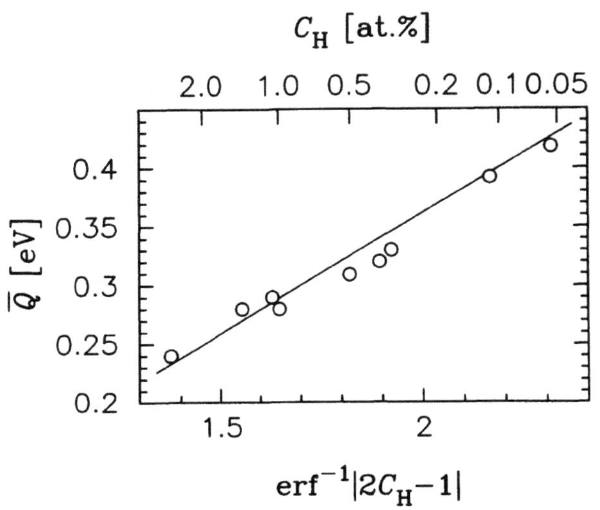

Fig. 10: Mean activation enthalpy of local reorientations, $\overline{Q}$, versus the inverse error function of $|2C_H - 1|$. The slope of the line corresponds to a 1/e width for the distribution of saddle-point enthalpies $\sigma_i = (0.18 \pm 0.04)\,\mathrm{eV}$ according to Eq. (9).

[7] shows that the mean activation enthalpy of H diffusion (see Fig. 9) is shifted to about 0.1 eV lower values for equivalent C_H in the nanocrystalline alloy, whereas the relation according to Eq. (9) yields the same slope. This approximately constant shift to lower enthalpies indicates a smaller difference in enthalpy between site and saddle point, i.e. lower potential barriers in the nanocrystalline $Fe_{90}Zr_{10}$ although the 1/e width of the distribution of sites is similar to that in the amorphous state.

5.3. LONG-RANGE DIFFUSION

The H degassing curves of nanocrystalline $Fe_{90}Zr_{10}$ ($\dot{T} = 5\,\mathrm{K/min}$) are shown in Fig. 11 for four different initial hydrogen concentrations. The high temperature part, i.e. low hydrogen concentrations, coincides well for all specimens, whereas the onset temperature of degassing shifts to lower temperatures with increasing hydrogen concentration, i.e. from 360 K down to 305 K. To separate the concentration dependence of the degassing kinetics, curves for subsequent initial concentrations are subtracted from each other, and, in addition, these differences are differentiated with respect to the degassing time (Fig. 12). In the curves obtained by this method the temperature of maximum degassing rate shifts to lower temperatures with increasing hydrogen

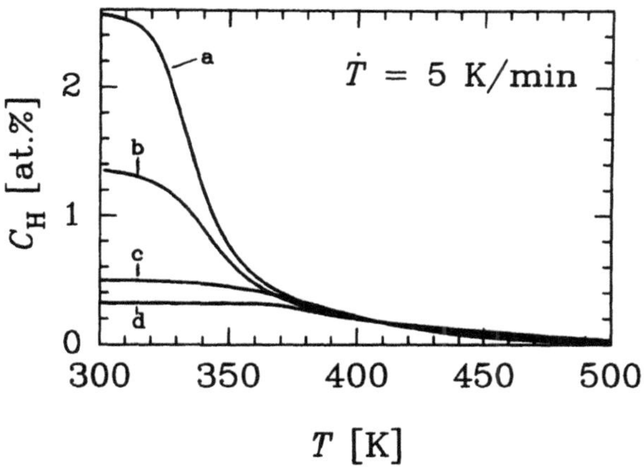

Fig. 11: H concentration, C_H, of nanocrystalline $Fe_{90}Zr_{10}$ during degassing in vacuum ($\dot{T} = 5$ K/min). Specimens with different initial H concentration are marked a to d.

concentration, which means that the fraction of hydrogen between the concentrations of curves a and b in Fig. 11, degasses easier than the fraction between b and c etc.

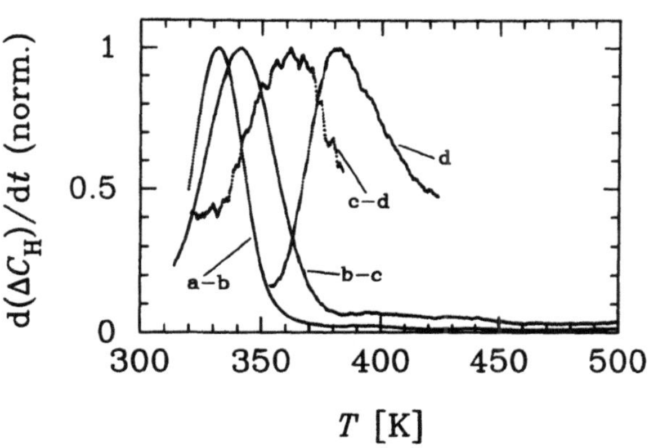

Fig. 12: Normalized degassing rate of H fractions, d(ΔC_H)/dt, obtained by subtracting the curves from Fig. 11 as marked a–b, b–c, c–d and d ($\dot{T} = 5$ K/min).

A temperature shift of the thermal degassing rate with the initial H concentration (Fig. 11), is again a clear indication of strongly concentration-dependent activation parameters of hydrogen diffusion. The maximum degassing rate of the different hydrogen fractions shown in Fig. 12 can be analysed according to the Kirchheim model by Eq. (10). This means that the a–b curve corresponds to an upper slice cut out of the distribution of sites, i.e. shallow sites, and the b–c curve to the next slice with deeper potential sites etc. In Fig. 13 the maximum temperatures are plotted as a function of $\mathrm{erf}^{-1}|\,2C_H - 1\,|$, where C_H are the average concentrations between a and b, b and c, etc., yielding a slope of $\sigma_T \approx 80$ K. Recently Maier and Kronmüller [8] determined the pre-exponential factor of H diffusion in nanocrystalline $Fe_{90}Zr_{10}$ to be $D_o \approx 10^6$ m^2/s, by TDS measurements with different heating rates. Applying

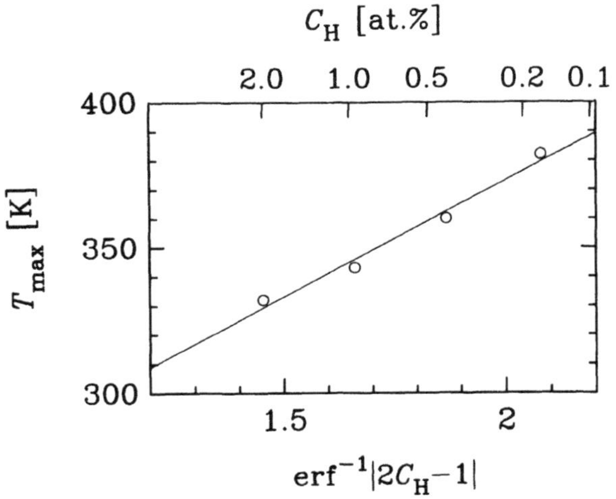

Fig. 13: Temperatures of maximum H degassing rate, T_{max}, from Fig. 12 versus the inverse error function of $|2C_H-1|$. The slope of the line corresponds to $\sigma_T \approx 80\,K$ according to Eq. (10).

Fick's laws of diffusion, and experimental parameters such as specimen size, heating rate and D_o, this distribution in temperature σ_T can be converted to a distribution of activation enthalpies $\sigma_i = (0.10 \pm 0.05)\,eV$, which again corresponds to the $1/e$ width of the interstitial site distribution.

6. Conclusions

The short-range and long-range diffusion of hydrogen in amorphous and nanocrystalline alloys shows a strong concentration dependence, which can be described by a potential with a distribution of site and saddle-point energies according to the Kirchheim model. In the nanocrystalline alloys this statistical potential is assigned to the interfacial regions possessing a structure similar to the amorphous state. The average activation enthalpy of hydrogen diffusion is lower in the nanocrystalline material than in the amorphous alloy, indicating lower potential barriers in the interfaces.

Acknowledgements

The author is very grateful to Prof. Dr. H. Kronmüller and Prof. Dr. R. Kirchheim for many helpful and stimulating discussions, and to Dr. E.H. Büchler and Dipl.-Phys. S. Zimmer for extensive discussions of their measurements.

References

[1] E. Adler, Z. Metallk. **56** (1965) 294.

[2] C.U. Maier, M. Hirscher, and H. Kronmüller, Phil. Mag. B **60** (1989) 627.

[3] M. Hirscher and H. Kronmüller, J. of Less-Common Met. **172-174** (1991) 658.

[4] H. Kronmüller, in: *Hydrogen in Metals I*, Eds. G. Alefeld and J. Völkl, Springer, Berlin - Heidelberg - New York, 1978, p.289.

[5] B. Hohler and H. Kronmüller, Phil. Mag. A **43** (1981) 1189 and **45** (1982) 607.

[6] M. Hirscher and H. Kronmüller, Rad. Effects and Defects in Solids **108** (1989) 359.

[7] A. Hofmann and H. Kronmüller, phys. stat. sol. (a) **104** (1987) 381 and 619.

[8] C.U. Maier and H. Kronmüller, J. of Less-Common Met. **172-174** (1991) 671.

[9] J. Mössinger and M. Hirscher, Z. phys. Chem. **183** (1994) 13.

[10] M. Vergnat, H. Chatbi, and G. Marchal, Appl. Phys. Lett. **64** (1994) 2084.

[11] E.H. Büchler, diploma thesis, Universität Stuttgart 1991.

[12] S. Zimmer, diploma thesis, Universität Stuttgart 1992.

[13] H. Kronmüller, *Nachwirkung in Ferromagnetika*, Springer, Berlin 1968.

[14] H. Kronmüller, in: *Vacancies and Interstitials in Metals*, Eds. A. Seeger, D. Schumacher, W. Schilling, and J. Diehl, North-Holland, Amsterdam, 1969, p.667.

[15] F. Walz, Appl. Phys. **3** (1974) 313 and phys. stat. sol. (a) **82** (1984) 179.

[16] G. Richter, Ann. Phys. **29** (1937) 605.

[17] J.R. Cost, J. Appl. Phys. **54** (1983) 2137.

[18] M. Abramowitz, and I. Stegun, *Handbook of Mathematical Functions* National Bureau of Standards 1972.

[19] R. Kirchheim, Acta metall. **30** (1982) 1069 and Prog. Mat. Sci. **32** (1988) 262.

[20] L.A. Jacobson and J. McKittrick, Mat. Sci. Eng. **R11** (1994) 355.

[21] R. Kirchheim, F. Sommer, and G. Schluckebier, Acta metall. **30** (1982) 1059.

[22] B.S. Berry and W.C. Pritchet, Scripta metall. **15** (1981) 637 and Z. phys. Chem. NF, **163** (1989) 381.

[23] H. Mizubayashi, H. Agari, and S. Okuda, Z. phys. Chem. NF, **163** (1989) 391.

[24] W. Ulfert and H. Kronmüller, Z. phys. Chem. NF, **163** (1989) 397.

[25] H. Gleiter and P. Marquardt, Z. Metallkund. **75** (1984) 263.

[26] Y. Yoshizawa, S. Oguma, and K. Yamauchi, J. Appl. Phys. **64** (1988) 6044.

[27] G. Herzer, IEEE Trans. Magn. **25** (1989) 3327, **26** (1990) 1397, and Mat. Sci. Eng. **A133** (1991) 1.

[28] R. Kirchheim, T. Mütschele, W. Kieninger, H. Gleiter, R. Birringer, and T.D. Koble, Mat. Sci. Eng. **99** (1989) 457.

Chapter 15

INTRODUCTION TO HARD MAGNETIC MATERIALS

K.H.J. Buschow
Philips Research Laboratories
5600 JA Eindhoven
The Netherlands

1. Introduction

Ferromagnetic materials are characterized by a parallel orientation of the participating magnetic moments below the Curie temperature. This parallel moment arrangement arises as a consequence of the exchange interaction between two magnetic spins, S_i and S_j, which is customarily represented by the Hamiltonian,

$$\underline{H}_{ex} = -2 J_{ij} \vec{S}_i \vec{S}_j = -2 J_{ij} (g-1)^2 \vec{J}_i \cdot \vec{J}_j, \tag{1}$$

where J_{ij} is the exchange constant and where we have used the relation $\vec{S} = (g-1)\vec{J}$. The orientation of a given moment at site i depends on the exchange interaction with the surrounding moments,

$$\underline{H}_i = -2 \sum J_{ij} (g-1)^2 \vec{J}_i \cdot \vec{J}_j. \tag{2}$$

If one makes the assumption that only the Z nearest neighbours contribute to the interaction one obtains with $J_{ij} = J_{ex}$,

$$\underline{H}_i = -2 Z J_{ex} (g-1)^2 \vec{J}_i \cdot \vec{J}_j. \tag{3}$$

If $J_{ex} > 0$ the lowest energy results if each moment, at site i, adopts a parallel orientation with respect to the moments at the nearest neighbour sites, j.

Each of the moments in the lattice carries a magnetic moment, $\mu = \mu_i = \mu_j = gJ\mu_B$. If there are N moments present, the macroscopic magnetization is $M = N\mu$. The exchange interaction of each moment with its neighbours can then be written in the form

$$\underline{H}_i = -\mu_o \vec{\mu} \cdot \vec{H}_m, \tag{4}$$

F. Grandjean et al. (eds.), Interstitial Intermetallic Alloys, 349–369.

where
$$\vec{H}_m = 2ZJ_{ex}(g-1)^2\vec{M}/\mu_o N g^2 \mu_B{}^2 = N_W\vec{M} \tag{5}$$

is the molecular field experienced by each of the individual moments, and N_W is the molecular field constant or Weiss field constant.

If one follows the above derivation step by step one finds that there is at no point any mention of the direction of the magnetic moments with respect to the crystallographic directions, whatever the magnitude and sign of the magnetic coupling constant, J_{ex}. This means that the exchange interaction is isotropic, and that the molecular field, H_m, and the macroscopic magnetization, M, have complete directional freedom. In fact, if the exchange interaction were the only interaction present it would lead to ideal soft magnetic materials. For applications of magnetic materials in permanent magnets such a property is undesirable because fixing the magnetization in a given direction is at a premium. This means that the ferromagnetic exchange interaction in permanent magnet materials has to be supplemented by a further mechanism which removes the directional freedom of the magnetization.

Modern magnet materials contain rare earth elements because of their strong magnetic anisotropy. The additional mechanism providing the magnetic anisotropy is due to the so-called crystal field interaction, to be described in detail below. Prior to the discussion of the origin of the magnetic anisotropy we will first deal with the magnetic anisotropy on a macroscopic scale by introducing macroscopic anisotropy constants and by describing a few prominent methods by which the magnetic anisotropy can be measured and characterized.

2. Macroscopic Description and Determination of Magnetic Anisotropy

A magnetic material exhibits magnetic anisotropy if its internal energy depends on the direction of the spontaneous magnetization. Phenomenologically the anisotropy energy, E_A, in a material with uniaxial (hexagonal and tetragonal) symmetry may be expressed by

$$E_A(\theta,\varphi) = K_1\sin^2\theta + K_2\sin^4\theta + K_3\sin^4\theta\cos^4\varphi..., \tag{6}$$

where K_1, K_2, and K_3 are the anisotropy constants and where the direction of the spontaneous magnetization relative to the single uniaxial (c-axis) direction) and the a-axis is given by the polar angles θ and φ, respectively (see Fig. 1). In most cases it is sufficient to consider only the K_1 and K_2 terms. The preferred magnetization direction will be along the c-axis in hexagonal or tetragonal structures if K_1 predominates and $K_1 > 0$. It will be perpendicular to the c-axis if $K_1 < 0$. When K_1 is not predominant, the preferred magnetization may point in other directions. In the following we will take only K_1 and K_2 into consideration. If

$$K_1 > 0 \text{ and } K_1 + K_2 > 0, \tag{7}$$

the lowest anisotropy energy is zero for $\theta=0$, whereas if $K_1 > 0$ and $2K_2 > -K_1$, the lowest anisotropy energy corresponds to a θ value given by

$$\sin^2\theta = -K_1/2K_2 \tag{8}$$

A diagram showing the preferred moment directions for different K_1 and K_2 values in a hexagonal crystal is given in Fig. 2.

The anisotropy fields, H_A, are commonly obtained by measuring magnetic polarization curves with the field parallel and perpendicular to the easy magnetization direction. The anisotropy field is then obtained as the intersection of the two magnetization curves mentioned. Illustrative examples of measurements of H_A obtained in this way for several permanent magnet materials by Strnat (1988) are shown in Fig. 3. Measurements of H_A can be helpful for obtaining an estimate of K_1. Suppose that the spontaneous polarization is J_s in tesla, and that in a small volume, ΔV, of the material the polarization is held in equilibrium by a field, H in A/m, normal to the preferred direction, such that J_s is inclined at an angle Θ and hence at $90°-\Theta$ to H. The magnetization force H then exerts a torque, $HJ_s\Delta V\cos\Theta$ in Nm, that tends to increase Θ. The torque tending to return J_s to the preferred direction is obtained by differentiating the expression for the crystal energy, i.e.

$$\frac{dE_A}{d\theta} = \frac{d}{d\theta}(K_1\sin^2\theta + K_2\sin^4\theta)\,\Delta V = (2K_1\sin\theta\cos\theta + 4K_2\sin^3\theta\cos\theta)\,\Delta V \tag{9}$$

Equating these torques leads to the relation

$$H = \frac{2K_1\sin\theta + 4K_2\sin^3\theta}{J_s} \tag{10}$$

The value of H that makes J_s parallel with the field is reached when $\sin\Theta = 1$. The anisotropy field, H_A, is then given by

$$H_A = (2K_1 + 4K_2)/J_s\ . \tag{11}$$

In some materials K_2 is negligible. In this case measurements of H_A are sufficient for the determination of K_1.

A frequently used method to determine K_1 and K_2 was developed by

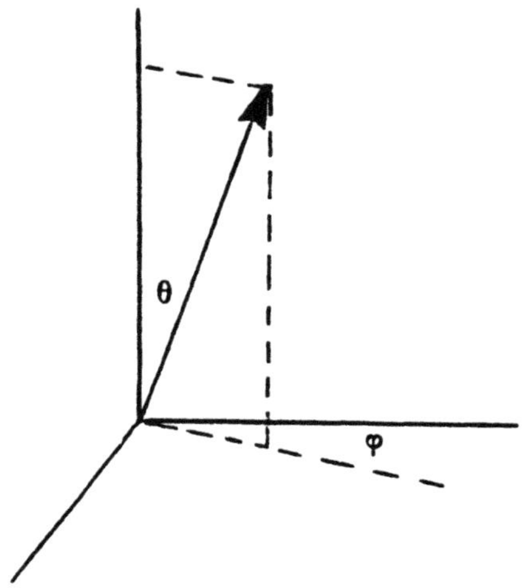

Fig. 1. Moment direction relative to the c-axis defining the angles θ and φ.

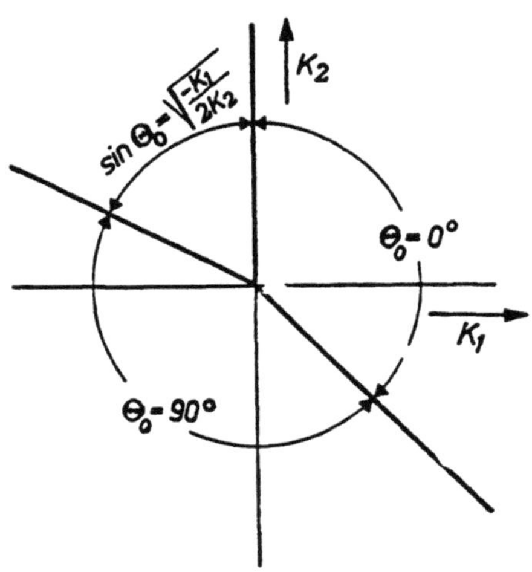

Fig. 2. Diagram showing the preferred moment directions for different K_1 and K_2 values in a hexagonal crystal (after Smit and Wijn, 1965).

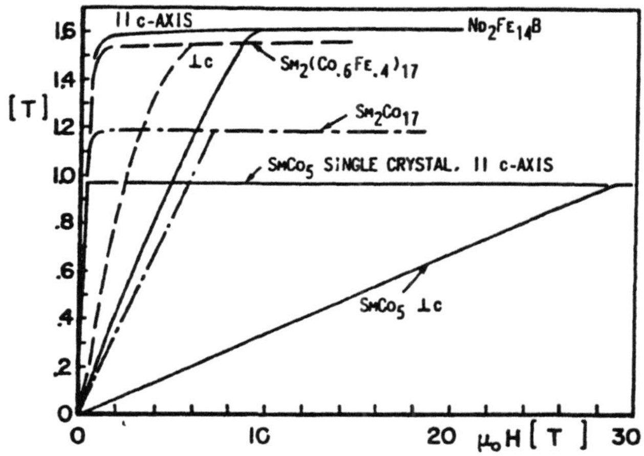

Fig. 3. Easy-axis and hard-axis magnetization curves of several high-anisotropy compounds on which practical permanent magnets are based. After Strnat (1988).

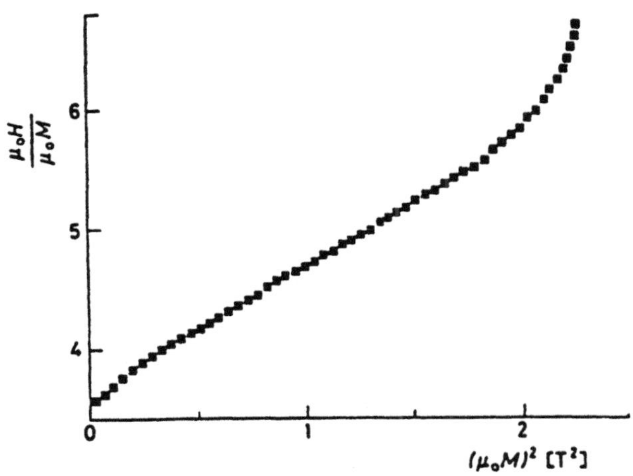

Fig. 4. Sucksmith-Thompson plot for $Nd_{15}Fe_{77}B_8$ measured at room temperature. After Durst and Kronmüller (1986).

Sucksmith and Thompson (1954) and is based on the relation

$$2K_1 / J_s^2 + (4K_2/J_s^4)J^2 = H/J, \tag{12}$$

which holds for the magnetization curve of a single crystal obtained in comparatively small fields applied perpendicular to the easy direction. Under such circumstances one may assume that the magnitude of the magnetization J_s does not change with field strength and hence $\sin\theta = J/J_s$. Substitution into relation (10) then leads to relation (12). When H/J is plotted versus J^2 the anisotropy constant K_1 in relation (12) may be derived from the horizontal intercept and the anisotropy constant K_2 from the slope of the straight line.

Substantial errors can arise from misalignment when this method is used for determining K_1 and K_2 from aligned powder samples. This misalignment leads to curvatures of the magnetization curve similar to those that would be produced by the effect of a larger value of K_2. Somewhat better in this respect is a method based on the Sucksmith-Thompson plot, as proposed by Ram and Gaunt (1983). In this modification, $\alpha^{-1}H/(J-J_R)$ is plotted versus $\alpha^2(J-J_R)^2$, where J_R is the remanence in the hard direction and where the factor $\alpha = (J_s-J_r)/J_s$ has been introduced to simulate perfect alignment of the powder particles. An example of a modified Sucksmith-Thompson plot, obtained on a single crystal of $Nd_2Fe_{14}B$ by Durst and Kronmüller (1986), is shown in Fig. 4. The values of K_2 and K_1 derived from the intercept and slope of this plot are equal to 1.5 MJ/m^2 and 3.9 MJ/m^2, respectively.

Other methods for determining the anisotropy constants make use of a torque magnetometer, by means of which it is possible to measure the torque, T, required to hold a crystal with its axes inclined at various known angles to an applied magnetic field. In the ideal case the measurements should be made with the sample cut in the shape of an oblate ellipsoid, but a thin disc is usually satisfactory, provided a field well in excess of H_A can be applied. The disc is rotated about an axis perpendicular to both its plane and the applied field. It is most important that the sample is circular and is mounted symmetrically about its centre, because otherwise spurious torques will be introduced. It is difficult to interpret the results if the applied field does not saturate the sample. For this reason the torque magnetometer is not usually used for investigating permanent materials based on rare earths that have very large anisotropies. Results obtained in this way for Gd_2Co_{17} by Franse et al. (1989) are shown in Fig. 5.

In contrast, the singular point detection (SPD) method can be used to study the anisotropy in permanent magnet materials having very high anisotropies (Asti and Rinaldi 1974). This method uses the singularity which shows up in the d^2J/dH^2 curve when the measurements are taken with the field perpendicular to the easy direction. In practice d^2J/dt^2 curves are measured using a sweep linear in time, in which H_A shows up as a cusp. The

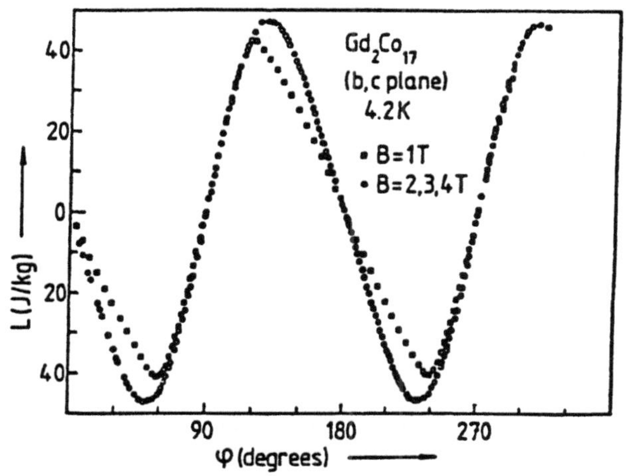

Fig. 5. The torque measured in the bc plane of Gd_2Co_{17} at 4.2 K for different applied magnetic fields. After Franse et al. (1989).

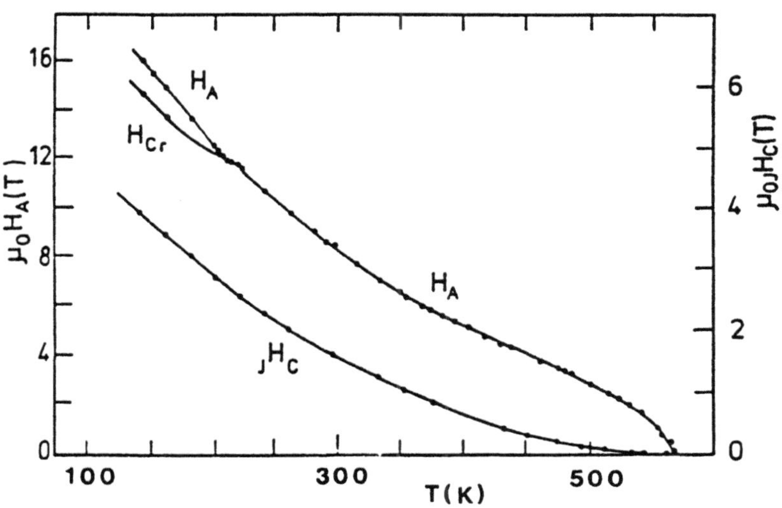

Fig. 6. Temperature dependence of the coercive field, $_JH_c$, the anisotropy field, H_A, and the critical field for FOMP transitions, H_{cr}, in a (Nd,Dy)-Fe-B magnet. After Grössinger et al. (1987).

method can be applied to polycrystalline materials and lends itself well to study the temperature dependence of H_A. Results obtained by Grössinger et al. (1987) on Dy-doped $Nd_2Fe_{14}B$ are shown in Fig. 6.

3. Crystal Field Theory

A substantial portion of the anisotropy in modern rare earth based magnet materials arises from the sublattice anisotropy of the rare earth component, R. The anisotropy of the 3d component is much weaker and, in some cases, such as $R_2Fe_{17}C$ and $R_2Fe_{17}N_3$, even has the wrong sign (easy plane magnetization) for applications in permanent magnets.

Generally one might say that the rare earth component in binary and ternary R-3d compounds is responsible for the magnetic anisotropy while the 3d component provides a sufficiently high magnetization and Curie temperature. In fact, the often huge anisotropy of the 4f systems can be regarded as the crucial parameter which is responsible for the excellent hard magnetic properties of some of these systems. The progress made in understanding the magnetocrystalline anisotropy in the R-3d systems is mainly based on crystal field theory in conjunction with extensive magnetic studies mostly performed on single crystals (see the reviews of Li and Coey 1991, Buschow 1988, 1991). Bandstructure calculation made on selected compounds (Coehoorn et al. 1990, 1991; Zhong and Ching 1989) have contributed much to the understanding of the origin of the anisotropy on an atomic scale.

3.1. QUANTUM MECHANICAL TREATMENT

In most compounds the magnetic atoms or ions form part of a crystalline lattice in which they are surrounded by other ions, the symmetry of the nearest neighbour coordination being determined by the crystal structure. In ionic crystals the metal ions are usually surrounded by negatively charged diamagnetic ions. In metallic systems the participating atoms carry an effective electric charge. This arises because they have donated all or at least a substantial portion of their valence electrons to the conduction band. The resultant positive ions are screened to some extent by the conduction electrons, making the effective charges smaller than the corresponding ionic charges. The electrostatic field experienced by the unpaired electrons of a given magnetic ion is called the crystal field or the ligand field. The neighbouring ions surrounding the atom with the unpaired electrons, are called the ligands. A typical situation, where the atom carrying the unpaired electrons is situated in a uniaxial crystal field, is shown in Fig. 7.

If J is the total angular momentum quantum number of the magnetic atom the 2J + 1 degeneracy of its ground state will be lifted in the presence

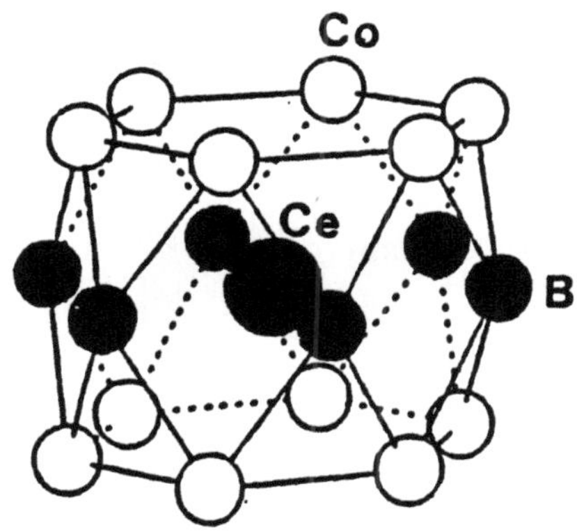

Fig. 7. Uniaxial crystal field produced by the effect of the neighbour atoms on the central atom.

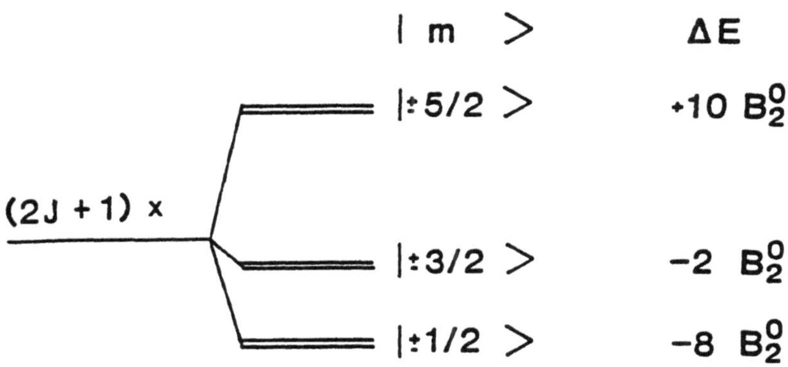

Fig. 8. Energy level scheme after removal of the 2J + 1 degeneracy for Ce^{3+} (J = 5/2) by the crystal field.

of a magnetic field or a crystal field. This will result in changes of the magnetic properties of the corresponding compound whenever crystal fields are present.

To obtain the magnetic properties it is necessary to solve the Hamiltonian of the crystal field interaction explicitly. The crystal field potential of the surrounding ions at the location of the kth unpaired electron of the magnetic ion, is

$$V_k(\vec{r}_k) = |e| \sum_j \frac{Z_j}{|\vec{R}_j - \vec{r}_k|} \tag{13}$$

where $|e|$ is the absolute value of the electron charge. The charge of the jth ligand ion is Z_j where Z_j can be either positive or negative. R_j and r_k are the positions of the jth ligand ion and the kth unpaired electron respectively. The summation is carried out over all ligand ions in the crystal.

In a more rigorous treatment the electric charges associated with the on-site valence electrons of the magnetic ion have to be included in the crystal field potential and the charges associated with the ligand atoms have to be included in the form of charge densities. The crystal field potential then takes the form of an integration in space over all on-site and off-site charge densities around V_k. We will return to this point later and use V_k as given above for introducing the operator equivalent method, without loss in generality.

The crystal field Hamiltonian of the magnetic ion is obtained from eq. (13) by summing over all unpaired electrons n_f

$$\mathcal{H}_c(\vec{r}) = -|e| \sum_{k=1}^{n_f} V_k(\vec{r}_k) \tag{14}$$

The Hamiltonian may be expanded in spherical harmonics because the charges of the CEF are outside the shell of the unpaired electrons, the 4f electrons in the case of rare earth atoms,

$$\mathcal{H}_c = \sum_{n=0}^{\infty} \sum_{m=-n}^{n} A_n^m \sum_{k=1}^{n_f} r_k^n \, Y_n^m(\theta_k, \varphi_k) \tag{15}$$

Here A_n^m are coefficients of this expansion. Their values depend on the crystal structure considered and determine the strength of the crystal field interaction. For instance, if the point charge model would be applicable, in which case the ions of the crystal are described by point charges located at the various crystallographic positions, the parameter A_n^m can be calculated by means of

$$A_n^m = -\frac{4\pi e^2}{2n+1}\sum_j \frac{Z_j}{R_j^{n+1}} Y_n^{m*}(\theta_j, \varphi_j) \tag{16}$$

where the summation again extends over all ligand charges Z_j and the corresponding ligand positions, R_j, θ_j, φ_j, in the crystal. Without going into detail about numerical computations of A_n^m in terms of point charges, we will keep the treatment general and consider them as numerical constants and focus our attention again on the Hamiltonian (15)

A relatively elegant form for this Hamiltonian can be obtained by using the Stevens' "Operator Equivalents" method. First the spherical harmonics $Y_n^m(\theta_k, \varphi_k)$ are expressed in Cartesian coordinates, f(x,y,z), after which x, y, and z are replaced by J_x, J_y, and J_z, respectively. In this way an operator is formed with the same transformation properties under rotation as the corresponding spherical harmonics. For instance

$$\sum_k Y_2^0 \rightarrow \sum_k (3z_k^2 - r_k^2) = \alpha_J\langle r^2\rangle[3J_z^2 - J(J+1)] = \alpha_J\langle r^2\rangle O_2^0 \tag{17}$$

$$\sum_k Y_2^2 \rightarrow \sum_k (x_k^2 - y_k^2) = \alpha_J\langle r^2\rangle[J_x^2 - J_y^2] = \alpha_J\langle r^2\rangle O_2^2 \tag{18}$$

where α_J is a constant and where $J_x^2 - J_y^2$ may be replaced by $\frac{1}{2}[J_+^2 + J_-^2]$. Equation 15 may now be rewritten as

$$\mathcal{H}_c = \sum_{n=0}^{\infty}\sum_{m=-n}^{n} A_n^m\, \theta_n^m \langle r^n\rangle O_n^m = \sum_{n=0}^{\infty}\sum_{m=-n}^{n} B_n^m O_n^m \tag{19}$$

For a magnetic ion with a given J value the operator equivalents, O_n^m, are known and a complete list of them and their relation to the spherical harmonics can be found in the review paper of Hutchings (1964). The quantities θ_n are the so-called reduced matrix elements that do not depend on the azimuthal quantum number m but do depend on J. Values of these quantities are also listed in Hutching 's paper. The latter constants are frequently indicated by α_J, β_J and γ_J for n = 2, 4, and 6, respectively.

Finally it can be shown that for f electrons $n \leq 6$. Furthermore n must be even because of the inversion symmetry of the CEF potential. This means that the above summation, for f electrons, contains only n = 2, 4, 6, because n = 0 gives an additive constant to the potential, which has no

Table 1. Effect of a second order crystal field perturbation on the (2J+1) manifold of $^2F_{5/2}(Ce^{3+})$.

$$H_c = B_2^0O_2^0 + B_2^2O_2^2 \qquad (1)$$

$$O_2^0 = 3J_z^2 - J(J+1) \qquad (2)$$

$$O_2^2 = J_x^2 - J_y^2 = \tfrac{1}{2}(J_+^2 + J_-^2) \qquad (3)$$

$$J_+ = J_x + iJ_y \qquad \text{raising operator} \qquad (4)$$

$$J_+ = J_x - iJ_y \qquad \text{lowering operator} \qquad (5)$$

$$\langle J, m|J_z|J, m\rangle = m \qquad (6)$$

$$\langle J, m|J^2|J, m\rangle = J(J+1) \qquad (7)$$

$$\langle J, m+1|J_+|J, m\rangle = [J(J+1) - m(m+1)]^{1/2} \qquad (8)$$

$$\langle J, m-1|J_-|J, m\rangle = [J(J+1) - m(m-1)]^{1/2} \qquad (9)$$

$$\begin{array}{c|cccccc|}
 & |5/2\rangle & |1/2\rangle & |-3/2\rangle & |-5/2\rangle & |-1/2\rangle & |3/2\rangle \\
\langle 5/2| & 10B_2^0 & \sqrt{10}B_2^2 & 0 & 0 & 0 & 0 \\
\langle 1/2| & \sqrt{10}B_2^2 & -8B_2^0 & 3\sqrt{2}B_2^2 & 0 & 0 & 0 \\
\langle -3/2| & 0 & 3\sqrt{2}B_2^2 & -2B_2^2 & 0 & 0 & 0 \\
\langle -5/2| & 0 & 0 & 0 & 10B_2^0 & \sqrt{10}B_2^2 & 0 \\
\langle -1/2| & 0 & 0 & 0 & \sqrt{10}B_2^2 & -8B_2^0 & 3\sqrt{2}B_2^2 \\
\langle 3/2| & 0 & 0 & 0 & 0 & 3\sqrt{2}B_2^2 & -2B_2^0
\end{array} \qquad (10)$$

Result: Three doublets of the type

$$\psi_p^+ = a_p|+5/2\rangle + b_p|+1/2\rangle + c_p|-3/2\rangle \qquad (11)$$

$$\psi_p^- = a_p|-5/2\rangle + b_p|-1/2\rangle + c_p|+3/2\rangle \qquad (12)$$

physical significance.

For crystal structures with uniaxial symmetry (tetragonal or hexagonal symmetry) it is sometimes sufficient to consider only the n = 2 terms and neglect the higher-order terms. In that case the crystal field Hamiltonian takes the relatively simple form

$$\mathcal{H}_c = \alpha_J \langle r^2 \rangle [A_2^0 O_2^0 + A_2^2 O_2^2] = B_2^0 O_2^0 + B_2^2 O_2^2 \tag{20}$$

In Table 1 an example is given of how the perturbation matrix may be obtained for the case J = 5/2. It is obvious in uniaxial systems to choose the c-axis as quantization axis or z-axis. The result is a lifting of the 2J+1 = 6 fold degeneracy of the ground state. The perturbation leads to three doublet states that are linear combinations of the states $|\pm5/2\rangle$, $|\pm3/2\rangle$ and $|\pm1/2\rangle$.

Several tetragonal structures have a fairly high symmetry which causes the B_2^2 term to be absent. It may be easily checked in Table 1 that the perturbation matrix (10) is then diagonal in m. The three resulting doublets are $|\pm5/2\rangle$, $|\pm3/2\rangle$ and $|\pm1/2\rangle$.

All the results described above follow from symmetry considerations. To obtain the relative energy of the three doublets one has to know the sign and magnitude of B_2^0. The energy scheme for $B_2^0 > 0$ is shown in Fig. 8. If the ligand charges are known accurately, the values of $B_2^0 = \alpha_J \langle r^2 \rangle A_2^0$ can be calculated if the point charge model is applicable. It is possible, however, to consider B_2^0 as a parameter that can be determined experimentally.

4. Crystal Field Induced Anisotropy

As will be discussed in more detail below, the magnetization in most magnetically ordered materials is not free to rotate but is linked to distinct directions. The latter directions are called the easy magnetization directions or equivalently, the preferred magnetization directions. Different compounds may have a different easy magnetization direction. In most cases, but not always, the easy magnetization direction coincides with one of the major crystallographic directions.

This section will show that the presence of a crystal field can be one of the possible origins of the anisotropy in the magnetization directions. In order to see this we will again consider a uniaxial crystal and assume that the crystal field interaction is sufficiently described by the $B_2^0 O_2^0$ term. Because we are discussing the situation in a magnetically ordered material we also have to take into account the strong internal magnetic field, H_m, as introduced in section 1.

The energy of the system is then described by a Hamiltonian containing the interaction of a given magnetic atom with the crystal field and the molecular field, such that

$$\underline{H}_{tot} = \underline{H}_c + \underline{H}_{ex} = B_2^o\, O_2^o - \mu_o \vec{\mu} \cdot \vec{H}_m. \tag{21}$$

The exchange interaction between the spin moments, as introduced in eq. 3, is isotropic. This means that it leads to the same energy for all directions, provided that the participating moments are collinear, i.e. parallel in a ferromagnet and antiparallel in an antiferromagnet. Thus the exchange interaction itself does not impose any restriction on the direction of H_m. The two magnetic structures shown in Fig. 9 have the same energy when one considers only the magnetic energy term, $\underline{H}_{ex} = \mu_o \vec{\mu} \cdot \vec{H}_m$.

The examples shown in Fig. 9 are ferromagnetic structures ($J_{nn} > 0$) and the same reasoning holds for antiferromagnetic structures ($J_{nn} < 0$) in which the moments are either parallel and antiparallel to c or are parallel and antiparallel to a direction perpendicular to c. Again, in these cases the two antiferromagnetic structures would have the same energy.

After inclusion of the $B_2^oO_2^o$ term in the Hamiltonian the energy becomes anisotropic with respect to the moment directions. This is illustrated by means of the two ferromagnetic structures shown in Fig. 9. We will assume that H_m is sufficiently large and that the exchange splitting of the $| \pm 5/2 \rangle$ level is much larger than the overall crystal field splitting, $| -5/2 \rangle$ being the ground state. The situation in Fig. 9 (A) corresponds to H_m // c or to H_m // z because for the crystal field we have chosen the z-direction along the uniaxial direction. The situation in Fig. 9 (B) corresponds to $H_m \perp$ c, so that we may write H_m // x. Rewriting the Hamiltonian (21) for both situations leads to

$$\underline{H}_{tot} = B_2^o \langle 3J_z^2 - J(J+1) \rangle + \mu_o g \mu_B \langle J_z \rangle H_{m,z} \tag{22}$$

and

$$\underline{H}_{tot} = B_2^o \langle 3J_z^2 - J(J+1) \rangle + \mu_o g \mu_B \langle J_x \rangle H_{m,x}, \tag{23}$$

where $| H_{m,z} | = | H_{m,x} | = H_m = N_W M$.

The Hamiltonian of eq. 22 is already in diagonal form. Because we have chosen H_m large enough, the ground state of eq. 22 is of course

$$| A \rangle = | -5/2 \rangle$$

One may easily obtain the ground state energy by calculating $\langle A | H_{tot} | A \rangle = \langle -5/2 | H_{tot} | -5/2 \rangle = (110/4)\, B_2^o - \mu_o g \mu_B H_m \cdot 5/2$.

In order to find the ground state energy for H_m // x, one has to diagonalize the Hamiltonian of eq. 23. This is a laborous procedure since the J_x

operator will admix all states differing by $\Delta m = \pm 1$ ($J_x = ½ [J_+ + J_-]$, see Table 1). It can be shown that the ground state wave function $|B>$ is of the type

$$|B> = a(|-5/2> - |+5/2>) + b(|-3/2> - |+3/2>) +$$
$$c(|-½> - |+½>)$$

We will not further investigate this wave function except to state that, because of the predominance of H_m, it corresponds to an expectation value $<J_x> = <B|J_x|B>$ almost equal to -5/2. In fact, almost the full moment is obtained, at 0 K, along the x-direction. This means that the magnetic energy contribution is almost equal for the two cases, the last term of eq. 22 and 23. In contrast, one may notice that $<B|J_z^2|B> = 0$ so that the crystal field contribution is strongly reduced when the moments point in the x-direction. The energies associated with the two Hamiltonians in eqs. 22 and 23 can now be written as

$$E_z = (114/4) \cdot B_2^\circ - 5/2\, \mu_o\, g\, \mu_B\, H_m \tag{24}$$

and

$$E_x = (35/4) \cdot B_2^\circ - 5/2\, \mu_o\, g\, \mu_B\, H_m. \tag{25}$$

It is clear from inspecting these two equations that E_z is lower than E_x for $B_2^\circ < 0$. For $B_2^\circ > 0$, and that the situation with the moments pointing along the x direction is energetically preferred. These results can be summarized by saying that, for a given crystal field ($B_2^0 < 0$ or $B_2^\circ > 0$), the 4f charge cloud adapts its orientation and shape in a way to minimize the electrostatic interaction with the crystal field. If the isotropic exchange fields experienced by the 4f moments are strong enough, one obtains the full moment $\mu = -g\mu_B m = g\mu_B J$, or at least a value very close. But the direction of this moment depends on the former energy minimization, i.e. on the sign of B_2°.

5. A Simplified View of 4f Electron Anisotropy

In the case of a simple uniaxial crystal field we found in Section 3 that the leading term of the crystal field interaction, E_1, is given by the expectation value of $B_2^\circ O_2^\circ$,

$$E_1 = B_2^\circ<O_2^\circ> = \alpha_J <r^2><3J_z^2 - J(J+1)> A_2^\circ. \tag{26}$$

In this Section we will show that the crystal field interaction expressed in eq. 24 can be looked upon in a different way, a way which provides a simple physical picture for this type of crystal field interaction. If the ground state at 0 K is $| J_z> = | -J>$, one has

$$E_1 = \alpha_J <r^2> (2J^2 - J) A_2^o, \tag{27}$$

where α_J is the second order Steven factor and $<r^2>$ is the 4f electron radial expectation value. We recall that A_2^o is the second order term of Y_2^o symmetry in the spherical harmonic expansion of the electrostatic crystal field potential. The latter quantity can be looked upon as the electric field gradient component of the crystal field. Equation 24 then represents the interaction of the axial quadrupole moment associated with the 4f charge cloud with the local electric field gradient. It is important to bear in mind that a nonzero interaction with an electric quadrupole moment requires an electric field gradient rather than an electric field. The shape of the 4f charge cloud resembles a discus when $\alpha_J > 0$ or a rugby ball when $\alpha_J < 0$. Examples of both types of charge clouds are shown in Fig. 10.

As was already mentioned above, the molecular field in a magnetically ordered compound is isotropic and $H_m = N_W M$ has the same strength in any direction if the exchange coupling between the moments is the only interaction present. Alternatively one might say that the magnetically ordered moments are free to rotate coherently into any direction. This directional freedom of the collinear system of moments is exploited by the interaction between the 4f quadrupole moment and the electric field gradient to minimize the energy expressed in eq. 24. If the crystal field is comparatively weak one may neglect any deformation of the 4f charge cloud and the aspherical 4f electron charge clouds, shown in Fig. 10, will simply orient themselves in the field gradient to yield the minimum energy situation.

It will be clear that, for a crystal structure with a given magnitude and sign of A_2^o, the minimum energy direction for the two types of shapes shown in Fig. 10 ($\alpha_J < 0$ and $\alpha_J > 0$) will be different. This also implies that the preferred moment direction for rare earth elements with $\alpha_J < 0$ and $\alpha_J > 0$ will be different. It may be shown from eq. 24 that the energy associated with the preferred moment orientation in a given crystal field A_2^o is proportional to $\alpha_J<r^2>(2J^2-J)$. Values of this quantity for several lanthanides have been included in Table 2.

6. Relation between Microscopic and Macroscopic Anisotropy Parameters

It can be shown that simple relations exist between the so-called crystal field parameters, A_n^m, that reflect the strength and symmetry of the crystal field and the macroscopic anisotropy constants, K_i introduced in Section 2. In the first order approximation the anisotropy constants, K_1 and K_2, are related to the crystal field parameters, A_n^m, via the relations (Rudowicz, 1985)

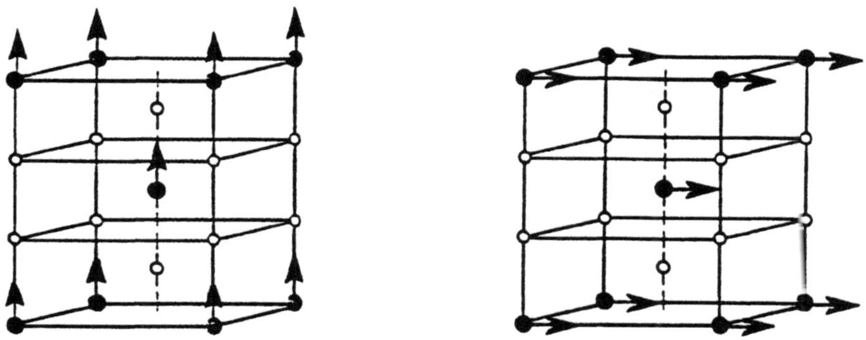

Fig. 9. Two energetically equivalent moment arrangements for the case of zero magnetocrystalline anisotropy.

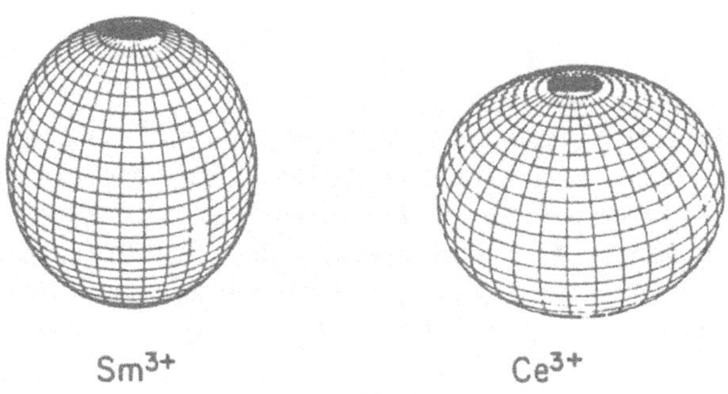

Fig. 10. Examples of 4f electron charge distribution for R^{3+} ions with $\alpha_J > 0$ (left) and $\alpha_J < 0$ (right).

$$K_1 = -\frac{3}{2}\alpha_J\langle r^2\rangle N_R A_2{}^0\langle O_2{}^0\rangle - 5N_R\beta_J\langle r^4\rangle A_4{}^0\langle O_4{}^0\rangle \,. \tag{28}$$

and

$$K_2 = \frac{35}{8}\beta_J\langle r^4\rangle N_R A_4{}^0\langle O_4{}^0\rangle \,. \tag{29}$$

where N_R is the rare earth atom concentration. The $<O_n{}^m>$ quantities are thermal averages of the Stevens operators, $O_n{}^m$, while $\alpha_J \equiv \theta_2{}^0$ and $\beta_J \equiv \theta_4{}^0$ are the second and fourth order Stevens constants introduced in Section 3.

These two latter quantities, α_J and β_J, reflect the shape of the 4f electron charge cloud densities, while the quantities $<r^n>$ are expectation values of r^n and describe the radial extent of the 4f wave function. For each of the rare earth elements the corresponding values of α_J, β_J, and $<r^n>$ have been accurately determined (Hutchings, 1964). In fact, the anisotropy constants, K_i, can be obtained rather straightforwardly by means of eqs. 28 and 29 if the values of the crystal field parameters, $A_n{}^m$, are available. Several experimental methods exist for determining the crystal field parameters, including measurement of the magnetization on single crystals in high magnetic fields as a function of temperature and field strength in various crystallographic directions and rare earth Mössbauer spectroscopy (see the reviews of Buschow 1991, Franse and Radwanski 1993, and Li and Coey 1990). Table 2 lists a few illustrative examples of values of $A_2{}^0$ and $A_4{}^0$ for several different rare earth ions.

7. Model Descriptions

In the previous sections we have kept the crystal field parameters as empirical parameters, their sign and magnitude being derived from experiment. For a rational optimization of the properties of rare earth based permanent magnet materials and, eventually, also for the search of novel materials, it seems desirable to have some understanding of the basic principles that determine the sign and magnitude of the crystal field parameters.

In Section 3 we mentioned that the coefficients, $A_n{}^m$, associated with the series expansion in spherical harmonics of the crystal field Hamiltonian 15 can be written in the point charge model in the form of eq. 16. In the particular case of $A_2{}^0$, after transformation into Cartesian coordinates, one obtains

$$A_2{}^0 = -\frac{|e|}{4}\Sigma\frac{Z_i\,(3\,z_i{}^2 - R_i{}^2)}{R_i{}^6} \tag{30}$$

where the summation is taken over all ligand charges, Z_i, located at a

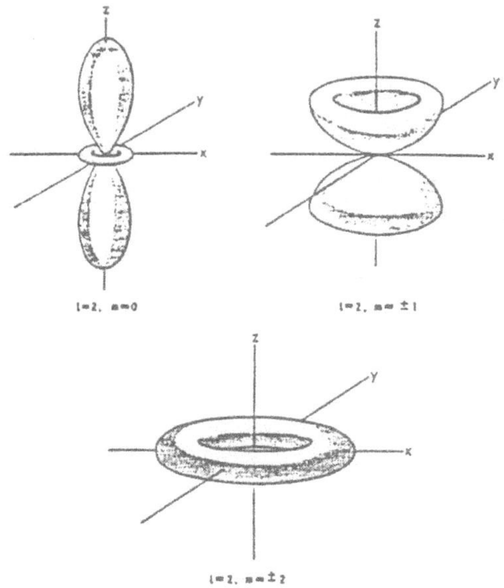

Fig. 11. Charge cloud asphericities of d electrons in a uniaxial crystal field.

Table 2. Values of the second order Stevens factor α_J and $\alpha_J<r^2><O_2^o>$ at 0 K, in units of a_o^2, where a is the Bohr radius, for several lanthanides.

R^{3+}	$\alpha_J\ 10^2$	$\alpha_J<r^2><O_2^o>$
Ce^{3+}	-5.714	-0.686
Pr^{3+}	-2.101	-0.713
Nd^{3+}	-0.643	-0.258
Sm^{3+}	+4.127	+0.398
Tb^{3+}	-1.010	-0.548
Dy^{3+}	-0.635	-0.521
Ho^{3+}	-0.222	-0.199
Er^{3+}	+0.254	+0.190
Tm^{3+}	+1.010	+0.454
Yb^{3+}	+3.175	+0.435

distance $R_j(x_j,y_j,z_j)$ from the central atom considered. Because, in a given crystal structure, the distances between a given atom and its surrounding atoms are exactly known it is possible to make a priori calculations of A_2^o which then can be compared with the experimental value.

The main problem associated with this approximation is the assumption that the ligand ions can be considered as point charges. In most cases the ligand ions have an extensive volume and the corresponding electrostatic field is not spherically symmetric. Also the magnitude of Z_j and, in some case, even the sign of Z_j is not known accurately. The only benefit one may derive from the point charge approximation is that it can be used to predict trends when comparing crystal field effects within a series of structurally similar compounds.

A special complication exists in intermetallic compounds of rare earth elements. This complication is due to the 5d (and 6p) valence electrons of the rare earth elements (Williams and Hirst 1969, Eagles 1975, Morin et al. 1974). When placed in the crystal lattice of an intermetallic compound, the charge cloud associated with these valence electrons will no longer be spherically symmetric but may become strongly aspherical. This may be illustrated by means of Fig. 11 which shows the orientation dependence of the d electron charge clouds in the forms appropriate for a uniaxial environment. Depending on the nature of the ligand atoms, the energy levels corresponding to the different forms in Fig. 11 will no longer be equally populated and will then produce an overall aspherical 5d charge cloud surrounding the 4f charge cloud. Because the 5d (and 6p) valence electrons are located on the same atom as the 4f electrons this on-site valence electron asphericity produces an electrostatic field that may be much larger than that due to the charges of the considerably more remote ligand atoms. It will be clear that results obtained by means of the point charge approximation are not expected to lead to the correct answer in these cases. Band structure calculations made for several type of intermetallic compounds have confirmed the important role of the on-site valence electron asphericities in determining the crystal field experienced by the 4f electrons (Coehoorn et al. 1990).

8. References

Asti G. and S. Rinaldi (1974), J. Appl. Phys. 45, 3600.

Buschow K.H.J. (1988) in "Ferromagnetic Materials Vol. 4" North Holland, Amsterdam. E.P. Wohlfarth and K.H.J. Buschow eds.

Buschow K.H.J. (1991), Rep. Prog. Phys. 54, 1123.

Coehoorn R. and K.H.J. Buschow (1991), J. Appl. Phys. 69, 5590.

Coehoorn R., K.H.J. Buschow, M.W. Dirken and R.C. Thiel (1990), Phys. Rev. B42, 4645.

Durst K.D. and H. Kronmüller (1986), J. Magn. Magn. Mater. 59, 86.

Eagles, D.M. Z. Phys. B21 (1975) 171.

Franse J.J.M. and R.J. Radwanski (1993) in Magnetic Materials Vol. 7, North Holland Amsterdam, K.H.J. Buschow Ed.

Franse J.J.M., S. Sinnema, R. Verhoef, F.R. de Boer and A. Menovsky (1989) in CEAM report p. 175, London 1989 I.V. Mitchel et al. eds.

Grössinger R., R. Krewenka,H.R. Kirchmayr, P. Naastepad and K.H.J. Buschow (1987), J. Less-Common Met. 134, L 17.

Hutchings M.T. (1964), Solid State Phys. 16, 227.

Li H.S. and J.M.D. Coey (1991) in Ferromagnetic Materials Vol. 6, North Holland, Amsterdam, K.H.J. Buschow Ed.

Morin P., J. Pierre, J. Rossad Mignod, K. Knorr and W. Drexel, Phys. Rev. B9 (1974) 4932.

Ram V.S. and P. Gaunt (1983), J. Appl. Phys. 54, 2872.

Rudowicz C. (1985), J. Phys. C 18, 1415.

Smit J. and H.P.J. Wijn (1965), Ferrites, Philips Techn. Libr. Eindhoven 1965.

Strnat K. (1988) in Ferromagnetic Materials Vol. 4, Amsterdam North Holland. E.P. Wohlfarth and K.H.J. Buschow eds.

Sucksmith W. and J.E. Thomson (1954) Proc. Roy. Soc. London A 225, 362.

Williams G. and Hirst L.L., Phys. Rev. 185 (1969) 407.

Zhong X.F. and W.Y. Ching (1989) Phys. Rev. B 39, 12018.

Chapter 16

PRODUCTION OF NITRIDES AND CARBIDES BY GAS-PHASE INTERSTITIAL MODIFICATION

R. SKOMSKI and S. BRENNAN
Physics Department
Trinity College (University of Dublin)
Dublin 2, Ireland

S. WIRTH
IfW Dresden
01171 Dresden, Germany

ABSTRACT. The modification of rare-earth intermetallics by interstitial nitrogen, carbon, and hydrogen is investigated. We describe gas-phase interstitial modification using N_2, NH_3, H_2, or hydrocarbon gases such as CH_4, and compare this method with alternative ways of producing interstitial materials, such as solid-phase interstitial modification. Starting from a simple lattice-gas model we investigate thermodynamics and diffusion kinetics of the interstitial modification reaction and discuss the site occupancy of the interstitial atoms. We put weight on the interstitial modification of Sm_2Fe_{17} by nitrogen and carbon, and discuss phenomena such as nitrogen overloading, macroscopic strain and interatomic interaction.

1. Introduction

1.1. PHYSICAL BACKGROUND

Interest in the behaviour of nitrogen and carbon in iron-rich rare-earth intermetallics was sparked by the discovery that the permanent magnetic properties of Sm_2Fe_{17} are considerably improved on nitrogenation [1]. From the point of view of iron magnetization, the dramatic effect of the interstitial atoms is a large increase in Curie temperature (from 116 °C to 476 °C for $Sm_2Fe_{17}N_3$) due to lattice expansion. Additionally, the interstitial atoms modify the crystal field at the rare-earth site so $Sm_2Fe_{17}N_3$ and $Sm_2Fe_{17}C_{3-x}$ exhibit easy-axis anisotropy, in contrast to the easy-plane anisotropy of the pure Sm_2Fe_{17} [1, 2].

The simplest way to conduct gas-phase nitrogenation is to heat finely ground R_2Fe_{17} (R = rare earth or yttrium) powder at about 450 °C in 1 bar molecular nitrogen. The interstitial modification process proceeds by thermally activated bulk diffusion, and, depending on the size of the particles, the reaction is completed after a few hours. Below about 350 °C diffusion kinetics are extremely sluggish, and, additionally, the nitrogen uptake tends to slow down due to the reduced surface activity of the particles. On the other hand, if the temperature is increased above about 550 °C a competing disproportionation reaction of the nitride intervenes [1, 3-5].

The nitrides have the same crystal symmetry as the parent compound, but the unit cell volume is expanded by about 6 %. In $Sm_2Fe_{17}N_3$ and $Sm_2Fe_{17}C_{3-\delta}$, the interstitial atoms occupy the large octahedral 9e sites nearly exclusively. These sites are coordinated by two rare-earth and four iron atoms, and the in-plane coordination of the rare-earth

F. Grandjean et al. (eds.), Interstitial Intermetallic Alloys, 371–409.

atoms is the main reason for the improved magnetocrystalline anisotropy of the Sm_2Fe_{17} nitrides and carbides. The nitrogen occupancy in $Sm_2Fe_{17}N_3$ was initially inferred from Sm-N bond lengths deduced from Sm L_{III} edge extended x-ray-absorption fine-structure (EXAFS) data [3, 6], but precise powder-neutron-diffraction studies on isostructural $Pr_2Fe_{17}N_3$ and $Nd_2Fe_{17}N_3$ have confirmed that nitrogen occupies the 9e sites in these nitrides [7]. Usually, the 9e sites are not fully occupied, hence the practice of writing the formulas as $Sm_2Fe_{17}Z_{3-\delta}$ where $\delta \approx 0.3$ for Z = N and $\delta \approx 0.8$ for Z = C. However, ***nonequilibrium*** methods such as ion implantation or use of flowing ammonia may populate other sites and yield nitrogen contents slightly larger than 3 [8, 9].

Main experiments to monitor gas-phase interstitial modification are thermopiezic analysis (TPA, Fig. 1) and X-ray diffraction. Typical constant-heating-rate TPA experiments are shown in Fig. 2. Hydrogenation of R_2Fe_{17} intermetallics (Fig. 2a) starts at about 100 °C, but the large entropy of the gas-phase causes the equilibrium hydrogen content to decrease with increasing temperature. Above about 600 °C the intermetallic lattice disproportionates into rare-earth hydride and α iron. The nitrogenation behaviour of R_2Fe_{17} intermetallics is illustrated in Fig. 2b. The main feature of the nitrogenation curves is that interstitial modification and disproportionation of the intermetallic lattice coincide: interstitial nitrides (and carbides with $\delta < 1$) are thermodynamically metastable with respect to a mixture of α iron and the corresponding rare-earth nitrides and carbides such as SmN. A standard method to investigate the phase structure of the nitrides and carbides is X-ray diffraction. Compared to the parent compounds, the X-ray diffraction lines of the carbides and nitrides are shifted towards smaller angles.

A key question is whether the quasi-equilibrium $Sm_2Fe_{17}N_x$ nitride ($0 \leq x < 3$) is a simple gas-solid solution with a continuous range of intermediate nitrogen contents or a two-phase mixture of nitrogen-poor (α) and nitrogen-rich (β) phases. There are two energies involved: the net reaction energy (solution energy) U_o and the interatomic interaction U_1. Phase segregation occurs below a critical temperature T_c and is due to attractive interaction U_1. Below T_c this interaction dominates and the interstitial atoms form macroscopic clusters, which shapes the quasi-equilibrium and non-equilibrium properties of the interstitial compounds. In the case of interstitial nitrides and carbides, the low diffusivity of the interstitial atoms at temperatures below about 400 °C complicates the experimental determination of the critical temperature, but there is much evidence that $Sm_2Fe_{17}N_x$ ($0 \leq x < 3$) is a single-phase system at typical nitrogenation temperatures [4, 10]. At room temperature the diffusivity of the nitrogen and carbon atoms is negligible, and the distribution of interstitial atoms remains quenched.

1.2. HISTORICAL REMARKS

1.2.1. *Hydrogen in Transition-Metal-Rich Rare-Earth Intermetallics*. Many transition-metal-rich rare-earth intermetallics such as $LaNi_5$, Sm_2Fe_{17}, or $Nd_2Fe_{14}B$ are able to absorb considerable amounts of hydrogen [12-18]. Apart from potential applications as hydrogen storage media, it is well-known that interstitial hydrogen influences the magnetic and electronic properties of the host lattice [17, 19]. In the case of iron-rich rare-earth intermetallics such as Sm_2Fe_{17}, the lattice expansion upon hydrogenation improves the fairly disappointing Curie temperature T_c of the parent compounds [15-17].

Still, the Curie temperature remains much smaller than that of pure iron, and the moderate increase in T_c is not accompanied by an improvement of the magnetocrystalline anisotropy. For permanent-magnetic applications, larger lattice expansions and a suitable modification of the electrostatic crystal field at the rare-earth site are required.

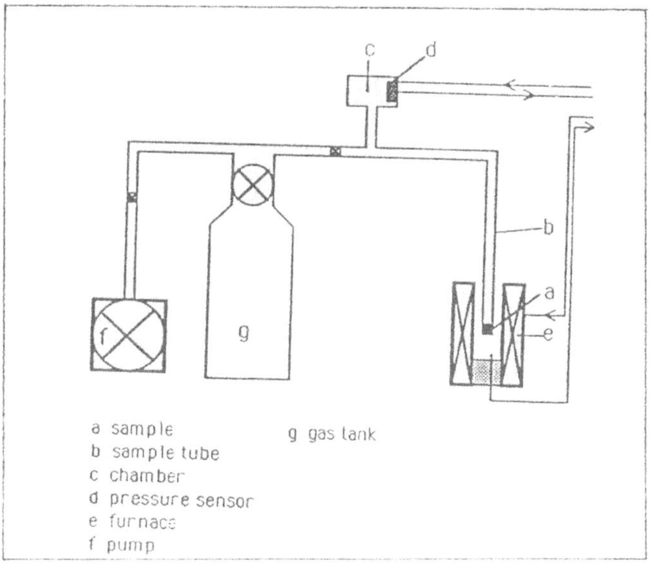

Fig. 1. Thermopiezic analyser [11].

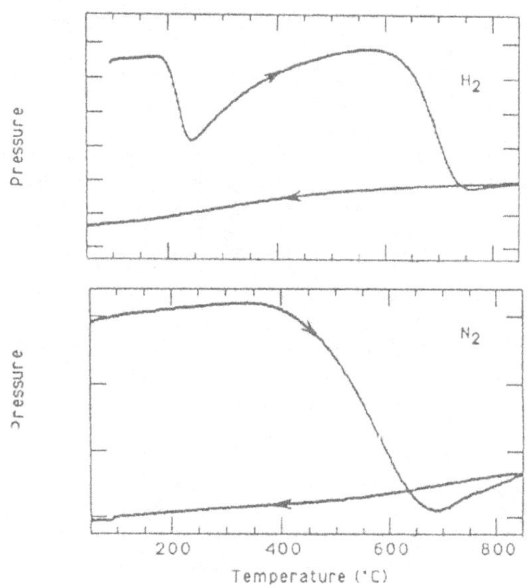

Fig. 2. Constant-heating-rate (10 K/min) thermopiezic analysis: Y_2Fe_{17} in H_2, Y_2Fe_{17} in N_2 [11].

1.2.2. *2p Elements and Metallurgy.* For a long time, solid-solid, liquid-solid, and gas-solid reactions have been used to improve the mechanical hardness of steel [20, 21]. These reactions yield complicated surface layers involving stoichiometrically well-defined iron borides, carbides and nitrides, so it is not meaningful to consider these structures as interstitial alloys. Still, metallurgical modification processes exhibit features which are relevant to interstitial modification as well. For instance, it is known that the use of ammonia instead of molecular nitrogen improves the reactivity of bcc iron [21] and other metallic materials [22]. On the other hand, nitrogen and carbon martensites [19, 21], and related materials such as the alleged giant-moment compound $Fe_{16}N_2$ [19, 21, 23, 24], can be interpreted as interstitially modified metals. Note that martensitic inhomogenities enhance not only the mechanical but also the magnetic hardness of steel: they act as pinning centres and improve the low coercivity of steel-based magnets [25].

It is possibe to introduce 2p atoms directly into the melt to produce ternary alloys such as $Nd_2Fe_{14}B$, $Dy_2Fe_{14}B$ or $Nd_2Fe_{14}C$ [26, 27]. In the case of 2-17 alloys, addition of carbon into the melt yields carbides such as $Sm_2Fe_{17}C_x$ with $x \approx 1$ [26, 28]. Unfortunately, this method cannot be used to produce interstitial nitrides, because the large electronegativity difference between nitrogen and rare earth atoms favours the formation of rare-earth nitrides RN.

1.2.3. *Gas-Phase and Solid-Solid Interstitial Modification.* In 1990, Coey and Sun [1] discovered that nitrogen could be introduced into Sm_2Fe_{17} from the gas phase to yield interstitial nitrides $Sm_2Fe_{17}N_x$ with $x \approx 3$. Since then, gas-phase interstitial modification using molecular nitrogen, ammonia, or hydrocarbon gases has been extended not only to other R_2T_{17} nitrides but also to carbides such as $Sm_2Fe_{17}C_x$ ($x > 2$) and 1-12 nitrides and carbides, such as $Sm(Fe_{11}Ti)N_x$ and $Sm(Fe_{11}Ti)C_x$ with $x \approx 1$ [2, 29-36] . Among these interstitially modified intermetallics, $Sm_2Fe_{17}N_x$ and $Sm_2Fe_{17}C_x$ and related materials such as $(\underline{Sm}Nd)_2(\underline{Fe}Co)_{17}N_3$ [37] are potential high-performance permanent magnets [29]. An alternative way to produce interstitial carbides $Sm_2Fe_{17}C_x$ ($x \geq 2$) is to introduce carbon directly from the solid phase [38].

From the point of view of thermodynamics, the term gas-phase interstitial modification stands for different types of reaction: it denotes (i) nitrogenation using molecular nitrogen or hydrogen [4, 10], (ii) nitrogenation using ammonia NH_3 [8, 9, 39], and (iii) carbonation using hydrocarbon gases such as CH_4 [38]. Ammonia nitrogenation is a complicated reaction, because it includes the catalytic dissociation of ammonia into hydrogen and nitrogen. In the case of gas-phase carbonation the presence of solid carbon (soot) at the metal surface indicates that the interstitial modification using hydrocarbon gases actually represents a derivate of the solid-solid interstitial modification [38].

2. Crystal Structure of Interstitial Nitrides and Carbides

Summary: Typically, interstitials in transition-metal-rich rare-earth intermetallics expand the lattice by a few volume percent but do not change the lattice structure of the parent compound. This volume expansion is the reason for the Curie-temperature increase of Sm_2Fe_{17} upon interstitial modification. The location of the nitrogen and carbon atoms on the large octahedral 9e sites in the rare-earth plane of the rhombohedral Th_2Zn_{17} structure is responsible for the improvement of the uniaxial magnetocrystalline anisotropy. In the tetragonal $ThMn_{12}$ structure, nitrogen and carbon occupy the octahedral 2b sites with *axial* coordination, and the uniaxial anisotropy of $Sm(Fe_{11}Ti)$ deteriorates upon nitrogenation.

2.1. CRYSTALLOGRAPHIC STRUCTURE OF THE PARENT COMPOUNDS

2.1.1. *$CaCu_5$ and Related Structures.* The hexagonal $CaCu_5$ structure (Fig. 3a) consists, roughly speaking, of alternating Cu and Cu-Ca planes. Important examples are the hydrogen storage alloy $LaNi_5$ and the permanent magnetic material $SmCo_5$. Beside the original $CaCu_5$ structure, there exists a whole zoo of 1-5 derivates obtained by systematically substituting transition-metal atoms for rare-earth atoms. Examples are the tetragonal $ThMn_{12}$ structure (Fig. 4), the hexagonal Th_2Ni_{17} structure (Fig. 3b), and the rhombohedral Th_2Zn_{17} structure (Fig. 3c). Note that the tetragonal $Nd_2Fe_{14}B$

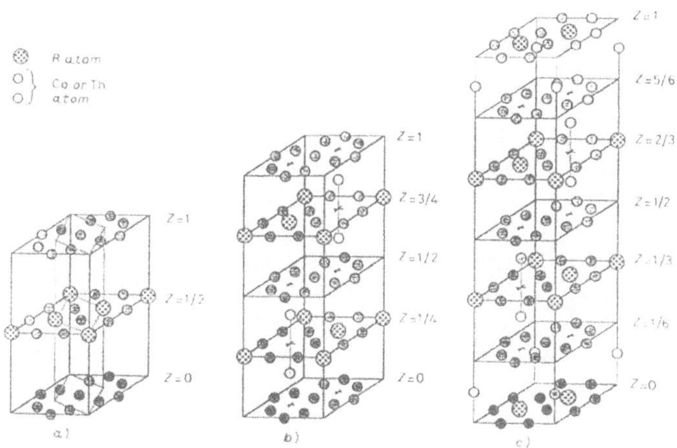

Fig. 3. The $CaCu_5$ structure and its 2-17 derivates: (a) $CaCu_5$, (b) Th_2Ni_{17}, and (c) Th_2Zn_{17} [42].

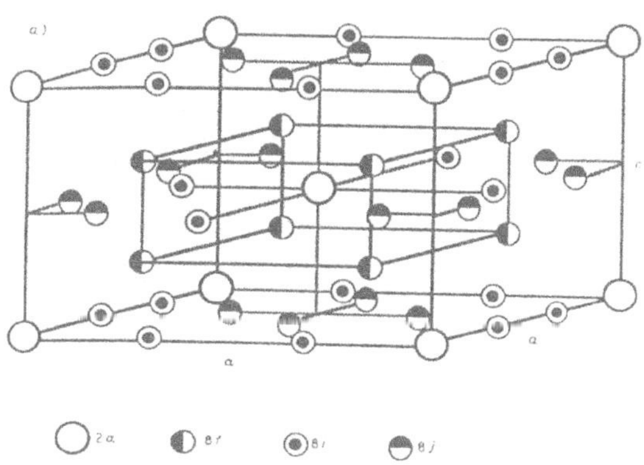

Fig. 4. The tetragonal $ThMn_{12}$ structure. The 2b site is cordinated by the 2a and 8j sites [40].

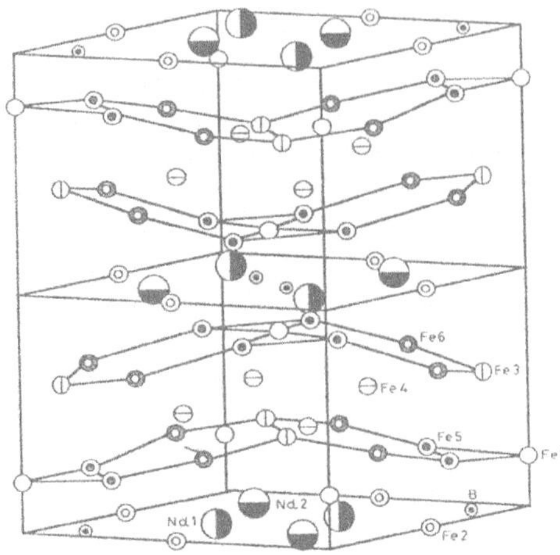

Fig. 5. The tetragonal $Nd_2Fe_{14}B$ structure [40].

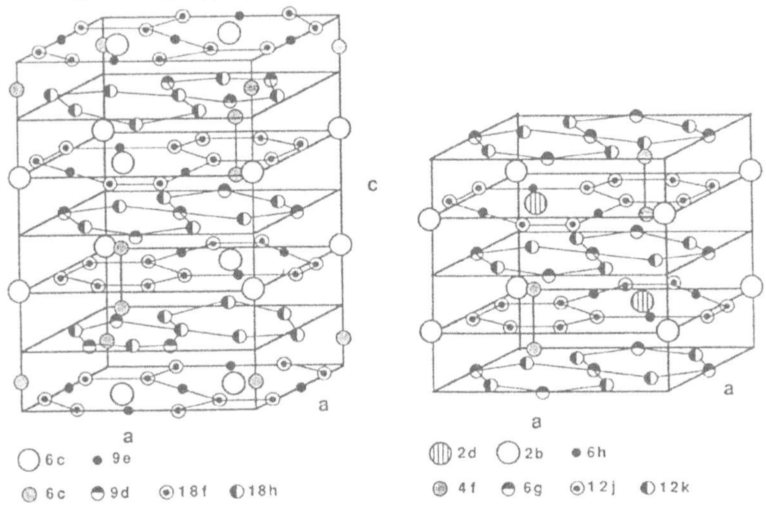

Fig. 6. Th_2Zn_{17} (left) and Th_2Ni_{17} (right) structures showing the octahedral interstitial 9c or 6h sites that may be occupied by C or N [44]. For further details see Fig. 8.

structure (Fig. 5) is not a $CaCu_5$ derivate. Many $CaCu_5$-based rare-earth transition-metal intermetallics, as well as $Nd_2Fe_{14}B$, are suitable for permanent-magnetic applications [40-43]. As a rule, the transition-metal atoms provide the necessary magnetization, while the rare-earth atoms are responsible for the easy-axis anisotropy required of a permanent magnet.

2.1.2. *Th_2Zn_{17} and Th_2Ni_{17} Intermetallics.* The Th_2Ni_{17} and Th_2Zn_{17} structures are closely related to each other, the differences arising from the stacking sequence of hexagonal planes along the crystallographic c axis (Fig. 3). From the point of view of magnetism, the main difference is that the hexagonal Th_2Ni_{17} structure has two nonequivalent rare-earth sites (2b and 2d), while there is only one rare-earth site (6c) in the rhombohedral Th_2Zn_{17} structure. As a rule, the structure of R_2Fe_{17} intermetallics depends on whether the rare earth is heavy or light. Compounds containing light rare earths crystallize in the rhombohedral Th_2Zn_{17} structure, and those containing heavy rare earth in the hexagonal Th_2Ni_{17} structure. It is worthwhile mentioning that both structure types may coexist for R = Gd, Tb and Dy. Yttrium, which is often used as a non-magnetic rare earth, behaves as a heavy rare earth. Note that the related permanent-magnetic material Sm_2Co_{17} is rhombohedral; iron substitution leads to a preferential occupation of 6c dumbell sites without much change in Curie temperature [45].

2.2. INTERSTITIAL MODIFICATION

2.2.1. *Metallic Elements.* Room-temperature α iron and the high-temperature phase δ iron crystallize in the bcc structure, while a large number of transition metals, such as nickel or palladium, form fcc crystals [19, 46, 47]. The dense-packed hcp and fcc structures differ only by their stacking sequences: hcp crystals such as a cobalt exhibit AB stacking in the (001) direction, as opposed to the fcc ABC stacking in the (111) direction [19]. A short description of these interstices is given in Table 1.

Hydrogen, carbon, nitrogen, and oxygen tend to occupy the large octahedral sites in the dense-packed fcc and hcp structures [19]. The composition of these interstitial compounds can be written as MZ_x (M = Ti, Ni, Cu, Zr, ... and Z = H, C, N, O), often with a small interstitial solubility $x \ll 1$ [19]. The preference of the octahedral site originates from the comparatively large (covalent) single-bond atomic radii $R_H = 0.30$ Å, $R_B = 0.88$ Å, $R_N = 0.70$ Å, $R_O = 0.66$ Å, and $R_F = 0.64$ Å [46]. For comparison, the metallic radii of iron and samarium are 1.24 Å and 1.80 Å, respectively (Fig. 7). On the other hand, carbon and nitrogen in α iron occupy the octahedral sites [19], which are *smaller* than the tetrahedral sites in the bcc structure. The reason for this apparent contradiction is the distorted symmetry of the octahedral sites in the bcc lattice: only two of the six surrounding metal atoms are close together, which leads to martensitic lattice distortion and makes the occupation of the octahedral site energetically favourable.

Table 1. Interstitial sites in bcc and close-packed structures.

Structure	Site	Number of sites per metal atom	Radius[a]	Radius (R = 1.25 Å)
bcc	octahedral	3	0.155 R	0.19 Å
bcc	tetrahedral	6	0.291 R	0.36 Å
fcc, hcp	octahedral	1	0.414 R	0.52 Å
fcc, hcp	tetrahedral	2	0.255 R	0.32 Å

[a] R is the metallic radius of the host-lattice atoms.

2.2.2. *Rare-Earth Intermetallics.*

Except for the lattice expansion, which is responsible for the increase in Curie temperature (cf. [27, 40, 48]), the structure of R_2Fe_{17} and RT_{12} intermetallics remains unchanged upon gas-phase interstitial modification [1, 49]. An exception are arc-melted carbides $Y_2Fe_{17}C_x$ ($x \leq 1.5$), where the crystal structure is hexagonal for $x < 1.0$ and rhombohedral for $x > 1.1$ [33]. Fig. 8 shows the rare-earth environment in intermetallics with the rhombohedral Th_2Zn_{17} structure. Nitrogen typically occupies the large 9e site, which is coordinated by two rare-earth and four iron atoms [3, 6, 7, 50]. The differences between the rhombohedral and hexagonal 2-17 structures are related to the stacking sequence and do not affect size and environment of the large octahedral 9e and 6h sites, respectively. Similiar steric arguments apply to the 2b sites in the tetragonal $ThMn_{12}$ structure, but here the axial coordination of the 2b interstices, as opposed to the in-plane one of the 9e and 6h sites, deteriorates the magnetocrystalline anistropy of the $Sm(\underline{Fe}M)_{12}N_{1-\delta}$ nitrides [51].

Hydrogen in Sm_2Fe_{17} occupies the 9e sites and up to two of the six 18g sites, which yields the nominal composition $Sm_2Fe_{17}H_{5-\delta}$ [17, 52]. There is some evidence that nonequilibrium gas-phase interstitial modification using ammonia yields overloaded nitrides $Sm_2Fe_{17}N_{3+\delta}$ ($\delta \leq 1$) where the excess nitrogen is very likely to occupy the 3b

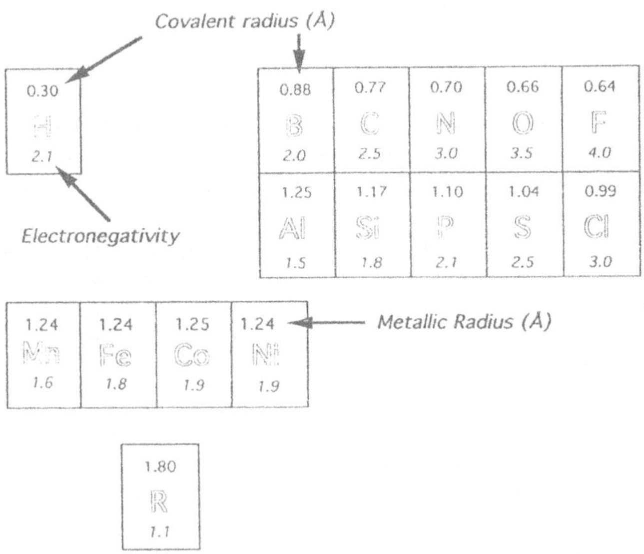

Fig. 7. Radii and electronegativities of selected elements [46].

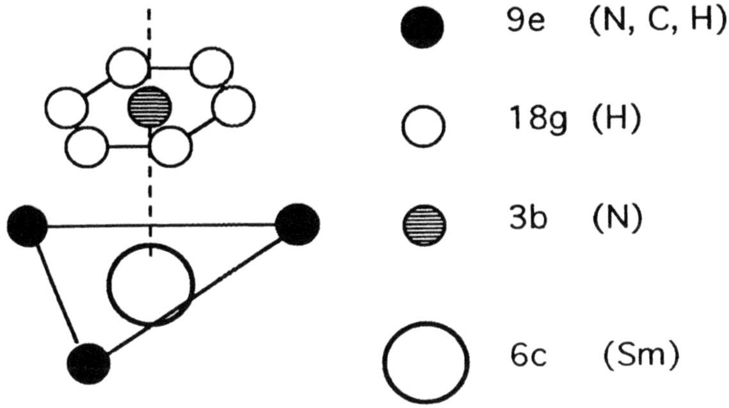

Fig. 8. The environment of the 6c Sm site in Sm_2Fe_{17} (cf. Fig. 6).

sites [8, 9]. However, it has not been possible to produce well-homogenized samples which clearly reveal the occupancy of the excess nitrogen (cf. [8, 9]).

Experimental lattice expansions are shown in Tables 2 and 3. The parameter $\xi = V(\Delta c/c - \Delta a/a)/\Delta V$ desribes the anisotropy of the lattice expansion: $\xi = 1$ points at pure c-axis expansion, $\xi = 0$ describes isotropic dilation, and $\xi = -0.5$ indicates in-plane expansion. In the case of $R_2Fe_{17}N_3$ nitrides, ξ is negative for light rare earths but positive for the heavy ones. This rule reflects, at least partly, the steric situation in 2-17 nitrides: the light rare earths are larger than the heavy ones (lanthanide contraction), and a comparatively large expansion perpendicular to the c axis is necessary to relieve the in-plane constraints. On the other hand, the preferential c-axis expansion in the case of overloaded nitrides ($x > 3$) speaks in favour of the 3b site occupancy.

Note that hydrogen in metals often, but not always, causes a lattice expansion $\Delta v \approx 2.9$ Å^3 per hydrogen atom [53, 54]. The volume expansion per nitrogen or carbon atom given in Tables 2 and 3 exceeds this value by a factor of order two.

Table 2. Structural properties of interstitially modified samarium and yttrium intermetallics.

	a (Å)	c (Å)	ΔV/V (%)	Δv (Å^3)[a]	ξ[b]	Ref.
Sm_2Fe_{17}	8.54	12.43				[49, 55, 56]
RHOMBOHEDRAL Th_2Zn_{17} STRUCTURE						
Sm_2Fe_{17}	8.55	12.45				[57]
$Sm_2Fe_{17}N_{1.0}$	8.62	12.51	2.1	5.6	- 0.16	[57]
$Sm_2Fe_{17}N_{2.1}$	8.73	12.69	6.7	8.3	- 0.02	[55]
$Sm_2Fe_{17}N_{2.3}$	8.73	12.64	6.3	7.1	- 0.09	[49]
$Sm_2Fe_{17}N_{2.5}$	8.73	12.64	6.3	6.6	- 0.09	[56]
$Sm_2Fe_{17}N_{2.6}$	8.71	12.61	5.1	5.2	- 0.11	[57]
$Sm_2Fe_{17}N_{2.9}$	8.74	12.70	7.0	6.3	- 0.02	[55]
$Sm_2Fe_{17}N_{3.7}$	8.75	12.83	8.3	6.0[c]	0.08	[9]
$Sm_2Fe_{17}C_{2.2}$	8.75	12.57	6.2	7.3	- 0.22	[56]
Y_2Fe_{17}	8.48	8.26				[49, 58]
HEXAGONAL Th_2Ni_{17} STRUCTURE						
$Y_2Fe_{17}N_{2.5}$	8.65	8.44	6.3	6.5	0.03	[58]
$Y_2Fe_{17}N_{2.6}$	8.65	8.44	6.3	6.2	0.03	[49]
$Y_2Fe_{17}C_{1.0}$	8.54	8.35	2.5	6.5	0.15	[33]
$Y_2Fe_{17}C_{2.2}$	8.66	8.40	6.1	7.1	- 0.07	[58]
$Y_2Fe_{17}H_{2.7}$	8.54	8.29	1.8	1.7	- 0.19	[58]
$Y(Fe_{11}Ti)$	8.50	4.80				[30]
TETRAGONAL $ThMn_{12}$ STRUCTURE						
$Y(Fe_{11}Ti)$	8.50	4.79				[44]
$Y(Fe_{11}Ti)N_{0.8}$	8.58	4.80	1.8	3.9	- 0.50	[30]
$Y(Fe_{11}Ti)N_{0.8}$	8.62	4.81	3.1	6.7	- 0.30	[44]
$Y(Fe_{11}Ti)C_{0.9}$	8.58	4.80	1.7	3.3	- 0.50	[30]
$Y(Fe_{11}Ti)H_{1.2}$	8.51	4.79	0.2	0.3	- 0.50	[44]
$Sm(Fe_{11}Ti)$	8.56	4.80				[44]
TETRAGONAL $ThMn_{12}$ STRUCTURE						
$Sm(Fe_{11}Ti)N_{0.8}$	8.64	4.84	2.8	6.2	- 0.04	[44]
$Sm(Fe_{11}Ti)C_{0.8}$	8.64	4.81	2.2	4.8	- 0.35	[44]
$Sm(Fe_{11}Ti)H_{1.2}$	8.55	4.80	-0.1	-0.2	- 0.50	[44]

[a] Δv is the volume expansion per interstitial atom [b] ξ is the anisotropy parameter of the lattice expansion [c] Estimate

Table 3 Structural properties of $R_2Fe_{17}N_x$ nitrides [49]

	Structure	a (Å)	c (Å)	ΔV/V (%)	Δv (Å^3)	ξ
Ce_2Fe_{17}	Th_2Zn_{17}	8 47	12 32			
$Ce_2Fe_{17}N_{2\,8}$	Th_2Zn_{17}	8 73	12 65	9 1	8 3	- 0 04
Pr_2Fe_{17}	Th_2Zn_{17}	8 57	12 42			
$Pr_2Fe_{17}N_{2\,5}$	Th_2Zn_{17}	8 77	12 64	6 6	6 9	- 0 09
Nd_2Fe_{17}	Th_2Zn_{17}	8 56	12 44			
$Nd_2Fe_{17}N_{2\,3}$	Th_2Zn_{17}	8 76	12 63	6 3	7 3	- 0 13
Sm_2Fe_{17}	Th_2Zn_{17}	8 54	12 43			
$Sm_2Fe_{17}N_{2\,3}$	Th_2Zn_{17}	8 73	12 64	6 3	7 1	0 09
Gd_2Fe_{17}	Th_2Zn_{17}	8 51	12 43			
$Gd_2Fe_{17}N_{2\,4}$	Th_2Zn_{17}	8 69	12 66	6 2	6 7	0 04
Tb_2Fe_{17}	Th_2Zn_{17}	8 45	12 41			
$Tb_2Fe_{17}N_{2\,3}$	Th_2Zn_{17}	8 66	12 66	7 2	8 0	- 0 07
Dy_2Fe_{17}	Th_2Ni_{17}	8 45	8 30			
$Dy_2Fe_{17}N_{2\,8}$	Th_2Ni_{17}	8 64	8 45	6 4	5 9	- 0 07
Ho_2Fe_{17}	Th_2Ni_{17}	8 44	8 28			
$Ho_2Fe_{17}N_{3\,0}$	Th_2Ni_{17}	8 62	8 45	6 5	5 5	- 0 01
Er_2Fe_{17}	Th_2Ni_{17}	8 42	8 27			
$Er_2Fe_{17}N_{2\,7}$	Th_2Ni_{17}	8 61	8 46	7 0	6 6	0 01
Tm_2Fe_{17}	Th_2Ni_{17}	8 40	8 28			
$Tm_2Fe_{17}N_{2\,7}$	Th_2Ni_{17}	8 58	8 47	6 7	6 3	0 02
Lu_2Fe_{17}	Th_2Ni_{17}	8 39	8 26			
$Lu_2Fe_{17}N_{2\,7}$	Th_2Ni_{17}	8 57	8 48	7 1	6 6	0 07
Y_2Fe_{17}	Th_2Ni_{17}	8 48	8 26			
$Y_2Fe_{17}N_{2\,6}$	Th_2Ni_{17}	8 65	8 44	6 3	6 2	0 03

3. Ideal Gas-Solid Equilibrium

Summary: Molecular gases such as N_2 or H_2 dissociate at the surface of the metallic particles and occupy well-defined interstitial sites inside the lattice. There are only two states per interstitial site (empty and occupied), so the temperature dependence of the equilibrium nitrogen concentration is reminiscent of a finite-temperature Fermi distribution. The pressure dependence of the nitrogen concentration arises from the large entropy of the gas phase and is calculated in terms of a configurational lattice-gas model. In the main, gas-solid interstitial modification reactions are entropically less favourable than solid-solid reactions. An exception is the 'non-equilibrium' NH_3 interstital modification reaction, where the maximum nitrogen content is $x \approx 4$, as opposed to $x \approx 3$ if N_2 is used.

3.1. BACKGROUND

3.1.1. *Lattice-Gas Interaction.* The simplest approach to understand the thermodynamics and statistics of gases in metals is to describe the interstitial atoms as a lattice gas (Fig. 9). Thermodynamical properties of lattices gases are largely independent of the detailed lattice structure if we assume that there are only two states per (isolated) site [59, 60]. In the case of interstitially modified metals, these two states may be interpreted as local concentrations $c_l = 0$ (empty) and $c_l = 1$ (occupied). Somewhat misleading, lattice-gas models where the condition $c_l \leq 1$ (site blocking) is the only interaction are called interaction-free. Interaction-free lattice gases have a number of pleasent properties: the

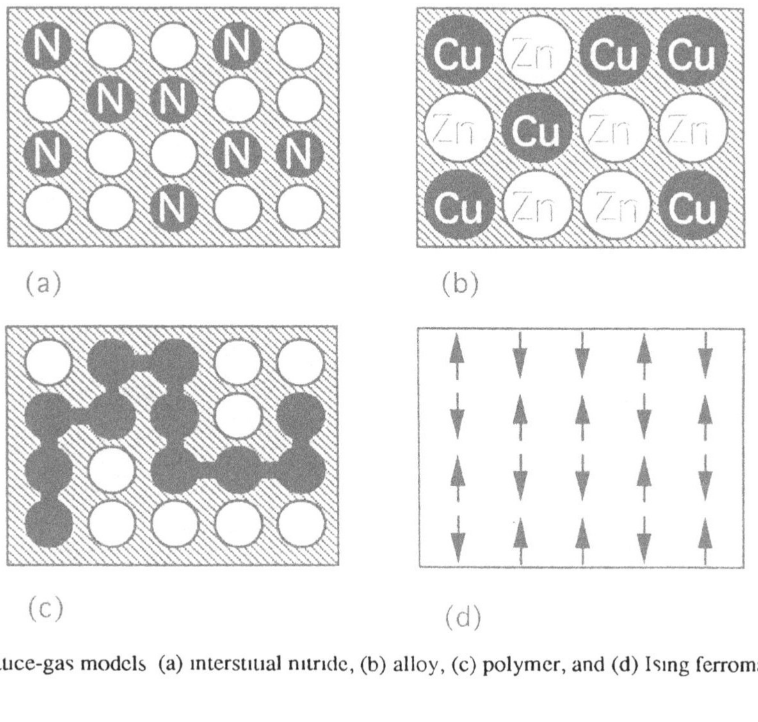

Fig 9 Lattice-gas models (a) interstitial nitride, (b) alloy, (c) polymer, and (d) Ising ferromagnet

equilibrium solubility <$c_i(T, p)$> can be calculated analytically, the reaction kinetics are governed by a simple diffusion equation, and concentration profiles of incompletely charged particles are continuous [4, 10, 61-63].

As we will discuss in Section 5, interaction effects dominate below a critical temperature T_c. There is much evidence that interstitially modified iron-rich rare-earth intermetallics such as $Sm_2Fe_{17}N_x$ behave as a gas-solid solution rather than a two-phase system with discontinuous interstitial concentration. In other words, the critical temperature T_c is lower than typical nitrogenation temperatures, and quenching fixes the distribution of the interstitial atoms at room temperature. For this reason it is appropriate to treat the interstitially modified intermetallics in question as gas-solid solutions. Deviations from the ideal gas-solid-solution behaviour will be discussed in Section 5.

3.1.2. *Microscopic Approaches*. First-principle calculations of interstitial solubilities are very difficult. The main contributions are (i) the electrostatic energy which reflects the difference in electronegativity, (ii) the covalent (or metallic) binding energy, and (iii) the elastic energy which is necessary to expand the lattice near the interstitial sites [64-69]. Despite much success in the description of systems such as hydrogen in metallic elements, it is often necessary to resort to phenomenological or semiphenomenological models [65, 70].

A simple approach, which has subsequently been made more sophisticated, is to consider the embedding of hydrogen and 2p atoms (B, N, C, O) in a homogeneous electron gas (jellium) [66, 67]. Interstitial modification of a homogeneous electron gas is energetically favourable in the case of alkali metals but unfavourable if the electron density reaches values typical of transition metals. Additionally, a fair amount of energy is required to break the atomic bonds in the elementary solids or gases [46].

Steric constraints, which are related to the repulsion of non-orthogonalized electronic states, influence the solubility of interstitial atoms. For hydrogen in metals there is a rule that interstitial modification is possible if the radius of the interstices exceeds 0.4 Å. Furthermore, the hydrogen atom separation must not be smaller than about 2.1 Å [54]. The radius of the 18g sites in Th_2Zn_{17} rare-earth intermetallics is of order 0.55 Å, so hydrogen occupancy is possible, but the too-small distances between the 18g sites (Fig. 10) lead to an incomplete filling of the 18g hexagon. The nominal composition $Sm_2Fe_{17}H_x$ with $x \leq 5$ and neutron-diffraction measurements [52] indicate

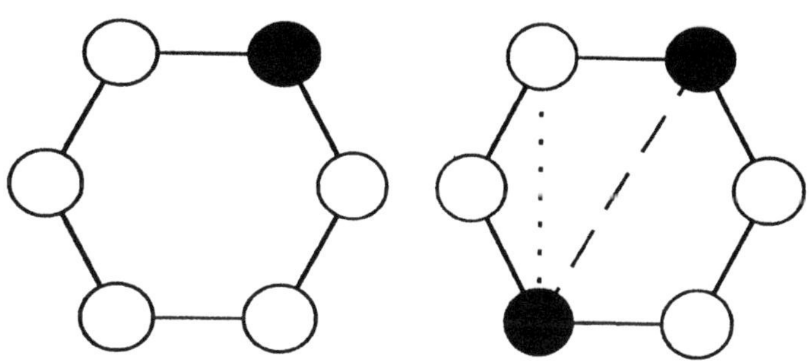

Fig. 10. The 18g hexagon of the Th_2Zn_{17} structure. The distances are 1.9 Å (dotted line) and 2.2 Å (dashed line).

the occupancy of the three 9e sites and up to two 18g sites per hexagon. The 9e sites are only slightly larger than the 18g sites (about 0.6 Å), but, as experiment shows, the neighbourhood of the strongly electronegative rare-earth atoms outweighs the energetically unfavourable steric configuration [10].

3.1.3. *Equation of State and Chemical Potential.* A widely used approach is to express the average equilibrium concentration $\langle c \rangle$ in terms of a chemical potential μ [60-62]. In the following, we will abandon this paradigm in favour of the thermodyamical variables T (temperature), N (number of interstitial atoms), and V (gas volume). In other words, we couple our isolated lattice-gas system to an external heat reservoir and replace the configurational entropy S by its thermodynamically conjugate variable T. This allows us to keep the more pictorial variables V and $N = \langle c \rangle n_s$ instead of p (gas pressure) and μ, respectively. Recall that the equation of state of a thermodynamic system remains *unchanged* if this replacement procedure is conducted properly. For instance, if the restriction to a fixed reaction-chamber volume were released than we would have to base the calculation on the enthalpy $H = U - pV$. A further advantage of the present approach is that all equations of state are obtained in terms of microscopic parameters such as binding energies. We are therefore able to replace the largely descriptive approach of chemical thermodynamics by a more physical view.

3.2. EQUILIBRIUM SOLUBILITY

3.2.1. *Lattice-Gas Statistics.* The lattice-gas model Fig. 11 consists of a solid with n_s (octahedral) sites in contact with a large but constant volume of gas V divided into $N_o = V/V_o$ cubic cells [4, 10]. We can interprete V_o as the cube of the atomic diameter of molecular nitrogen, about 1 Å^3. A particular feature of lattice-gas models such as that shown in Fig. 11 is that the concentration of interstitial atoms in the metal does not depend on translational and vibrational degrees of freedom (configurational approach). As a matter of fact, both experiment and theoretical arguments show the limited influence of motional degrees of freedom. For instance, replacing the energy $U = mv^2/2$ of an ideal gas atom by $U = 0$ does not affect the equation of state $pV = k_B T$.

Let us first consider the case of a single atom in the model Fig. 11, where the energy of the atom in the metal be $\mathbb{H} = U_o$, but $\mathbb{H} = 0$ in the 'gas phase'. The probability p_i of finding the atom on a particular site is now given by the Boltzmann factor

$$p_i = \frac{1}{Z} e^{-\beta \mathbb{H}_i} \tag{3.1}$$

with $\beta = 1/k_B T$. The partition function reads

$$Z = \sum e^{-\beta \mathbb{H}} \tag{3.2}$$

where the summation includes all $\Omega = N_o + n_s$ possible single-atom configurations. There are n_s interstitial sites, so the probability of finding the atom in the metal is $\langle c \rangle_{eq} = n_s \exp(-\beta U_o)/\{N_o + n_s \exp(-\beta U_o)\}$ or

$$\langle c \rangle_{eq} = \frac{1}{1 + \frac{N_o}{n_s} \exp(\beta U_o)} \tag{3.3}$$

We see that interstitial occupancy is favourable for (i) high gas pressure (N_0 small), and (ii) exothermic reactions ($U_0 < 0$) at (iii) low temperature The temperature dependence of the concentration <c> is reminiscent of the quantum-mechanical Fermi-Dirac distribution [47, 69], which reflects the binary lattice-gas occupancy Note that the derivation of Eq (3 3) is purely microscopic phenomenological quantities such as chemical potentials are not involved On the other hand, the restriction to a single atom is an oversimplification, and it is difficult to specify thermodynamical quantities such as pressure

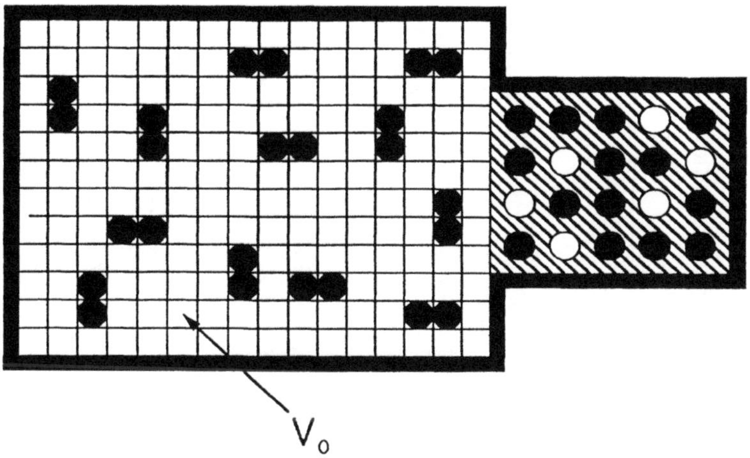

Fig 11 Schematic illustration of the lattice gas model, showing nitrogen (black circles) in the gas phase (left hand side) and in solid solution in the intermetallic compound (right hand side) Note the existence of unoccupied octahedral 9e sites

3 2 2 *Partition Function and Equation of State* The generalization of Eq (3 3) to molecular gases (n atoms) is slightly more complicated [4, 10] First, we have to redefine the net reaction energy (energy of solution) U_0 as the energy difference per atom refering to the *molecular* state Next, the partition function has to be calculated from the number $N(\mu)$ of configurations with a fixed number $\mu \leq n$ of interstitial atoms (Fig 11)

$$Z = \sum_{\mu = 0}^{n_s} N(\mu)\, e^{\beta\mu U_o} \qquad (3\ 4)$$

The calculation of $N(\mu)$ and Z requires some bookkeeping but is straight forward [4]

Once Z is known, the equation of state can be calculated analogously to Eq (3 3) or as a derivative of Z (see Appendix) The result for diatomic gases is [4, 10]

$$<c>_{eq} = \frac{1}{1 + \sqrt{\frac{p_o}{p}\, e^{\beta U_o}}} \qquad (3\ 5)$$

with $p = n\,k_BT/2$ and the pressure $p_o = 3k_BT/V_o \approx 150$ kbar. Typical curves for the nitrogenation of Sm_2Fe_{17} in N_2 are shown in Fig. 12. Based on long-time isothermal absorption experiments the net reaction energy $U_o = -57 \pm 5$ kJ/mole [4, 10] has been obtained (cf. Table 6 in Section 4).

In the high-temperature limit $k_BT \gg |U_o|$, and for $U_o \geq 0$ at all temperatures, Eq. (3.5) reduces to Sieverts' law $\langle c \rangle = (p/p_o)^{1/2} \exp(-U_o/k_BT)$ [19, 71]. Typical examples of endothermic gas-solid systems ($U_o > 0$) are hydrogen and nitrogen in α iron [19]. The low but finite endothermic solubility is due to the thermal activation, which 'pushes' gas atoms into the lattice, and *increases* with temperature.

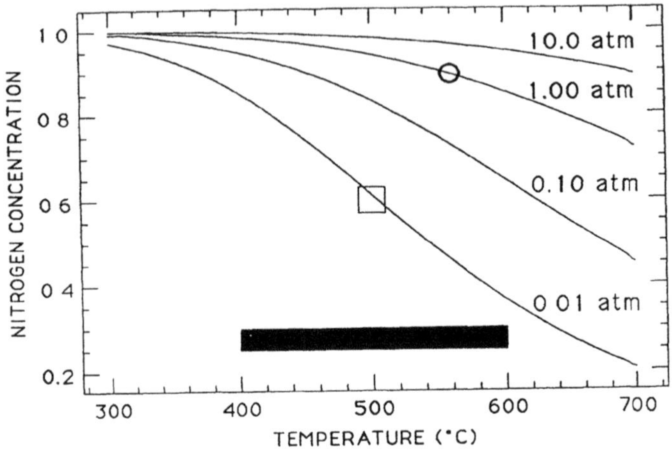

Fig. 12. Equilibrium nitrogen concentrations in Sm_2Fe_{17} as function of temperature and pressure, calculated from Eq. (3 5) The black bar shows the temperature region where interstitial modification may be carried out.

3.2.3. *Solid-Solid Reaction.* Apart from molecular gases, atomic gases and solids are conceivable sources for interstitial modification. It has been shown [38] that interstitial carbides can be produced from a mixture of rare-earth intermetallics, in the form of powder, and solid carbon, in the form of graphite (Fig.13). Both reaction conditions and lattice expansions, $\Delta V/V \geq 5$ vol.%, are similiar to those of gas-phase interstitial modification using hydrocarbon gases. In fact, optical observation — soot at the particles' surface — suggests that the reaction of rare-earth intermetallics with hydrocarbon gases is ultimately a solid-solid reaction.

The entropy of solid carbon is negligibly small, which yields a pressure-independent equilibrium carbon content. The temperature dependence of the carbon concentration reads

$$\langle c \rangle_{eq} = \frac{1}{1 + e^{\beta U_o}} \tag{3.6}$$

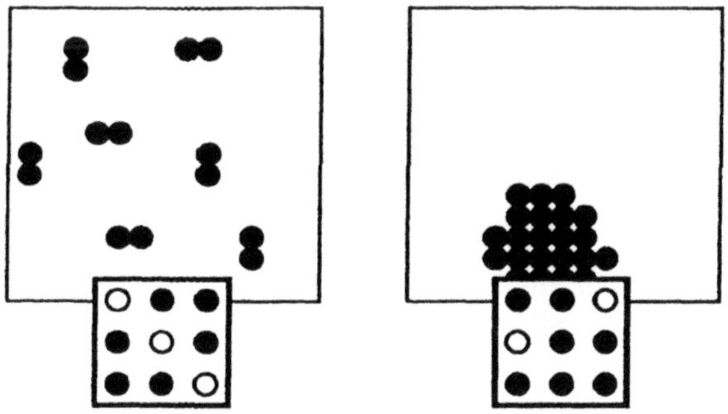

Fig. 13. N_2 gas-phase (left) and solid-phase (right) interstitial modification.

so the assumption of a weakly negative reaction energy U_o = - 5 kJ/mole yields x = 3 $\langle c \rangle_{eq}$ = 2.13 at 400 °C and x = 1.75 at 1500 °C. Gas-phase interstitial modification yields carbon contents of order x = 2.2, while melted carbides [26] are characterized by x ≤ 1.5.

3.2.4. *Gas Mixtures.* It is likely that hydrogen improves the particle surface reactivity and creates microcracks, so it has been proposed to conduct gas-phase inerstitial modification in a mixture of hydrogen and nitrogen gases (cf. Section 4). From the point of view of equilibrium thermodynamics, however, there is no reason to assume that hydrogen influences the gas-phase nitrogenation reaction. As it can be seen from Fig. 2, the solubility of hydrogen in Sm_2Fe_{17} is very low above 400 °C, so the 'substitution' of interstitial hydrogen by nitrogen is a rare event. Upon heating, the hydrogen atoms leave the lattice before the nitrogen atoms enter the lattice. An estimate based on Eq. (3.5) and Table 6 in Section 4 is that less than about 1% of the nitrogen atoms have to be replaced by hydrogen, and at present it it not possible to detect such small amounts of hydrogen quantitatively.

3.3. OVERLOADED NITRIDES

3.3.1. *Maximum Nitrogen Content.* There are reports dealing with possible nitrogen contents x > 3.0 in 2-17 and x > 1.0 in 1-12 nitrides, which are incompatible with an exclusive occupation of the large R_2Fe_4 octahedra in the Th_2Zn_{17} and $ThMn_{12}$ structures (see e.g. [39, 55, 72]). $Sm_2Fe_{17}N_x$ nitrides with x ≤ 6 are reported in [39], but no structural data to situate the excess nitrogen are given, and the observed volume expansion (≈ 7.5%) appears to be inconsistent with an interstitial nitrogen occupancy very much greater than three. In [55], ammonia treatment is believed to yield $Sm_2Fe_{17}N_x$ with nitrogen contents as high as x = 8.2. Here again, neither volume expansion (7.7%) nor structural data support a nitrogen content much larger than three. Note that the complete disproportionation of Sm_2Fe_{17} into SmN and Fe_4N requires 6.25 nitrogen atoms per formula unit, which falls in the range of nitrogen contents claimed to be possible. In the case of RT_{12} intermetallics, nitrides $RFe_{11}TiN_x$ with nitrogen contents x ≤ 1.4 have been reported [72], but the lattice expansion $\Delta V/V \leq$ 2.9% is not much greater than that observed for x ≈ 0.8 (cf. Table 2).

3.3.2. *Non-Equilibrium Interstitial Modification Using NH_3.* All nitrides discussed in the preceeding section were procuced by gas-phase interstitial modification using ammonia. Even the first attempts to produce $Sm_2Fe_{17}N_x$ [1] were based on an ammonia treatment, but shortly afterwards it became clear that molecular nitrogen can be used equally well. As a matter of fact, nitrogen overloading deteriorates the permanent magnetic performance of $Sm_2Fe_{17}N_x$ [9, 39, 55], and there is no point in artifically increasing the nitrogen content beyond x = 3. Furthermore, the characterization and investigation of samples produced by gas-phase interstitial modifaction using molecular nitrogen is much more transparent. The special interest in the gas-phase interstitial modification using ammonia is related to (i) the scientifically interesting reaction kinetics and (ii) the creation of microcracks in ammonia-treated powders which may facilitate industrial-scale interstitial modification.

As discussed in Section 1, ammonia treatment can be used to produce metallic nitrides. The reason for this tendency is the comparatively low entropy of gaseous ammonia: despite the endothermic decomposition of ammonia, the entropy gain upon formation of 1.5 H_2 molecules per NH_3 molecule makes the catalytic decomposition of ammonia thermodynamically favourable. At 500 °C and 1 bar the concentration of ammonia in *equilibrium* with molecular hydrogen and nitrogen is of order 1%, and the modification reaction is actually conducted in a mixture of nitrogen and hydrogen gases. However, due to high activation energies the catalytic decomposition of ammonia proceeds sluggishly, and the N_2-H_2/NH_3 subsystem remains for a long time in a non-equilibrium state. In this regime, nitrogen absorption includes the decomposition of ammonia and is therefore thermodynamically favourable, even if it involves energetically unfavourable site occupancy.

A way to describe the reaction with ammonia is to replace Eq. (3.5) by the quasi-equilibrium equation

$$\langle c\rangle_{eq} = \frac{1}{1 + \sqrt{\dfrac{p_H{}^3}{p_{NH_3}{}^2 p_c}\, e^{\beta U'_o}}} \qquad (3.7)$$

where U'_o is the net reaction energy for the non-9e site and $p_c \approx 70$ kbar [8, 9]. Eq. (3.7) shows that the use of ammonia is equivalent to a very high nitrogen pressure. In the case of a pure ammonia atmosphere, $p_H = 0$, we obtain $\langle c\rangle_{eq} = 1$. Based on Eq. (3.7) an estimate $U'_o = 5 \pm 10$ kJ/mole has been obtained [9].

Catalytic non-equilibrium nitrogenation using ammonia yields volume expansions up to 8.3% [8, 9]. Assuming that there is a volume expansion of about 2% per nitrogen atom and formula unit, this expansion corresponds to the nominal composition $Sm_2Fe_{17}N_4$ [8, 9].There is some evidence [9], but no final proof, that the fourth nitrogen atom occupies the 3b site at the centre of the 18g hexagon (Fig. 8): (i) there is exactly one 3b site per formula unit, (ii) nitrogen overloading leads to a preferential expansion in the crystallographic c direction, and (iii) the measured decrease in coercivity (cf. [39, 55]) can be explained by the presence of nitrogen atoms at 3b sites.

4. Diffusion

Summary: Above the critical temperature, gas-phase interstitial modification proceeds by thermally activated diffusion. In the case of spherical Sm_2Fe_{17} particles, the nitrogen concentration profile c(**r**) is given by the average concentration <c(**r**)>, which, in turn, depends on particle rardius, temperature, time, and activation energy. The profile does *not* depend on parameters such as the net reaction energy (energy of solution). After the initial stage of reaction, where the reaction velocity obeys a square-root law, the reaction slows down, and the long-time regime is characterized by exponential or power-law behaviour, depending on the particle size distribution.

4.1. DIFFUSION EQUATION

4.1.1. *Random-Walk Statistics.* A pictorial way of dealing with diffusion is to decompose the diffusion process into random hopping events (Fig. 14). This randomness is characteristic of diffusive processes, and the term diffusion should not be used to denote largely deterministic phenomena such as nitrogen transport along microcracks.

Let us first derive the interaction-free one-particle diffusion equation. The extension to higher dimensionalities and particle numbers is straightforward, while interaction effects have to be discussed separately (Section 5). The probability $P_i(t) = P(x_i,t)$ of finding the interstitial atom at the site x_i obeys the rate equation

$$P_i(t+\tau_o) - P_i(t) = W_{i,i+1}\, P_{i+1}(t) + W_{i,i-1}\, P_{i-1}(t) - W_{i+1,i}\, P_i(t) - W_{i-1,i}\, P_i(t) \tag{4.1}$$

where τ_o is a small time interval and W_{ij} the transition probability $W(x_j \rightarrow x_i)$. In our case, $W_{ij} = W_{ji} = \exp(-E_a/k_BT)$ where the activation energy E_a describes thermally activated hopping.

If the x and t dependence of P(x, t) is sufficiently weak (mesoscopic limit) then series expansion of Eq. (4.1) yields

$$\frac{\partial P(x,t)}{\partial t} = \Gamma_o\, a^2 \exp(-E_a/k_BT) \frac{\partial^2 P(x,t)}{\partial x^2} \tag{4.2}$$

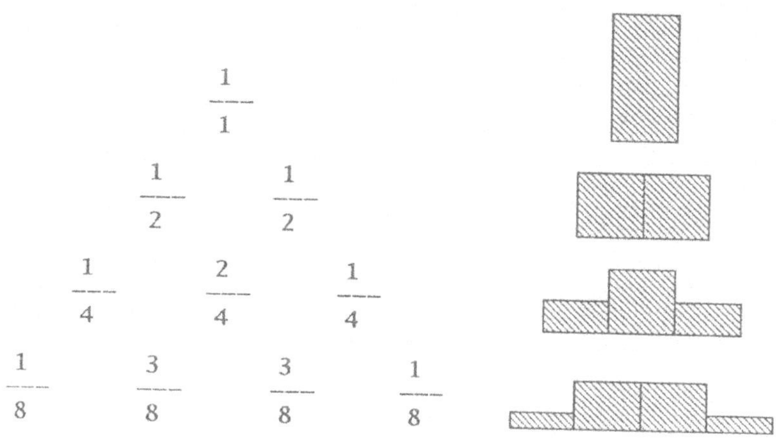

Fig. 14. Pascal's triangle which can be used to visualise ond dimentional diffusion.

where $\Gamma_o = 1/\tau_o$ and $P_{i\pm1} = P(x_i \pm a)$. Generalization to three dimensions and putting $P(\mathbf{r},t) = \langle c(\mathbf{r}, t)\rangle$ yields the diffusion equation

$$\frac{\partial \langle c(\mathbf{r}, t)\rangle}{\partial t} = D\, \nabla^2 \langle c(\mathbf{r}, t)\rangle \tag{4.3}$$

where we have put $D = D_o \exp(-E_a/k_BT)$ with $D_o = \Gamma_o\, a^2$. As it will be shown in Section 5, the diffusion constant of interaction-free systems does *not* depend on the interstitial concentration.

4.1.2. *Diffusion Constants*. Table 4 shows experimental values D_o and E_a for nitrogen in Sm_2Fe_{17} obtained from long-time isothermal experiments. Typical microscopic times are of order $\tau_o = 10^{-14}$ s, so D_o should be of order $\Gamma_o a^2 \approx$ mm^2/s. Gas-phase carbonation is usually conducted under conditions which are typical of nitrogenation, so the diffusion constants of nitrogen and carbon in iron-rich rare-earth intermetallics must be comparable [10]. For the sake of comparison, Table 5 shows typical parameters for the diffusion of 2p atoms in metals.

Table 4. Experimental diffusion constants for nitrogen in Sm_2Fe_{17}.

E_a kJ/mole	D_o mm^2/s	Ref.
78	0.000195	[3]
133 ± 5	1.02	[4, 10]
100 - 163	-	[73]

Table 5. Diffusivity of 2p atoms in bcc metals [19].

System	E_a kJ/mole	D_o mm^2/s
C in Fe	80.3	0.39
N in Fe	76.8	0.49
O in Fe	97.5	3.7
C in Nb	138.1	0.40
N in Nb	146.0	0.86
O in Nb	112.6	2.1

4.1.3 *Tunneling Effects*. Below a cross-over temperature T_o, the leading diffusion mechanism is quantum tunneling. For hydrogen in metals T_o is of order 200 K, but the values for deuterium and tritium are significantly lower [74]. To estimate T_o for nitrogen we use the quasi-empirical relation (cf. [75])

$$E_a = c_a E \sqrt[3]{Z^2}\, a_o^3 \tag{4.4}$$

Here $c_a \approx 3$ is a numerical factor, E denotes Young's modulus, and Z is the atomic number of the diffusing atom. The exponent 2/3 originates from the R^2 dependence of the elastic energy on the deformation and the approximate relation $R \sim Z^{1/3}$.

To describe the tunneling process we use the Gamov factor

$$\tau' = \tau_o \exp\left(\sqrt{\frac{8\, m\, E_a\, d_B}{h}}\right) \tag{4.5}$$

where d_B is the thickness of the energy barrier. Putting $\tau' = \tau_o$ (see Section 4.1.1 and 4.1.2) we obtain the cross-over temperature

$$T_o = \sqrt{\frac{c_a\, a_o^3\, z^{2/3}\, h^2\, E}{8\, k_B^2\, m\, d_B^2}} \tag{4.6}$$

which is $T_o \approx 170$ K and $T_o \approx 90$ K for nitrogen in Sm_2Fe_{17}. Although the approximate overall dependence $T_o \sim 1/m^{1/6}$ is much weaker than the simple isotope law $T_o \sim 1/m^{1/2}$, for practical purposes the restriction to thermally activated nitrogen diffusion is justified.

4.2. INITIAL STAGE OF NITROGENATION

4.2.1. *Diffusion Mechanisms*. The equilibrium state $\langle c(\mathbf{r})\rangle = \langle c\rangle_{eq}$ is achieved by solid-state diffusion from the surface of the intermetallic particles. Bulk diffusion, however, presupposes a sufficiently-high nitrogen flux through the particle surface. In practice, microcrack formation [39] and low surface activity may influence the nitrogenation behaviour. A characteristical feature of these mechanisms is a linear time dependence of the average nitrogen content, as opposed to the square-root dependence of diffusion-controlled reactions (Section 4.3).

Possible reasons for microcrack formation and surface destruction are mechanical stress and strain caused by the inhomogenuous lattice expansion (Section 6), exothermic heat released during nitrogenation, and the use of ammonia which implies a large effective nitrogen pressure.

In practice, it is appropriate to descibe the influence of microcracks as a reduction of the average particle size. This effective particle size may decrease in the initial stage of nitrogenation, but ultimately the reaction proceeds by thermally activated bulk diffusion in geometrically well-defined small grains.

4.2.2 *Surface Barriers*. Unlike nonlinear algebraical equations, linear differential equations such as the diffusion equation Eq. (4.3) have always a trivial solution $\langle c(\mathbf{r}, t)\rangle$ = const., and we have to formulate physically reasonable boundary conditions. Often it is assumed that the concentration $\langle c(\mathbf{R},t)\rangle$ at the particle surface is sufficiently high to for sufficiently-high reaction temperatures, but below about 350 °C the flux through the surface is limited by an energy barrier arising from surface deactivation (Fig. 15). In the surface-controlled low-temperature regime we have to use the boundary condition $\mathbf{j}(R, t) = j_o\, \mathbf{e}_r$. This boundary condition yields parabolic concentration profiles $c(r, t) \sim j_o (r^2 + 6Dt - 3R^2/5)$ and a linear time dependence of the average nitrogen content. From the suface-bulk crossover behaviour one can deduce surface activation energies E_s (Table 6). Note that these energies, as well as the corresponding crossover temperatures, are real-structure quantities which strongly depend on the previous powder treatment.

Table 6. Diffusion and reaction parameters deduced from thermopiezic absorption experiments.

System	E_s kJ/mole	E_a kJ/mole	D_o mm^2/s	U_o kJ/mole
N in Sm_2Fe_{17}	53 ± 15	133 ± 10	1.0	- 57 ± 10
H in Sm_2Fe_{17}	81 ± 15	31 ± 15	0.4	- 26[a] ± 10
N in $NdFe_{11}Ti$[b]	61 ± 15	158 ± 15	6.5	- 51 ± 10
H in $NdFe_{11}Ti$	58 ± 15	45 ± 15	0.5	—

[a]Average, including 9e and 18g sites. An estimate for Y_2Fe_{17} is U_o = - 35 ± 15 kJ/mole. [b]$NdFe_{11}TiN$ is less stable than $Sm_2Fe_{17}N_3$.

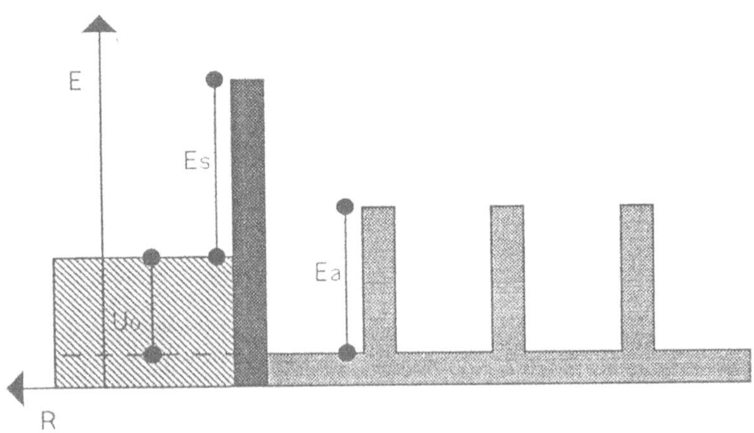

Fig. 15. Lattice-gas energies. E_s is the surface activation energy, E_a the (bulk) activation energy, and U_o the ent reaction energy. The dashed area represents the gas phase.

4.3. NITROGENATION PROCESS

4.3.1. *Concentration Profiles.* In the bulk diffusion regime the concentration $<c(\mathbf{r},t)>$ is found by solving the diffusion equation Eq. (4.3) subject to the boundary condition $<c(\mathbf{r}_s,t)> = <c>_{eq}$ at the particle surface.

For spherical particles the diffusion problem can be solved analytically [75], and we obtain the local concentration

$$<c(r,t)> = <c>_{eq}\left(1 + \frac{2\,R}{\pi\,r}\sum_{m=1}^{\infty}\frac{(-1)^m}{m}\sin\left(\frac{m\,\pi\,r}{R}\right)\exp\left(-\frac{m^2\pi^2Dt}{R^2}\right)\right) \tag{4.7}$$

where R is the radius of the particle. Eq. (4.7) yields smooth, bathtub-like profiles for intermediate nitrogen contents. Volume integration of Eq. (4.7) yields

$$<c(r,t)> = <c>_{eq}\left(1 - \frac{6}{\pi^2}\sum_{m=1}^{\infty}\frac{1}{m^2}\exp\left(-\frac{m^2\pi^2Dt}{R^2}\right)\right) \tag{4.8}$$

In the initial stage of reaction, the nitrogen absorption rate is proportional to the total powder surface area, and the average concentration reads [4]

$$<c(r,t)> = 6\,<c>_{eq}\sqrt{\frac{D\,t}{\pi\,R^2}} \tag{4.9}$$

The long-time behaviour of Eq. (4.8) is given by

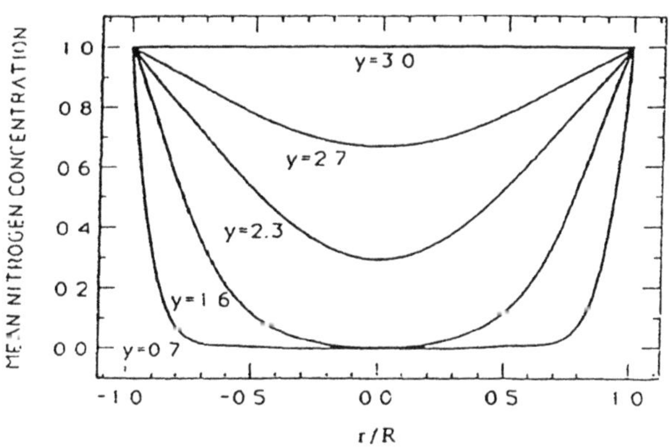

Fig. 16. Nitrogen profiles for spherical particles. The equilibrium nitrogen content y is taken as 3.0.

$$<c(r,t)> = <c>_{eq}\left(1 - \frac{6}{\pi^2}\exp\left(-\pi^2 Dt/R^2\right)\right) \qquad (4.10)$$

In the short-time limit the influence of particle shape and particle-size distribution reduces to a renormalization of the particle radius R, while the long-time behaviour strongly depends on the particle-size distribution [4].

Table 7. Diffusion constants deduced from D_o = 1.02×10^{-6} mm^2/s and E_a = 133 kJ/mole [4] [10].

Temperature °C	D $\mu m^2/h$	D m^2/s
400	0.17	4.38×10^{-17}
425	0.41	1.13×10^{-16}
450	0.90	2.50×10^{-16}
475	1.88	5.24×10^{-16}
500	3.76	1.05×10^{-15}
525	7.20	2.00×10^{-15}
550	13.24	3.68×10^{-15}
575	23.48	6.52×10^{-15}
600	40.31	1.12×10^{-14}

Table 8. Reduced nitrogenation times t/t_o. The data refer to the average interstitial concentration $<c>$ and to the nitrogen concentration c(0) at the particle centre (**R** = 0), respectively.

Condition	Illustrative Comment	t/t_o
$<c>$ = 10 %	initial stage	0.001
$<c>$ = 50 %	half-empty particle	0.03
c(0) = 10 %	nitrogen reaches particle centre	0.07
c(0) = 50 %	half-empty particle centre	0.14
$<c>$ = 90 %	nearly full nitrogenation	0.18
c(0) = 90 %	nearly homogeneous nitrogenation	0.30

4.2.4. *Reaction Times*. The nomograph Fig. 17 gives a rough description of the nitrogenation reaction. A slightly more sophisticated procedure is based on Tables 7 and 8. Let us first approximate our particles by spheres of radius R. In the case of a broad size distribution we have to consider all fractions separately, because the nitrogenation time depends strongly on R. Now we calculate the reference time $t_o = R^2/D$, where D is given in Table 7, and obtain the relevant reaction time $t = t/t_o\ R^2/D$ from Table 8. It is remarkable that there is a time factor 10 if we compare 50% nitrogenation with nearly homogeneous nitrogenation.

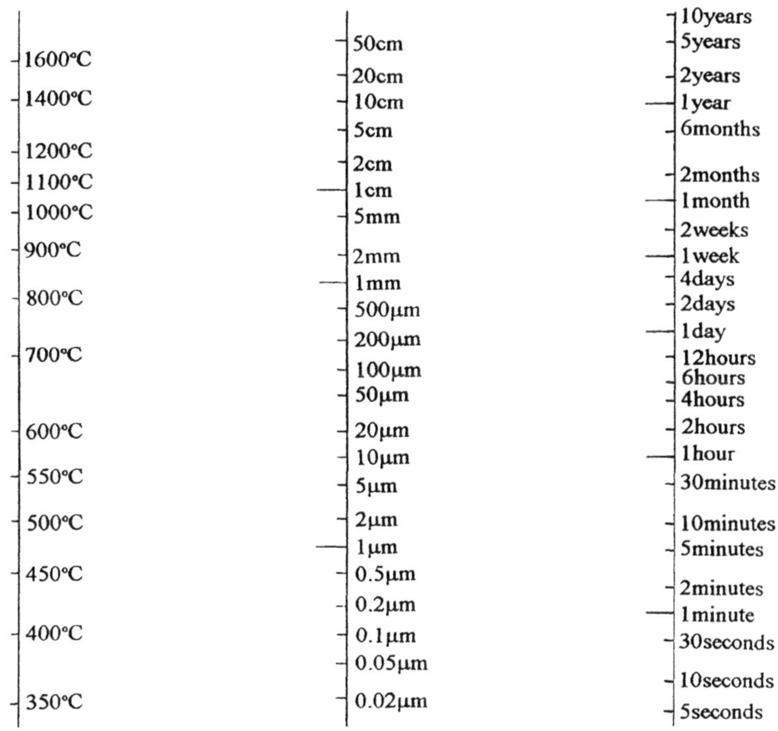

Fig. 17. Nitrogenation nomograph. A straight horizontal line connects possible Sm_2Fe_{17} particle radii, nitrogenation temperatures T, and nitrogenation times t for 90% nitrogenation. To avoid disproportionation during the reaction (Section 6), the straight line has to lie below the black circle.

5. Interaction Effects

Summary Ideal gas-solid systems are characterized by the absence of any interaction which goes beyond molecular dissociation, gas-lattice binding and site repulsion Interaction between interstitial atoms on different sites leads to essential deviations from the ideal gas-solid behaviour If this interaction is attractive, as it is the case for long-range elastic interaction, then the intersitial atoms tend to form clusters, and phase segregation into low- and high-concentration phases occurs Above a critical temperature T_c, which depends on the number of available interstitial sites, these clusters are destroyed by thermal activation From the point of view of kinetics, above T_c the diffusivity of the interstitial atoms is large enough to overcome the interaction forces which keep the clusters together

5 1 PHENOMENOLOGY OF INTERACTING LATTICE GASES

5 1 1 *Experimental Situation* A key question is whether the quasi-equilibrium nitrides are a simple gas-solid solution with a continuous range of intermediate nitrogen contents or a two-phase mixture of nitrogen-poor (α) and nitrogen-rich (β) phases [4, 5, 10] Fig 18 shows a schematic phase diagram of gas-metal systems [18, 60, 62, 76] Spinodal phase segregation into α and β phases occurs below a critical temperature T_c A typical value is T_c = 565 K (292 °C) for palladium hydride [19, 62]

Room-temperature R_2Fe_{17} nitrides and carbides, quenched from typical reaction temperatures T ≥ 400 °C, are solid solutions [4, 10] The experimental evidence can be summarized as follows [4, 10, 77] (i) by X-ray diffraction (Fig 19) it is possible to observe intermediate nitrogen contents, which are typical of gas solid solutions [18], (ii) samples with intermediate nitrogen contents exhibit intermediate Curie temperatures [37], (iii) homogenization of partly nitrided grains yields intermediate nitrogen concentrations [57], (iv) NMR measurements demonstrate variable rare-earth environments [78], and (v) domain size observations [57, 79] can be explained by intermediate anisotropy constants only

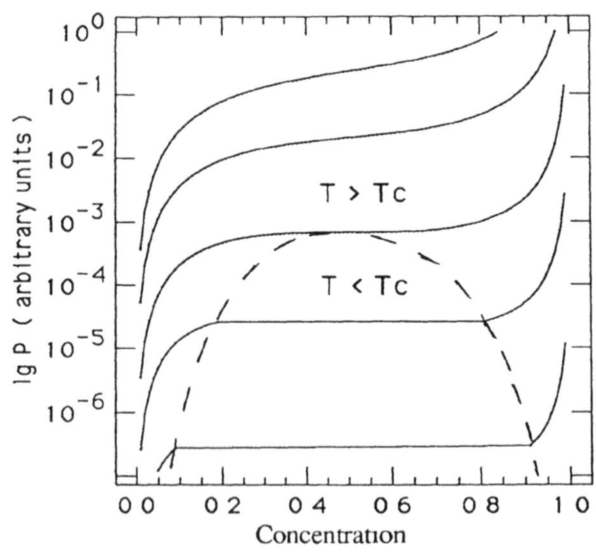

Fig 18 Schematic gas-metal phase diagram

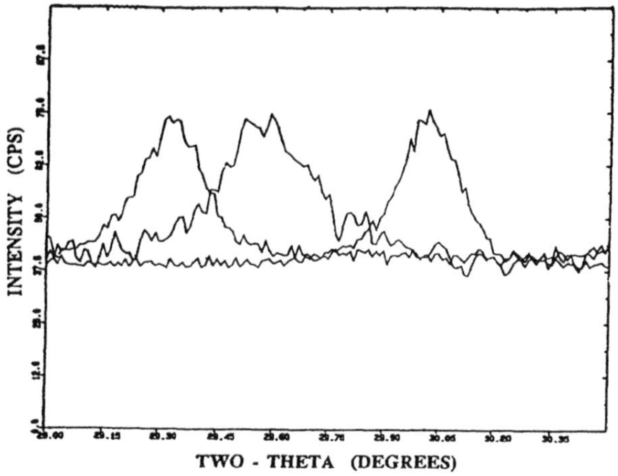

Fig. 19. (113) X-ray diffraction lines for different long-time nitrided $Sm_2Fe_{17}N_x$ samples. From right to left: pure Sm_2Fe_{17}, partly nitride $Sm_2Fe_{17}N_x$, and fully nitrided $Sm_2Fe_{17}N_{3-\delta}$ [10]. Note that it has been possible to prepare quasi-continuous sets of well-homogenized nitrides with nitrogen contents $0 \leq x \leq 3$ [77].

Table 9. Single-barrier jumping times for hydrogen (E_a = 45 kJ/mole) and nitrogen (E_a = 133 kJ/mole) in Sm_2Fe_{17}.

T	Hydride	Nitride
400 °C	31 ns	0.21 ms
200 °C	0.93 μs	4.9 s
RT	1.1 ms	160 a

From the point of view of non-equilibrium statistics, the room-temperature nitrides and carbides in question are quasi-equilibrium compounds. Table 9 shows jumping times deduced from $\tau = \tau_o \exp(E_a/k_B T)$ with $\tau_o = 10^{14}$ s. At room temperature, the mobility of nitrogen in Sm_2Fe_{17} is extremely low, and it is difficult to investigate the 'low-temperature' Sm-Fe-N phase diagram.

5.2. MEAN-FIELD APPROXIMATION

5.2.1. *Model and Results.* Let us start with the Hamiltonian (see appendix)

$$\mathbb{H} = -\frac{1}{2} \sum_{i,j=1}^{N} A_{ij}\, c_i\, c_j \; + \; \sum_{i=1}^{N} \mu_i\, c_i \qquad (5.1)$$

where A_{ij} describes the interaction between interstitial atoms located at $\mathbf{r}_i$ and $\mathbf{r}_j$. $\mu_i = U_o + \Delta\mu$ is the chemical potential of the interstitial atoms. Using the analogy to the spin-1/2 Ising model (Appendix) we obtain the mean-field equation of state

$$\langle c \rangle_{eq} = \frac{1}{1 + e^{\beta(\mu - zAc)}} \tag{5.2}$$

In the limit $A_{ij} = 0$ this result must reduce to Eq. (3.5), which reproduces the well-known ideal-gas expression $\mu = U_o - 0.5\, k_B T \ln(p/p_o)$. The critical properties of Eq. (5.2) are obtained by a simple replacement procedure (see Appendix). The 'H = 0' limit yields the critical pressure $p_c = p_o \exp(-4) \exp(2U_o/k_B T_c)$. Finally, the critical temperature reads $T_c = zA/4k_B$.

5.2.2. *Discussion.* A subtle question is to what extent the van't Hoff relation [26]

$$\ln p[N_2] = \Delta H_f/3RT - \Delta S_f/3R \tag{5.3}$$

is able to serve as an equation of state. ($p[N_2]$ denotes the gas pressure measured in bar). In particular, we have to find out (i) how the phenomenological quantities enthalpy of formation ΔH_f and entropy of formation ΔS_f relate to the phase diagram Fig. 18, and (ii) whether Eq. (5.3) is equivalent to Eq.(5.2).

First, we have to bear in mind that Eq. (5.3) refers to the *two-phase* region below T_c [26]. In this region, we can apply familiar phase-transitions concepts [61, 80], and it turns out that Eq. (5.3) is the coexistence curve which separates nitrogen-poor α and nitrogen-rich β phases (Fig. 20). In the case of the lattice-gas model Eq. (5.1), this coexistence line reads $\ln (p/p_o) = (2U/k_B - 4T_c)/T$. However, at present there is no indication that nitrogenation at about 500 °C yields two-phase nitrides, so that the concentration-independent relation Eq. (5.3) gives only a crude extrapolative description of the nitrogenation process.

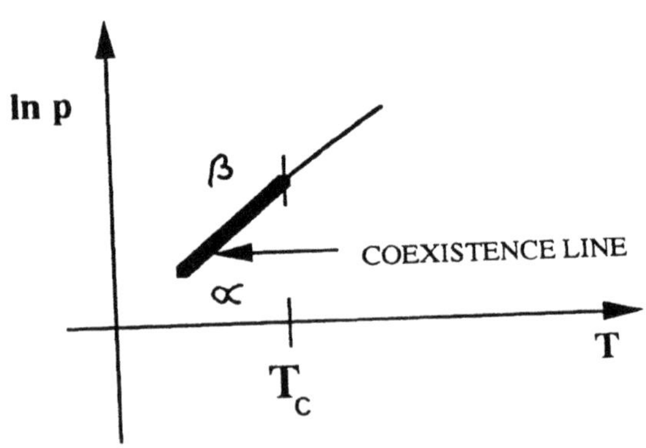

Fig. 20. Alternative presentation of the phase diagram Eq. (18). The gas-liquid analogon to this curve is given by the symbolic replacement 'α → gas' and 'β → liquid'.

A more subtle problem is that Eq. (5.3) is based on the Maxwell construction of the coexistence curve (cf. [80, 81]). In fact, there is no mathematical motivation for the Maxwell rule, and the 'nucleation-based' delay rule is equally compatible with the theory of phase transitions [80, 81]. In the case of gases in metals, the Maxwell rule presupposes the destruction of the host lattice (incoherency), while plain interstitial modification obeys the delay rule [18]. Illustratively speaking, long-range elastic strain in the coherent lattice inhibits the nucleation of relaxed nitrogen-rich regions and leads to bulk hysteresis effects below T_C.

5.3. MICROSCOPIC ORIGIN OF INTERACTION EFFECTS

5.3.1. *Electronic Interaction.* In a wider sense, electronic interaction means the influence of the local electronic environment and includes all electrostatic and steric effects which go beyond the simple site-blocking hard-core repulsion. As a rule, electronic interaction involves the orthogonalization of quantum-mechanical wave functions and is repulsive (cf. Fig. 10). Repulsive interaction means $A_{ij} < 0$ in Eq. (5.1) which is incompatible with a positive critical temperature T_C. Positive critical temperatures are due to *attractive* interaction. Furthermore, due to the strong screening of impurity charges in metals [47] this interaction is short-ranged, while experimental evidence favours long-range interaction [18, 76].

5.3.2. *Elastic interaction.* There is much theoretical and experimental evidence that the α-β phase transition in gas-metal systems is mainly due to long-range eleastic interaction [18, 76]. Fig. 21 gives an illustrative idea of this mechanism. Once the first interstitial atom has entered the lattice it enlarges the interstitial holes in its neighbourhood. The occupancy of these pre-expanded interstices is energetically more favourable, and the interstitial atoms tend to form clusters (phase segregation). Below T_C these clusters are thermodynamically stable, but above T_C they are destroyed by thermal excitation.

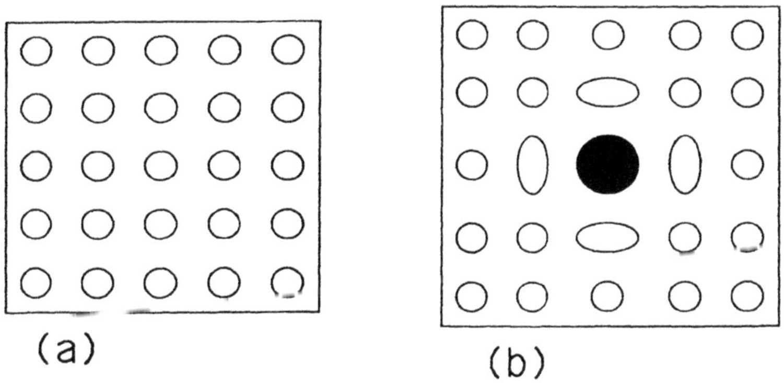

Fig. 21. Elastic interaction and phase segregation.

The critical temperature increases with the number of available interstitial sites [18], because short distances between interstices enhance the 'hole enlargement'. The small fraction 3/(2 + 17) of nitrogen atoms per metal atom in $Sm_2Fe_{17}N_3$ — as compared to 5/(1 + 5) hydrogen atoms per metal atom in $LaNi_5H_5$ — explains the comparatively low critical temperature of nitrogen in Sm_2Fe_{17}, even if the volume expansion per interstitial atom is larger in the nitrides than in the hydrides. A corresponding estimate of the critical temperature of the Sm_2Fe_{17}-nitrogen system is $T_c = 400$ K ± 50% (cf. [10]). Due to the increased number of available interstices, the critical temperature of overloaded nitrides is likely to be higher than 400 K.

It is worthwhile mentioning that this attractive interaction presupposes free particle surfaces. In the case of clamped surfaces ($\Delta V/V = 0$) the size of neighbouring interstitial holes is actually reduced and the interaction becomes repulsive [18, 76].

5.4. DIFFUSIVITY OF INTERACTING INTERSTITIALS

5.4.1. *Concentration Dependence of the Diffusion Constant.* The discrete master equation Eq. (4.1) describes the motion of a single interstitial atom — one can imagine that interstitial diffusion is affected by the total number of interstitial atoms in the lattice. Surprisingly, site blocking does not influence the bulk (or: chemical) diffusion constant [63]. Only the tracer diffusion coefficient, which describes, let's say, the motion of single radioactively marked atom, is bound to vanish for $\langle c(\mathbf{r})\rangle = 1$. To explain this paradox we multiply the transition rates $W_{ij} = W(x_j \rightarrow x_i)$ in Eq. (4.1) by a factor $1 - P_i$ which assures $W_{ij} = 0$ if the final site is already occupied. This replacement creates quadratic terms in Eq. (4.1), but it turns out that all these quadratic terms cancel each other, and the interaction-free diffusion constant is independent of concentration.

5.4.2. *Concentration Profiles.* Interatomic interaction A_{ij} makes the bulk diffusion constant concentration-dependent [63, 76]. The mean-field result D for the diffusion constant is [63]

$$D = D_F\left(1 - \frac{1}{4}\langle c\rangle(1 - \langle c\rangle)\,\frac{T_c}{T}\right) \tag{5.4}$$

where D_F denotes the interaction-free diffusion constant. At this point it is worthwhile emphasizing that the diffusion constants of interacting lattice gases, as well as the tendendency towards phase segregation, do *not* depend on the net reaction energy U_0. Let us next consider intermediate nitrogen concentrations $0 < \langle c\rangle < 1$. For $\langle c\rangle = 0.5$, Eq. (5.4) reduces to $D = D_F\,(1 - T_c/T)$. At T_c this equation yields $D = 0$ (critical slowing down [82]), while $T < T_c$ implies $D < 0$. Negative diffusion constants describe the formation of phase boundaries whose thickness is given by the correlation length ξ (see Appendix). In other words, original concentration inhomogenities are not smoothed but enhanced. Recall, however, that the existence of phases and phase boundaries is an *equilibrium* property. Insufficient nitrogenation (cf. Table 8), surface effects [39, 83], which are not related to q. (5.4), and non-equilibrium ammonia treatment may mimic two-phase behaviour, but careful long-time nitrogenation using N_2 or NH_3 reveals the single-phase character of the quasi-equlilibrium nitrides [77].

6. Properties of Particles with Intermediate Nitrogen Content

Summary: During nitrogenation, inhomogeneous concentration profiles cause mechanical stress and strain which influences both processing and magnetic properties. Stress and strain facilitate the disproportionation of $Sm_2Fe_{17}N_3$ into SmN and bcc iron, but the rate-determining step of the disproportionation reaction is iron diffusion. The magnetic properties of incompletely nitrided $Sm_2Fe_{17}N_x$ grains are affected by the magnetically soft core.

6.1. STRESS AND STRAIN

6.1.1. *Linearized Theory.* Inhomogeneous concentration profiles (Fig. 16) cause inhomogeneous stress and strain. The fully nitrided outer 'shell' of the particle cannot expand freely because it is connected to the un-nitrided inner core. Elastic stress and strain in spherical particles have been calculated in [4]. Starting point is Hooke's law

$$\sigma_{ik} = \frac{E}{1+\nu}(\varepsilon_{ik} - \varepsilon_{ik}^{H}) + \frac{\nu E}{(1+\nu)(1-2\nu)}\delta_{ik}\,\mathrm{Tr}(\varepsilon_{ik} - \varepsilon_{ik}^{H}) \qquad (6.1)$$

where E and ν are Young's modulus and Poisson's number, respectively. $\varepsilon_{ik}^{H} = \delta_{ik}\,\varepsilon_o$ $\langle c(\mathbf{r})\rangle$ denotes the stress-free reference expansion ($\varepsilon_o = \Delta V/3V \approx 2\%$).

The strain which minimizes the elastic energy calculated from Eq. (6.1) reads

$$\varepsilon_{ik}(\mathbf{r}) = \int_0^R G^{\varepsilon}{}_{ik}(\mathbf{r},r')\,\langle c(r')\rangle\,dr' \qquad (6.2)$$

Using the $G^{\varepsilon}{}_{ik}(\mathbf{r},r')$ given in [4] the lateral and radial strain profiles are calculated from $\langle c(r)\rangle$.

6.1.2. *Strain Profiles.* Fig. 22 shows strain profiles for a partly nitrided grain. It is remarkable that (i) there is a finite strain at the particle centre even if there is (nearly) no nitrogen there, and (ii) the radial expansion at the particle surface is as high as 3%. The

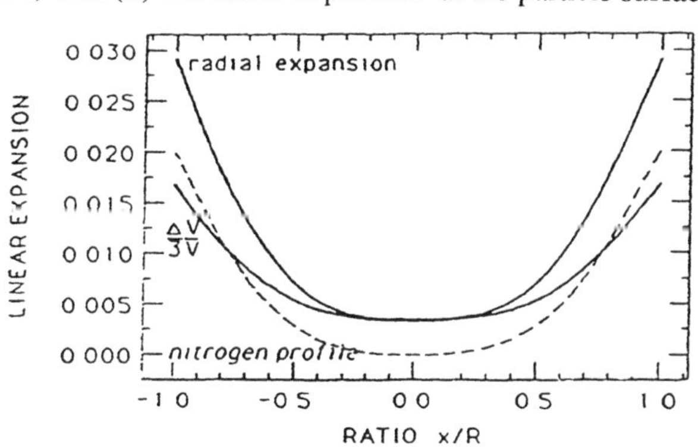

Fig. 22 Radial and volume expansion for a spherical grain of $Sm_2Fe_{17}N_{1.6}$.

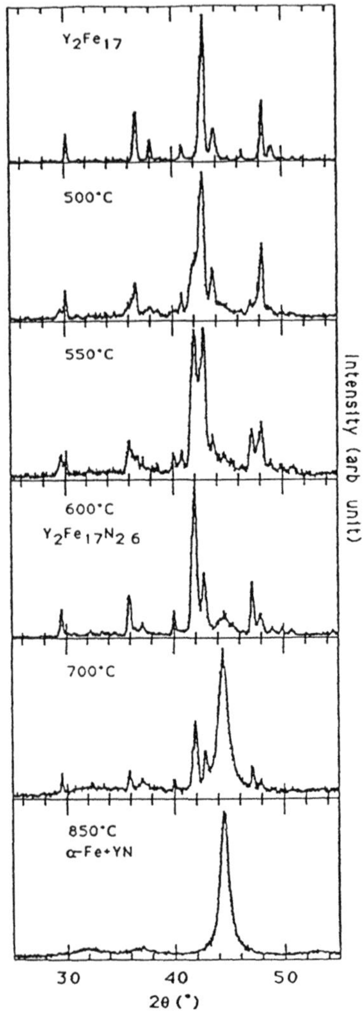

Fig 23 X-ray diffraction patterns of Y_2Fe_{17} heated at a constant heating rate 10 K/min [85] The main bcc-iron peak at 44 7° refeclects the disproportionation of the lattice Arrhenius analysis of the X-ray diffraction patterns yields an activation energy of about 200 kJ/mole, which falls in the range of activation energies typical of iron diffusion [75]

high excess strain and the acompanying mechanical stress indicate that range of validity of Hooke's law is actually left and non-linear phenomena, including lattice disproportionation, take over.

It is worthwhile mentioning that Eq. (6.2) describes a purely macroscopic strain which is only loosely related to the enlargement of interstitial holes shown in Fig. 21.

6.2. MAGNETIC PROPERTIES AND STABILITY

6.2.1. *Magnetic Performance.* On the whole, the magnetic properties of partly nitrided $Sm_2Fe_{17}N_x$ grains are inferior to those of fully nitrided particles. The main reason for this is the low coercivity of partly nitrided grains. The maximum coercivity in a rare-earth permanent magnet is determined by the crystal-field parameter A_2^0, which depends linearly on $\langle c(\mathbf{r})\rangle$ [51]. The Sm_2Fe_{17} parent compound exhibits easy-plane anisotropy with negligibly small coercivity, but the potential coercivity of the fully nitrided intermetallic is very high. However, coercivity is subjected to a 'weakest link in the chain' principle (nucleation), so low A_2^0 values at the particle centre drastically deteriorate the permanent magnetic performance [84].

On the other hand, it is known that the Curie temperature of iron-rich rare-earth intermetallics strongly increases with lattice expansion [1, 27, 48]. Fig. 22 shows a finite volume expansion of about 1.1% at the particle centre. This implies a Curie-temperature increase at the particle centre, even if there is no nitrogen there.

6.2.2. *Stability.* The formation of the ionic compound SmN is energetically more favourable than the interstitial modification reaction. This means that $Sm_2Fe_{17}N_3$ is a metastable quasi-equilibrium compound which has to be treated 'carefully' to avoid the disproportionation of the intermetallic lattice. Various aspects of this problem have been dealt with in Sections 3.3.2, 4.2.1, and 6.1.2, but the main, rate-determining disproportionation step is the diffusion of iron atoms. As it can be seen from Fig. 23, the disproportionation rate strongly increases with temperature. It is therefore impossible, for instance, to produce sintered $Sm_2Fe_{17}N_3$ permanent magnets.

The disproportion behaviour of $Sm_2Fe_{17}N_3$, based on an assumed activation energy $E_a(Fe) = 200$ kJ/mole, is included in Fig. 17. The straight line which connects T, R, and t must lie below the black circle to avoid disproportionation during nitrogenation. For practical purposes, i.e. establishing a time scale of about 100 years, well-nitrided $Sm_2Fe_{17}N_{3-\delta}$ can be considered as stable below 250 °C [5].

Conclusions.

Interstitially modified intermetallics such as $Sm_2Fe_{17}N_3$ are solid solutions at typical nitrogenation temperatures T = 400-500 °C and can be regarded as archetypical lattice-gas systems. Due to the low diffusivity of nitrogen below 400 °C is has not been possible to measure critical temperatures. In reality, quenching fixes the solid-solution structure at room temperature. This behaviour is independent of whether the interstitial 2p atoms are introduced using molecular nitrogen, ammonia, hydrocarbon gases, or solid carbon.

Apart from secondary factors such as surface activity or microcrack formation, the interstitial modification reaction proceeds by thermally activated bulk diffusion. The ultimate concentration profiles are smooth and bathtub-like, as opposed to two-phase systems where the concentration profiles are discontinuous. Using the experimentally determined diffusion parameters E_a = 133 kJ/mole and D_o =1.0 mm^2/s it is possible to predict the nitrogenation behaviour of spherical Sm_2Fe_{17} grains.

$Sm_2Fe_{17}N_3$ easily disproportionates at temperatures $T \geq 600$ °C. This disproportionation may be enhanced for instance by mechanical stress and strain but the rate determining step is iron diffusion. Below 250 °C the nitride $Sm_2Fe_{17}N_3$ can be regarded as quasi-stable.

Acknowledgements

The authors are indebted to J. M. D. Coey for stimulating discussions. This work forms part of the BRITE/EURAM program of the European Commission.

Appendix: The Ising model

Definition. Many physical systems can be described in terms of local phase-space variables $s_i = s(\mathbf{r}_i) = \pm 1$ (see Table A1 and Fig. 9). Models with this property are called spin-1/2 Ising models [59, 60, 86-88]. The Hamiltonian of the model reads

$$\mathbb{H} = -\frac{1}{2}\sum_{i,j=1}^{N} J_{ij}\, s_i\, s_j - \sum_{i=1}^{N} H_i\, s_i \qquad \text{(A.1)}$$

where the matrix J_{ij} describes the interaction between the N spins; the external magnetic field H_i serves to the align the spins in z direction. Note that the relation $[\mathbb{H}, s] = [\mathbb{H}, \sigma_z] = 0$ implies the equivalence the classical and quantum-mechanical Ising models. In the case of ferromagnetic coupling, $J_{ij} > 0$ and a second-order phase transition is observed at the critical point $T = T_c$ and $H_i = 0$. The ferromagnetic phase diagram is shown in Fig. 24: below T_c a small applied field $H_i = \pm H$ causes the macroscopically large equilibrium magnetization to switch, while above T_c this change is continuous (paramagnetism).

The equilibrium average of any quantity B is given by

$$\langle B \rangle = \frac{1}{Z}\sum_{s_i} B(s_i)\, e^{-\beta\, \mathbb{H}(s_i)} \qquad \text{(A.2)}$$

with $\beta = 1/k_BT$. With the Hamiltonian Eq. (A.1) the partition function reads

$$Z = \sum_{s_i} \exp\left(\frac{1}{2}\beta\, \Sigma_{ij}\, J_{ij}\, s_i\, s_j + \beta\, \Sigma_i\, H_i\, s_i\right) \qquad \text{(A.3)}$$

Once the partition function is known, equilibrium averages are obtained by differentiation. For instance, evaluating $\partial Z/\partial H_i$ yields $\langle s_i \rangle = k_BT\, Z^{-1}\, \partial Z/\partial H_i$.

Mean-Field Approximation. In practice, it is very difficult to calculate Z. At present, there is no exact solution in three dimensions, though considerable progress has been made in understanding the behaviour near the critical point [86, 87, 89]. In one dimension $T_c = 0$, while the famous d = 2 Onsager result predicts a finite Curie temperature.

Table A1. Ising systems.

system	interaction	phase space	external field
magnet	exchange	magnetization s_i	magnetic field H_i
gas-liquid	atomar	local density $\rho(\mathbf{r}_i)$	pressure p
gas-solid	elastic	concentration c_i	pressure p
binary alloy	exchange	concentration c_i	chemical potential μ

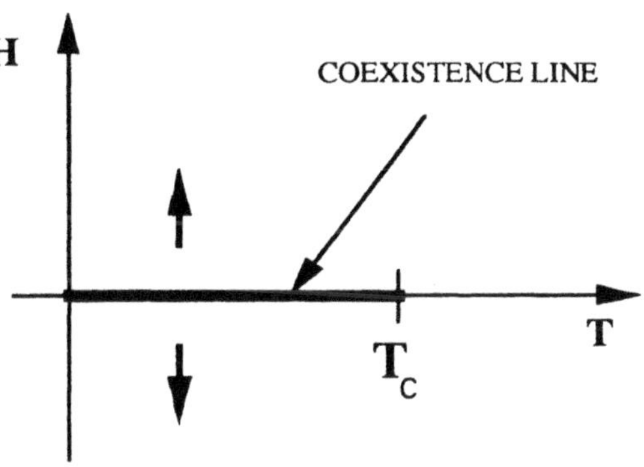

Fig. 24. Phase diagram of an Ising ferromagnet. From the point of view of lattice-gas statistics, this diagram is equivvalent to Fig. 20.

To simplify matter, let us assume that the interaction can be mapped onto a local field h_i which yields the mean-field Hamiltonian $\mathbb{H}_{MF} = - \sum h_i s_i$. Utilizing the Bogol'ubov inequality $F \leq F_0 - <\mathbb{H} - \mathbb{H}_0>_0$ we obtain $h_i = H_i + \sum_j J_{ij} <s_j>$ [86], and the partition function reads $Z = 2 \cosh(\beta H_i + \beta \sum_j J_{ij} <s_j>)$. In the case of a homogeneous system $H_i = H$ and $\sum_j J_{ij} = z J$, where J is the coupling between neighbouring spins and z the number of neighbours, and the magnetization $m = <s_i>$ obeys

$$m = \tanh(\beta H + z \beta J m) \tag{A.4}$$

Series expansion of this equation with respect to the small parameters H and m yields $T_c = z J_o/k_B$ and $m(H = 0) \alpha (T_c - T)^{1/2}$. Renormalization-group analysis [87, 89] reveals that the mean-field exponent 1/2 is true for d ≥ 4 only - and, additionally, in the case of long-range interaction [18, 76]. The exact value for short-range interaction is 0.33, but here we will restrict ourselves to the mean-field result. Note that the long-range character of the elastic interaction (Section 5.3) speaks in favour of the mean-field exponent [18].

The same applies to the behaviour of other quantities such as the correlation length ξ, which scales as $|T - T_c|^{-1/2}$.

To realize the site occupancy $c_i = \{0, 1\}$ instead of the spin projection $s_i = \{-1, +1\}$ we have to put $s_i = 2\, c_i - 1$. Comparing the resulting expression with Eq. (5.1) we obtain $A_{ij} = 4\, J_{ij}$ and $\mu_i = 2\, \Sigma_j\, J_{ij} - 2\, H_i$. Furthermore, with $m = 2 <c>_{eq} - 1$, $\Sigma_j A_{ij} = z\, A$, $\mu_i = \mu$, and Eq. (A.4) we obtain Eq. (5.2). The critical concentration $<c>_{eq,c} = 0.5$ is deduced from the fact that m = 0 at the coexistence line.

References

[1] J. M. D. Coey and H. Sun, J. Magn. Magn. Mater. **87**, L251 (1990).

[2] J. M. D. Coey, H. Sun, Y. Otani, and D. P. F. Hurley, J. Magn. Magn. Mater. **98**, 76 (1991).

[3] J. M. D. Coey, J. F. Lawler, H. Sun, and J. E. M. Allan, J. Appl. Phys. **69**, 3007 (1991).

[4] R. Skomski and J. M. D. Coey, J. Appl. Phys. **73**, 7602 (1993).

[5] R. Skomski and J. M. D. Coey, Journal of Materials Engineering and Performance **2** (1993) 241.

[6] T. W. Capehart, R. K. Mishra, and F. E. Pinkerton, Appl. Phys. Lett. **58**, 1395 (1991).

[7] O. Isnard, S. Miraglia, J. L. Soubeyroux, J. Pannetier, and D. Fruchart, Phys. Rev. B **45**, 2920 (1992).

[8] R. Skomski, S. Brennan, and J. M. D. Coey, phys. stat. sol. (a) **139**, K11 (1993).

[9] S. Brennan, R. Skomski and J. M. D. Coey, EMMA'93 Kosice, IEEE Trans. Magn. (in press).

[10] J. M. D. Coey, R. Skomski, and S. Wirth, IEEE Trans. Magn. **28**, 2332 (1992).

[11] H. Sun, *PhD Thesis*, Trinity College, Dublin 1992.

[12] G. Sandrock, S. Suda, and L. Schlapbach, "Applications" in *Hydrogen in Intermetallic Compounds II*, L. Schlapbach, ed., Springer, Berlin, 1992, p.197.

[13] J. H. N. van Vucht, F. A. Kuipers, and H. C. A. M. Bruning, Philips Res. Repts. **25**, 133 (1970).

[14] J. M. Cadogan and J. M. D. Coey, Appl. Phys. Lett. **48**, 442 (1986).

[15] X.-Zh. Wang, K. Donelly, J. M. D. Coey, B. Chevalier, J. Etourneau, and T. Berlureau, J. Mater. Sci. **23**, 329 (1989).

[16] B.-P. Hu and J. M. D. Coey, J. Less-Common Met. **142**, 295 (1988).

[17] D. Fruchart and S. Miraglia, J. Appl. Phys. **69**, 5578 (1991).

[18] H. Wagner and H. Horner, Adv. Phys. **23**, 587 (1974).

[19] J. D. Fast, *Gases in Metals*, MacMillan, London, 1976.

[20] M. Merkel and K.-H. Thomas, *Technische Stoffe*, Fachbuchverlag, Leipzig, 1988.

[21] K. H. Jack, Proc. R. Soc. A **208**, 200, 216 (1951).

[22] M. Katsura and H. Serizawa, J. Alloys and Compounds **196**, 191 (1993).

[23] J. M. D. Coey, K. O'Donnell, Q.-N. Qi, E. Touchais, and K. H. Jack, J. Phys.: Condens Matter **6**, L23 (1994).

[24] T. K. Kim and M. Takahashi, Appl. Phys. Lett. **20**, 492 (1972).

[25] J. E. Evetts, ed., *Concise Encyclopedia of Magnetic and Superconducting Materials*, Pergamon, Oxford, 1992.
[26] K. H. J. Buschow, Rep. Prog. Phys. **54**, 1123 (1991).
[27] J. F. Herbst, Rev. Mod. Phys. **63** , 819 (1991).
[28] D. B. de Mooij and K. H. J. Buschow, J. Less-Common Met. **142**, 349 (1988).
[29] J. M. D. Coey, Physica Scripta **T39**, 21 (1991).
[30] Qi-Nian Qi, Y. P. Li, and J. M. D. Coey, J. Phys.: Condens. Matter **4**, 8209 (1992).
[31] Y.-C. Yang, X.-D. Zhang, S.-L. Ge, Q. Pan, L.-S. Kong, H.-L.Li, J.-L. Yang, B.-S. Zhang, Y.-F. Ding, and C.-T. Ye, J. Appl. Phys. **70**, 6001 (1991).
[32] J. M. D. Coey, H. Sun, and D. P. F. Hurley, J. Magn. Magn. Mater. **101**, 310 (1991).
[33] H. Sun, B.-P. Hu, H.-S. Li, and J. M. D. Coey, Solid State Comm. **74**, 727 (1990).
[34] Y.-P. Li and J. M. D. Coey, Solid State Comm. **81**, 447 (1992).
[35] D. P. F. Hurley and J. M. D. Coey, J. Magn. Magn. Mater. **99**, 229 (1991).
[36] D. P. F. Hurley and J. M. D. Coey, J. Phys.: Condens. Matter **4**, 5573 (1992).
[37] M. Katter, J. Wecker, C. Kuhrt, L. Schultz, X. C. Kou, and R. Grössinger, J. Magn. Magn. Mater. **111** (1992) 293.
[38] R. Skomski, C. Murray, S. Brennan, and J. M. D. Coey, J. Appl. Phys. **73**, 6940 (1993).
[39] T. Iriyama, K. Kobayashi, T. Fukuda, H. Kato, and Y. Nakagawa, IEEE Trans. Magn. **28**, 2326 (1992).
[40] J. M. D. Coey, "Rare-Earth - Iron Permanent Magnets", in: *Current Trends in the Physics of Materials*, G. F. Chiarotti, F. Fumi, and M. P. Tosi, eds., North-Holland, Amsterdam, 1990, p. 265.
[41] K. H. J. Buschow, Materials Science Report, **1**, 1 (1986).
[42] J. M. D. Coey, "Intermetallic Compounds and Crystal-Field Interaction", in: *Science and Technology of Nanostructured Materials*, G. C. Hadjipanayis and G. A. Prinz, eds., Plenum Press, New York, 1991, p. 439.
[43] M. Sagawa, S. Hirosawa, H. Yamamoto, S. Fujimura, Y. and Masuura, Jpn. J. Appl. Phys. **26**, 785 (1987).
[44] H.-S. Li and J. M. D. Coey, "Magnetic Properties of Ternary Rare-Earth Transition-Metal Compounds", in: *Handbook of Magnetic Materials VI*, K.H. J. Buschow, ed., Elsevier, Amsterdam, 1991, p. 1.
[45] J. F. Herbst, J. J. Croat, R. W. Lee, and W. B. Yelon, J. Appl. Phys. **53**, 250 (1982).
[46] J. Emsley, *The Elements*, University Press, Oxford, 1989.
[47] N. W. Ashcroft and N. D. Mermin, *Solid State Physics*, Holt, New York, 1976.
[48] M. Brouha and K. H. J. Buschow, J. Appl. Phys. **44**, 1813 (1973).
[49] H. Sun, J. M. D. Coey, Y. Otani, and D. P. F. Hurley, J. Phys.: Condens. Matter **2**, 6465 (1990).
[50] R. M. Ibberson, O. Moze, T. H. Jacobs, and K. H. J. Buschow, J. Phys.: Condens. Matter **3**, 1219 (1991).
[51] R. Skomski, M. D. Kuz'min, and J. M. D. Coey, J. Appl. Phys. **73**, 6934 (1993).
[52] O. Isnard, J. L. Soubeyroux, S. Miraglia, D. Fruchart, L. M. Garcia, and J. Bartolomé, Physica B **180&181**, 629 (1992).
[53] H. Peisl, "Lattice Strains due to Hydrogen in Metals" in *Hydrogen in Metals I*,

G. Alefeld and J. Völkl, eds., Springer, Berlin, 1978, p.53.
[54] K. Yvon and P. Fischer, "Crystal and Magnetic Structures of Ternary Metal Hydrides: A Comprehensive Review" in *Hydrogen in Intermetallic Compounds I*, L. Schlapbach, ed., Springer, Berlin, 1988, p.88.
[55] Y.-N. Wei, K. Sun, Y.-B. Fen, J.-X. Zhang, B.-P. Hu, Y.-Zh. Wang, X.-L. Rao, and G.-Ch. Liu, J. Alloys and Compounds **194**, 9 (1993).
[56] J. M. D. Coey and D. P. F. Hurley, J. Magn. Magn. Mater. **104**, 1098 (1992).
[57] T. Mukai and T. Fujimoto, J. Magn. Magn. Mat. **103**, 165 (1992).
[58] Q.-N. Qi, H. Sun, R. Skomski, and J. M. D. Coey, Phys. Rev. B **45**, 12278 (1992).
[59] T. D. Lee and C. N. Yang, Phys. Rev. **87**, 410 (1952).
[60] G. Alefeld, phys. stat. sol. **32**, 67 (1969).
[61] T. B. Flanagan and W. A. Oates, "Thermodynamics of Intermetallic Compound-Hydrogen Systems" in *Hydrogen in Intermetallic Compounds I*, L. Schlapbach, ed., Springer, Berlin, 1988, p.49.
[62] E. Wicke and H. Brodowsky, "Hydrogen in Palladium and Palladium Alloys" in *Hydrogen in Metals II*, G. Alefeld and J. Völkl, eds., Springer, Berlin, 1978, p.73.
[63] R. Kutner, K. Binder and K. W. Kehr, Phys. Rev. B **26**, 2967 (1982).
[64] L. Pauling, *The Nature of the Chemical Bond*, Cornell, New York, 1960.
[65] R. Griessen and T. Riesterer, "Heat of Formation Models" in *Hydrogen in Intermetallic Compounds I*, L. Schlapbach, ed., Springer, Berlin, 1988, p.49.
[66] P. Nordlander, J. K. Nørskov, and F. Besenbacher, J. Phys. F: Met. Phys. **16**, 1161 (1986).
[67] J. K. Nørskov, Phys. Rev. B **26**, 2875 (1982).
[68] J. Friedel, Adv. Phys. **3**, 446 (1954).
[69] N. F. Mott and H. Jones, *The Theory of the Properties of Metals and Alloys*, University Press, Oxford, 1936.
[70] A. R. Miedema, P. F. de Chatel, and F. R. de Boer, Physica **100B**, 1 (1980).
[71] G. Alefeld and J. Völkl, in *Hydrogen in Metals I*, Springer, Berlin, 1978, p.1.
[72] L. Cao, L.-Sh. Kong, and B.-G. Shen, phys. stat. sol. (a) **134**, K69 (1992).
[73] H.-H. Uchida, H. Uchida, T. Yanagisawa, S. Kise, T. Suzuki, Y. Matsumura, U. Koike, K. Kamada, T. Kurino, and H. Kaneko, J. Alloys and Compounds **196**, 71 (1993).
[74] L. L. Dhawan and S. Prakash, J. Phys. F **14**, 2329 (1984).
[75] B. S. Bochstein, *Diffusion in Metals*, Metallurgiya, Moscow, 1978.
[76] H. Wagner, "Elastic Interaction and Phase Transition in Coherent Metal-Hydrogen Alloys" in *Hydrogen in Metals I*, G. Alefeld and J. Völkl, eds., Springer, Berlin, 1978, p.5.
[77] S. Brennan, R. Skomski, and J. M. D. Coey, to be published.
[78] C. Kapusta, R. J. Zhou, M. Rosenberg, P. C. Riedi, and K. H. J. Buschow, J. Alloys Compounds **178**, 139 (1992).
[79] K.-H. Müller, P. A. P. Wendhausen, D. Eckert, and A. Hanstein, in: *7th International Symposium Magn. Anisotropy and Coercivity in RE-TM Alloys*, Canberra 1992, p. 34; P. A. P. Wendhausen, J. Zawadzki *et al.*, to be published.
[80] D. L. Goodstein, *States of Matter*, Dover, New York,1985.
[81] H. Triebel, *Analysis und mathematische Physik*, Teubner, Leipzig, 1984.
[82] J. Völkl and G. Alefeld, "Diffusion of Hydrogen in Metals" in *Hydrogen in Metals I*, G. Alefeld and J. Völkl, eds., Springer, Berlin, 1978, p.321.

[83] C. C. Colucci, S. Gama, and F. A. O. Cabral, IEEE Trans. Magn. **28**, 2578 (1992).
[84] R. Skomski, K.-H. Müller, P. A. P. Wendhausen, and J. M. D. Coey, J. Appl. Phys. **73**, 6047 (1993).
[85] J. M. D. Coey and Y. Otani, J. Mag. Soc. Japan **15**, 677 (1991).
[86] J. M. Yeomans, *Statistical Mechanics of Phase Transitions*, University Press, Oxford, 1992.
[87] K. H. Fischer and J. A. Hertz, Spin Glasses, University Press, Cambridge 1991.
[88] S. G. Brush, Rev. Mod. Phys. **39**, 883 (1967).
[89] C. Itzykson and J.-M. Drouffe, *Statistical Field Theory I*, University Press, Cambridge, 1989.

Chapter 17

ELECTRONIC STRUCTURE AND PROPERTIES OF PERMANENT-MAGNET MATERIALS

S. S. JASWAL
Department of Physics and Astronomy
and Center for Materials Research and Analysis
University of Nebraska — Lincoln
Lincoln, Nebraska 68588-0111
U.S.A.

ABSTRACT. Band structure theory used in electronic-structure calculations is reviewed briefly. The various properties of the recently discovered classes of permanent-magnet materials derived from the band calculations and their comparison with the photoemission and other experimental data are discussed.

1. Introduction

Almost all the observed properties of atoms, molecules and solids are determined by their valence electrons. Thus a good knowledge of the valence-electron states is essential to understand the physical properties of these systems. The discrete quantum states in atoms become bands of states in solids due to the overlap of valence electrons of neighboring atoms. The valence bands are completely occupied in insulators and semiconductors but only partially occupied in metals. Normally the valence bands of a metal contain equal number of electrons with up and down spin such that the net spin angular momentum and hence the spin moment is zero. Since the orbital angular momentum of an electron is quenched in most of the metals, there are very few metals in the periodic table that have useful magnetic properties. The magnetism of some of the transition metals such as Fe, Co and Ni is determined by their 3d energy bands. This chapter deals with the derivation of the main properties of the permanent-magnet materials from the electronic structure calculations. After a brief introduction to band theory, the electronic structure and properties of the recently discovered classes of permanent-magnet materials are discussed.

2. Band Theory

Band theory is a procedure to solve the many-electron wave equation

$$H\Psi = E\Psi , \qquad (1)$$

F. Grandjean et al. (eds.), Interstitial Intermetallic Alloys, 411–432.

where H and ψ are the many-body Hamiltonian and wave function, respectively, and E is the total energy. Since Eq. (1) cannot be solved exactly, a variational procedure is used to simplify it to an effective one-body problem. This is achieved by expressing ψ as a linear combination of the products of one-body functions ϕ_i. The simplest approximation is the Hartree approximation in which

$$\Psi = \prod_i \phi_i, \qquad (2)$$

i.e., the many-body function ψ is a simple product of one-body function ϕ_i without any correlation effects. The next is the Hartrec-Fock (H-F) approximation in which ψ is antisymmetrized to satisfy the Pauli principle, i.e., ψ is a Slater determinant whose order is equal to the number of electrons.

The effective one-body H-F equations become

$$\left[-\nabla^2 + V(r) + \sum_j \int \frac{|\phi_j(r')|^2}{|r-r'|} d^3r' - \sum_j^{\text{11spin}} \int \frac{\phi_j^*(r')\phi_i(r')\phi_j(r)}{\phi_i(r)|r-r'|} d^3r'\right]\phi_i(r) = \varepsilon_i \phi_i(r), \qquad (3)$$

where the first and second terms on the left hand side are the kinetic energy and the Coulomb potential due to the nuclie respectively giving the one-body Hamiltonian, the third term is the Coulomb term due to the electrons and the last term is the exchange term arising from all the electrons with the same spin as ϕ_i. As can be seen, the exchange term is a complicated non-local term making Eq. (3) a set of non-linear equations difficult to solve. Without the exchange term Eq. (3) corresponds to the Hartree approximation. Using plane waves for the ϕ_i, Slater derived a local density approximation to the exchange term

$$V_{ex}^{Slater} \propto \rho(r)^{1/3}, \qquad (4)$$

where $\rho(r)$ is the particle density.

It should be remembered that the anti-symmetrized wave function allows the correlation among the electrons of the same spin only. Coulomb correlation effects are included by going beyond the H-F approximation.

We now discuss the physical significance of ϕ_i and ε_i, the solutions of Eq. (3). The ε_i are the Lagrange multipliers introduced by the variational procedure mentioned above. Koopmann's theorem gives the physical significance of ε_i. Assuming that ϕ_i do not change if one of the particles is removed from a large system, $-\varepsilon_i$ is the energy required to remove that particle. This assumption is quite reasonable except for highly correlated systems. The particle density is related to ϕ_i as follows:

$$\rho(r) = \sum_i |\phi_i(r)|^2. \qquad (5)$$

Kohn and Sham [1] used the density functional theory of Hohenberg and Kohn [2] to derive the effective one-body Hamiltonian similar to Eq. (3). They arrived at the following local density approximation to the exchange-correlation term in the Hamiltonian

$$V_{xc}^{k-s} = \frac{4}{3}\varepsilon_{xc}, \quad (6)$$

where ε_{xc} is the exchange-correlation energy per particle of the free electron gas. This is related to the Slater term as follows:

$$V_{xc}^{k-s} = \frac{2}{3}V_{xc}^{Slater}. \tag{7}$$

Gáspár independently derived Eq. (6) [3]. The density functional theory, with a different form of the exchange-correlation term in the Hamiltonian, is now universally used in band-structure calculations.

The electronic structure calculations are variational solutions of Eq. (3) with an appropriate choice of basis functions for the valence electrons. This is accomplished by choosing ϕ to be a linear combination of the Bloch functions x_i made from the various basis functions, i.e.,

$$\phi = \sum_i c_i x_i. \tag{8}$$

Minimization of

$$\varepsilon = \frac{\langle\phi|H|\phi\rangle}{\langle\phi|\phi\rangle} \tag{9}$$

with respect to c_i leads to a set of linear equations

$$\left(H_{ij} - \varepsilon S_{ij}\right)c_j = o, \tag{10}$$

where H_{ij} and S_{ij} are the Hamiltonian and overlap matrices with respect to x_i.

3. Exchange Interactions and Magnetic Ordering

We see from Eq. (3) that the exchange interactions lower the energy of a system of fermions. Thus the exchange term alone favors ferromagnetic ordering of all the electrons. However, the Pauli principle would force some of the electrons in such an arrangement to occupy higher energy states thereby increasing the kinetic energy of the system. It is the competition between these two terms that determines the magnetic order in a given system. In the simple metals, where s-p conduction bands are broad, there is no magnetic ordering due to the dominance of the kinetic energy term. In some of the transition metals, the d-bands are narrow enough that the exchange term dominates leading to the magnetic order.

A quantitative relationship that takes into account the two terms mentioned above is the Stoner criterion for magnetic ordering [4]:

$$IN(\varepsilon_f) > 1, \tag{11}$$

where I is the Stoner parameter determined by the exchange interactions and $N(\varepsilon_f)$ is the paramagnetic density of states (DOS) at the Fermi energy reflecting the shape of the bands near the Fermi energy.

Magnetic properties are very sensitive to the local atomic structure. An excellent example of this is the compound VPd_3 [5]. The elemental metals V and Pd are non-magnetic. Fig. 1 shows the real VPd_3 ($TiAl_3$) and a hypothetical VPd_3 (Cu_3Au) structure which is quite similar. DOS at the Fermi energy for the real structure is low (Fig. 2) and the Stoner criterion for magnetic ordering is not satisfied. This is in agreement with the experimental results. However, DOS at the Fermi energy is so large for the hypothetical structure (Fig. 3) that the Stoner criterion is easily satisfied. Finally, the magnetic properties of some of the transition metal impurities are known to be very sensitive to the matrix in which they are located.

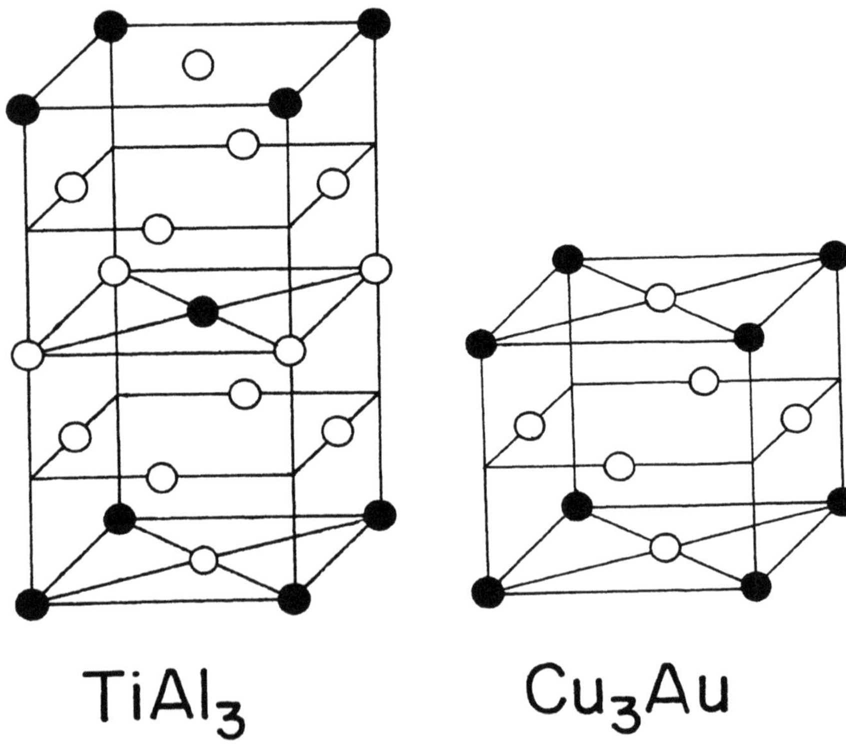

Fig. 1. Unit cells for $TiAl_3$ and Cu_3Au structures.

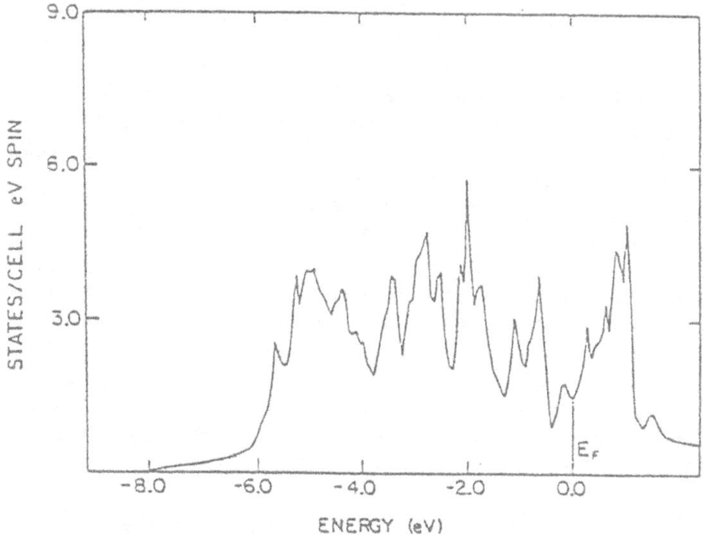

Fig. 2. The density of states for paramagnetic VPd_3 in the $TiAl_3$ structure. The energy zero is at the Fermi level.

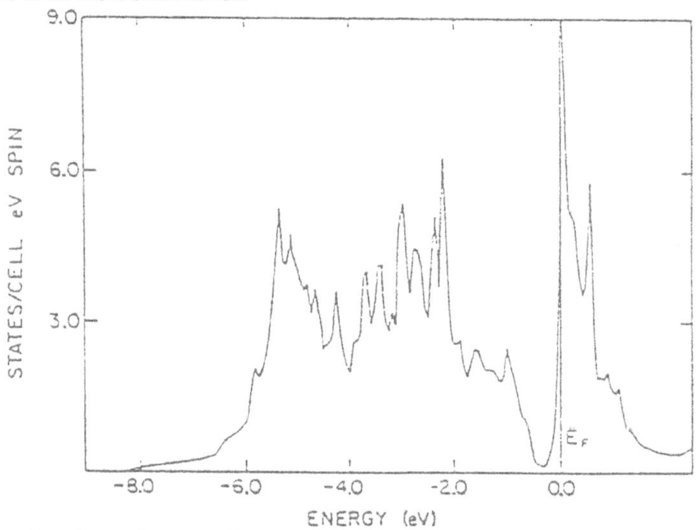

Fig. 3. The density of states for paramagnetic VPd_3 in the Cu_3Au structure. The energy zero is at the Fermi level.

The best way to study the magnetic properties of a system theoretically is to perform self-consistent spin-polarized electronic-structure calculations on it.

4. Self-Consistent Spin-polarized Electronic Structure Calculations

The problem is to solve the set of equations given by Eq. (10) self-consistently. It is almost universal in condensed matter physics now that the Hamiltonian is treated in the local density approximation of the density functional theory. The best set of basis functions is given by the augmented plane wave method in which the functions are atomic-like near the nuclei and plane waves in the interstitial regions [6]. Unfortunately, this method is very cumbersome and therefore not suitable for permanent-magnet materials with many atoms per unit cell. One of the most efficient methods for such a system is the linear-muffin-tin-orbitals (LMTO) method due to Andersen [7,8]. In this method the Hankel functions in the interstitial region are continuously and differentially augmented inside each sphere by atomic-like functions linearized about a suitably chosen energy for each band. Thus Eq. (10) is now linear in energy based on a very realistic set of basis functions. Using the atomic charge densities as the starting point, the LMTO method converges rapidly to self-consistent results. For spin-polarized results, the calculations are carried out both for spin-up and spin-down charge densities.

The self-consistent electronic structure is then used to derive the various magnetic properties. For example, the magnetization is given by the difference in the spin-up and spin-down charge densities. The magnetic moment at a given site is the difference in the site-projected spin-up and spin-down charge densities.

The anisotropy study requires fully relativistic total-energy calculations with magnetization pointed in different directions. Since this requires the inclusion of spin-orbit interactions which are very small compared to exchange interactions, anisotropy calculations are extremely difficult and time consuming computationally.

Mohn-Wohlfarth spin-fluctuation theory can be used to study variations in the Curie temperature with volume changes introduced by interstitials such as C or N in permanent-magnet materials [9].

5. Electronic Structure and Properties of Hard-Magnet Materials

A good permanent magnet must have a large saturation magnetization, high Curie temperature, and large magnetic anisotropy for high coercivity. Fe and Co have reasonably large magnetic moments [μ(Fe) = 2.12 μ_B, μ(Co) = 1.56 μ_B] and high Curie temperatures [T_c(Fe) = 1043K, T_c(Co) = 1388K] but very low coercivities. On the other hand, many of the rare-earth (R) metals have large moments and single-ion anisotropies, but very low Curie temperatures. Thus, the researchers have been trying to exploit the good properties of the metals in these two classes by studying their compounds which are rich in Fe and or Co.

It is only the light R metals that are of interest in this problem because they couple ferromagnetically to Fe or Co. The ferromagnetic coupling can be understood as follows: The non-4f valence states of an R metal are similar to those of an early transition metal and hence lie on the average higher than Fe or Co majority d-band. Thus they are closer to the minority Fe or Co d-band and their hybridization with each other through exchange interaction induces non-4f spin moment on the R metal which is opposite to that of the magnetic atom. The induced moment on the R metal couples ferromagnetically with the 4f spin moment $\vec{S}$ through exchange interactions. Thus the 4f $\vec{S}$ is opposite to that of Fe or Co. Since the total angular momentum $\vec{J}$ is anti-parallel to $\vec{S}$ for light R metals due to a Hund's rule, $\vec{J}$ is parallel to the spin of Fe or Co leading to a ferromagnetic coupling between the two.

$Nd_2Fe_{14}B$ is the latest excellent entry to the permanent-magnetic field [10]. We give below a brief discussion of the electronic structure and properties of this and other promising classes of permanent-magnet materials. Unless stated otherwise, the LMTO calculations are in the local-density and scalar relativistic approximations.

5.1 $R_2Fe_{14}B$(2-14) CLASS OF PERMANENT MAGNETS

These compounds are tetragonal with 68 atoms per primitive cell (Fig. 4) and the space-group symmetry $P4_2/mnm$ [10]. We reported the first electronic-structure results by photoemission and self-consistent spin-polarized calculations [11,12]. Now the 4f states of the rare-earths are quite localized and highly correlated. As a result they are not described well by the standard band structure calculations. One tries to understand their contributions to magnetization and anisotropy from atomic like calculations. Thus the band structure calculations are normally carried out on a prototype rare-earth compound such as $Y_2Fe_{14}B$ or the 4f states are treated as core states. Since the compounds under study are Fe- or Co-rich, this procedure is fairly successful in giving the desired magnetic properties.

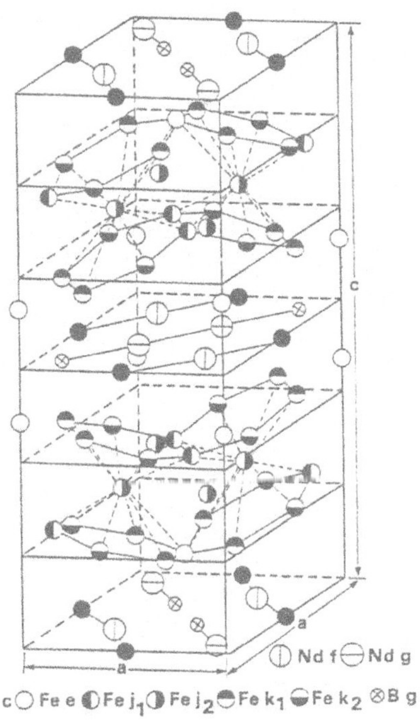

Fig. 4. Tetragonal unit cell of $Nd_2Fe_{14}B$.

As an example, we show in Fig. 5 the partial spin-polarized density of states (DOS) for $Y_2Fe_{14}B$. As expected, Fe d-band dominates the spectrum. The average exchange splitting of ~2.1 eV is similar to that of pure Fe. The total (spin summed) DOS of $Y_2Fe_{14}B$ multiplied with the zero-temperature Fermi function and broadened with a Gaussian of 0.3 eV width is compared with the HeI photoemission spectrum of $Nd_2Fe_{14}B$ with the best resolution in Fig. 6. This comparison is justified because the 4f photoionization cross section at HeI photon energy of 21.2 eV is negligible compared to that of the other valence electrons. Aside from small energy shifts in the peak positions, it can be seen that the overall agreement is very good. Our results show that all compounds of the form $R_2Fe_{14}B$ have essentially identical electronic structures except for the 4f levels. We also found that the photoemission spectrum of $Nd_2Fe_{14}B$ in the paramagnetic state at 623K (T_c = 585K) is quite similar to that of the ferromagnetic state. Thus the short-range magnetic order persists in these compounds above T_c as in the fluctuating-band picture [13].

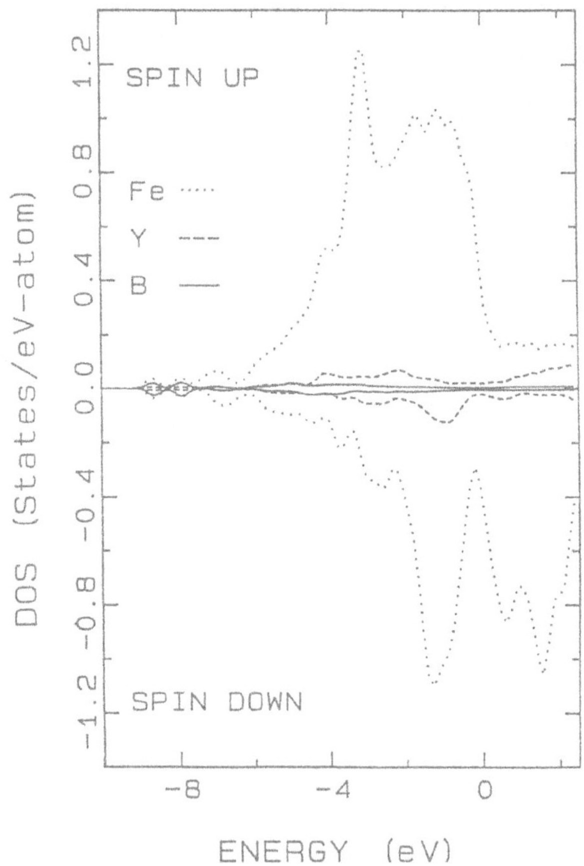

Fig. 5. Partial spin-polarized density of states for Fe, Y and B in $Y_2Fe_{14}B$.

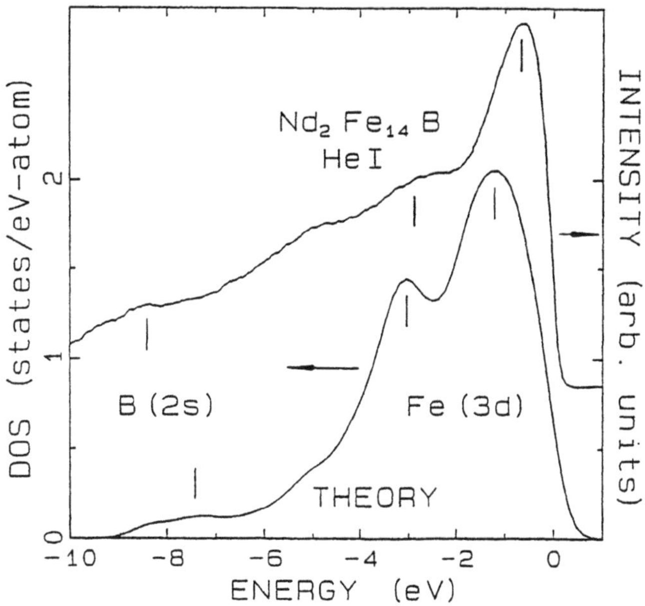

Fig. 6. Comparison of the broadened DOS for $Y_2Fe_{14}B$ with the $Nd_2Fe_{14}B$ photoemission spectrum (HeI). The Nd (4f) levels are not seen in the photoemission spectrum at HeI photon energies (see text).

The calculated magnetic moments for different Fe sites are in good agreement with the experimental data as shown in Table I.

Table I. Magnetic moments (in μ_B) for different sites in $Y_2Fe_{14}B$ and $Nd_2Fe_{14}B$

	$Y_2Fe_{14}B$		$Nd_2Fe_{14}B$	
	Calc.	Expt.[a]	Calc.	Expt.[a]
Y or Nd (f)	-0.64	—	-0.55[b]	—
Y or Nd (g)	-0.60	—	-0.52[b]	—
Fe (k_1)	2.08	2.07-2.25	2.12	2.08-2.60
Fe (k_2)	2.15	2.23-2.32	2.18	2.16-2.60
Fe (j_1)	2.07	2.10-2.40	2.12	2.06-2.30
Fe (j_2)	2.74	2.43-2.80	2.74	2.43-2.85
Fe (e)	2.06	2.03-2.28	2.13	2.00-2.28
Fe (c)	2.53	1.24-2.25	2.59	1.97-2.75
B (g)	-0.19	—	-0.20	—

[a]R. Fruchart *et al.*, J. Phys. F 17, 483 (1987); H. Onodera *et al.*, J. Magn. Magn. Mater. 68, 15 (1987); J.M. Friedt *et al.*, J. Phys. F. 16 651 (1986); H.M. Van Noort *et al.*, J. Less-Common Metals 115, 155 (1986); D. Givord *et al.*, J. Appl. Phys. 57, 4100 (1985).

[b]Without the 4f electron contributions.

Finally, the anisotropy arises from magneto-static dipole-dipole and spin-orbit interactions. It is normally assumed that the dipolar contribution to the intrinsic anisotropy is zero as given by macroscopic theory for uniform magnetization. This is true only for simple highly symmetric lattices (cubic, hcp) where the local field correction is isotropic. We have calculated the dipolar contributions to the intrinsic anisotropy (K_u^d) in several 2-14 compounds which favor the easy axis to lie in the plane perpendicular to the c-axis [14]. The spin-orbit contribution in these systems can be divided into 4f single-ion term (K_u^{si}) and the universal non-4f term (K_u^{so}). Since K_u^{si} is zero in R = Y, Gd, and Lu compounds, the calculated K_u^d along with the experimental data are used to deduce K_u^{so} for them. Then K_u^d and K_u^{so} together with the experimental data lead to an estimate of K_u^{si} for several 2-14 compounds. The results are given in Table II. As expected, the single-ion contribution dominates over the other two terms in these compounds. Regarding the single-ion anisotropy in 2-14 compounds, Coehoorn and Buschow concluded that the second-order crystal field parameter A_2^0 is determined primarily by the asphericity of the valence electron density of the rare earth [15].

Table II. Calculated and experimental anisotropy energy densities in 10^7 ergs/cm^3

	K_u^d	$K_u^{expt\,a}$ (4K)	K_u^{so}	K_u^{si}
$Y_2Fe_{14}B$	-0.13	1.14	1.27	—
$Gd_2Fe_{14}B$	-0.56	0.70	1.26	—
$Lu_2Fe_{14}B$	-0.13	1.17	1.30	—
$Nd_2Fe_{14}B$	0.0	12.5	1.3	11.2
$Tb_2Fe_{14}B$	-0.76	7.9	1.3	7.4
$Dy_2Fe_{14}B$	-0.87	3.86	1.3	3.43

[a]Ref. 10.

5.2 PERMANENT-MAGNET MATERIALS OF R_2Fe_{17}(2-17) STRUCTURE

Since the discovery of $Nd_2Fe_{14}B$ as a hard magnet, ternary compounds R-Fe(Co)-M have been actively studied where M is a metalloid atom such as B, C or N. Compounds of R_2Fe_{17} class have large magnetization values but low Curie temperatures. However, Sun *et al.* found that absorption of N in these compounds almost doubles their Curie temperatures and changes the anisotropy of Sm_2Fe_{17} from planar to uniaxial [16]. This discovery has produced considerable research activity in the permanent-magnet field. $R_2Fe_{17}N_x$ compounds can be prepared with x values typically up to 2.6 and generally having the hexagonal Th_2Ni_{17} structure for the heavy rare earths, and the rhombohedral Th_2Zn_{17} structure for the light rare earths. The two structures are quite similar as shown in Fig. 7.

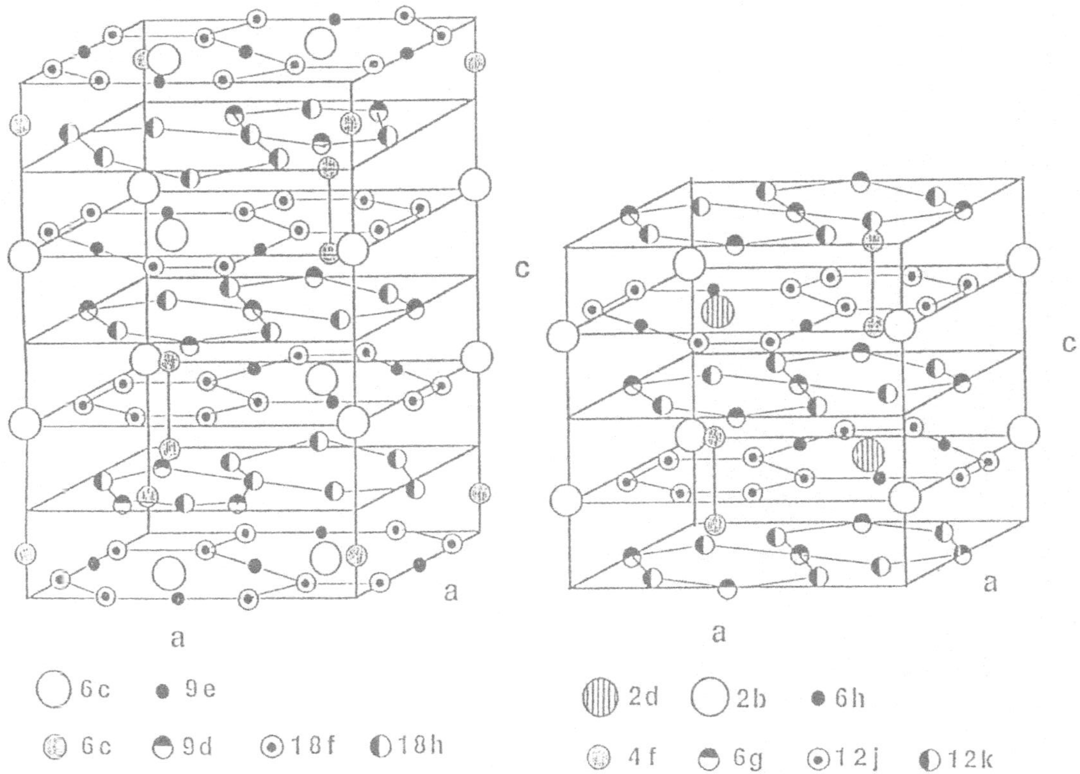

Fig. 7. Unit cells of rhombohedral Th_2Zn_{17} (left) and hexagonal Th_2Ni_{17} (right) structures.

We have studied the electronic structure of these compounds by photoemission and band structure calculations [17-19]. The band structures of the R_2Fe_{17} and $R_2Fe_{17}N_x$ compounds were investigated by focusing on Y_2Fe_{17} and $Y_2Fe_{17}N_3$ in the experimentally observed hexagonal structure. The choice of the rare earth prototype Y was made for the reasons given in the last section. To study the effect of nitrogenation the compounds Y_2Fe_{17} and $Y_2Fe_{17}N_3$ were studied with 38 and 44 atoms per primitive cell respectively. The N atoms were placed at the experimentally found h sites in the hexagonal structure which are equivalent to the sites occupied by the majority of the N atoms in the rhombohedral structure.

The spin-polarized DOS for Y_2Fe_{17} and $Y_2Fe_{17}N_3$ are shown in Fig. 8. As expected, the DOS are dominated by the Fe d-band near the Fermi energy. The structure around 6 eV in $Y_2Fe_{17}N_3$ is due to the N 2p states. The structure due to the Fe d bands in $Y_2Fe_{17}N_3$ DOS is shifted to the higher binding energy due to the reduction in overlap upon nitrogenation. This leads to a decrease in the values of both the up- and down-spin DOS values at the Fermi energy.

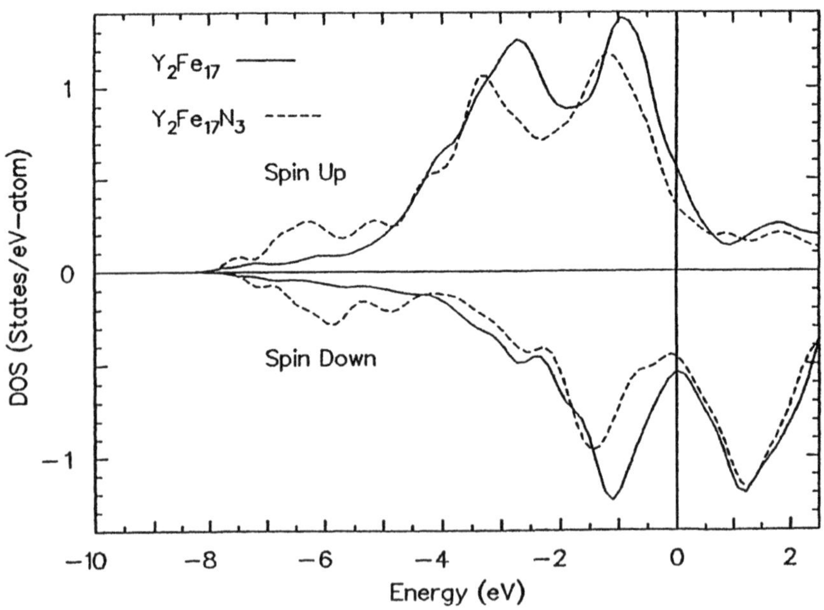

Fig. 8. Spin-polarized DOS for Y_2Fe_{17} and $Y_2Fe_{17}N_3$. The zero of the energy represents the Fermi energy.

The total calculated DOS for Y_2Fe_{17} and $Y_2Fe_{17}N_3$ are compared with the photoemission data for various N concentrations in Sm_2Fe_{17} in Fig. 9. The surface contaminants C and O are below the detection limit of Auger electron spectroscopy. The overall agreement between the calculated DOS and experimental data is good. The subtle difference between the data for the clean and nitrided compounds near the Fermi energy is shown in Fig. 10. The lines are drawn through the data points to serve as a guide to the eye. The systematic shift to higher binding energy by about 0.05 eV in the nitride data is clear in the expanded binding energy scale in Fig. 10. This is in agreement with the corresponding result for the calculated DOS.

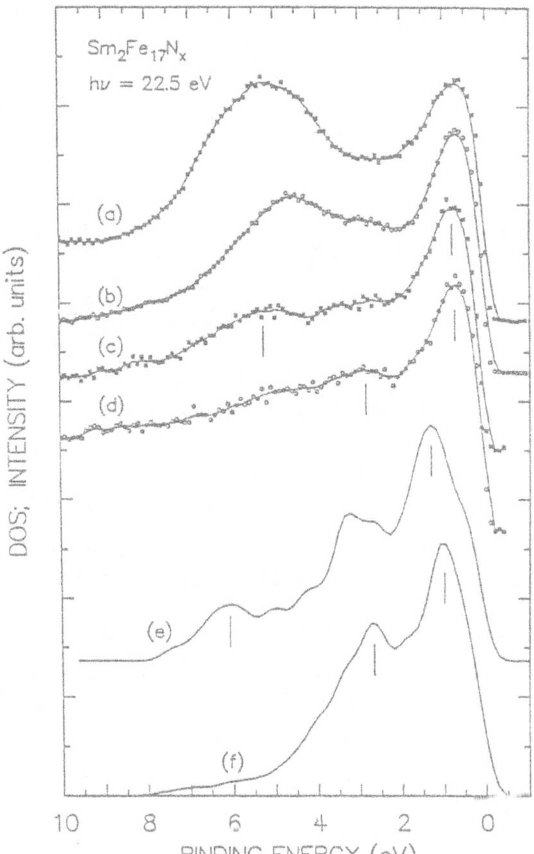

Fig. 9. Room-temperature photoemission spectra and calculated DOS for clean and nitrided 2-17 compound. (a) Surface data after sputtering with N with x = 9.8. (b) Data for N sputtered and heated surface with x=3.2. (c) Data of the subsurface with x = 2.6 obtained by argon sputtering the N rich surface layer of (b) and annealing at 350°c. (d) Data before nitriding. (e) Calculated DOS for $Y_2Fe_{17}N_3$. (f) Calculated DOS for Y_2Fe_{17}.

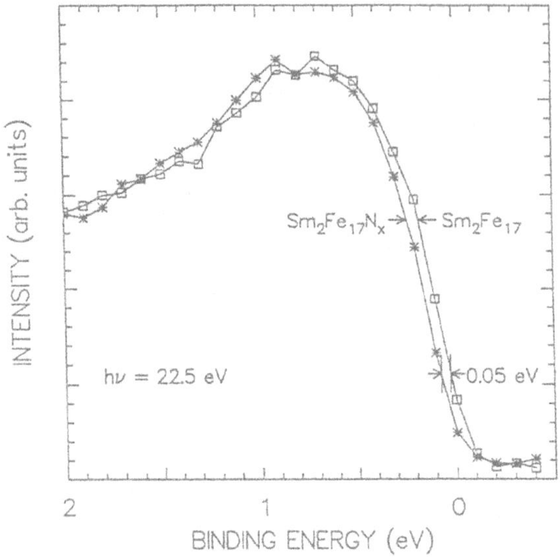

Fig. 10. Expanded view of the data in (c) ($Sm_2Fe_{17}N_{2.6}$) and (d) (Sm_2Fe_{17}) of Fig. 9. The lines connect the data points to serve as a guide to the eye.

The magnetic moments for the different atomic sites in the two compounds are compared in Table III. The expansion of the lattice due to N lowers the overlap among the 6g and 12k Fe atoms in the Fe only planes (z = 0, 1/2) and hence their magnetic moments go up by about 30 percent. Since the 12j Fe atoms in the nitride have overlaps with the N atoms, their moments go down by about 12 percent. The 4f-site Fe moment is the least effected, with only a small increase in its value. The calculated moments are in reasonable agreement with the Mössbauer measurements [20]. Once again Y has small induced negative moments due to the reasons given before.

Table III. Magnetic moments (in μ_B) for different sites in R_2Fe_{17} and $R_2Fe_{17}N_x$ compounds.

	Calculated Values		Experimental Values*	
Site	Y_2Fe_{17}	$Y_2Fe_{17}N_3$	Lu_2Fe_{17}	$Lu_2Fe_{17}N_x$
R(2b)	-0.47	-0.20	—	—
R(2d)	-0.45	-0.45	—	—
Fe(4f)	2.53	2.65	2.47	2.69
Fe(6g)	1.92	2.53	2.23	2.51
Fe(12j)	2.25	2.01	2.07	2.35
Fe(12k)	2.00	2.57	1.87	2.17
N(6h)	—	-0.04	—	—

*From Mössbauer results of Ref. 20.

It is natural to ask if the increase in T_C upon nitrogenation can be understood theoretically. According to the spin-fluctuation theory of Mohn and Wohlfarth [9] T_C is given by

$$T_c \propto M_0^2/\chi_0$$

where M_0 is the zero-temperature magnetic moment per Fe atom and the enhanced susceptibility χ_0 is given by

$$\chi_0^{-1} = \left[1/\left[2N_\uparrow(\varepsilon_f)\right] + 1/\left[2N_\downarrow(\varepsilon_f)\right] - I\right]/2\mu_B^2$$

$N_\uparrow(\varepsilon_f)$ and $N_\downarrow(\varepsilon_f)$ are the up- and down-spin DOS at the Fermi energy and I is the Stoner parameter. These parameters as obtained from the band structure results are listed in Table IV. They give $R = T_C(Y_2Fe_{17}N_3)/T_C(Y_2Fe_{17}) = 2.34$. The experimental data for x = 2.6 give R = 694/325 = 2.14 [16]. An increase in M_o and a substantial decrease in $N_\uparrow(E_f)$ upon nitrogenation are responsible for almost doubling T_C. This good agreement between experiment and theory concerning the increase of T_C indicates that the spin-fluctuation theory is a reasonable model for the magnetism of this class of compounds. It should be mentioned that the absolute value of the Curie temperature is not correctly predicted by this theory using our band-structure parameters.

Table IV. Band Structure Parameters.

	Density of States States/(eV-Fe Atom)		Magnetic Moment (μB/Fe Atom)	Stoner Parameter (eV)
	Up	Down		
Y_2Fe_{17}	0.663	0.545	2.11	0.95
$Y_2Fe_{17}N_3$	0.365	0.536	2.34	0.93

Coehoorn and Daalderop calculated the crystal field parameter A_2^0 for Gd_2Fe_{17} and $Gd_2Fe_{17}N_3$ for the single-ion anisotropy of the rare earth [21]. They found that A_2^0 is essentially unaffected by a 5 percent volume expansion, but changes significantly upon nitrogenation. This implies that it is the bonding of the metalloid atom and not the volume expansion that changes the single-ion anisotropy in 2-17 compounds.

In summary, electronic structure calculations show the nitrogenation of R_2Fe_{17} compounds produces the following changes in their magnetic properties:

i) The magnetic moments of the Fe atoms in the vicinity of N atoms decrease due to their overlap with the N atoms. The magnetic moments of the Fe atoms away from N atoms increase due to the decrease in their overlap caused by the lattice expansion.

ii) The increase in magnetization and the decrease in DOS at the Fermi energy caused by the volume expansion explain, through spin-fluctuation theory, the large increase in T_c observed experimentally.

iii) It is the bonding between N and the rare earth and not the lattice expansion that produces significant changes in the single-ion anisotropy parameter A_2^0.

5.3 PERMANENT-MAGNET MATERIALS OF $ThMn_{12}$(1-12) STRUCTURE

The history of research for this class of compounds is somewhat similar to that of 2-17 compounds discussed in the last section. A binary compound in which Fe replaces Mn in $ThMn_{12}$ structure is not stable. However, a fairly large class of ternary compounds $RFe_{12-x}T_x$ exists where T is a transition metal [22]. Once again, this class of materials has borderline hard-magnet properties which improve considerably upon nitrogenation [23]. One of the most promising compounds in this class is $NdFe_{11}Ti$. The addition of N enhances its Curie temperature by about 30% and $NdFe_{11}TiN_x$ has uniaxial anisotropy. We have studied the electronic structure and properties of several of these compounds and, as an example, discuss here the results for $NdFe_{11}TiN_x$ [24-27].

The $ThMn_{12}$ structure is body-centered tetragonal (space group I4/mmm) as shown in Fig. 11. The stabilizing element, such as Ti in $NdFe_{11}Ti$, replaces one of the 8i Fe atoms [28]. Apparently N partially occupies the empty 2b sites and for $x = 0.5$, half the 2b sites on the average will be occupied. Thus $NdFe_{11}TiN_x$ is a disordered compound with Ti and N randomly occupying 8i and 2b sites, respectively. This randomness is simulated in electronic structure calculations by a supercell containing four primitive cells of $ThMn_{12}$ structure. The tetragonal supercell is $a \times a \times 2c$ and contains 52 atoms without N, and 54 or 56 atoms with N for $x = 0.5$ or 1.0, respectively. The four Ti atoms are permuted among the four equivalent i sites in different $ThMn_{12}$ primitive cells and the N atoms occupy every other b site in $NdFe_{11}TiN_{0.5}$.

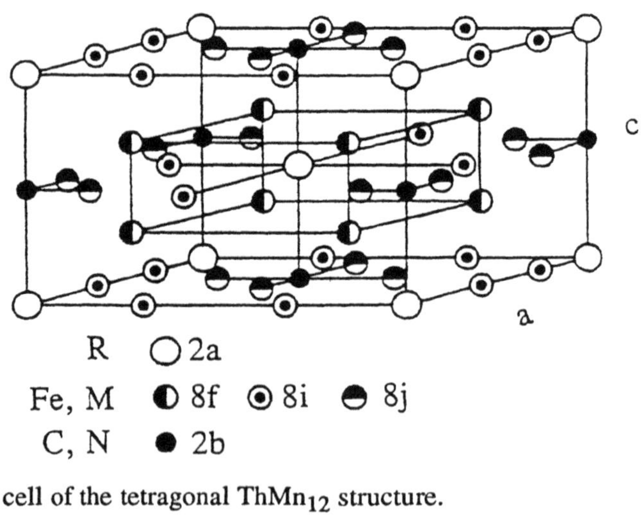

Fig. 11. Unit cell of the tetragonal $ThMn_{12}$ structure.

The spin-polarized partial DOS for $NdFe_{11}Ti$ and $NdFe_{11}TiN_{0.5}$ are shown in Figs. 12 and 13, respectively. As expected, DOS are dominated by the Fe d states near the Fermi energy. The results for the two systems are quite similar with N making a very small contribution. The DOS at the Fermi energy for both spins are slightly lower for $NdFe_{11}TiN_{0.5}$ compared to those for $NdFe_{11}Ti$. The total DOS are in good agreement with the photoemission data as shown in Fig. 14.

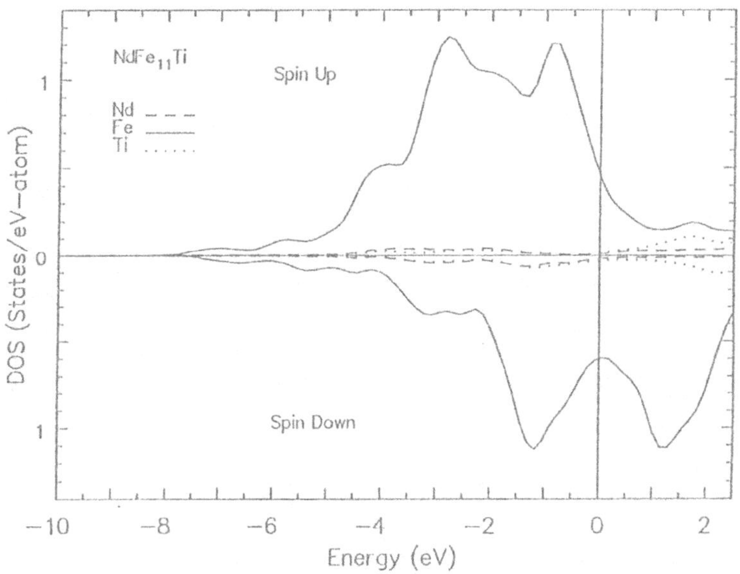

Fig. 12. Spin-polarized partial DOS for $NdFe_{11}Ti$. Nd 4f electrons are polarized and treated as core electrons.

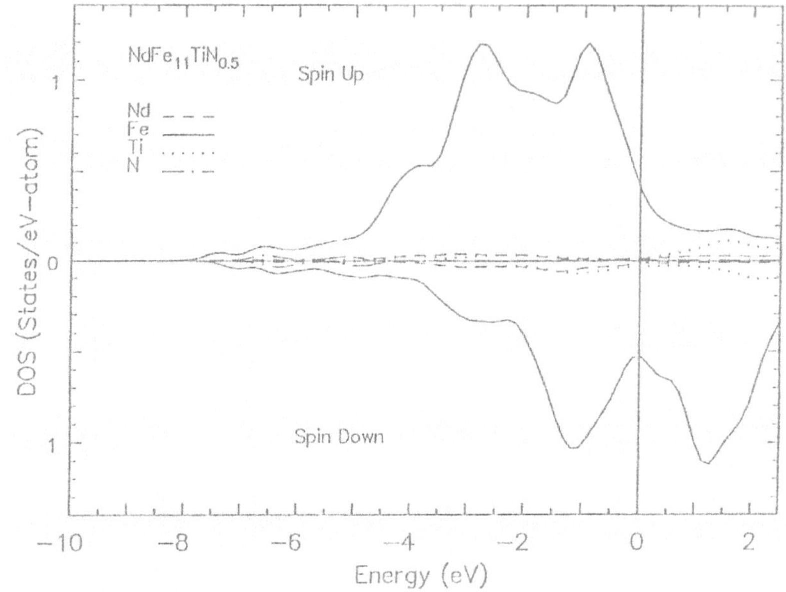

Fig. 13. Spin-polarized partial DOS for $NdFe_{11}TiN_{0.5}$. Nd 4f electrons are polarized and treated as core electrons.

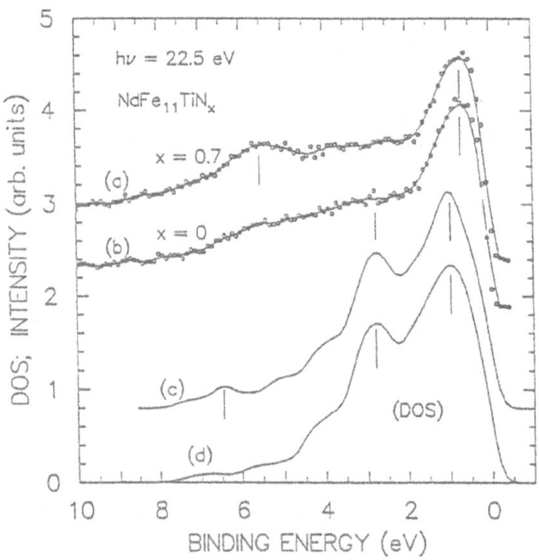

Fig. 14. (a) and (b) Photoemission spectra. (c) Calculated DOS for $NdFe_{11}TiN_{0.5}$. (d) Calculated DOS for $NdFe_{11}Ti$. The zero of the energy is at the Fermi level.

The calculated magnetic moments are compared with the experimental data in Table V. There is good agreement between experiment and theory for the 8f sites, but not for 8i and 8j sites. The origin of the disagreement for the i and j sites may be the shortcomings of our model in representing the disorder due to Ti and N. These results are similar to those of Sakuma for the Y compounds [30].

Table V. Magnetic moments (in μ_B) for different sites in $NdFe_{11}Ti$ and $NdFe_{11}TiN_{0.5}$ compounds.

Site	$NdFe_{11}Ti$ Theory[a]	$NdFe_{11}Ti$ Experiment[b] (Mössbauer)	$NdFe_{11}TiN_{0.5}$ Theory[a]	$NdFe_{11}TiN_{0.5}$ Experiment[c] (Neutron)
Nd (2a)	-0.44[d]		-0.36[d]	
Fe (8f)	1.67	1.67	1.93	2.18
Fe (8i)	2.46	1.97	2.55	2.28
Ti (8i)	-0.86		-0.89	
Fe (8j)	2.32	1.79	2.14	2.80
N (2b)			-0.02	

[a]Since the structure is disordered, these are the average moments at the original $ThMn_{12}$ sites.
[b]Ref. 29.
[c]Ref. 28 (The data for $YFe_{11}TiN_{0.5}$).
[d]Without the 4f contribution.

The band-structure data required to compare the Curie temperatures in spin-fluctuations theory discussed in the last section are given in Table VI. They give

$$R \equiv T_c(NdFe_{11}TiN_{0.5})/Tc(NdFe_{11}Ti) = 1.32.$$

The corresponding experimental value of R =723/551 = 1.31 [23]. The almost perfect agreement is not to be taken too seriously. The Curie-temperature enhancement here is smaller than that in 2-17 compounds due to smaller volume expansion.

Table VI. Band structure parameters for the Curie-temperature calculation.

	Density of States (States/eV-Fe Atom) Up	Density of States (States/eV-Fe Atom) Down	Magnetic Moment (μ_B/Fe Atom)	Stoner Parameter (eV)
$NdFe_{11}Ti$	0.553	0.706	2.12	0.90
$NdFe_{11}TiN_{0.5}$	0.50	0.646	2.17	0.88

Finally, the anisotropy of these systems is primarily due to the single-ion contribution of the rare-earth ion Nd. The anisotropy constant

$$K_1 = -\frac{3}{2}\alpha_j \langle r_f^2 \rangle [2J^2 - J] A_2^0,$$

where α_j (= -0.006 for Nd) is the second-order Stevens coefficient, $< r_f^2 >$ is the quantum-mechanical average of the square of the f-electron radial distance, the total angular momentum quantum number J = 4.5 for the ground state of Nd 4f electrons and A_2^0 is the first non-zero crystal-field parameter.

Following Coehoorn we write

$$A_2^0 = A_2^0(\text{val}) + A_2^0(\text{lat})$$

where A_2^0 (val) is the contribution of the valence electrons of Nd and A_2^0 (lat) is due to the charge distribution of the rest of the lattice [31]. It has been shown by several authors that A_2^0 (val) is the dominant crystal-field term [31-33]. (This means that the point-ion model of anisotropy is unphysical.) Therefore, we ignore the variation of A_2^0 (lat) in our study of the effect of nitrogenation on the anisotropy. Coehoorn approximates the p and d subshell contributions to A_2^0 (val) as follows:

$$A_2^0(\text{val,p}) = \frac{e^2}{4\pi\varepsilon_0}\frac{1}{5}S_p\Delta n_p$$

$$\text{with } \Delta n_p = n_z - \frac{1}{2}(n_x + n_y),$$

$$A_2^0(\text{val,d}) = \frac{e^2}{4\pi\varepsilon_0}\frac{1}{7}S_d\Delta n_d$$

$$\text{with } \Delta n_d = \frac{1}{2}(n_{xz} + n_{yz}) + n_{z^2} - n_{x^2-y^2} - n_{xy}$$

The n_i are the occupation numbers for the various p and d orbitals and the S_i are the overlap integrals of the squares of the radial parts of the valence and 4f orbitals of Nd.

We calculate Δn_p and Δn_d to study the nature of the anisotropy as a function of the nitrogenation and the results are given in Table VII. The last row in Table VII corresponds to the b sites being fully occupied by N. Experimentally the N concentration is usually not well defined. We see that Δn_p and Δn_d have the same sign for a given N concentration. The negative (positive) values of Δn_p and Δn_d make K_1 negative (positive) implying planar (c-axis) anisotropy. The positive values of Δn_p and Δn_d mean that the valence charge distribution is extended in the c direction, forcing the oblate axis and hence the magnetic moment of the 4f electrons of Nd in the same direction. By similar reasoning, the negative values of Δn_p and Δn_d lead to planar anisotropy. We see from Table VII that the increasing N concentration changes the anisotropy from planar to c-axis configuration, a result in qualitative agreement with the experimental data. In the last case, there is enough hybridization of the valence bands of Nd and N alternating along the c-axis to make Δn_p and Δn_d positive.

Table VII. Anisotropy parameters Δn_p and Δn_d as functions of nitrogenation

			Anisotropy	
	Δn_p	Δn_d	Theory	Experiment[a] (10K)
$NdFe_{11}Ti$	-0.023	-0.093	planar	cone
$NdFe_{11}TiN_{0.5}$	-0.015	-0.045	planar	c-axis[b]
$NdFe_{11}TiN$	0.849	0.676	c-axis	

[a]Ref. 25.
[b]Experimentally the N concentration is not well defined.

In summary, the calculated DOS is in good agreement with the photoemission data for $NdFe_{11}TiN_x$. The agreement between theory and experiment for magnetization in 1-12 compounds is not as good as in 2-14 and 2-17 compounds. This may be related to the disorder in 1-12 compounds. The spin-fluctuation theory, in combination with the band-structure results predicts the enhancement in Curie temperature on nitrogenation in very good agreement with the experimental data. This enhancement is similar to, but smaller than that in 2-17 compounds. This is because the volume increase upon nitrogenation is smaller in 1-12 (about 2%) than that in 2-17 (about 6%) compounds. Thus the interatomic exchange coupling responsible for the Curie temperature does not increase as much in 1-12 as in 2-17 compounds. The single-ion anisotropy, based on the Nd valence electron distribution, changes from planar to c-axis configuration with increasing N concentration in qualitative agreement with the experimental data.

This work is supported by the U.S. Department of Energy under Grant No. DE-FG02-86ER45262, National Science Foundation (ESPCoR) and the Cornell National Supercomputing Facility.

References

1. W. Kohn and L.J. Sham, Phys. Rev. 140, A1133 (1965).
2. P. Hohenberg and W. Kohn, Phys. Rev. 136, B864 (1964).
3. R. Gáspár, Acta Phys. Soc. Acad. Sci. Hung. 3, 263 (1954).
4. E.C. Stoner, Proc. Roy. Soc. A165, 372 (1938).
5. D.K. Misemer, S. Auluck, D.J. Sellmyer, S.S. Jaswal and A.J. Arko, Phys. Rev. B31, 3356 (1985).
6. T. Loucks, *Augmented PLANE Wave Method* (Benjamin, New York, 1967).
7. O.K. Andersen, Phys. Rev. B12, 3060 (1975).
8. H.L. Skriver, *The LMTO Method*, vol 41 of Springer Series in Solid State Sciences (Springer-Verlag, New York, 1984).
9. P. Mohn and E.P. Wohlforth, J. Phys. F 16, 2421 (1987).

10. J.F. Herbst, Rev. Mod. Phys. 63, 819(1991).
11. D.J. Sellmyer, M.A. Engelhardt, S.S. Jaswal and A.J. Arko, Phys. Rev. Lett 60, 2077 (1988).
12. S.S. Jaswal, Phys. Rev. B 41, 9697 (1990).
13. See the reviews in *Metallic Magnetism,* edited by H. Capellmann, *Topics in Current Physics Vol. 42* (Springer-Verlag, New York, 1987).
14. S.S. Jaswal and A.A. Kusov, J. Magn. Magn. Mater 109, L151 (1992).
15. R. Coehoorn and K.H.J. Buschow, J. Appl. Phys. 69, 5590 (1991).
16. H. Sun, J.M.D. Coey, Y. Otani and D.P.R. Hurley, J. Phys.: Condens. Matter 2, 6465 (1990).
17. S.S. Jaswal, W.B. Yelon, G.C. Hadjipanayis, Y.Z.Wang and D.J. Sellmyer, Phys. Rev. Lett. 67, 644 (1991).
18. S.S.Jaswal, IEEE Trans. Magn. 28, 2322 (1992).
19. J.P. Woods, A.S. Fernando, S.S. Jaswal, B.M. Patterson, D. Welipitiya and D.J. Sellmyer, J. Appl. Phys. 73, 6913 (1993).
20. G. Zouganelis, M. Anagnostou, and D. Niarchos, Solid State Commun. 77, 11 (1991).
21. R. Coehoorn and G.H.O. Daalderop, J. Magn. Magn. Mat. 104-107, 1081 (1992).
22. K.H.J. Buschow, J. Appl. Phys. 63, 3130 (1988).
23. G.C. Hadjipanayis, Z.X. Zang, W. Gong and E.W. Singleton, Rare-Earth Workshop, Australia (1992).
24. S.S. Jaswal, Y.G. Ren, and D.J. Sellmyer, J. Appl. Phys. 67, 4564 (1990).
25. S.S. Jaswal, J. Appl. Phys. 69, 5703 (1991).
26. A.S. Fernando, J.P. Woods, S.S. Jaswal, B.M. Patterson, D. Welipitiya, A.S. Nazareth and D.J. Sellmyer, J. Appl. Phys. 73, 6919 (1993).
27. S.S. Jaswal, Phys. Rev. B48, 6156 (1993).
28. Y. Yang, X.Zhang, L. Kang, Q. Pan, S. Ge, J. Yang, Y. Ding, B. Zhang, C. Ye and L. Yin, Solid State Commun. 78, 313 (1991).
29. B-P. Hu, H-S. Li, J.P. Gavigem and J.M.D. Coey, J. Phys.: Condens. Matter 1, 755 (1989).
30. A. Sakuma, J. Phys. Soc. Jpn. 61, 4119 (1992).
31. R. Coehoorn, in *Supermagnets, Hard Magnetic Materials,* Edited by G.J. Long and F. Grandjean (Kluwer Academic, Dordrecht, 1991), p. 133.
32. Z.Q. Gu and W.Y. Ching, Phys. Rev. B36, 8530 (1987).
33. K. Hummler and M. Fähnle, Phys. Rev. B45, 3161 (1992).

Chapter 18

APPLICATIONS OF NEUTRON SCATTERING TO INTERSTITIAL ALLOYS

W. B. YELON
University of Missouri
Research Reactor
Columbia, MO 65211

1. Introduction

There is an extensive literature on the effect of hydrogen (and deuterium) in metals as well as a variety of studies of the effects of other interstitial species. It is not possible to comprehensively review this literature, but references to some of the review papers will be given. In addition, examples will be derived from many sources, in order to give a flavor for the type of problem successfully investigated with neutrons. These will cover the full range of techniques discussed in a previous chapter. It will be clear that the versatility of the neutron probe has led to a broad range of investigation, which has had a profound impact on our understanding of these technologically important systems. These examples may suggest experiments to help clarify outstanding questions.

2. Structure Studies

As previously discussed, neutron scattering takes advantage of the relatively large scattering of light elements to determine their locations. In addition, the powder refinement methods allow an accurate determination of the lattice and magnetic changes accompanying the incorporation of interstitial species.

2.A. DETERMINATION OF INTERSTITIAL ATOM LOCATION

2.A.1. *Hydrides and deuterides.* Because of the intense interest in hydrogen storage systems and problems arising from hydrogen embrittlement, a great deal of attention has focussed on a few prototypical systems. The first neutron study to locate these atoms dates from 1948 [1]. For these studies D is used more often than H, due to the large incoherent scattering contributed by H atoms. This leads to a worsened signal-to-noise ratio compared to deuterium. It is generally assumed that, apart from mass mediated effects, which may change, for example, a phase transition temperature, the hydride and deuterides are otherwise identical.

2.A.1.a. Nb-Hydrides. As an example of prototypical systems, Nb, V and Ta hydrides have been extensively studied. For a complete description of the structural work, see Skold et al

F. Grandjean et al. (eds.), Interstitial Intermetallic Alloys, 433–461.

[2]. The phase diagram for NbH_x is shown in Figure 1 [3]. As a function of temperature and hydrogen concentration, a large number of phases are observed. The α and α' phases are bcc metals with disordered H positions, with the H atoms most likely located on tetrahedral sites. In the β phase with x ≈ 1, the metal lattice is fc orthorhombic, with a and b approximately equal to $\sqrt{2}(a_c)$, the cubic lattice constant. In this phase, the H atoms are found at (1/4,1/4,1/4), (1/4,3/4,3/4) + translations of the space group Cccm. A variety of other ordered phases are indicated on the phase diagram. These can be thought of as distorted CaF_2 structures with only about 1/2 of the fluorite type H sites occupied. The differences in these structures, then, correspond to different arrangements of hydrogen ordering and accompanying lattice distortions. The δ phase, at high H content, has sufficient H to cause equal distortion in all directions and therefore returns to cubic symmetry.

2.A.1.b. Ti-Hydrides. Ti-Al alloys are candidates for structural components for advanced aeronautical vehicles, including the National Aerospace Plane, because of their light weight, good strength and relatively good high temperature performance. The National Aerospace Plane is expected to have hydrogen fueled engines, and, therefore, understanding of the behavior of these alloys exposed to low concentrations of hydrogen at high temperature is essential. A combined transmission electron microscopy and neutron diffraction study was carried out on a number of alloys exposed to different environments. Depending upon the alloy composition, the hydrogen may be dissolved in the α phase, leading to small lattice expansion [4]. In several cases, novel hydride phases are observed, and one of them was completely characterized (Figure 2). In pure α_2 alloy, Ti_3Al, exposure to hydrogen leads to the formation of a hydride with stoichiometry Ti_3AlH [5]. This phase has the symmetry Pm3m, with Ti atoms at the face centers, Al at the origin, and the H atom at the body center (1/2,1/2,1/2). It is interesting that H and Ti have almost identical scattering lengths, and

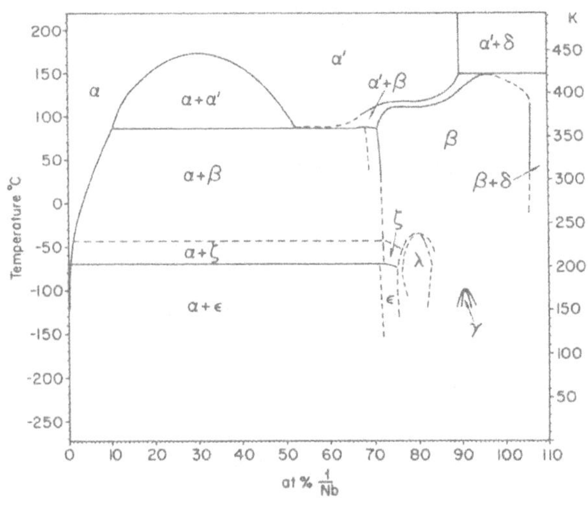

Figure 1. Phase diagram for the Nb-H system from Schober and Wenzl [3].

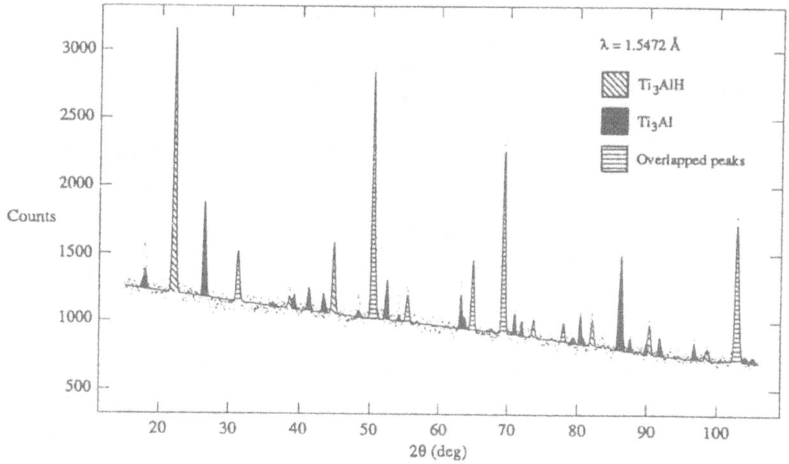

Figure 2. Refined Neutron data for hydrogenated Ti_3Al showing peaks from the parent phase and the hydride phase [4].

therefore are indistinguishable in the refinement. The bond lengths between the atoms, however, allow unambiguous identification. The hydride phase is found to be heavily faulted along the (111) planes, the direction in which metal layers alternate with hydrogen planes. The presence of the hydrogen presumably allows slip deformations between these planes, leading to the formation of this cubic phase from the hexagonal parent alloy.

2.A.1.c. Rare-earth transition metal compounds. $LaNi_5$ and related compounds are well known for their ability to absorb (and release) hydrogen at densities greater than is achieved in the liquid. There have been numerous studies of the location of the hydrogen in these compounds. Since many of these compounds are also magnetic, neutron diffraction has also been brought to bear on the changes in magnetism. The results of a study on $LaNi_{4.5}Al_{0.5}D_{4.5}$ [6] are shown in Tables I and II and Figure 3. The refinement was originally carried out in two different space groups, P6/mmm, the space group of the parent phase, and P31m, a lower symmetry group proposed in the literature. Of the five possible D sites in the P31m group, only two were found to be significantly occupied, and these are quite close to positions of the P6/mmm structure. Likewise, additional Ni/Al positions were refined to be nearly vacant, and the higher symmetry model gave agreement factors only slightly worse than the fully refined lower symmetry model. However, these crystallographic agreement factors (R) were found to be quite shallow, as a function of atom position, which may account for the dispute over the correct space group assignment.

Table I. Refinement results for $LaNi_{4.5}Al_{0.5}D_{4.5}$ at 298 K [6]

atom	P6/mmm site	positional parameters	(a)	(b)
La	1a	x,y,z	0.0	0.0
Ni	2c	z	0.0	0.980(9)
Ni	3g	x	0.5	0.485(5)
		z	0.5	0.495(5)
Ni	6l	x	0.290	0.270*
		y	0.580	0.540*
		z	0.0	0.02*
Ni	2e	z	0.310	0.290*
Al	3g	x	0.5	0.485(5)
		z	0.5	0.495(5)
D	6m	x	0.139	0.155(20)
		y	0.861	0.836(18)
		z	0.5	0.508(16)
D	6i	x	0.5	0.490(9)
		z	0.120	0.090(7)
R-factor (profile, weighted)			4.11	4.53

Refined composition
P6/mmm — $La_{0.91}Ni_{4.50}Al_{0.50}D_{4.43}$
P31m — $La_{0.90}Ni_{4.50}Al_{0.50}D_{4.72}$

(a) P6/mmm symmetry with two deuterium sites.
(b) P31m symmetry with two deuterium sites.
* Occupation of these sites too small to obtain meaningful standard deviations.

Table II. Dependence of R factors on atom coordinates in $LaNi_{4.5}Al_{0.5}D_{4.5}$ at 298 K

Refinement No.	1	2	3	4	5	6	7*	8
z (2c Ni site)	0.910	0.925	0.940	0.955	0.970	0.985	0.000	0.015
x (3g Ni-Al site)	0.470	0.475	0.480	0.485	0.490	0.495	0.500	0.505
z (3g Ni-Al site)	0.4550	0.4625	0.4700	0.4775	0.4850	0.4925	0.5000	0.5075
z (6m D site)	0.590	0.575	0.560	0.545	0.530	0.515	0.500	0.485
x (6i D site)	0.4550	0.4625	0.4700	0.4775	0.4850	0.4925	0.5000	0.5075
R-factor**(298K)	12.56	9.80	7.82	5.97	5.12	5.20	5.19	9.97
R-factor**(77K)	13.15	10.04	7.27	6.30	5.28	5.20	5.31	5.96

*These values correspond to P6/mmm symmetry.
**These are the weighted profile R-factors expressed in percent.

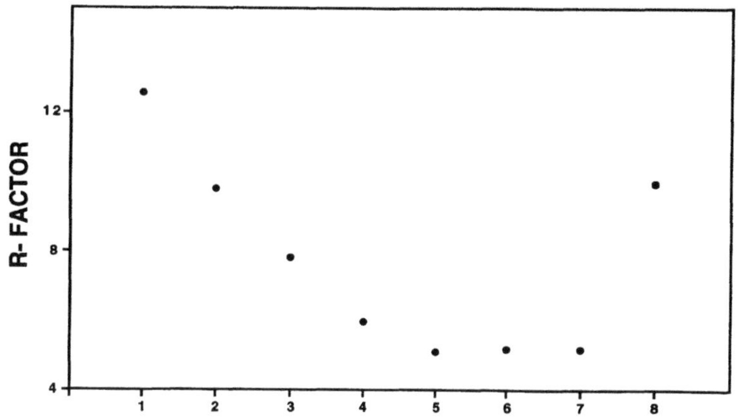

Figure 3. Crystallographic agreement (R) factor for $LaNi_{4.5}Al_{0.5}D_{4.5}$ as a function of the position parameters shown in Table II [6].

2.A.2. *Nitrides and Carbides*. The recent discovery of the enhancement in Curie point of numerous R-Fe compounds, through the incorporation of N or C into the lattice, has led to a serious interest in interstitial compounds of this type. Some controversy continues over the location of these interstitial atoms in some cases.

2.A.2.a. $\underline{R_2Fe_{17}N_x}$. The R_2Fe_{17} compounds are found to form in two structure types, rhombohedral and hexagonal. The former structure (Figure 4) is found to be stoichiometric, with a perfectly regular stacking of the Fe-Fe dumbbells, which replace the rare-earth in the $CaCu_5$ structure according to the formula

$$3(RFe_5) - R + 2Fe = R_2Fe_{17} ,$$

while in the latter there can be considerable disorder in the stacking, with some R atoms on the ideal Fe dumbbell sites and vice-versa. This can lead to a considerable stoichiometry range. The disorder significantly complicates any neutron diffraction analysis of the interstitial site occupancy. For the rhombohedral form, (space group R-3m) however, it is consistently reported that the N (or C) atoms occupy the 9e sites, leading to a maximum x = 3 [7-9]. While there have been several reports of 18h occupancy, this site is thought to be too small to accommodate a N atom, although it is known to be occupied in the hydride [10]. Studies which report occupancy of this site do not report a lattice expansion commensurate with the lattice strain expected to result from its filling. Thus, a consensus appears to reject this result. It is important to note that the incorporation of either C or N causes the sample to

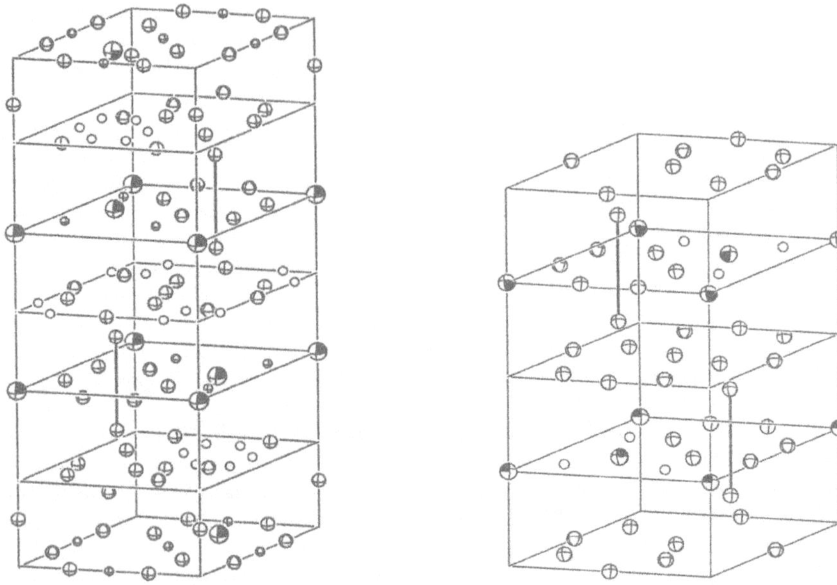

Figure 4. Structure of the rhombohedral and hexagonal R_2Fe_{17} phases showing the possible interstitial sites. The largest symbols represent the R atoms, the smallest symbols are the interstitial atoms, while the remainder are Fe atoms. For the hydrides, both the octahedral and tetrahedral interstitial sites are (partially) occupied. For the nitrides and carbides only the octahedral site, which is seen in the basal layer on the left hand figure, is occupied.

shatter and the Bragg peaks are broadened (Figure 5). This may be the result of particle size and strain effects, as well as compositional inhomogeneities, due to incomplete interstitial incorporation. The broadening may compromise the quality of the neutron refinements, which may be the cause of the discrepancies in the reported 18h occupancy. In the hexagonal R_2Fe_{17} compounds, it is clear that N occupies the equivalent site to the 9e site in the rhombohedral phase. For that case, however, one must refine the parent material, in order to model the disorder, and then constrain the metal occupancies in the nitride to those refined for the parent. Fully nitrided compounds are expanded by 6-7%, while carbided compounds have somewhat smaller expansions.

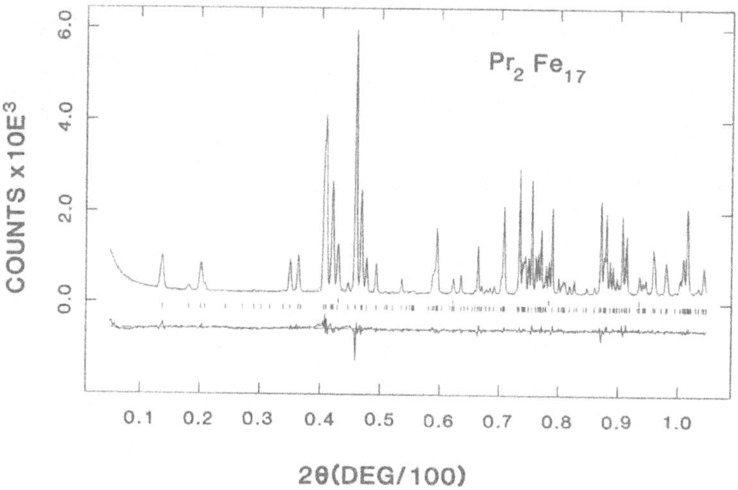

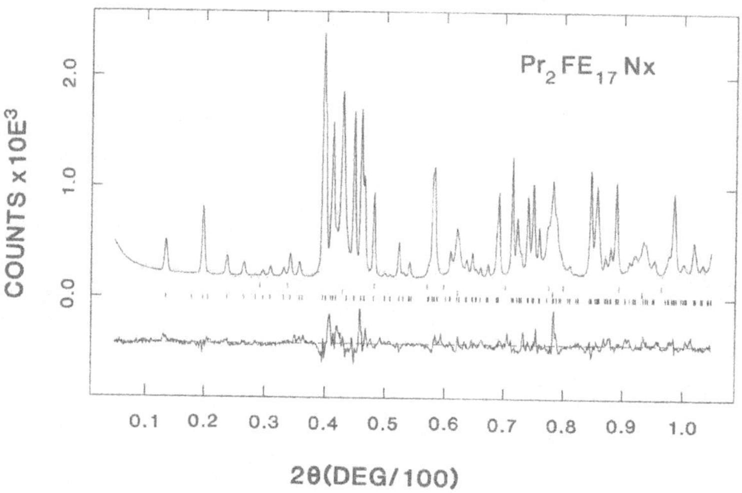

Figure 5. Neutron diffraction diagrams for Pr_2Fe_{17} and its nitride. The shift of peaks to lower angle, due to lattice expansion, and peak broadening are both apparent in the nitride diagram. A prominent α-Fe peak is visible in the nitride.

2.A.2.b. $\underline{RT_{12}}$. The RT_{12} phase forms with up to about 11 Fe atoms per formula unit; the remainder consisting of Mo, Ti, Al, V or other elements (Figure 6). In these compounds, interstitial N has been found to substantially increase the Curie temperature, as well as to change the magnetic anisotropy, from basal to axial, for R = Nd. Carbon can also be incorporated into the lattice. Either species is found at the edge center (0,0,1/2), and the lattice expansion and Curie temperature are proportional to the interstitial population (Table III) [11]. The substitutional species are found at different Fe sites, depending upon the particular substituent, but this appears to have little, or no effect on the interstitial incorporation.

Table III. Refinement Results for $RFe_{12-x}T_xN_y$

		$YFe_{12-x}V_xN_y$	$NdFe_{12-x}Mo_x$ Sample 1	$NdFe_{12-x}Mo_xN_y$ Sample 2	$NdFe_{12-x}Mo_xN_y$ Sample 1 + N
RE(a)					
(x=y=z=0, n=1)					
Fe(j)	n	0.87(2)	1	1	1
(y=1/2, x=0)	x	0.2728(8)	0.2765(8)	0.2743(5)	0.2766(4)
Fe(i)	n	1	0.43(5)	0.57(4)	0.47(5)
(y=0, z=0)	x	0.3632(7)	0.3590(3)	0.3607(4)	0.3604(4)
Fe(f)	n	1	1	1	1
(x=y=z=1/4)					
N(b)	n	0.44(3)	0	0.48(3)	0.834(3)
(x=y=0, z=1/2)					
a (Å)		8.5436(22)	8.6155(5)	8.6593(5)	8.6878(6)
c (Å)		4.7834(13)	4.8188(3)	4.8295(4)	4.8835(4)
Vol. (Å3)		349.16(26)	357.68(5)	362.13(4)	368.60(7)

n is the fractional site occupancy; values in parentheses indicate statistical uncertainties in least significant digits. If no estimate is given, parameter was not refined. Broad peaks for the V containing sample give an unrealistically large Fe fraction on the j site.

2.B. AMORPHOUS HYDRIDES

Studies of crystalline hydrides have, for the most part, used the deuterides in order to eliminate the large incoherent background which would be observed with H. Studies of amorphous hydrides make further use of the difference between these two isotopes, since each will give a different contribution to the coherent scattering. An example is an early study of amorphous Si and its hydrides [11]. The large angle scattering for these three samples is shown in Figure 7. The scattering intensity from a binary system is given by

$$I(k) = b_\alpha^2 C_\alpha^2 S_{\alpha\alpha}(k) + Q b_\alpha b_\beta C_\alpha C_\beta S_{\alpha\beta}(k) + b_\beta^2 C_\beta^2 S_{\beta\beta}(k)$$

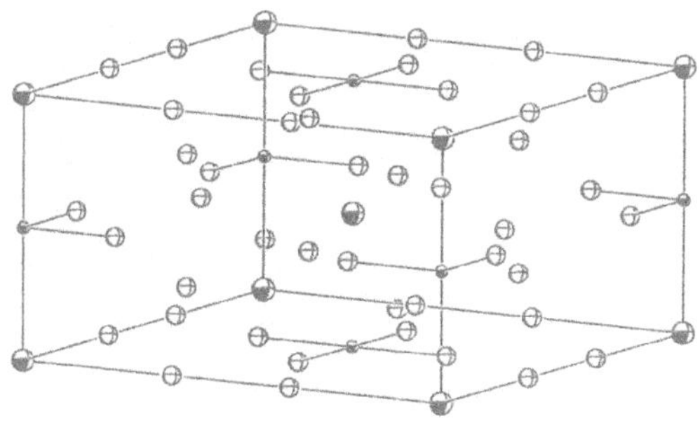

Figure 6. Structure of tetragonal $RFe_{12-x}T_xN_y$. The symbols are the same as in Figure 4.

where α and β refer to the species in the system, b_α and b_β are their respective scattering lengths, C is the concentration and S is the structure factor. If the concentration, C, is small, then the third term may be neglected. The Si-H and Si-D partial structure factors $S_{\alpha\beta}$ can be obtained by taking the differences in the coherent structure factors of α-Si and the hydrogenated and deuterated samples (Figure 8). Aside from the lowest angles where the background subtraction is problematical, the three curves lie on top of each other, indicating that the difference is meaningful, and that the Si structure is not modified by the incorporation of hydrogen. Although the q range of the data is too limited to allow the pair distribution function of the hydrogen to be calculated directly, the data can be compared with computer simulation. This model shows that about half of the hydrogen is incorporated in the bulk, while the remainder is on surfaces.

The Si-H system is one in which the metal structure is not altered by the hydrogen. If this was not the case, then it would be possible to examine the hydrogen partial structure factor as determined above, and to directly measure the metal structure factor by using a mixture of hydrogen and deuterium at the concentration which gives no coherent scattering (64% H - 36% D). This may then be compared with the partial structure factor obtained by subtracting the H partial structure factor from the total. For nitrides, a similar scheme may be used, although there is no null mixture with the two isotopes available. For carbon the difference in scattering length for the two isotopes is too small to be useful, but it is also possible, in many cases to make use of the metal isotopes for the same result.

2.C. CHANGES ACCOMPANYING INTERSTITIAL INSERTION

The examples, thus far, have focussed on the location of the interstitial atoms. The host lattice is, of course, affected in various ways by the incorporation of the atoms. X-ray diffraction is capable of measuring the lattice expansion and changes in symmetry, but cannot, in general, determine the structural details when single crystals are unavailable, as is usually the case. Thus, Rietveld refinement of neutron data has become the main tool for these studies.

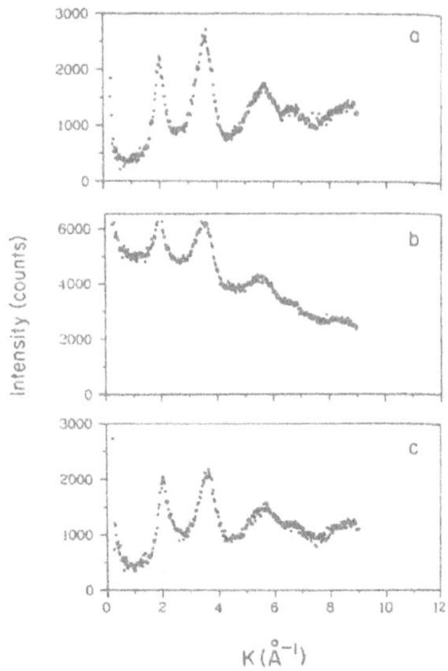

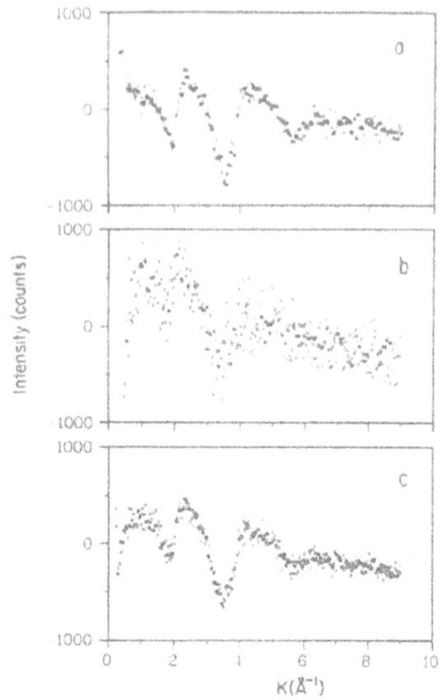

Figure 7. Neutron scattering intensities for a) pure α-Si, b) hydrogenated α-Si, and c) deuterated α-Si, as a function of momentum transfer K [12].

Figure 8. Differences of the coherent structure factors of the three samples shown in Figure 7: a) SiD-Si, b) SiD-SiH, c) Si-SiH. Apart from noise, the three difference are identical indicating that the Si-Si distribution function is the same for all samples [12].

2.C.1. *Lattice Distortions*. In a metal such as Nb, incorporation of an interstitial atom may change the crystal symmetry and expand the lattice, and, thus, the bonding between metal atoms is likely to change in a uniform way. For more complex systems, such as the R_2Fe_{17} structure, the changes may be more complex. This is especially true when the bonding of the interstitial species shows a preference for a particular type of neighbor, or when other interactions, such as the magnetic exchange, are altered by the lattice expansion. Table IV shows the change in average Fe-metal bond lengths at each Fe site in $Pr_2Fe_{17}N_{2.6}$ [7]. These range from a maximum expansion of 5.1% to a contraction of -0.3%. Similar changes are seen in the Nd compound [8]. Obviously, any attempt to model this compound, using the x-ray determined changes in lattice parameter, and assuming uniform shifts (i.e. that the positional coordinates are unchanged), is likely to lead to incorrect conclusions. A careful study of the temperature dependent magnetostriction would help to determine if these changes are due to magnetic interactions or to interstitial atom bonding effects which modify, in turn, the bonding of the nearby metal atoms. To date, such studies have not been carried out.

Table IV. Lattice Parameters and Average Bond Lenths in Pr_2Fe_{17} and $Pr_2Fe_{17}N_{2.6}$. Also given are the relative changes in $Nd_2Fe_{17}N_x$ [8].

Parameter			Pr_2Fe_{17}	$Pr_2Fe_{17}N_{2.6}$	% change	% change Nd
Lattice parameters	a, Å		8.621(4)	8.8223(5)	2.3	2.4
	c, Å		12.518(6)	12.7079(9)	1.5	1.8
	c/a		1.452	1.440	-0.8	-0.6
	V, Å3		805.0(1)	856.6(1)	6.4	6.8
Average bond distance, Å	Fe, 6c to	Fe, 6c	2.412	2.458	1.9	0.9
		Fe, 9d	2.640	2.698	2.2	2.6
		Fe, 18f	2.754	2.774	0.7	0.1
		Fe, 18h	2.652	2.669	0.6	0.6
		Pr, 6c	3.111	3.103	-0.3	2.7
	Fe, 9d to	Fe, 18f	2.450	2.487	1.5	1.6
		Fe, 18h	2.474	2.505	1.2	1.1
		Pr, 6c	3.342	3.375	1.0	2.5
	Fe, 18f to	Fe, 18f	2.475	2.487	0.5	-0.1
		Fe, 18h	2.620	2.681	2.3	2.7
		Pr, 6c	3.096	3.193	3.1	3.7
		N, 9e	—	1.924	—	—
	Fe, 18h to	Fe, 18h	2.540	2.669	5.1	5.6
		Pr, 6c	3.201	3.284	2.6	2.5
		N, 9e	—	1.946	–	—

2.C.2. *Phase Transformations*. As was the case for Nb, many systems undergo phase transformations as the interstitial atoms are incorporated. The lattice type may be defined by the metal atoms, which can be observed with x-rays, but the space group may be determined only by taking into account the position and occupancy of the interstitial atoms. While entropy calculations suggested that $LaNi_{4.5}Al_{0.5}$ would transform to a lower symmetry space group upon hydrogen absorption, the neutron data, discussed above, failed to support such a conclusion. In fact, many of the rare-earth transition metal compounds do not change symmetry, but merely expand. There are counter examples, however. The R_6T_{23} phase can absorb about 22 hydrogen atoms per formula unit. At room temperature the structure remains in the fcc structure, with 4 distinct sites populated by D. At low temperature, however, the material undergoes a transformation to a primitive tetragonal symmetry, with half the volume (+2.2%) of the cubic cell [13]. The refinement results for this compound are shown in Table V. There are twelve populated D sites. Some of these are fully occupied, while others show only partial occupancy. The refinement is based on a random distribution of interstitial atoms, but it is likely that the actual distribution is non-random, in the sense that a D atom on a specific location may prevent others from populating nearby sites. This complication also affects the bonding, but only the average structure is refined.

Table V. Atomic Parameters of $Ho_6Mn_{23}D_{23}$ at 9 K in the P4/mmm Structure, N is the site occupancy. Sites are labeled according to the International Tables for X-ray Crystallography [13].

fcc sites	Tetragonal site	*N*	Refined coordinates *x*	*y*	*z*	Moment (μ_B/atom)	θ* (degrees)
Ho e	Ho g	2	0.00	0.00	0.0201	3.4	51.4
	h	2	0.5	0.5	0.304	3.4	128.6
	j	4	0.222	0.222	0.0	3.5	100.1
	k	4	0.299	0.299	0.5	3.5	79.9
Mn b	Mn b	1	0.0	0.0	0.5	3.5	29.5
	c	1	0.5	0.5	0.0	3.5	150.5
Mn d	Mn e	2	0.0	0.5	0.05	2.3	67.7
	f	2	0.5	0.0	0.0	2.3	112.3
	r_1	8	0.246	0.246	0.762	0.0	—
Mn f_1	Mn s_1	8	0.329	0.0	0.161	2.8	117.5
	t1	8	0.103	0.5	0.312	2.8	62.5
Mn f_2	Mn s_2	8	0.256	0.0	0.399	1.5	30.3
	t_2	8	0.266	0.5	0.150	1.5	149.7
D a	D a	1	0.0	0.0	0.0		
	d	1	0.5	0.5	0.5		
D f3	D s_3	3.2	0.152	0.0	0.059		
	t_3	8	0.270	0.5	0.387		
D j	D p	3.2	0.191	0.449	0.0		
	q	2.6	0.199	0.995	0.5		
	r_2	8	0.144	0.144	0.326		
D j	D r_3	0	—	—	—		
	r_4	4	0.335	0.335	0.059		
	r_5	0	—	—	—		
D k	D s_4	1.4	0.238	0.0	0.117		
	t_4	1.4	0.192	0.5	0.458		
	u_1	0.4	0.188	0.134	0.162		
	u_2	8	0.277	0.380	0.322		

Cell parameters a = 8.981 Å, c = 12.638 Å $\chi = 1.9$
R-factor (weighted profile) = 5.71% R-factor (nuclear)= 5.73%
R-factor (expected) = 3.02% R-factor (magnetic)= 13.3%

*θ is the angle between the magnetic moment and the c axis.

An unusual situation is encountered for $Nd_2Fe_{17-x}Si_x$. As discussed earlier, the parent phase, Nd_2Fe_{17}, does not change symmetry with interstitial atom absorption. When C is incorporated into this compound from the gas phase, however, the structure depends upon the subsequent annealing. If the sample is annealed below about 700°C, it maintains the rhombohedral structure and the C is found in the 9e site, as reported for N or C in Nd_2Fe_{17}. The uncarbided material is stable, up to its melting point, but when the carbided material is heated above 700°C, it transforms to the $BaCd_{11}$ structure plus other phases [14]. The same result was found by LeRoy, who attempted to form the $BaCd_{11}$ structure from the same constituents [15]. Without C, he obtained only the R_2Fe_{17} structure. Small additions of C to the melt led to the formation of the $BaCd_{11}$ phase. Apparently, the presence of C stabilizes the phase, which does not like to form with high Fe concentrations [16]. However, it is appropriate to consider this as an interstitial phase, since it can form without the C atom (with other elements), and, in this case, can form with a variable C content. The direct synthesis produced samples with roughly 25% occupancy of the C site, while the solid state transformation from the R_2Fe_{17} precursor resulted in nearly stoichiometric occupancy of the C site, for a final composition of $NdFe_{10}SiC_{1.94}$. Not surprisingly, the lattice parameters of this sample were increased compared to the sample prepared by direct synthesis.

2.C.3. *Magnetic Effects*. Interstitial atoms may significantly alter the interactions in metallic alloys through a number of different routes. Electron transfer, especially for the hydrides, may significantly change the band structure and induce or suppress magnetic ordering. For example, Y_6Mn_{23} is a ferrimagnet with a Curie temperature of 486 K, while its hydride is a Pauli paramagnet. Conversely, Th_6Mn_{23}, which is an isostructural compound, is a paramagnet at room temperature, which becomes ferrimagnetic with H insertion. Changes in the lattice parameters may significantly alter the magnetic exchange interactions, as is predicted from the Bethe-Slater curves. Changes in the c/a ratio may also affect the single ion anisotropy and lead to changes in the easy axis of magnetization. Many of the compounds of interest are metastable, decomposing at relatively low temperature to RN_x and α-Fe. Because of the large scattering amplitude of Fe and its simple crystal structure, leading to a few intense reflections, this decomposition is seen with neutrons at a much earlier stage than with x-rays, and *in situ* studies of the nitrogenation have been be carried out in order to optimize the process parameters [17].

Lattice expansion effects have been seen in a variety of compounds. The $NdFe_{10}SiC_x$ system is a good example [14-15]. With x =0.5 this compound has a Curie point of 410 K. When the C site is fully occupied, T_c increases to 505 K. Because the magnetic interactions are fairly long range, and every Fe atom has numerous Fe neighbors, the phase transition is sharp, despite the fact that the exchange must vary between cells in which the C site is occupied and those which are vacant. In contrast, the introduction of H into $ErFe_2$ is thought to create a random anisotropy, due to the incomplete and random filling of the sites in the available tetrahedral sites [18]. The Er sublattice magnetization is significantly reduced (Figure 9) as H is incorporated, presumably due to the creation of a "fan" like wandering of the easy axis. The net Er moment, measured by neutron scattering, is, thus, strongly reduced. At the same time, the Fe moments remain largely unchanged, since they are not substantially affected by the random rare earth anisotropy. In this compound, however, the Fe-Fe exchange is decreased, leading to a decrease in the Curie temperature.

An important anisotropy change is observed in $NdFe_{12-x}T_x$ upon nitrogenation. These compounds have Curie points above 500 K, making them potentially useful for permanent magnet applications, but the Nd compound has basal anisotropy. Incorporation of N or C

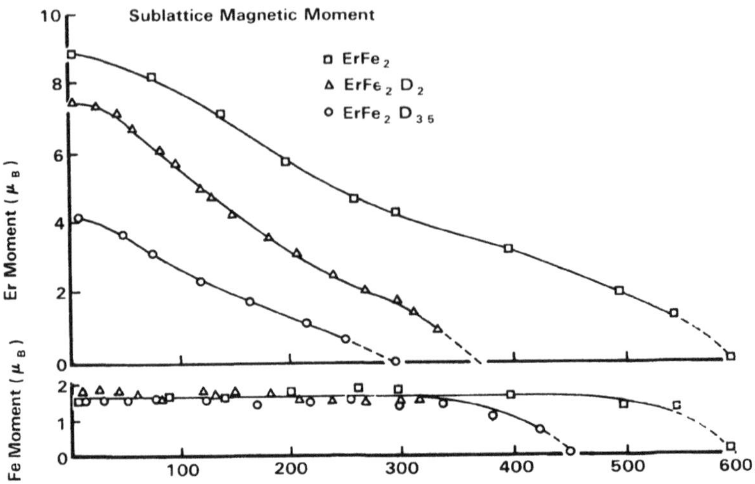

Figure 9. Site magnetizations of Er and Fe in the compounds $ErFe_2D_2$ and $ErFe_2D_{3.5}$. The reduction of the Er moment compared to $ErFe_2$ is suggested to arise from randomization of the crystal field [18].

not only increases the Curie point to over 700 K, but switches the anisotropy from basal to axial. This change is easily seen with neutron diffraction. These materials have not received as much attention as $Sm_2Fe_{17}N_x$, due to their apparently low remanences. The neutron data suggest that the remanences are significantly higher than the reported bulk data, perhaps due to problems of alignment of the nitrided powders, and these compounds may be of considerable technological interest. Because of the unique ability of the neutron to determine the magnitude and direction (within the limitations of polycrystalline averaging) of the site specific magnetic moments, it is truly an indispensable tool in the study of magnetic changes induced by interstitial atom incorporation. The interpretation of these changes, however, is still a challenge for theory. Even when the changes can largely be inferred from bulk measurements, the details of bonding are crucial data for any band theory calculations, the results of which must be compared to both the macroscopic and microscopic measurements of moment and easy axis.

2.C.4. *Induced Amorphization.* Hydrogen induced amorphization is another subject of interest in this volume. Neutron scattering can play a significant role in following this process. A recent study was carried out at the University of Missouri. A small stainless steel cell was made to contain the sample and gas which would evolve in an uncontained system.

$ErFe_2D_{3.2}$ was studied as a function of temperature [19]. At approximately 90°C, a reversible order-disorder transition from the room temperature rhombohedral structure, associated with H ordering, to a cubic phase in which the "$ErFe_3$" interstitials are randomly occupied by the H atoms [20]. The transition agrees well with Differential Scanning Calorimetry (DSC) observations. A second DSC signal, observed near 200°C is not accompanied by any structural change and is probably the Curie point of this compound. Between 220 and 240°C the Bragg peaks disappear in an irreversible transition (Figure 10). Although the amorphization temperature is significantly lower than measured by DSC, the latter temperature may be shifted by a time delay in the scanning experiments caused by the activation energy of the process. It is believed that the amorphization is driven by short range clustering of the Fe and Er atoms through the creation of energetically more favorable sites for the H atoms. The stainless steel peaks, from the sample container, were excluded in the analysis of these data, but contributed to the background and the loss of statistical accuracy. It should be possible to carry out such studies in specially fabricated V cells, free from Bragg

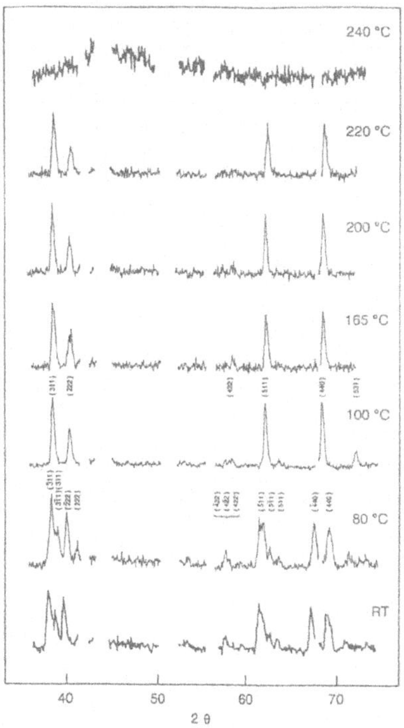

Figure 10. Neutron powder diagrams for $ErFe_2D_{3.2}$ as a function of temperature, showing the transformation from tetragonal to cubic at about 100°C and hydrogen induced amorphization at about 230°C [19].

peaks and to obtain greater detail about the interstitial atom locations. As previously discussed, a study with H and D (separately) would also provide the information about the hydrogen partial structure factor.

2.D. DEFECT STUDIES

Diffuse neutron scattering can be used to extract information about the location of interstitial atoms, even at low concentrations, as well as to provide information about the local symmetry of the strain field. Bauer has measured the diffuse scattering from $NbD_{0.0256}$ over an extended scattering vector range [21]. The data are then compared to models in which both the location of the D atoms, and the tetragonality of the sites are modelled. The data (Figure 11) are found to be consistent with tetrahedral site occupancy with a cubic strain field. While there is no obvious reason why the strain field should have this symmetry, it is suggested that the long range interactions play a critical role.

Compounds such as VC_x can also be considered as interstitial alloys, as the C content can be varied over a considerable range. Since V has a near zero coherent scattering length, scattering from this compound will be due almost entirely to the long and short range ordering of the C atoms. Diffuse neutron scattering studies of $VC_{0.75}$ [22] have been used to determine the Warren short range order parameters (Table VI) for comparison with theory and with earlier electron diffraction measurements [23]. These results, while in better agreement with theory than the electron diffraction data, still show some discrepancies. It was concluded that it might have been better to use Nb or Ta, rather than V, since the last contributes significant background, while the others contribute only to the Bragg intensities, which are ignored in any case.

Table VI. Warren Short Range Order Parameters for V_4C_3 from Neutron and Electron Diffraction Studies

α_n	**Neutron Determination** [22]	**Electron Diffraction** [23]
α_1	-0.17 ± .005	-0.17
α_2	-0.20 ± .01	-0.30
α_3	+0.15	+0.18
α_4	-0.03	+0.01
α_5	+0.04	+0.06
α_6	-0.08	-0.11
α_7	-0.05	-0.10
α_8	+0.05	+0.18

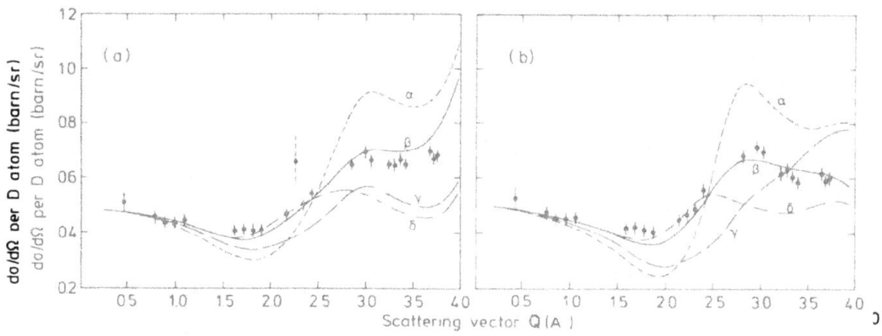

Figure 11. Diffuse elastic neutron scattering cross sections for D in Nb for two directions in reciprocal space, measured in an $NbD_{0.0256}$ single crystal at 300K. The symbols are the experimental data; the curves are calculated for different defect bonding models. The best agreement is observed for a cubic force field [21].

3. Small Angle Scattering

Small angle neutron scattering (SANS) can be used to investigate any type of inhomogeneity which leads to a variation in scattering length density on a length scale from about 2 to 500 nm. For x-ray scattering the contrast change produced by H is usually negligible, but the substantial scattering amplitudes of H or D leads to significant contrast change with neutrons. In many cases the change in coherent scattering amplitude would favor the use of H, but the large incoherent scattering of H leads to high background and compromises the subtraction of reference and sample data. Thus, D is most often used. If the interstitial species is uniformly distributed through the bulk, no new signal will be seen at small angles. However, if the atoms concentrate at defects, grain boundaries, or voids, changes in signal will be seen relative to an uncharged sample. Likewise, if incorporation of the interstitial atom leads to decomposition and phase separation, intensity will be observed, provided there is a component in the size range probed.

3.A. DEFECTS AND GRAIN BOUNDARIES

As has been the case so far, systems such as Pd and Nb have received attention due to their relative simplicity as models for more general behaviors. Studies carried out on deformed single crystals and polycrystalline rods compared the scattering with and without dissolved D in Pd [24]. The scattering from deuterium trapped at dislocations and at grain boundaries is distinguished by the Q ranges (where $Q = 4\pi\sin\theta/\lambda$); the D at grain boundaries contributes to the scattering only at low Q (< 0.025 $Å^{-1}$), and thus the high angle region may be used to determine the dislocation effects (Figure 12). In samples with of the order of 6000 ppm D, it is found that about 2 deuterons per Å are trapped along the dislocations, which, in this case,

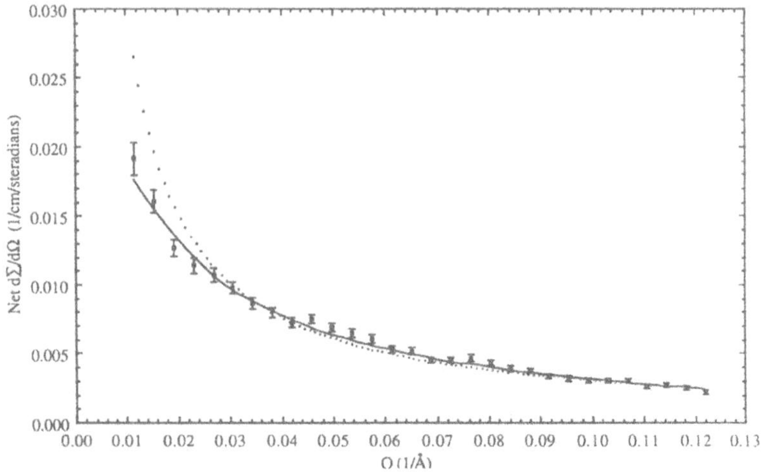

Figure 12. Net small angle scattering signal for Pd-D, for a single crystal cycled through the $\alpha/\beta/\alpha$ transition. Data are fitted to a model with the D atoms localized along dislocation lines about 100 Å long [24].

could be modeled as rods 50-100 Å long. The intensity of the scattering was roughly proportional to the fraction of the material which transformed in the $\alpha/\beta/\alpha$ cycle. In samples which were annealed to reduce the dislocation density, the low angle scattering revealed excess scattering due to about 0.4 deuterons per $Å^2$ at grain boundaries. The trapping of "interstitial" species in these locations should be a reminder that knowledge of the total H or D content may not adequately describe the state of the sample. These trapped atoms must be considered as excess; not incorporated into the crystal lattice under inspection.

3.B. VOID SCATTERING

The case of amorphous Si, previously considered [12], is another example of excess H located outside the bulk structure. The low angle scattering is modeled by isolated isotropic "particles" of average radius of gyration, R_G, leading to a scattered intensity given by

$$I = I_o \exp\left(\frac{-k^2 R_G^2}{3}\right) \cdot$$

For the samples considered, the hydrogenated material yielded R_G = 270 Å. Unexpectedly, the contrast was larger for the hydrogenated samples than for the deuterated samples. This suggests that the scattering is strongly affected by the H on the surface, which gives great contrast to the bulk, rather than by the contrast between the surface of the void and the interior of the void. Simulations of this situation suggest that this is, in fact, the case for voids smaller than about 800 Å diameter, as was determined here. The combination of absolute intensity measurements for both the SANS measurements and the large angle scattering leads to the results that about 50% of the "interstitial" atoms in this system are located within the voids, and do not contribute to the amorphous scattering pattern.

3.C. PHASE SEPARATION

Phase transitions related to concentration fluctuations of the "lattice gas" of H atoms in metals can be studied with neutrons. Munzing [25] measured the critical scattering in $NbD_{0.31}$ single crystals as a function of temperature. The scattering is expected to show a Curie-Weiss behavior,

$$\frac{d\sigma}{d\Omega} = b_D^2 \frac{T}{T - T_s} \eta(Q) \,,$$

where $\eta(\mathbf{Q})$ is related to the lattice strains caused by the compositional fluctuations. Only after the quasielastic-elastic scattering was separated from the inelastic scattering was the expected behavior observed (Figure 13). Because the spinodal decomposition is interrupted by incoherent phase separation, above T_s, the data must be extrapolated to the pure spinodal temperature. The spinodal temperature was found to be direction dependent, and well below the incoherent phase separation, indicating the strong influence of lattice strains on coherent compositional fluctuations.

4. Quasielastic Scattering

An excellent review of quasielastic neutron scattering can be found in the book by Bee [26]. The focus of the book is on orientational transitions but hydrogen in metals is briefly considered and the reader can also get more information about theory and experimental details than can be given here. Inelastic and quasielastic scattering from hydrogen in metals has been reviewed a number of times. A recent reference by Springer and Richter [27] can be found in *Methods in Experimental Physics*. The majority of the studies concern hydrogen, rather than D, due to the large incoherent scattering used to probe the single particle motions.

The earliest study of hydrogen diffusion in metals was performed on a powder sample of α-PdH_x [28]. The full width at half maximum of the quasielastic scattering was measured as a function of temperature and Q and compared with the model of Chudley and Elliot [29], based on diffusion between only one type of site. The results (Figure 14) are in good agreement with theory and show that the proton diffuses between near neighbor octahedral sites. This experiment was repeated using a single crystal sample [30]. While the earlier experiment was subject to some interpretation, due to the averaging over different crystallographic directions in the powder, the latter study unambiguously shows that the system is described by octahedral-octahedral jumps. At high temperature, the data are less well described by the simple model, which was subsequently modified to include the possibility of double jumps. This model is justified by the assumption that the interstitial creates significant lattice strains which do not completely relax after the hydrogen has left the interstitial site. This lowers the barrier to the subsequent jump, which occurs on a very short

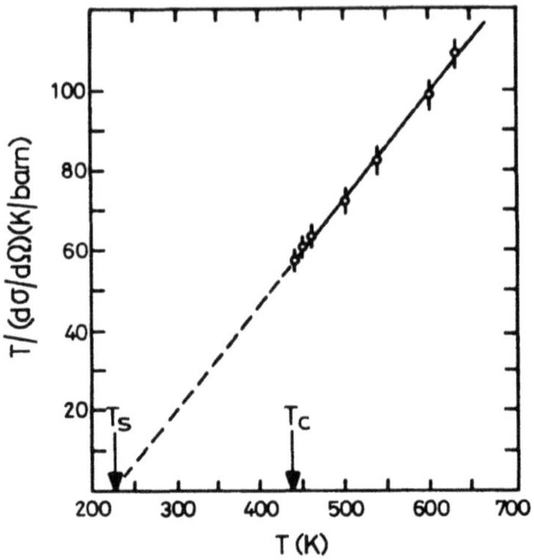

Figure 13. The inverse elastic and quasielastic scattering intensity multiplied by the temperature T versus T above the incoherent decomposition temperature T_c for $NbD_{0.31}$. The extrapolated curve shows a coherent spinodal temperature well below T_c [25].

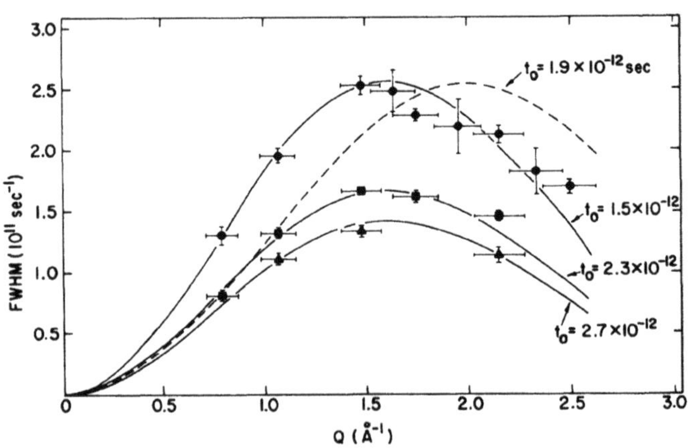

Figure 14. Full width at half maximum of the quasielastic neutron intensity in PdH_x in the α-phase at three temperatures (582K, 630K, 704K) [28]. The data are well described by the Chudley-Elliot model [29] for octahedral-octahedral jumps with residence times indicated.

time scale. This process becomes easier as the temperature is raised, accounting for the deviations from the earlier model.

The diffusion of hydrogen may be significantly modified by the presence of other trapped impurities such as N or O. At small Q a single Lorentzian linewidth is observed in the scattering, corresponding to the self-diffusion constant, modified by the trapping

$$\Gamma = Q^2D = Q^2D_f\,\tau_1/(\tau_0 + \tau_1)\ .$$

At large Q, however, two Lorentzian components appear. One has a width of approximately $1/\tau_1$, the time the hydrogen atom spends at the trap site, while the other is approximately that of the freely diffusing atom. This can be further modified by the presence of multiple trap sites.

The single trap site situation was investigated in NbH_yN_c, as a function of concentration and temperature [31]. The effects of the additional traps are clearly seen (Figure 15).

The asymptotic quasielastic linewidth $\Gamma = DQ^2$ has often been used to determine the self-diffusion constant, although great care must be taken to assure that all components of the quasielastic scattering are observed. In $LaNi_5H_x$ an interesting effect is seen. As x increases above 5 the diffusion constant increases, contrary to the expectations of blocking [32]. This is attributed to the presence of different sites in the lattice. The deeper sites fill first, and then shallower, more mobile sites fill. H atoms on these shallow sites are, then, able to diffuse more rapidly (Figure 16).

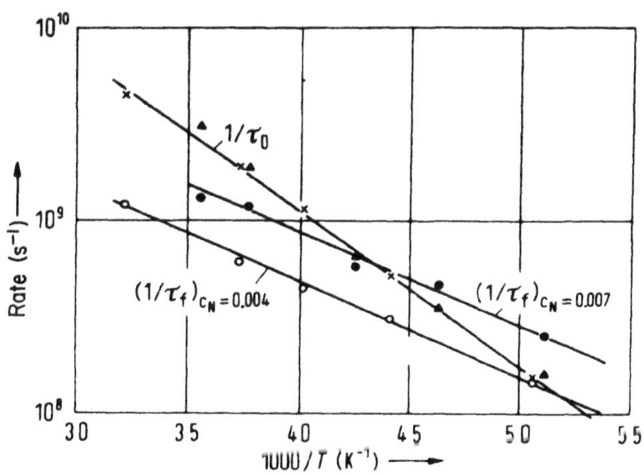

Figure 15. Mean escape rate $1/\tau_0$ from the traps and mean trapping rate $1/\tau_0$ as a function of temperature for hydrogen diffusion in $NbN_cH_{0.004}$ [31].

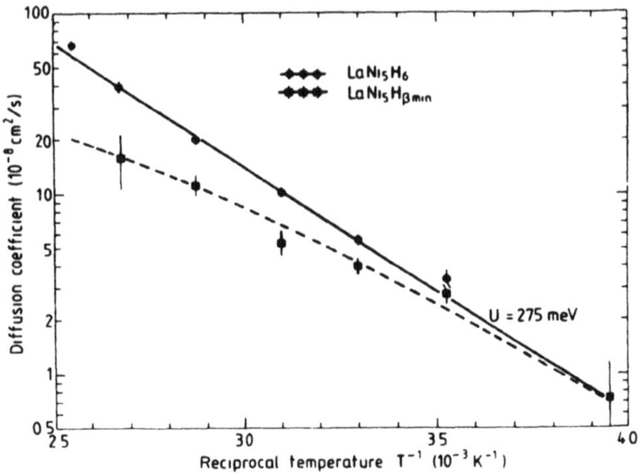

Figure 16. Self-diffusion constant for hydrogen in $LaNi_5H_x$ from quasielastic neutron scattering, for two hydrogen concentrations. At higher concentration the diffusion rate increases [32].

5. Inelastic Scattering

In a large percentage of the cases discussed above, measurements could be performed on polycrystalline samples, accepting the limitations imposed by the averaging over different directions in reciprocal space. Only in certain, fairly simple cases, such as hydrides of Nb, Ta, etc. are single crystals of interstitial compounds available. This puts an unfortunate limitation of the use of inelastic scattering, for, in many cases of interest, it is essential to have detailed knowledge of **Q**. Even when single crystals of the parent phase can be prepared, the strong lattice expansion, and interstitially induced phase transitions will reduce the specimen to powder. Since many of the interstitial alloys of interest are also thermodynamically unstable, attempts to grow single crystals from the melt are unlikely to be successful. Thus, the contribution of inelastic scattering has been more limited than its potential. Nevertheless, when single crystals are available, or when the excitations of interest are nearly dispersionless, inelastic neutron scattering can be applied.

5.A. PHONON DISPERSION

Unlike the previous examples, phonon studies of hydrogen in metal have used the coherent scattering of D, which is comparable to that of the metal atoms in the lattice, and for which the background is low. The phonon dispersion is a measure of the force constants between the various atoms in the crystal. If the D-D force constants are negligible, then the optic branches will be essentially flat, and their energies will be determined by the metal-D force constants.

The acoustic branches will be modified, both by the M-D force constants and by the changes in M-M force constants due to the lattice expansion and possible band effects. The case of $PdD_{0.63}$ has been investigated by Rowe et al. (Figure 17) [33]. It is seen that the optic branches show significant dispersion, indicating that the D-D interactions can not be neglected, and that the acoustic branches are softened relative to the pure metal phonon spectrum. Apparently, the lattice expansion dominates the other effects and reduces the strength of the lattice coupling. This is consistent with the normally observed reduction in the transition temperature between low and high temperature phases observed in many interstitial systems. In addition to the phonon softening, the phonon groups are broadened, due to the random distribution of D atoms in the interstitial sites.

Changes in the phonon dispersion of $NbD_{0.85}$ show the disappearance or reduction in many of the anomalous features of the pure metal (Figure 18) [34,35]. These data were collected at high temperature to assure that the D atoms were in solid solution and distributed randomly in the tetrahedral sites. The effects of the deuterium are thought to arise from changes in the electronic structure.

5.B. VIBRATIONAL MODES

In most hydride systems the hydrogen isotopes can be regarded as independent three-dimensional oscillators in a deep well. The frequencies can be modeled using an

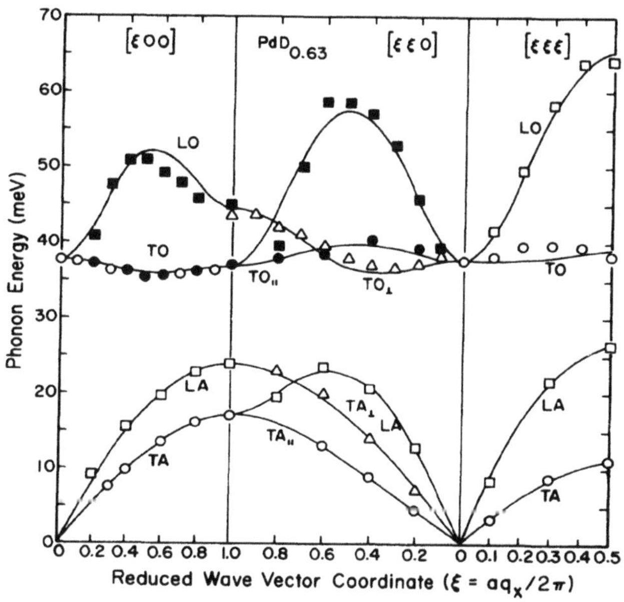

Figure 17. Phonon dispersion curves for $PdD_{0.63}$ at 150K [33].

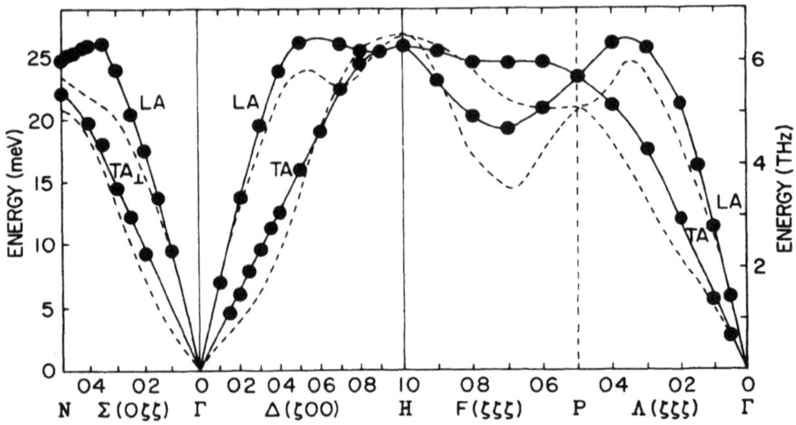

Figure 18. The accoustic branches of the phonon dispersion curves for $NbD_{0.85}$ at 160°C [34]. Also shown (dashed lines) are the dispersion curves for pure Nb [35].

effective single-particle potential. The shape of this potential can be extracted from the higher harmonics of the vibrational spectrum and from the isotope shifts. Three fundamental frequencies describe the independent oscillator in Nb and related metals. If the interstitial is in a tetrahedral site with tetragonal symmetry, two of the three frequencies are degenerate. Eckert [36] considered in detail the anharmonicity, which can be determined from the higher harmonics. This includes the anharmonicity of the z vibration and the anharmonic coupling between the z and x,y vibrations. In addition, deviations from tetragonal symmetry will lead to a lifting of the degeneracy, but the anharmonic shifts may still be determined. The isotope shifts are connected to both types of anharmonicity and the measured shifts can be used to confirm the parameters inferred from the higher harmonics. Until the advent of spallation sources, it was generally not possible to measure more than the fundamental frequency for hydrogen vibrations. With the higher energy neutrons now available, more comprehensive data can be obtained. Figure 19 shows the isotope shifts in NbH_x [37], while Figure 20 shows both the fundamentals and the first harmonic in $NbD_{0.85}$ [38]. Table VII, from the review by Springer and Richter [27], summarizes the parameters extracted from a number of studies on Nb and Ta hydrides. The reader is advised to refer to this work for details of the model and for full references.

The studies on the pure metals have been extended to systems with interstitial traps and substitutional impurities. In the former case, little change is seen compared to the pure metals, while in the latter there are changes both in frequency and in lineshape, indicating that the impurity acts as a strong trap and locally causes the symmetry to deviate from the tetragonal environment of the pure metal.

While considerable information can be extracted from the H isotopes, similar studies have not been performed with N or C interstitials. Their larger masses cause some of the assumptions leading to the independent oscillator model to break down, and the interpretation of results, in the complex crystal structures in which these interstitial atoms are found, is likely to be quite problematical.

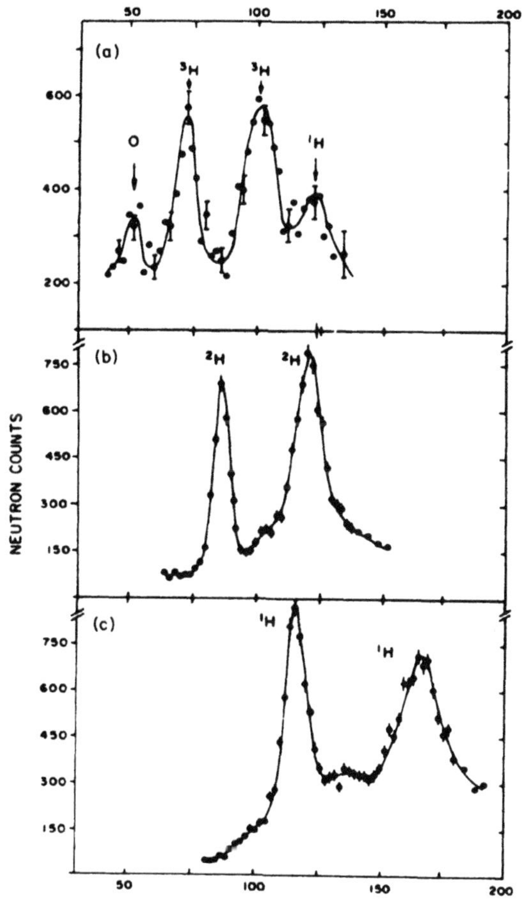

Figure 19. Inelastic scattered intensity for Nb hydrides: a) $NbT_{0.2}$, b) $NbD_{0.72}$, and c) $NbH_{0.32}$ at room temperature [37].

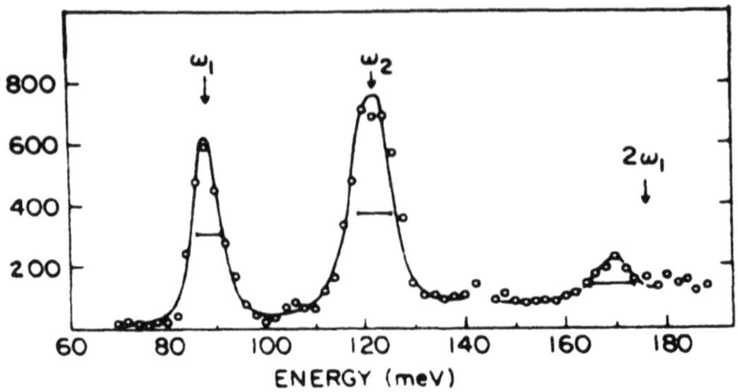

Figure 20. Inelastic scattered intensity for $NbD_{0.85}$ at 10K. The horizontal bars give the experimental resolution. The arrow at $2\omega_1$ shows the expected position of the second harmonic peak for an harmonic potential [38].

5.C. MAGNETIC EXCITATIONS

In principle, the magnon dispersion curves of interstitial compounds will give detailed information about the changes in magnetic coupling induced by the insertion. In addition to the unavailability of single crystals, already mentioned, there are several obstacles to such a study. First, most of the systems of interest have numerous magnetic atoms and the determination of the exchange constants is quite complex. Secondly, the incomplete filling of interstitial sites leads to a broadening of the excitation spectra due to the local variation of the exchange interactions. For these reasons such studies have not been performed.

One type of study, however, is possible with polycrystalline material. The crystal field levels of the rare earth ions may be investigated [39], especially in time-of-flight experiments, and the changes with interstitial concentration determined. These studies may help to shed some light on the enhancement effects recently observed for the nitrides and carbides of the binary rare-earth Fe compounds.

6. Conclusions

It can been seen, from the examples given, and from the extensive literature, that neutron scattering has and will continue to play a central role in our understanding of interstitial alloys. The scope of neutron scattering, already large for conventional alloy systems, is further expanded by the unique ability of the neutron to observe light atoms, and by the contrast variation available through the use of different isotopes, especially hydrogen and deuterium. The neutron serves both as a probe of the coherent processes (structure and lattice dynamics) and of the localized motions of interstitial species (diffusion and vibrational spectroscopies). No other probe can provide the breadth of capability of the neutron, and it is sensible to think about neutron characterization of any new system, if one is interested in developing a detailed understanding of the mechanisms responsible for the bulk properties of that system. Unfortunately, the relative dearth of neutron sources, because of permanent

24. B. J. Heuser, J.S. King, G.C. Summerfeld, F. Boue and J.E. Epperson, Acta Metall. Mater. **39**, 2815 (1991).
25. W. Munzing, Dr.rer.nat. Dissertation, Technische Universitate Munchen, Germany (1977) reproduced from G. Kostroz, *Treatise on Materials Science and Technology*, Vol. **15**, Academic Press, New York, (1979).
26. M. Bee, *Quasielastic Neutron Scattering*, Adam Hilger, Bristol and Philadelphia, (1988).
27. T. Springer and D. Richter, in *Methods of Experimental Physics*, eds. K. Sköld and D. L. Price, Vol. **23**, Part A, Academic Press,, San Diego (1987), pp. 131-186.
28. K. Skold and G. Nelin, J. Phys. Chem Solids **28**, 2369 (1967).
29. C.T. Chudley and R.J. Elliot, Proc. Phys. Soc. London,**77**, 353 (1961).
30. J.M. Rowe, J.J. Rush, L.A. de Graff, and G.A. Ferguson, Phys. Rev. Letters **29**, 1250 (1972).
31. D. Richter and T. Springer, Phys. Rev. **B18**, 126 (1978).
32. D. Richter, R. Hemplemann and L.A. Vinhas, J. Less-Common Met. **88**, 353 (1982).
33. J.M. Rowe, J.J. Rush, and H.E. Flowtow, Phys. Rev. **B9**, 5039 (1974).
34. S.M. Shapiro, Y. Noda, T.O. Brun, J.F. Miller, H.K. Birnbaum and T. Kajitani, Phys. Rev. Letters **41**, 1051 (1978).
35. Y. Nakagawa and A.D.B. Woods, Phys. Rev. Letters **11**, 271 (1963).
36. J. Eckert, J.A. Goldstone, D. Tonks, and D. Richter, Phys. Rev. **B27**, 1980 (1983).
37. J.J. Rush, A. Magerl, J.M. Rowe, J.M. Harris and J.L. Provo, Phys. Rev. **B24**, 4903 (1983).
38. D. Richter and S.M. Shapiro, Phys. Rev **B22**, 599 (1980).
39. See, for example, M. Loewenhaupf, I. Sosnowska, A. Taylor and R. Osborn, J. Appl. Phys. **69**, 5593 (1991).

Chapter 19

MÖSSBAUER SPECTROSCOPIC STUDIES OF INTERSTITIAL INTERMETALLIC COMPOUNDS

Fernande Grandjean
Institut de Physique, B5
Université de Liège
B-4000 Sart-Tilman, Belgium

and

Gary J. Long
Department of Chemistry
University of Missouri-Rolla
Rolla, MO 65401-0249, USA

1. Introduction

The complex room temperature Mössbauer spectra of Nd_2Fe_{17} and $Nd_2Fe_{17}N_{2.6}$, [1] shown in Figure 1, may seem rather ominous to the reader who is not familiar with the technique of Mössbauer spectroscopy. However, this complexity contains much useful information about the structural and magnetic properties of these materials. This chapter, which will show how this useful information may be extracted from such complex spectra, will introduce the basic principles of iron-57 Mössbauer spectroscopy and will stress the resolution achieved with this isotope. Then it will discuss the fundamental hyperfine parameters as measured by Mössbauer spectroscopy in rather more simple materials than Nd_2Fe_{17} and $Nd_2Fe_{17}N_{2.6}$. A latter section will cover the computer analysis techniques which may be used to extract the maximum amount of information from complex spectra such as those shown in Figure 1. Finally, the last section will review the application of iron-57 Mössbauer spectroscopy to interstitial materials. Some pertinent studies with the gadolinium-155 Mössbauer isotope will also be discussed.

2. Basic Principles of Mössbauer Spectroscopy

Mössbauer spectroscopy is the recoil-free emission and resonant absorption of γ-rays without the loss of energy to the lattice. Because of the recoil-free nature of the emission process, the energy distribution of the emitted Mössbauer γ-radiation has a Lorentzian lineshape, centered at E_γ, with the natural linewidth of the transition.[2] The theoretical resolution of the technique is therefore determined by the natural linewidth of the nuclear transition or the lifetime of the excited state of the nuclear transition.[3] In the remainder of this section, we will concentrate on iron-57, which is the most common isotope for Mössbauer spectroscopy and is especially useful because so many permanent magnets and interstitial materials contain iron.

F. Grandjean et al. (eds.), Interstitial Intermetallic Alloys, 463–496.

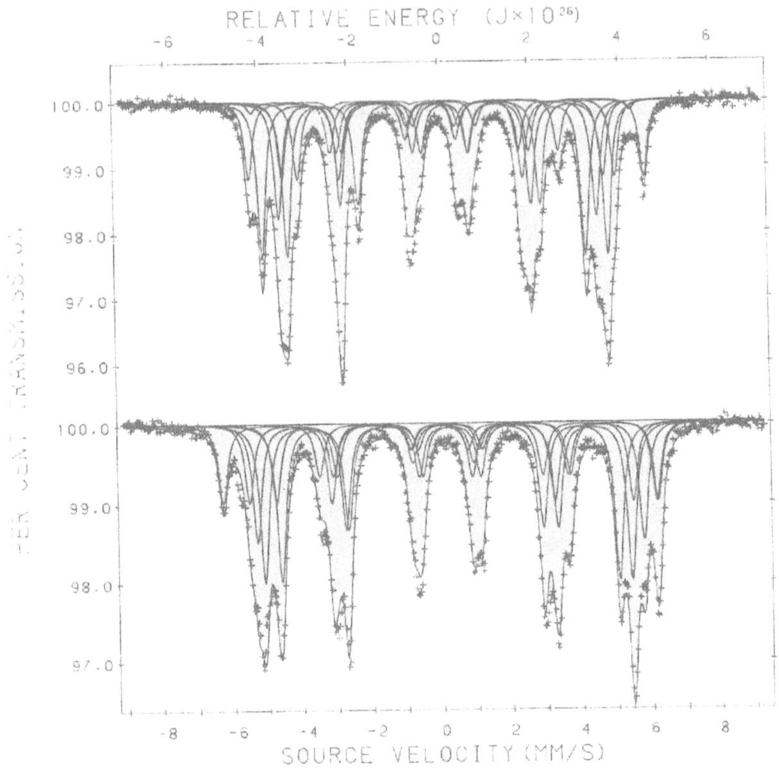

Figure 1. The iron-57 Mössbauer effect spectra of Nd_2Fe_{17}, top, and $Nd_2Fe_{17}N_{2.6}$, bottom, obtained at 78K.

A Mössbauer spectrum is obtained as illustrated schematically in Figure 2a. The γ-radiation is emitted from a source containing the radioactive parent isotope, cobalt-57, which decays to the excited state of the iron-57 Mössbauer isotope. This excited state decays by recoil-free emission of the Mössbauer γ-radiation with an energy, E_γ, of 14.4keV or 2.304×10^{-15}J, as shown in Figure 2b. The transmitted intensity of this radiation through an absorber containing iron-57 atoms is measured as a function of the γ-ray energy which is varied by Doppler shifting the source relative to the absorber. This explains why the energy scale in Figure 1 is given in mm/s and why energies are measured in mm/s by Mössbauer spectroscopists. The vertical scale of a Mössbauer spectrum is given in percent transmission or percent absorption and an important quantity for the spectroscopist is the absorption area of the spectrum. This area is a measure of the recoil-free fraction of the absorber, which is the fraction of nuclei undergoing recoil-free resonant absorption.[2,4]

For iron-57 Mössbauer spectroscopy, the natural linewidth of the 14.4keV transition is 4.6×10^{-9}eV or 0.095mm/s. Because the spectrum is obtained as the convolution of the emission and absorption lines, the smallest observable linewidth is 9.2×10^{-9}eV, or 0.19mm/s. Because of source and instrumental broadening, the typical observed linewidths are between 0.23 and 0.27mm/s. Hence, an energy difference of about

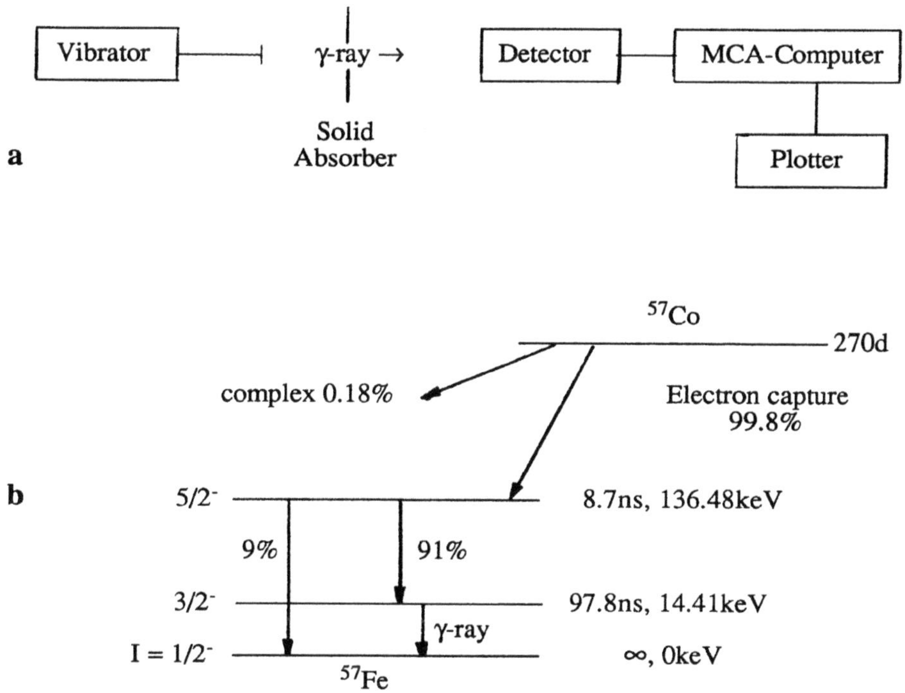

Figure 2. Schematic Mössbauer effect spectrometer, a, and the cobalt-57 nuclear decay scheme, b.

10^{-8}eV or 1.6×10^{-27}J may be observed. The hyperfine interactions between electrons and nuclei are of the order of 10^{-8}eV and hence may be observed in the Mössbauer spectra. These interactions are discussed in the next section. Finally, because of the narrow observed linewidth, iron-57 Mössbauer spectroscopy has a remarkably high relative resolution of 10^{-12}, the ratio of the linewidth to the transition energy.

Iron-57 is the most frequently used Mössbauer isotope because well resolved spectra can be obtained even at room temperature. However, many other isotopes [4] may be used and some examples will be discussed in the section devoted to the R_2Fe_{17} compounds. Because of intrinsically broader lines, the high resolution of iron-57 Mössbauer spectroscopy is often not achieved with other isotopes. In the case of gadolinium-155, for which an example is given in Section 8, the natural linewidth is 7.5×10^{-8}eV, or 0.25mm/s, and hence the smallest observable Mössbauer spectral linewidth is 0.50mm/s, and the theoretical resolution is thus 2×10^{-12}. The source linewidth is typically 0.35mm/s and hence typical observed linewidths are 0.70mm/s which gives a resolution of 3×10^{-12}.

3. Hyperfine Interactions

The interactions between the Mössbauer nucleus and its surrounding electronic environment are of a magnetic and electrostatic nature and are known as the hyperfine interactions. The interaction between the nuclear magnetic dipole moment and the magnetic induction generated by the electrons at the nucleus gives rise to the magnetic hyperfine splitting in the Mössbauer spectra, as is observed in magnetic compounds. The electrostatic interaction between the nucleus and the electric charges in its environment results in the isomer shift and the quadrupole interaction in the Mössbauer spectra. Because this chapter deals with the application of Mössbauer spectroscopy to magnetic materials, we will first discuss the magnetic hyperfine interaction, then the isomer shift, and finally the quadrupole splitting.

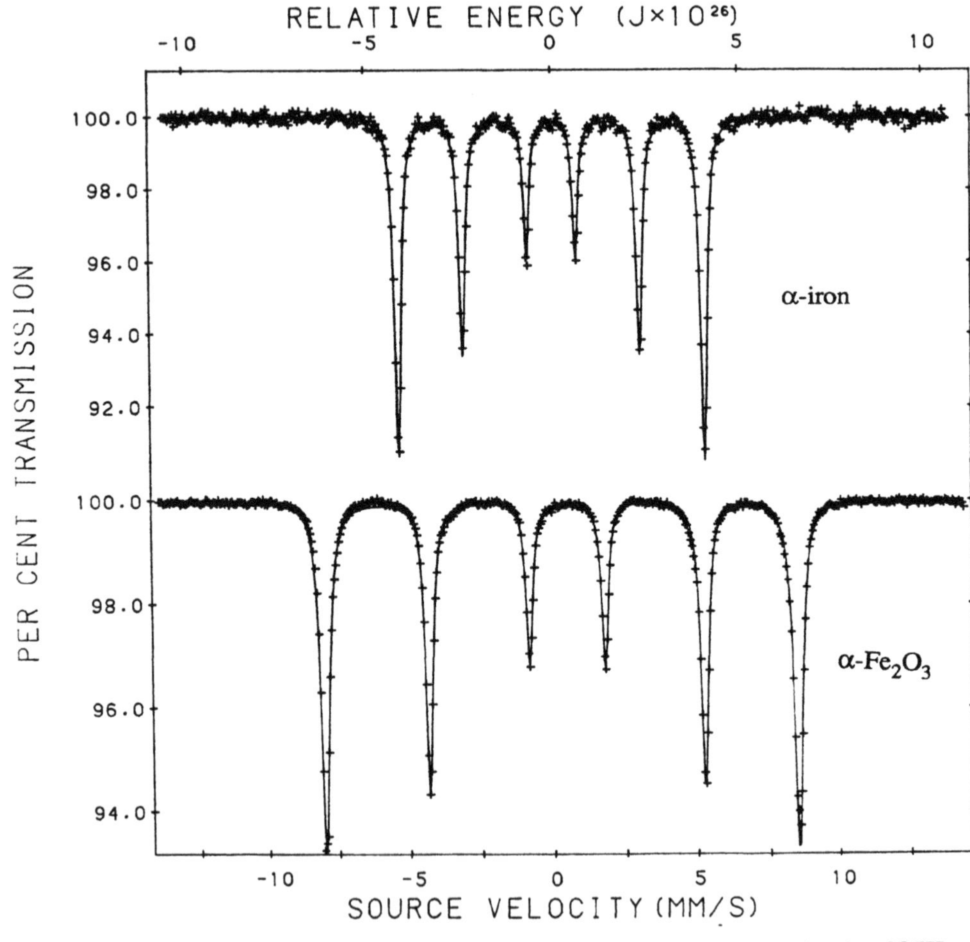

Figure 3. The Mössbauer effect spectra of α-iron and α-Fe_2O_3 obtained at 295K.

3.1. Magnetic Hyperfine Interaction

Figure 3 shows the effect of the magnetic hyperfine interaction in the rather simple Mössbauer spectra of α-iron, which has a body centered cubic structure, and α-Fe_2O_3, which has a hexagonal close packed structure. The most obvious characteristic of these spectra is the presence of six absorption lines. Indeed, the Zeeman interaction between the nuclear magnetic dipole moment of the iron nucleus and the magnetic induction generated at the nucleus by the electrons completely removes the degeneracy of the nuclear ground and excited states as is shown in the left two portions of Figure 4. The ground state of iron-57 has a nuclear spin of 1/2 and hence splits into two m_I states, whereas the excited state has a nuclear spin of 3/2 and hence splits into four m_I states. Because the 14.4keV Mössbauer transition of iron-57 is a pure magnetic dipole transition, for which the allowed transitions require $\Delta m_I = 0, \pm 1$, only six out of the eight possible transitions are observed, as is shown in Figure 3. In a random powder sample or an unmagnetized foil, the relative area [4] of these six lines is in the ratio 3:2:1:1:2:3, as observed in Figure 3. For a non-random powder or a magnetized foil, the area ratio is given by 3:x:1:1:x:3, where x is determined by the angle between the γ-ray direction and the magnetic induction. For instance, in a foil magnetized in its plane, the ratio is 3:4:1:1:4:3, whereas if the foil is magnetized normal to its plane, and hence parallel to the γ-ray direction, the ratio is 3:0:1:1:0:3.

The main information obtained from a magnetic Mössbauer spectrum, as shown in Figure 3, is the overall splitting which is proportional to the magnetic induction experienced by the nucleus below the Curie temperature of a ferromagnetic material or below the Néel temperature of an antiferromagnetic material. This magnetic induction is usually referred to as the internal hyperfine field. In the case of α-iron, the room temperature splitting is 10.6mm/s and corresponds to a magnetic induction or hyperfine field of 330kOe. In the case of α-Fe_2O_3, the room temperature splitting is 16.6mm/s and corresponds [5] to a magnetic induction or hyperfine field of 517.5kOe. A comparison of these splittings with the typical iron-57 experimental linewidth of 0.25mm/s indicates that a hyperfine field of 10kOe corresponds to approximately one experimental linewidth. In other words the magnetic hyperfine splitting is often 30 to 50 times the experimental linewidth, a factor which is very fortunate for the study of permanent magnets. If the hyperfine field is above 100kOe, a resolved magnetic sextet spectrum may be observed. However, smaller internal fields can easily be observed and an accurate value of the field obtained through computer fitting techniques.

The range of hyperfine fields observed in various iron magnetic compounds is rather broad, but several classes may be distinguished. Iron and iron-based steels exhibit hyperfine fields in the range of 300 to 350kOe.[6,7] Thus, small variations in the hyperfine fields with alloy composition may be measured. For metallic compounds the hyperfine field is to a first approximation proportional to the magnetic moment carried by the iron atom with a proportionality constant of 125 to 150kOe per Bohr magneton.[8] High-spin iron(III) compounds exhibit hyperfine fields in the range of 450 to 550kOe and different iron(III) compounds can often be distinguished by their splittings. For instance, two important octahedral iron(III) containing minerals, goethite, α-FeOOH,[9] and hematite, α-Fe_2O_3,[5] have saturation hyperfine fields of 506 and 541.7kOe, respectively. High-spin iron(II) compounds exhibit substantially smaller hyperfine fields in the range of 200 to 250kOe.

In compounds where iron is present on more than one crystallographic site, the different sites may often be distinguished as different components in a magnetic Mössbauer spectrum. If the recoil-free fractions of the different sites are equal, each component has a relative absorption area which is proportional to the relative iron population of the crystallographic site.

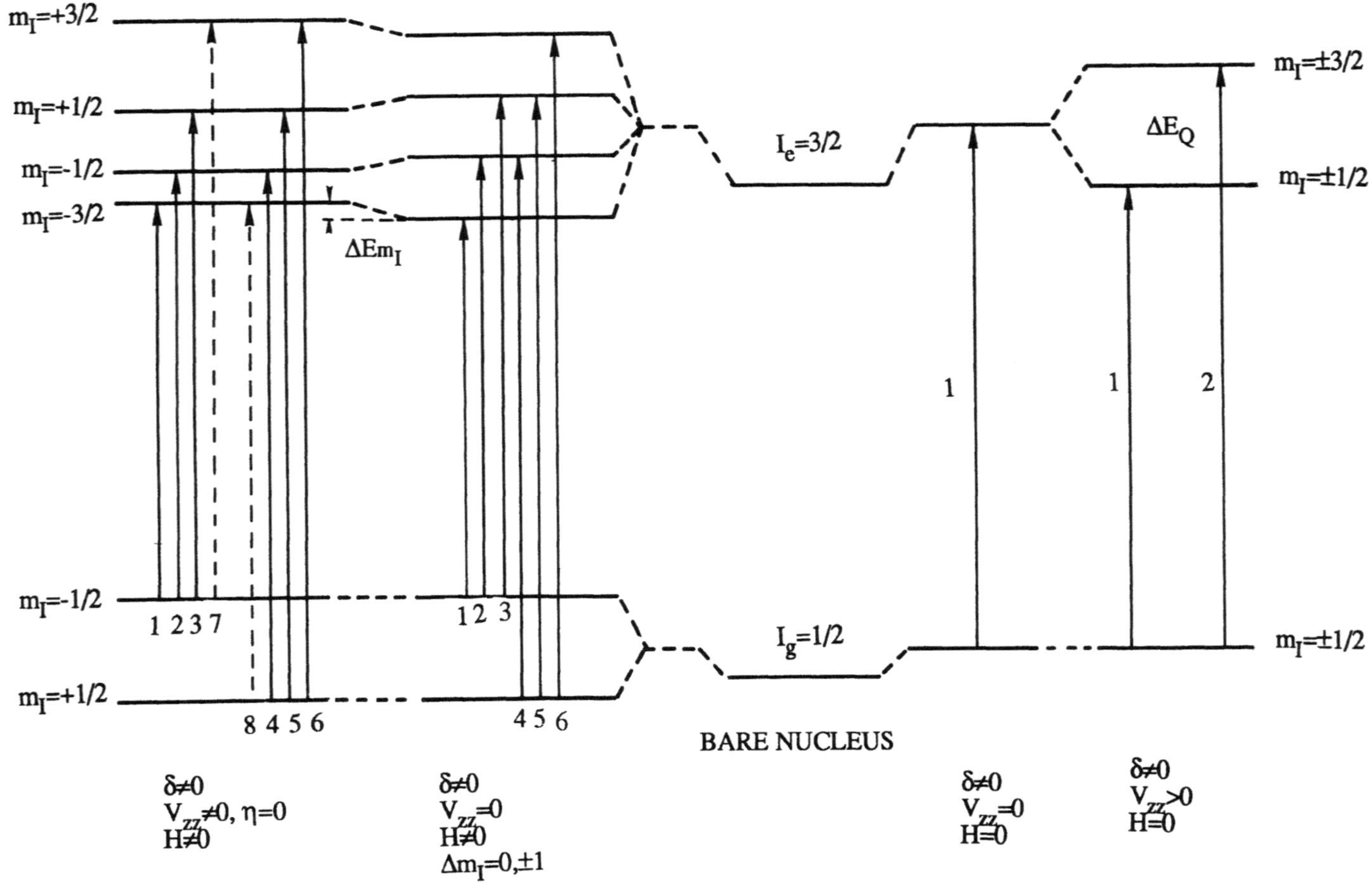

Figure 4. Nuclear energy level diagram for iron-57 showing the different hyperfine interactions.

In the oxidic spinels, such as $CoFe_2O_4$ and $NiFe_2O_4$, the iron(III) ions occupy two different crystallographic sites, the A site, in which the iron ion is in a tetrahedral environment of four oxygen ions, and the B site, in which the iron ion is in an octahedral environment of six oxygen ions. The A site is characterized by a saturation hyperfine field of about 500kOe, whereas the B site is characterized by a slightly larger saturation hyperfine field of 530 to 550kOe.[10] The difference of 30 to 50kOe in the hyperfine fields is quite easily observed in the Mössbauer spectra of spinels. In compounds where iron atoms occupy more than two sites, the Mössbauer spectra become complex and more difficult to analyze because of the overlap of various absorption lines, as is illustrated in Figure 1 for Nd_2Fe_{17} and $Nd_2Fe_{17}N_{2.6}$. In order to analyze such complex spectra and to obtain the hyperfine fields, information about the crystallographic and magnetic structure, the isomer shift, and the quadrupole splitting is very helpful and often essential for a complete analysis.

So far, we have discussed only the hyperfine field obtained from the typical six-line Zeeman pattern in a Mössbauer spectrum. In all the above spectra, there is at least one additional hyperfine parameter, the isomer shift, and sometimes a second hyperfine parameter, the quadrupole interaction, which must be considered. In the simple six-line spectra shown in Figure 3, the isomer shift manifests itself as a shift of the center of the spectrum relative to zero velocity. In the more complex spectra of Figure 1, the differing isomer shifts for each component in the spectra are very helpful in the analysis of the spectra.

3.2. Isomer Shift

The electronic charge density at the nucleus gives rise to shifts in the energy of the nuclear states, as is shown in Figure 4.[2] As the size of the nucleus changes during the transition from its excited state to its ground state, the nuclear transition energy is shifted by an energy which is proportional to the difference in the square of the nuclear radii in the two states. Because the Mössbauer atoms in the source and the absorber are usually present in different chemical environments, the Mössbauer absorption line is shifted by the isomer shift, δ, which is proportional to the difference in the electronic charge density at the nucleus in the absorber and in the source. Hence the isomer shift is a function of both a nuclear term and an electronic term. Fortunately the nuclear term is fixed for a given isotope, such as iron-57, and in practice the isomer shift is directly related to the electronic environment in an absorber relative to that of a standard.

The isomer shift of an absorber is usually given relative to a standard reference material, typically room temperature α-iron, which defines a zero velocity as well as a velocity scale. If the Mössbauer spectrum is a single line spectrum, the isomer shift is measured by the shift of the absorption line from zero velocity. If the Mössbauer spectrum is a doublet, as discussed in the next section and shown in Figure 5, the isomer shift is the shift of the mean of the two peak positions. If the Mössbauer spectrum is a sextet, as shown in Figure 3, the isomer shift is obtained either by computing the average of the eight possible line positions, shown in Figure 4, or, for a random absorber, by computing the area weighted average of the six observed line positions.

Because only the *s*-electrons have a non-zero probability density at the nucleus, the total electron density at the nucleus is dominated by the number of *s*-electrons in the atom. Hence, in an iron atom, which has the $[Ar]3d^64s^2$ electronic configuration, the total electron density at the nucleus is made up of contributions from the *1s, 2s, 3s*, and *4s* electrons. Although the remaining electrons have a zero probability density at the nucleus, they can shield the *s*-electrons and effectively change the *s*-electron density at the nucleus. Thus the *s*-electron densities at the nucleus for the iron(II) and iron(III) free ions, with the $3d^6$ and $3d^5$ valence electronic configurations, are different and their

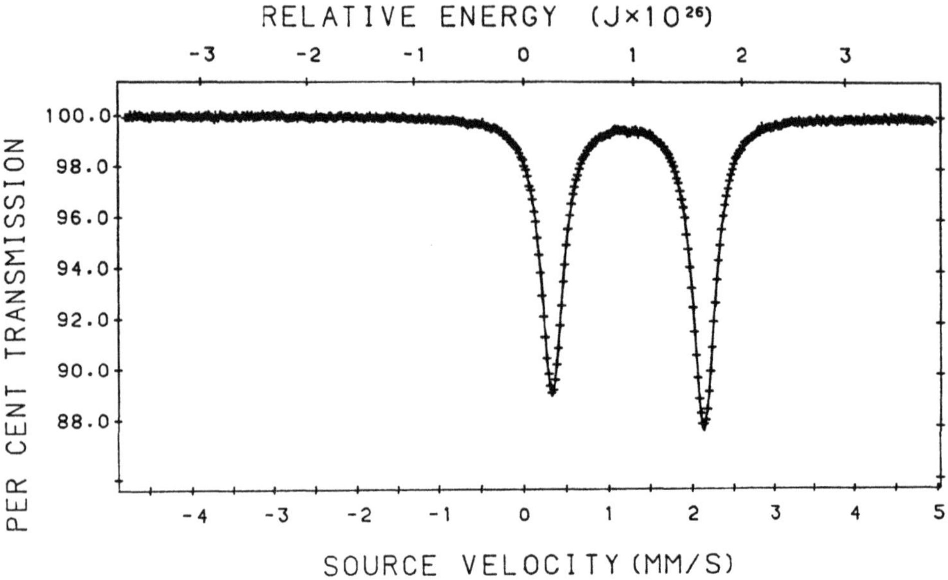

Figure 5. The Mössbauer effect spectrum of siderite, $FeCO_3$, obtained at 295K.

isomer shifts are different. The iron(II) free ion has an isomer shift which is larger than the iron(III) free ion by approximately 0.85mm/s.[11] This difference corresponds to approximately three to four times the linewidth and is easily observed.

In practice high-spin iron(II) compounds have room temperature isomer shifts in the range of 0.8 to 1.2mm/s, low-spin iron(II) compounds in the range of 0.4 to 0.6mm/s, high-spin iron(III) compounds in the range of 0.2 to 0.5mm/s, and metallic iron compounds and alloys in the range of –0.1 to 0.2mm/s. Hence, it is easy to distinguish high-spin iron(II) from the other iron configurations. Further, differences in isomer shift as small as 0.01mm/s may be obtained from the computer analysis of carefully measured Mössbauer spectra. For instance, the difference in iron(III) isomer shift between the octahedral and the tetrahedral site in spinels is usually in the range 0.09 to 0.14mm/s [10] and is easily observed. In R_2Fe_{17} compounds, the isomer shifts vary from –0.10 to 0.3mm/s and are found in the metallic iron range. The reliability of the observed isomer shifts may be checked through their temperature dependence. Indeed, as expected from the second-order Doppler shift, the iron-57 isomer shift in a highly ionic material should decrease linearly with increasing temperature, with a theoretical slope of -7.3×10^{-4}mms^{-1}K^{-1}.[12,13] Such a behavior is observed in the R_2Fe_{17} permanent magnetic materials and is discussed in Section 6.

3.3. Quadrupole Interaction

An asymmetric distribution of the charges surrounding the Mössbauer nucleus, in combination with a non-spherical distribution of the charge within the nucleus, gives rise to the quadrupole interaction. However this interaction only partly removes the degeneracy of the nuclear levels, as is shown in Figure 4.[2] The resulting Mössbauer spectrum of a random powder sample is a symmetric doublet, as is shown in Figure 5.

The obvious parameter for such a doublet is the splitting between the two lines, ΔE_Q, given by the expression

$$\Delta E_Q = (1/2)eQV_{zz}(1 + \eta^2/3)^{1/2}, \quad (1)$$

where Q is the nuclear quadrupole moment, V_{zz} is the principal component of the electric field gradient tensor at the nucleus, and $\eta = (V_{xx}-V_{yy})/V_{zz}$ is the asymmetry parameter of the electric field gradient tensor if the inequalities $0 \leq \eta \leq 1$ and $|V_{zz}| \geq |V_{yy}| \geq |V_{xx}|$ hold. It is not possible to obtain V_{zz}, η, or the sign of the quadrupole interaction from an iron-57 quadrupole doublet Mössbauer spectrum of a random powder. However, the sign of the interaction may be obtained from single crystal [4] or from applied field studies. In contrast, V_{zz}, η, and the sign may be obtained from the magnetic spectrum, as will be shown below. The quadrupole splitting measures the asymmetry of the electric charge distribution around the iron nucleus. The larger is the distortion from cubic symmetry, the larger is the quadrupole splitting, ΔE_Q.

The electric field gradient is made up of two contributions.[14] The first is a valence or electronic term, q_{val}, which can result from the inequivalent electronic population of the five iron 3d orbitals, such as is the case in the $t_{2g}^4e_g^2$ electronic configuration in a distorted octahedral high-spin iron(II) compound. The second contribution is the lattice contribution, q_{lat}, which results if the nucleus is in a non-cubic lattice. This would be the case in a distorted octahedral high-spin iron(II) compound. For a high-spin iron(III) compound, the valence contribution, q_{val}, is essentially zero because of the half-filled $3d^5$ valence shell electronic configuration, and any observed quadrupole splitting results predominantly from the lattice contribution. High-spin iron(II) compounds can have quadrupole splittings in the range of zero to 3.5mm/s, but more typically are found in the range of 1.5 to 2.5mm/s. High-spin iron(III) compounds have smaller quadrupole splittings typically in the range of zero to 1.0mm/s.

3.4. Combined Quadrupole and Magnetic Hyperfine Interaction

Most magnetic Mössbauer spectra result from a combined quadrupole and magnetic hyperfine interaction. The α-iron spectrum shown in Figure 3 has six lines which are symmetrically positioned relative to the center of the spectrum. Such a symmetry in the spectrum indicates the absence of a quadrupole interaction, as would be expected for the iron nuclei in the cubic lattice of α-iron. A close examination of the spectrum of α-Fe_2O_3, shown in Figure 3, indicates that the two highest velocity lines are closer together than the two lowest velocity lines. This asymmetry in the spectrum results from a combined quadrupole and magnetic hyperfine interaction. In the presence of both quadrupole and magnetic hyperfine interactions, the degeneracy of the nuclear levels is removed, as is shown in Figure 4. The splittings between the nuclear excited state sublevels are now no longer equal and the absorption lines are not equally spaced in energy. The energy level diagram shown in Figure 4 is valid for materials in which the perturbation by the quadrupole interaction is smaller than the perturbation by the magnetic interaction. This condition is often met in iron(III) compounds where the quadrupole interaction is usually small. It is certainly satisfied in α-Fe_2O_3, where the quadrupole interaction is 0.2mm/s, a value much smaller than the total magnetic splitting of 16mm/s. The next section will discuss how an estimate of the quadrupole interaction may be obtained from a magnetic spectrum.

4. Analysis of Complex Magnetic Spectra

4.1. General Considerations

The determination of the hyperfine parameters from Mössbauer spectra resulting from a combined quadrupole and magnetic hyperfine interaction is not straightforward because of the vector character of the hyperfine field and the tensor character of the electric field gradient. Usually, as is shown in Figure 6, the orientation of the magnetic hyperfine field is given relative to the principal axes of the electric field gradient tensor.

If the asymmetry parameter, η, of the electric field gradient tensor is zero, i.e. for an axial system in which $V_{xx} = V_{yy}$, the angle, θ, defines the orientation of the magnetic hyperfine field relative to V_{zz}, the principal axis of the electric field gradient tensor. If the quadrupole interaction can be treated as a small perturbation of the magnetic interaction, then it can be shown that the m_I levels of the nuclear excited states, see Figure 4, are shifted by the amount

$$\Delta E_{m_I} = (-1)^{|m_I|+1/2} \frac{eQV_{zz}}{4} \left(\frac{3\cos^2\theta - 1}{2} \right), \qquad (2)$$

where e is the electron charge, Q, the iron-57 nuclear quadrupole moment, V_{zz}, the principal component of the electric field gradient, and θ is defined as shown in Figure 6. From the shifts given by Equation 2, it is easy to compute the difference in the splitting between lines 1 and 2 and the splitting between lines 5 and 6 in a magnetic sextet. This difference is referred to as the quadrupole shift, ε or QS, and is given by the equation,

$$\varepsilon = (1/2)eQV_{zz}(3\cos^2\theta - 1). \qquad (3)$$

Thus, if the hyperfine field is parallel to the principal axis of the electric field gradient tensor, θ is zero, and ε is equal to eQV_{zz}, i.e., twice the paramagnetic quadrupole splitting. If θ is 90°, the hyperfine field is normal to the principal axis of the electric field gradient tensor and ε is equal to $-eQV_{zz}/2$, i.e., the absolute value of the para-

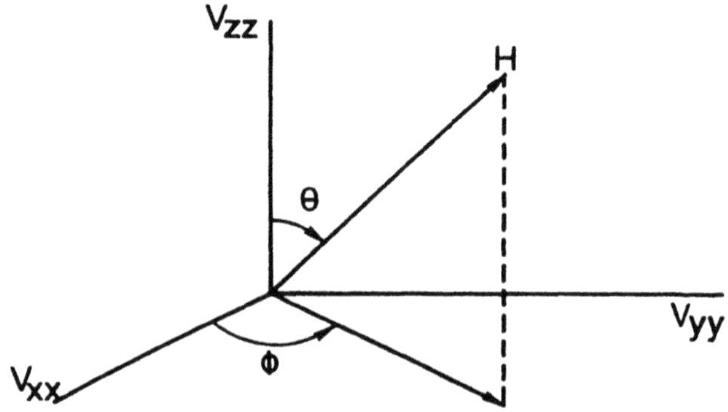

Figure 6. Orientation of the magnetic hyperfine field in the axes of the electric field gradient tensor.

magnetic quadrupole splitting. If θ is 54.7°, the 'magic angle', ε is zero and the spectrum is symmetric as would be observed for a pure magnetic interaction. In this special case, the quadrupole interaction cannot be obtained from the magnetic spectrum.

In conclusion, if Equation 3 applies, the measurement of the difference in the splitting of the 1, 2 and 5, 6 line positions gives an estimate of the quadrupole interaction. If the paramagnetic quadrupole interaction is known, then conclusions may be drawn about the angle θ, and the orientation of the magnetic field relative to the axes of the electric field gradient tensor. If the orientation of these axes is known relative to the crystallographic axes of the compound, as for instance from a calculation of the electric field gradient tensor, then the magnetic field can be oriented in the crystallographic axes. Or, if the orientation of the magnetic field is known relative to the crystallographic axes, as for instance from the orientation of the magnetic moments obtained from neutron diffraction measurements, then the electric field gradient axes may be oriented relative to the crystallographic axes.

The techniques described above to obtain the hyperfine parameters use the line positions in the magnetic Mössbauer spectra. The usual procedure is to fit the line positions with a computer program based on a least-square minimization method, in agreement with both the energy levels given in Figure 4 and Equation 2. If the first-order approximation is not acceptable because the quadrupole interaction is comparable with the magnetic interaction, another, more complex, approach to fit the spectra and obtain the hyperfine parameters must be used. The basic idea is to compute a theoretical spectrum from a first guess of the hyperfine parameters and then to adjust this theoretical spectrum by adjusting the hyperfine parameters until agreement is obtained with the experimental spectrum.

In order to compute the iron-57 theoretical spectrum, the Hamiltonians describing the combined quadrupole and magnetic interaction for the nuclear excited state and the magnetic interaction for the nuclear ground state must be written and solved for their eigenvalues and eigenvectors. The Hamiltonians are given by the expressions

$$\mathcal{H}_g = C \begin{bmatrix} -\cos\theta & -\sin\theta\, e^{-i\phi} \\ -\sin\theta\, e^{i\phi} & \cos\theta \end{bmatrix}, \tag{5}$$

$$\mathcal{H}_e = \begin{bmatrix} -3A\cos\theta + B & -3^{1/2}A\sin\theta\, e^{-i\phi} & \eta B/3^{1/2} & 0 \\ -3^{1/2}A\sin\theta\, e^{i\phi} & -A\cos\theta - B & -2A\sin\theta\, e^{-i\phi} & \eta B/3^{1/2} \\ \eta B/3^{1/2} & -2A\sin\theta\, e^{i\phi} & A\cos\theta - B & -3^{1/2}A\sin\theta\, e^{-i\phi} \\ 0 & \eta B/3^{1/2} & -3^{1/2}A\sin\theta\, e^{i\phi} & 3A\cos\theta + B \end{bmatrix}, \tag{6}$$

where A is $hg_eH/4\pi$, B is $eQV_{zz}/4$, C is $hg_gH/4\pi$, and g_e and g_g are the nuclear g-factors for the excited and ground states. The eigenvalues give the energies of the levels in Figure 4 and from these values, the energies of the transitions are easily computed. The eigenvectors are used to compute the relative areas of the transitions. The observed spectral areas are obtained by multiplying the relative areas by the experimental absorption area given in (mm/s)(%effect). For each transition, a Lorentzian line with linewidth, Γ, and with the proper area is computed. The spectrum is then the sum of all the Lorentzian lines. Although Equations 5 and 6 are written specifically for iron-57, similar equations would apply for other I=1/2 to 3/2 nuclear transitions.

For the simple case in which η is zero, Equation 2 may be obtained from the eigenvalues of the matrix given in Equation 6. In the most complex case, where η is non-zero and ϕ is different from zero, the matrices are complex and consequently, the calculation of the eigenvalues and eigenvectors is very time-consuming and spectra are seldom fit under these conditions. Nevertheless, this is not a real limitation because the effects of a non-zero η and a non-zero ϕ are small and more highly influence the intensities of the lines than their energies. A typical procedure for fitting a complex Mössbauer spectrum is to obtain a first guess of the hyperfine parameters from a first-order fit with Equation 2. The first guess is used to compute a theoretical spectrum and then the hyperfine parameters are adjusted to give the best fit to the experimental spectrum. Obviously, the first guess for the hyperfine field may be rather accurate, but the values of eQV_{zz}, η, and θ, are all contained in the quadrupole shift, ε, and are difficult to determine. Thus, all available information must be used to obtain a good first approximate for θ. In the initial fits, η may be taken as zero and then adjusted in subsequent fits.

4.2. Application to the Mössbauer-effect Spectra of R_2Fe_{17} and $R_2Fe_{17}N_x$

In R_2Fe_{17} and $R_2Fe_{17}N_x$ there are four crystallographically inequivalent iron sites and, as discussed in Section 3.1, one would expect that only four sextets would be necessary to fit the Mössbauer spectra of these compounds. However, this is the case only if the magnetization and, hence, the hyperfine field lies along the crystallographic c-axis of the unit cell. This results because, for the point symmetry of the iron sites, the angle, θ, between the principal axis of the electric field gradient and the hyperfine field is unchanged by the crystallographic symmetry operations and each iron site is characterized by a single θ value. Figure 7 demonstrates this for the 6c and 18f sites. For the 6c site, the principal axis of the electric field gradient lies along the c-axis and thus, when the magnetization is axial, the θ value is zero. For the 18f site, the principal axis lies in the basal plane, and could be along [100], [010], or a more general direction. Then if the magnetization is axial, the θ value is 90°. This is the case for $Sm_2Fe_{17}N_3$ as will be discussed in Section 6.

If the magnetization lies along [010] in the basal plane, the four crystallographically inequivalent iron sites are further subdivided into seven magnetically inequivalent sites. Indeed, as shown in Figure 7 for the 6c site, if the magnetization lies along [010], the angle θ is 90° and there is only one θ value. However, for the 18f site, if the magnetization lies along [010], V_{zz} is either parallel to the magnetization and θ is zero, or V_{zz} makes an angle of ±60° with the magnetization and θ is ±60°. The crystallographically equivalent 18f sites are thus subdivided into two magnetically inequivalent sites with relative populations of one to two. A similar reduction in the magnetic degeneracy, as compared to the crystallographic degeneracy, also occurs for the 9d and 18h sites in the R_2Fe_{17} and $R_2Fe_{17}N_x$ compounds with basal anisotropy. Hence, if the magnetization lies along one of the basal axes, seven sextets with relative areas of 6:6:3:12:6:12:6 are required to fit the 6c, 9d, 18f, and 18h components in the Mössbauer spectra.

We have discussed in detail the case where the magnetization lies along [010]. If the magnetization lies along [100], the number of sextets required to fit the spectra is also seven, but the relative intensities of a pair of crystallographically equivalent iron sites is reversed. This is illustrated by the differences observed in the Mössbauer spectra of Pr_2Fe_{17} and Nd_2Fe_{17}, as will be discussed in Section 6.

Because of the large numbers of sextets required to fit the Mössbauer spectra of the R_2Fe_{17} and $R_2Fe_{17}N_x$ compounds, and because of the large hyperfine magnetic fields observed in these compounds, their Mössbauer spectra have been fit with a first-order

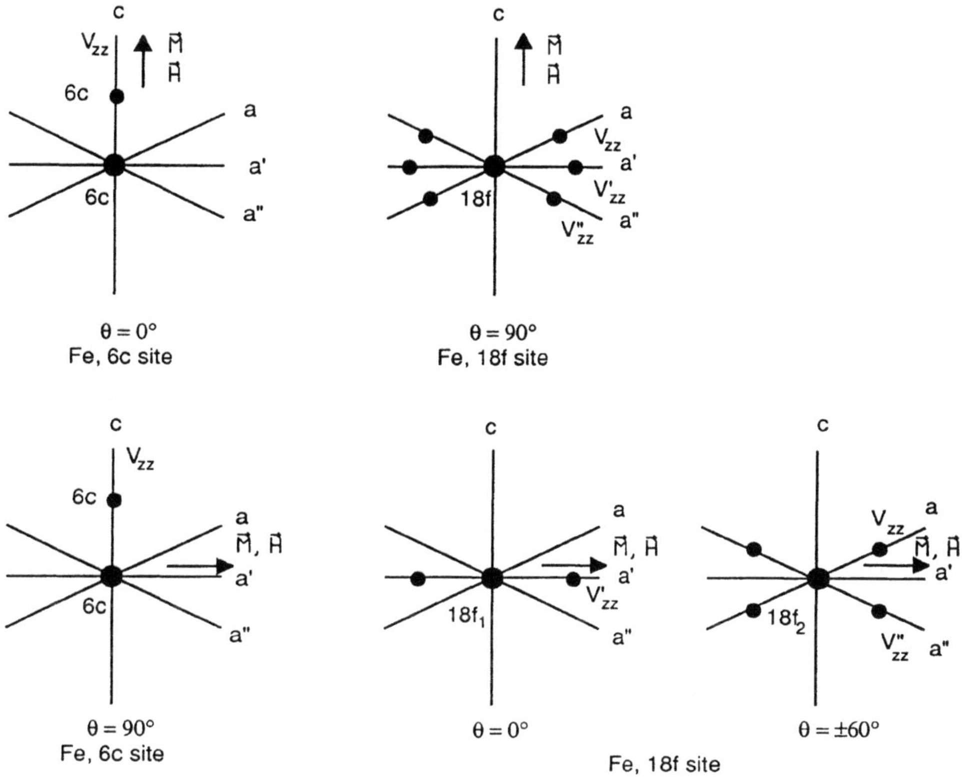

Figure 7. The orientations of the magnetization and hyperfine field, arrow, and the principal axis of the electric field gradient tensor, V_{zz}, in $R_2Fe_{17}N_x$ at the 6c and 18f sites for axial magnetization, top, and basal magnetization, bottom.

approximation in which the quadrupole interaction is treated as a small perturbation of the magnetic interaction, as was described in Section 4.1. In this approximation, each sextet is defined by its isomer shift, its hyperfine field, and its quadrupole shift. Hence, for seven sextets, there are 21 variable hyperfine parameters. Fortunately, this number of variable parameters can be reduced. There is no reason for crystallographically equivalent iron sites to have different isomer shifts, and hence, the isomer shifts of pairs of crystallographically equivalent sites are constrained to be equal. The number of hyperfine parameters is thus reduced to 18. The hyperfine fields of crystallographically equivalent sites are expected to be slightly different, the difference corresponding to the anisotropic contribution to the hyperfine field, i.e., the orbital and dipolar contributions, which are typically [15-17] of the order of 10 to 30kOe. The quadrupole shifts of crystallographically equivalent sites are of course different, but not independent, as they depend upon θ and eQV_{zz} as given by Equation 3.

In addition to the 18 hyperfine parameters, one spectral absorption area, one linewidth, and the baseline must also be fit. The total number of adjustable parameters is thus 21, but in practice no more than 16 parameters are simultaneously adjusted. It may seem that this is a large number of parameters with which one could always find an acceptable fit. However, all the constraints imposed by the crystallographic structure and the requirement of physically reasonable parameters as a function of temperature, nitrogenation, and rare-earth atom place many constraints on the fits, fits which often yield quite unique models, especially when the linewidth is small.

5. Wigner-Seitz Cell Analysis of the Site Environments

Because the model used to analyze the Mössbauer effect spectra of R_2Fe_{17} and $R_2Fe_{17}N_x$ is extensively based on the Wigner-Seitz cell volume [18] of each of their crystallographic sites, this section presents some results which are common to all compounds and which form a common basis for the fits. The Wigner-Seitz near-neighbor environment of each of the crystallographic sites in $R_2Fe_{17}N_3$ is given in Table I. The only difference between $R_2Fe_{17}N_3$ and R_2Fe_{17} is the presence of the nitrogen near neighbors at the rare-earth 6c and iron 18f and 18h sites. Fortunately for Mössbauer spectral work, the four crystallographic iron sites have different numbers of iron near neighbors. Because it has been observed [19-21] in other permanent magnetic materials that the hyperfine field increases with the number of iron near neighbors, the hyperfine fields in the R_2Fe_{17} compounds are expected to increase in the sequence, 18h < 9d ≅ 18f < 6c. As will be shown in Section 6, this sequence is observed.

The Wigner-Seitz cell volume of each iron site in R_2Fe_{17} and $R_2Fe_{17}N_3$, where R is Pr, Nd, and Th, is given in Table II. The crystallographic data used to obtain these values were taken from various sources.[22-27] As expected, the unit cell expansion observed upon nitrogenation of the R_2Fe_{17} compounds is reflected in the individual Wigner-Seitz cell volumes, but there are clear differences between the different sites. The 18f site shows the smallest increase, the 6c site shows an intermediate increase, and the 9d and 18h sites show large increases.

The isomer shift of an iron site is related to the Wigner-Seitz cell volume because it is proportional to the *s*-electron density at the nucleus. A larger Wigner-Seitz cell volume leads to a more extended *s*-electron radial distribution, a smaller *s*-electron density at the nucleus, and hence, a larger isomer shift. This relationship between the isomer shift and the Wigner-Seitz cell volume is extensively used in the assignment of the Mössbauer spectral components, as will be discussed below.

TABLE I. Near-neighbor Environments in $R_2Fe_{17}N_3$.

	Site Environment[a]						Near-neighbors			
Site	R, 6c	Fe, 6c	Fe, 9d	Fe, 18f	Fe, 18h	N, 9e	Fe	R	N	Total
R, 6c	1	1	3	6	9	3	19	1	3	23
Fe, 6c	1	1	3	6	3	0	13	1	0	14
Fe, 9d	2	2	0	4	4	0	10	2	0	12
Fe, 18f	2	2	2	2	4	1	10	2	1	13
Fe, 18h	3	1	2	4	2	1	9	3	1	13
N, 9e	2	0	0	2	2	0	4	2	0	6

[a]As determined from the Wigner-Seitz cell.

TABLE II. Wigner-Seitz Cell Volumes[a] for Several R_2Fe_{17} and $R_2Fe_{17}N_3$ Compounds.

Compound	6c	9d	18f	18h
Pr_2Fe_{17}	12.53	11.43	11.94	12.25
$Pr_2Fe_{17}N_3$	12.91	11.93	11.95	12.66
increase	0.38	0.50	0.01	0.41
% increase	3.0	4.4	0.1	3.4
Nd_2Fe_{17}	12.33	11.23	11.70	11.99
$Nd_2Fe_{17}N_3$	12.76	11.84	11.74	12.43
increse	0.43	0.61	0.04	0.44
% increase	3.5	5.4	0.3	3.7
Th_2Fe_{17}	11.97	11.09	11.61	12.19
$Th_2Fe_{17}N_3$	12.87	12.01	11.87	12.62
increase	0.90	0.92	0.26	0.43
% increase	7.5	8.3	2.2	3.5

[a]Room temperature volumes in Å^3 calculated by using twelve coordinate metallic radii of 0.92, 1.26, 1.82, 1.82, and 1.80Å, for nitrogen, iron, praseodymium, neodymium, and thorium, respectively.

6. Iron-57 Mössbauer-effect Studies of R_2Fe_{17} and $R_2Fe_{17}N_x$

The simplest iron-57 Mössbauer spectrum to be discussed is that of $Sm_2Fe_{17}N_{2.6}$ which is shown in Figure 8. Indeed, for this compound the magnetization,[28] and hence, the hyperfine field at each crystallograhic site is parallel to the c-axis of the unit cell. As explained in Section 4.2, because of the point symmetry of each site there is only one θ value and thus only one quadrupole shift for each site. As a result, each crystallographic site corresponds to one sextet with a relative area equal to its crystallographic degeneracy. The spectrum in Figure 8 has been analyzed [29] with four sextets with relative areas in the ratio 6:9:18:18, corresponding to the 6c, 9d, 18f, and 18h sites, plus one sextet due to an iron nitride impurity with a relative area of 20%. The 85K hyperfine parameters resulting from this fit are given in Table III. The assignment of the four sextets to the four crystallographic sites is first based on their relative areas and then, for the 18f and 18h sextets upon the isomer shifts, hyperfine fields, and Wigner-Seitz cell volumes. In general, the larger the Wigner-Seitz cell volume of a site, the larger is its isomer shift. Further the larger the number of iron near-neighbors, the larger is its hyperfine field. These general rules form the basis for the assigment of the sextets to the different crystallographic sites, as will be illustrated below.

In the R_2Fe_{17} and $R_2Fe_{17}N_x$ compounds, where R is Pr, Nd, and Ho, and in Sm_2Fe_{17}, the magnetization,[28,30] and hence the hyperfine field at each crystallographic site, lies along one of the basal axes of the unit cell. As explained in Section 4.2, because of the crystallographic point symmetry, the 9d, 18f, and 18h, sites are divided into two groups of magnetically inequivalent sites with relative areas of one to two, whereas the 6c site remains magnetically unique. Hence, seven sextets, constrained as described in Section 4.2, are used to fit the spectra. Figures 9 and 10 show the 85K Mössbauer spectra of several R_2Fe_{17} and $R_2Fe_{17}N_x$ compounds. The hyperfine fields deduced from these constrained fits are given in Tables III and IV.

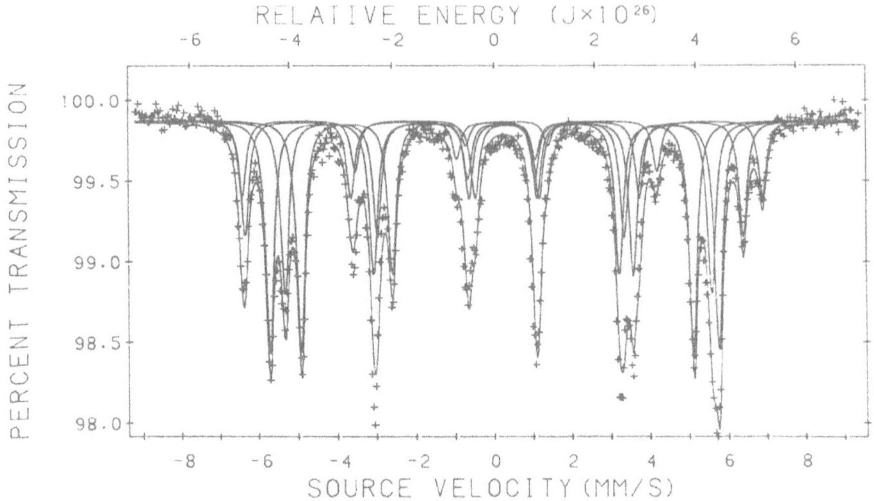

Figure 8. The Mössbauer effect spectrum of $Sm_2Fe_{17}N_{2.6}$ obtained at 85K.

The assignment of the various sextets to the crystallographic sites is based, first, upon their relative intensities and, second, upon their isomer shifts and their changes upon nitrogenation. In Section 5, we have given the near neighbor environment and Wigner-Seitz cell volume for each site. Because the isomer shift is related to the *s*-electron density at the nucleus, it is sensitive to the Wigner-Seitz cell volume. As indicated above, a larger Wigner-Seitz cell volume leads to a more positive iron-57

TABLE III. Mössbauer Spectral Hyperfine Parameters Measured at 85K for $R_2Fe_{17}N_x$.

	R	6c	$9d_6$	$9d_3$	$18f_{12}$	$18f_6$	$18h_{12}$	$18h_6$	Wt. Av.
H,	Pr	359	371	380	315	325	327	292	331.9
kOe	Nd	362	383	371	300	318	328	334	333.8
	Sm	413	395	-	355	-	311	-	353.0
	Ho	375	393	382	313	333	341	349	346.9
δ,[a]	Pr	0.300	0.025	0.025	0.140	0.140	0.190	0.190	0.156
mm/s	Nd	0.290	-0.002	-0.002	0.175	0.175	0.195	0.195	0.164
	Sm	0.285	0.045	0.045	0.144	0.144	0.215	0.215	0.168
	Ho	0.270	-0.010	-0.010	0.150	0.150	0.180	0.180	0.146
QS,[b]	Pr	-0.36	0.09	-0.70	0.30	-0.34	0.30	0.01	-
mm/s	Nd	-0.38	-0.18	0.36	0.10	0.38	-0.20	0.38	-
	Sm	-0.09	-0.02	-	0.44	-	-0.21	-	-
	Ho	-0.34	-0.36	0.34	0.06	0.33	-0.16	0.38	-

[a]Relative to room temperature α-iron foil. [b]Quadrupole shift values, except for the 6c site, for which QS is the quadrupole interaction, $eQV_{zz}/2$, with θ equal to 90° for Pr, Nd, and Ho, and 0° for Sm.

isomer shift. In addition to this volume dependence, the presence of a nitrogen near-neighbor at a given site is expected to further increase the isomer shift because of the higher electronegativity of nitrogen as compared to iron. Both of these effects are clearly illustrated in Figure 11 for Nd_2Fe_{17} and $Nd_2Fe_{17}N_{2.6}$. For Nd_2Fe_{17}, the sequences of the isomer shifts and the Wigner-Seitz cell volumes are identical. Upon nitrogenation, the isomer shift of the 6c and 9d sites increase because the Wigner-Seitz cell volumes increase, the isomer shift of the 18f site increases because of the presence of a nitrogen near neighbor, and the isomer shift of the 18h site increases both because of the increase in Wigner-Seitz cell volume and the presence of a nitrogen near neighbor. Similar plots result for Pr_2Fe_{17}, Sm_2Fe_{17}, and Ho_2Fe_{17} and their nitrides.[22, 29] The internal consistency of the analysis of the Mössbauer spectra of the R_2Fe_{17} and $R_2Fe_{17}N_x$ compounds for several rare-earth atoms is illustrated for the isomer shifts in Figure 12.

The above assignment of the sextets, based on the isomer shifts, leads to a very reasonable assignment of the hyperfine fields. In the R_2Fe_{17} compounds of Pr, Nd, Sm, and Ho, the hyperfine fields are directly related to the number of iron near neighbors as determined from the Wigner-Seitz cell analysis, see Section 5. The 6c site, with 13 iron near neighbors has the largest field, the 18h site with 9 iron near neighbors has the smallest field, and the 9d and 18f sites with 10 iron near neighbors each, have intermediate fields, see Table IV. For the $R_2Fe_{17}N_x$ compounds of Pr, Nd, Sm, and Ho, the hyperfine fields are not directly related to the number of iron near neigbors. However, the changes in hyperfine field upon nitrogenation can be understood on the basis of the unit cell expansion, and, further, the changes correlate well with the increases in magnetic moment measured by neutron diffraction or determined through band structure calculations. Table V gives the site average hyperfine fields, the measured and calculated moments, and their changes upon nitrogenation for several R_2Fe_{17} compounds. For comparison, we have included, in Table V, Y_2Fe_{17} and Th_2Fe_{17} which are not discussed in detail herein because of their additional complexity.[31,32] For Y_2Fe_{17}, which has the hexagonal Th_2Ni_{17} structure, the crystallographic sites have been relabelled with the equivalent rhombohedral site notation.

TABLE IV. Mössbauer Spectral Hyperfine Parameters Measured at 85K for R_2Fe_{17}.

	R	6c	$9d_6$	$9d_3$	$18f_{12}$	$18f_6$	$18h_{12}$	$18h_6$	Wt. Av.
H, kOe	Pr	343	282	306	298	264	266	263	286.2
	Nd	351	301	275	285	315	271	267	292.3
	Sm	354	304	288	286	324	275	281	298.0
	Ho	360	316	287	292	326	283	274	302.3
δ,[a] mm/s	Pr	0.240	-0.095	-0.095	0.065	0.065	0.040	0.040	0.049
	Nd	0.250	-0.070	-0.070	0.060	0.060	0.064	0.064	0.061
	Sm	0.245	0.080	0.080	0.050	0.050	0.060	0.060	0.054
	Ho	0.240	-0.070	-0.070	0.050	0.050	0.020	0.020	0.041
QS,[b] mm/s	Pr	-0.03	0.14	-0.36	0.06	0.58	0.16	-0.62	-
	Nd	-0.05	-0.10	0.34	0.44	-0.10	-0.42	0.48	-
	Sm	-0.13	-0.34	0.06	0.42	-0.09	-0.33	0.69	-
	Ho	-0.07	-0.21	0.36	0.50	-0.08	-0.38	0.55	-

[a]Relative to room temperature α-iron foil. [b]Quadrupole shift values, except for the 6c site, for which QS is the quadrupole interaction, $eQV_{zz}/2$, with θ equal to 90°.

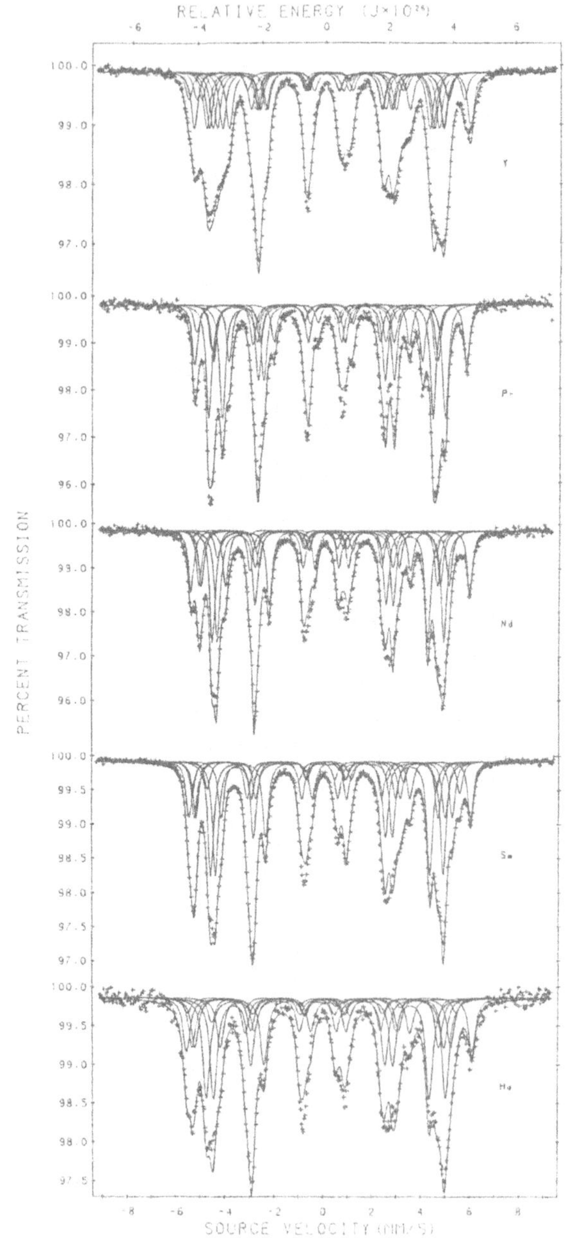

Figure 9. The Mössbauer effect spectra of R_2Fe_{17}, where R is Y, Pr, Nd, Sm, and Ho, obtained at 85K.

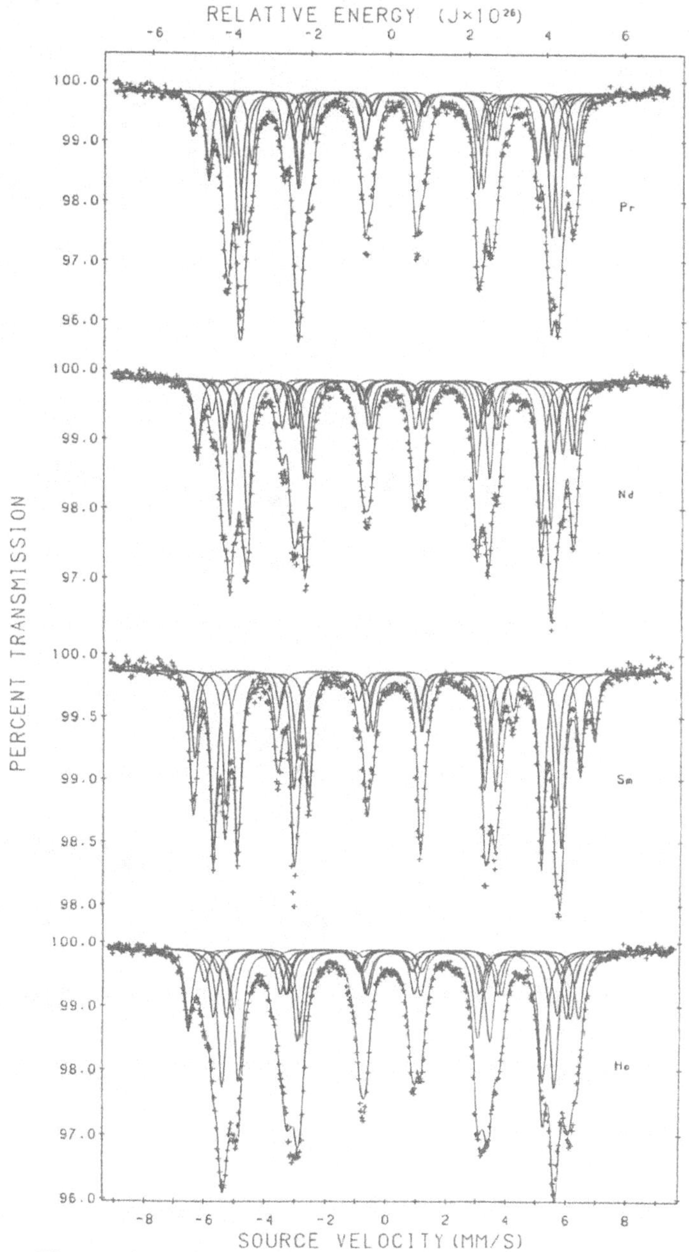

Figure 10. The Mössbauer effect spectra of $R_2Fe_{17}N_x$, where R is Pr, Nd, Sm, and Ho, obtained at 85K.

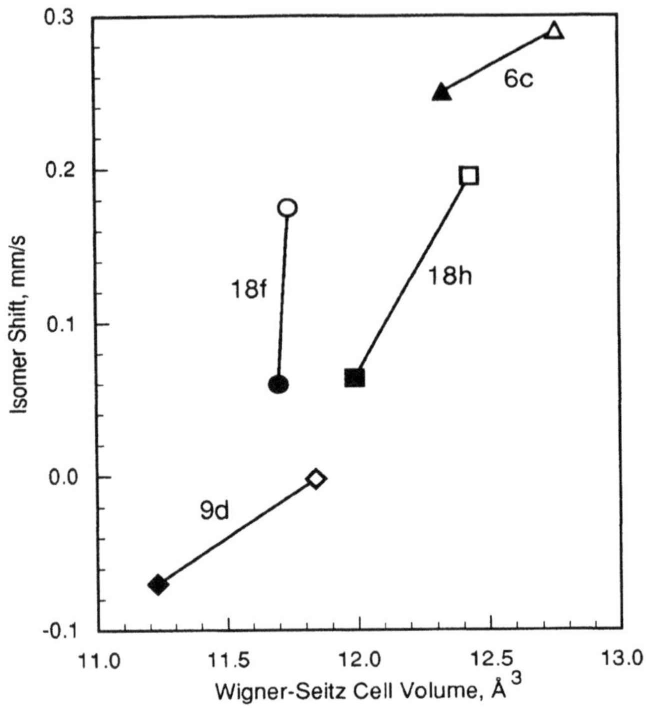

Figure 11. The correlation between the Wigner-Seitz cell volume and the 85K isomer shift for each site in going from Nd_2Fe_{17}, solid symbols, to $Nd_2Fe_{17}N_{2.6}$, open symbols.

For all the rare earth compounds studied, the hyperfine field at the 9d site shows a large increase upon nitrogenation. A similar large increase in the magnetic moment of the 9d site is both found by neutron diffraction studies [25,26] of $Th_2Fe_{17}N_{2.6}$ and predicted by band structure calculations for the nitrides of Nd_2Fe_{17},[33] Gd_2Fe_{17},[34] and Y_2Fe_{17}.[16] Except for Sm_2Fe_{17}, the hyperfine field of the 18h site also shows a substantial increase upon nitrogenation. A similar substantial increase of the 18h magnetic moment is both found from neutron diffraction studies of $Th_2Fe_{17}N_{2.6}$ [25,26] and predicted by band structure calculations for the nitrides of Gd_2Fe_{17},[34] and Y_2Fe_{17}.[16] The increases in the 9d and 18h hyperfine fields also correlate well with the large increases in Wigner-Seitz cell volumes of these sites upon nitrogenation. For all the rare-earth compounds studied, the 6c and 18f hyperfine fields experience small increases upon nitrogenation.

In the case of $Y_2Fe_{17}N_3$, the hyperfine fields, measured at 85K may be compared with the calculated [35] hyperfine fields shown in Figure 13. Rather typically, and for some obscure reason, the calculated hyperfine fields in intermetallic compounds are often ca. 50kOe smaller than the measured hyperfine fields. However, the observed sequence, 6g > 4f > 12k > 12j, is well reproduced by the calculations. This is not the case for an alternative model used to analyze the Mössbauer spectra of $Y_2Fe_{17}N_3$.[36] Figure 14 shows the change in the weighted average hyperfine field with rare earth in the R_2Fe_{17} and $R_2Fe_{17}N_x$ series.

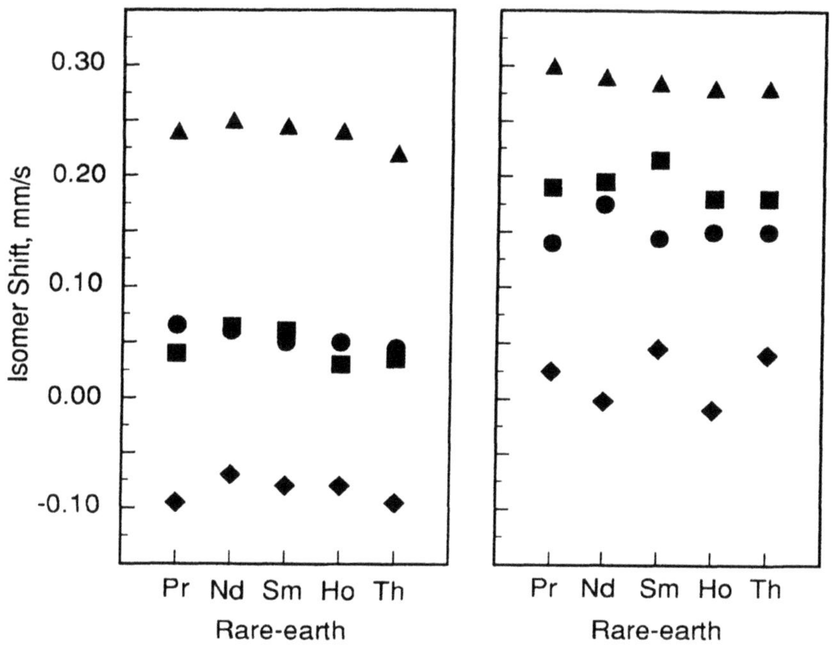

Figure 12. The 85K iron-57 isomer shifts for the 6c, ▲, the 9d, ◆, the 18f, ●, and the 18h, ■, sites in R_2Fe_{17}, left, and $R_2Fe_{17}N_x$, right, for different rare-earth atoms.

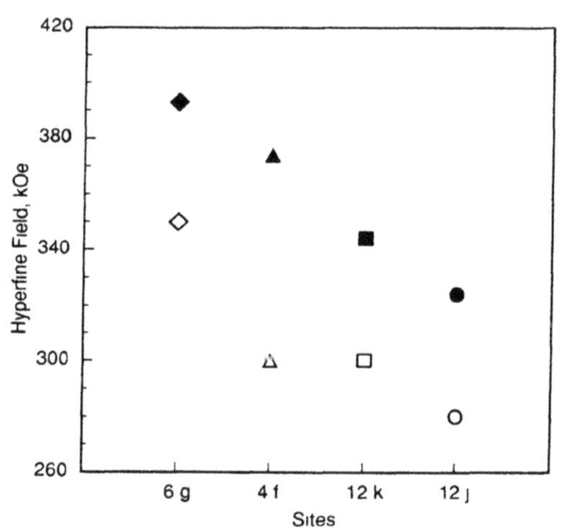

Figure 13. The observed hyperfine fields [31] in $Y_2Fe_{17}N_{2.6}$, solid symbols, and calculated fields, open symbols, [35] in $Y_2Fe_{17}N_3$.

TABLE V. Hyperfine Fields and Related Quantities in R_2Fe_{17} and $R_2Fe_{17}N_x$.

	T, K	Compound	Ref.	6c	9d	18f	18h	Wt. Av.
H, kOe[a]	85	Y_2Fe_{17}	31	350	326	284	279	297
	85	$Y_2Fe_{17}N_{2.6}$	31	374	392	324	344	349
ΔH				24	66	40	65	52
H, kOe	calc	Y_2Fe_{17}	35	339	314	295	271	295
	calc	$Y_2Fe_{17}N_3$	35	290	350	280	300	301
ΔH				−49	36	−15	29	6
μ, μ_B	calc	Y_2Fe_{17}	16	2.53	1.92	2.25	2.00	2.14
	calc	$Y_2Fe_{17}N_3$	16	2.65	2.53	2.01	2.57	2.37
Δμ				0.08	0.61	−0.24	0.57	0.23
H, kOe[a]	85	Pr_2Fe_{17}	22	343	290	287	265	286
	85	$Pr_2Fe_{17}N_{2.6}$	22	359	374	318	306	332
ΔH				16	84	31	41	46
μ, μ_B[b]	2	Pr_2Fe_{17}	26	2.17	2.17	2.17	2.17	2.17
	295	$Pr_2Fe_{17}N_{2.9}$	26	2.45	2.45	2.45	2.45	2.45
Δμ				0.28	0.28	0.28	0.28	0.28
H, kOe[a]	85	Nd_2Fe_{17}	1	351	292	295	270	292
	85	$Nd_2Fe_{17}N_{2.6}$	1	362	379	306	330	334
ΔH				11	87	11	60	42
μ, μ_B[b]	295	Nd_2Fe_{17}	23	1.85	1.58	1.56	1.47	1.57
	295	$Nd_2Fe_{17}N_{2.4}$	24	1.90	1.50	1.50	1.50	1.55
Δμ				0.05	−0.08	−0.06	0.03	−0.02
μ, μ_B	calc	Nd_2Fe_{17}	33	2.68	1.93	2.02	2.34	2.19
	calc	$Nd_2Fe_{17}N_3$	33	2.48	2.66	1.51	2.19	2.07
Δμ				−0.20	0.73	−0.51	−0.15	−0.12
H, kOe[a]	85	Sm_2Fe_{17}	29	354	299	299	277	298
	85	$Sm_2Fe_{17}N_{2.6}$	29	413	395	355	311	353
ΔH				59	96	56	34	55
H, kOe[a]	85	Ho_2Fe_{17}	29	360	306	303	280	302
	85	$Ho_2Fe_{17}N_{2.6}$	29	375	389	320	344	347
ΔH				15	83	17	64	45
μ, μ_B	calc	Gd_2Fe_{17}	34	2.38	2.15	2.29	2.06	2.19
	calc	$Gd_2Fe_{17}N_3$	34	2.43	2.41	2.15	2.44	2.33
Δμ				0.05	0.26	−0.14	0.38	0.14
H, kOe[a]	85	Th_2Fe_{17}	32	313	261	256	235	256
	85	$Th_2Fe_{17}N_{2.6}$	32	357	354	327	331	337
ΔH				44	93	71	96	81
μ, μ_B[b]	85	Th_2Fe_{17}	25	2.81	2.14	1.83	1.74	1.97
	85	$Th_2Fe_{17}N_{2.4}$	26	3.26	2.74	2.19	2.15	2.40
Δμ				0.45	0.60	0.36	0.41	0.43

[a]Mössbauer spectral results. [b]Neutron diffraction results.

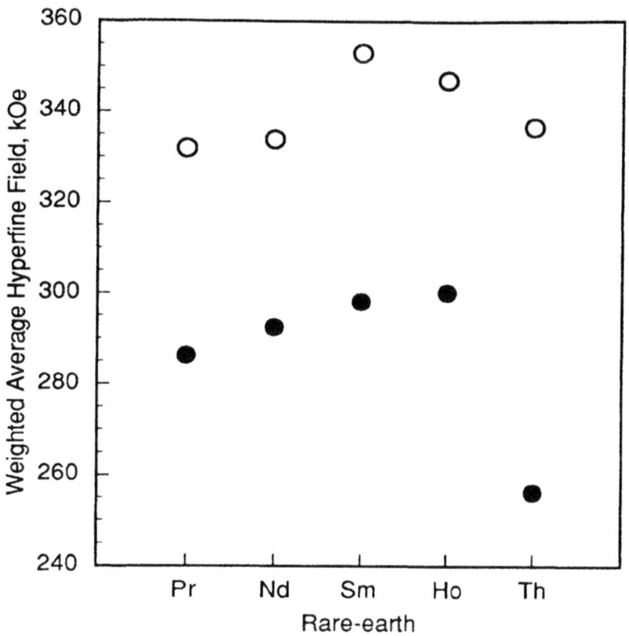

Figure 14. The 85K weighted average hyperfine fields in R_2Fe_{17}, solid symbols, and $R_2Fe_{17}N_x$, open symbols, as a function of rare-earth atom.

Useful information can also be deduced from the quadrupole shifts determined for the different sites. For a pair of magnetically inequivalent, but crystallographically equivalent sites there is usually a positive and a negative value of the quadrupole shift. The change in sign is due to the different θ values, one above and one below the magic angle of 54.7°. More interesting perhaps is the change in sign of the quadrupole shift in going from the Pr to the Nd compounds. This change results from the change in the orientation of the magnetization in the basal plane. In Pr_2Fe_{17} the magnetization lies along the b-axis, whereas in Nd_2Fe_{17} it lies along the a-axis, as has already been noted by Gubbens et al.[30,37] Hence, the Mössbauer spectra reveal the orientation of the magnetization in the basal plane. For $Y_2Fe_{17}N_{2.6}$ [31] and $Th_2Fe_{17}N_{2.6}$,[32] the magnetization takes on a general direction and the analysis of the Mössbauer spectra requires ten sextets, a complication which is not discussed in this chapter.

Figure 15 shows the temperature dependence of the hyperfine fields at the four crystallographic sites in Nd_2Fe_{17} and $Nd_2Fe_{17}N_{2.6}$. Similar plots result for Pr_2Fe_{17}, Sm_2Fe_{17}, and Th_2Fe_{17}, and their nitrides. As expected, the hyperfine fields decrease with increasing temperature, but the most apparent feature is the difference in the decrease in the fields between Nd_2Fe_{17} and $Nd_2Fe_{17}N_{2.6}$, a difference which results because of the increase in Curie temperature upon nitrogenation.

Another way to check the validity of the model for the Mössbauer spectra of the R_2Fe_{17} and $R_2Fe_{17}N_x$ compounds is to study the temperature dependence of the spectra and, in particular, the temperature dependence of the isomer shifts and hyperfine fields. Figures 16 and 17 show the Mössbauer spectra of Nd_2Fe_{17} and $Nd_2Fe_{17}N_{2.6}$ obtained at various temperatures and Figure 18 shows the corresponding temperature dependence of

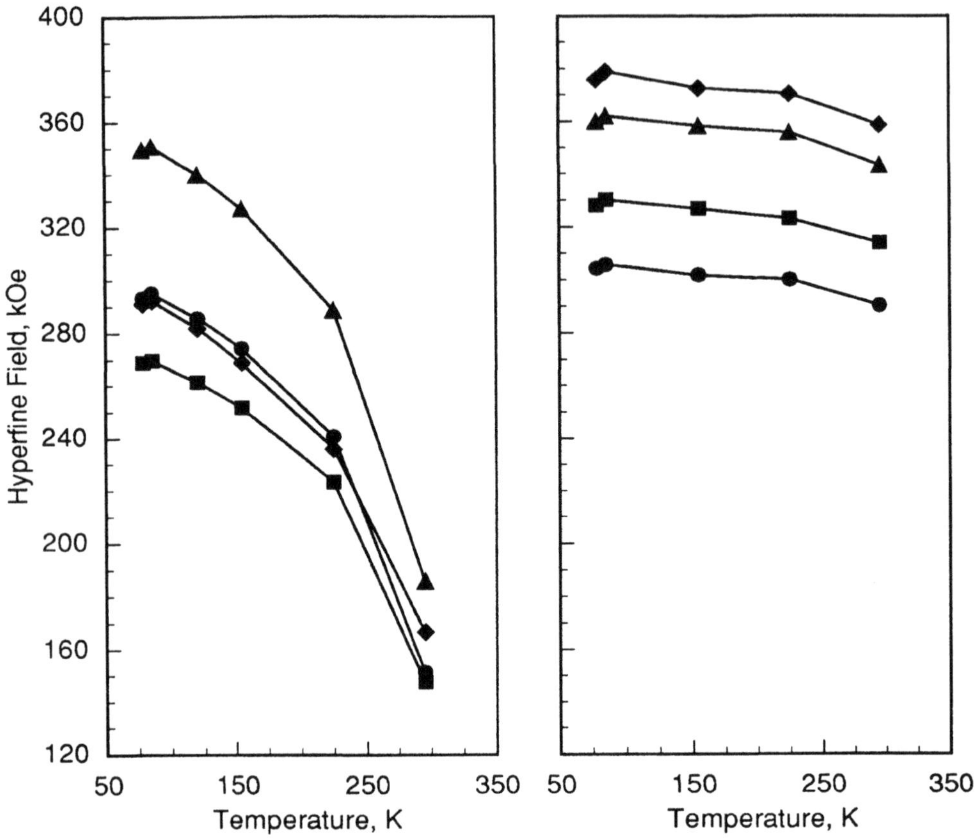

Figure 15. The temperature dependence of the iron-57 Mössbauer effect hyperfine fields of the 6c, ▲, 9d, ◆, 18f, ●, and 18h, ■, sites in Nd_2Fe_{17}, left, and $Nd_2Fe_{17}N_{2.6}$, right.

the isomer shifts. Similar figures have been published [22] for Pr_2Fe_{17} and $Pr_2Fe_{17}N_{2.6}$ and found for Sm_2Fe_{17} and Th_2Fe_{17} and their nitrides. As explained in Section 3.2, the approximately linear decrease of the isomer shift with increasing temperature is due to the second-order Doppler shift. The slopes for the four crystallographic sites vary between -6.3×10^{-4} and -4.1×10^{-4} $mms^{-1}K^{-1}$ for Nd_2Fe_{17} and -5.3×10^{-4} and $-4.5\times10^{-4}mms^{-1}K^{-1}$ for $Nd_2Fe_{17}N_{2.6}$. The small differences between sites indicate small differences in their bonding. For the Pr, Nd, Sm, and Th compounds, the slopes observed for the nitrided compounds are systematically larger than those observed for the unnitrided compounds. As a result, the effective recoil masses [12] deduced from these slopes are larger in the nitrided compounds, presumably because of the added covalency of the bonding with the nitrogen atom, a covalency which has also been revealed by band structure calculations.[16]

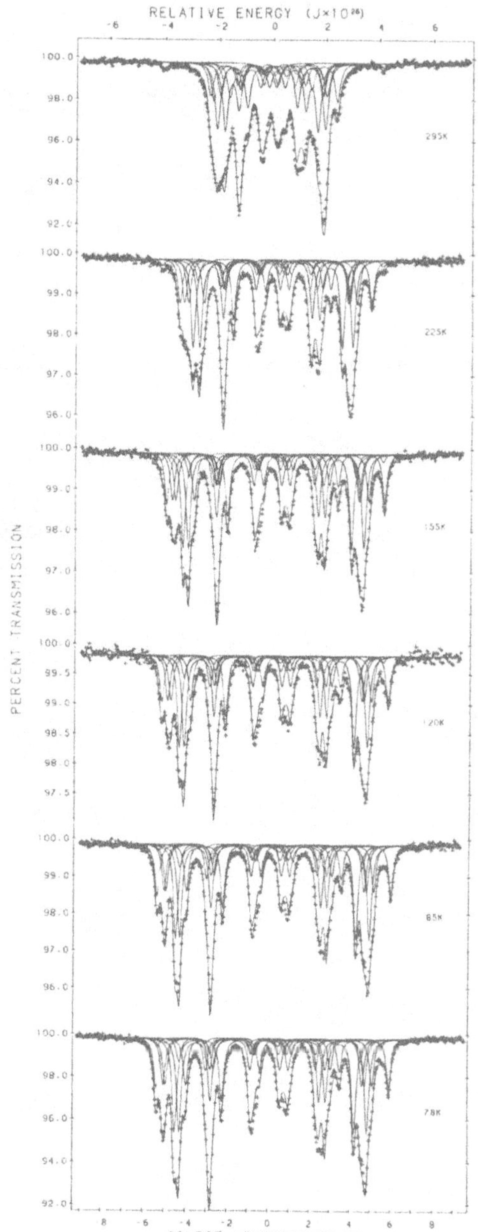

Figure 16. The Mössbauer effect spectra of Nd_2Fe_{17} obtained at the indicated temperatures.

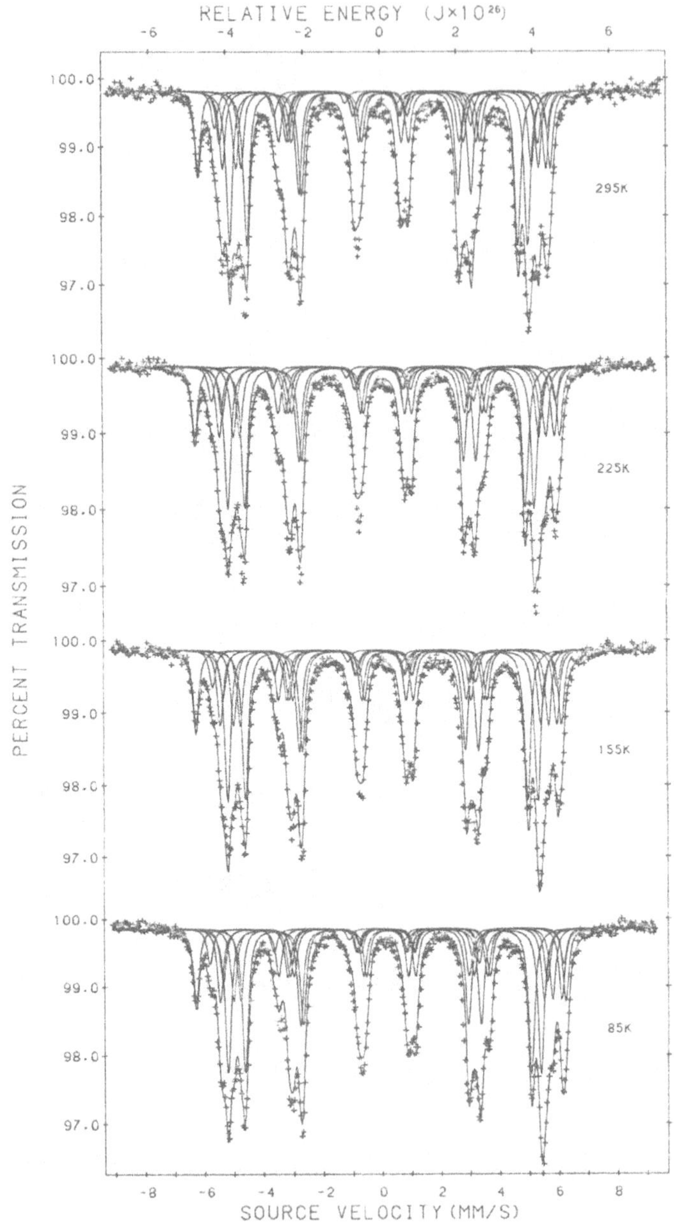

Figure 17. The Mössbauer effect spectra of $Nd_2Fe_{17}N_{2.6}$ obtained at the indicated temperatures.

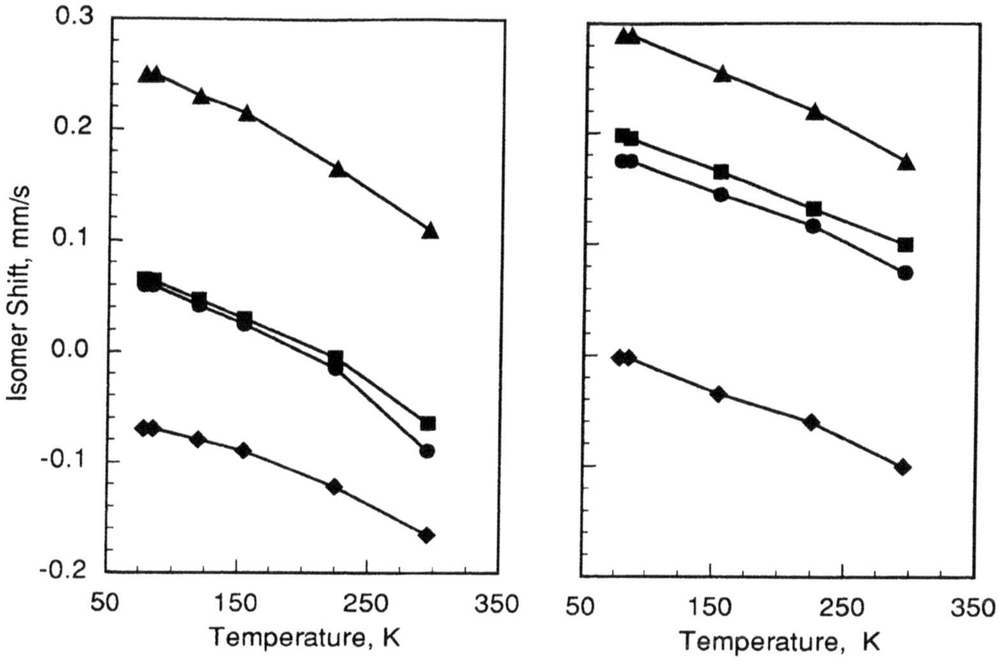

Figure 18. The temperature dependence of the iron-57 Mössbauer effect isomer shifts of the 6c, ▲, 9d, ◆, 18f, ●, and 18h, ■, sites in Nd_2Fe_{17}, left, and $Nd_2Fe_{17}N_{2.6}$, right.

7. Critical Review of Spectral Models

It is useful to compare the model and its results, described in Section 6, with other published work. Most authors [37-40] use six or seven sextets to represent the inequivalent magnetic iron sites in the R_2Fe_{17} compounds and their nitrides. In addition to these seven sextets, Hu et al. [38] and Rupp and Wiesinger [39] introduce a central doublet of uncertain origin. Furthermore, Gubbens et al.[37] and Hu et al.[38] introduce an additional sextet which is assigned to α-iron, but its relative intensity is not reported. Because often only some of the hyperfine parameters are reported by these authors, it is difficult to determine whether the constraint of equal isomer shifts for the pairs of crystallographically equivalent sites has been used in their fits. The positions of the lines indicated by the bars in the spectra of Nd_2Fe_{17} and its hydride in the figures of Rupp and Wiesinger [39] indicate that this constraint was not used. From the published work of Gubbens et al.[37] and Hu et al.[38] it is impossible to determine which constraints were used.

Another main problem in analyzing complex Mössbauer effect spectra resides in line broadening. It is crucial that great care be taken to avoid experimental line broadening, a broadening which decreases resolution and thus obscures the details of the hyperfine structure. Indeed a broadened spectrum is easier to fit than a sharp spectrum, but less spectral information is obtained, and often no reliable conclusions can be drawn from broadened spectra. All the spectra which are discussed in Section 6, were fit with a single linewidth for all the sextets, a linewidth always found to be in the range of 0.24 to 0.29mm/s. These values compare well with the linewidth of 0.24 to 0.26mm/s observed with α-iron at room temperature on the same spectrometer.

Hu et al.[38] used different linewidths for the outer, middle, and inner pairs of lines in a given sextet; however, they do not indicate whether different linewidths were used for the different sextets. There is no physical basis for differing linewidths for the different sextets in R_2Fe_{17}. In $R_2Fe_{17}N_x$, because of the non-stoichiometric nitrogen content, the iron on the 18h and 18f sites can have either zero or one nitrogen near neighbor. This variation in their near-neighbor environments may lead to a small distribution of hyperfine parameters and, hence, a broadening of the absorption lines of these sites. When the linewidths of the 18f and 18h sites are varied in the spectra of $Nd_2Fe_{17}N_{2.6}$, they are found to be larger by 0.007 and 0.002mm/s, respectively, than those of the 6c and 9d sites. Because such a small increase was observed, we did not allow the linewidths of the 18f and 18h sites to vary from those of the 6c and 9d sites in subsequent fits. This procedure has been confirmed by a recent Mössbauer spectral study [41] of $Nd_2Fe_{17}N_3$ which gave results virtually identical to those shown in Figures 1 and 17 and given in Table III for $Nd_2Fe_{17}N_{2.6}$.

The experimental data published by Hu et al.[38] reveal broad lines, and the authors quote linewiths in the range of 0.28 to 0.36mm/s. For Y_2Fe_{17} and its carbide and nitride, Qi et al.[36] report linewidths in the range of 0.15 to 0.36mm/s. Chen et al.[40] report a unique linewidth of 0.34mm/s for their spectra of Y_2Fe_{17} and Sm_2Fe_{17} and their interstitial derivatives. A linewidth smaller than the natural linewidth of 0.195mm/s is completely unrealistic and a broadening of 50% of the typical calibration linewidth is unacceptable if a detailed discussion of the hyperfine parameters is to be undertaken.

8. Rare-earth Mössbauer Effect Studies

Rare-earth Mössbauer effect spectra of R_2Fe_{17} and their interstitials are also very helpful in understanding their microscopic properties. Three rare earths are discussed in this section, gadolinium-155,[42,43] erbium-166,[44] and thulium-169.[45] The usual hyperfine parameters, magnetic hyperfine field, quadrupole splitting and isomer shift, have been measured. The isomer shift is sensitive to the bonding of the rare-earth atom and hence, is expected to change with the introduction of interstitial atoms, which are near neighbors of the rare-earth atom as indicated in Table I. The magnetic hyperfine field at the rare earth is the sum of four contributions, the core polarization by the local 4f moment, H_{cp}, the conduction electron contribution due to the self polarization by the 4f moment at the probe, H_{sp}, and the transferred fields, $H_{tr}(Fe)$ and $H_{tr}(R)$, resulting from the surrounding iron and rare-earth moments, respectively. H_{cp} is generally well determined for a saturated moment rare-earth atom. $H_{tr}(Fe)$ at the rare-earth atom has been measured by nuclear magnetic resonance studies [46] of the non-magnetic yttrium sites in Y_2Fe_{17} and is –204kOe. The iron transferred field at the other rare-earth atoms is obtained by multiplying by the ratio of the hyperfine coupling constants of the specific rare-earth and yttrium. Hence the sum of H_{sp} and $H_{tr}(R)$ can be obtained from the experimentally observed Mössbauer hyperfine field.

The most interesting parameter is the quadrupole splitting because it contains information on the second order crystal field parameter, A_2^0, associated with the rare-earth atom. Because the sign of A_2^0 is related to the rare-earth magnetic anisotropy, the easy magnetization direction of the rare-earth sublattice can be inferred. In SI units,[47] the relationship [48] between the principal component of the electric field gradient, V_{zz}, and A_2^0 is

$$A_2^0 = -eV_{zz}/4, \tag{7}$$

if the aspherical charge density which produces the electric field gradient and the 4f crystal field are outside the 4f orbital. In Equation 7, V_{zz} is given in V/m^2 and e is the electronic charge in coulomb and thus the SI unit for A_2^0 is J/m^2. Because A_2^0 is usually given in units of K/a_0^2, a value of V_{zz} in V/m^2 must be multiplied by 0.3211×10^{-16} in order to obtain A_2^0 in K/a_0^2. Because the 4f charge density extends over a much larger range than the nuclear charge density, shielding effects must be included in Equation 7, in the form of a shielding factor, $(1-\sigma_2)$. Experimental and calculated [49] values of σ_2 are in the range of 0.4 to 0.7. Further, the shielding of point charges by core electrons given by the Sternheimer antishielding factor, γ_∞, must be included in Equation 7, which then becomes

$$A_2^0 = -\frac{(1-\sigma_2)\, e\, V_{zz}}{(1-\gamma_\infty)\; 4}. \tag{8}$$

The $(1-\gamma_\infty)/(1-\sigma_2)$ term has been measured in many rare-earth compounds and found [48] to be between 120 and 300 for gadolinium. Band structure calculations [48] show that Equation 8 is an empirical relationship, which seems to be valid for a series of similar compounds in which the 6p and 5d orbitals of the rare-earth atom show similar asphericities.

In conclusion, although the exact value of A_2^0 obtained from V_{zz} values with Equation 8, depends upon the rather arbitrarily chosen value for the shielding factors, its sign is clearly determined by the sign of V_{zz}. Further, for a series of compounds, a proportionality between A_2^0 and V_{zz} is observed and thus, the comparison of V_{zz} values within a series of compounds is meaningful.

The magnetic anisotropy of the R_2Fe_{17} compounds and their interstitial derivatives is primarily determined by the rare-earth magnetic anisotropy, whose first-order anisotropy constant, K_1, is given by

$$K_1 = -(3/2)\, \alpha_J <r^2>_{4f} A_2^0\, (3<J_z^2> - J(J+1)), \tag{9}$$

where α_J is the second order Stevens coefficient and $<r^2>_{4f}$ is the expectation value for the 4f orbitals. Both values are tabulated in the literature.[50-52] If K_1 is positive, the easy direction of magnetization is expected, at least at low temperature, to be parallel to the c-axis, whereas, if K_1 is negative, the easy direction of magnetization is expected to be in the basal plane.

Gadolinium-155 Mössbauer spectroscopy is a very attractive technique for the study of Gd_2Fe_{17} and its interstitial derivatives because it yields good spectral resolution [49] and permits the measurement of the quadrupole interaction at the gadolinium site. Figures 19 and 20 show the gadolinium-155 Mössbauer spectra of Gd_2Fe_{17},[43] $Gd_2Fe_{17}H_x$,[43] and $Gd_2Fe_{17}N_x$[42] at 4.2K and Table VI gives the corresponding hyperfine parameters. As explained above, the most interesting parameter is the quadrupole splitting, ΔE_Q, from which values of A_2^0 have been obtained by using a proportionality constant between A_2^0 and V_{zz} which was different for Gd_2Fe_{17} and its hydrides [43] and for Gd_2Fe_{17} and its carbide and nitride.[42] Even though the exact values of A_2^0 may not be significant, the changes with hydrogen, nitrogen, or carbon are significant. It is clear that A_2^0 becomes less negative as the hydrogen content increases. Hence, the rare-earth anisotropy decreases with increasing hydrogen concentration. In contrast, A_2^0 becomes progressively more negative as carbon and nitrogen are inserted into Gd_2Fe_{17}, indicating that the rare-earth anisotropy increases. The different influence of hydrogen, nitrogen, or carbon on the rare-earth sublattice anisotropy is surprising and not well understood at this time.

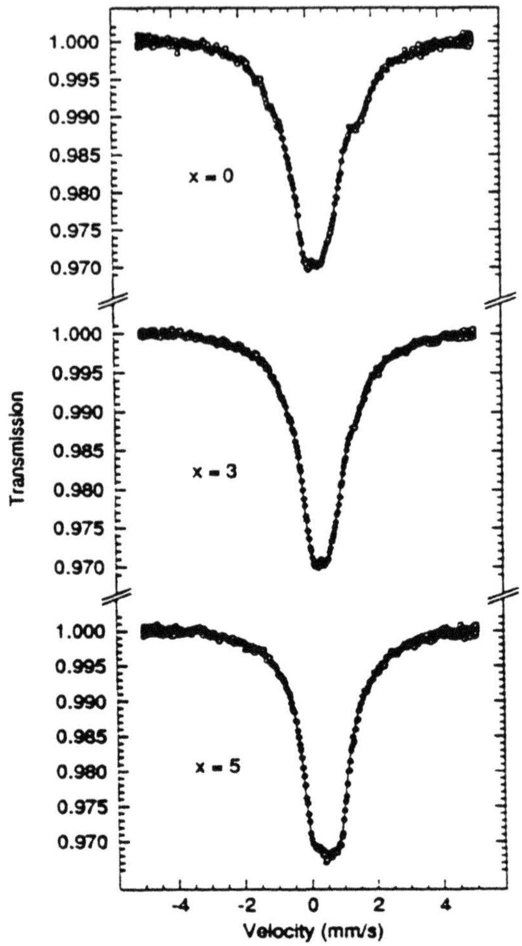

Figure 19. The Gadolinium-155 Mössbauer effect spectra of $Gd_2Fe_{17}H_x$, with x=0, 3, and 5, measured at 4.2K. (Reproduced with permission from ref. 43.)

As is shown in Table VI, the magnetic hyperfine field measured at the gadolinium site decreases when hydrogen, carbon, or nitrogen is introduced into Gd_2Fe_{17}. As explained above, the field is the sum of four contributions. The core polarization field, H_{cp}, for saturated gadolinium moments is –340kOe, and the transferred field due to the iron moments, $H_{tr}(Fe)$, obtained from the field measured in Y_2Fe_{17}, is +473kOe. Hence, from the observed field in Gd_2Fe_{17}, a value of +70kOe is deduced for the sum of H_{sp} and $H_{tr}(Gd)$. Upon hydrogenation, the hyperfine field decreases because H_{sp} decreases, as was observed earlier for $R_2Fe_{14}BH_x$ interstitial compounds.[53] The larger decrease in the hyperfine field observed upon the insertion of carbon and nitrogen may be due to a decrease in the transferred field resulting from the iron moments, moments which are now further from the gadolinium atoms because of the lattice expansion.

TABLE VI. Rare-earth Hyperfine and Crystal Field Parameters.

Gadolinium-155	Gd_2Fe_{17}	$Gd_2Fe_{17}H_3$	$Gd_2Fe_{17}H_5$	$Gd_2Fe_{17}C_{1.2}$	$Gd_2Fe_{17}N_3$
δ, mm/s	0.25(1)	0.37	0.42	0.34	0.47
H, kOe	+206(4)	+164(3)	+170(3)	+106.9	+47
ΔE_Q, mm/s	+1.81	+1.20	0		
V_{zz}, $\times 10^{21}$V/m^2	+4.1(1)	+2.66	0	+9.3	+12.6
A_2^0, K/a_0^2	–351, –200	–233	0	–428	–580
Reference	42, 43	43	43	55	42
Erbium-166[a]	Er_2Fe_{17}			$Er_2Fe_{17}C_x$	$Er_2Fe_{17}N_x$
H, kOe	84.1			82.4	78.1
ΔE_Q, mm/s	8.4			4.9	3.7
A_2^0, K/a_0^2	–50			–290	–400
Thulium-169[b]	Tm_2Fe_{17}			$Tm_2Fe_{17}C_x$	$Tm_2Fe_{17}N_x$
A_2^0, K/a_0^2	90			–300	–300

[a]Data obtained from ref. 44. [b]Data obtained from ref. 45.

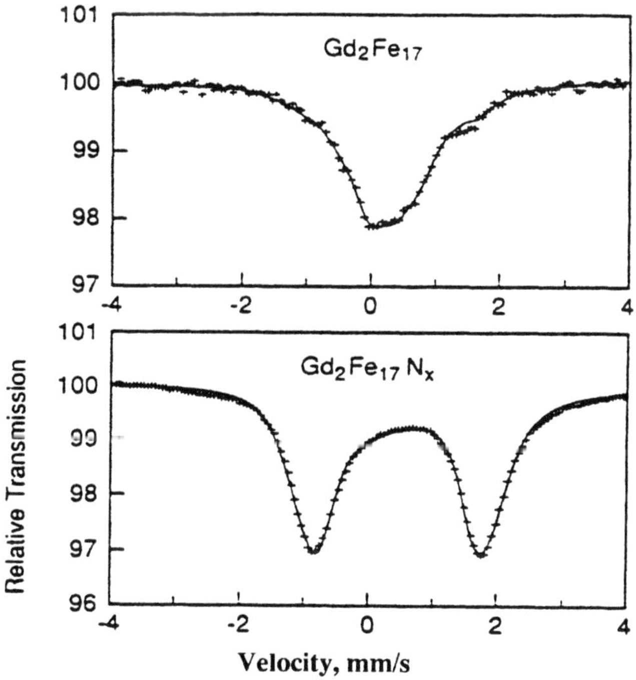

Figure 20. The Gadolinium-155 Mössbauer effect spectra of Gd_2Fe_{17} and $Gd_2Fe_{17}N_x$, measured at 4.2K. (Reproduced with permission from ref. 42.)

As is indicated in Table VI, the isomer shift increases and hence, the *s*-electron density at the nucleus decreases, with the introduction of hydrogen, carbon, or nitrogen. This occurs because of a charge transfer from the rare-earth 6s orbitals to the hydrogen, carbon, or nitrogen atom.[54] The largest increase in the isomer shift is observed for $Gd_2Fe_{17}N_3$ and probably results from a combination of charge transfer and volume expansion.

Erbium-166 and thulium-169 Mössbauer spectra of Er_2Fe_{17} and Tm_2Fe_{17} and their carbides and nitrides have also been studied.[44,45] The parameters obtained from these spectra are included in Table VI. Again, the most interesting parameter is the quadrupole splitting, from which the A_2^0 parameter was obtained by assuming a proportionality constant between the two parameters. A_2^0 clearly becomes progressively more negative as carbon or nitrogen are inserted into Er_2Fe_{17} and Tm_2Fe_{17}. This behavior indicates, in agreement with the results for gadolinium, that the rare-earth anisotropy increases as carbon or nitrogen are inserted into the lattice.

Acknowledgments

The authors would like to thank Dr. K. H. J. Bushow for providing numerous samples and for many helpful discussions during the course of this work, and Mr. Sanjay Mishra for his experimental help. The authors acknowledge, with thanks, NATO for a cooperative scientific research grant (92-1160) and the Division of Materials Research of the US National Science Foundation, for grant DMR-9214271. GJL would like to thank the Commission for Educational Exchange between the United States of America, Belgium, and Luxembourg for a Fulbright Research Fellowship and the 'Fonds National de la Recherche Scientifique,' Belgium, for a visiting professorship at the Universsity of Liège, during the 1993-1994 academic year.

References

[1] G. J. Long, O. A. Pringle, F. Grandjean, and K. H. J. Buschow, J. Appl. Phys. **72,** 4845 (1992).

[2] S. Mørup, "Mössbauer Spectroscopy and its Application to Studies of Time-dependent Phenomena," in *The Time Domain in Surface and Structural Dynamics,* G. J. Long and F. Grandjean, eds., Kluwer Academic Publishers, Dordrecht (1988), p. 271.

[3] F. Grandjean, "Mössbauer Spectral Lineshapes in the Presence of Electronic State Relaxation," in *The Time Domain in Surface and Structural Dynamics,* G. J. Long and F. Grandjean, eds., Kluwer Academic Publishers, Dordrecht (1988), p. 287.

[4] N. N. Greenwood and T. C. Gibb, *Mössbauer Spectroscopy*, Chapman and Hall, London (1981).

[5] E. Murad and J. H. Johnston, "Iron Oxides and Oxyhydroxides," in *Mössbauer Spectroscopy Applied to Inorganic Chemistry*, Vol. 2, G. J. Long, ed., Plenum Press, New York (1989) p. 507.

[6] S. M. Dubiel, Z. Zurek, and M. Przybylski, "Sulfidation-induced Changes in Iron-Aluminum and Iron-Silicon Alloys," in *Industrial Applications of the Mössbauer Effect*, G. J. Long and J. G. Stevens, eds., Plenum Press, New York (1986) p. 217.

[7] B. Fultz and J. W. Morris, Jr., "Hyperfine Fields in Fe-Ni-X Alloys and Their Application to a Study of Tempering of 9Ni Steel," in *Industrial Applications of the Mössbauer Effect*, G. J. Long and J. G. Stevens, eds., Plenum Press, New York (1986) p. 237.

[8] A. J. Freeman, Phys. Rev. **130**, 888 (1963).
[9] C. Wivel and S. Mørup, J. Phys. E **14**, 605 (1981).
[10] R. E. Vandenberghe and E. De Grave, "Mössbauer Effect Studies of Oxidic Spinels," in *Mössbauer Spectroscopy Applied to Inorganic Chemistry*, Vol. 3, G. J. Long and F. Grandjean, eds., Plenum Press, New York (1989) p. 59.
[11] G. K. Shenoy, "Mössbauer Effect Isomer Shifts," in *Mössbauer Spectroscopy Applied to Inorganic Chemistry*, Vol. 1, G. J. Long, ed., Plenum Press, New York (1989) p. 57.
[12] R. H. Herber, "Structure, Bonding, and the Mössbauer Lattice Temperature," in *Chemical Mössbauer Spectroscopy*, R. H. Herber, ed., Plenum Press, New York (1989) p. 199.
[13] B. Kolk, "Studies of Dynamical Properties of Solids with the Mössbauer Effect," in *Dynamic Properties of Solids*, G. K. Horton and A. A. Maradudin, eds., North-Holland, Amsterdam (1984) p. 5.
[14] G. J. Long, "Basic Concepts of Mössbauer Spectroscopy," in *Mössbauer Spectroscopy Applied to Inorganic Chemistry*, Vol. 1, G. J. Long, ed., Plenum Press, New York (1989) p. 7.
[15] M. T. Averbuch-Pouchot, R. Chevalier, J. Deportes, B. Kebe, and R. Lemaire, J. Magn. Magn. Mater. **68**, 190 (1987).
[16] S. S. Jaswal, IEEE Trans. Magn. **28**, 2322 (1992).
[17] H. Kawakami, T. Hihara, Y. Koi, and T. Wakiyama, J. Phys. Soc. Japan **33**, 1591 (1972).
[18] L. Gelato, J. Appl. Cryst. **14**, 141 (1981).
[19] F. Grandjean, G. J. Long, D. E. Tharp, O. A. Pringle, and W. J. James, J. Physique, Coll. **49**, C8-581 (1988).
[20] F. Grandjean, G. J. Long, O. A. Pringle, and J. Fu, Hyperfine Interactions **62**, 131 (1990).
[21] G. J. Long, R. Kulasekere, O. A. Pringle, F. Grandjean, and K. H. J. Buschow, J. Magn. Magn. Mater. **117**, 239 (1992).
[22] G. J. Long, O. A. Pringle, F. Grandjean, W. B. Yelon, and K. H. J. Buschow, J. Appl. Phys. **74**, 504 (1993).
[23] J. F. Herbst, J. J. Croat, R. W. Lee, and W. B. Yelon, J. Appl. Phys. **53**, 250 (1982).
[24] S. Miraglia, J. L. Soubeyroux, C. Kolbeck, O. Isnard, and D. Fruchart, J. Less-Common Met. **171**, 51 (1991).
[25] K. H. J. Buschow, J. Appl. Phys. **42**, 3433 (1971).
[26] O. Isnard, J. L. Soubeyroux, D. Fruchart, T. H. Jacobs, and K. H. J. Buschow, J. Alloys Compounds **186**, 135 (1992); O. Isnard, S. Miraglia, J. L. Soubeyroux, D. Fruchart, and J. Pannetier, Phys. Rev. B **45**, 2920 (1992).
[27] T. H. Jacobs, Doctoral Dissertation, University of Leiden, the Netherlands (1992).
[28] H. Sun, J. M. D. Coey, Y. Otani, and D. P. F. Hurley, J. Phys. Condens. Matter **2**, 6465 (1990).
[29] G. J. Long, S. Mishra, O. A. Pringle, F. Grandjean, and K. H. J. Buschow, J. Appl. Phys. **75**, 5994 (1994).
[30] P. C. M. Gubbens, Doctoral Dissertation, Delft University of Technology, Delft, the Netherlands (1977).
[31] F. Grandjean, G. J. Long, O. A. Pringle, and K. H. J. Buschow, Hyperfine Interactions (1994), in press.
[32] G. J. Long, O. A. Pringle, F. Grandjean, T. H. Jacobs, and K. H. J. Buschow, J. Appl. Phys. **75**, 2598 (1994).
[33] Z. Gu and W. Lai, J. Appl. Phys. **71**, 3911 (1992).

[34] R. Coehoorn and G. H. O. Daalderop, J. Magn. Magn. Mater. **104-107**, 1081 (1992).
[35] T. Beuerle and M. Fähnle, Phys. Stat. Sol. (b) **174**, 257 (1992).
[36] Q. Qi, H. Sun, R. Skomski, and J. M. D. Coey, Phys. Rev. B **45**, 12278 (1992).
[37] P. C. M. Gubbens, J. J. Van Loef, and K. H. J. Buschow, J. Phys. Colloq. **35**, C6-617 (1974).
[38] B. P. Hu, H. S. Li, H. Sun, and J. M. D. Coey, J. Phys. Condens. Matter **3**, 3983 (1991).
[39] B. Rupp and G. Wiesinger, J. Magn. Magn. Mater. **71**, 269 (1988).
[40] X. Chen, D. H. Ryan, Z. Altounian, and L. X. Liao, J. Appl. Phys. **73**, 6038 (1993).
[41] F. Grandjean, G. J. Long, S. Mishra, O. A. Pringle, O. Isnard, S. Miraglia, and D. Fruchart, Hyperfine Interactions, submitted for publication, July 1994.
[42] M. W. Dirken, R. C. Thiel, R. Coehoorn, T. H. Jacobs, and K. H. J. Buschow, J. Magn. Magn. Mater. **94**, L15 (1991).
[43] O. Isnard, P. Vulliet, A. Blaise, J. P. Sanchez, S. Miraglia, and D. Fruchart, J. Magn. Magn. Mater. **131**, 83 (1994).
[44] P. C. M. Gubbens, A. A. Moolenaar, G. J. Boender, A. M. van der Kraan, J. Magn. Magn. Mater. **97**, 69 (1991).
[45] P. C. M. Gubbens, A. A. Moolenaar, T. H. Jacobs, and K. H. J. Buschow, J. Magn. Magn. Mater. **104-107**, 1113 (1992).
[46] A. Oppelt and K. H. J. Buschow, J. Phys. F **3**, L212 (1973).
[47] The reader should be careful when reading the literature because some authors [41-43] use a relation between A_2^0 and V_{zz} which is not written in SI units.
[48] R. Coehoorn, "Electronic Structure Calculations for Rare Earth-Transition Metal Compounds," in *Supermagnets, Hard Magnetic Materials*, G. J. Long and F. Grandjean, eds., Kluwer Academic Publishers, Dordrecht (1990) p. 133.
[49] G. Czjzek, "Mössbauer Spectroscopy of New Materials Containing Gadolinium," in *Mössbauer Spectroscopy Applied to Magnetism and Materials Science*, Vol. 1, G. J. Long and F. Grandjean, eds., Plenum Press, New York (1993) p. 373.
[50] K. W. H. Stevens, Proc. Phys. Soc. (London) **A65**, 209 (1952).
[51] B. Bleaney and K. W. H. Stevens, Rept. Prog. Phys. **16**, 108 (1953).
[52] A. J. Freeman and J. P. Desclaux, J. Magn. Magn. Mater. **12**, 11 (1979).
[53] J. M. Friedt, A. Vasquez, J. P. Sanchez, P. L'Héritier, and R. Fruchart, J. Phys. F **16**, 651 (1986); J. P. Sanchez, J. M. Friedt, A. Vasquez, P. L'Héritier, and R. Fruchart, Solid State Commun. **57**, 309 (1986).
[54] G. K. Shenoy, B. Schuttler, P. J. Viccaro, and D. Niarchos, J. Less-Common Met. **94**, 37 (1983).
[55] M. W. Dirken, R. C. Thiel, L. J. de Jongh, T. H. Jacobs, and K. H. J. Buschow, J. Less-Common Met. **155**, 339 (1989).

Chapter 20

NMR STUDIES OF INTERMETALLICS AND INTERSTITIAL SOLUTIONS CONTAINING H, C and N

CZ. KAPUSTA[&] and P.C. RIEDI[#]
[&]*Department of Solid State Physics*
Faculty of Physics and Nuclear Techniques
University of Mining and Metallurgy
30-059 Cracow, Poland
[#]*Department of Physics and Astronomy*
University of St.Andrews
St.Andrews, Fife
KY 16 9SS,Scotland, UK

ABSTRACT. A survey is given of the NMR work on the rare earth - 3d metal compounds containing interstitial hydrogen, carbon and nitrogen. In particular the resolution and usefulness of NMR for studying the variation of hyperfine parameters caused by interstitial elements is stressed. Information on the site occupation and its influence on magnetic hyperfine fields and electric field gradients is analysed. The variation of magnetic moments and the rare earth contributions to the crystal electric field induced magnetocrystalline anisotropy in these materials are also discussed.

1. Introduction

Novel material science usually deals with multi-element materials that often contain several sublattices. Most of their intrinsic parameters relevant to applications are products of individual site contributions. Thus, a knowledge of individual site properties is essential if the desired intrinsic parameters are to be obtained. This concerns also magnetic properties and, in particular, the rare earth based permanent magnet materials. A very promising method for modification of the intrinsic properties of materials involves the introduction of light elements such as hydrogen, carbon or nitrogen on the interstitial sites in the crystal structure [1,2]. A huge increase of the magnetocrystalline anisotropy and a doubling of the Curie temperature can be obtained.

Nuclear magnetic resonance (NMR) is one of the few experimental techniques which allows the study of individual site properties such as magnetic moments, contributions to the magnetocrystalline anisotropy and the site environment. Information on these quantities is obtained from the hyperfine parameters reflecting the interaction of the spin and quadrupole moment of a nucleus with magnetic fields and electric field gradients (EFG) produced by electron shells. The occupation of unfilled electronic shells is usually different between inequivalent sites or atomic positions in a molecule. It corresponds to the appearance of different electron spin and orbital polarisations in an applied magnetic or exchange molecular field and results in a difference in magnetic fields and EFGs between the sites or atomic positions. Measurements of nuclear relaxation rates can be used for studying molecular and lattice dynamics, as well as the diffusion of atoms or magnetic domain wall motions.

F. Grandjean et al. (eds.), Interstitial Intermetallic Alloys, 497–520.

Nuclei with spin I different from zero can be used as probes for local magnetic fields. Those with I>1/2, having nonvanishing nuclear quadrupole moments, can also probe EFG. The majority of isotopes in the periodic table can serve as NMR probes. For information on the basic principles of NMR see Abragam [3] and Riedi [4] - this volume. The requirements of the experimental technique and the information obtained depend on the materials investigated. For example, features of classical NMR in diamagnetic materials [1,5] are different from those of NMR in metals [6], which in turn differ from the frequency swept spin-echo NMR in strongly magnetic materials [7,8,9,10].

NMR has been widely applied in research on metals [6] and their hydrides (see review [11] - this volume). It has proved to be also very useful for studying the recently discovered permanent magnet materials containing interstitial carbon and nitrogen [12,13,14,15]. In this paper we present a survey of NMR results on the $RE_2TM_{17}A_x$ series (RE = rare earth and yttrium; TM = Fe, Co; A = H, C, N) and $NdFe_{11}Ti$. For a comparison of the influence of the interstitial element, the NMR results on the Laves phase YFe_2H_x and $GdMn_2H_x$ are shown. An example for permanent magnet boride $La_2Fe_{14}B$ is also presented. All the results shown were obtained on polycrystalline samples at 4.2 K

2. NMR in strongly magnetic materials

The characteristic features of NMR in strongly magnetic materials are related to the presence of internal fields resulting from magnetic order. Thus, the resonant condition can be fulfilled at zero applied magnetic field making use of the effective internal field with its magnetic induction denoted as B_e. For the quadrupole interaction much smaller than the magnetic one the resonant frequencies in the first order approximation are given by the relation:

$$\nu_{m,m-1} = | \gamma_n B_e/2\pi + P_q(2m-1)| \qquad (1),$$

where m runs in integer steps from 1-I to I. The spectrum of a nucleus with I>1/2 has the form of 2I equidistant lines.

In a NMR experiment at zero applied field the internal magnetic field B_e at a nucleus can be expressed as follows:

$$\mathbf{B_e = B_{loc} + B_{hf}} \qquad (2).$$

B_{loc} is the local field consisting of the Lorentz field and the dipolar field B_{dip}, which is usually the dominant contribution to B_{loc}. B_{dip}, however, rarely exceeds 1 T. The hyperfine field B_{hf} is usually the dominant contribution and B_e is sometimes called synonymously "hyperfine field" and abbreviated as HFF. B_{hf} can be expressed as a sum:

$$\mathbf{B_{hf} = B_s + B_{orb} + B_n} \qquad (3).$$

B_s is the self polarisation term originating from core and conduction electron polarisation by the spin moment of the parent atom. B_{orb} is the orbital field produced by nonvanishing orbital moments of the 4f electrons at the rare earth nuclei and of the 3d electrons at the 3d transition metals nuclei. B_n is the polarisation contribution from neighbouring magnetic atoms often called the "transferred" HFF. B_{orb} is dominant for the non S-state rare earths and can be comparable with B_s for the 3d metals. The contributions to HFF are related to the partial moments through the hyperfine coupling and can be written as $B_i = A_i \times \mu_i$, where i stands for "s", "orb" and "n" indexes.

The quadrupole parameter can be expressed as:

$$P_q = -3eQ_nV_{zz}/4hI(2I-1) \qquad (4)$$

where Q_n is the nuclear quadrupole moment and V_{zz} is the EFG component with maximum absolute value. From the quadrupole splittings $\Delta\nu_q$ derived from the line separations of the quadrupole spectra the corresponding values of the electric field gradient (EFG) component along the hyperfine field Vii can be derived using the formula:

$$|V_{ii}| = 2I(2I-1)h\,\Delta\nu_q / 3e|Q_n| \qquad (5).$$

The EFG at the nucleus of a non-S state rare earth originates mainly from a nonspherical distribution of the 4f electron density of the parent ion as well as from the asphericity of the 6p and 5d electron density of the parent ion due to the presence of neighbouring atoms in the lattice [16]. Thus, the diagonal EFG component along the hyperfine field direction may be written as:

$$V_{ii} = V_{ii}(4f) + V_{ii}(latt) \qquad (6).$$

In the compounds where RE-TM exchange interaction is much stronger than the crystal electric field, e.g. in $RE_2TM_{17}A_x$, the rare earth preserves its fully polarised ground state irrespective of the direction of magnetisation. Thus, $V_{ii}(4f)$ is independent of the magnetic moment direction and has its maximum value denoted as $V_{zz}(4f)$, which is proportional to the quantum number J_z. Also B_{orb} preserves its maximum value.

The whole spectrum of a nucleus very often covers a range of a few hundred Tesla, so the most useful technique is frequency swept spin-echo NMR at zero applied field. A standard two-pulse sequence of radiofrequency (rf) pulses is usually applied and the amplitude of the nuclear spin-echo appearing after the second pulse at the distance equal to the pulse separation is measured. An external magnetic field can be applied for saturation of the sample or determination of the origin of an unknown signal. Recent progress in construction of broadband rf electronics allowed the construction of an untuned computer controlled frequency swept spectrometer (see review by Riedi - this volume [4]) and made the NMR research in magnetic materials much more effective.

The presence of magnetic domain structure in ordered materials leads to enhancement of the rf field and lowers the pulse power required to form a spin echo. However, NMR signals corresponding to different magnetic regions appear and very often overlap, which brings about some difficulties in the interpretation of the spectra. For studying anisotropic properties such as orbital contributions to magnetic moments and EFGs it is essential to be able to obtain a signal corresponding to a well defined direction of local magnetisation with respect to the crystallographic axes. In the subsequent paragraphs we will briefly outline the way of solving this problem.

The enhancement effect is caused by oscillations of the electronic magnetisation excited by the rf pulses. A rf field enhanced through the hyperfine coupling appears at the nucleus. The effect is usually much larger in domain walls than in domains so the signals from domain walls are usually observed. The rf enhancement in highly anisotropic materials also varies across the domain wall and is larger at the domain wall centre (DWC) than at the domain wall edge (DWE). Thus, the DWC signals can be distinguished from the DWE ones on the basis of their dependence on the rf pulse power. The maximum of the former corresponds usually to a much lower pulse power than for the latter. Thus, by increasing pulse power we can scan through the domain wall. For large enough power we can study DWE regions adjacent to domain interiors, where the directions of magnetic moments are nearly parallel and are representative of moments direction in the domain interior.

Another possibility to distinguish between the DWC and DWE signals are the nuclear relaxation rates. Since the relaxation of the DWC signal is much faster than the DWE one, a measurement at a large pulse separation removes the DWC signal but leaves the DWE one, e.g. the La resonance in $La_2Fe_{14}B$.

A feature of zero applied field NMR worth noting is that the spectrum depends on the mutual directions of B_e, which usually follows the local magnetisation, and the principal axes of the EFG tensor, which are rigidly coupled to the crystal lattice. The spectra of polycrystalline samples are therefore equivalent to those of single crystals.

3. $RE_2TM_{17}A_x$

The host compounds crystallise in the hexagonal structure for Y and in the rhombohedral structure for light rare earths. A building segment for unit cells of both structures is presented in Fig.1. Both structures contain four inequivalent TM sites 12i, 12k, 6g, 4f in the hexagonal and 18h, 18f, 9d, 6c in the rhombohedral compound. RE elements occupy a single site 6c in the rhombohedral and two sites 2b and 2d in the hexagonal structure.

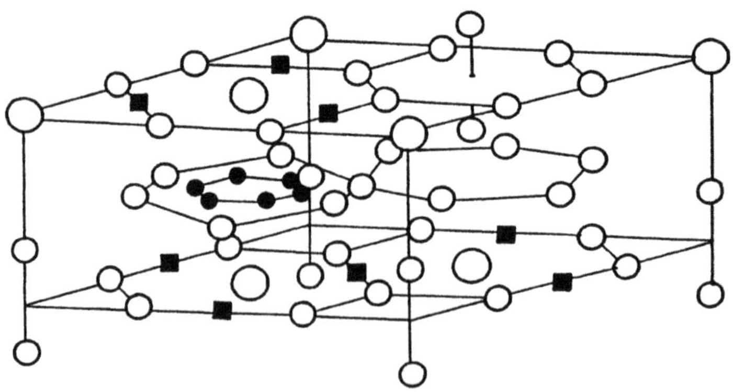

Fig. 1. A building block of the $RE_2TM_{17}A_x$. Large empty circles - RE, small empty circles - TM, filled circles - A 9e (6h) sites, filled cubes - A 18g (12i) sites

Neutron diffraction experiments shown that the C atoms occupy randomly the interstitial positions 9e (6h) [1]. The hydrogen atoms locate at 9e and 18g (6h and 12i) sites [17]. Nitrogen locates almost entirely at 6h (9e) interstitial positions [18], but recently the succesful preparation of a nitride containing nitrogen at the 18g (12i) sites was reported [19]. This result has not however been generally accepted. The number of 9e (6h) and 18g (12i) nearest neighbour (NN) positions to a RE site is 3 and 6 respectively.

3.1. ^{89}Y SPECTRA

3.1.1. $Y_2Fe_{17}C_x$. The NMR spectra measured on the samples prepared by co-melting appropriate amounts of the constituents are shown in Fig.2 [12]. Three well resolved satellite lines in addition to the ^{89}Y (I = 1/2 - no quadrupole splitting) line of the host compound [20] were observed. The lines at 42, 37, 29 and 17 MHz were assigned to the Y sites with 0, 1, 2 and 3 C atoms respectively. A comparison of the corresponding B_e values shows that for the 3C configuration B_e is reduced by 60% with respect to the value in Y_2Fe_{17}. As B_e on yttrium is mostly of transferred hyperfine origin, it indicates a large C NN influence on the valence electron distribution at the RE site.

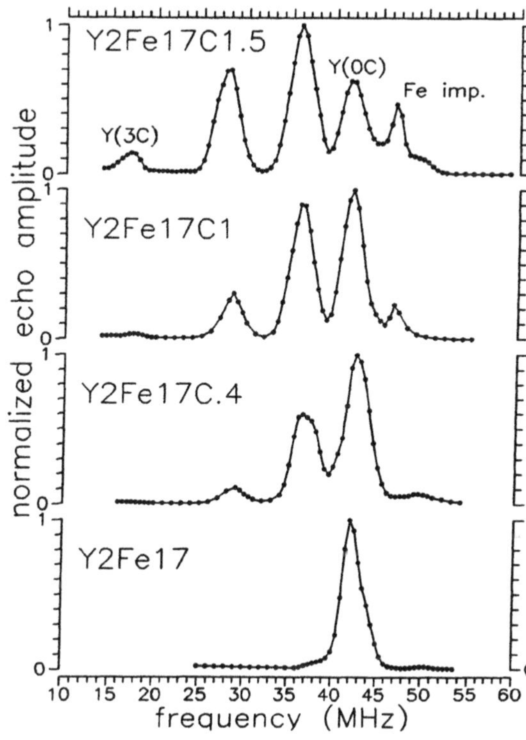

Fig.2. The ^{89}Y NMR spin-echo spectra of $Y_2Fe_{17}C_x$ at 4.2 K.

The line at 46 MHz appearing in the carbides has the broad echo shape characteristic of a narrow line, and is attributed to the ^{57}Fe signal from a spurious α-Fe phase present in the samples (see Fig.2).

The ^{57}Fe signals from the Fe sites in the structure are overlapped by the much stronger signal of ^{89}Y . However, the Fe resonance line at about 50 MHz is detected in all the samples.

3.1.2. $Y_2Fe_{17}N_x$. Four samples prepared by heating the Y_2Fe_{17} powder in an atmosphere of N_2 were measured [12]. The amount of absorbed nitrogen x was estimated from the pressure difference before and after the reaction for the first three samples. With increasing reaction time between subsequent samples the approximate values of x = 0.5, 1.2 and 2.5 were obtained. However the nitrogen distribution is non-uniform as discussed below.

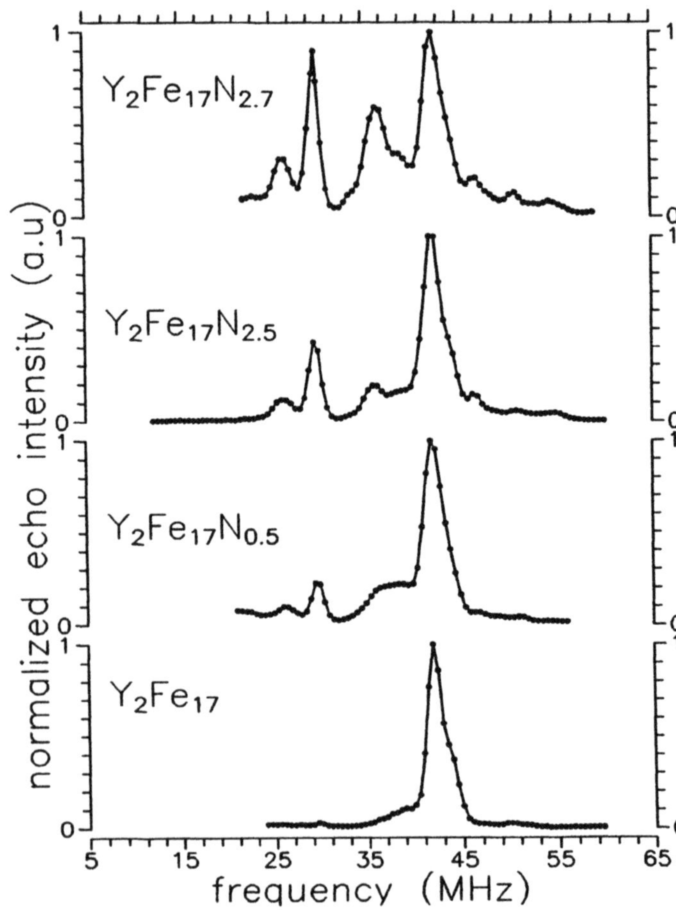

Fig.3. The ^{89}Y NMR spin-echo spectra of $Y_2Fe_{17}N_x$ at 4.2 K.

The ^{89}Y spectra shown in Fig.3 consist of four resonance lines: the main line at 42 MHz and the lines at 36, 30 and 26 MHz respectively. In a similar way to those in carbon containing compounds, the 36, 30 and 26 MHz lines are assigned to Y atoms with 1N, 2N and 3N atoms as NNs respectively. The set including 1N, 2N and 3N lines increases its intensity with x with respect to the 0N line. The intensity ratio within the set however remains unchanged which reflects the growth of the nitrided outer shells of grains. As X-ray diffraction shows that the samples contain both the expanded nitride phase and the unnitrided one, most of the 0N intensity can be attributed to the unnitrided cores of the grains. The intensity of the 2N line is the largest and the relative intensities within the set of 1N, 2N and 3N lines are approximately constant with x. This indicates that the distribution of nitrogen is not random, but that there is a preference for 2N neighbours around a Y atom. It also shows that during the growth of the nitrided outer shell the nitrogen concentration is approximately constant and corresponds to a value of x in the chemical formula slightly larger than 2, which is smaller than that deduced from mass enhancement or nitrogen pressure.

The reduction of the Y HFF per nitrogen NN atom decreases with the increasing number of N neighbours. In the carbon containing compounds an opposite tendency is observed.

3.1.3. $Y_2Fe_{17}CN_x$. The ^{89}Y spectrum presented in Fig.4 [12] consist of 6 lines at 42, 36, 30, 25, 20 and 15 MHz respectively. The results are interpreted consistently with those for carbides and nitrides. Thus, the 36 MHz line is assigned to Y atoms either with 1C or with 1N as NN, the 30 MHz line to either 2C or 2N as NN and 25 MHz line to either 3N or 2N+1C as NN. The 20 MHz line is interpreted as caused by 2C+1N and the 15 MHz line by 3C configuration. The environements with 2C and 1N+1C give a contribution to the shoulder at about 27 MHz. The calculated lowering of the Y HFF in 2C+1N and 1C+2N environements by simple summation of contributions per one atom in 3N and 3C configurations of $Y_2Fe_{17}C$ and $Y_2Fe_{17}N_x$ provides the expected resonance lines at 19.5 MHz and 22 MHz respectively, very close to that observed experimentally. On this basis an independent influence of C and N on Y HFF in mixed environements in this compound is concluded.

3.1.4. $Y_2Fe_{17}H_x$. Only one Y resonance line was detected, Fig.5 [12]. The line broadens from a 2.7 MHz linewidth for x=0 to 4.0 MHz and 4.2 MHz for x=1.6 and 5.5 respectively. It shifts from 42.5 to 41.5 and 40.9 MHz for x = 0, 1.6 and 5.5 respectively. From the above results no local influence but only a slight global influence of hydrogen on Y HFF is deduced. The Y-H distance is 2.5 Å for the hydrogen 6h sites and 2.35 Å for the 12i ones. As it was recently found [17] the 6h sites are filled for x<3 and a maximum of 2 of the 12i sites are occupied for x=5. The value of x equal to 5.5 given in [12] now appears to be overestimated by 10%.

3.1.5 $Y_2Fe_{17-y}Co_yC_x$. For y = 17 the samples of nominal carbon content x = 0 and 0.8 were measured [15]. The main line at 18.5 MHz (B =8.9 T) as well as a weaker line at 21 MHz (B_e =10.1 T) appearing in Y_2Co_{17} sample (Fig.6) according to literature data [20] are assigned to Y resonance in rhombohedral and hexagonal structures respectively. The additional weak line observed in the C containing sample at 15.7 MHz (B_e =7.5 T) is attributed to Y environements with one carbon NN. The corresponding relative decrease

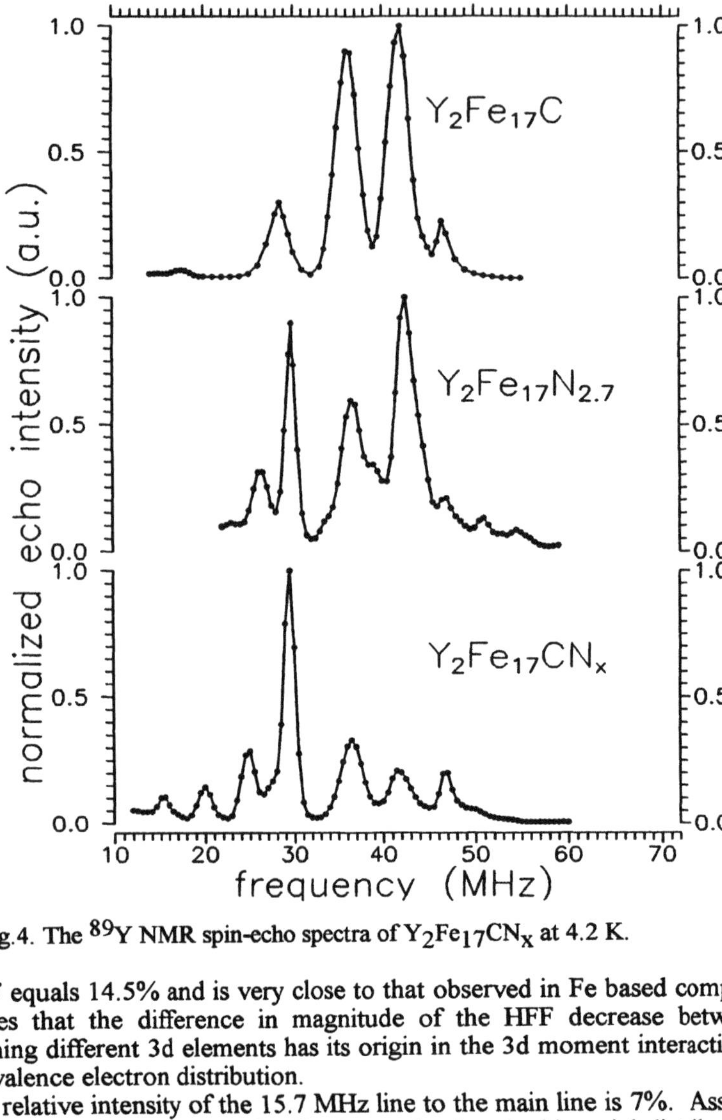

Fig.4. The ^{89}Y NMR spin-echo spectra of $Y_2Fe_{17}CN_x$ at 4.2 K.

of HFF equals 14.5% and is very close to that observed in Fe based compound (13%). It indicates that the difference in magnitude of the HFF decrease between compounds containing different 3d elements has its origin in the 3d moment interaction rather than in the Y valence electron distribution.

The relative intensity of the 15.7 MHz line to the main line is 7%. Assuming a random distribution of C atoms over the 6h sites and using the binomial distribution function we get the real concentration of interstitial carbon x=0.07, which is an order of magnitude smaller than the nominal (as weighted) x=0.8. It means that only a slight amount of carbon can be introduced into the Y_2Co_{17} lattice by co-melting.

In a similar way to that for $Y_2Co_{17}C_x$ the real carbon concentrations in the samples $Y_2Fe_6Co_{11}C_{0.8}$ and $Y_2Fe_3Co_{14}C_{0.8}$ amounting to x=0.8 and x=0.5 respectively were estimated [21]. These values can be treated as solubility limits for the compounds

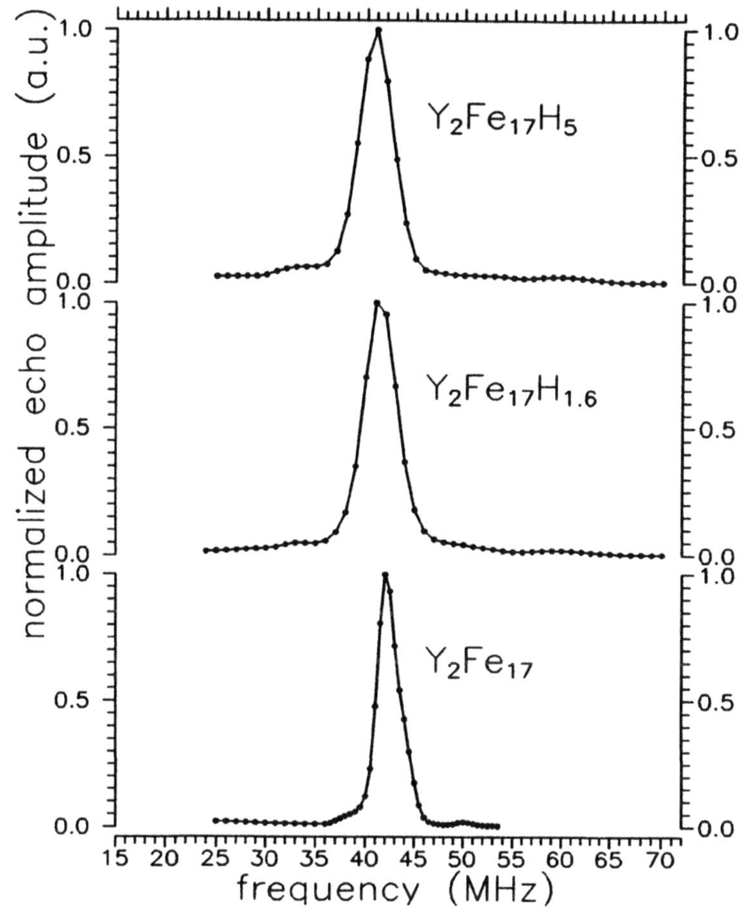

Fig.5. The ^{89}Y NMR spin-echo spectra of $Y_2Fe_{17}H_x$ at 4.2 K.

prepared from the melt. With increasing Co concentration the solubility of carbon decreases from x=1.5 for Y_2Fe_{17} down to x=0.07 for Y_2Co_{17}, Fig.7.

3.1.6. $Y_2Co_{17}N_x$. A similar evolution of the ^{89}Y spectra with x to that in the iron based compound is observed [22]. The lines, however, are broader. The real concentration x of the interstitial nitrogen is deduced to be approximately two.

3.1.7. $Y_2Co_{17}H_x$. For the measured sample with x = 3.3, similarly to $Y_2Fe_{17}H_x$, no local influence of hydrogen on the yttrium hyperfine field is observed [22]. The resonance line broadens and remains at the same position as in Y_2Co_{17}.

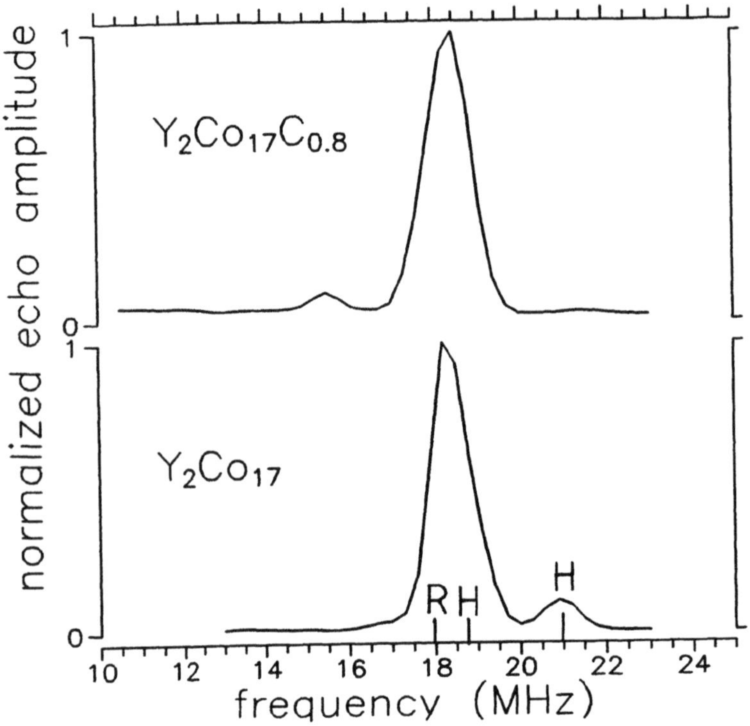

Fig.6. The ^{89}Y NMR spin-echo spectra of $Y_2Co_{17}C_x$ at 4.2 K.

3.2. ^{57}Fe SPECTRA

The ^{57}Fe signal in $Y_2Fe_{17}A_x$ not overlapped by a much stronger Y signal is observed at 50 MHz (B_e = 36.2 T), Fig.2 [12]. The position of this line in carbides is independent of x and no other Fe signals at higher frequencies appear. This line is assigned to the 4f (6c) dumb-bell site.

In the nitrogen containing compound an additional Fe resonance line at 55 MHz (B_e = 40.0 T) appears in good agreement with Mössbauer results [13], while the 50 MHz line still remains at the same frequency. A significant decrease of the NMR enhancement factor indicates a considerable influence of nitrogen on local magnetic anisotropies of Fe sites in contrast to carbon and hydrogen.

The ^{57}Fe spectra in $Lu_2Fe_{17}A_x$ are reported in [14]. Polycrystalline samples prepared in the same way as for yttrium containing compounds were measured. In the real structure of Lu_2Fe_{17} a statistical distribution of the Lu atoms and the Fe dumb-bells (4f) occurs. This leads to a splitting of the Fe sites making the comparison of results with

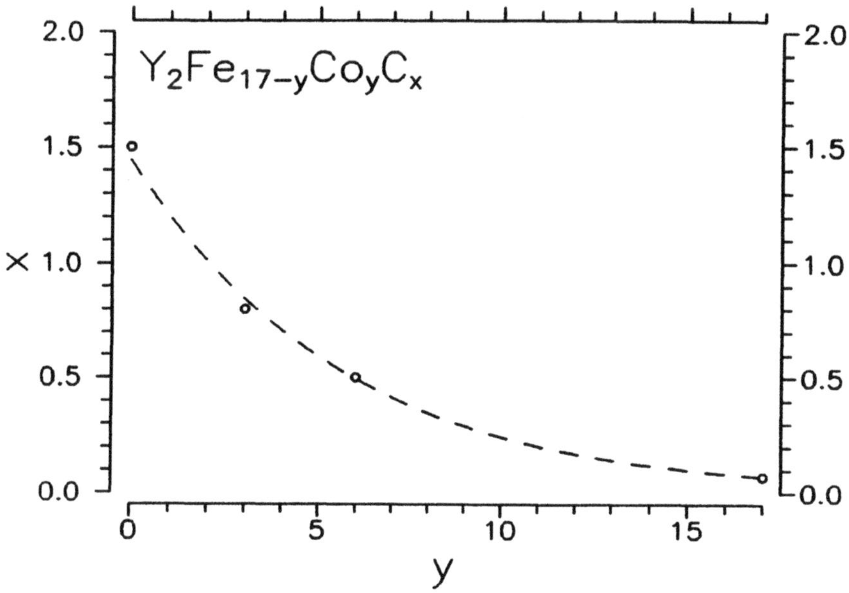

Fig.7. The real concentration of interstitial C in $Y_2Fe_{17-y}Co_yC_x$. The dashed line shows the approximate carbon solubility limit.

those for the Y-containing compounds difficult. The average HFF increases more in the nitride than for the carbide, but decreases for the hydride.

Two competing mechanisms have to be taken into account in analysing the Fe HFF changes: the magnetovolume effect and the local influence of the interstitial atom via bonding effects. The former mechanism leads to an increase of magnetic moments (and HFFs) with the lattice expansion, whereas the latter causes a decrease of them. Thus, in the hydride the local influence appears to be dominant, in contrast to the carbide and nitride.

3.3. ^{59}Co SPECTRA

The strongest changes of the spectrum as compared to the host Y_2Co_{17} are observed in the nitride [22]. For the carbide and hydride additional resonance lines to those of the host Y_2Co_{17} are observed. They are assigned to the HFF changes caused by the neighbouring H and C atoms. Much larger changes of the HFF in the nitride can be attributed to a change of the easy magnetization direction (EMD) from the basal plane for Y_2Co_{17} to the c-axis in the nitride. This means that there is a large influence of nitrogen on the orbital moment of the cobalt 3d electron shell, leading to changes of the spin-orbit interaction anisotropies of the individual cobalt sites. As the carbide contains a small amount of interstitial carbon of x=0.07, the results cannot be compared quantitatively with those for the nitride. A strong influence of nitrogen on the Fe anisotropy is also observed in $Y_2Fe_{17}N_x$.

3.4. ^{147}Sm AND ^{149}Sm SPECTRA

3.4.1 $Sm_2Fe_{17}N_x$. The host compound Sm_2Fe_{17} possesses a planar anisotropy due to the anisotropy of the Fe sublattice which overcomes the Sm sublattice preference to align along the c-axis. The introduction of nitrogen gives rise to a dramatic increase of the crystal electric field (CEF) potential at the Sm site, leading to a strong enhancement of the anisotropy of the samarium sublattice. This results in the large uniaxial magnetic anisotropy of $Sm_2Fe_{17}N_x$. Two groups of samples prepared in two ways described in [23] (A) and [24] (B) were measured in [25] and [13] respectively. The samples (A) were prepared as fine powders and intermediate concentrations of nitrogen $0<x<3$ were succesfully obtained.

The ^{147}Sm spectra of $Sm_2Fe_{17}N_x$ are shown in Fig.8. The ^{149}Sm spectra are fully consistent with them. The observed sets of quadrupole septets (nuclear spin for both Sm isotopes I = 7/2) are assigned to Sm 6c sites with different numbers of nitrogen neighbours. The samples (A) have linewidths 5 times larger than those of (B), indicating that they have much larger local inhomogeneities.

It was confirmed in an NMR experiment under applied magnetic field that the observed signal in $Sm_2Fe_{17}N_x$ is of DWE origin. The corresponding directions of the Sm moments are perpendicular to the c-axis for Sm_2Fe_{17}, parallel to the c-axis for x > 1.2 and tilted from the c-axis for x = 0.4. As the main septet for x = 0.4 is assigned to Sm with 0N the change of its quadrupole splitting, relative to Sm_2Fe_{17}, can be assigned to the change of easy magnetisation direction (EMD). The small septet is attributed to Sm with 2N neighbour and the septet in the fully nitrided sample (B) to Sm with 3N neighbours. A deviation of the intensities of the septets from those expected for a random distribution of nitrogen can be attributed here to large differences in the individual site anisotropies and a smearing of the signal of the low symmetric Sm sites.

The slight change of HFF with magnetic moment direction for the 0N site indicates that Sm preserves its fully polarised state. As the lattice EFG at the 0N site is axially symmetric with its V_{zz} along the c-axis, Laplace's equation can be applied i.e.:

$$V_{cc}(\text{latt}) = -2\, V_{pp}(\text{latt}) \qquad (7),$$

where c and p stand for c-axis- and c-plane components. Thus, using (6) from the quadrupole splitting for Sm_2Fe_{17} and for 0N septet in the sample with x = 1.2 the $V_{zz}(4f)$ and $V_{cc}(\text{latt})$ were obtained. The slightly smaller quadrupole splitting of the 0N septet for x = 0.4 than for x = 1.2 indicates a deviation of EMD from the c-axis. The angle between the c-axis and EMD for x = 0.4 is estimated to be 26°. The change of EMD between Sm_2Fe_{17} and $Sm_2Fe_{17}C_{0.5}$ is reflected in the ^{147}Sm spectra [26] in the same way as for the nitrides.

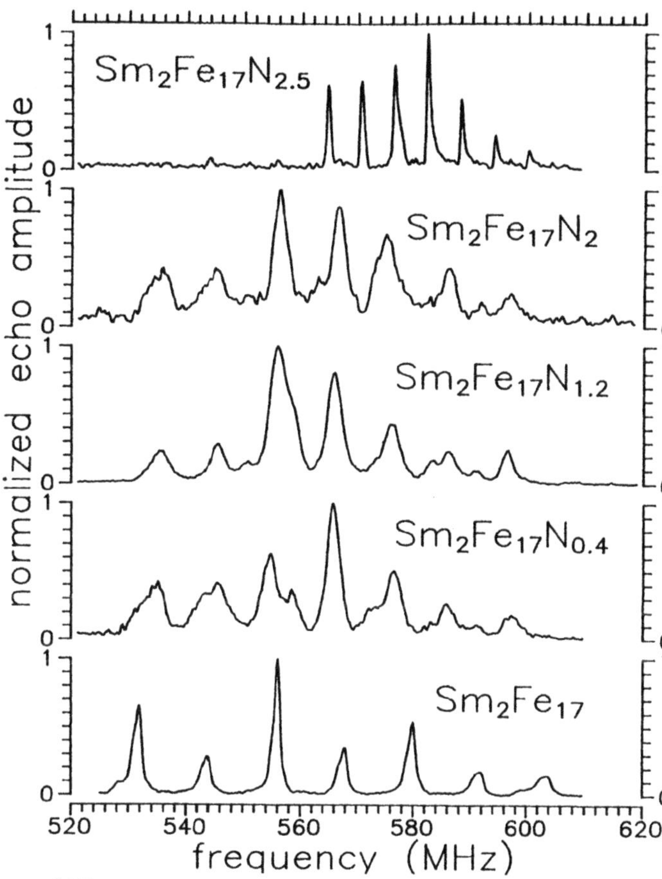

Fig.8 The ^{147}Sm NMR spin-echo spectra of $Sm_2Fe_{17}N_x$ at 4.2 K.

For 0N sites the V_{cc}(latt) value of 26×10^{20} V/m^2 was obtained, and for the 3N environements V_{cc}(latt) as large as 141×10^{20} V/m^2 is measured. The latter value bearing an error margin of 5% is close to that obtained from Gd Mössbauer measurements for $Gd_2Fe_{17}N_x$, $V_{cc}(latt) = 126\times10^{20}$ V/m^2 [27].

For $V_{zz}(4f)$ the value of -249×10^{20} V/m^2 is obtained, about 20% larger than the absolute value of 211×10^{20} V/m^2 derived from an atomic beam experiment [28], indicating a contraction of the 4f shell in the compound as compared to a free atom.

The lowest order term of the CEF interaction corresponds to the interaction of the quadrupole moment of the 4f electron shell with the EFG produced mainly by the asphericity of the 6p and 5d shells of the parent atom resulting from the presence of neighbours in the lattice [16]. The CEF coefficient A_2^0 used in the CEF formalism is proportional to this EFG. A lack of a strict relation between EFG at nucleus and EFG exerted by the 4f shell means that only an approximate relation can be used:

$$A_2^0 = - D/4 \; e \; V_{cc}(latt) \qquad \textbf{(8)}.$$

Taking the value of 1/320 for the coefficient D as obtained from a comparison of a Moessbauer spectroscopy study with bulk magnetic data for Gd_2Fe_{17} and other RE-3d intermetallics [29], the CEF coefficients A_2^0 for the Sm sites with different numbers of nitrogen neighbours in the compound can be determined. They change from -66 Ka_0^{-2} for 0N to -358 Ka_0^{-2} for 3N atoms as nearest neighbours to the Sm site, where a_0 is the Bohr radius. The value of -242 Ka_0^{-2} reported recently for A_2^0 in $Sm_2Fe_{17}N_3$ from extrapolation of bulk anisotropy data [30] is about 30% smaller than the value of A_2^0 for Sm with 3N derived from the nuclear magnetic resonance spectra.

3.4.1 $Sm_2Fe_{17}H_x$

The ^{147}Sm NMR spin echo spectra of the $Sm_2Fe_{17}H_x$ (x = 0, 2.9, 4.6) series at 4.2K are reported in [31]. Powder samples prepared in the same way as for the Y_2Fe_{17} hydrides were measured.

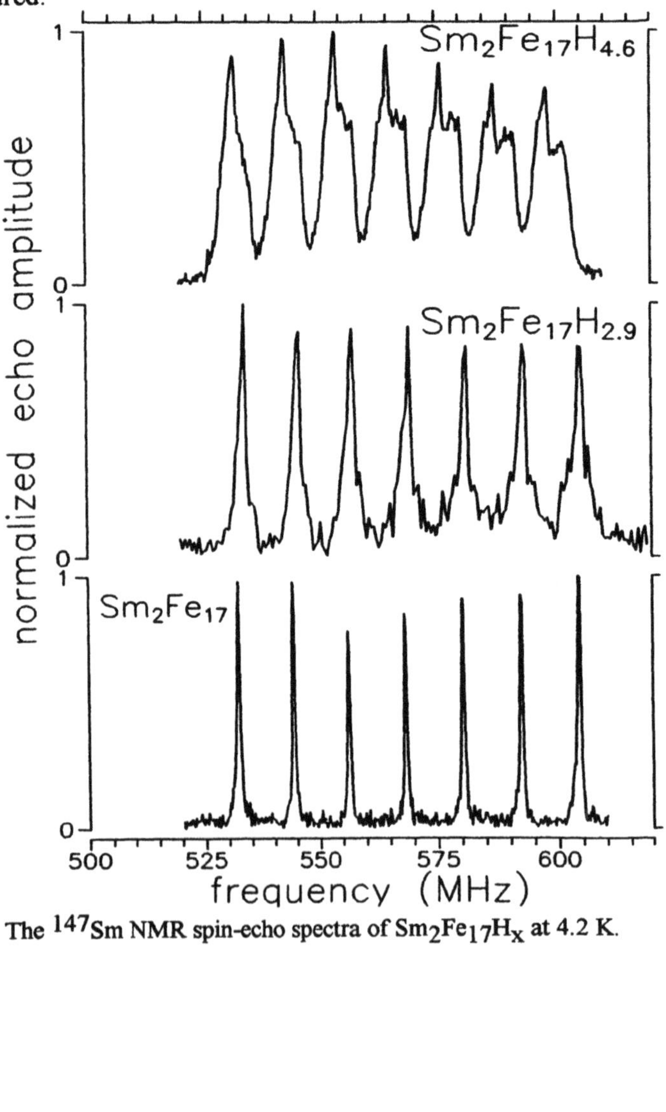

Fig.9. The ^{147}Sm NMR spin-echo spectra of $Sm_2Fe_{17}H_x$ at 4.2 K.

As was shown in the neutron diffraction study on the isostructural rhombohedral $Nd_2Fe_{17}H_x$ [17], hydrogen enters the octahedral 9e and the tetrahedral 18g sites. Hydrogen atoms fill the 9e sites first, up to their full occupation corresponding to the formula $Sm_2Fe_{17}H_3$, and subsequently enter the 18g sites. The maximum concentration of hydrogen in the compound corresponds to the formula $Sm_2Fe_{17}H_5$, i.e. the maximum number of H atoms occupying the 18g sites is two per formula unit.

The ^{147}Sm spectra of Sm_2Fe_{17} and $Sm_2Fe_{17}H_{2.9}$ (Fig.9) consist of a single septet of very similar central line positions and line separations between both samples. The quadrupole splitting and the HFF obtained from the spectra in the same way as for $Sm_2Fe_{17}N_x$ amount to 11.9 MHz and 323.7 T (central frequency 569.1 MHz) for $Sm_2Fe_{17}H_{2.9}$, which is very close to the values of 12.0 MHz and 323.1 T (central frequency 568.8 MHz) for the host Sm_2Fe_{17}. For the $Sm_2Fe_{17}H_{4.6}$ two overlapping septets with different quadrupole splittings and HFFs appear, which can be explained by a partial occupation of the 18g sites. At such a concentration all the 9e sites in number 3 per formula unit are filled, and the remaining 1.6 H/f.u. are distributed over the two 18g sites per formula unit available to hydrogen [32]. Using the binomial distribution for 1.6 H atoms randomly distributed over these two available 18g sites, the probabilities of finding 0, 1 or 2 hydrogen at 18g sites as the nearest neighbours to a Sm atom are 4%, 32% and 64% respectively. The last two values are close to the relative intensities of the two septets in the NMR spectrum of $Sm_2Fe_{17}H_{4.6}$. Thus, the septet with larger intensity can be attributed to the Sm sites with 3H atoms at the 9e and 2H atoms at the 18g NN sites. The septet with the smaller intensity can consequently be assigned to the Sm sites with 3H atoms at the 9e sites and 1H atom at the 18g sites. As the Sm sites with 3H atoms at the 9e sites and 0H atoms at the 18g sites have an order of magnitude smaller population, the corresponding septet in the spectrum is not resolved.

A significant influence of H 18g and a slight influence of H 9e on the lattice EFG can be explained in terms of the strength of the bonding effects being strongly distance dependant. According to the above results their influence on the RE 5d and 6p electron population is visible already for the Sm-H(18g) distance of 2.35 Å but is negligible for the Sm-H(9e) separation of 2.5 Å.

3.5. ^{143}Nd AND ^{145}Nd SPECTRA

3.5.1 $Nd_2Fe_{17}N_x$. The rhombohedral Nd_2Fe_{17} sample and its nitride with nominal x close to 3 prepared as described elsewhere [24] were measured. The ^{145}Nd spectra obtained are presented in Fig.10 [33]. The spectrum of Nd_2Fe_{17} consists of a quadrupole septet (nuclear spin of ^{145}Nd $I=7/2$) corresponding to the single Nd crystallographic site in the structure. A single septet with a broad shoulder at the low frequency side is also obtained for the nitride and assigned to the Nd sites with three nitrogen NN.

Similarly to the Sm based series [13], the value of B_e for the Nd sites with nitrogen neighbours is higher than that of the non-nitrided sample. As for Sm and Nd B_{orb} is dominant and antiparallel to the transferred HFF, the effect is consistent with that for the Y-based compounds.

The planar anisotropy of the Nd and other rare earths with the second order Stevens factor $\alpha_J < 0$ strengthens with nitriding. Thus, from the difference of quadrupole splittings between Nd_2Fe_{17} and the nitride the EFG change caused by the 3N neighbours $\Delta V_{pp}{}^{3N}(\text{latt})$ is directly obtained. The corresponding $\Delta V_{cc}{}^{3N}(\text{latt})$ amounts to 111.2×

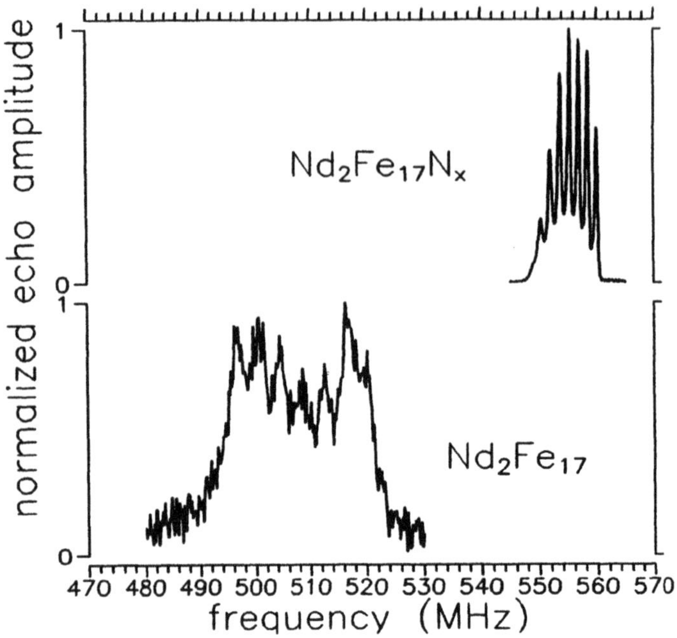

Fig.10. The ^{145}Nd NMR spin-echo spectra of $Nd_2Fe_{17}N_x$ at 4.2 K.

10^{20} V/m^2, very close to that of 115×10^{20} V/m^2 obtained in the Sm based nitride [25]. This indicates that the change of $A_2{}^0$ caused by 3N neighbours is only weakly dependent on the RE type in these compounds.

4. $NdFe_{11}TiN_x$

Introduction of nitrogen to the interstitial sites in the structure of $NdFe_{11}Ti$ [34,35] leads to a great improvement of its permanent magnetic properties, similarly to that in Sm_2Fe_{17}. The EMD of $NdFe_{11}Ti$ switches from the c-plane to the c-axis with nitriding.

In the tetragonal $ThMn_{12}$ type crystal structure of $REFe_{11}Ti$ RE atoms occupy a single 2a site, Fe enters the three inequivalent sites 8i, 8j and 8f, whereas Ti is distributed over the Fe 8i sites. The neutron diffraction study on the isostructural $YFe_{11}TiN_x$ [36] showed that the nitrogen atoms enter octahedral 2b holes. The RE site has two nearest neighbour 2b sites: one above and one below it, along the c-axis direction.

The samples of $NdFe_{11}Ti$ and its nitride prepared as described elsewhere [37,38] were measured [39]. The ^{145}Nd spectra are shown in Fig.11. The ^{143}Nd spectra fully consistent with them were also obtained.

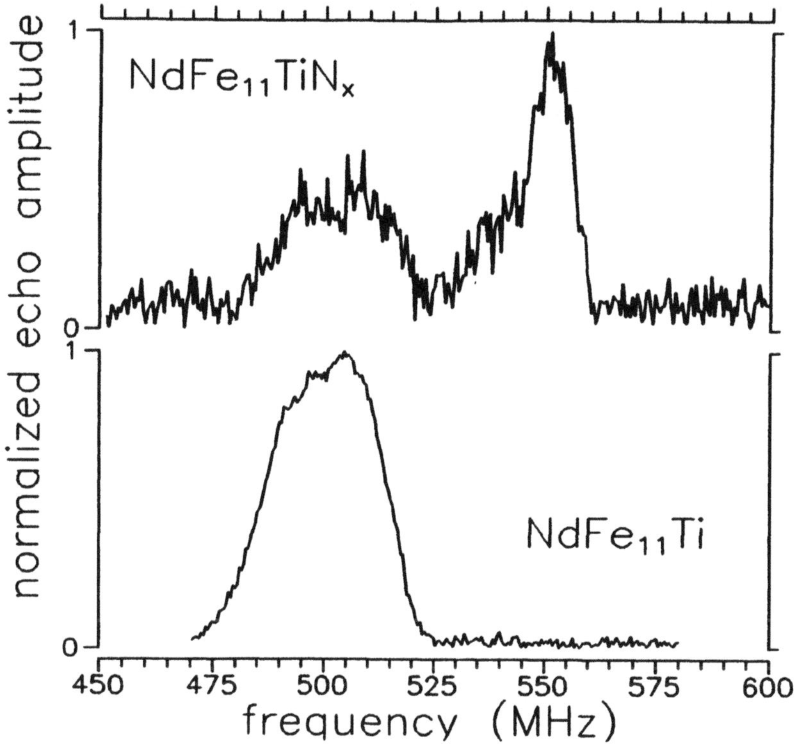

Fig.11. The ^{145}Nd NMR spin-echo spectra of $NdFe_{11}TiN_x$ at 4.2 K.

The spectra of $NdFe_{11}Ti$ for both isotopes consists of a single broad line corresponding to the single Nd 2a crystallographic site in the compound. A lack of the resolved quadrupole septet patterns (nuclear spin of ^{145}Nd and ^{143}Nd I=7/2) can be explained by the presence of Nd environments with various numbers of Ti neighbours resulting in a broad distribution of the quadrupole splitting. Assuming the quadrupole interaction is the main contribution to the total linewidth of the spectrum as in $Nd_2Fe_{17}N_x$ [33], the spectra were fitted with septets of equidistant lines. The values of the quadrupole splittings of 5.2 MHz for ^{145}Nd and 9.6 MHz for ^{143}Nd are obtained. Central frequencies amount to 499 MHz for ^{145}Nd and 803 MHz for ^{143}Nd, which correspond to the HFF at Nd nuclei of 347 T and is close to that of 353.4 T in Nd_2Fe_{17}.

An additional narrow line appearing in the spectra of the nitride at higher frequencies for both isotopes is assigned to the Nd sites with nitrogen at the nearest neighbour 2b sites. Its width corresponds to quadrupole splittings of 1.3 MHz and 3.1 MHz for ^{145}Nd and ^{143}Nd respectively. Central frequencies amounting to 550 MHz for ^{145}Nd and 883

MHz for ^{143}Nd were derived, which correspond to a HFF of 382 T, close to that of 385.7 T obtained for the 3N sites in the Nd_2Fe_{17} nitride. The low frequency lines are similar to those of the host compound with central frequency corresponding to HFF of 348 T and quadrupole splitting of 4.4 MHz for ^{145}Nd.

Assuming a statistical distribution of nitrogen over the two NN sites to a Nd atom, one would expect for the nitrogen content x of 0.5 three resonance lines corresponding to 0, 1 and 2 N neighbours with maximum intensity for the 1N line. As the nitriding process is similar to that for the RE_2Fe_{17} nitrides we also have a similar situation with nitrogen rich outer shells of the grains and the inner cores containing less nitrogen. Such a situation results in an increased number of the Nd atoms with no nitrogen neighbours inside grains and with the two N neighbours in the outer shells leading to a depletion of the environments with one nitrogen.

According to this interpretation we can attribute the upper line in the nitride to the 2N environments and the lower line to 0N sites. A low frequency shoulder of the upper line can possibly be assigned to the 1N sites. A similar effect of increased intensities of the lines for the zero and maximum nitrogen number environments was observed for the $Sm_2Fe_{17}N_x$ series [25]. Also the HFF for the Nd sites with N neighbours is higher than that for those with no nitrogen neighbours.

Similarly to $RE_2Fe_{17}N_x$ the reduction of the quadrupole splitting by nitrogen neighbours is attributed to an increase of the magnitude of lattice EFG. From the difference of EFGs between 0N and 2N lines the change of the lattice EFG component along the c–axis caused by the two N neighbours $\Delta V_{cc}{}^{2N}(\text{latt}) = -73\times10^{20}$ V/m^2 was derived. The corresponding contribution to $A_2{}^0$ from the two nitrogen neighbours amounts to 185 $Ka_0{}^{-2}$, which gives 92.5 $Ka_0{}^{-2}$ per one N neighbour. This is very similar in absolute value to that of -94 $Ka_0{}^{-2}$ obtained for $Nd_2Fe_{17}N_x$. As the N atoms in $NdFe_{11}Ti$ are slightly closer to the RE site, a larger influence on the EFG is expected. The error margin, however is as large as 30% and thus EFG and $A_2{}^0$ are possibly underestimated.

On the basis of the above obtained results it can be concluded that the lowest order CEF anisotropy energy term contributions as calculated per one nitrogen neighbour in $Nd_2Fe_{17}N_x$ and $NdFe_{11}TiN_x$ are similar in absolute value, but opposite in sign. The difference in sign is consistent with the "opposite" location of the N neighbours; along the c-axis in $REFe_{11}TiN_x$ and in the c-plane in $RE_2Fe_{17}N_x$ and agrees with the bulk magnetic anisotropy data.

5. $GdMn_2H_x$

The ^{55}Mn spectra for $GdMn_2H_x$ (x = 0, 0.5, 1, 2, 3.4, 3.5 - cubic Laves phase and 4.3 - rhombohedral) have been obtained [40]. Their features are very similar to those for YMn_2H_x [41] The host compound $GdMn_2$ is a canted antiferromagnet (antiferromagnetic order within the Mn sublattice vs. canted alignment of Gd moments) with a Neel temperature about 100 K [42]. Hydrogen uptake raises the ordering temperature. Also a stepwise increase of magnetisation with hydrogen content at x about 3 is observed [43]. From a neutron diffraction experiment on the isostructural YMn_2H_x it was found that hydrogen occupies exclusively the 96g tetrahedral sites having as the NNs 2Y and 2Mn atoms [44].

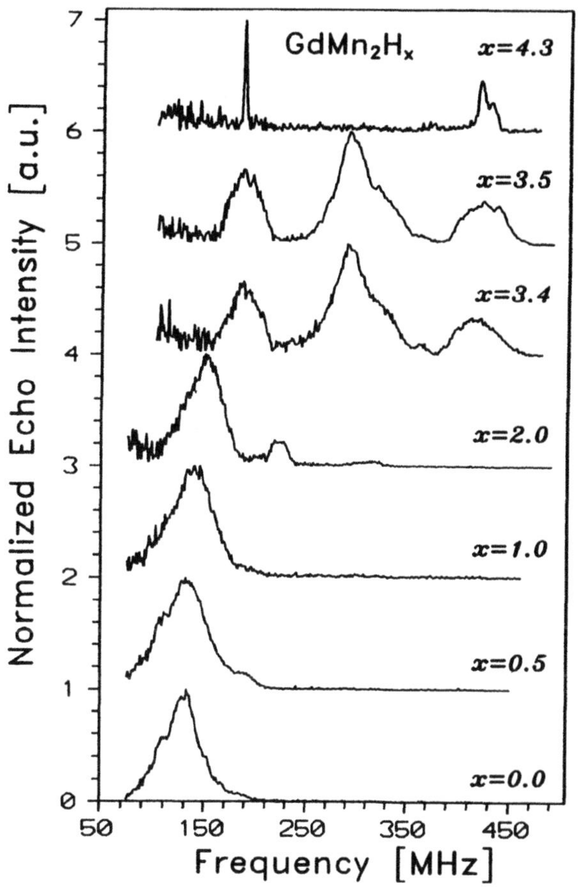

Fig. 12. The ^{55}Mn NMR spin-echo spectra of $GdMn_2H_x$ at 4.2 K.

For the host $GdMn_2$ a single broad line is observed indicating a single Mn magnetic moment in the compound - Fig. 12. The line moves slightly to higher fields with increasing x and weak satellite lines appear for x = 2. In the spectra for x = 3.4 and 3.5 three strong lines are observed with much different HFFs, similarly to that in YMn_2H_x [41]. This suggests that the Mn moments can have three values. Large differences of HFF indicate, that a possible reason for the effect is the change of the orbital contribution to the Mn magnetic moment caused by the presence of hydrogen NNs.

The two narrow lines obtained for antiferromagnetic rhombohedral $GdMn_2H_{4.3}$ are assigned to the two Mn crystallographic sites in the rhombohedral structure.

6. YFe_2H_x

The ^{89}Y spectra of the cubic Laves phase ferromagnet and its deuteride $YFe_2D_{3.5}$ were measured [45]. In YFe_2 a single narrow line at 46 MHz corresponding to the single Y site in the compound was obtained (Fig.13), in agreement with the literature data [46]. Five satellite lines at lower frequencies down to 8 MHz were detected in the deuteride. These satellites can be assigned to the yttrium sites with 1, 2, 3, 4 and 5 hydrogen atoms as NNs. A huge reduction of HFF by 87% is found for Y sites with 5H NNs. It indicates that the hydrogen NN, which is at a distance of 2.1 Å from the Y atom, i.e. closer than the sum of atomic radii of Y and H, significantly influences the yttrium valence electron distribution.

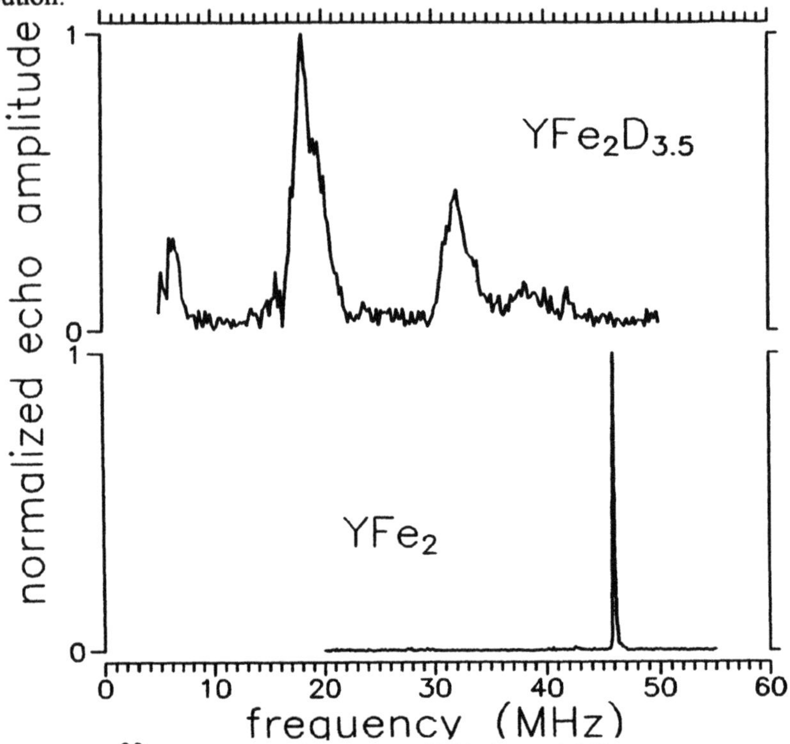

Fig.13. The ^{89}Y NMR spin-echo spectra of YFe_2D_x at 4.2 K.

7. $La_2Fe_{14}B$

The ^{139}La spectra of $La_2Fe_{14}B$, the compound isostructural with the high performance permanent magnet material $Nd_2Fe_{14}B$, are reported in [47]. As La has the $4f^0$ configuration, the 4f electron contribution to the HFF and EFG vanishes, giving an opportunity for an accurate determination of the lattice EFG and transferred HFF at both crystallographic sites 4f and 4g in the compound. From Gd Mössbauer measurements

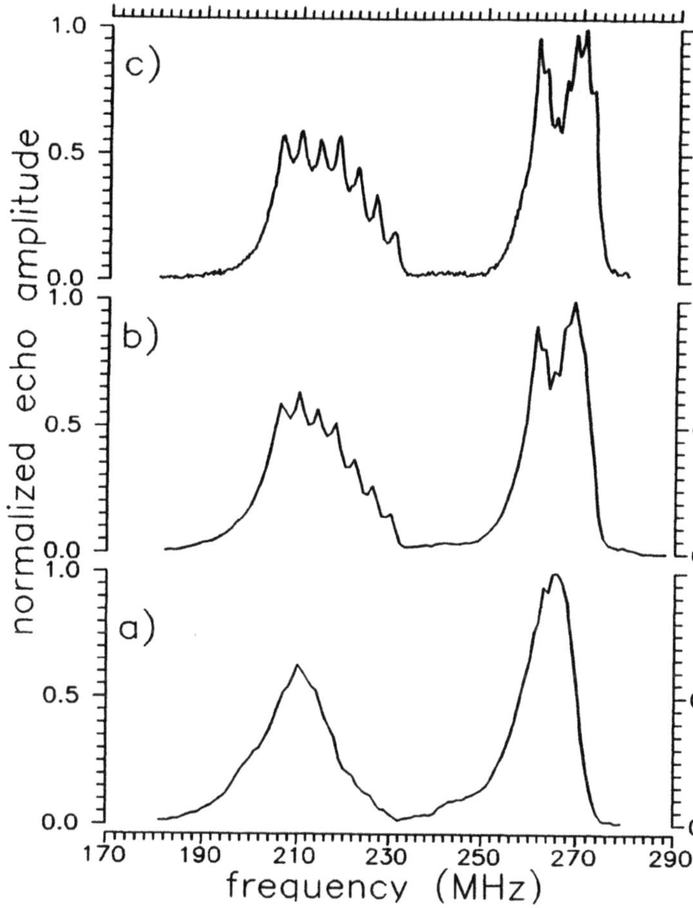

Fig. 14. The ^{139}La NMR spin-echo spectra of $La_2Fe_{14}B$ at 4.2 K for the pulse sequences 0.2μs/τ/0.4μs. a) τ = 15μs, b) τ = 200μs and c) τ = 500μs.

[48] and theoretical calculations for $Gd_2Fe_{14}B$ [49] very similar V_{cc}(latt) and $A_2{}^0$ values for both sites were found, in contrast to the Sm and Nd NMR in $Sm_2Fe_{14}B$ and $Nd_2Fe_{14}B$ where both values differ by a factor of more than two [50].

The spectra obtained for different pulse separation are presented in Fig. 14. The spectrum obtained for the pulse separation τ = 15 μs shows no resolved quadrupole pattern of the two broad resonant lines. As indicated in Sec.2 in order to remove DWC signals large pulse separations were used. For τ = 200 μs one gets two resolved quadrupole septets and the resolution further improves for τ = 500 μs. As the EMD for $La_2Fe_{14}B$ is the c-axis, the quadrupole splitting corresponds directly to V_{cc}(latt). Following the theoretical calculations for the "transferred" HFF [51], the upper septet is assigned to the 4f site and the lower to the 4g one. The corresponding B_e amounts to 36.1 T for the 4g and 44.1 T for the 4f site.

The obtained values of the quadrupole splittings amounting to 4.05 MHz for the 4g- and 1.93 MHz for the 4f site indicate that V_{cc}(latt) and thus A_2^0 for the 4g site is more than two times larger than for the 4f one.

8. Conclusions

The HFFs and EFGs at the individual sites in the RE-TM intermetallics and their interstitial solutions containing hydrogen, carbon and nitrogen are obtained with good accuracy by means of frequency swept spin-echo NMR.

The local influence of the neighbouring interstitial element on these parameters at the RE sites is related to the change of the RE valence electron distribution. For the 3d metals the 3d orbital moment is also influenced. The influence depends on the site distance and the kind of element.

For hydrogen at a distance of 2.5 Ä from the RE site no influence on either HFF or EFG is observed. For a distance of 2.35 Ä a measurable influence on the EFG appears with a slight influence on the HFF, and for a distance of 2.1 Ä a strong influence on the HFF is observed. A strong influence of N and C on both the EFG and HFF is already observed at a distance of 2.5 Ä, indicating that the local influence becomes effective at a distance approximately equal to the sum of the atomic radii. This suggests that the interstitial atom influence arises from bonding effects.

The different influence of interstitial H, C and N on the Fe HFF is assigned to the competitive influence of the magnetovolume effect and the local influences of the C and N nearest neighbours. Large changes of Co and Fe anisotropies caused by nitrogen are attributed to its influence on the 3d electron orbital moments, and thus, on the spin-orbit interaction anisotropies of the Fe and Co sites.

With increasing content of interstitial hydrogen the Mn magnetic moment takes up to three values in $GdMn_2H_x$, which is attributed to a change of the orbital contribution to the Mn moment caused by the neighbouring H atoms.

A large increase in the magnitude of the lattice EFG with number of N neighbours is related to the changes of the CEF coefficient A_2^0, corresponding to a big rise of the CEF induced magnetocrystalline anisotropy with nitriding.

Fully polarised Sm and Nd states, irrespective of the EMD direction are found in $RE_2Fe_{17}A_x$ and in $NdFe_{11}TiN_x$. The 4f electron contribution to EFG on Sm in the compounds is smaller by 20% than that in a free atom, reflecting a contraction of the 4f shell in the solid as compared to a free atom.

The lattice EFG at the RE site 4g in $La_2Fe_{14}B$ was found to be larger by a factor of two than that of the 4f site, indicating different lowest order term contributions to the magnetocrystalline anisotropy from these sites.

Information on the real concentration of carbon and nitrogen at the interstitial sites in the materials prepared in different ways has been obtained. On this basis hints on preparation procedures can be given.

ACKNOWLEDGEMENTS. The experimental work surveyed in this paper was done in St. Andrews, Bochum and Cracow with the financial support of SERC (UK), the Alexander von

Humboldt Foundation (Germany) and the State Committee for Scientific Research (Poland) - grant 203409101 We are indebted to Prof. K.H.J.Buschow for the loan of samples and helpful discussions. Cz.K. acknowledges collaboration with Prof. M. Rosenberg and Prof. H. Figiel.

REFERENCES

1. D.B.de Mooij and K.H.J.Buschow, J.Less-Comm.Met. 142 (1988) 349
2. Hong Sun, J.M.D.Coey, Y.Otani and D.P.F.Hurley, J.Phys.: Condens.Matter 2 (1990) 6465.
3. A.Abragam, The Principles of Nuclear Magnetism, Oxford, Clarendon Press, 1961.
4. P.C.Riedi, Fundamentals of Nuclear Magnetic Resonance, this volume.
5. E.Fukushima and S.B.W.Roeder, Experimental Pulse NMR, Addison-Wesley Publ.Comp.Inc., Reading, Massachusetts, 1981.
6. J.Winter, Magnetic Resonance in Metals, Oxford, Clarendon Press, 1971.
7. A.M.Portis and R.H.Lindquist, in Magnetism (Eds G.T.Rado and H.Suhl, Vol.IIA, Academic Press, New York, 1965) p.357.
8. M.A.H.McCausland and I.S.Mackenzie, Adv. in Phys. 28, (1979); also: Nuclear Magnetic Resonance in Rare Earth Metals, Taylor & Francis Ltd, London 1980.
9. P.C.Riedi, in Magnetism in Solids (Eds. A.P.Cracknell and R.A.Vaughan, SUSSP Edinburgh, 1981), p. 445.
10. H.Figiel, Magn.Res.Rev., 16 (1991) 101.
11. P.C.Riedi, An Introduction to the Study of Hydrogen Motion in Metals by NMR, this volume.
12. Cz.Kapusta, M.Rosenberg, J.Żukrowski, H.Figiel, T.H.Jacobs and K.H.J.Buschow, J.Less-Common Met. 171, 101 (1991).
13. Cz.Kapusta, R.J.Zhou, M.Rosenberg, P.C.Riedi and K.H.J.Buschow, J.All.Comp. 178 (1992) 139.
14. Cz.Kapusta, M.Rosenberg, H.Figiel, T.H.Jacobs and K.H.J.Buschow, J.Magn.Magn. Mater. 104-107 (1992) 1331.
15. Cz.Kapusta, M.Rosenberg, K.V.Rao, Hang Zheng-he, T.H.Jacobs and K.H.J.Buschow, J.Less-Common Met. 169, L5 (1991).
16. R.Coehoorn and K.H.J.Buschow, J.Appl.Phys. 69 (1991) 5590.
17. O.Isnard, S.Miraglia, J.LSoubeyroux and D.Fruchart, Sol.St.Commun. 81 (1992) 13.
18. O.Isnard, S.Miraglia, J.L.Soubeyroux and D.Fruchart, J. Alloys Comp., 190 (1992) 129.
19. Q.W.Yan, P.L.Zhang, Y.N.Wei, K.Sun, B.P.Hu, Y.Z.Wang, G.C.Liu, C.Gau and Y.F.Cheng, Phys.Rev. B 48 (1993) 2878.
20. H.Figiel, A.Oppelt, E.Dormann and K.H.J.Buschow, Phys.Stat.Sol. (a) 36, 275 (1976).
21. Cz. Kapusta, G.Stoch, H.O.Koths, M.Rosenberg and K.H.J.Buschow, submitted to the Conference ICM'94, Warsaw.
22. Cz.Kapusta, M.Rosenberg, J.Żukrowski, G.Stoch, H.Figiel, T.H.Jacobs and K.H.J.Buschow, J.All.Comp. 182 (1992) 331
23. M.Katter, J.Wecker, L.Schultz and R.Groessinger, J.Magn.Magn.Mater. 92 (1990) L14.
24. K.H.J.Buschow, R.Coehoorn, D.B.de Mooij, K.de Waard and T.H.Jacobs, J.Magn.Magn.Mater. 92 (1990) L35.
25. Cz.Kapusta, M.Rosenberg, P.C.Riedi, M.Katter and L.Schultz, J. Magn. Magn. Magn. Mater. 134 (1994) 106.
26. Cz.Kapusta, J.S.Lord, P.C.Riedi and K.H.J.Buschow, to be published.
27. M.W.Dirken, R.C.Thiel, R.Coehoorn, T.H.Jacobs and K.H.J.Buschow, J. Magn. Magn. Mater. 94 (1991) 415.

28. B.Bleaney, in Magnetic Properties of Rare Earth Metals, ed. J.R.Elliot (Plenum NY 1972) Chap.8.
29. P.C.M.Gubbens, A.M.van der Kraan and K.H.J.Buschow, Hyp.Int. 53 (1989) 37.
30. Hong-Shuo Li and J.M.D. Coey, J. Magn. Magn. Mater. 115 (1992) 152.
31. Cz.Kapusta, J.S.Lord and P.C.Riedi, submitted to the Conference ICM'94, Warsaw
32. Ch.N.Christodoulou and T.Takeshita, J.Alloys Comp. 198 (1993) 1.
33. Cz.Kapusta, J.S.Lord, P.C.Riedi, K.H.J.Buschow, submitted to the 8th International Symposium on Magnetic Anisotropy and Coercivity in Rare Earth - Transition Metal Alloys, Birmingham '94.
34. Y-C.Yang, X-D.Zhang, L-S.Kong, Q.Pan, S-L.Ge, Appl.Phys.Lett. 58 (1991) 2042.
35. J.M.D.Coey, Y.Otani, H.Sun, D.P.Hurley, J.Magn.Soc.Japan 15 (1991) 769.
36. Y-C.Yang, X-D.Zhang, L-S.Kong, Q.Pan, S-L.Ge, J-L.Yang, Y-F.Ding, B-S.Zhang, C-T.Ye, L.Jin, Sol.St.Commun. 78 (1991) 313.
37. K.H.J.Buschow, J.Appl.Phys. 63 (1988) 3130.
38. X.C.Kou, T.S.Zhao, R.Groessinger, H.R.Kirchmayr, X.Li, F.R.de Boer, Phys.Rev. B 47 (1993) 3231.
39. Cz.Kapusta, H.Figiel, X.C.Kou, G.Wiesinger, J.S.Lord, P.C.Riedi and K.H.J.Buschow, submitted to the Conference ICM'94, Warsaw.
40. Cz.Kapusta, J.Przewoznik, J.Zukrowski, H.Figiel, J.S.Lord, P.C.Riedi, V.Paul-Boncour, A.Percheron-Guegan, submitted to the Conference ICM'94, Warsaw.
41. Cz.Kapusta, J.Przewoznik, J.Zukrowski, N.Spiridis, H.Figiel and K.Krop, Hyp.Int. 59 (1990) 353.
42. H.Wada, K.Yoshimura, M.Shiga, T.Goto and Y.Nakamura, J.Phys.Soc.Jpn., 54 (1985) 3543.
43. J.Przewoznik, private communication
44. M.Latroche, private communication
45. Cz.Kapusta, M.Latroche, A.Percheron-Guegan, J.S.Lord and P.C.Riedi, unpublished data.
46. A.Oppelt and K.H.J.Buschow, J.Phys. F: Met.Phys. 3 (1973) L212
47. Cz.Kapusta, submitted to the Conference ICM'94, Warsaw
48. M.Boge, G.Czjzek, D.Givord, C.Jeandey, H.S.Li and J.L.Oddou, J.Phys F: Met.Phys. 16 (1986) L67.
49. R.Coehoorn, in "Supermagnets, Hard Magnetic Materials", Eds. G.J.Long and F.Grandjean, Kluwer, Dordrecht, 1990.
50. Cz.Kapusta, H.Figiel, G.Stoch, J.S.Lord and P.C.Riedi, IEEE Trans.Magn. 29 (1993) 2893
51. R.Coehoorn and K.H.J.Buschow, J.Magn.Magn.Mater 118 (1993) 175.

Chapter 21

STRUCTURAL CHANGES OF Fe_2Ce ALLOYS BY HYDROGEN ABSORPTION

E. H. BÜCHLER, M. HIRSCHER and H. KRONMÜLLER
Max-Planck-Institut für Metallforschung
Institut für Physik
Heisenbergstr. 1
D-70569 Stuttgart
Federal Republic of Germany

ABSTRACT: Most C-15 Laves phase compounds of rare earth (RE) and transition metals absorb hydrogen (H) readily while changing their atomistic structure, as indicated by X-ray diffraction (XRD), differential scanning calorimetry (DSC) and thermal H desorption measurements. The local H diffusion process in ferromagnetic compounds can be investigated by means of magnetic susceptibility and magnetic after-effect (MAE) measurements, respectively. The H atom can therefore be used as a probe for studying the microstructure and the amorphization process on an atomistic level.

In this work the amorphization kinetics of the Fe_2Ce Laves phase – a representative of the Fe_2RE Laves phase compounds – was studied, employing the above mentioned measurement techniques. Additionally, thermal hydrogen desorption was used to determine H concentrations and to investigate long-range H diffusion processes.

Fe_2Ce can easily be amorphized ($p_{H_2} = 5 \cdot 10^3$ to $1 \cdot 10^5$ Pa, T <400 K), whereby the irreversible hydrogen-induced amorphization takes place in a narrow phase boundary which migrates from the surface into the sample. Furthermore, cyclic charging of amorphous Fe_2Ce below its crystallization temperature reveals a degradation in H-storage capacity from 53 at.% to 18 at.% after only three cycles. This degradation is attributed to a *hydrogen-induced* phase separation into crystalline CeH_2 and α-Fe which is enhanced by the high mobility of Fe atoms.

These findings may be helpful for the improvement of metal-hydride storage systems, of which Fe_2Ce provides a model system of extreme degradation characteristics.

1. Introduction

In 1983 Yeh and Samwer observed the amorphization of crystalline Zr_3Rh when exposing this compound to a hydrogen atmosphere [1, 2]. H-induced amorphization had already been reported for rare earth intermetallic compounds, but usually studied

F. Grandjean et al. (eds.), Interstitial Intermetallic Alloys, 521–539.

from the point of view of changes in their magnetic properties [3, 4, 5, 6, 7, 8]. By that time industrial production of amorphous alloys had started to increase rapidly, especially for the use as soft magnetic materials. The standard way of preparing amorphous alloys is by rapid quenching from the melt, where cooling rates between 10^4 and 10^8 K/min are necessary [9]. This rather expensive method can only be applied to a restricted number of alloys, so interest turned towards H-induced amorphization as a means of achieving large scale production of amorphous compounds. It therefore became necessary to understand the kinetics and mechanism of the H-induced amorphization process and it was soon established that the high mobility of the H atom in an otherwise thermodynamically metastable compound played the key role. A good summary of the amorphization mechanism is given in [10].

Investigations with H pressures of up to 5 MPa revealed that nearly all M_2RE (M=transition metal, RE=rare earth) Laves phase compounds show H-induced amorphization [11, 12]. But H pressures of 5 MPa are not absolutely necessary. For example, H-induced amorphization in Ni_2Dy and Ni_2Ce can already be observed at a hydrogen pressure of only 10^5 Pa [13, 14]. Fe_2Ce can even be amorphized below $p_{H_2} = 5 \cdot 10^3$ Pa, as demonstrated by this work.

Hydrogen absorption measurements during the H-induced amorphization process revealed hydrogen concentrations above 100 at.%, that is one metal atom for each hydrogen atom, in the amorphous hydride [15, 16, 17, 18]. In light of these high values, interest turned towards using the M_2RE Laves phase compounds and their amorphous hydrides as alloys for cyclic hydrogen storage. These metal hydrides are starting to play an ever increasing role in the field of stationary and mobile energy storage systems. Although already well advanced, these alloys still show degradation, having therefore only a restricted life time when cyclically charged with hydrogen.

The purpose of the present investigation was to study the amorphization and degradation process employing Fe_2Ce as a representative for Fe_2RE Laves phase compounds. Furthermore, Fe_2Ce provides a good model for the more complex, commercially available metal-hydride systems.

2. Experimental Details

2.1 SPECIMEN PREPARATION

The Fe_2Ce compound was prepared by induction melting in argon atmosphere using high purity metals (Fe: 99.99 at.%, Ce: 99.9 at.%). Afterwards the ingot was homogenized in an evacuated quartz-glass tube for 5 days at 600 °C. The ingot was wrapped in tantalum foil to prevent contamination from the quartz glass. The homogenization led to a single Fe_2Ce phase component of the cubic (C15) Laves phase structure as indicated by X-ray diffraction analysis (XRD), Fig.3a. The chemical composition of Fe (66 ± 1) at.% and Ce (34 ± 1) at.% was determined by EDX analysis in the scanning electron microscope (SEM). Within the error limits no traces of silicon or tantalum could be detected.

2.2 HYDROGEN-CHARGING

About 30 g of crystalline Fe_2Ce were exposed to a pure H atmosphere (99.999 at.%) at a H pressure of $p_{H_2}=10^5$ Pa. As the H-absorption process is exothermic for Fe_2Ce, the H filled quartz tube containing the specimen had to be cooled down several times in a dewar filled with ice water so as to prevent heating above 80°C. Any formation of additional phases or decomposition of the hydride could thus be avoided. After an exposure time of several minutes the crystalline sample started to pulverize, indicating a very fast hydride formation causing large internal stresses. To ensure a complete hydride formation through the whole sample bulk, the exposure time was prolonged to a maximum of 5 days.

A reduction in the speed of H absorption could be achieved by exposing the crystalline Fe_2Ce Laves phase to an argon atmosphere diluted with hydrogen ($Ar_{95}H_5$ at $p=10^5$ Pa), equivalent to a partial H pressure of $p_{H_2}=5\cdot10^3$ Pa. The temperature was slightly raised to 60°C in order to activate the H absorption process. After 2 days of exposure in this diluted atmosphere only the outside of the sample was found to have pulverized, indicating that hydride formation had solely occurred in the surface area of an otherwise crystalline Fe_2Ce sample. This sample consisting of two chemically and structurally different phases will, in the following, be referred to as heterogenous Fe_2Ce.

2.3 STRUCTURAL ANALYSIS

Structural and chemical analysis of crystalline and amorphous Fe_2Ce were carried out using different analyzing techniques:

XRD measurements on uncharged and charged Fe_2Ce samples employing CuK_α radiation gives insight into the structural changes which take place during hydrogen absorption. Changes in the position and the appearance of new XRD maxima are attributed to phase transitions. The method of differential scanning calorimetric analysis (DSC) is applied for a more thorough investigation of the phase transitions in the amorphous Fe_2Ce hydride. The (pulverized) sample is heated at a constant heating rate of 5 K/min and any heat absorption or desorption registered. Maxima in the DSC spectrum correspond to endothermic phase transitions, while minima are attributed to exothermic phase transitions.

2.4 THERMAL HYDROGEN DESORPTION

The measurement of the thermal hydrogen desorption rate is an excellent method for studying long-range H diffusion. Additionally, the hydrogen concentration of H-charged samples can be determined. Only a small mass of around 3 mg was needed for heating up in a previously evacuated chamber at a constant heating rate. The thermally desorbed hydrogen leads to a rise in the chamber pressure which is measured by a membrane pressure gauge. A cooling trap, located between the chamber and the pressure gauge, is kept at a constant temperature of 25 K. This trap ensures that only hydrogen contributes to the pressure rise; gases like O_2 and N_2 (except helium) are completely frozen out. Calibration with a known volume filled with pure hydrogen reveals that a pressure rise of 1 Pa is equivalent to $1.57 \cdot 10^{17}$ hydrogen atoms being desorbed. For a known specimen mass the H concentration C_{H} can thus readily be determined. It has to be mentioned however, that C_{H} is based only on the amount of hydrogen which has thermally desorbed up to T_{A}=750 K, the temperature limit of the apparatus. Any hydrogen which is so strongly bound that it is not removed at 750 K cannot be detected.

By plotting the H desorption rate $\mathrm{d}C_{\mathrm{H}}/\mathrm{d}t$ against temperature, different processes contributing to the desorption kinetics can readily be recognized. All thermal hydrogen desorption measurements are carried out with a constant heating rate of 5 K/min, which allows a direct comparison of the H desorption spectra with those of the DSC analysis.

2.5 MAGNETIC AFTER-EFFECT

The microscopic mechanism of the magnetic after-effect is based on the dependence of the magnetic interaction energy between the directions of spontaneous magnetization within a domain wall and the symmetry axis of anisotropic hydrogen interstitial configurations [19]. Owing to thermally activated jumps, the reorientation of H atoms lowers this interaction energy, leading to a decrease of the initial susceptibility $\chi(t)$, i.e. the slope of the virgin magnetization curve. In our case, $\chi(t)$ is measured using an AC method, where the specimen is centered in a coil which forms part of a LC oscillator [20, 21]. The applied oscillating magnetic field is kept well below the coercive field H_c of the sample to ensure a measurement in the Rayleigh region of the magnetization curve.

Isothermal relaxation curves of the reluctivity

$$r(t) = \frac{1}{\chi(t)} \tag{1}$$

are measured after each demagnetization of the specimen at different temperatures. The experimental results are represented as isochronal relaxation curves of the reluctivity amplitude between the times t_1 and t_2, defined as

$$\frac{\Delta r(t_1, t_2, T)}{r(t_1, T)} = \frac{r(t_2, T) - r(t_1, T)}{r(t_1, T)}. \tag{2}$$

In the measurements presented, $t_1 = 0.5$ s and t_2 varies from 1.5 s to 179.5 s. The ratio $\Delta r/r$ represents the relative reluctivity amplitude of the relaxation. Since $\chi(t)$ decreases with increasing time, it follows that $r(t_2, T) > r(t_1, T)$ and $\Delta r/r$ corresponds to positive quantities.

For thermally activated reorientation of anisotropic defects, the dependence of the reluctivity $r(t, T)$ may be written as

$$r(t, T) = r(0, T) + (r(\infty, T) - r(0, T))G(t), \tag{3}$$

where $r(0, T)$ and $r(\infty, T)$ is the reluctivity at t=0 and $t = \infty$, respectively. In the case of a Debye process, the MAE function obeys an exponential law,

$$G(t) = 1 - \exp(-t/\tau). \tag{4}$$

For a satisfactory description of the magnetic relaxation spectra a superposition of Debye processes is necessary. Therefore $G(t)$ will depend on the distribution $p(\tau)$ of the relaxation times,

$$G(t) = \int_{\tau_1}^{\tau_2} p(\tau)[1 - \exp(-t/\tau)]\mathrm{d}\tau. \tag{5}$$

In our case, the magnetic relaxation processes are described by a box-type distribution in the logarithm of the relaxation time τ between a lower limit τ_1 and an upper limit τ_2 [22]. The relaxation time, $\tau_1 < \tau < \tau_2$, obeys the Arrhenius equation

$$\tau = \tau_o \cdot \exp(Q/kT). \tag{6}$$

Here τ_o denotes the pre-exponential factor and the activation enthalpy follows a box-type distribution with $Q_1 \leq Q \leq Q_2$. The width of the enthalpy box is $\Delta Q = Q_2 - Q_1$. In the case of n such boxes with an equal width ΔQ, the MAE function is given by

$$G(t) = 1 + \sum_{i}^{n} p_i \cdot \frac{\{\mathrm{Ei}(-t/\tau_i) - \mathrm{Ei}(-t/\tau_{i+1})\}}{\ln(\tau_2/\tau_1)}, \tag{7}$$

where Ei denotes the exponential integral

$$\mathrm{Ei}(-x) = -\int_{x}^{\infty} \frac{e^{-t}}{t}\,\mathrm{dt}. \tag{8}$$

The weights of the boxes, p_i, are normalized to one, $\sum_i p_i = 1$. For a satisfactory description of a broad relaxation maximum, a superposition of ten such box-type distributions with one common pre-exponential factor τ_0 is required.

3. Experimental Results

3.1 STRUCTURE AND PHASE TRANSITIONS

The XRD measurement on crystalline, mechanically powdered Fe_2Ce in Fig.3a shows sharp reflections at angles corresponding to the Bragg conditions for the cubic C15 Laves phase structure. The absence of any further peaks in the XRD pattern confirms the result of the SEM (EDX analysis) where no other phase than the cubic C15 Fe_2Ce Laves phase could be detected. During hydrogen absorption the Fe_2Ce Laves phase turns into fine powder with an average particle size of 200 μm. A SEM picture (Fig.1) of one of these powder particles reveals a number of cracks which have led to the pulverization of the Fe_2Ce Laves phase. Their origin lies in the high amount of absorbed hydrogen, causing stress in the Fe_2Ce lattice.

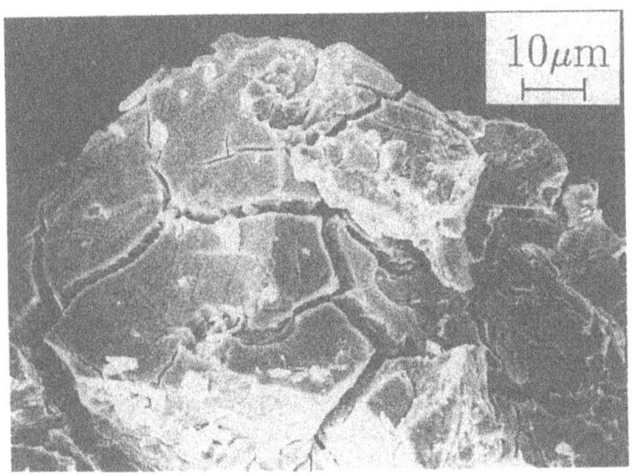

Fig. 1: Scanning electron microscope (SEM) picture of a Fe_2Ce hydride particle after pulverization in hydrogen atmosphere

In the XRD spectrum of the Fe_2Ce hydride powder (Fig.3b) no sharp reflexes are visible, indicating the total absence of a long-range atomic order. This absence is characteristic for the disordered structure of amorphous alloys.

Fig. 2: Electron diffraction pattern in the TEM. The broad diffraction rings are typical for amorphous alloys.

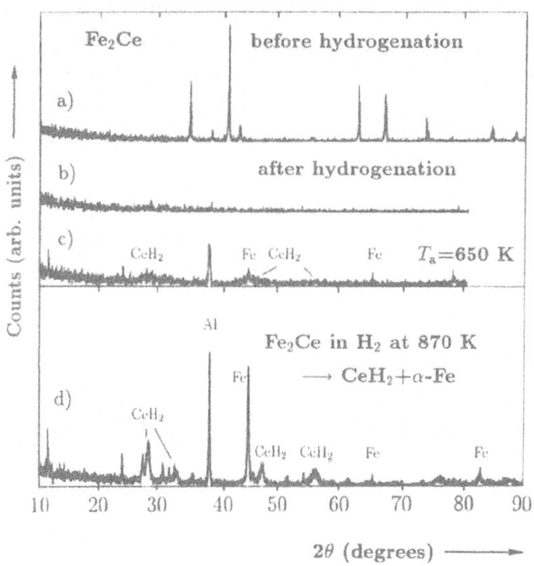

Fig. 3: X-ray diffraction analysis of (a) crystalline Fe_2Ce, (b) hydrogen-charged, amorphous Fe_2Ce, (c) amorphous Fe_2Ce hydride after annealing at T_a=650 K in a He atmosphere and (d) Fe_2Ce after annealing at T_a=870 K in a pure hydrogen atmosphere. An aluminium specimen holder was used, leading to a diffraction reflections at 38° in all four spectra.

Furthermore, the electron diffraction in the TEM (Fig.2) shows broad diffraction rings usually found in amorphous alloys. After heating the amorphous Fe_2Ce hydride up to 650 K under a Helium atmosphere small XRD maxima become visible at diffraction angles characteristic of crystalline Ce hydride and α-Fe (Fig.3c). When an amorphous Fe_2Ce sample is annealed at 870 K, these reflections become much sharper (Fig.3d). It is therefore possible to assign these reflections to both crystalline CeH_2 and α-Fe phases, clearly indicating a complete phase separation.

The DSC analysis of amorphous Fe_2Ce hydride is shown in Fig.4, measured at a heating rate of 5 K/min under an argon atmosphere. The broad maximum between 400 K and 480 K corresponds to the endothermic hydrogen desorption from amorphous Fe_2Ce. At 570 K and 630 K two separate exothermic reactions are observed, indicating the formation of crystalline CeH_2 and α-Fe, respectively. The formation of Ce hydride at 560 K marks the beginning of crystallization of amorphous Fe_2Ce. This value was determined by the intersection of the steepest slope of the second minimum with the base line. The endothermic reaction at 1080 K marks the melting of any surplus Ce formed during crystallization. At 1180 K the peritectic reaction of Fe_2Ce into $Fe_{17}Ce_2$ and liquid FeCe is observed.

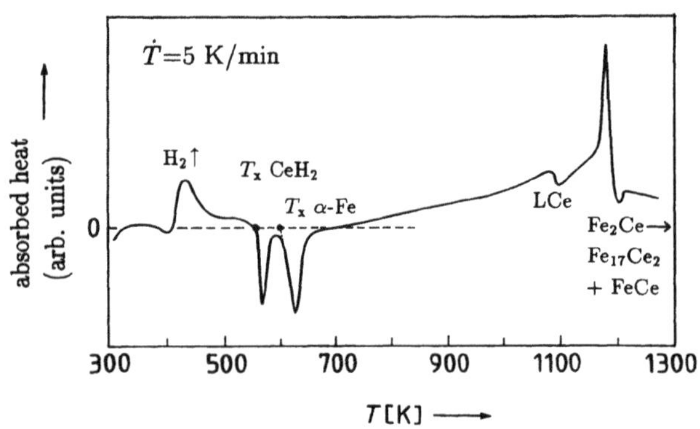

Fig. 4: DSC analysis of H charged, amorphous Fe_2Ce. The maximum between 400 K and 480 K indicates H desorption. The first minimum at 570 K is attributed to the formation of Ce_2H ($T_{cryst.}$=560 K), the second at 630 K to the formation of α-Fe.

3.2 THERMAL HYDROGEN DESORPTION

Fig.5 shows the rate of hydrogen desorption from amorphous Fe_2Ce hydride measured with a heating rate of 5 K/min. A broad maximum between 400 K and 520 K is observed, showing a double peak at 440 K and 475 K. It is attributed to hydrogen desorption from the amorphous Fe_2Ce hydride. Temperature wise it falls in line with the broad DSC maximum in Fig.4. In addition, the two desorption peaks indicate two sites with different binding energies for hydrogen in amorphous Fe_2Ce.

Their identification will be discussed in 3.3. The two steps in the desorption rate at 550 K and 610 K correspond to the formation of CeH_2 and α-Fe, respectively. Their formation can also be observed in the DSC spectrum at the same temperatures.

After H-induced amorphization the hydrogen concentration in the amorphous Fe_2Ce hydride was found to be C_H^{amorph}=53 at.%. This value was determined by measuring the total amount of desorbed hydrogen up to a maximum temperature of 750 K. It does not account for any hydrogen retained during the formation of CeH_2. The hydrogen concentration achieved in amorphous Fe_2Ce hydride phase is independent of hydrogen charging pressure.

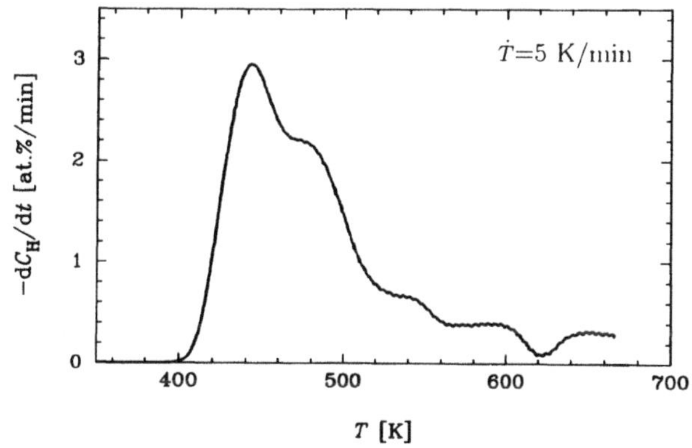

Fig. 5: Thermal hydrogen desorption spectra of hydrogen charged, amorphous Fe_2Ce.

In the case of crystalline, uncharged Fe_2Ce the hydrogen concentration is much smaller, C_H=10^{-2} at.%. Some hydrogen absorption during specimen preparation must have taken place, despite the fact that the crystalline Fe_2Ce sample was not exposed to any hydrogen.

3.3 MAGNETIC SUSCEPTIBILITY

Fig.6 shows the temperature dependence of the magnetic susceptibility for crystalline and amorphous Fe_2Ce. In the case of crystalline Fe_2Ce, the susceptibility $\chi(T)$ increases with temperature before dropping sharply. This drop is attributed to the magnetic phase transition from ferro- to the paramagnetic state of the crystalline Fe_2Ce Laves phase at a Curie temperature of T_c=230 K. The sharp transition in $\chi(T)$ at T_c is in agreement with literature values and the absence of any ferromagnetic component in $\chi(T)$ above 230 K confirms the presence of a single crystalline Fe_2Ce phase without any additional Fe-rich phases. The small minimum in $\chi(T)$ at around 60 K originates from the H-relaxation in Fe_2Ce which is discussed in 3.4.

When Fe_2Ce is totally amorphized, T_c shows shifts up to 380 K (C_H=53 at.%).

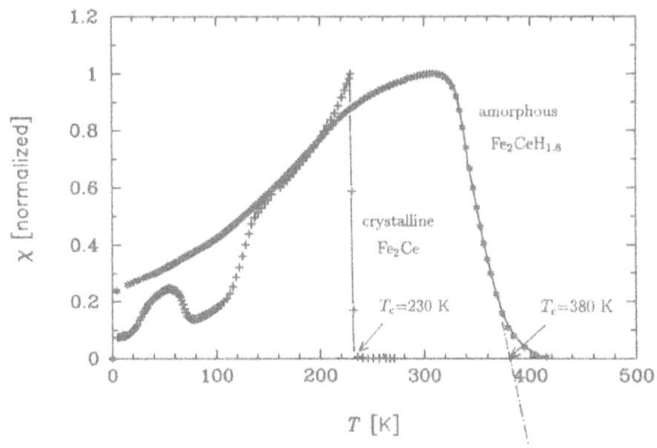

Fig. 6: Temperature dependence of the initial magnetic susceptibility $\chi(T)$ of crystalline and amorphous Fe_2Ce.

This rather drastic rise in T_c is attributed to a local rearrangement of the metal atoms induced by local H diffusion. The dominant role in the ferromagnetic behaviour of Fe_2Ce is played by the Fe atoms, because any change in their inter-atomic distances will cause a change in the ferromagnetic exchange, and thus in T_c [5, 7].

In the case of amorphous Fe_2Ce hydride, annealing is the only way in which hydrogen concentrations lower than C_H=53 at.% can be achieved. The influence of different annealing temperatures T_a, hence different hydrogen concentrations C_H^{amorph}, on the magnetic susceptibility is shown in Fig.7.

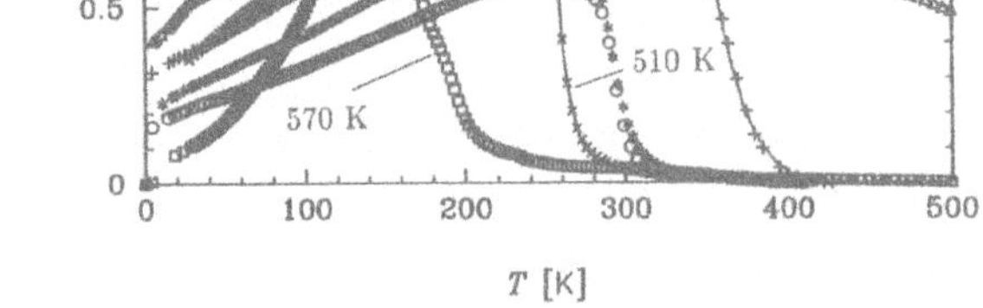

Fig. 7: Temperature dependence of the initial magnetic susceptibility $\chi(T)$ of amorphous Fe_2Ce after annealing at different temperatures T_a. C_H^{amorph} decreases with increasing T_a.

The temperature corresponding to the sharp drop in $\chi(T)$ decreases with increasing annealing temperature T_a. This may be explained by a lattice relaxation brought about by hydrogen desorption, which in turn leads to a decrease in the ferromagnetic exchange coupling and thus T_c. When amorphous Fe_2Ce is heated up to and above T_{cryst}=560 K, the previously observed sharp decrease in $\chi(T)$ changes into a very broad flank between 150 K and 250 K. Now the Curie temperature shows a whole spectrum of values between 190 K and 240 K. The spectrum in T_c indicates a broad distribution of distances between the Fe atoms in the amorphous matrix. This means that considerable changes in the positions of Fe atoms must have taken place, despite the fact that crystallization of amorphous Fe_2Ce commences primarily with the formation of CeH_2.

When $T_a > 650$ K, ferromagnetic behaviour is observed at temperatures up to 500 K, which is brought about by the ferromagnetic α-Fe phase. The appearance of this phase indicates the complete crystallization of the amorphous Fe_2Ce hydride.

A summary of Fig.7 is given in Fig.8. Here the Curie temperature is plotted against T_a, reflecting the strong correlation between the Fe distribution (T_c) and annealing temperature T_a, or C_H. It was possible to determine C_H for each T_c explicitly by measuring the amount of hydrogen being thermally desorbed up to T_a.

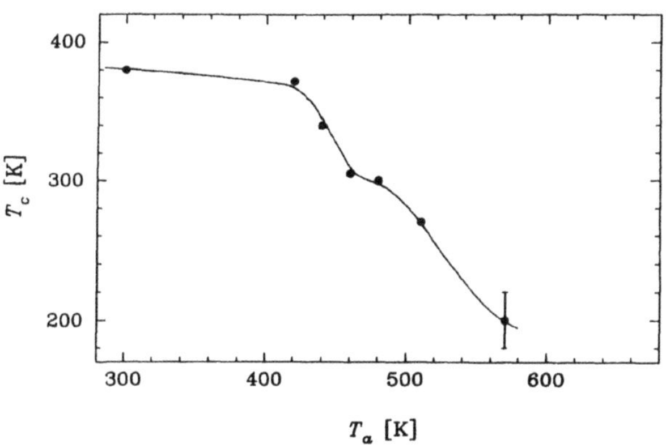

Fig. 8: Curie temperature T_c of hydrogen-charged, amorphous Fe_2Ce in dependence of annealing temperature T_a.

The direct dependence of T_c on C_H is shown in Fig.9. As the remaining amount of hydrogen in the Fe_2Ce hydride is reduced, T_c decreases in three steps. T_c decreases rather rapidly from 380 K to 300 K between C_H=53 at.% and 38 at.%. Here the hydrogen can only originate from Ce_3Fe tetrahedra (or tetrahedra with a higher Fe content), as T_c is strongly affected by a change in the distance between the Fe atoms. This means that the first desorption maximum in Fig.5 between C_H=53 at.% and 38 at.% has to be assigned to hydrogen atoms leaving tetrahedra containing Fe atoms.

Hydrogen desorption between 38 at.% and 19 at.% leads to the second desorption maximum at 475 K in the desorption spectrum (Fig.5). Although the same amount of hydrogen is desorbed as during the first maximum, T_c decreases by a mere 20 K. This means that the desorbed hydrogen originates only from Ce_4 tetrahedra, as any change in their size has a negligible influence on T_c. Between 19 at.% and 8 at.% T_c starts to decreases more rapidly again. The presence of a spectrum in T_c at 8 at.% indicates the start of crystallization and formation of ferromagnetic α-Fe.

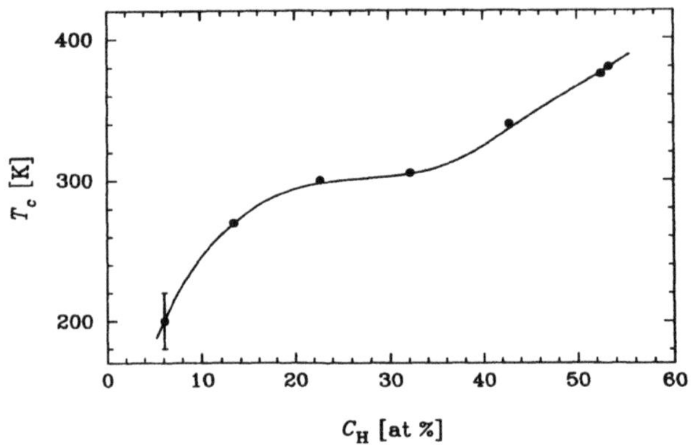

Fig. 9: Curie temperature T_c of hydrogen-charged, amorphous Fe_2Ce as a function of hydrogen concentration C_H.

The depopulation behaviour of Ce_3Fe and Ce_4 tetrahedral sites in amorphous Fe_2Ce described above is also observed in other Laves phase compounds. Vergnat [23], for example, observed a 'stepwise' depopulation of different tetrahedral sites in Fe_2Gd. Here the desorption temperature for hydrogen from different tetrahedral sites increases with the number of Gd atoms making up the tetrahedron. The same is true for amorphous NiZr, CuTi, and CuZr compounds: at low hydrogen concentrations the Zr_4 and Ti_4 tetrahedra are filled prior to Zr_3Ni, Ti_3Cu and Zr_3Cu tetrahedra [24, 25, 26]. Hydrogen desorption, on the other hand, takes place in the reverse order.

3.4 MAGNETIC AFTER-EFFECT

Fig.10 shows the relaxation spectra of hydrogen in crystalline Fe_2Ce obtained with the measurement of the magnetic after-effect. The symbols represent the experimental data (t_1=0.5 s, t_2=1.5, 4.5, 9.5, 19.5, 29.5, 89.5, 119.5, 179.5 s) and the solid lines correspond to the numerical fit of the 9 isochronal curves employing equations (3) and (6). The hydrogen concentration is less than 0.01 at.%.

The relaxation maximum at 55 K and 65 K can only be observed when the crystalline structure is present. Its amplitude decreases with decreasing hydrogen concentration. Furthermore, the maxima of the different isochronal curves show a shift in temperature, which is typical for a thermally activated relaxation process. The MAE-relaxation maximum must therefore be attributed to the thermal hydrogen relaxation process in the cubic (C15) Fe_2Ce Laves phase.

Detailed MAE analysis performed on crystalline Fe_2Zr reveal a relaxation maximum indicating H jump processes only between Fe_2Zr_2 tetrahedral sites [27]. The same can be assumed for crystalline Fe_2Ce, where the hydrogen relaxation process occurs only between Fe_2Ce_2 tetrahedral sites. Additionally the hydrogen concentration of 0.01 at.% is far too low for any Ce_3Fe sites to be occupied. The MAE relaxation maximum in Fig.10 must therefore be solely attributed to H jumps between Fe_2Ce_2 tetrahedral sites in the Fe_2Ce Laves phase.

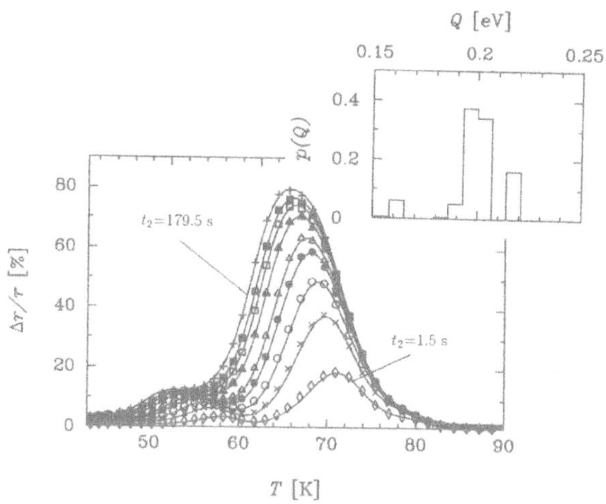

Fig. 10: H-induced MAE relaxation maximum in crystalline Fe_2Ce ($C_H < 0.01$ at.%). Experimental data are represented by symbols and the solid lines correspond to the numerical fit. The insert shows the distribution of activation enthalpies. The pre-exponential factor τ_0 has a value of 10^{-14} s.

A numerical evaluation of the MAE spectra with the time law for orientational MAE reveals an activation enthalpy of $\bar{Q}_H=0.2$ eV and a pre-exponential factor of $\tau_0=1\cdot10^{-14}$ s for the H jump process. The enthalpy distribution necessary for a satisfactory numerical fit of the MAE is depicted in the insert of Fig.10.

After hydrogen absorption at a hydrogen pressure of $p_{H_2}=5\cdot10^3$ Pa, an additional, very broad MAE relaxation maximum at 135 K is observed (Fig.11), indicating the formation of an additional phase. This maximum shows a shift towards higher temperatures when the hydrogen concentration in the amorphous Fe_2Ce hydride is lowered, a behaviour typical of amorphous alloys (Fig.12). It must therefore be attributed to the thermally activated H jump processes in the amorphous Fe_2Ce hydride. A numerical fit reveals $\bar{Q}_H=0.33$ eV and $\tau_0=10^{-14}$ s for the H jump activation parameters in this phase.

Thermal H desorption measurements in connection with magnetic susceptibility measurements of amorphous Fe_2Ce have revealed that only Ce_3Fe and Ce_4 tetrahedral sites are occupied by hydrogen in the amorphous Fe_2Ce hydride phase (see Figures 5 and 9 in parts 3.2 and 3.3). The broadness of the enthalpy spectrum in Fig.11 can be explained by the energetically different relaxation processes which take place between these energetically different sites. The statistical distribution of atoms in the amorphous Fe_2Ce alloy leads to an additional broadening of the enthalpy spectrum.

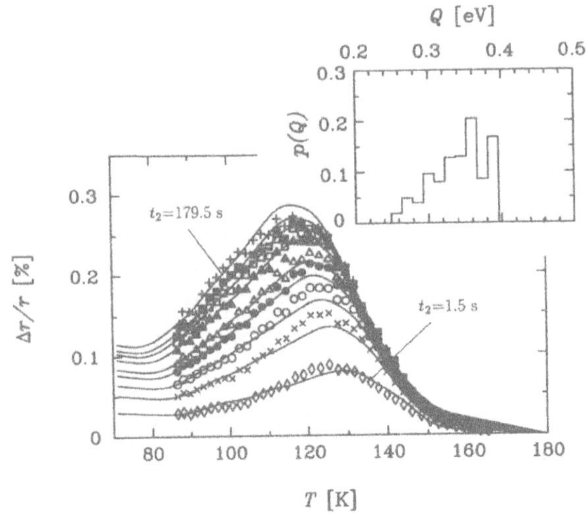

Fig. 11: H-induced MAE relaxation maximum in amorphous Fe_2Ce hydride (C_H^{amorph}=53 at.%). Isochrones and fit as in Fig.10.

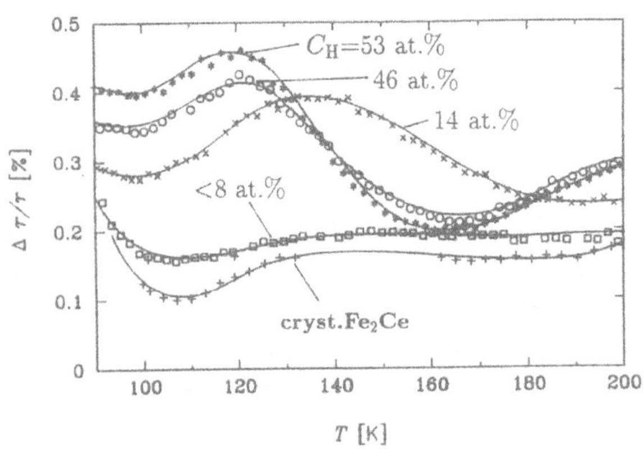

Fig. 12: H-induced MAE relaxation maximum in amorphous Fe_2Ce hydride at different H concentrations. Only isochrones with t_2=179.5 s are shown.

MAE measurements on totally amorphous Fe_2Ce show a complete disappearance of the 65 K relaxation maximum, whereas the 135 K maximum is still preserved. The disappearance of the low temperature maximum is the direct result of the total absence of crystalline Fe_2Ce, and hence of Fe_2Ce_2 tetrahedral sites.

3.5 CYCLIC CHARGING

In order to investigate reversibility, amorphous Fe_2Ce was cyclically charged with hydrogen up to three times. Care was taken not to exceed its crystallization temperature of T_{cryst}=560 K during cycling. Every charging cycle was performed at 50°C under pure hydrogen (99.999 at.%, $p_{H_2}=10^5$ Pa). Before and after each H charging the magnetic susceptibility of the amorphous sample was measured. From these measurements T_c could be obtained. With the aid of T_c in connection with Fig.9 it was possible to determine C_H^{amorph} of the *amorphous hydride phase* without having to heat the bulk sample above T_{cryst}. Additionally, measurements of the magnetic after-effect (T_{max}, $\Delta r/r$) were made in order to detect any changes in the H relaxation in the amorphous hydride.

The desorbed amount of hydrogen from the cycled sample could be obtained by thermally desorbing a small portion taken from the bulk sample after each H charge. Despite the fact that the bulk sample was never heated above 400 K, a desorption spectra reaching up to 700 K could thus be obtained. The reduction in mass of the bulk sample after three cycles amounts to only 3 weight % and is therefore negligible. The values for the MAE relaxation temperature T_{max} and amplitude $\Delta r/r$, Curie temperature T_c with corresponding hydrogen concentration C_H^{amorph} and desorbed hydrogen concentration C_H^{desorp} are summarized in Table 1.

Although cyclic charging was performed at temperatures well below the crystallization temperature of amorphous Fe_2Ce, the hydrogen concentration C_H^{desorp} decreases rapidly from 53 at.% to 18 at.% after just three charging cycles. However, this drastic degradation is not reflected in the hydrogen concentration concerning the amorphous Fe_2Ce phase: Here the value for C_H^{amorph} drops from 53 at.% to 48 at.%, 23 at.% and 43 at.% respectively, but can always be increased again to 53 at.% by hydrogen charging. The fact that the whole sample shows degradation whereas the amorphous phase remains fully rechargeable indicates a phase separation into a non-hydrogen absorbing phase and a hydrogen absorbing phase. The volume of the non-hydrogen absorbing phase increases with every charging cycle, leading to the drastic decline in C_H^{desorp}.

Fig.13 shows the susceptibility of the bulk sample before (b, d, f) and after (a, c, e, g) hydrogen charging. After only two charging cycles, a second phase with a higher Curie temperature than that of amorphous Fe_2Ce is formed, indicating the formation of the α-Fe phase.

In Fig.14 the desorption spectra after each hydrogen absorption step (a, c, d) are shown. The broad double maximum moves to higher temperatures with every charging cycle, revealing a faster decrease in the first maximum's amplitude in comparison to that of the second maximum. As the first maximum is related to hydrogen desorbing out of Ce_3Fe tetrahedral sites, its rapid decrease indicates a change or even total destruction of these sites. The second maximum, related to Ce_4 sites, is still present after three charging cycles. This means that Ce_4 tetrahedra are much more stable than Ce_3Fe tetrahedra in view of these data.

The shift of the desorption step at 610 K towards lower temperatures indicates a decrease in crystallization temperatures for the α-Fe phase. This means that the tendency for amorphous Fe_2Ce to crystallize increases with every charge and discharge cycle.

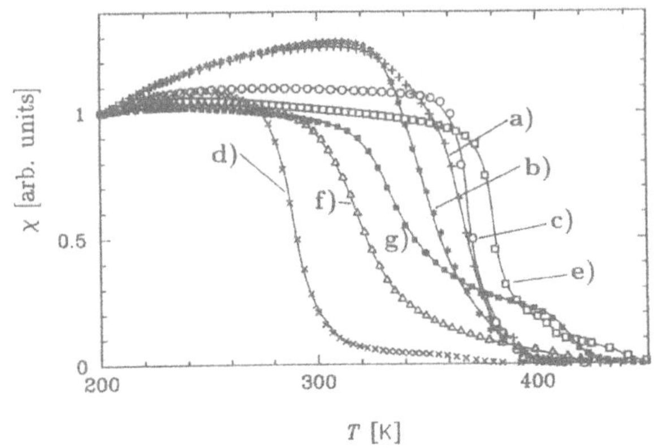

Fig. 13: Temperature dependence of the initial susceptibility χ for cyclically charged amorphous Fe_2Ce. a) to g) correspond to states listed in Table 1.

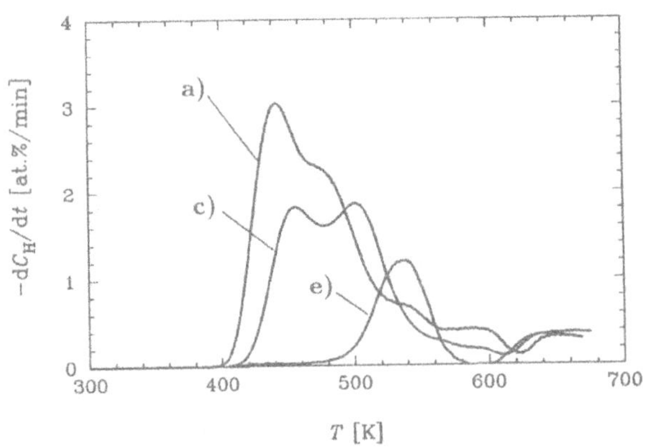

Fig. 14: Thermal H desorption obtained during cyclic charging (heating rate 5 K/min). a), c) and e) correspond to 1st, 2nd and 3rd H charging (see Table 1).

	Initial Compound: crystalline Fe_2Ce	MAE		χ		Desorption
		T_{max} [K]	$\Delta r/r$ [%]	T_c [K]	C_H^{amorph} [at.%]	C_H^{desorp} [at.%]
a	1st H charge t=24 hrs	133	0.3	380	53	53
b	1st H discharge T_{max}=420 K	134	0.24	370	48	-
c	2nd H charge t=20 hrs	130	0.2	380	53	38
d	2nd H discharge T_{max}=480 K	135	0.16	300	23	-
e	3rd H charge t=60 hrs	132	0.14	390	54	18
f	3rd H discharge T_{max}=480 K	135	0.1	340	43	-
g	4th H charge t=20 hrs	-	0.8	370	50	-

Table 1: Experimental data of the MAE (T_{max}, $\Delta r/r$) and magnetic susceptibility measurements (T_c). With T_c and Fig.9 the hydrogen concentration C_H^{amorph} in the amorphous Fe_2Ce phase could be obtained. The hydrogen concentration C_H^{desorp} was determined by thermal hydrogen desorption of a small amount from the bulk sample. H charging took place under pure hydrogen atmosphere at 50°C.

4. Summary and Conclusion

Previous XRD and DSC measurements have shown that Fe_2Ce, a representative of the rare earth and transition metal cubic (C15) Laves phase compounds, can be amorphized easily when exposed to a high-pressure hydrogen (H) atmosphere (p_{H_2}=5 MPa). In order to slow down the amorphization process, much lower H-charging pressures between $p_{H_2} = 5 \cdot 10^3$ Pa and $1 \cdot 10^5$ Pa were employed in this work, making the study of the amorphization kinetics much easier. In addition to standard XRD and DSC analysis, the method of thermal hydrogen desorption was applied, giving information on concentration and desorption characteristics of hydrogen in crystalline and amorphous Fe_2Ce. With the aid of the magnetic after-effect (MAE) it was possible to investigate the local hydrogen diffusion processes. Herein H is employed as a probe in order to study different microstructures and the amorphization process on an atomistic level. The investigations show that for low H concentrations ($C_{\rm H} \sim 0.01$ at.%), the Fe_2Ce Laves phase is preserved. MAE measurements reveal $Q_{\rm H}$=0.2 eV and $\tau_o = 10^{-14}$ s as activation parameters for the local H diffusion in this phase. These activation parameters are solely attributed to H jumps between Fe_2Ce_2 tetrahedral sites in the C15 structure.

Stronger H charging of Fe_2Ce at a H pressure of $p_{\rm H_2} = 5 \cdot 10^3$ Pa (T <400 K) reveals an additional broad MAE relaxation maximum at 135 K and an increase in the Curie temperature from 230 K to 380 K. The 135 K MAE maximum shows the typical H concentration dependence for amorphous alloys, indicating the formation of the amorphous Fe_2Ce hydride phase. Susceptibility measurements on this heterogenous Fe_2Ce sample, i.e. crystalline and partially amorphized, show that the irreversible hydrogen induced amorphization takes place in a narrow phase-boundary which migrates from the surface into the sample during H absorption.

At H pressures of $p_{\rm H_2} = 1 \cdot 10^5$ Pa (T <400 K) Fe_2Ce becomes completely amorphized. Independent of charging pressure, the H concentration of the amorphous hydride phase is found to be $C_{\rm H}^{amorph}$=53 at.%. This phase decomposes into CeH_2 and α-Fe under H desorption when heated above 610 K.

The occurrence of two desorption maxima well below the decomposition temperature in combination with susceptibility measurements indicate, that only Ce_4 and Ce_3Fe tetrahedral sites are occupied by H in amorphous Fe_2Ce. The activation parameters for local H jumps between these sites are $\bar{Q}_{\rm H}$=0.33 eV and $\tau_o = 10^{-14}$ s, whereas occupation of Fe_2Ce_2 tetrahedra could be not observed.

Cyclical charging and discharging of fully amorphized Fe_2Ce well below its crystallization temperature $T_{\rm cryst}$=560 K leads to a reduction in the storage capacity from $C_{\rm H}^{max}$=53 at.% to 18 at.% after only three charging cycles. This degradation is attributed to a *hydrogen-induced* phase separation into crystalline CeH_2 and α-Fe, which occurs via enhancement of the short range diffusivity of Fe atoms. Although the amorphous Fe_2Ce hydride phase can be recharged up to $C_{\rm H}^{amorph}$=53 at.% after each cycle, its volume fraction diminishes gradually with cycling.

References

[1] Yeh X.L., Samwer K., Johnson W.L., *Applied Physics Letters*, **42**, 242, (1983).

[2] Samwer K., in *Hydrogen in Disordered and Amorphous Solids*, edited by G.Bambakids and R.C. Bowman, Jr., New York: Plenum, 173, (1986).

[3] Buschow K.H.J., *Solid State Communications*, **19**, 421 (1976).

[4] Buschow K.H.J., *Physica B*, **86-88**, 79 (1977).

[5] Diepen A.M., Buschow K.H.J., *Solid State Communications*, **22**, 113 (1977).

[6] Malik, S.K., Wallace, W.E., *Solid State Communications*, **24**, 283 (1977).

[7] Deriagin A.V., Kazakov A.A., Kudrevatykh N.V., Moskalev V.N., Mushnikov N.V., Teret'yev S.V., *Phys. Met. Metall*, **60** Nr. 2, 81 (1985).

[8] Klose F., et al, *Zeitschrift für Physik*, **90**, 79 (1993).

[9] Cahn, R.W., in *Physical Metallurgy* edited by R.W. Cahn and P. Haasen, North Holland (1983).

[10] Chung U-I., Kim Y.-G., Lee J.-Y., *Philosophical Magazine B*, **63**, 1119 (1991).

[11] Aoki K., Yanagitani T., Li X.-G., Masumoto T., *Material Science Enginering*, **97**, 35 (1988).

[12] Aoki K., Yamamoto T., Masumoto T., *Scripta Metallurgica*, **21**, 27 (1987).

[13] Cohen R.L., West K.W., Oliver F., Buschow K.H.J., *Physical Review B*, **21**, 21 (1980).

[14] Bancour V.P., Lartigue C., Guegan A.P., Achard J.C., Pannetier J., *Journal of Less-Common Metals*, **35**, 301, (1988).

[15] Aoki K., Yamamoto T., Satoh Y., Fuchamichi K., Masumoto T., *Acta Metallurgica*, **35**, 2465 (1987b).

[16] Aoki K., Li X-G., Masumoto T., *Acta Metall. Mater.*, **40**, (1992).

[17] Christodoulou C.N., Takeshita T., *Journal of Alloys and Compounds*, **194**, 31 (1993).

[18] Kim Y.-G., Lee J.-Y., *Journal of Alloys and Compounds*, **191**, 243 (1993).

[19] Kronmüller H., *Nachwirkung in Ferromagnetika*, Springer-Verlag, Berlin (1968).

[20] Walz F., *Applied Physics*, **3**, 313 (1974).

[21] Walz F., *Phys. Stat. Sol. A*, **82**, 179 (1984).

[22] Richter G., *Annalen der Physik*, **29**, 605 (1959).

[23] Vergnat M., Chatbi H., Marchal G., *Applied Physics Letters*, **64**, 2084 (1994).

[24] Mizubayashi H., Arai H. und Okuda, S., *Zeitschrift für physikalische Chemie, Neue Folge*, **163**, 391 (1989).

[25] Mizubayashi H., Naruse T. und Okuda S., *Journal of Less-Common Materials*, **172-174**, 908 (1991).

[26] Harris J.H. and Curtin W.A., *Zeitschrift für physikalische Chemie, Neue Folge*, **163**, 315 (1989).

[27] Maier C.U., Kronmüller H., *J. Phys.: Condens. Matter*, **4**, 4409, (1992).

Chapter 22

MAGNETIC PROPERTIES OF THE RARE-EARTH TRANSITION-METAL COMPOUNDS AND THEIR MODIFICATION BY HYDROGENATION

J. BARTOLOME
Instituto de Ciencia de Materiales de Aragón
CSIC-Universidad de Zaragoza
Facultad de Ciencias
E-50009 Zaragoza
Spain

ABSTRACT. The interatomic magnetic interactions and the crystal field interation acting on the rare-earth, which give rise to the intrinsic magnetic properties of the rare-earth transition-metal intermetallic compounds $R_2Fe_{14}Z$; Z = B and C, R_2Fe_{17} and $R(Fe_{12-x}M_x)$; M = Ti, V, Mo on which some of the new permanent magnets are based, are reviewed in the first section. The parameters which describe these interactions have been found to depend on interatomic distances and nearest neighbor geometry. In the second section the location of hydrogen on interstitial sites, the modification of the Curie temperature, magnetization, and magnetocrystalline anisotropy are discussed in connection to the interactions described above.

1. Introduction

Permanent magnets are materials which form components in many appliances, for example in motors. The improvement of their magnetic force to weight ratio is one of the main goals in the quest for reduced energy consumption. In a vehicle this is obvious, because a reduction in its total weight implies a direct reduction in fuel consumption. Tailoring the magnet properties of a material to demand is also a desirable feature as it will optimize its function. In the last decade new families of intermetallic compounds have provided materials with both capacities, in particular the $R_2Fe_{14}B$, R_2Fe_{17} and $R(Fe_{12-x}M_x)$ (R = rare earth, M = V, Ti, Mo, Cr, Si, W, and Rh) compounds. The inclusion of interstitial atoms, H, N, and C, has been demonstrated to be a very fruitful development of the previous series. Because of the vast selection of materials to be discussed, this chapter is limited to general trends with examples drawn from the above mentioned series of compounds. In the first section the magnetic properties of the parent compounds are described in terms of their crystal field and magnetic exchange interactions. The second section describes the modification in these properties due to hydrogen uptake.

2. Magnetic properties of the parent compounds

The new magnets are Fe-based intermetallic compounds. The richness of Fe as a magnetic element reduces the price as compared to Co, while the proper R atom helps in achieving high magnetic anisotropy. The prime performance parameter is the energy product, $(BH)_{max}$, which requires a high coercivity and magnetization. The coercivity is mainly related to the extrinsic characteristics of the material, especially its microstructure, whereas magnetization is related to intrinsic characteristics. We shall focus on the latter.

F. Grandjean et al. (eds.), Interstitial Intermetallic Alloys, 541–559.

In the general case the magnetic constituent atoms, R and Fe, form at least two sublattices. In the R atom the unpaired spins are well localized in the 4f shell and its ground state is given by Hund's rule. The atoms surrounding the R atom create a crystal field, CF, which splits the R ground state multiplet and may give rise to a non-quenching of its orbital moment and, thus, to strong single ion anisotropy. In contrast, the Fe electrons are partially localized and partially itinerant. The relevant interatomic interactions will be the Fe-Fe, Fe-R, and, R-R coupling, in the order of decreasing importance. The interactions give rise to magnetic ordering at the Curie temperature, T_C, and their detailed interplay, together with the R anisotropy at different temperatures, produce the observed different magnetic phases and the spin reorientation transitions, SRT, between these phases.

2.1. Fe-Fe INTERACTIONS

The Fe-Fe interaction is ferromagnetic and is described within the simple Stoner model of magnetism of metals. It may be studied in compounds where R is non-magnetic, such as Y. Recently, it has been shown that the main limitation of the Stoner model, namely that the predicted T_C is too high, can be overcome by considering both the single-particle excitations of the Stoner model and the collective spin-fluctuations of itinerant electron magnetism. Mohn and Wohlfahrt (1987) have proposed a simple model, renormalizing the Landau type development of the Stoner free-energy in terms of the parallel and transverse components of the locally fluctuating magnetic moments. They predicted that the experimental T_C should obey the relation,

$$\frac{T_C^2}{T_{SF}^2} + \frac{T_C}{T_{SF}}\frac{T_C^{s2}}{T_{SF}^2} - \frac{T_C^{s2}}{T_{SF}^2} = 0, \tag{1}$$

where T_C^S is the Stoner Curie temperature, which is 1508 or 612K for $Y_2Fe_{14}B$ or Y_2Fe_{17}, respectively. In this equation, T_{SF} is the characteristic spin-fluctuation temperature, given by $T_{SF} = M_0^2 / 10k_B\chi_o$, where M_o is the equilibrium magnetization at T = 0K and χ_o is the exchange enhanced susceptibility at equilibrium. By substituting M_o and χ_o, determined from experiment, the values T_{SF} = 674 and 615K were deduced for $Y_2Fe_{14}B$ and Y_2Fe_{17}, respectively. The T_C of $Y_2Fe_{14}B$ and Y_2Fe_{17} agree very well with the T_C/T_{SF} versus T_C^S/T_{SF} curve, see Fig. 1. From the proximity of T_C to one extreme or the other on this curve one concludes that Y_2Fe_{17} is near the Stoner limit, whereas $Y_2Fe_{14}B$ has a stronger spin-fluctuation. Moreover, the heat capacity anomaly at the Curie magnetic ordering temperature for $Y_2Fe_{14}B$ could be interpreted within the Mohn and Wohlfart model, Fig. 2, by means of the analytic expressions of the magnetic specific heat found for a simplified model of spin fluctuations (Mohn and Hilscher 1989). The theoretical curve depends only on the parameter $t_C = T_C/T_C^S = 0.38$, which was determined independently (Luis et al. 1994).

The magnetic moments on the Fe sites depend in detail on the interatomic distances with the nearest neighbors. They may be determined by neutron diffraction and, in some cases, very reasonable predictions are provided by band structure calculations. For example, for $Y_2Fe_{14}B$ the six different Fe moments, corresponding to the different sites (Givord et al., 1985), are reproduced to within 10% by a tight-binding d-band model (Inoue and Shimizu, 1987). They range between 1.95 and 2.8μ_B. For Y_2Fe_{17} the local moments have values that range between 1.91 and 2.41μ_B, as obtained from band structure calculations (Beuerle et al., 1991), a trend which is followed by the experimental hyperfine fields detected by Mössbauer spectroscopy (Gubbens et al., 1987). In contrast, in $Y(Fe_{10}M_2)$ and

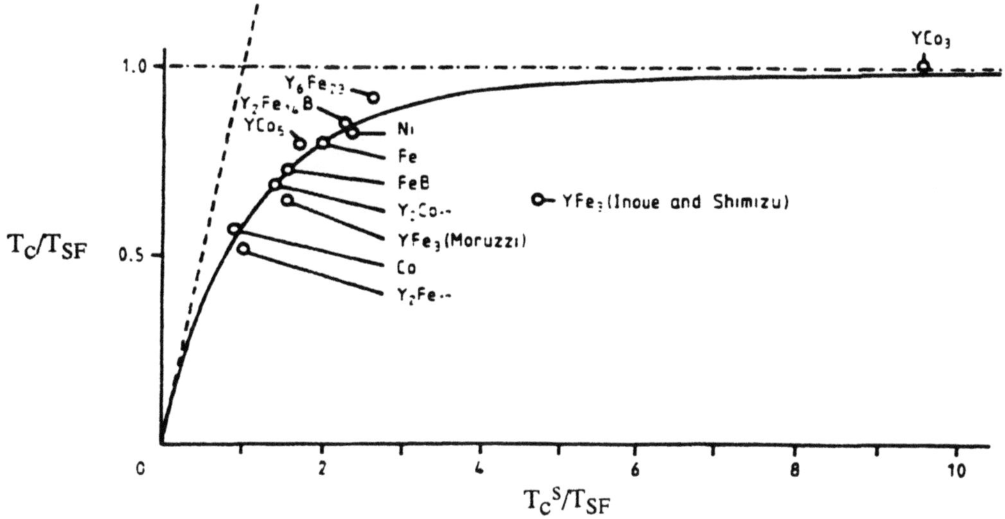

Figure 1. Graphical representation of Eq. 1. with data corresponding to various compounds including $Y_2Fe_{14}B$, and Y_2Fe_{17}. (- -) Stoner Model limit. (- · -) Pure fluctuation behavior. (Mohn and Wohlfarth, 1987)

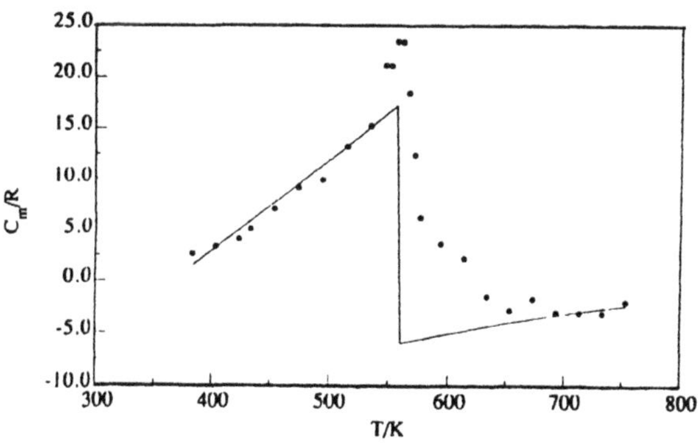

Figure 2. $C_p(T)$ of $Y_2Fe_{14}B$. (-) prediction with Mohn and Wohlfarth spin-fluctuation theory (Luis et al., 1994).

$Y(Fe_{11}M)$, with M = V,Ti, or Mo, the Fe moments are systematically smaller. For example, in $Y(Fe_{10}V_2)$ they range from 1.16 to 1.95μ_B, as determined from neutron diffraction studies (Helmholdt et al., 1988). A review of the dependence of the Fe moments on interatomic distances has been recently published (Isnard and Fruchart, 1994a). From the above description it is rather clear that the Fe-Fe interaction should be described by the Stoner parameter with the effective exchange integral, I, in a band model. Indeed, a value of I = 0.925 eV has been given for Y_2Fe_{17} (Coehoorn, 1989).

However, it is far more common to approximate the Fe-Fe interaction with a mean-field model, MF, such that,

$$H_{Fe\text{-}Fe} = -\,2\, z_{Fe\text{-}Fe}\, J_{Fe\text{-}Fe}\, \overline{S}_{Fe} \langle \overline{S}_{Fe} \rangle = -\; g\, \mu_B\, \overline{S}_{Fe}\, B_{MF}, \tag{2}$$

where $z_{Fe\text{-}Fe}$ is the number of Fe nearest neighbors, $J_{Fe\text{-}Fe}$ is the effective Fe-Fe interaction coupling constant, and $\overline{S}_{Fe}$ is the effective Fe spin. In the MF model $B_{MF} = n_{FeFe}M_{Fe}$, where M_{Fe} is the Fe sublattice magnetization and n_{FeFe} is the molecular field interaction parameter. B_{MF} is 390T for Y_2Fe_{17}, 590T for $Y_2Fe_{14}B$, and 622T for $Y(Fe_{11}Ti)$. It is important to note that B_{MF} decreases with increasing magnetic coordination number of the interacting Fe spins, in contrast to the expected linear dependence on $z_{Fe\text{-}Fe}$. From the simple relation, $n_{FeFe}=3k_BT_C/\mu_{Fe}{}^2$, the dependence of n_{FeFe} on the local coordination number of the Fe, c_n, has been estimated and represented in Fig. 3 (Gavigan et al., 1988). It is very instructive to note that there is a regular decrease for increasing c_n for compounds with c_n ranging between 6 for RFe_2 and 10.3 for R_2Fe_{17}, a decrease which cannot be ascribed only to differences in the interatomic distances. From band structure calculations (Kübler, 1981) it has been proposed both that ferromagnetic interactions are associated with open structures with a low Fe coordination number, and that the interactions will be strongly reduced for close packed structures. In conclusion, it is the local coordination number, c_n, and geometry of the nearest Fe moments which dominate the Fe-Fe interactions.

The Fe-Fe interaction gives rise to the so called "weak" ferromagnetism, which occurs when both the majority and minority spin sub-bands remain partially filled. In a simple phenomenological model the anisotropy energy is given by,

$$E_a^{Fe} = K_1^{Fe} \sin^2\theta, \tag{3}$$

where θ is the angle between the magnetization direction and the c-axis. The magnetic anisotropy induced by this interaction depends on the families of compounds studied; it is axial ($K_1 > 0$) for $Y_2Fe_{14}B$ and $Y(Fe_{11}Ti)$, whereas it is planar ($K_1 < 0$) for Y_2Fe_{17}. For Y_2Fe_{17} (Deportes et al., 1986) and $Y(Fe_{11}Ti)$ (Hu et al., 1989) the magnetocrystalline anisotropy constant, K_1, increases with decreasing temperature in very much the same fashion as for a model of localized moments at low temperatures, Fig. 4, for which

$$K_1(T) = K_1(0)\left[\frac{M(T)}{M(0)}\right]^3. \tag{4}$$

In contrast, in $Y_2Fe_{14}B$, $K_1(T)$ has an abnormal maximum at 310K, which is related to a giant volume magnetostriction, and a plateau at lower temperatures (Hirosawa, 1987).

2.2. Fe-R INTERACTIONS

The presence of a magnetic rare earth serves the purpose of inducing a strong anisotropy, transmitting the crystal field induced local anisotropy to the Fe sublattice by means of the

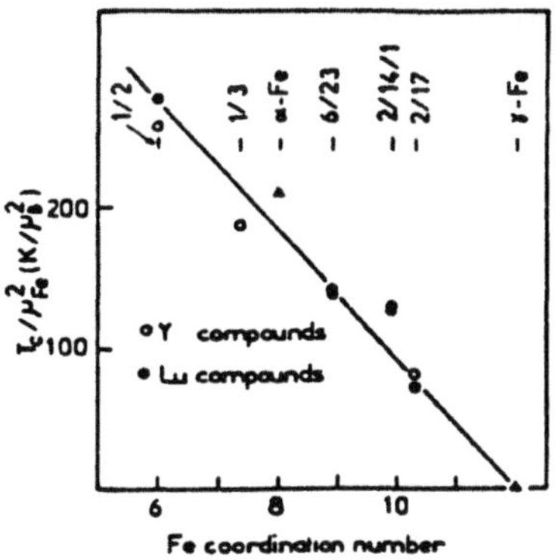

Figure 3. T_c/μ^2_{Fe} in Y-Fe and Lu-Fe compounds as a function of the mean local coordination of Fe atoms. (Gavigan et al., 1988)

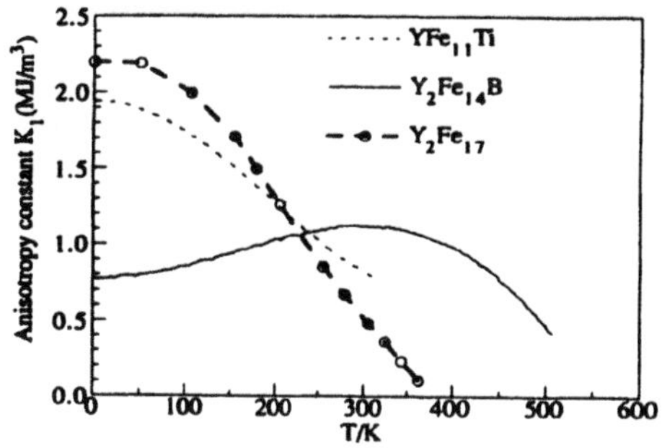

Figure 4. Temperature dependence of the Fe sublattice anisotropy constant K_1. (-) Hirosawa et al 87), (- -) (Hu et al., 1989), (o - o) (Deportes et al., 1986) (For Y_2Fe_{17}, $K_1<0$).

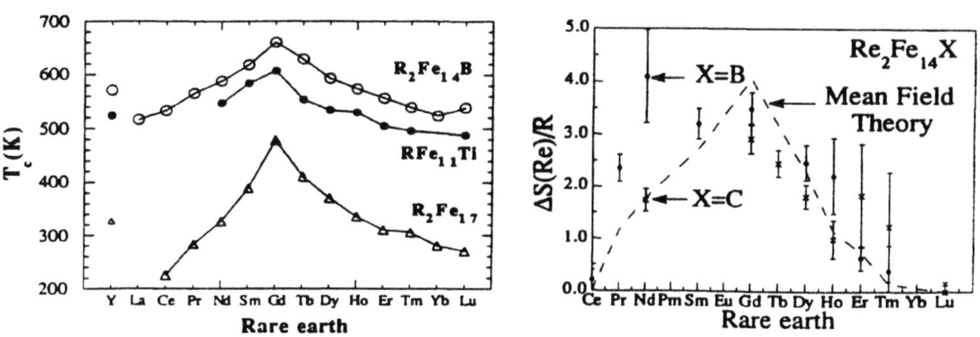

Figure 5. a) T_c of the formula compounds. b) Contribution of the R sublattice to the magnetic entropy in the $R_2Fe_{14}B$ series, compared to Mean Field theoretical predictions (Luis et al., 1994).

Fe-R interaction. This interaction is caused, according to the model proposed by Campbell (1972), initially by hybridization of the 3d electrons of the T atom with the 5d electrons of R, and then by means of the intraatomic coupling between the 5d and 4f electrons. Thus the 5d electrons mediate the 4f-3d interaction. In fact, recently the antiferromagnetic character of the 3d-5d interaction was observed by means of X-ray magnetic circular dichroism experiments on the $R_2Fe_{14}B$ compounds (Giorgetti et al., 1993, Chaboy et al., 1994, Isnard et al., 1994b). In contrast, the 4f-5d coupling favours parallel alignment of the 5d and 4f spins (Brooks et al., 1991). The net effect is an antiferromagnetic 4f-3d spin interaction. For light R, where J = L - S, the Fe and R total moments are ferromagnetically coupled, whereas for the heavy R, where J = L + S, they are antiferromagnetically coupled. Thus for permanent magnet materials, the best candidates are based on the light rare earths.

The simplest description of ferrimagnetism is in terms of the Nééł mean field two sublattice model. The mean field acting on the R moment is given by

$$B_{MF}^{R} = n_{RT} M_T + n_{RR} M_R = B_{MF}^{RT} + B_{MF}^{RR}, \tag{5}$$

where M_T is the transition metal sublattice magnetization, n_{RT} is the rare earth-transition metal molecular field parameter, M_R, is the rare earth sublattice magnetization, and n_{RR} is the rare earth-rare earth coupling constant. The coupling constant, n_{RT}, is related to the interatomic spin-spin interaction, J_{RT}, which causes the coupling, through the relation,

$$n_{RT} = \frac{J_{RT}\, z_{RT}(g_R - 1)}{N_T\, \mu_B^2\, g_R}, \tag{6}$$

where z_{RT} is the number of nearest neighbor transition metal atoms, N_T is the number of transition metal atoms per formula unit, and g_R is the gyromagnetic factor for R. The ordering temperature increase, relative to the non-magnetic rare earth compound, is given, according to the mean field relation, by

$$T_c = \frac{1}{2}\left\{T_T + T_R + \left[(T_T - T_R)^2 + 4\, T_{RT}^2\right]^{1/2}\right\}. \tag{7}$$

The characteristic temperatures representing the contributions to T_C, arising from the T-T, R-R, and R-T interactions, are

$$T_T = n_{TT}\, C_T\,,\ T_R = n_{RR}\, C_R\,,\ \text{and}\ T_{RT} = n_{RT}\left[C_R C_T\right]^{1/2},$$

where C_R and C_T are the Curie constants of the R and T sublattices. For a given series of compounds, T_C clearly follows the de Gennes factor, $G=(g_J -1)^2 J(J+1)$, see Fig. 5a. It is interesting to note that the entropy, subtended by the magnetic ordering heat capacity anomaly at the Curie point, also follows the de Gennes trend (Luis et al., 1994) as is shown in Fig. 5b.

The J_{RT} interaction decreases linearly with increasing concentration of the T atoms, as is shown for a series of Er-Fe compounds in Fig. 6 (Liu et al., 1994), while the transition metal magnetic moments, μ_T, increase with increasing concentration of T. This dependence is thought to be due to reduction of the 3d-5d hybridization with increasing T concentration (Brooks and Johansson, 1993). Within the lanthanide series, J_{RT} decreases with increasing atomic number, rapidly for light R and slowly for heavy R, as deduced from high field magnetization measurements (Liu et al., 1994), and from Curie temperature measurements (Belorizky et al., 1987) according to the relation,

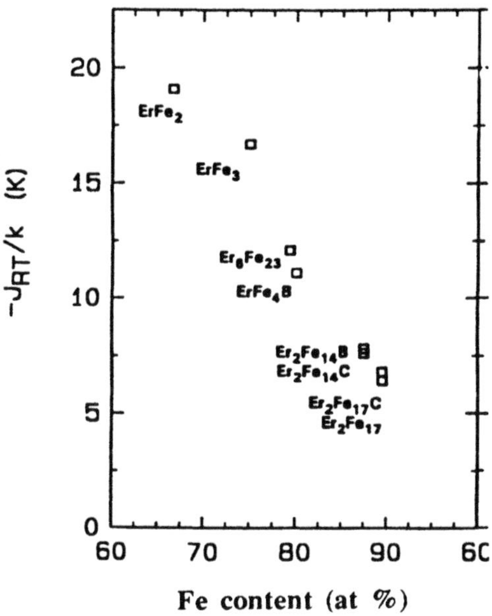

Figure 6. Fe concentration dependence of the magnetic-coupling J_{ErFe} of various Er-Fe compounds (Liu et al., 1994).

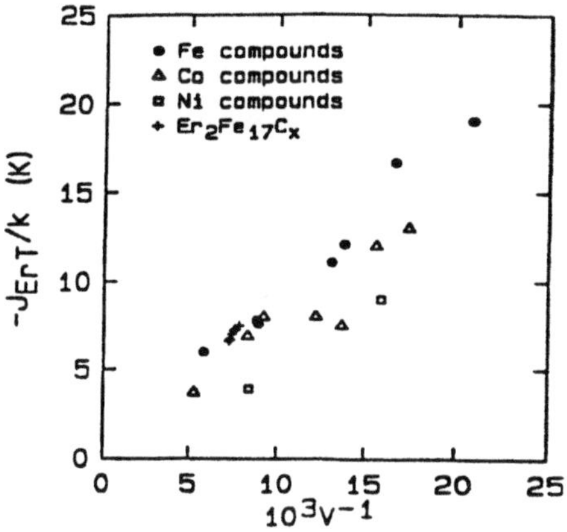

Figure 7. $J_{R\text{-}T}$ of Er-3d compounds *vs* reciprocal values of the normalized volume per formula unit (Liu et al., 1994).

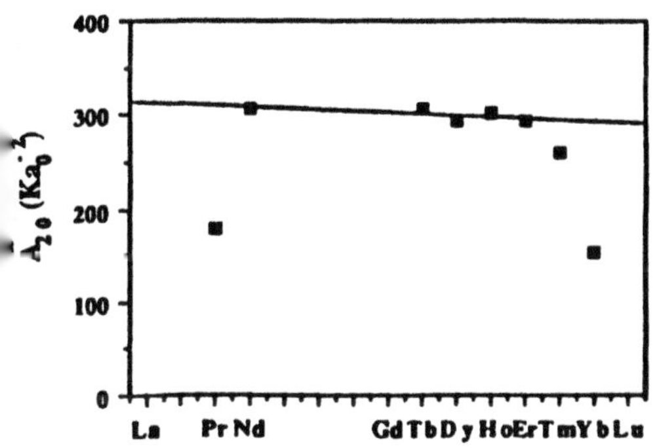

'igure 8. Variation of A_2^0 across the $_2Fe_{14}B$ series (Coey et al., 1989)

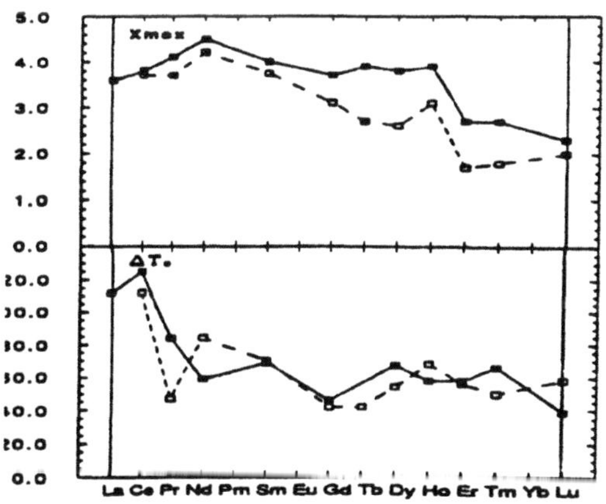

Figure 9. x_{max} and increment of T_c for the $R_2Fe_{14}XH_x$ series, with X=B (■) and C (□) (Bartolomé et al., 1991)

$$n_{RT} = \left[\frac{(T_c\text{-}T_T)(T_c\text{-}T_R)}{C_R C_T}\right]^{1/2}, \tag{8}$$

where C_R and C_T are the respective Curie constants. The approximation $T_R = 0$ is valid because the R-R interactions are weak, as will be discussed below.

Two mechanisms have been invoked to account for the J_{RT} dependence on the R atom. On one hand, across the lanthanide series, the atom radius decreases with increasing atomic number, and the interatomic distances decrease accordingly, thus favouring an increase in J_{RT}. On the other hand, the 4f radius decreases more rapidly than the 5d radius, thus decreasing the 4f-5d overlap and decreasing the 4f-5d-3d interaction. Finally, an empirical relation has been found between J_{RT} and the reciprocal volume of the R-Fe intermetallics (Liu et al., 1994), Fig. 7. This probably reflects the variation in the 3d-5d hybridization which increases with decreasing volume.

2.3 R-R INTERACTIONS

R-R interactions play a secondary role in Fe intermetallic compounds because they are at least an order of magnitude lower than the Fe-R interactions. However, some general comments are pertinent. The mechanism involved in this interaction has been proposed to be either the RKKY interaction, occuring via the conduction electrons, or indirect exchange occuring in two stages, first an intraatomic 4f-5d interaction, and second, via the overlap of the 5d rare earth orbitals with the 4s orbitals of the 3d atoms. Both types of interaction may be modelled with a Heisenberg Hamiltonian involving the R spins,

$$H_{RR} = -\sum_{ij} J_{RR}\, \bar{S}_{Ri} \cdot \bar{S}_{Rj} = -\sum_{ij} J_{RR}(g_R-1)^2 \bar{J}_{Ri} \cdot \bar{J}_{Rj}, \tag{9}$$

where the de Gennes factor is introduced to relate S to the J operators. It has been shown that, for some R-Al and Zn compounds, J_{RR} decreases strongly across the light rare earths, while it decreases less rapidly, or remains constant in the heavy rare earths (Belorizky et al., 1988). For the $R_2Fe_{14}B$ compounds the only reliable value is the B_{MF}^{RR} = 54T value found for $Gd_2Fe_{14}B$, which proves the relatively low value of the interaction as compared to the B_{MF}^{RFe} = 270 T (Liu et al. 1994).

2.4. CRYSTAL FIELD INTERACTION

The single-ion magnetic anisotropy of the R atoms is essentially created by the electronic crystal field produced by the neighboring atoms. The simplest approach gives

$$H_{CF} = \sum_{n,m} \theta_n \langle r^n \rangle A_n^m O_n^m = \sum_{n,m} B_n^m O_n^m, \tag{10}$$

where θ_n are the Stevens factors, referred to as α_J, β_J, and γ_J, for n = 2, 4, and 6, respectively, $\langle r^n \rangle$ is the Hartree-Fock radial integral, A_n^m is the crystal field parameter, and O_n^m is the Stevens operator. The B_n^m parameters yield an alternative, equivalent, description.

For a given series, A_n^m may be expected to be independent of the R atom. However, this statement is not strictly true. In Fig. 8 the variation of A_2^0 across the lanthanide series in the $R_2Fe_{14}B$ compounds (Coey et al., 1989) is shown, in which Pr and Yb, for example, depart from the general, slightly decreasing trend. In the $R(Fe_{11}Ti)$ compounds the differences are even larger, A_2^0 varies from -32 for Dy to -143Ka_o^{-2} for Sm. More relevant is the change of sign that occurs for the series of different stoichiometry. Li and Coey

(1991) have systematized the experimental data for A_2^0 and shown that its sign is negative for all compounds studied, including $R(Fe_{12-x}M_x)$ and R_2Fe_{17}, except for the $R_2Fe_{14}Z$ series, in which it is positive. By departing from a simple point-charge model calculation, to band structure calculations, the dependence of A_2^0 on the valence electrons has been shown to depend on the charge density at the edge of the atomic Wigner-Seitz cell described by the Miedema parameter, n_{ws}. It is found that $A_2^0 > 0$ if the neighbors with the highest n_{ws} values are on the z-axis, and $A_2^0 < 0$ if they are in the x-y plane around the R-atom (Coehoorn, 1990a).

The anisotropy will depend on the values and signs of the α_J, β_J, and γ_J coefficients. The R anisotropy energy is given by symmetry adapted functions of the angles determining the direction of magnetization with respect to the crystal. For tetragonal symmetry, up to the fourth order, it is given by

$$E_a^R = K_1 \sin^2\theta + K_2 \sin^4\theta + K_3 \sin^4\theta \cos^2\varphi \sin^2\varphi. \qquad (11)$$

In this expression the K_i coefficients are related to the A_n^m coefficients, in particular K_1 depends essentially on A_2^0, while K_2 depends on A_4^0 (Bartolomé, 1990). For the lanthanides, α_J is positive for Sm, Er, Tm, and Yb, and negative for Pr, Nd, Tb, Dy, and Ho, and zero for Gd. Thus uniaxial anisotropy occurs when α_J and A_2^0 are of opposite signs, and basal anisotropy occurs when they are of the same sign. $Nd_2Fe_{14}B$ provides the best example of high uniaxial anisotropy at room temperature by having $\alpha_J < 0$ with $A_2^0 > 0$. The total anisotropy energy is simply obtained by adding expressions (3) and (11), where the Fe sublattice only affects the first term. Because the K_i coefficients are temperature dependent through the exchange coupling and thermal population of the crystal field states, a change of sign may occur, giving rise to spin reorientation transitions, transitions which will be discussed in another chapter.

3. Modification of the magnetic properties by hydrogenation

The interstitial inclusion of atoms in the lattice of R-3d intermetallic compounds leads frequently to drastic modifications in the physical properties of the parent compounds. The mechanisms invoked may be structural, because of the modification of interatomic distances, or electronic, because the variation of distances, the creation of new s-orbitals states by the extra electrons supplied by hydrogen, and the supply of extra electrons which may modify the Fermi energy. Now that we know how the different interactions give rise to the magnetic properties of the three R-Fe series of interest, we shall proceed by describing the modified, or new, magnetic properties induced by hydrogen uptake.

3.1. HYDROGEN INSERTION

For the three series, hydrogenation is achieved by an exothermal reaction in an autoclave at low hydrogen pressures and temperatures. Their stability depends on the series; in the $R_2Fe_{14}BH_x$ series (L'Héritier et al., 1984), hydrogen is depleted at 450K (Österreicher and Österreicher, 1984), while in $R_2Fe_{17}H_x$ hydrogen is depleted at 670K (Isnard et al., 1992b) and in $R(Fe_{11}M)$ at 400K (Obbade et al., 1988). The effect of H uptake is to expand the unit cell without modifying the space group of the compound. The H atoms occupy interstitial sites following some general rules, a) the Fe metal has a small affinity for H, so the occupied sites will contain only one hydrogen, b) the available size size per hydrogen-atom should be of a radius larger than 0.4Å (Westlake, 1983), c) if one site is occupied, the nearest sites are probably empty because of H-H repulsion.

With this filling scheme one can explain the observed occupied sites in the $R_2Fe_{14}BH_x$ compounds (Ferreira et al., 1985), namely four tetrahedral sites, one with 3R-1Fe coordination, and three with 2R-2Fe coordination (Fruchart et al., 1988b). From neutron diffraction experiments on deuterated compounds, it has been found that the ratio of occupancy of the 3R-1Fe (8j) versus the 2R-2Fe ($16k_1$) sites is 0.6 and 2 for the Y and Ce compounds (light R) and Er and Ho compounds (heavy R), respectively (Obbade et al., 1991). Hence, the $16k_1$ site is preferred when the cell dimensions provide sufficient space. However, when, because of the lanthanide contraction, the volume of this site decreases, the preferred site is the 8j site.

The maximum hydrogen uptake, x_{max}, in $R_2Fe_{14}ZH_x$, Z = B, C, depends on the rare-earth element involved. The Nd compound has the highest x_{max} among the light R compounds, and the Ho compound among the heavy R compounds. In Fig. 9 the values range from 1.8 for $Er_2Fe_{14}CH_x$ to a maximum of 4.6 for $Nd_2Fe_{14}BH_x$ (Fruchart et al., 1988a, Bartolomé et al., 1991). In all cases x_{max} is smaller in the carbide than in the boride compounds. A peculiar step-like decrease is observed in going from the Ho to the Er compound, Fig. 9, (Bartolomé et al., 1991) in both boride and carbide compounds. The amount of hydrogen absorbed is consistent with the above site filling model. However, the peculiar double maximum shape of x_{max} across the lanthanide series led to three proposed criteria to explain the observed trends in both series, Fig. 10, (Fruchart et al 1988a). First, the 2R-2Fe $16k_1$ site participates in the absorption independently of geometrical factors, yielding a constant uptake component. Second, the other available sites are filled depending on the dimensions of the unit cell, and thus on the R atomic radii. Thus, a component in x which decreases along the lanthanide series is expected. Third, the excess filling may again be tentatively ascribed to a 'magnetic' bonding effect, because it follows the trend which is proportional to the J values of the R atoms.

In both the rhombohedral and hexagonal R_2Fe_{17} compounds the occupied sites are of two types, a 2R-4Fe site of distorted octahedral coordination, and a second 2R-2Fe site of tetrahedral coordination, yielding a maximum content of five hydrogen atoms per formula unit (Isnard et al., 1990, 1992a). The octahedral sites seem to be preferred, their occupancy rate is higher, they form a regular network in the crystal with the same symmetry as graphite (rhombohedral or hexagonal) and permits three interstitial H atoms to be absorbed. Unlike in the $R_2Fe_{14}B$ compounds, x_{max} does not show the double maximum feature as a function of R, however, the H uptake is lower for heavy R compounds. This trend has been related to the larger size of the octahedral hole in the rhombohedral light R compounds (Isnard et al., 1990). Hence, less elastic energy is involved in its deformation. Neutron diffraction studies, as a function of temperature have shown that the octahedral sites are more stable, because they begin to be depleted only when the tetrahedral sites are fully empty (Isnard et al., 1992b). In a recent review, (Christodoulóu, 1993) a comprehensive systematization of the occupancy of the different sites is given. In particular, it is shown that, because of competing effects between repulsive H-H forces and the attractive electrostatic potential for the localization of H, only two of the six possible tetrahedral sites found in the Fe hexagons, are occupied by hydrogen. As the hole size decreases along the lanthanide series, just one site may be occupied, thus explaining the maximum uptake of x = 4 for $Lu_2Fe_{17}H_4$ (Isnard 1994c).

In the $R(Fe_{10.5}Mo_{1.5})H$ compounds, the H atoms have been located at the 8j octahedral sites with two R atom and 4 Fe atom near neighbors, thus allowing for just one H per formula unit, as observed by Tomey et al., (1994a and b). The same holds for $Tm(Fe_{11}Ti)D_{0.7}$ and $Tb(Fe_{11}Ti)D$ (Isnard, 1994c).

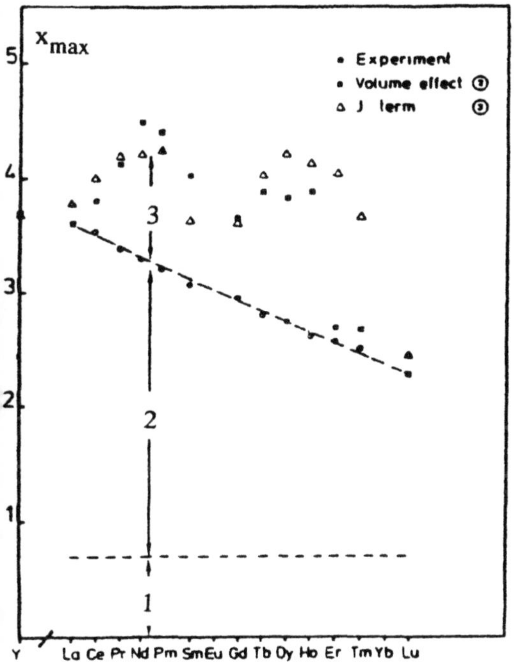

Figure 10. x_{max} for the $R_2Fe_{14}B$ series. (1) volume and electronegativity effect , (2) $r^{3+}(R)$ hole size variation, (3) Rare-earth J value (Fruchart et al., 1988 a).

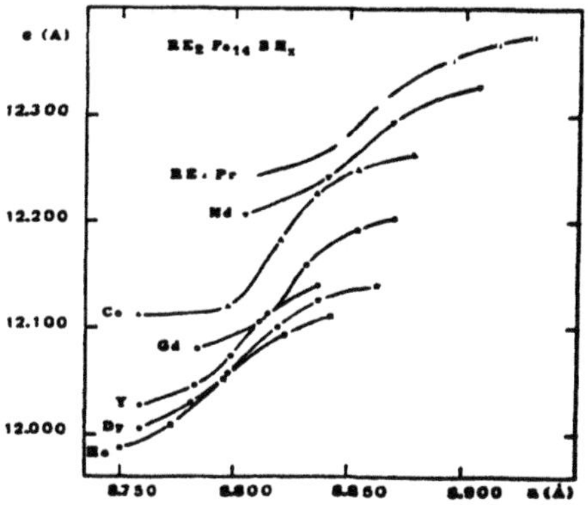

Figure 11. The c-parameter *vs* a-parameter in the $R_2Fe_{14}BH_x$ series (0<x<5) (Fruchart et al., 1990)

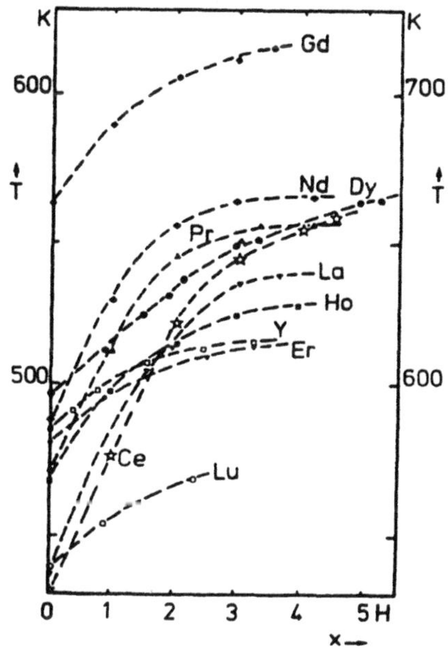

Figure 12. T_c of the $R_2Fe_{14}BH_x$ series. Ce is refered to the right scale. (Fruchart et al., 1988b)

3.2. STRUCTURE MODIFICATIONS

The net effect of hydrogen uptake is the expansion of the unit cell, in general without a modification of the space group. However, the actual deformation may be isotropic or anisotropic, depending on the series of compounds and on the H uptake sequence. The expansion of the lattice produces a modification of the interatomic distances and, consequently, of the magnetic interactions.

In the $R_2Fe_{14}ZH_x$ compounds, the relative c versus a unit cell expansion is anisotropic, and reflects the two-step process of H absorption, Fig. 11. For the initial hydrogenation, the a dimension expands more rapidly, a change which is related to filling the 8j sites first. For x > 1 the c dimension increases more rapidly, and is related to the filling of the $16k_1$ sites (Fruchart et al., 1990). However, this interpretation is contradicted by neutron diffraction data which show that, for the light-R compounds, the $16k_1$ sites fill first (Obbade et al., 1991). In any case, the interatomic distances increase, in particular the Fe-Fe distances within the hexagons forming the layers in this structure increase from the anomalously short value of 2.33 to 2.44 Å upon hydrogenation from x=0 to 3.1 in $Ho_2Fe_{14}BH_x$ (Fruchart and Miraglia, 1991). In contrast, the Fe-Ho distances show little change.

For the $R_2Fe_{17}H_x$ ($x \approx 5$) compounds, the volume increases by 4%, in a rather isotropic way, although the expansion is slightly larger in the a-b plane (±1.5%) than in the c direction (≈0.68%). Regarding the Fe-Fe distances, the situation is different from that of the $R_2Fe_{14}Z$ compounds, because the R_2Fe_{17} structure derives from the hypothetical RFe_5 structure in which some of the R atoms are substituted by Fe-Fe dumbbells. The Fe-Fe distance in the dumbbells is anomalously short at ≅ 2.39Å for the Pr and Nd compounds. The effect of hydrogenation on these dumbbell sites is indirect, because they are not nearest neighbors of the sites occupied by H. However, because of the cell expansion upon hydrogenation, these stressed dumbbells relax and have longer Fe-Fe distances. This increase is thought to play an important role in modifying the compounds magnetic properties. The expansion in the a-b plane also increases somewhat the Fe-Fe distances in the hexagons (Isnard et al., 1992c).

The expansion in the $R(Fe_{11}M)H_x$ compounds is very isotropic, and smaller than in the previous compounds, as would be expected for compounds with less hydrogen content. These compounds are related to the previous ones in that they contain Fe-Fe pairs with very short interatomic distances. Hydrogen uptake is expected to act in a similar way as in the $R_2Fe_{17}H_x$ series (Obbade et al., 1988). Only an average increase in the Fe-Fe distances of ca. 0.4 % can be estimated from neutron diffraction experiments. The same increase in Fe-Fe distances upon hydrogenation has been observed for $R(Fe_{10.5}M_{1.5})H$ compounds (Tomey et al., 1994a and b).

3.3.INCREASE OF THE CURIE TEMPERATURE

A general feature observed in all hydrogenated compounds is that H uptake induces a pronounced increase in T_C. In fact, this is one of the motivations for introducing interstitial atoms into the new families of Fe-rich intermetallic permanent magnets, because their common drawback for many applications is a low T_C.

For the $R_2Fe_{14}Z$; Z = B, C series, the increase in T_C is shown in Fig. 9 (Bartolomé et al., 1991) where a definite reduction in ΔT_C is observed across the lanthanide series. As is shown in Fig. 12, the increase of T_C is not linear in x. For x < 2.5, T_C increases sharply with increasing x, and for x > 2.5 T_C becomes constant (Fruchart et al., 1988b). Moreover, the initial slope clearly divides the compounds into two groups, for light R the slope is larger than for heavy R. This difference is probably related to the different site filling preference, as was discussed in Section 3.1 for the $R_2Fe_{14}B$ compounds. The increase in T_C has beenrelated first, to the increase in the Fe-Fe distances described above, and second, to the increase in the magnetization caused by an increase of a few percent in the average Fe moments, as observed for example in $Y_2Fe_{14}B$ (Andreev et al., 1986). The increase in the distances increases the ferromagnetic character of the Fe-Fe interaction because, according to the Slater-Nééel theory, the Fe-Fe ferromagnetic interaction decreases with decreasing distance, even becoming antiferromagnetic. The increase in the magnetization results from the increase of the absolute Fe moments, and is observed in the increase of the hyperfine field at the different Fe sites, measured by Mössbauer spectroscopy (Fig. 13) (Ferreira et al., 1985, Coey et al., 1986) The cause of this increase in Fe magnetic moment is still under discussion (Herbst, 1991). Two reasons are invoked, a reduction in the 5d-3d hybridization giving rise to an increased Fe moment or a screening due to H. However, as indicated in Section 2.1, we may also consider that, because T_C is strongly related to the spin fluctuations, where increasing fluctuations imply lower T_C, the increase in J_{FeFe} caused by H uptake could reduce the fluctuations and thus increase T_C.

For the magnetic $R_2Fe_{14}B$ compounds, an estimate of the modification of the molecular field acting on the R sublattice, due to hydrogenation, can be deduced from,

$$T_{RFe}(x) = T_c(x)\,[1-T_{Fe}(x) / T_c(x)]^{1/2}, \qquad (12)$$

where $T_{Fe}(x)$ is the corresponding Curie temperature of the non-magnetic La compound with the same x. Because T_{RFe} is directly proportional to n_{RFe}, the x dependence of T_{RFe} yields a rough indication of the molecular field variation. For all compounds it shows an increase for x = 1 and a subsequent decrease for x > 1 (Fruchart, 1988a). From magnetization measurements, the reduction in the Gd-Fe interaction was estimated to be 40% for x = 3.4 (Bartashevitch and Andreev, 1990), which results in far too low an interaction.

In R_2Fe_{17} the hydrogen uptake produces an increase in T_C of about 30 to 60%. In fact, as in the pure compounds, the short Fe-Fe dumbbell distances yield exceptionally low T_C values. Thus it appears that the relaxation of these 'compressed' dumbbells gives rise to a significantly enhanced T_C. This effect, compared to the effect of C and N substitution, is shown in Fig. 14 (Fruchart et al., 1992). Although an increase in the magnetic moment has been observed in the $Nd_2Fe_{17}H_x$ series (Rupp and Wiesinger, 1988) and with other R (Chevalier et al., 1988), it is argued by comparison with C and N interstitial compounds and by theoretical band-structure calculations, that the main cause of the T_C increase is the increase in relative volume caused by H uptake (Qi et al., 1992).

The $R(Fe_{12-x}M_x)$ compounds also increase their T_C upon hydrogenation, although to a lower extent, due to their lower H uptake. Nevertheless, an increase of ca. 20% has been observed for various members of this series, an increase which is correlated with a small increase in the Fe sublattice magnetization in $YFe_{12-x}M_x$; M = Ti and V (Obbade et al., 1988). The interpretation of the increase in T_C is identical to the R_2Fe_{17} increase because of their structural similarity.

3.4. MAGNETIC ANISOTROPY

The absortion of hydrogen by the Fe-rich intermetallic compounds produces, in general, a modification of their magnetic anisotropy. The hydrogenation process is reversible and the magnetic properties are restored after the desorption of H. It is for this reason that hydrogen decrepitation (HD) and hydrogenation disproportionation desorption recombination (HDDR), are important intermediate processes for obtaining fine particles in the manufacture of $R_2Fe_{14}B$ magnets. Tailoring the exact magnetic performance by interstitial substitution has been envisaged as possible by means of hydrogenation. However, stability problems may favour achieving the same purpose by means of other atomic substitutions.

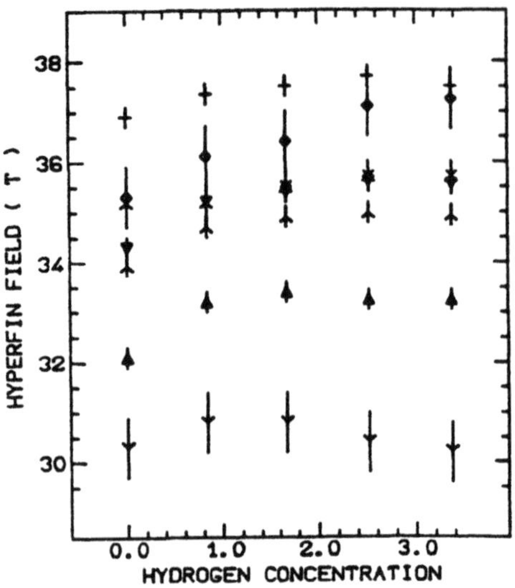

Figure 13. B_{hf} at the six sites of $Y_2Fe_{14}BH_x$ at 4.2 K, from Mössbauer spectroscopy (Coey et al., 1985).

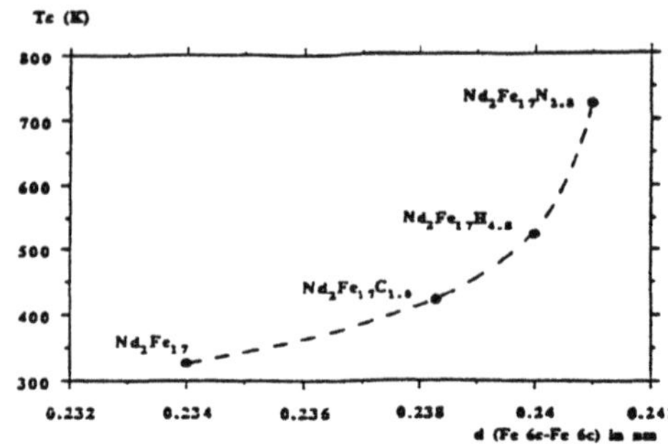

Figure 14. Increase of T_c vs the Fe-Fe interatomic distances at the 6c positions of the Nd_2Fe_{17} Z_x , Z=C, H and N series (Fruchart et al 92).

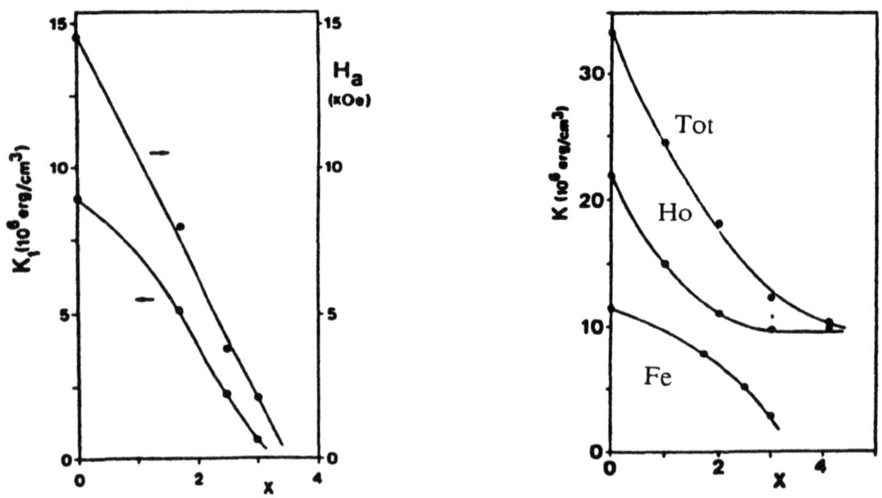

Figure 15. a) Anisotropy field H_a and K_1 dependence on x, at 78 K, for $Y_2Fe_{14}BH_x$. b) Decomposition of the total K_1 for $Ho_2Fe_{14}B$ (Tot), in the Fe sublattice contribution as deduced from the $Y_2Fe_{14}BH_x$ (Fe), and the deduced Ho sublattice contribution (Ho) (Pareti et al., 1988).

The anisotropy field, H_a, has been determined traditionally by means of magnetization measurements as a function of applied field. However, because of the high fields required these determinations are subject to error. A recent technique, the singular point detection method (SPD), has helped to overcome this problem (Pareti et al., 1988). These authors studied the $R_2Fe_{14}BH_x$ series, which became paradigmatic in the study of the anisotropy modifications.

First the influence of hydrogenation on the Fe sublattice is studied in the $Y_2Fe_{14}BH_x$ system. The anisotropy field decreases sharply and linearly with x to an extrapolated intercept for $K_1 = 0$ at $x = 4.5$, (i.e., becoming isotropic). We have scanned the $3 < x < 5$ region with the standard magnetization technique, and have found that the K_1 decrease tapers off to a small, but non-zero, asymptotic value (Kuz'min et al., 1994). Of course, this reduction of H_a can be described alternatively as a reduction of K_1. Both H_a and K_1 are shown in Fig. 15a. It is noteworthy that the anomalous temperature dependence of H_a and K_1, mentioned in Section 2.1, remains upon hydrogenation. In fact, the temperature of the maximum seems to shift towards higher values because of the increase in T_C. This is believed to indicate that the local anisotropies of the different Fe sites have opposite signs. Thus, the anomalous increase in K_1 with increasing temperature is caused by the temperature dependence of the competing Fe sublattice magnetizations. Because the H is located in preferential sites, its effect on the neighbouring Fe atoms should be larger, thus modifying the temperature dependence. However, the competing anisotropies do not compensate completely and the anomaly remains (Bartashevich and Andreev, 1989). In contrast, in the $Th_2Fe_{14}BH_x$ system, in which Th is not magnetic, $Th_2Fe_{14}B$ does not show the anomaly nor does the hydrogenated compound. Moreover, the K_1 decrease upon H uptake is less pronounced than in the $R_2Fe_{14}BH_x$ compounds, with R = Y, La, and Lu. This is believed to indicate that, in the Th compound the Fe local anisotropies do not compete, whereas they do compete in the other R compounds (Andreev and Bartashevich, 1990).

Hydrogenation does not modify the R net moment appreciably, however, in the magnetic rare-earth compounds the anisotropy field also decreases sharply upon hydrogenation, the decrease originates partly in the Fe sublattice, as was shown above. Then the reduction in anisotropy due to the R atom is estimated by subtraction of K_1^{Fe} as determined in $Y_2Fe_{14}BH_x$ for the same x content, see Fig.15b (Pareti et al., 1988). The result for Ho, Nd, and Tm is that the total anisotropy, $K^R = K_1^R + 2K_2^R$, decreases strongly for $x < 2.5$, and remains rather constant for higher values of x. This could indicate that the H affects the crystal field parameters in two ways. For the initial uptake, the effect is strong and, for $x > 1$ it tends to increase at a lower rate and to saturate. The actual decomposition of K into the contributions due to K_1 and K_2 has been attempted, by assuming that the quadrupole interaction observed at room temperature by Mössbauer spectroscopy at the R site is related to B_2^0 and, consequently, to K_1. The quadrupole interaction was found to increase for increasing x, as deduced from Dy 161 Mössbauer experiments (Ferreira et al., 1985), an increase which corresponds to a decrease in the lattice contribution, which has the opposite sign and is proportional to B_2^0. These results subtracted from the known valence contribution, $e^2qQ(Dy)=140mm/s$, and normalised to unity at $x = 0$, follow the same decrease upon hydrogenation as B_2^0. Does K_2 increase to compensate for this decrease and give rise to a constant K. This will be discussed, together with the temperature dependence of the K_i parameters, in Chapter 23 in relation to the spin reorientation transitions present in these compounds.

The magnetization at room temperature increases sharply with H uptake in all compounds of the $R_2Fe_{17}H_x$ series. However, at 5K the magnetization increases only slightly. The first effect is due to the dramatic increase in T_C that H uptake produces in these compounds (Isnard et al., 1990). It has also been observed that hydrogenation has the effect of reducing the Fe sublattice anisotropy, while maintaining the basal orientation of the easy magnetization direction (Isnard et al., 1994a). Mössbauer experiments performed on $Gd_2Fe_{17}H_x$ have demonstrated that the electric field gradient, V_{zz}, at the Gd nucleus decreases with H uptake. In so far as this parameter is related to B^0_2, and B^0_2 is the main contribution to K_1, the R single ion anisotropy decreases upon hydrogenation (Isnard et al., 1994b). This behaviour is opposite to the strong increase in K^R_1 produced by C or N uptake on these compounds.

In contrast with the above examples, in $R(Fe_{11}Ti)H_x$, with R = Y and Sm, the anisotropy field, H_a, increases with H uptake. A possible correlation with the reduction of the a/c ratio upon hydrogenation was proposed, in contrast to the increase in the a/c ratio observed in the $R_2Fe_{14}B$ compounds (Zhang and Wallace, 1989). We have observed a similar increase in $R(Fe_{10.5}Mo_{1.5})H$ for R = Y and Er (Tomey et al., 1994a, 1994b). The existence of at least three different Fe sites, with local anisotropies of opposite signs, may again explain the increase in anisotropy, if hydrogen reduces the anisotropy of the Fe contributing the negative K_1 component.

As a summary of the influence of H, C, and N interstitials on the magnetic properties we reproduce Table 1 by Fruchart et al. (1992). In conclusion, hydrogenation has been found to be an important processing step for the preparation of $R_2Fe_{14}Z$ based magnets and has also paved the way in the search for other interstitial atoms, such as N, which have augmented the panoplia of interesting magnetic materials.

Table 1. Main modifications in the magnetic properties as observed after an initial charging.

Series	$R_2Fe_{14}X$	R_2Fe_{17}		$R(Fe_{11-x}X_x)$	
Interstitial	H	H	C,N	H	C,N
T_C	II	II	III	II	III
M_s	I	I or C	I	II	
H_a	D or C	D or C	III	Ior C	

D= decrease, C= constant, I=increase, II=large increase and III=very large increase.

Acknowledgements

Extensive discussions with D. Fruchart are acknowledged. The LEA "MANES" has aided the Grenoble-Zaragoza cooperation. This work has been performed under the CEE CEAM III project, and has been financially supported by CICYT Project MAT93-0240-C04-04.

References for Section 2 (Parent compounds)

Bartolomé J. (1990) in "Supermagnets, Hard Magnetic Materials," Eds. G.J. Long, and F. Grandjean (Kluwer) p. 261.

Belorizky E., M.A. Fremy, J.P. Gavigan, D. Givord, and H.S. Li (1987) J. Appl. Phys. 61, 3971-3.

Belorizky E., J.P. Gavigan, D. Givord, and H.S. Li (1988) Europhysics Lett. 5, 349.

Beurle T., P. Braun, and M. Fähnle (1991) J. Magn. Magn. Mater. 94, L11.

Brooks M.S.S., L. Nordström, and B. Johansson (1991) J. Phys: Condens. Matter 3, 3393.

Brooks M.S.S. and B. Johansson (1993), in "Magnetic Materials," Vol. 7, (Elsevier Science Publ., Amsterdam) K.H.J. Buschow Ed., p. 139.
Campbell I.A. (1972) J. Phys. F 2, L47.
Chaboy J., A. Marcelli, L.M. García, J. Bartolomé, M.D. Kuz'min, H. Maruyama, K. Kobayashi, H. Kawata, and T. Iwazumi (1994) to be published.
Coehoorn R. (1989) Phys. Rev. B39, 13072.
Coehoorn R. (1990a) "Supermagnets, Hard Magnetic Materials," Eds. G.J. Long and F. Grandjean (Kluwer) p. 133.
Coehoorn R. (1990b) Phys. Rev. B41, 11790.
Coey J.M.D., H.S. Li, J.P. Gavigan, J.M. Cadogan, and B. P. Hu (1989) "CEAM report". Ed. I.V. Mitchell (Elsevier, New York), p. 76-97.
Deportes J., B. Kebe, and R. Lemaire (1986) J. Magn. Magn. Mater. 54-57, 1089-1090.
Gavigan J.P., D. Givord, H.S. Li, and J. Voiron (1988) Physica B 149, 345-351.
Giorgetti C., S. Pizzini, E. Dartyge, A. Fontaine, F. Baudelet, C. Brouder, G. Krill, S. Miraglia, D. Fruchart, and J.P. Kappler (1993) Phys. Rev. B 48, 12732.
Givord D., H.S. Li, and F. Tasset (1985) J. Appl. Phys. 57, 4100.
Gubbens P.C.M., A.M. van der Kraan, J.J. van Loef, and K.H.J. Buschow (1987) J. Magn. Magn. Mater. 67, 255.
Helmholdt R.B., J.J.M. Vleggar, and K.H.J. Buschow (1988) J. Less-Common Met. 138, L11.
Hirosawa S., K. Tokuhara, H. Yamamoto, S. Fujimura, M. Sagawa, and H. Yamaguchi (1987) J. Appl. Phys. 61, 3571.
Hu B.P., H.S. Li, J.P. Gavigan, and J.M.D. Coey (1989) J. Phys: Condens. Matter 1, 755.
Hu B.P., H.S. Li, J.P. Gavigan, and J.M.D. Coey (1990) Phys. Rev. B 41, 2221.
Inoue J. and M. Shimizu (1986) J. Phys. F 16, L157.
Isnard O. and D. Fruchart (1994a) J. Alloys Comp. 205, 1-15.
Isnard O., S. Miraglia, D. Fruchart, C. Giorgetti, S. Pizzini, E. Dartyge, and G. Krill (1994b) Phys. Rev. B 49, 21.
Kaneko T., M. Yamada, K. Ohashi, Y. Tawara, R. Osugi, H. Yoshida, G. Kido, and Nakagawa (1989) Proc. 10th Int. Work. on Rare Earth Magnets, 1, 191.
Kübler J. (1981) Phys. Lett. A81, 81.
Liu J.P., F.R. de Boer, P.F. de Châtel, R. Coehoorn, and K.H.J. Buschow (1994) J. Magn. Magn. Mater. 132, 159-179.
Luis F., P. Infante, J. Bartolomé, R. Burriel, C. Piqué, R. Ibarra, and K.H.J. Buschow (1994) J. Magn. Magn. Mat., submitted.
Mohn P. and E.P. Wohlfarth (1987) J. Phys. F: Met. Phys. 17, 2421-2430.
Mohn P. and G. Hilscher (1989) J. Phys. Rev. B 40, 9126.

References for Section 3 (Hydrides).

Andreev A.V., A.V. Deryagin, N.V. Kudrevatikh, M.V. Mushnikov, V.A. Reimer, and S.V.T. Rev (1986) Sov. Phys. JETP 63, 608.
Andreev A.V. and M.I. Bartashevitch (1990) J. Less-Common Metals 167, 107-111.
Bartashevitch M.I. and A.V. Andreev (1989) Sov. Phys. JETP 69, 1192-1195.
Bartashevitch M.I. and A.V. Andreev (1990) Physica B 162, 52-56.
Bartolomé J. , F. Luis, D. Fruchart, S. Miraglia, S. Obbade, and K. H. J. Buschow (1991) J. Magn. Magn. Mater. 101, 411-413.
Chevalier B, Etourneau, and J.M.D. Coey (1988) "CEAM report". Ed. I.V. Mitchell (Elsevier, New York) p. 214.
Christodoulou Ch. and T. Takeshita (1993) J. Alloys and Compounds 198, 1-24.

Coey J.M.D., A. Yaouanc, and D. Fruchart (1986) Solid. State. Comm. 58, 413-416.
Ferreira L.P., R. Guillien, P. Vulliet, A. Yaouanc, D. Fruchart, P. Wolfers, P. L'Héritier, and R. Fruchart (1985) J. Magn. Magn. Mater. 53, 145-152.
Fruchart D., L. Pontonnier, F. Vaillant, J. Bartolomé, J.M. Fernandez, J.A. Puertolas, C. Rillo, J.R. Regnard, A. Yaouanc, R. Fruchart, P. L'Héritier (1988a) IEEE Trans. Magn, 24, 1641.
Fruchart D., P. Wolfers, S. Miraglia, L. Pontonnier, F. Vaillant, H. Vincent, D. Le Roux, A. Yaouanc, P. Dalmas de Reotier, P. l'Héritier, and R. Fruchart (1988b) "CEAM report" Ed. I.V. Mitchell (Elsevier, New York) p. 214
Fruchart D., S. Miraglia, S. Obbade, P. Ezekwenna, and P.L'Héritier (1990) J. Magn. Magn. Mater. 83, 291-292.
Fruchart D. and S. Miraglia (1991) J. Appl. Phys 69, 5578.
Fruchart D., O. Isnard, S. Miraglia, S. Obbade, C. Rillo, and J.L. Soubeyroux (1992) Physica B 180 & 181, 632-634.
Isnard O., S. Miraglia, J.L. Soubeyroux, D. Fruchart, and A. Stergiou (1990) J. Less-Common Metals 162, 273-284.
Isnard O., S. Miraglia, J.L. Soubeyroux, and D. Fruchart (1992a) Solid State Comm. 81, 13-19.
Isnard O., J.L. Soubeyroux, S. Miraglia, D. Fruchart, L.M. García, and J. Bartolomé (1992b) Physica B 180-181, 629-631.
Isnard O., S. Miraglia, D. Fruchart, and J. Deportes (1992c) J. Magn. Magn. Mat 103, 157-169
Isnard O., S. Miraglia, D. Fruchart, and M. Guillot (1994a) J. Magn. Magn. Mat. (in press).
Isnard O., P. Vuillet, A. Blaise, J.P. Sanchez, S. Miraglia, and D. Fruchart (1994b) J. Magn. Magn. Mater. 131, 83-89.
Isnard O. (1994c) Thesis, Grenoble
Kuz'min M.D., L.M. García, I. Plaza, J. Bartolomé, D. Fruchart, and K.H.J. Buschow. (1994) to be published in J. Magn. Magn. Mat.
L'Héritier Ph., P. Chaudouët, R. Madar, A. Rouault, J.P. Sénateur, and R. Fruchart. (1984) C.R. Acad. Sci. (Paris) 299, 849.
Obbade S., S. Miraglia, D. Fruchart, M. Pre, P. L'Héritier, and A. Barlet (1988) C. R. Acad. Sci. Paris T 307, serie II 889.
Obbade S., S. Miraglia, D. Fruchart, and P. L'Héritier (1989) Z. Physicalische Chemie
Obbade S., S. Miraglia, P. Wolfers, J.L. Soubeyroux, D. Fruchart, F. Lera, C. Rillo, B. Malaman, and G. le Caër (1991) J. Less common Metals. 171, 71.
Österreicher K. and H. Österreicher (1984) Phys. Stat. Sol. (a) 85, K61.
Pareti L, O. Moze, D. Fruchart, Ph. L'Heritier, and A Yaouanc (1988) J. Less-Common Met. 142, 187-194.
Qi Qi-nian, H. Sun, R. Skomski, and J.M.D. Coey (1992) Phys. Rev. 45, 12278-12286.
Rupp B. and Wiesinger G. J. (1988) J. Magn. Magn. & Mater. 71,262-278.
Tomey E., D. Fruchart, J.L. Soubeyroux, and D. Gignoux (1994a) EMMA 94. To be published in IEEE Trans. Magn.
Tomey E, M. Bacmann, D. Fruchart, S. Miraglia, J.L. Soubeyroux, D. Gignoux, and E. Palacios (1994b) EMMA 94. To be published in IEEE Trans. Magn.
Westlake D.G. (1983) J. Less Common Metals 90, 1, 91-251.
Zhang L.Y. and W.E. Wallace (1989) J. Less-Common Metals 371-376.

Reviews

Burzo E., and H.R. Kirchmayr (1989) "Handbook on the Physics and Chemistry of Rare Earths," Vol. 12, Ed. K.A. Gschneider and L. Eyring, p. 71.
Buschow K.H.J., P.C.P. Bouten, and A.R. Miedema (1982) Rep. Prog. Phys. 45, 937.
Buschow K.H.J. (1988) "Ferromagnetic Materials" Vol. 4, (Elsevier. P.C.) Eds. E.P. Wohlfarth and K.H.J. Buschow, Chapter 1, p. 1.
Buschow K.H.J. (1991) Rep. Prog. Phys. 54, 1123-1213.
Franse J.J.M. and R.J. Radwanski (1991) "Handbook of Magnetic Materials," Vol. 7, (Elsevier. P.C.) Edit. K.H.J. Buschow, Chapter 5, 307.
Herbst J.F. (1991) Reviews of Modern Physics., 63, 819-898.
Li H.S. and J.M.D. Coey (1991) "Handbook of Magnetic Materials," Vol. 6, (Elsevier. P.C.) Ed. K.H.J. Buschow Chapter 1, p. 1.
Wiesinger G. and G. Hilsher (1991) "Handbook of Magnetic Materials," Vol. 6, (Elsevier. P.C.) Ed. K.H.J. Buschow, Chapter 6, p. 511.

Chapter 23

INTERSTITIAL NITROGEN, CARBON, AND HYDROGEN: MODIFICATION OF MAGNETIC AND ELECTRONIC PROPERTIES

R. SKOMSKI
Physics Department
Trinity College (University of Dublin)
Dublin 2, Ireland

ABSTRACT. It is shown how the electronic and magnetic properties of rare-earth transition-metal intermetallics are modified by interstitial atoms. On an introductory level and starting from illustrative physical ideas we relate band structure, statistics, and electrostatic crystal field effects to the intrinsic properties magnetic moment, Curie temperature, and magnetocrystalline anisotropy. We discuss how the permanent magnetic properties of Sm_2Fe_{17} and other intermetallics are improved upon interstitial modification, and put particular weight on recent developments such as spin-fluctuation theory and screening of crystal-field charges.

1. Introduction

1.1. SCIENTIFIC BACKGROUND

The improvement of permanent magnetic properties due to interstitial modification is one of the most captivating trends in modern materials science. Considerable interest in the electronic and magnetic properties of interstitially modified rare-earth intermetallics was stirred up by the discovery that nitrogen introduced from the gas phase drastically improves the permanent magnetic properties of Sm_2Fe_{17} (Coey and Sun 1990 [1]). Compared to the parent compound Sm_2Fe_{17}, the Curie temperature increases from 116 °C to 476 °C, the magnetization is enhanced by about 50%, and the magnetocrystalline anisotropy changes from easy-plane to easy-axis anisotropy (Table 1). This means that the interstitial nitride $Sm_2Fe_{17}N_3$ represents a promising magnetic material, in striking contrast to Sm_2Fe_{17} which is not suitable for permanent magnetic appplications.

Since then, gas-phase interstitial modification involving molecular nitrogen has been extended to other hexagonal and rhombohedral 2:17 intermetallics, 1:12 compounds such as $R(Fe_{11}Ti)N_{1-\delta}$ [2, 3], and interstitial carbides $R_2Fe_{17}C_{3-\delta}$ and $RT_{12}C_{1-\delta}$ [4 - 6]. Further quasi-compounds are obtained by transition-metal substitution, e.g. Co for Fe in 2:17 nitrides [7, 8].

An overall improvement of permanent magnetic properties is not always observed. A good example is nitrogen in $SmFe_{11}Ti$. Curie temperature and magnetization of $SmFe_{11}Ti$ improve upon nitrogenation, but the anisotropy drastically deteriorates and changes from easy-axis to easy-plane (Table 1). On the other hand,The permanent magnetic properties of the carbides tend to be slightly worse than those of the isostructural nitrides [9], and for hydrogen, which has long been known as an interstitial in rare-earth transition-metal intermetallics, changes of the magnetic properties are much less pronounced.

F. Grandjean et al. (eds.), Interstitial Intermetallic Alloys, 561–598.

Table 1. Magnetic properties of selected samarium-iron intermetallics.

	T_c °C	M_o(RT) T	B_a(RT) T	
Sm_2Fe_{17}	116	1.00	< 0[a]	[10]
$Sm_2Fe_{17}N_{3-\delta}$[c]	476	1.54	20	[9]
$Sm_2Fe_{17}C_{3-\delta}$	395	1.43	15	[11]
$Sm(Fe_{11}Ti)$	311	1.15	10	[12]
$Sm(Fe_{11}Ti)N_{1-\delta}$	496	1.28[b]	< 0[a]	[2, 13]

[a] $B_a \leq 0$ means easy-plane magnetization, which excludes permanent magnetic applications. [b] Deduced from T_c. [c] δ indicates nearly complete occupancy.

1.2. MAGNETIC MATERIALS

Interstitial permanent magnets are not only a scientifically interesting class of materials but also a major potential application of interstitial modification. There are myriads of permanent magnets around us: apart from table magnets and toys we can find them in wrist watches, refrigerators, TV sets, electronic typewriters, and, abundantly, in cars. To start with, let us shortly recall the main features of permanent magnetism.

(i) *Magnetic Moment.* Isolated electrons always have a magnetic moment μ, but most solids are non-magnetic due to double-occupied ↑↓ orbitals (chemical interaction). Solid-state magnetism is caused by the partially filled ('unfilled') inner electron shells of transition-metal atoms [14]. As a rule of thumb, the magnetization of iron and cobalt is largest. In the case of rare-earth atoms, the large volume of the lanthanide atoms (about three times that of iron or cobalt) leads to an unfavourable dilution of the magnetization.

(ii) *Curie temperature.* Thermal excitation destabilizes the microscopic spin alignment, so a coupling mechanism must be found which assures spin alignment at and above room temperature. Due to the weakness of magnetostatic interaction, this coupling has to be of electrostatic origin (exchange interaction). Experience shows that sufficiently large amounts of iron or cobalt are a necessary but not sufficient condition to obtain a high Curie temperature.

(iii) *Anisotropy.* The main job of permanent magnets is to do mechanical work. As opposed to soft magnetic materials, where the magnetization is allowed to rotate freely, permanent magnets require a mechanism which keeps the magnetization in a desired direction. The anisotropy of modern permanent magnets is largely provided by the rare-earth sublattice.

(iv) *Coercivity.* Magnetostatic interaction favours the formation of magnetic domains with zero net magnetization, so the metastable magnetized state must be stabilized by a suitable metallurgical microstructure. This coercivity requirement is not related to the electronic structure of the permanent magnet and will be dealt with in other chapters of this volume.

Modern permanent magnetic materials such as $Sm_2Fe_{17}N_3$, $Nd_2Fe_{14}B$, or $SmCo_5$ [1, 15 - 19] have the typical structure $R_aT_bZ_x$ (Z = B, C, N). As already mentioned, the 3d transition-metal atoms (T) are responsible for magnetic moment and Curie temperature, while the rare-earth atoms (R) provide the necessary anisotropy. The function of the non-magnetic 2p atoms is (i) to modify the electronic structure of the parent compound in order to improve mgnetization and Curie temperature, (ii) to improve the anisotropy by changing the electrostatic crystal field, and (iii) to stabilize the intermetallic lattice, as it is the case in $Nd_2Fe_{14}B$ and other $R_2T_{14}Z$ compounds. Here we are not concerned with metallurgical problems and exclude the stabilization function from our consideration. In other words, we assume that the 2p atoms are introduced from the gas or solid phase without changing the lattice structure or forming a two-phase system [20, 21].

2. Moment and Magnetization

Summary: Metallic magnetism is due the existence of partially filled inner shells in transition-metal atoms. These shells are fairly well separated from each other, which suppresses the formation of non-magnetic electron pairs. Spontaneous magnetization occurs if the gain in electrostatic energy upon moment formation outweighs the accompanying increase in kinetic energy. The enhancement of the magnetic moment in nitrides and carbides is caused by the lattice expansion. Chemical interaction is a secondary effect and tends to reduce the moment.

2.1. MAGNETISM OF IONS AND SOLIDS

2.1.1. *Relativistic Origin of Magnetism* Let us first remember that magnetism is a relativistic phenomenon. 'Relativistic' means that a physical fact can be described by space-time symmetric equations. The simplest example is the relativistic propagation of light, $x^2 + y^2 + z^2 = c^2t^2$, which can be written as $\mathbf{x}^+\mathbf{x} = 0$ with $\mathbf{x} = (x, y, z, ct)$ and $\mathbf{x}^+ = (x, y, z, -ct)$.

The Schrödinger equation mixes first and second derivatives with respect to $\mathbf{r} = (x, y, z)$ and ct, and is therefore non-relativistic. To describe magnetism, we have to go beyond the Schrödinger equation (Dirac equation). Fortunately, in solids the velocity v of electrons is small compared to the velocity of light c, and we can use a series expansion to simplify the Dirac equation [22, 23]. The small parameter of the theory is Sommerfeld's fine-structure constant $\alpha = e^2/4\pi\varepsilon_o\hbar c$. Strictly speaking, $\alpha \approx 1/137$ has to be replaced by $Z\alpha$, where Z is the effective nuclear charge acting on the electron, because core electrons have a higher velocity than valence electrons. Electrons in inner shells, such as 4f electrons, feel a large nuclear charge and exhibit comparatively pronounced relativistic properties. As we will see, this is the reason for the large anisotropy of the rare-earth sublattice, which characterizes modern permanent magnets.

In the case of a single isolated electron in the electrostatic potential of a point charge and an external magnetic field the Hamiltonian reads [23]

$$\mathbb{H} = mc^2 + \frac{1}{2m}\mathbb{p}^2 + \frac{1}{8m^3c^2}\mathbb{p}^4 - \frac{Ze^2}{4\pi\varepsilon_o|\mathbf{r}|} + e\,V_{ex} - \mu_B\,(\mathbb{l} + 2\,\mathbb{s})\mathbf{H} + \frac{\hbar^2Ze^2}{8\pi\varepsilon_o m^2c^2|\mathbf{r}|^3}\,\mathbb{l}\,\mathbb{s} \tag{2.1}$$

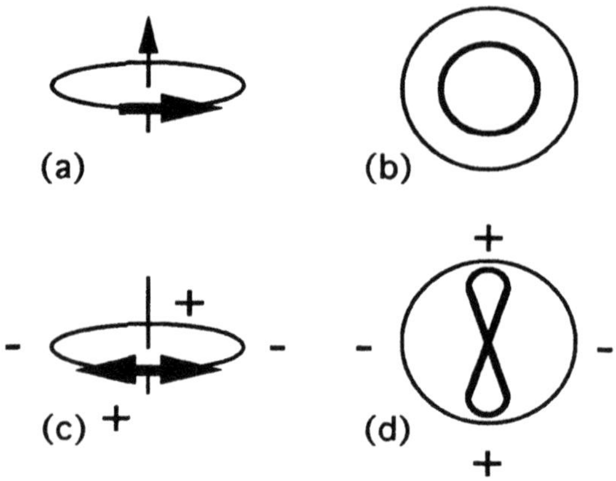

Fig. 1. (a) and (b) show a circular-current electron with 'running' angular wave function $\Psi = \Psi_o \exp(i\phi)$ and rotationally invariant electron density $\rho(\phi) = \Psi^*\Psi$. (c) and (d) symbolize the same electron in a strong crystal field. Spin-orbit coupling favours $<L_z> \neq 0$, i.e. a free-running electron, but this trend is outweighed by the low electrostatic energy of the standing-wave configuration with $\Psi=\Psi_o \cos\phi$ and $\rho(\phi) = \Psi_o^2 (1 + \cos 2\phi) / 2$. Calculation of $<L_z> = <\Psi|i\hbar\partial/\partial\phi|\Psi>$ yields $<L_z> = 0$ for the oscillating electron yields (quenching).

Here $\mathbf{l}$ and $\mathbf{s}$ are the vector operators of the orbital and spin momentum, respectively, **H** is the external magnetic field, and V_{ex} an external electrostatic potential. In this expression, the only zeroth order term is the large, but constant and therefore uninteresting rest energy mc^2. The kinetic energy $\mathbf{p}^2/2m$ and the Coulomb energy $Ze^2/4\pi\varepsilon_o|\mathbf{r}|$ turn out to be of order $(\alpha Z)^2$. The remainder in Eq. (2.2) scales as $(\alpha Z)^4$. The $\mathbf{p}^4$ term is the lowest-order relativistic correction to the kinetic energy, as can be seen from the series expansion of the famous result $E_{kin} = m_{eff}\, c^2$ with $m_{eff} = m/(1-v^2/c^2)^{1/2}$. The Zeeman energy couples $\mathbf{l}$ and $\mathbf{s}$ to the external magnetic field **H**, whereas the $\mathbf{l}\,\mathbf{s}$ term (spin-orbit coupling) describes the alignment of the electron in the magnetic field generated by its own motion.

From Eq. (2.1) we can draw a number of conclusions:

(i) Magnetic interaction is very weak compared to elementary-particle interaction, where typical energies are of order MeV, and electrostatic interaction, where the energies are of order 1 eV. An estimate of the magnitude of magnetic interaction is $E/k_B = \alpha^2 \cdot eV/k_B \approx 0.6$ K. To maintain magnetic order at and above room temperature, an auxiliary electrostatic mechanism is required.

(ii) In solids, the orbital motion of the electrons is subject to the electrostatic potential V_{ex} of the crystal. This potential obeys the overall symmetry of the lattice, so the spin-orbit coupling causes the magnetic energy to depend on the crystallographic orientation of the magnetization (magnetocrystalline anisotropy).

(iii) There is a competition between electrostatic crystal field and spin-orbit coupling. In the case of 3d metals, the magnetic electrons are delocalized and feel a nearly unscreened crystal field. For energetical reasons they are compelled to adapt their charge density to the charge distribution of the surrounding electrons and metal ions (Fig. 1). On the other hand, the effective nuclear charge and the spin-orbit coupling of delocalized electrons are small. This means that the electrostatic crystal field dominates the spin-orbit coupling: orbital moment and spin behave independently (quenching), and the magnetocrystalline anisotropy is comparatively low.

2.1.2. *Magnetism in Solids*. Isolated electrons and many ions have a magnetic moment which can be aligned by an external magnetic field. In solids, however, chemical interaction tends to destroy the magnetic moment. The ultimate reason for this is the Pauli principle. Phase-space is divided into cells of size $\Delta r\, \Delta p \approx h$ which are either single-occupied, with the possible spin directions $\uparrow$ (spin up) and $\downarrow$ (spin down), or double-occupied ($\uparrow\downarrow$) with zero net magnetization. If electrons are put together in a solid then the smallness of Δr, which reflects the reduced interatomic distances, must be compensated by a large Δp. This mechanism causes the kinetic energy to increase, and the electrons prefer to condense in low-lying double-occupied orbitals. On the other hand, electrons in $\uparrow\downarrow$ orbitals feel a strong Coulomb repulsion, which works against double occupation. Coulomb repulsion and increase in kinetic energy mechanisms are competitive, and the outcome of this competition determines whether a metallic moment is formed [23, 24]. In simple metals, all valence electrons are delocalized and the trend to reduce the kinetic energy dominates. A magnetic moment, however, is observed in transition metals and their compounds, such as Fe, CrO_2, $ZrZn_2$, Nd, or UH_2. Transition metals have unfilled inner shells, and the properties of the electrons in these shells are reminiscent of isolated atoms.

The behaviour of the well-localized 4f electrons in rare-earth atoms is determined by Coulomb repulsion. Alternatively, 4f electrons are located deep inside the atom so that the gain in kinetic energy due to interatomic hopping is negligible. Magnetically, rare-earth atoms behave as ions, typically R^{3+}, and exhibit a well conserved moment. Unfortunately, the large volume of rare-earth atoms ($R_a \approx 1.8$ Å) dilutes the moment μ so that the magnetization $M_o = \mu/V$ is only moderate.

The most important group of transition-metal elements are those of the iron group (3d elements). 3d electrons are delocalized (itinerant), but nevertheless exhibit an electron density $\rho(\mathbf{r})$ which is similiar to that of ionic orbitals. To find out which 3d elements and intermetallics are magnetic we have to investigate the band structure of these metals.

2.2. ITINERANT MAGNETISM

2.2.1. *Homogeneous Hartree-Fock Electron Gas*. As first pointed out by Bloch [25], the low-density homogeneous electron gas in Hartree-Fock approximation exhibits ferromagnetism. In this approximation, plane waves are used to construct a Slater determinant and to calculate the energy as a function of the Fermi wave vector k_F. The T = 0 energy of the non-magnetic state is [26]

$$E_o = N_{el}\left(\frac{3}{5}\frac{\hbar^2 k_F^2}{2m} - \frac{3e^2 k_F}{16\pi^2\varepsilon_o}\right) \tag{2.2}$$

The k_F^2 term in this equation gives the kinetic energy, while the term linear in k_F describes the (electrostatic) exchange interaction.

To describe magnetism, this relation has to be generalized to $E = 0.5\, E_o(k_{F\uparrow}) + 0.5\, E_o(k_{F\downarrow})$. Here $k_{F\uparrow}$ and $k_{F\downarrow}$ are the spin-up and spin-down Fermi wave vectors, respectively (Fig.2). For instance, the fully spin-polarized state is described by $k_{F\uparrow} = (2)^{1/3}$ and $k_{F\downarrow} = 0$ [23, 26, 27]. In the particularily instructive case of small polarization, $s = (N_\uparrow - N_\downarrow)/(N_\uparrow + N_\downarrow) \ll 1$, the energy reads

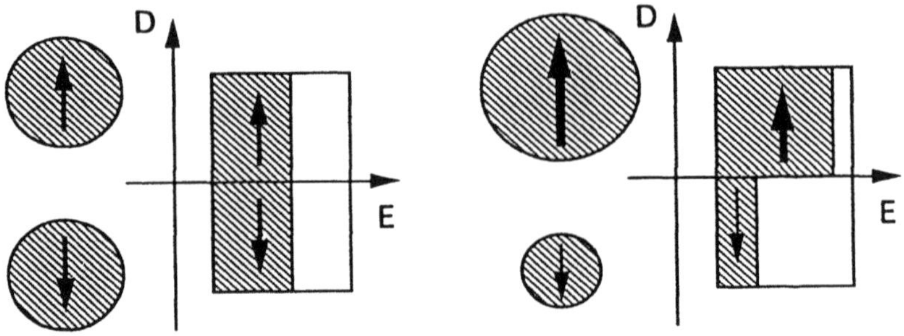

Fig. 2. Spin polarization caused by an external magnetic field or, alternatively, by exchange interaction (Stoner theory). The schematic figure illustrates the transfer of spin-down (minority) electrons to the spin-up (majority) band. The size of the Fermi spheres depends on the total number of spins in the spin-up and spin-down bands. Note that the exchange field is not included in this figure, so the spin-up and spin-down Fermi energies are different. A more common representation is used in Fig. 5.

$$E = E_o + (N_\uparrow + N_\downarrow)\left\{\frac{1}{4}\left(\frac{1}{D(E_F)} - I\right)s^2 - \mu_o\,\mu_B\,H\,s\right\} \tag{2.3}$$

Here $\mu_B = 9.274\ 10^{-24}$ A/m^2 is the Bohr magneton and $D(E_F) = 3/4(\hbar^2 k_F^2/2m)$ the free-electron density of states per spin and electron. The Stoner parameter $I = e^2k_F/6\pi^2\varepsilon_o$ describes the exchange interaction of the homogeneous electron gas.

Minimizing the energy Eq. (2.3) with respect to s we find that an external magnetic field H generates a magnetization $M = \mu_B s N_{el}/V = \chi H$ where χ is the 'exchange-enhanced' Pauli susceptibility

$$\chi = \frac{N_{el}}{V}\,\frac{2\mu_o\mu_B^2 D(E_F)}{1 - ID(E_F)} \tag{2.4}$$

For I = 0 this equation reproduces the ordinary Pauli susceptibility $\chi = \alpha^2 k_F a_o/\pi$. At $ID(E_F) = 1$, the susceptibility diverges and the paramagnetic state becomes unstable with respect to infinitesimal distortions, so ferromagnetism is predicted for $ID(E_F) > 1$. The same condition is obtained from Eq. (2.3) if we require the spontaneously magnetized state to be energetically more favourable than the non-magnetic one.

Expressing I and $D(E_F)$ in terms of k_F we see that Eq. (2.3) predicts ferromagnetism for $\pi k_F a_o < 1$. According to this result, simple bcc and fcc metals with atomic radii larger than 2.81 Å and 2.89 Å, respectively, should be ferromagnetic. As a matter of fact, alkali metals are enhanced paramagnets [28, 29], but the atomic radii of the ferromagnetic 3d elements iron, cobalt and nickel are much smaller: $R_{Fe} = 1.24$ Å, $R_{Co} = 1.25$ Å, and $R_{Ni} = 1.25$ Å [30]. Apart from this, Eq. (2.2) represents a ***high-density*** approximation, which hardly applies to metals with atomic radii approaching 3 Å [23, 26].

2.2.2. *Stoner Theory*. Despite these shortcomings, the criterion $ID(E_F) > 0$ (the Stoner criterion) gives an illustrative explanation of ferromagnetism in metals, and we will see that this simple picture can be resurrected if physically more reasonable values of $D(E_F)$ and I are used. In particular, we have to adjust $D(E_F)$ to calculated or measured densities of states of 3d metals and alloys.

3d electrons are close to localization so that the interatomic interaction, and with it the width of the 3d band, are fairly small. On the other hand, the 3d band has to be able to accommodate 10 electrons per atom, in contrast to only two electrons in the broad 4s band. This means that the 3d density of states is much larger than predicted from the free-electron theory, and the condition $ID(E_F) > 1$ can be satisfied even if the atomic radii are much smaller than about 3 Å. Remember that the orbital moment of 3d electrons is largely quenched, so the restriction to the spin moment is a physically reasonable assumption.

Putting $k_F = 1$ Å^{-1} in Eq. (2.3) we find that the Stoner parameter I for the homogeneous electron gas is of order 3 eV. It can be shown, for instance by comparing the results of sophisticated band-structure calculations with Eq. (2.3) [16], that the Stoner parameter of 3d metals and alloys is of order 1 eV (Fig. 3). This means that the Stoner theory predicts ferromagnetism for 3d elements and alloys where the density of states is larger than about 1 eV^{-1} (Table 2 and Fig. 3).

It is worth mentioning that a satisfied Stoner criterion $ID(E_F) > 1$ does not necessarily yield full exchange splitting. Examples of weak ferromagnets are iron and many iron intermetallics where interaction is not able to fill the spin-up band. The exchange splitting can be improved, however, by alloying with a strong ferromagnet ($Fe_{65}Co_{35}$ has a magnetization of 2.43 T !), or by interstitial modification. On the other hand, complete exchange coupling does not necessarily yield a high Curie temperature: there are magnetic systems which are strong from the point of view of ground-state magnetism but weak from the thermodynamic point of view (see Section 3).

2.3. INTERSTITIAL MODIFICATION AND MAGNETIC MOMENT

2.3.1. *Interstitial Modification and Density of States*. Interstitial modification influences band structure and magnetic moment. A simple picture is given by the theory of chemical bonding: caused by the electronegativity difference between host lattice and interstitial atoms there is an electron transfer. Small atoms with high valency, such as nitrogen, are electronegative and attract screening electrons. These electrons occupy low-lying, well-localized orbitals, which form narrow peaks in the density of states (Fig. 4e). Rather electropositive interstitials, such as hydrogen, in a sufficiently electronegative host loose their valence electrons and fill the holes in the transition-metal bands (Fig. 4f). The archetypical example of this mechanism is palladium hydride PdH_x [32, 33]. For $x \geq 0.6$ the palladium 4d band is filled (Fig. 4c) and the Pauli susceptibility, which is enhanced in pure palladium, reduces to a value typical of simple metals. Note that the hydrogen uptake is exothermic for $x < 0.6$ but endothermic for $x > 0.6$: the filling of the broad s band (Fig. 4c) is energetically very unfavourable.

Anionic modification (low-lying orbitals) and 'protonic' modification (band filling) are idealized limits [34, 35]. Even in palladium there are only 0.36 holes per atom in the 4d band, i.e. only 60% of the hydrogen electrons actually fill the 4d band. On the other hand, hydrogen in rare-earth dihydrides, such as PrH_2, behaves anionically [35]. Note that the Pauling electronegativity of palladium, 1.5, is larger than that of the rare-earths, which is of order 1.1 [30].

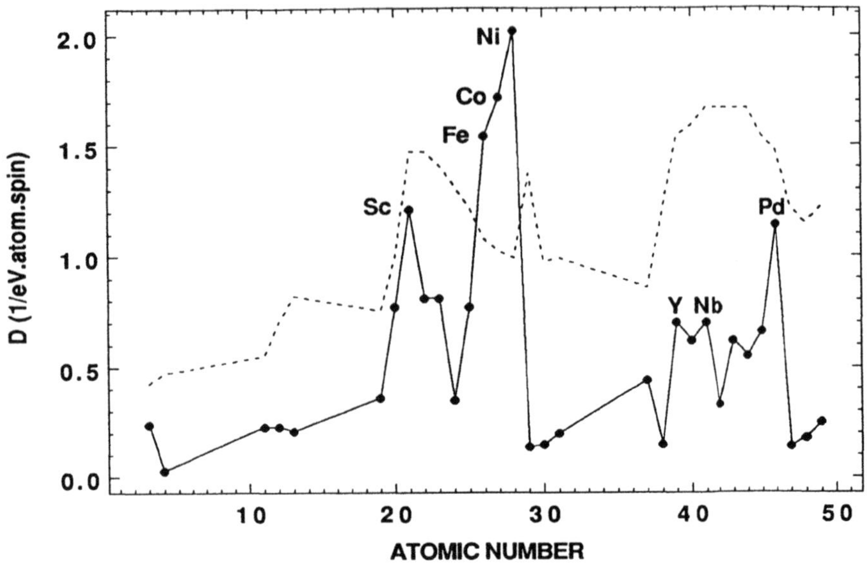

Fig. 3. Calculated densities of states at the Fermi level (solid line) and densities of states above which ferromagnetism occurs (dashed line). The elements iron, cobalt, and nickel satisfy the Stoner criterion $ID(E_F) > 1$ [28, 31].

Table 2. Ferromagnetic 3d elements.[a]

	μ	$\mu_0 M_0$ (RT)	T_c	$D(E_F)$	I
	μ_B	T	K	1/eV	eV
Fe	2.217	2.16	1044	1.54	0.93
Co	1.753	1.76	1360	1.72	0.99
Ni	0.616	0.61	627	2.02	1.01

[a]Data from [15, 31].

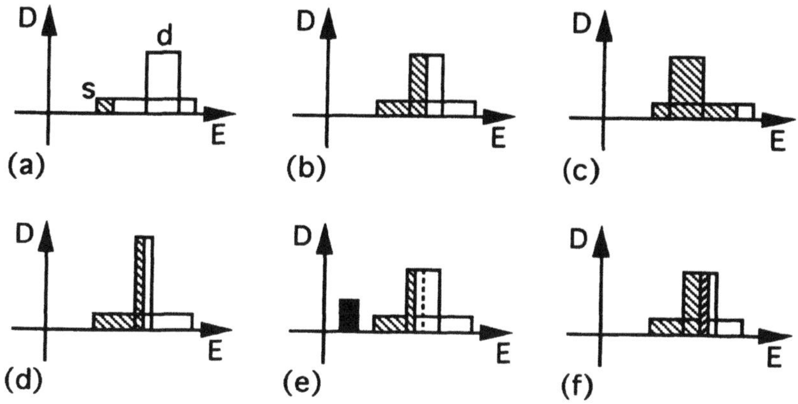

Fig. 4. Schematic band structures of (a) metals with empty inner shell, (b) enhanced paramagnets, (c) metals with completely filled inner shell, and (d) magnetic 3d metals (before exchange splitting). The main difference between enhanced Pauli paramagnets and ferromagnets is the density of states at the Fermi level. The anionic and 'protonic' limits of interstitial modifiaction are illustrated in (e) and (f), respectively [32, 36].

2.3.2. *Iron-Rich Rare-Earth Intermetallics*. Band structure and magnetic moment of transition-metal-rich rare-earth intermetallics and their interstitially modified derivates have been subject to major scientific interest (see e.g. [4, 16, 37-43]). An example of a calculated band structure (Y_2Fe_{17} and $Y_2Fe_{17}N_3$) is shown in Figure 5 [40]. Note that the different rare earths, including the non-magnetic 'rare earth' yttrium, have very similiar chemical properties, so the band structures of isostructural rare-earth intermetallics are fairly universal.

Somewhat simplifying, the band structures of interstially modified iron-rich rare-earth intermetallics and their parent compounds reveal the following tendencies [37, 44, 45]:

(i) Due to the volume expansion of about 6%, the magnetic moment increases by about 15%. This fact very well agrees with the general trend that large atomic distances favour ferromagnetism by reducing the overlap of the 3d wave functions. This overlap reduction narrows the 3d band and enhances the tendency towards ferromagnetism. It is worthwhile mentioning that 2:17 nitrides are nearly strong ferromagnets: $Y_2Fe_{17}N_3$ misses strong ferromagnetism by only 3% [44].

(ii) Chemical interaction tends to reduce the magnetic moment by formation of covalent bonds. These hybridization effects are larger in carbides than in nitrides — nitrogen electrons tend to occupy well-localized low-lying states which do not interfere very much with the 3d band. As a consequence, the moment of the carbides is smaller by about 5% compared to the nitrides.

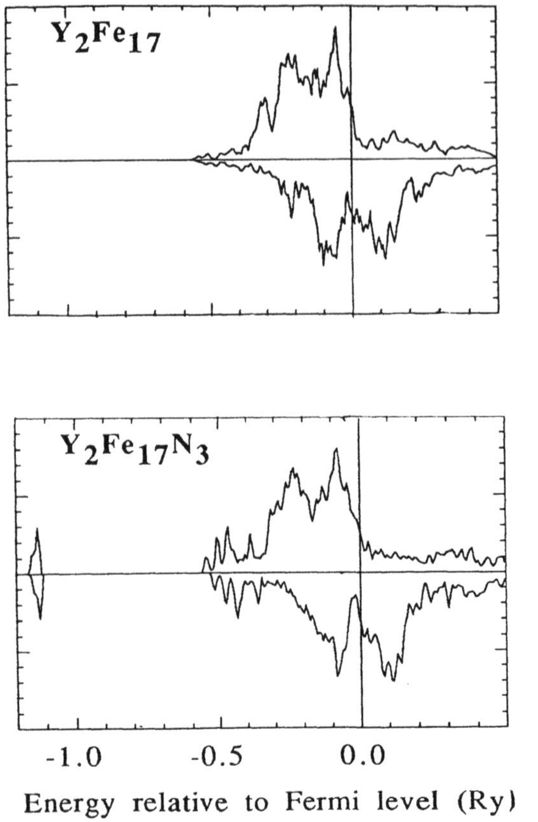

Fig. 5. Result of LMTO band-structure calculations [40]: (a) Y_2Fe_{17} and (b) $Y_2Fe_{17}N_3$. Note the separation of the spin-up and spin-down bands in bonding (low energy) and antibonding (high energy) subbands, separated by valleys. The molecular-field energy is included in the Hamiltionian, so the spin-up and spin-down bands have the same Fermi level (exchange splitting).

(iii) There is a fair amount of hybridization in 2:17 hydrides. The volume expansion is only 2%, so the magnetic moment of the hydrides is actually lower than that of the parent compounds. On the other hand, hybridization in hypothetical fluorides $R_2Fe_{17}F_3$ is predicted to be negligible [45]. Presupposing that the hypothetical fluoride has a volume expansion of about 6%, we can summarize our findings as $\mu(H) < \mu(C) < \mu(N) < \mu_{hyp}(F)$.

(iv) Compared to the parent compound, the room-temperature magnetization of $SmFe_{17}N_3$ is enhanced by about 50%. This cannot be ascribed to the the 15% increase in magnetic moment, particularly since the 6% volume increase dilutes the magnetization $M = \mu/V$. In fact, the large increase in room-temperature magnetization is mainly due to the increased Curie temperature of the nitride.

On the whole, there is a reasonable agreement between (i) different band structure calculations and (ii) theory and experiment [44]. The detailed description of correlation phenomena, which goes beyond the Hartree-Fock approximation, is still a problem [16,

24], but these effects don't affect the numerical exactness of the calculated moments very much. In conclusion, there is a basic, and largely quantitative, understanding of the magnetic moment in interstitially modified rare-earth intermetallics.

3. Curie Temperature

Summary: Compared to other intrinsic properties, in particular magnetization and anisotropy, the Curie temperature of interstitially modified rare-earth intermetallics is least understood. Phenomenologically, the increase in Curie temperature is due to the lattice expansion of about 6 vol.%, but it has not been possible to calculate the Curie temperature from first principles. There are two opposite limits to explain finite-temperature magnetic ordering: non-metallic local-moment magnetism, and metallic band magnetism. Metallic 3d ferromagnetism represents an *intermediate* case. To give a tentative description of the iron-rich parent compounds the concept of quasi-weak ferromagnetism is used, while interstitial nitrides such as $Sm_2Fe_{17}N_3$ are very close to the local-moment picture.

3.1. BACKGROUND

3.1.1. *Curie Temperature and Electrostatic Interaction.* As discussed in Section 2, magnetic interaction is very weak. Typical magnetic dipole fields are of order $\mu_0 H \approx 1$ T while Bohr's magneton, which gives the order of magnitude of magnetic interaction, can be written as $\mu_B/k_B = 0.672$ K/T. This means, even low-temperature thermal excitations ($T \approx 4.2$ K) are able to destroy magnetic order. To assure macroscopic magnetism at and above room temperature, a much stronger coupling mechanism must be found.

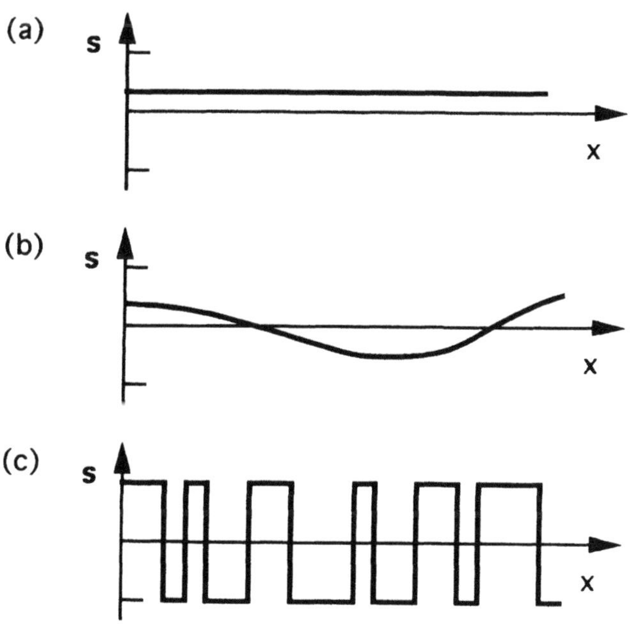

Fig. 6. Schematic spin structure of a ferromagnet near T_C: (a) itinerant ferro-magnet, (b) intermediate region, and (c) local-moment ferromagnet.

Table 3. Intrinsic magnetic properties of selected yttrium intermetallics.

	T_c K	μ μ_B/f.u.	m_T μ_B/at.	K_1 (4.2 K)[a] MJ/m^3	Ref.
Y_2Fe_{17}	325	32.8	1.98	-2.5	[12, 44]
Y_2Fe_{17}		35.4	2.11		[37, 39]
$Y_2Fe_{17}N_{2.5}$	694	38.1	2.29	-1.3	[12, 44]
$Y_2Fe_{17}N_3$		39.6	2.34		[37, 39]
$Y_2Fe_{17}C_{1.0}$	510	31.6	2.08		[6]
$Y_2Fe_{17}C_{2.2}$	660	36.1	2.17		[44]
$Y_2Fe_{17}H_{2.7}$	475	31.8	1.92		[44]
$Y(Fe_{11}Ti)$	524	19.0	1.76	2.0	[4, 13]
$Y(Fe_{11}Ti)N_{0.8}$	742	21.7	2.01	1.3	[4]
$Y(Fe_{11}Ti)C_{0.9}$	678	21.6	2.00	1.4	[4]
$Y_2Fe_{14}B$	586	30.5	2.18	1.1	[15, 17]

[a] See Section 4.

Electrostatic interaction provides this coupling: the Pauli principle forbids the occupation of electronic orbitals by two electrons with the same spin, so the electrostatic interaction becomes spin dependent. Under certain circumstances , the energy of the aligned (ferromagnetic) state is lower than that of the antiferromagnetic state (cf. Section 2). This reference to exchange interaction, however, reveals only *one* feature of the thermal behaviour of metallic ferromagnets. As we will see, even the simplest metal, the free electron gas, is able to surpress states with inhomogeneus spin orientation - a mechanism which does not involve exchange interaction.

3.1.2. *Localized and Delocalized Excitations*. Fig. 6 gives a schematic idea of ferromagnetic spin structures near T_c. In the metallic limit, Fig. 6a, the mesoscopic local magnetization $s(\mathbf{r}) = M_z(\mathbf{r})/M_o$ is realized by ideally delocalized Bloch functions $\Psi_k(\uparrow)$ and $\Psi_k(\downarrow)$. At T_c, both magnetization $M = M_o <s(\mathbf{r})>$ and moment $\mu = M_o<s(\mathbf{r})^2>^{1/2}$ vanish. A theory which describes delocalized excitations is the temperature-dependent Stoner theory [29].

Fig. 6c shows the non-metallic local-moment limit. At and even above T_c, there are well-defined ionic moments $\mu = M_o$. Below T_c, these moments order ferromagnetically without changing their magnitude. The reason for the stability of local moments are the strong electron correlations in non-metals and the corresponding Hund's-rule electron structure. Any decrease of the magnetic moment implies the double occupation ($\uparrow\downarrow$) of another orbital, which is energetically unfavourable due to intra-atomic Coulomb repulsion. A well-known local-moment approach is the mean-field theory of the spin-1/2 Ising model [46, 47].

Real metallic ferromagnets represent an intermediate case (Fig. 6b). At T_c, the moment of ferromagnetic materials is reduced but finite. In the case of very weak itinerant ferromagnets, such as $ZrZn_2$ with $T_c = 22$ K, long-wavelength spin-fluctuations dominate and the picture is very close to Fig. 6a [48, 49]. On the other hand, in strong and nearly strong ferromagnets, such as iron or $Sm_2Fe_{17}N_3$, short-wavelength spin-fluctuations dominate and the moment at T_c is only slightly reduced [16, 24, 50]. This means, the situation found in strong ferromagnets is rather close to that shown in Fig. 6c, and the local-moment approach represent a physically reasonable starting approximation. Theories which deal with the intermediate case Fig. 6b are generally referred to as spin-fluctuation theories. In the following subsections we will discuss some basic features of these theories.

3.1.3. *Itinerant Picture.* Let us remember the situation shown in Fig. 2, where a zero-temperature molecular field, described by the Stoner parameter I, transfers electrons from the spin-down to the spin-up band. At finite temperature, thermal excitations will redistribute the electrons - and reduce the magnetic moment - by thermally smearing out the Fermi surface.

To calculate the Stoner Curie temperature T_{cs}, at which these Stoner excitations destroy ferromagnetism, one can use a Sommerfeld expansion involving derivatives of the density of states at the Fermi level [26, 49]. For our purposes, however, it is sufficient to use the estimate $T_{cs} = I\mu/4\mu_B k_B$ which is obtained by assuming that a single peak in the density of states dominates the behaviour of the magnet [29]. Using the data given in Table 2, we obtain T_{cs} = 5709 °C for iron, which is much larger than the observed Curie temperature T_c = 771 °C. There are more sophisticated calculations which yield somewhat lower values of T_{cs} (cf. [51] and references therein), but none is able to reproduce the observed Curie temperatures.

A simple explanation of this failure is the fact that a lower limit of $k_B T_{cs}$ is given by the 'fine structure' of the peaks and valleys which form the density of states. We cannot expect, for instance, that a Curie temperature decrease from 500 K to 50 K is accompanied by a fine graining of the density of states by a factor 10. There must exist a mechanism which yields low Curie temperatures without referring to the Stoner theory.

3.1.4. *Local Moment Picture.* The comparatively high excitation energies of the Stoner theory are due to the delocalized character of the underlying Bloch states. Imagine a burning candle at one end of an iron spoon - the Stoner theory requires the magnetization to change uniformly, including the 'cold' end of the rod! Localized excitations are much more effective in destroying magnetization, even if the magnetic ground state is delocalized.

Let us first discuss the highly idealized limit Fig. 6(c). The archetypical approach to rationalize the Curie temperature of local-moment systems is the spin-1/2 Ising model [46, 52]. The starting point is the Ising Hamiltonian

$$\mathbb{H} = -\frac{1}{2} \sum_{<ij>} J_{ij}\, s_i\, s_j \;-\; H \sum_i s_i \qquad (3.1)$$

where H is the (reduced) external magnetic field and $J_{ij} = J_o$ the coupling between next neighbours. The mean-field treatment of this Hamiltonian (cf. Appendix B) yields the Curie temperature $T_c = z\,J_o/k_B$, where z is the number of next neighbours. In fact, critical fluctuations near T_c do not only change the critical exponents but also reduce the Curi temperature. A numerical result is $T_c = 0.764\, z\, J_o/k_B$ for a simple cubic lattice (z = 6) [46, 47], while the prefactor in dense-packed lattices is more close to one. Similiar arguments apply to more sophisticated Ising and Heisenberg systems.

In spite of its disadvantages, mean-field analysis represents a standard method of evaluating experimental Curie-temperature data (see e.g. [15, 17, 52-54]). For instance, it allows to investigate how the leading 3d Curie-temperature contribution is modified by the rare-earth sublattice.

3.2. SPIN-FLUCTUATION THEORY

3.2.1. *Selfconsistent Renormalization of Spin Fluctuations.* A theory which describes spin fluctuations in *very weak* itinerant ferromagnets, such as $ZrZn_2$ with $T_c = 22$ K, is the self-consistent renormalization of spin fluctuations (Murata and Doniach, 1972 [48]). This approach, which is very close to the itinerant limit Fig. 6(a), treats the spin fluctuations $s_i = s(\mathbf{r}_i)$ as contineous long-wavelength excitations [24, 48, 50]. In the simplest case, the Hamiltonian reads [48]

$$\mathbb{H} = \int \big(A\,[\nabla s(\mathbf{r})]^2 + \eta(s(\mathbf{r})) \big)\, d\mathbf{r} \qquad (3.2)$$

where A is a coupling constant and $\eta(s(\mathbf{r}))$ a non-linear function which describes the energy of the homogeneously spin-polarized system (see Section 3.3).

The thermal behaviour of the system Eq. (3.2) is given by the free energy

$$F = -\,k_B T \ln\left(\int e^{-H/k_B T}\, Ds(\mathbf{r})\right) \qquad (3.3)$$

where the functional integral $\int ...\, Ds(\mathbf{r})$ includes all microscopic states (continuous spin configurations) $s(\mathbf{r})$. The calculation of the free energy is easy if Eq. (3.3) represents a Gaussian functional integral, i.e. if $\eta = 0.5\, a_{eff}\, s^2$ (Appendix A). In this case, the Curie temperature is obtained from the 'trivial' stability condition $a_{eff} = 0$, but a_{eff} has to be calculated from $\eta(s)$ in a selfconsistent manner.

For simplicity, let's assume $\eta = -\,0.5\, a\, s^2 + 0.25\, b\, s^4$, which is a widely used ansatz in statistical physics. Applying the Bogol'ubov inequality to this function (Appendix B), we obtain $a_{eff} = -\,a + 3\, b \langle s^2 \rangle$ so that T_c is given by $a = 3\, b \langle s^2 \rangle$. The squared moment $\mu^2 = \langle s^2 \rangle$ is calculated in Appendix A, and the Curie temperature finally reads [48]

$$T_c = \frac{2\,\pi^2\, A\, a}{3\, b\, k_B\, k_o} \qquad (3.4)$$

The cut-off wave vector k_o [48, 55] is necessary to match the dimension of the coupling constant, which is measured in J/m. An upper limit of k_o is given by the lattice spacing.

Depending on the strength of the coupling and the number of **k** modes involved, spin fluctuations can be very effective in destroying magnetic order. In particular, Eq. (3.4) yields Curie temperatures which are much lower than those predicted from the Stoner theory, where $k_o = 0$. There are more sophisticated extensions [55, 56] of this theory with similiar results, but in any case the validity of continuous spin-fluctuation theories is resticted to very weak itinerant ferromagnets close to the limit Fig. 6a.

3.2.2. *Nearly Strong Ferromagnets.* Most permanent magnets are strong or nearly strong ferromagnets, and formulas such as Eq. (3.4) cannot be used to calculate their Curie temperatures. In fact, the incorporation of localized as well as itinerant features of the d electrons has been a long-standing problem [17, 24, 57-63]. Apart from the quantum-mechanical treatment, which is usually based on Hubbard [24, 59] or Hartree-Fock [60, 61] approaches, Curie-temperature calculations are a statistical problem.

The simplest models which give some non-trivial insight into the statistics of spin fluctuations near the local-moment limit are Ising models — due to $[\mathbb{H}, \sigma_z] = 0$ the quantum-mechanical and classical limits of Ising-type models coincide. Let us start from the spin-1 Ising Hamiltonian

$$\mathbb{H} = -\frac{1}{2} \sum_{<ij>} J_{ij}\, s_i\, s_j - U_D \sum_i s_i^2 - H \sum_i s_i \tag{3.5}$$

with $s_i = 0, \pm 1$. The parameter U_D describes the formation of local moments. If U_D is very large then states with $s_i^2 = 1$ are energetically more favourable than states with $s_i^2 = 0$. Per definition, $U_D = 0$ in the spin-1/2 Ising model Eq. (3.1): due to $s_i^2 = 1$ there the inclusion of diagonal interaction would merely yield a physically irrelevant shift of the zero-point energy.

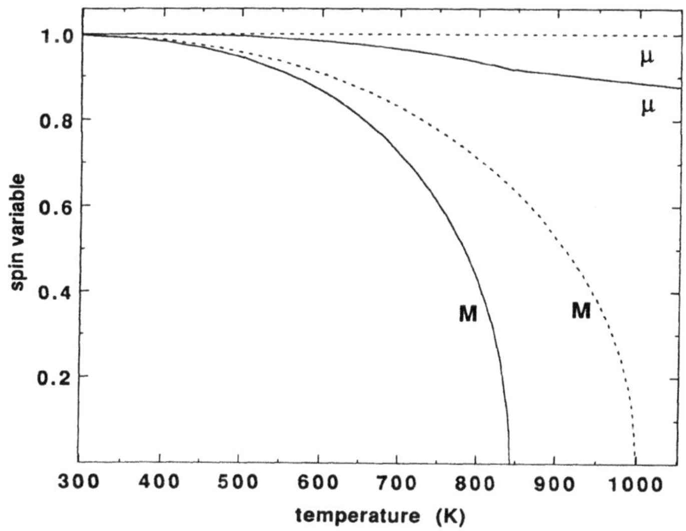

Fig. 7. Magnetic moment μ and magnetization M as a function of temperature (cf. Appendix B). (dashed lines) local-moment limit with $zJ_o/k_B = 1000$ K, and (solid lines) intermediate regime with $zJ_o/k_B = 1000$ K and $U_D/k_B = 1000$ K. With decreasing U_D the graphs μ(T) and M(T) become increasingly more similiar (weak ferromagnetism).

An oversimplified mean-field approach is to use the identity $s_i^2 = 2<s_i> s_i + (s_i - <s_i>)^2 - <s_i>^2$ and to neglect (i) the correlation term $(s_i - <s_i>)^2$ and (ii) the term $<s_i>^2$, which does not explicitly depend on s_i. This procedure is actually equivalent to the customary mean-field approach used to treat the Hamiltonian Eq. (3.1). Still, it greatly overestimates the Curie temperature if $U_D \gg J_{ij}$. As in the Stoner theory, the presupposed but physically unreasonable homogenity mimics a strong ferromagnetic interaction and a high Curie temperature.

It should be emphasized, however, that this failure arises from the sloppy treatment of the diagonal interaction and not from the simplicity of the statistical mean-field approximation. In fact, a proper mean-field treatment of Eq. (3.5) produces Onsager-like reaction fields (cf. [24]) and leads to resonable Curie temperatures (Fig. 7 and Appendix B).

3.3. CURIE TEMPERATURE AND INTERSTITIAL MODIFICATION

3.3.1. *Experimental Results.* Table 4 shows Curie temperatures and room-temperature magnetizations of selected $R_2Fe_{17}Z_x$ intermetallics. On the whole, the Curie temperature increases by about 100%, and the room-temperature magnetization is enhanced by about 50%. The main reason for the increase in room-temperature magnetization is the enhanced Curie temperature; the moment increase (Section 2) is of secondary importance. However, only $Sm_2Fe_{17}N_x$ and $Sm_2Fe_{17}C_x$ exhibit the easy-axis anisotropy required of a permanent magnetic material.

The energy product, the main permanent-magnetic figure of merit, obeys $(BH)_{max} \leq \mu_o M_o^2/4$, so $Sm_2Fe_{17}N_3$ with $\mu_o M_o \approx 1.54$ T is nearly as good a potential high-performance permanent magnet as the present room-temperature record holder $Nd_2Fe_{14}B$ with $\mu_o M_o \approx 1.61$ T [17, 18]. In the technically important region T > 150 °C, however, the low Curie temperature $T_c = 312$ °C of $Nd_2Fe_{14}B$ takes effect and $Sm_2Fe_{17}N_3$ is actually the potentially best permanent magnetic material (cf. [17, 64]). Finally, pronounced high-temperature applications are covered by Sm-Co magnets with Curie temperatures of order 1000 K.

As opposed to cobalt-rich rare-earth intermetallics, whose Curie temperatures approach the value of pure cobalt, the Curie temperatures of iron-rich rare-earth intermetallics are much lower than that of pure iron (Tables 2 and 3). For instance, $Nd_2Fe_{14}B$ consists largely of iron, which is very favourable from the point of view of raw material prices and magnetic moment, but its Curie temperature is much lower than expected from the volume fraction of iron. On the other hand, the low Curie temperatures of iron-rich rare-earth intermetallics are accompanied by a strong negative pressure dependence dT_c/dp [65], which leads us to the conclusion that the Curie temperature increase upon interstitial modification originates from the the volume expansion [1, 2].

At present, there is no detailed, quantitative understanding of Curie-temperature trends in transition-metal-rich rare-earth intermetallics. Simple ferromagnets, in particular elementary iron and nickel, have attracted much attention, and a large number of approaches, theories, and approximations have been used to study their finite-temperature magnetic properties (see e.g. [17, 24, 49, 50, 57-63]). These investigations have yielded much insight into phenomena such as finite-temperature moment formation or short-range spin fluctuations, but it is unlikely that this ponderous hierarchy of approaches will survive the confrontation with the properties of more complex ferromagnets such as $Sm_2Fe_{17}N_3$ and $Nd_2Fe_{14}B$.

Table 4. Curie temperature and magnetization of selected 2:17 compounds.

	T_c °C	M_o (RT) T	ΔT_c K	$\Delta T_c/T_c$ [a] %	Ref.
Sm_2Fe_{17}	116	1.00			[10]
Sm_2Fe_{17}	125	1.17			[11]
$Sm_2Fe_{17}N_{2.3}$	476	1.34	360	93	[10]
$Sm_2Fe_{17}N_{2.3}$	476	1.54	360	93	[12]
$Sm_2Fe_{17}N_{2.5}$	476	1.55	351	88	[11]
$Sm_2Fe_{17}N_{2.5}$	476	1.54	360	93	[9]
$Sm_2Fe_{17}C_{1.1}$	279	1.24	163	42	[12]
$Sm_2Fe_{17}C_{2.2}$	395	1.43	270	68	[11]
Ce_2Fe_{17}	- 32	0.00			[10]
$Ce_2Fe_{17}N_{2.8}$	440	1.52	472	196	[10]
Nd_2Fe_{17}	57	0.76			[10]
$Nd_2Fe_{17}N_{2.3}$	459	1.69[b]	402	122	[10]
Gd_2Fe_{17}	204	0.47			[10]
$Gd_2Fe_{17}N_{2.4}$	485	1.33	281	59	[10]
Dy_2Fe_{17}	94	0.52			[10]
$Dy_2Fe_{17}N_{2.8}$	452	1.15	358	98	[10]
Er_2Fe_{17}	23	0.34			[10]
$Er_2Fe_{17}N_{2.7}$	424	1.36	401	135	[10]

[a]Increase referring to the parent compound. [b]1.60 T in [8].

In the following subsections we analyse the predictive power of various approaches. The underlying reference point is a possible first-principle guidance to the improvement of metallic Curie temperatures.

3.3.2. *Itinerant Coupling*. Let us first discuss to what extent the Curie temperature of interstitial materials can be predicted from the density of states at the Fermi level, for instance as an extrapolation of the long-wavelength spin fluctuation theory. To answer this question we have to investigate the microscopic origin of the parameters which describe the behaviour of very weak itinerant ferromagnets. These parameters, such as A, a and b in Eqns. (3.4) and (B.9), are related to the band structure [48, 49, 55]. In the case of a Hartree-Fock free electron gas the lowest-order terms of the series expansion read [23]

$$\mathbb{H} = \frac{N_T}{2V} \int \left(\frac{1}{24\, D(E_F)\, k_F^2} (\nabla s)^2 + \frac{1}{2} \left(\frac{1}{D(E_F)} - I \right) s^2 + \frac{1}{4} B\, s^4 \right) d\mathbf{r} \quad (3.6)$$

To adapt the free-electron result to the usual definition of D(E) as a density of states per transition-metal atom and spin we have introduced the (fictitious) number of transition-metal atoms $N_T = V/V_T$ and the (fictitious) moment per transition-metal atom $s = \mu_T/\mu_B$. The homogeneous second-order term is equivalent to that in Eq. (2.3), while the $(\nabla s)^2$ term describes the coupling between free-electron spins. This coupling is called inverse **k**-dependent spin susceptibility [23, 49, 55] and gives the energy which has to be paid to create a state with inhomogeous spin density. Imagine a free electron with $|\mathbf{k}| \approx k_F$ whose wave function is modulated by an external magnetic field $H(\mathbf{r}) \propto \cos(\mathbf{qr})$. Using $\mathbb{H} = \hbar^2\nabla^2/2m$ we find that these modulations yield an energy contribution of order $\hbar^2q^2/2m = \hbar^2k_F{}^2/2m \times q^2/k_F{}^2$, which, due to $E_F \propto D(E_F)^{-1}$, reproduces the $(\nabla s)^2$ term in Eq. (3.6). As it can be seen from Eqns. (3.2) and (3.4), the coupling term in Eq. (3.6) favours a high Curie temperature if $D(E_F)$ is small. However, if this were the only consideration then the increase of D(E) upon lattice expansion would *reduce* the Curie temperature. In fact, the behaviour of T_c is largely determined by the energy density η; the coupling constant $A \propto J$ serves as a common prefactor which fixes the order of magnitude of T_c.

Higher-order corrections and more complicated band structures involve derivatives such as $D'(E_F)$ (see e.g. [49, 55]). The procedure is straightforward - except the 'minor' fact that the determination of derivatives of D(E) from numerical band-structure data is fairly inaccurate. Still, it is not possible to extend band structure approach to strong and nearly strong ferromagnets. There are two reasons: first, the wave vectors involved are not small compared to k_F [48, 55], and secondly, the series expansion on which the approach is based does not converge properly (Appendix C). In a word, we cannot expect that an extrapolation of the long-wavelength spin-fluctuation theory yields quantitative Curie temperature predictions.

This general result explains, for instance, the failure [66] of the $\mathbf{k} = 0$ Mohn-Wohlfarth approach [51]. For $T_c \ll T_{cs}$ the Mohn-Wohlfarth theory predicts $T_c \approx T_{sf}$ with $k_BT_{sf} = 0.05\ (\mu/\mu_B)^2\ \{0.5\ D_\uparrow(E_F)^{-1} + 0.5\ D_\downarrow(E_F)^{-1} - I\}$. In the case of bcc iron, this equation yields $T_c \approx 4000$ K if $D_\uparrow$ and $D_\downarrow$ are obtained from modern band-structure calculations (cf. [16, 40, 66]). The only way to solve this contradiction is to make the physically unreasonble assumption that the magnetic moment of iron vanishes at T_c. The Curie temperature is then given by the Stoner temperature T_{cs}, which, in turn, has to be strikingly low to reproduce the observed Curie temperature. Similiar arguments refer to rare-earth intermetallics and interstitial nitrides and carbides [66]. Though widely used to discuss Curie temperature data of rare-earth intermetallics and their interstitially modified derivates [4, 16, 37, 39, 42, 51, 67], the Mohn-Wohlfarth theory suffers all shortcomings of the very-weak-itinerant limit and is not able to reproduce observed Curie temperatures in a quantitatively satisfactory manner.

3.3.3. *Exchange Coupling and Curie Temperature*. Apart from the **k**-dependent spin susceptibility of the free-electron gas, the simplest coupling mechanism is the hydrogen-molecule exchange. If $|a\rangle$ and $|b\rangle$ denote two suitable one-electron states, then the ferromagnetic exchange splitting reads $J = 2\ \{\langle a\, b\,|\, V\,|\, b\, a\rangle - S \langle a\, b\,|\, V\,|\, a\, b\rangle\}/(1 - S^2)$ with $S = |\langle a\,|\, b\rangle|^2$ [29].

It is tempting to explain magnetization and Curie temperature in metals by referring to sign and magnitude of J. For instance, let us explain the low Curie temperature of Sm_2Fe_{17} by assuming that the average interaction J is just going to become negative (antiferromagnetic). If J increases with the average iron-iron distance then a comparatively small lattice expansion is sufficient to cause a considerable relative increase in J and T_c [12, 13, 14, 15, 17]. This argument refers, in particular, to the iron atoms on the 'dumbell' 6c and 4f sites in the Th_2Zn_{17} and Th_2Ni_{17} structures, respectively. R_2T_{17} intermetallics can be interpreted as the result of a hypothetical substitution reaction $3\ RT_5 + 2\ T \leftrightarrow R_2T_{17} + R$, where the two dumbell iron atoms have to substitute on the rare-earth site. In fact, the Curie temperature $T_c = 1150$ K of $Nd_2Fe_2Co_{15}$, where iron occupies the 6c dumbell sites [68], is not much smaller than the Curie temperature $T_c = 1180$ K of the isostructural compound Nd_2Co_{17}. Linear interpolation between Nd_2Fe_{17} and Nd_2Co_{17} yields $T_c = 1080$ K for $Nd_2Fe_2Co_{15}$, while antiferromagnetic dumbell contributions would reveal themselves by an even lower T_c.

Similiar arguments apply to other attempts to explain Curie temperature trends in iron-rich rare-earth intermetallics by distance dependent exchange integrals [17, 52]. However, these shortcomings do not exclude a phenomenological description of Curie-temperature trends by exchange parameters [15, 53] or generalized Néel-Slater curves Fig. (8).

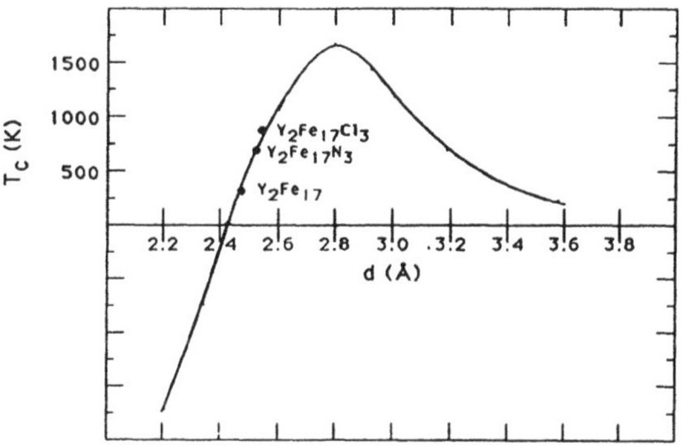

Fig. 8. Modified Néel-Slater curve for Y_2Fe_{17} interstitial compounds [44]. The curve, which is a semiempirical fit based on the idea of a distance-dependent exchange interaction, predicts that the hypothetical compound $Y_2Fe_{17}Cl_3$ has a Curie temperature of about 900 K.

Microscopically, the orthogonality of Bloch functions (S = 0) and the repulsive interaction $V = e^2/4\pi\varepsilon_0|\mathbf{r}_1 - \mathbf{r}_2|$ yields $J > 0$, and, as emphasized by Slater [69], Heisenberg exchange in metals is always *ferromagnetic*. A simple example is the homogeneous electron gas Eq. (2.2) with the exchange term generating ferromagnetism but not antiferromagnetism if the Stoner criterion is satisfied.

From a rather general point of view, this dilemma is solved by breaking the Bloch symmetry of the electronic states (see e.g. [50]). This can be done, for instance, by strong Coulomb interaction. The simplest approach is to neglect exchange altogether and to restrict oneself to the Hubbard interaction $V = U \sum n_{i\uparrow} n_{i\downarrow}$, with U being the intratomic Coulomb repulsion and $n_{i\uparrow}$ and $n_{i\downarrow}$ the occupation number operators of the lattice site i [23, 24]. With $n_i = n_{i\uparrow} + n_{i\downarrow}$ and $s_{zi} = n_{i\uparrow} - n_{i\downarrow}$ the interaction reads $V = U \sum(n_i^2 - s_{zi}^2)/4$ [24, 70], and we see that states with a large (squared) magnetic moment $\mu^2 = s_{zi}^2$ are energetically favourable. Still, an exact treatment of this simple but nevertheless non-trivial many-body problem is very difficult. Furthermore, it is difficult to quantify how the exclusion of exchange interaction [49] affects the predictions of the Hubbard model.

3.3.4. *RKKY interaction*. As we have seen (Section 3.2), thermal excitations are able to break Bloch symmetry by causing spin fluctuations. In the simplest case, a single spin is switched and the corresponding change in exchange potential is treated as a lattice impurity [50, 57]. It is even possible to map non-random many-body Hamiltonians onto non-interacting random-field problems (Hubbard-Stratonovich transformation [24, 71]).

Related to this approach, but not equivalent, is the RKKY coupling between rare-earth ions [23, 72, 73]. Consider an itinerant electron gas with a magnetic impurity, i.e. itinerant spin-up and spin-down electrons feel a different exchange potential $V_{\pm}(\mathbf{r})$. The impurity potential breaks the Bloch symmetry, but in the case of a weak potential $V_{\pm}(\mathbf{r})$ we can use perturbation theory to calculate the perturbed wave functions $\Psi_{\mathbf{k}\pm}(\mathbf{r}) = \Psi_{\mathbf{k}\pm}{}^{o}(\mathbf{r}) + \Delta\Psi_{\mathbf{k}\pm}(\mathbf{r})$ and the perturbed spin-up and spin-down densties. Of course, only waves with $|\mathbf{k}| \leq k_F$ are involved in dealing with the impurity, so $\Delta\Psi_{\mathbf{k}\pm}(\mathbf{r})$ contains oscillations down to $\lambda \approx 2\pi/k_F$. Simply speaking, these Friedel oscillations are a rudimentary equivalent of the shell structure of atomic matter.

In the case of point impurities in a free electron gas this mechanism introduces the so-called RKKY interaction [23]

$$J(\mathbf{r}) = J_o \frac{\sin(2k_F r) - 2k_F r \cos(2k_F r)}{(2k_F r)^4} \tag{3.7}$$

The oscillatory behaviour of $J(\mathbf{r})$ corresponds to a distance dependent alternation between ferromagnetism and antiferromagnetism, and predicts Curie temperature trends similiar to those observed in interstitially modified rare-earth intermetallics.

Note that the assumptions of a localized impurity and a free electron gas are *not* essential - the basic feature of this approach is a weak but spin-dependent potential $V_{\pm}(\mathbf{r})$. The analysis of this problem (see e.g. [72, 73]) reveals that the coupling $J(\mathbf{r}, \mathbf{r}')$ is proportional to $\int \mathrm{Im}\, \{G_o(\mathbf{r}, \mathbf{r}'; E)\, G_o(\mathbf{r}', \mathbf{r}\,; E)\}\, dE$, where $G_o(\mathbf{r}, \mathbf{r}'; E)$ is the single-electron Green function of the unperturbed problem. The single-particle Green function, as well as the positions of the nodes of $J(\mathbf{r})$, can be calculated from the wave functions $\Psi_{\mathbf{k}\pm}{}^{o}(\mathbf{r})$ [23] and depend on the number of itinerant electrons per atom.

3.3.5. Curie Temperature and Band Structure. The band structure is known for many rare-earth intermetallics (Section 2), and it would be useful to have a tool which predicts the Curie temperature from, let's say, the density of states. In fact, band-structure data serve as input in many Curie-temperature calculations [24, 59, 60, 73-75], but the numerical character of these calculations makes it difficult to extract physical information from the calculated Curie temperatures.

To some extent, the local-environment approach of Kakehashi [74, 75] represents an exception. In this theory, a self-consistent effective-medium approach is used to calculate finite-temperature magnetic properties of amorphous transition-metal alloys. By variation of the input density of states it is shown that the Curie temperature is largest if the paramagnetic Fermi level lies in the (antibonding) main peak of the density of states. Without reference to exchange integrals this rule relates the occurrence of ferromagnetism to the Curie tempeature. For instance, bcc iron, in contrast to fcc iron, has a high Curie temperature because the main bcc peak is located at a comparatively low energy. On the other hand, nickel, which has a nearly-filled 3d band, requires a high-lying main peak to develop its maximum Curie temperature.

Of course, as it has been emphasized in [74], this argumentation is related to the Stoner theory, which is inapplicable to finite-temperature magnetism. Furthermore, it is difficult to see why Sm_2Fe_{17}, whose moment is not much smaller than that of $Sm_2Fe_{17}N_3$, should be a very weak itinerant ferromagnet. In the following we will show that complete exchange splitting (strong ferromagnetism) and very weak itinerant ferromagnetism do not exclude each other ('quasi-weak' ferromagnetism).

Let us start from the (paramagnetic) density of states D(E), the Stoner parameter I, and an exchange parameter J which we assume to be fairly independent of small changes in the next-neighbour distances. For the sake of simplicity, we require that the theory reproduces (i) the Stoner theory at T = 0 and (ii) the local-moment mean-field result $k_B T_c = J \mu^2/\mu_B^2$ for $I = \infty$, where μ is the ground-state moment per transition-metal atom.

A Hamiltonian which satisfies these requirements is $\mathbb{H}_{MF} = \sum_i \mathbb{H}(s_i)$ with

$$\mathbb{H}(s) = -\frac{1}{4}(I - 2J)\, s^2 - J \langle s \rangle s - \mu_B H s - \eta(s) \tag{3.8}$$

where

$$\eta(s) = \int_0^{E_{F\uparrow}(s)} D(E)\, E\, dE + \int_0^{E_{F\downarrow}(s)} D(E)\, E\, dE \tag{3.9}$$

is the energy density of the Stoner theory. Starting from the partition function $Z = \int \exp(-\mathbb{H}(s)/k_B T)\, ds$ it is now straightforward to derive the implicit relation $k_B T_c = J \langle s^2(T_c) \rangle$. As expected, due to $\langle s^2 \rangle \leq \mu^2/\mu_B^2$ spin fluctuations always reduce the Curie temperature.

To discuss the possibility of quasi-weak ferromagnetism we use the rectangular density of states $D(E) = n/2W$ where W and n are the band width and the maximum number of antibonding electrons, respectively. The number $n' \leq n$ of antibonding electrons actually present in the band determines the complete-exchange-splitting moment μ_s. The Curie temperature is now given by

$$\frac{I - 2W/n}{2J} = \frac{\mu_s}{\mu_B} \frac{e^{\mu^2 (I - 2W/n - 2J)/4 k_B T_c \mu_B^2}}{\int_0^{\mu_s/\mu_B} e^{s^2 (I - 2W/n - 2J)/4 k_B T_c}\, ds} \tag{3.10}$$

Taking $\mu_s = 2\,\mu_B$, $I = 1$ eV, $k_B J = 300$ K, and $n = 5$ we obtain $\mu = 2\,\mu_B$ and $T_c = 644$ K for $W = 2.1$ eV as compared to $\mu = 2\,\mu_B$ and $T_c = 363$ K for $W = 2.4$ eV. We see that Eq. (3.10) predicts a considerable Curie-temperature reduction without any change in ground-state magnetization and exchange interaction.

In the limit $W \approx 0$, i.e. strong ferromagnetism (Fig. 4d and solid line in Fig.11), Eq. (3.10) predicts $k_B T_c = J \mu_s^2 (1 - 4J/D)/\mu_B^2$, thus $T_c = 1076$ K for the assumed parameter set. On the other hand, ferromagnetism breaks down if the band width, i.e. the inverse density of states, exceeds a critical value W_c. In the quasi-weak regime close to W_c (Fig. 4b and pointed line in Fig.11) the Curie temperature scales as $T_c \propto 1/\ln((W_c - W)/W_c)$. Using Heine's result [76] that the 3d band width scales as $V^{-5/3}$ we expect a volume change of 6% to yield an effective band narrowing of 10%. The corresponding Curie-temperature change is compatible with our numerical estimate, but Fig. 5 reveals that it is actually very difficult to define the peak widths from band-structure densities of states.

It is worthwhile mentioning that Eqs. (3.8) and (3.10) represent a direct but non-trivial generalization of the Murata-Doniach theory [48]. The main difference consists in the treatment of singularities in the density of states (Appendix C). On the other hand,

Eq. (3.9) is *not* used to exploit finite-temperature results of the Stoner theory [76-80]; it only serves as a tool to describe the ground state properties of the magnet.

4. Crystal Field and Anisotropy

Summary: The magnetocrystalline anisotropy of rare-earth permanent magnets and their interstitially modified derivates is mainly provided by the rare-earth sublattice. Interstitial atoms modify the electrostatic crystal field around the rare-earth atoms and cause the 4f electron distributions to change their orientation. Depending on whether the 4f orbitals are prolate or oblate, the change in magnetocrystalline anisotropy can have either sign. In the case of $Sm_2Fe_{17}N_3$, the large increase in magnetocrystalline anisotropy is a combined effect of the prolate shape of the Sm 4f shell, the negative crystal-field charge of nitrogen, and the in-plane coordination of the interstial 9e site.

4.1. BACKGROUND

4.1.1. *Phenomenological Free Energy.* It is convenient to express the macroscopic free energy as a phenomenological series such as

$$\frac{F}{V} = K_1 \sin^2\theta + K_2 \sin^4\theta + K_3 \sin^6\theta - \mu_o H M_o \cos(\theta - \Phi) \tag{4.1}$$

where θ is the angle between crystallographic c axis and magnetization, and Φ the angle between external magnetic field and c axis. M_o is the saturation magnetization, H the external magnetic field, and the coefficients K_m are called anisotropy constants. Note that the detailed structure of this series depends on the overall crystal symmetry.

Table 5. Crystal-field related properties of permanent magnets.[a]

	B_a(RT)[b]	K_1 (RT)	K_2(RT)	$A_2{}^0$
	T	MJ/m^3	MJ/m^3	K/$a_o{}^2$
$Sm_2Fe_{17}N_x$	20	8.5	1.5	-180
$Sm_2Fe_{17}C$	5.3			-130
$Sm_2Fe_{17}C_{2.2}$	15	7.4	0.74	-200
$Sm(Fe_{11}Ti)$	10			-140
$SmCo_5$	30	17		-200
Sm_2Co_{17}	6.5	3.3		- 80
$Nd_2Fe_{14}B$	9	4.4	0.66	300

[a] Data from [8, 9, 12, 19, 64, 97]. [b] The most common definition of the anisotropy field is $B_a = (2 K_1 + 4 K_2 + 6 K_3)/M_o$.

Depending on the structure type, many intermetallics exhibit additional *in-plane* contributions [15, 81, 82]: $K_3' \sin^6\theta \cos6\phi$ in hexagonal $CaCu_5$ or Th_2Ni_{17} compounds as well as in rhombohedral Th_2Zn_{17} compounds, and $K_2' \sin^4\theta \cos4\phi + K_3' \sin^6\theta \cos4\phi$ in tetragonal $ThMn_{12}$ or $Nd_2Fe_{14}B$ compounds. Furthermore, in the case of in-plane anisotropy the field dependent Zeeman term in Eq. (4.1) is more complicated. Typically, the 4f anisotropy contribution dominates in rare-earth intermetallics, and free-energy series is limited to K_k contributions with $k \leq 3$, but extemely strong crystal fields (quenching) cause higher-order contributions.

Often the lowest-order anisotropy contribution $K_1 \sin^2\theta$ dominates at and above room temperature. As a crude explanation, this is due to the distance dependence of the crystal-field interaction (Section 4.3.1) and the more pronounced temperature dependence of the higher-order anisotropy constants (Section 4.2.2). In the case of samarium compounds the $J = 5/2$ ground state yields $K_3 = 0$ and $K_3' = 0$, but in reality there are small K_3 and K_3' contributions due to temperature or crystal-field induced admixture of $J = 7/2$ states (J mixing) [15, 81, 82, 83].

In good permanent magnets K_1 is large and positive, which fixes the magnetization direction parallel to the crystallographic c axis. If K_1 is too small then the magnetization deviates from the crystallographic c axis, which yields complicated spin structures or in-plane magnetization. Table 5 shows anisotropy-related data of selected rare-earth permanent magnets.

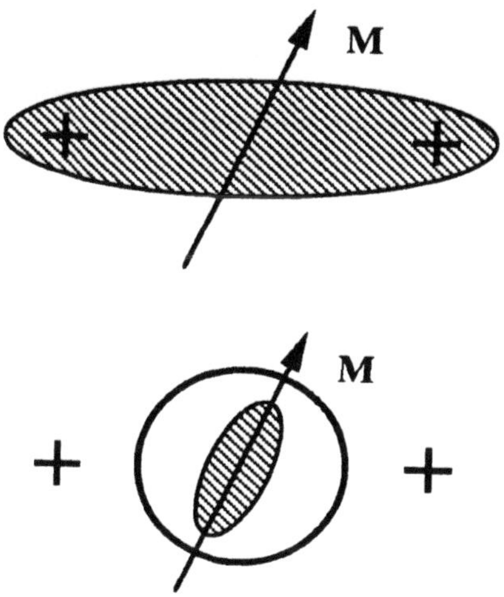

Fig. 9. Microscopic origin of magneto - crystalline anisotropy: (above) The orbital motion of delocalized 3d atoms is quenched by the strong crystal-field interaction, while the spin-orbit interaction is not strong enough to provide the coupling between orbital motion (lattice) and magnetization. This is the reason for the low anisotropy of the 3d sublattice. (below) The large spin-orbit coupling in rare-earth ions yields a firm connection between 4f charge cloud and rare-earth magnetization. The crystal field is partly screened but nevertheless causes a strong coupling between magnetization and lattice.

4.1.2. *Microscopic Origin of Magnetocrystalline Anisotropy.* Magnetocrystalline anisotropy arises from spin-orbit coupling. The electrons move in the lattice and try to avoid regions with high electron density and energetically unfavourable electrostatic potential. Of course, the electrostatic crystal field reflects the lattice symmetry, so energetically favourable orbitals yield, via spin-orbit coupling, a preferential magnetization direction.

The magnetocrystalline anisotropy of pure 3d compounds is comparatively small. This is due to the quenching of the orbital motion. Figure 9 gives a schematic idea of this effect. The orbital motion of the delocalized 3d electrons is dominated by the electrostatic crystal field (quenching), and the comparatively weak spin-orbit coupling is not able to induce an appreciable orbital moment to interact with. Strictly speaking, the quenching is not complete, and there remains a small orbital moment, typically a few tenths of a Bohr magneton, which couples to the spin and yields some 3d sublattice anisotropy.

4.2. CRYSTAL-FIELD THEORY

4.2.1. *Multipole Expansion.* It is worth mentioning that the ultimate way of dealing with magneto-crystalline anisotropy is to solve the many-body problem called solid-state physics. Still, imagine a giant computer which provides this solution in form of myriades of wave functions and Slater determinants — then the problem would be to extract physically meaningful information from this torrent of data.

Let us make the following, widely used approximations: (i) we neglect the transition-metal contribution, so the magnetocrystalline anisotropy is given by rare-earth sublattice, (ii) the rare-earth atoms do not interact (single-ion approximation), (iii) we restrict ourselves to electrostatic interaction and neglect exchange effects (cf. [84]), (iv) the spin-orbit interaction is strong enough to yield a rigid coupling between rare-earth moment and 4f electron cloud, and (v) the 4f electrons are localized well inside the rare-earth atom so that the crystal field is not able to deform the charge distribution of the 4f ground-state multiplet .

These conditions mean that we can use the unperturbed 4f wave functions ψ_l to calculate the crystal-field energy $\Sigma_l\langle\psi_l|V|\psi_l\rangle$ in lowest order perturbation theory. Alternatively speaking, we treat the rare-earth ion as rigid charge distribution $\rho_{4f}(\mathbf{r}) = -e\, n_{4f}(\mathbf{r})$ whose energy is given by the electrostatic interaction with the density $\rho(\mathbf{r})$ of the non-4f charges present in the lattice:

$$V_{CF} = -\frac{e}{4\pi\varepsilon_o}\int\frac{n_{4f}(\mathbf{r}')\,\rho(\mathbf{r})}{|\mathbf{r}-\mathbf{r}'|}\,d\mathbf{r}\,d\mathbf{r}' \tag{4.2}$$

To solve the integral Eqn. (4.2) it is convenient to make an electrostatic multipole expansion. The lowest-order terms of this expansion are

$$n_{4f}(\mathbf{r}) = \frac{1}{\sqrt{4\,\pi}}\,n_o\,f(|\mathbf{r}|) + \sqrt{\frac{5}{16\,\pi}}\,(3\cos^2\Theta - 1)\,n_2\,f(|\mathbf{r}|) \tag{4.3}$$

where $f(|\mathbf{r}|)$ is the radial 4f charge density and Θ the angle between $\mathbf{r}$ and the symmetry axis of the 4f charge cloud. The physically meaningless prefactors are the normalization constants of the spherical harmonics and simplify the mathematical treatment of the crystal-field interaction. The spherical 'monopole' term (index o) does not affect the magnetocrystalline anisotropy and can be dropped. The 4f quadrupole moment n_2, which must not be confused with the nuclear quadrupole moment, indicates whether the 4f electron cloud is prolate ($n_2 > 0$) or oblate ($n_2 < 0$). In a given crystal environment the second-order crystal-field contribution is proportional to n_2, so ions with $n_2 > 0$ and $n_2 < 0$, respectively, have opposite effects on the anisotropy constant K_1.

Quadropole moments, as well hexadecapole moments and 64-pole moments, are calculated from the 4f ground-state wave function (cf. [83, 85]). Table 6 gives the shape of the rare-earth ions in their Hund's-rule ground states. The sign of the quadrupole moment exhibits an easy-to-remember symmetry:

Table 6. Properties of tripositive rare-earth ions.[a]

		Shape [b]	Ground state and l_z: 3 2 1 0 -1 -2 -3	S	L $(= \Sigma l_z)$	J	R_{4f} [c] (Å)
57	La		| | | | | | | |	0	0	0	-
58	Ce		| ↓| | | | | | |	1/2	3	5/2	0.580
59	Pr		| ↓| ↓| | | | | |	1	5	4	0.551
60	Nd		| ↓| ↓| ↓| | | | |	3/2	6	9/2	0.529
61	Pm		| ↓| ↓| ↓| ↓| | | |	2	6	4	0.512
62	Sm		| ↓| ↓| ↓| ↓| ↓| | |	5/2	5	5/2	0.497
63	Eu		| ↓| ↓| ↓| ↓| ↓| ↓| |	3	3	0	0.490
64	Gd		| ↓| ↓| ↓| ↓| ↓| ↓| ↓|	7/2	0	7/2	0.477
65	Tb		|↑↓|↑ |↑ |↑ |↑ |↑ |↑ |	3	3	6	0.461
66	Dy		|↑↓|↑↓|↑ |↑ |↑ |↑ |↑ |	5/2	5	15/2	0.451
67	Ho		|↑↓|↑↓|↑↓|↑ |↑ |↑ |↑ |	2	6	8	0.441
68	Er		|↑↓|↑↓|↑↓|↑↓|↑ |↑ |↑ |	3/2	6	15/2	0.432
69	Tm		|↑↓|↑↓|↑↓|↑↓|↑↓|↑ |↑ |	1	5	6	0.423
70	Yb		|↑↓|↑↓|↑↓|↑↓|↑↓|↑↓|↑ |	1/2	3	7/2	0.414
71	Lu		|↑↓|↑↓|↑↓|↑↓|↑↓|↑↓|↑↓|	0	0	0	0.402

[a] Yttrium is often regarded as a 'non-magnetic' rare-earth with zero quadrupole moment.
[b] Quadrupole moment n_2 [c] $R_{4f} = (<r^2>_{4f})^{1/2}$.

(i) the quadrupole moment of empty, half-filled or filled shells is zero (sphericity), (ii) electrons added to an empty or half-filled shell occupy states with large orbital moment, which corresponds to oblate 'circular current' orbitals and (iii) the removal of electrons from a filled or half-filled shell corresponds to the substraction of an oblate charge distribution from a spherical one, which leaves a prolate remainder.

Putting Eq. (4.3) in Eq. (4.2) reveals that the second-order crystal-field contribution separates into n_2 and an integral which contains all information about the electrostatic crystal field.

4.2.2. *Crystal-Field Energy*. The multipole approach described in the preceeding section is more illustrative, and physically more general, than the widely used point-charge approach based on an expansion of the electrostatic potential (see Section 4.3.1). For historical reasons it is customary to use a notation originating from the point-charge model [15, 85, 86]. For instance,

$$n_2 = \sqrt{\frac{5}{16\pi}}\, \alpha_J \langle O_2^0\rangle \langle r^2\rangle_{4f} \tag{4.4}$$

where $\langle O_2^0\rangle$ is the Stevens operator equivalent, $\alpha_J = \theta_2$ is the second order Stevens coefficient, and $\langle r^2\rangle_{4f}$ the average squared 4f-shell radius (Table 6). Note that $\mathrm{sgn}(\alpha_J) = \mathrm{sgn}(n_2)$, so prolate ions have $\alpha_J > 0$.

Equations (4.1) and (4.2) establish the relation between the macroscopic anisotropy constants and the microscopic electron structure. Including the hexadecapole moment we obtain [15, 81-86]

$$K_1 = -\frac{1}{V_R}\left(\frac{3}{2}\alpha_J A_2^0 \langle O_2^0\rangle \langle r^2\rangle_{4f} + 5\,\beta_J A_4^0 \langle O_4^0\rangle \langle r^4\rangle_{4f}\right) \tag{4.5}$$

and

$$K_2 = -\frac{1}{V_R}\frac{35}{8}\beta_J A_4^0 \langle O_4^0\rangle \langle r^4\rangle_{4f} \tag{4.6}$$

Here V_R is the crystal volume per rare-earth atom, $\langle O_4^0\rangle$ the fourth-order operator equivalent, $\beta_J = \theta_4$ the fourth-order Stevens coefficient, and $\langle r^4\rangle_{4f}$ the radial average of r^4. The quantities α_J, β_J, $\langle r^2\rangle_{4f}$, and $\langle r^4\rangle_{4f}$, as well as the zero-temperature values of $\langle O_2^0\rangle$ and $\langle O_4^0\rangle$ are essentially known for all rare-earths and describe the multipole properties of the rare-earth ions. The information about the intermetallic lattice is contained in the crystal-field parameters A_2^0 and A_4^0

$$A_2^0 = -\frac{5e}{64\pi^2\varepsilon_0\langle r^2\rangle_{4f}}\int \frac{\rho(\mathbf{r}')\,(3\cos^2\Theta - 1)\, f(|\mathbf{r}|)}{|\mathbf{r}' - \mathbf{r}|}\, d\mathbf{r}'\, d\mathbf{r} \tag{4.7a}$$

and

$$A_4^0 = -\frac{9e}{1024\pi^2\varepsilon_0\langle r^4\rangle_{4f}}\int \frac{\rho(\mathbf{r}')\,(35\cos^4\Theta - 30\cos^2\Theta + 3)\, f(|\mathbf{r}|)}{|\mathbf{r}' - \mathbf{r}|}\, d\mathbf{r}'\, d\mathbf{r} \tag{4.7b}$$

Note that the parametrization of crystal fields is not standardized; the notation implied in Eq. (4.7) is widely used in literature (cf. [82, 84, 85]).

4.2.3. *Temperature Dependence.* To realize macroscopic anisotropy, the highly anisotropic rare-earth sublattice must be coupled to the transition-metal sublattice. The interaction energy between the two sublattices scales as $E_{RT} = \mu_B n_{RT} M_T$ where n_{RT} describes the exchange coupling between rare-earth and transition-metal sublattices, and M_T is the magnetization of the transition-metal sublattice. With the typical value $n_{RT} \approx 300\ \mu_o$ [13, 53] we find that E_{RT}/k_B is of order 300 K. This indicates moderately strong inter-sublattice coupling, and thermal excitations are very effective in destroying anisotropy. Illustratively speaking, due to thermal excitation the orientation of the 4f orbitals becomes random and the asphericity of the 4f shell vanishes. Compared to this mechanism, the temperature variation of the crystal field is negligible.

In more detail, there are 2J + 1 possible magnetization orientations m = J, J - 1, ... - J per rare-earth ion, and the statistical problem reduces to the calculation of the average multipole moments $\langle n_k \rangle = Z^{-1} \Sigma_m n_k(m) \exp(-c\, m\, E_{RT}/k_B T)$. Here c is a numerical factor which describes the spacing between the energy levels: $c = 2\, S/(J + \eta)$, where $\eta = 1$ for the first half of the rare-earth series, and $\eta = 0$ for the second one including Gd.

Table 7. Low-temperature multipole plateaus.[a]

	c	$T_p(n_2)$ K	$T_p(n_4)$ K	$T_p(n_6)$ K
Nd	0.55	116	71	57
Sm	1.43	232	156	-
Dy	0.67	194	104	81
Tm	0.33	83	48	38

[a] T_p is the temperature at which the lowest-order correction to n_k or $\langle O_k^0 \rangle$ reaches 10% of the zero temperature value. For simplicity we assume $\mu_o M_T = 2$ T and $n_{RT} = 300\ \mu_o$.

Due to the finite energy spacing, the multipole moments $\langle n_k \rangle$ approach a plateau at low temperatures [82]. The functional structure of c, reveals that this plateau is most pronounced for rare earths with large S or small J (Table 7). Samarium has a comparatively large c, which favourably influences the $Sm_2Fe_{17}N_3$ anisotropy at low and moderate temperatures [64]. Still, it should not be sweeped under the carpet that a detailed and comprehensive investigation of the temperature dependence of the anisotropy remains a demanding experimental and theoretical task.

Compared to the quadrupole moment, higher-order multipole moments exhibit a more pronounced temperature dependence [15, 81, 82]. Higher-order multipole moments describe, for instance, whether a prolate orbital looks like a bone or rather like a stretched snake with a rabbit in her stomach, and thermal excitations are extremely effective in smearing out such 'details' of shape, in particular at higher temperatures.

4.3. CRYSTAL FIELD AND INTERSTITIAL MODIFICATION

4.3.1. *Point-Charge Analysis of Interstitial Magnets.* The archetypical approach to deal with the integrals Eq. (4.7) is to assume that the crystal-field generating charges $\rho(\mathbf{r})$ are sufficiently far away from the rare-earth ion so that $\rho(\mathbf{r})$ can be replaced by $\rho(\mathbf{r}) = Q\,\delta(\mathbf{r} - \mathbf{R})$. Here Q and **R** are the charge and the position of the crystal-field generating atom, respectively. Note that Eq. (4.7) obeys the superposition principle [84] so we can restrict ourselves to the case of a single atom; the total crystal field is obtained by adding up all single-atom contributions. After some calculations we obtain

$$A_2{}^0(\mathrm{p.c.}) = -\frac{eQ}{4\pi\varepsilon_o R^3}\,\frac{1}{4}\,(3\cos^2\Theta - 1) \qquad (4.8a)$$

and

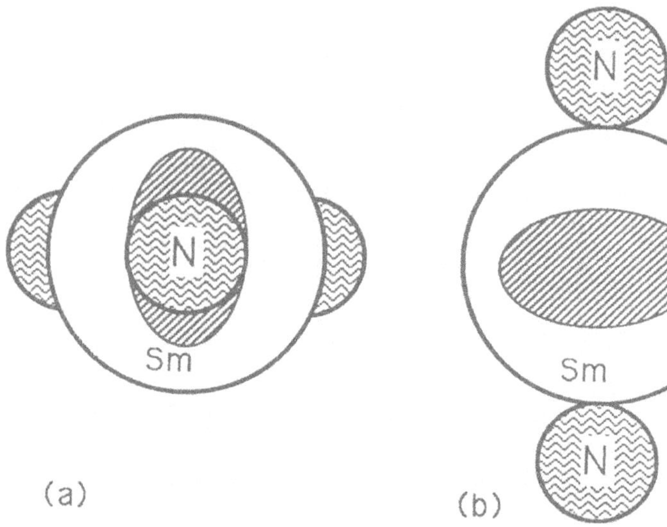

Fig. 10. The coordination of interstitial nitrogen in (a) $Sm_2Fe_{17}N_3$ and (b) $Sm(Fe_{11}Ti)N$. The Sm^{3+} ion is prolate, so the electrostatic repulsion between 4f shell and the negatively charged nitrogen atoms yields easy-axis behaviour in the 2:17 nitride but easy-plane magnetization in the 1:12 nitride [87, 88].

$$A_4{}^0(\mathrm{p.c.}) = -\frac{eQ}{4\pi\varepsilon_o R^5}\,\frac{1}{256}\,(35\cos^4\Theta - 30\cos^2\Theta + 3) \qquad (4.8b)$$

An equivalent way to derive Eq. (4.8) is to assume that the electrostatic potential at the rare-earth site obeys $\nabla^2 V(\mathbf{r}) = 0$ [84, 85].

As a matter of fact, Eq. (4.8) is not a good approximation to describe metallic magnetism. The point-charge model neglects the screening of localized charges by conduction electrons, and physically reasonable point charges yield much-too-high crystal-field parameters (see Section 4.2.2). Simply speaking, there is no point in describing delocalized electrons as point charges, and explaining measured magnetization curves and crystal-field parameters in terms of fictious point charges is equivalent to a mere reparametrization of the crystal field [84].

Nevertheless, Eq. (4.8) gives a qualitative explanation of anisotropy trends in interstitial magnets [2, 9, 54, 87]:

(i) The strong distance dependence of the crystal-field parameters means that the leading crystal-field contributions are those of the neighbouring atoms. This refers in particular to higher order crystal-field parameters.

(ii) The crystal-field parameters, and with it the magnetocrystalline anisotropy, depend on the coordination angle Θ between crystallographic c axis and point-charge position. Moving a crystal-field generating charge from $\Theta = 0$ to $\Theta = 90°$ changes the sign of its crystal-field contribution (See Fig. 10).

(iii) If sign and position of the point charge are known, we can predict whether K_1 increases upon interstitial modification (Eqs. (4.5) and (4.8a)). To simplify the confusing binary algebra we can use the 'sign rule' $\mathrm{sgn}(\Delta K_1) = \mathrm{sgn}(Q)\,\mathrm{sgn}(\alpha_J)\,\mathrm{sgn}\{\Theta\}$, were $\mathrm{sgn}\{0\} = 1$ and $\mathrm{sgn}\{90°\} = -1$.

First, let us assume that the charge of interstitial nitrogen is negative — most lattice atoms and interstitials turn out to bear *negative* effective point charges [84]. An exception is hydrogen which tends to be neutral, or even positive in a sufficiently electronegative environment [35, 89], but there is no reason to believe that nitrogen atoms loose strongly-bound electrons to a electropositive lattice (cf. Section 4.3.2). In the case of samarium compounds, $\mathrm{sgn}(\alpha_J) = 1$ implies that we have to look for interstitial sites with in-plane coordination $\Theta = 90°$ to obtain useful permanent magnetic properties. This is precisely the case for the 9e sites in Sm_2Fe_{17}, and the anisotropy changes from easy-plane in Sm_2Fe_{17} to strong easy-axis in $Sm_2Fe_{17}N_3$ (Fig. 10).

There are other rare earths with $\mathrm{sgn}(\alpha_J) = 1$, but in these cases the positive anisotropy contribution due to nitrogen is not large enough to overcome the easy-plane anisotropy of the parent compound and to surpress spin-reorientation transitions caused by higher-order anisotropy contributions [2, 9, 10, 12]. The 2b site in $SmFe_{11}Ti$ has axial coordination with $\Theta = 0$, and the presence of nitrogen deteriorates the favourable easy-axis anisotropy of the parent compound (see Fig. 10 and [9, 54, 82]). In turn, the unfavourable anisotropy of $RFe_{11}Ti$ compounds with $\alpha_J(R) < 0$ improves upon nitrogenation, but the overall intrinsic properties of 1:12 nitrides and carbides remain inferior to those of $Sm_2Fe_{17}N_x$ and $Sm_2Fe_{17}C_x$.

4.3.2. *Screening*. It is difficult to extract direct physical information from Eq. (4.7). A tool to reduce the six-fold integrals Eq. (4.7) to simple three-dimensional ones are crystal-field weight functions $W_k(r)$. For instance [87]

$$W_2(R) = \frac{e}{16\pi\varepsilon_o \langle r^2\rangle_{4f}} \left(\int_0^R \frac{\xi^4}{R^3} f(\xi)\, d\xi + \int_r^\infty \frac{R^2}{\xi} f(\xi)\, d\xi \right) \tag{4.9}$$

so that $A_2^0 = -\int (3\cos^2\Theta - 1)\, W_2(r)\, \rho(\mathbf{R})\, d\mathbf{R}$. Apart from the 'trivial' coordination factor $3\cos^2\Theta - 1$, $W_2(r)$ gives the A_2^0 contribution of a (selfconsistently) determined charge element $\rho(\mathbf{R})\, d\mathbf{R}$. It turns out [87, 90] that $W_2(R)$ reproduces the point-charge behaviour at large distances R but becomes zero at $R = 0$, where the point-charge model predicts the crystal field to diverge. This disagreement is due to the fact that the point-charge model neglects the finite size of the rare-earth ion (charge penetration).

Still, the determination of the crystal-field parameters is a complicated task, particularly since the distribution of non-4f charges $\rho(\mathbf{r})$ is not known. A possible way is to calculate the crystal-field parameters numerically using $\rho(\mathbf{r})$ values obtained from band structure calculations [16, 91-93]. It turns out that these calculations reduce the giant predictions of the point-charge model to a physically reasonable order of magnitude.

A different approach is to use the selfconsistent Thomas-Fermi theory [23, 26, 33, 94] to calculate $\rho(\mathbf{r})$ and A_2^0. In this theory [90, 95], the interstitial atoms are assumed to yield a weak perturbation of a homogenous electron gas, which is described by the inverse Thomas-Fermi screening length q. The main results of this theory, which does not contain freely adjustable parameters, are:

(i) The distance dependence of the crystal-field parameters A_m^n is given by exp(-qR)/R, as opposed to the prediction $\exp(-qR)/R^{m+1}$ for a screened point charge. Note that the semiempirical bonding-charge model [96] predicts $\exp(-qR) = \exp(-qR)/R^0$.

(ii) A prefactor 1/(2m+1)! recovers the intrinsic hierarchy $A_6^n \ll A_4^n \ll A_2^n$.

(iii) Even in the limit of very large distances, the crystal-field parameters depend on the size of the rare-earth ion. For instance , $A_2^0 \propto (1 + q^2 \langle r^4\rangle_{4f} / 14\langle r^2\rangle_{4f})$ or, in hydrogenic approximation, $A_2^0 \propto (1 + 0.105\, q^2 \langle r^2\rangle_{4f})$.

(iv) The crystal-field charge, which is the Thomas-Fermi analogue to the point charge, reads $Q_{cf} = \int \exp(-\, q\mathbf{r}\mathbf{R}/R)\, \rho_L(\mathbf{R} + \mathbf{r})\, d\mathbf{r}$ where $\rho_L(\mathbf{r})$ describes the nucleus of the crystal-field generating atoms and the electrons which are *localized* around this atom. Tentative calculations of this integral show that the q-dependent crystal-field charge Q_{cf} is negative, except hydrogen and (perhaps) boron where it can have either sign [90].

Let us finally discuss a few numerical data. In [82], the interstitial crystal-field contribution in terms of the equation $\Delta A_2^0 = 0.5\, N_I\, (3\cos^2\Theta - 1)\, A'_2$ where N_I is the number of interstitials per rare-earth atom and A'_2 the intrinsic crystal-field contribution per interstitial atom. This 'superposition analysis' [84], which goes beyond the point-charge model, yields the estimates $A'_2 = 200 \pm 60\ Ka_o^2$ for nitrogen in Sm_2Fe_{17} and $A'_2 = 270 \pm 60\ Ka_o^2$ for nitrogen in $SmFe_{11}Ti$. Both values are similiar, which confirms the relevance of our tentative crystal-field analysis and justifies, to some extent, the simple picture presented in Section 4.2.1. Equivalent conclusions refer to R_2Fe_{17} carbides, where the linear dependence of A_2^0 on N_I has been investigated [96, 97].

5. Conclusions

The magnetic and electronic properties of interstitially modified transition-metal-rich rare-earth intermetallics are a scientifically fascinating and technologically important subject. The intrinsic magnetic properties of $Sm_2Fe_{17}N_3$ stand comparison with any other permanent magnetic material and demonstrate how interstitial modification can be used to improve material properties in a systematic manner.

Nitrogen and carbon in R_2T_{17} and RT_{12} intermetallics yield a moderate increase in magnetic moment, which is mainly due to the lattice expansion upon interstitial modification. Chemical effects are of secondary importance and try to reduce the magnetic moment. On the whole, the magnetic moment of interstitially modified rare-earth intermetallics is fairly well understood.

The increase in room-temperature magnetization mainly originates from the Curie temperature enhancement. From a rather phenomenological point of view, the Curie-temperature increase is caused by the lattice expansion on interstitial modification. The microscopic interpretation of this behaviour is much more difficult. The ground state of the intermetallic nitrides is delocalized, but for excitations relevant at T_c the local-moment description is a better starting approximation. Apart from the determination of low-lying excitations, which is quite a demanding quantum-mechanical task, the problem is in the statistical treatment of these excitations. Simple models are able to give non-trivial insight into finite temperature magnetism, but there is still a long way to a comprehensive understanding of Curie temperature trends in 3d ferromagnets.

Interstitial atoms modify the magnetocrystalline anisotropy by changing the electrostatic crystal field acting on the rare-earth 4f electrons. Models as simple as the point-charge model give a qualitative explanation of these changes, and the inclusion of charge penetration yields a physically reasonable, semi-quantitative description of the crystal-field interaction. Still, a detailed quantitative understanding has not yet been achieved. Open questions are, for instance, the influence of exchange interaction and 'non-linear' effects such as the deformation of the rare-earth 4f shell due to very strong crystal fields.

Acknowledgements

The author is indebted to J. M. D. Coey, S. Brennan, and Q.-N. Qi for stimulating discussions and help in details. This work forms part of the BRITE/EURAM program of the European Commission.

Appendices

APPENDIX A: GAUSSIAN FUNCTIONAL INTEGRALS

From the mathematical point of view, functional integrals (see Section 3.2) are a continuous generalization of multiple integrals and can be written as

$$\int f(s(\mathbf{r}))\, Ds(\mathbf{r}) = \iint \ldots \int f(s_1, s_2, \ldots, s_N)\; ds_1 ds_2 \ldots ds_N\,, \qquad N = \infty \qquad (A.1)$$

if we denote the value of the spin variable s at a given point by $s(\mathbf{r}_i) = s_i$.

Once F is known, thermodynamical observables are obtained in form of *functional derivatives*. For instance, the (reduced) magnetization, reads $\langle s(\mathbf{r})\rangle = - \delta F/\delta h(\mathbf{r})$, the susceptibility

$$\chi(\mathbf{r}, \mathbf{r}') = - \frac{\delta^2 F}{\delta h(\mathbf{r}) \delta h(\mathbf{r}')} \qquad (A.2)$$

and correlation function $\langle s(\mathbf{r})\, s(\mathbf{r}')\rangle = \langle s(\mathbf{r})\rangle \langle s(\mathbf{r}')\rangle + T\, \chi(\mathbf{r}, \mathbf{r}')$ where $h(\mathbf{r})$ denotes the local magnetic field. The derivation of these formulas is trivial if we bear in mind that the functional derivative is the continuous generalization of the partial derivative. Replacing $\delta \ldots /\delta h(\mathbf{r})$ by $\partial \ldots /\partial h_i$ reproduces the well-known results of the Ising model [46].

Functional integrals can be calculated explicitly if the exponent - $\mathbb{H}/k_B T$ is a quadratic expression (Gaussian functional integrals)

$$\mathbb{H} = \frac{1}{2} \int J(\mathbf{r} - \mathbf{r}')\, s(\mathbf{r})\, s(\mathbf{r}')\, d\mathbf{r}\, d\mathbf{r}' - \int h(\mathbf{r})\, s(\mathbf{r})\, d\mathbf{r} \tag{A.3}$$

To do this we remember that quadratic expressions diagonalize by a a suitable unitary transformation. As a consequence of the diagonalization the functional integral factorizes and we can use the identity

$$\int_{-\infty}^{\infty} e^{-0.5\,\alpha\, x^2 + \gamma\, x}\, dx = \sqrt{\frac{2\pi}{\alpha}}\; e^{0.5\,\gamma^2/\alpha} \tag{A.4}$$

to realize the integration. After some elementary calculation we obtain

$$F = -\frac{1}{2} \int \chi(\mathbf{r}, \mathbf{r}')\, h(\mathbf{r})\, h(\mathbf{r}')\, d\mathbf{r}\, d\mathbf{r}' \tag{A.5}$$

where the susceptibility $\chi(\mathbf{r}, \mathbf{r}')$ satisfies $\int \chi(\mathbf{r}, \mathbf{r}')\, J(\mathbf{r}', \mathbf{r}'')\, d\mathbf{r}' = \delta(\mathbf{r} - \mathbf{r}'')$. The average magnetization $\langle s(\mathbf{R})\rangle$ is given by the equation of state $\langle s(\mathbf{R})\rangle = \int \chi(\mathbf{R}, \mathbf{r})\, h(\mathbf{r})\,)d\mathbf{r}$.

In the paramagnetic regime, i.e. for $T > T_c$, a small symmetry-breaking magnetic field $h(\mathbf{r})$ is not very effective in magnetizing the sample and $\langle s(\mathbf{R})\rangle \approx 0$. At $T \le T_c$, however, the susceptibility diverges and an infinitesimally small field generates a macroscopically large spontaneous magnetization. An alternative formulation of this criterion is that spontaneous magnetization starts if the lowest eigenvalue of the operator $J(\mathbf{r}, \mathbf{r}')$ equals zero.

An important case is the quasi-local interaction [48, 49]

$$\int J(\mathbf{r} - \mathbf{r}')\, s(\mathbf{r})\, s(\mathbf{r}')\, d\mathbf{r}\, d\mathbf{r}' = \int \left(A\, [\nabla\, s(\mathbf{r})]^2 + a_{eff}\, s^2(\mathbf{r}) \right) d\mathbf{r} \tag{A.6}$$

where the spin-spin coupling is described by an effective exchange stiffness A. This expression is diagonalized by Fourier transformation. The eigenvalues are $J_{\mathbf{k}} = a_{eff} + A\, \mathbf{k}^2$, so the system becomes unstable at $\mathbf{k} = 0$ and $a = 0$. After some calculation [48] the average squared magnetization, i.e. the squared moment [24], is obtained

$$\langle s^2(\mathbf{r})\rangle = \frac{k_B\, T}{2\, \pi^2} \int_0^{k_o} \frac{k^2\, dk}{a_{eff} + A\, k^2} \tag{A.7}$$

Here k_o is a cut-off length which has to be determined externally. The most interesting point about Eq. (A.7) is that $\langle s^2(\mathbf{r})\rangle$ remains finite at T_c while $\langle s(\mathbf{r})\rangle$ vanishes. Putting a = 0 in Eq. (A.7) we obtain

$$\langle s^2(\mathbf{r}, T = T_c)\rangle = \frac{k_B\, T_c\, k_o}{2\, \pi^2\, A} \tag{A.8}$$

APPENDIX B: THE BOGOL'UBOV INEQUALITY

Variational Free Energy. The calculation of the free energy $F = - k_BT \ln Z$ of a physical system can be very complicated. To treat approximations in a systematic way, it would be nice to have a variational principle similiar to Ritz's procedure in quantum mechanics.

Such a variational free energy is given the equation

$$F \leq F_t = F_o + \langle \mathbb{H} - \mathbb{H}_o \rangle_o \qquad \text{(B.1)}$$

where $\mathbb{H}_o$ is an approximate Hamiltonian and $\langle ... \rangle_o$ the statistical average in terms of $\mathbb{H}_o$. Minimization of F_t with respect to suitable variational parameters in $\mathbb{H}_o$ yields an optimum free energy.

Let us first prove Eq. (B.1), which is known as Bogol'ubov inequality, Peierls inequality, or Feynman inequality [24, 46, 48, 56]. With $\mathbb{H} = \mathbb{H}_o + (\mathbb{H} - \mathbb{H}_o)$, $Z = \sum \exp(-\mathbb{H}/k_BT)$, $Z_o = \sum \exp(-\mathbb{H}_o/k_BT)$ and $\langle ... \rangle_o = Z_o^{-1} \sum ... \exp(-\mathbb{H}_o/k_BT)$ we obtain

$$F = F_o - k_BT \ln \left\langle e^{-(\mathbb{H} - \mathbb{H}_o)/k_BT} \right\rangle_o \qquad \text{(B.2)}$$

and

$$F_t - F = T \ \ln \left\langle e^{\{\langle(\mathbb{H} - \mathbb{H}_o)\rangle_o - (\mathbb{H} - \mathbb{H}_o)\}/k_BT} \right\rangle_o \qquad \text{(B.3)}$$

Now the inequality $\exp(x) \geq 1 + x$ and the mathematical properties of the logarithmic function yield $F_t - F \geq 0$, which is the sought for proof.

Conventional Mean-Field Theory. The application of the trial Hamiltonian

$$\mathbb{H}_o = - h \sum_i s_i \qquad \text{(B.4)}$$

to the spin-1/2 Ising model yields the well-known selfconsistent mean-field equation $\langle s \rangle = \tanh(z J_o \langle s \rangle / k_BT)$ and the Curie temperature $k_BT = z J_o$ [46].

Situation changes, however, if $s_i \neq \pm 1$ is permitted. This is the case, for instance, in the spin-1 Ising model Eq. (3.5) with $s_i = 0, \pm 1$. Applying Eq. (B.4) and the Bogol'ubov inequality to the Hamiltonian Eq. (3.5), we obtain, after some calculation, the set of selfconsistent equations

$$\langle s \rangle = \frac{\sinh(h/k_BT)}{\frac{1}{2} + \cosh(h/k_BT)} \quad \text{and} \quad h = H + z J_o \langle s \rangle + U_D \langle s \rangle \frac{1 - \langle s^2 \rangle}{\langle s^2 \rangle - \langle s \rangle^2} \qquad \text{(B.5)}$$

In the case of very large U_D the behaviour near T_c is characterized by $\langle s^2 \rangle \approx 1$ and $\langle s \rangle \approx 0$, so $h = H + z J_o \langle s \rangle + U_D (1 - \langle s^2 \rangle) \langle s \rangle$. We see that the effect of U_D is reduced by a factor of order $(1 - \langle s^2 \rangle) \ll 1$. This reduction, which limits the influence of U_D on the Curie temperature, is an example of an Onsager reaction field (see e.g. [24, 93]).

In the case of the spin-1 Ising model we can go even further [58]. Using the trial Hamiltonian

$$\mathbb{H}_o = - u \sum_i s_i^2 - h \sum_i s_i \qquad \text{(B.6)}$$

yields $u = U_M$ and $h = H + z\,J_o <s>$, we obtain the selfconsistent equation of state

$$<s> = \frac{\sinh((H + zJ_o<s>)/k_BT)}{\frac{w}{2}\, e^{-U_M/k_BT} + \cosh((H + zJ_o<s>)/k_BT)} \tag{B.7}$$

where the weight parameter w, which can can be used to mimic details of the underlying band structure, describes multiple occupancy of the $s_i = 0$ state.

Figure 7 shows the local-moment limit $U_D = \infty$ and $w = 4$, which reproduces the customary spin-1/2 Ising model, and a solution with finite U_D. The Curie temperature is obtained by putting $<s> = 0$ in Eq. (B.7). We see that the Curie temperature is only slightly reduced compared to the local moment case, which indicates that the local-moment description works reasonably well in the case of strong and nearly strong ferromagnets. Note that w does not affect the qualitative behaviour of Eq. (B.6).

Renormalization of Gaussian models. Let us finally consider the contineous Hamiltonian

$$\mathbb{H} = \frac{1}{2}\int J(\mathbf{r} - \mathbf{r}')\, s(\mathbf{r})\, s(\mathbf{r}')\, d\mathbf{r}\, d\mathbf{r}' + \int \eta(s(\mathbf{r}))\, d\mathbf{r} - \int h(\mathbf{r})\, s(\mathbf{r})\, d\mathbf{r} \tag{B.8}$$

with

$$\eta(s(\mathbf{r})) = -\frac{1}{2}\, a\, s^2(\mathbf{r}) + \frac{1}{4}\, b\, s^4(\mathbf{r}) + + \frac{1}{6}\, c\, s^6(\mathbf{r}) + \ldots \tag{B.9}$$

To obtain an analytical solution of the functional integral (I.1), we have to replace the energy density Eq. (B.8) by $\eta(s(\mathbf{r})) = a_{eff}\, s^2(\mathbf{r})/2$, where a_{eff} is a trial parameter [48, 49]. Application of the Bogol'ubov inequality to this trial energy density and the complete Hamitonian Eqns. (B.7) and (B.8) yields (cf. [48, 49])

$$a_{eff} = a - 3 <s^2> b - 15 <s^4> c \tag{B.10}$$

In other words, we have to replace s^4 by $\gamma <s^2> s^2$, where the factor $\gamma = 6$.

Note that the Eq. (B.9) represents a scalar result, i.e. $\mathbf{s}(\mathbf{r}) = s(\mathbf{r})\, \mathbf{e}_z$. In D spin dimensions Eq. (B.10) reads $a_{eff} = a - f_D <s^2> b + \ldots$ with $f_D = 1 + 2/D$. In particular, $f_1 = 3$ and $f_3 = 5/3$ (cf. [55]).

APPENDIX C: NON-ANALYTIC POINTS AND MAGNETIC ENERGY

It is well-known that long-wavelength spin-fluctuation theory does not apply to strong and nearly strong ferromagnets where the condition $|\mathbf{k}| \ll k_F$ is violated [48, 55]. Apart from this, there is a statistical reason which prohibits the extrapolation of Eq. (3.7) to strong ferromagnets. Let us first note that it is possible to express the parameters B and b in Eqns. (3.7) and (B.8), respectively, by derivatives of the density of states [49, 55]. In the case of a more complex band structure we could use these parameters and the renormalization procedure (Section 3.2.1) to calculate the Curie temperature Eq. (3.4). Unfortunately, the metallic density of states D(E) and with it the magnetic energy $\mathbb{H}$ involve nonanalytic points such as Van Hove singularities, and it is questionable whether a series expansion such as Eq. (3.7) is physically meaningful.

To investigate the influence of non-analytic points we use the theorem

$$a_{eff} = \frac{(2\pi)^{-1/2}}{<s^2>^{5/2}} \int_{-\infty}^{\infty} \{ s^2 - <s^2> \} \, \eta(s) \, e^{-s^2/(2<s^2>)} \, ds \qquad (C.1)$$

which is obtained by applying the Bogol'bov inequality (Appendix B) directly to the Hamiltonian Eq. (3.2). This theorem can be used to derive Eq. (B.9), but it works equally well if $\eta(s)$ is a non-analytic function, which cannot be expanded in a Taylor series. In real magnets, there exists an upper limit $s \leq s_{max}$, where s_{max} denotes the reduced moment of the *fully* spin-polarized magnet. This implies $\eta(s > s_{max}) = \infty$ and, as it can be seen by analysing Eq. (C.1), $a_{eff} = \infty$ unless $s_{max} = \infty$. In very weak itinerant ferromagnets, the smallness of μ_T yields $s_{max} \gg \mu_T/\mu_B$, and the condition $s_{max} = \infty$ is approximately satisfied. In the case of strong and nearly strong ferromagnets, however, this 'excuse' does not work, and it is not possible to deduce physically reasonable values a_{eff} from $\eta(s)$ or from the density of states.

To illustrate these findings, let us look at two simple model ferromagnets. For a rectangular band, the Stoner energy is shown in Fig. 11. The derivatives of $D(E_F)$, and with it the higher coefficients of the Taylor series, vanish while the graph clearly reveals a minimum at $s = s_{max}$. The Hartree-Fock free electron gas behaves similiarly [23, 27] - except the fact that all even derivatives are negative, which is even more bizarre.

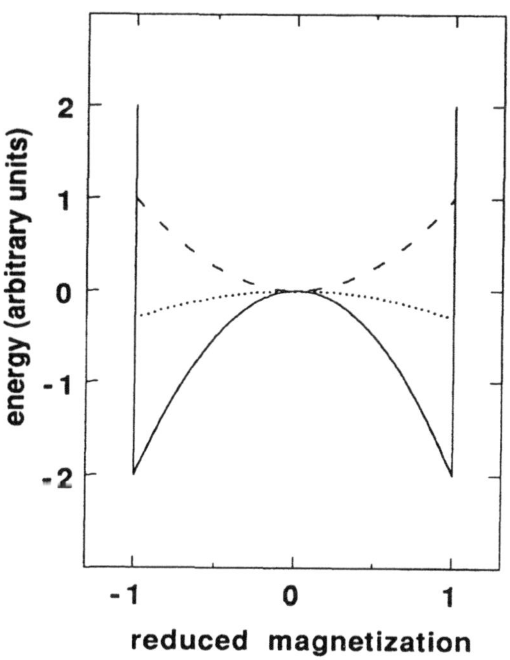

Fig. 11. Stoner criterion and local magnetic energy of a rectangular-band magnet: (solid line) strong ferromagnet, (pointed line) quasi-weak ferromagnet, and (dashed line) paramagnet. The picture reveals the non-analytic behaviour at $s = \pm s_{max}$, which cannot be treated in terms of a series expansion.

References

[1] J. M. D. Coey and H. Sun, J. Magn. Magn. Mater. **87**, L251 (1990).
[2] J. M. D. Coey, H. Sun, and D. P. F. Hurley, J. Magn. Magn. Mater. **101**, 310 (1991).
[3] Y.-C. Yang, X.-D. Zhang, S.-L. Ge, Q. Pan, L.-S. Kong, H.-L.Li, J.-L. Yang, B.-S. Zhang, Y.-F. Ding, and C.-T. Ye, J. Appl. Phys. **70**, 6001 (1991).
[4] Qi-Nian Qi, Y. P. Li, and J. M. D. Coey, J. Phys.: Condens. Matter **4**, 8209 (1992).
[5] J. M. D. Coey, H. Sun, Y. Otani, and D. P. F. Hurley, J. Magn. Magn. Mater. **98**, 76 (1991).
[6] H. Sun, B.-P. Hu, H.-S. Li, and J. M. D. Coey, Solid State Comm. **74**, 727 (1990).
[7] D. P. F. Hurley and J. M. D. Coey, J. Magn. Magn. Mater. **99**, 229 (1991).
[8] M. Katter, J. Wecker, C. Kuhrt, L. Schultz, X. C. Kou, and R. Grössinger, J. Magn. Magn. Mater. **111,** 293 (1992).
[9] J. M. D. Coey, Physica Scripta **T39**, 21 (1991).
[10] H. Sun, J. M. D. Coey, Y. Otani, and D. P. F. Hurley, Phys.: Condens. Matter **2**, 6465 (1990).
[11] J. M. D. Coey and D. P. F. Hurley, J. Magn. Magn. Mater. **104**, 1098 (1992).
[12] J. M. D. Coey and Y. Otani, J. Mag. Soc. Japan **15**, 677 (1991).
[13] H.-S. Li and J. M. D. Coey, "Magnetic Properties of Ternary Rare-Earth Transition-Metal Compounds", in: *Handbook of Magnetic Materials VI*, K. H. J. Buschow, ed., Elsevier, Amsterdam, 1991, p. 1.
[14] J. C. Slater, Phys. Rev. **36**, 57 (1930).
[15] J. M. D. Coey, "Rare-Earth - Iron Permanent Magnets", in: *Current Trends in the Physics of Materials*, G. F. Chiarotti, F. Fumi, and M. P. Tosi, eds., North-Holland, Amsterdam, 1990, p. 265.
[16] R. Coehoorn, "Electronic Structure Calculations for Rare-Earth Transition Metal Compounds" in *Supermagnets, Hard Magnetic Materials*, G. J. Long and F. Grandjean, eds., Kluwer, Dordrecht, 1991, p. 133.
[17] J. F. Herbst, Rev. Mod. Phys. **63** , 819 (1991).
[18] M. Sagawa, S. Hirosawa. H. Yamamoto, S. Fujimura, Y. and Masuura, Jpn. J. Appl. Phys. **26**, 785 (1987).
[19] K. H. J. Buschow, Materials Science Report, **1**, 1 (1986).
[20] R. Skomski and J. M. D. Coey, J. Appl. Phys. **73**, 7602 (1993).
[21] J. M. D. Coey, R. Skomski, and S. Wirth, IEEE Trans. Magn. **28**, 1992 (2332).
[22] J. S. Griffith, *The Theory of Transition-Metal Ions*, University Press, Cambridge, 1964.
[23] W. Jones and N. H. March, *Theoretical Solid-State Physics I*, Wiley & Sons, London, 1973.
[24] P. Fulde, *Electron Correlations in Molecules and Solids*, Springer, Berlin, 1991.
[25] F. Bloch, Z. Phys. **57**, 545 (1929).
[26] N. W. Ashcroft and N. D. Mermin, *Solid State Physics*, Holt, New York, 1976.
[27] S. Blügel, in *24. IFF-Ferienkurs*, IFF, Jülich, 1993, ch. 6.
[28] J. F. Janak, Phys. Rev. B **16**, 255 (1977).
[29] H. Ibach and H. Lüth, *Solid-State Physics*, Springer, Berlin, 1993.
[30] J. Emsley, *The Elements*, Clarendon, Oxford, 1989.
[31] R. Zeller, in *24. IFF-Ferienkurs*, IFF, Jülich, 1993, ch. 18.

[32] J. D. Fast, *Gases in Metals*, MacMillan, London, 1976.
[33] N. F. Mott and H. Jones, *The Theory of the Properties of Metals and Alloys*, Clarendon, Oxford, 1936.
[34] A. C. Switendick, "The Change in Electronic Properties on Hydrogen Alloying and Hydride Formation" in *Hydrogen in Metals I*, G. Alefeld and J. Völkl, eds., Springer, Berlin, 1978, p.101.
[35] W. E. Wallace, "Magnetic Properties of Metal Hydrides and Hydrogenated Intermetallic Compounds" in *Hydrogen in Metals I*, G. Alefeld and J. Völkl, eds., Springer, Berlin, 1978, p.169.
[36] J. Friedel, Adv. Phys. **3**, 446 (1954).
[37] S. S. Jaswal, IEEE Trans. Magn. **28**, 2322 (1992).
[38] Y. P. Li, H.-S. Li, and J. M. D. Coey, phys. stat. sol. (b) **166**, K107 (1991).
[39] S. S. Jaswal, W. B. Yelon, G. C. Hadjipanayis, Y. Z. Wang, and D. J. Sellmyer, Phys. Rev. Lett. **67**, 644 (1991).
[40] Q.-N. Qi and J. M. D. Coey, to be published.
[41] Y.-P. Li and J. M. D. Coey, Solid State Comm. **81**, 447 (1992).
[42] T. Beuerle and M. Fähnle, J. Magn. Magn. Mat. **110**, L29 (1992).
[43] T. Beuerle, P. Braun, and M. Fähnle, J. Magn. Magn. Mater. **94**, L11 (1991).
[44] Q.-N. Qi, H. Sun, R. Skomski, and J. M. D. Coey, Phys. Rev. B **45**, 12278 (1992).
[45] M. Fähnle and T.Beuerle, phys. stat. sol. (b) **177**, K95 (1993).
[46] J. M. Yeomans, *Statistical Mechanics of Phase Transitions*, Clarendon, Oxford, 1992.
[47] C. Itzykson and J.-M. Drouffe, *Statistical Field Theory I*, University Press, Cambridge, 1989.
[48] K. K. Murata and S. Doniach, Phys. Rev. Lett. **29**, 285 (1972).
[49] M. Shimizu, Rep. Prog. Phys. **44**, 329 (1981).
[50] T. Moriya, J. Magn. Magn. Mater. **100**, 261 (1991) .
[51] P. Mohn and E. P. Wohlfarth, J. Phys. F: Met. Phys. **17**, 2421 (1987).
[52] J. P. Gavigan, D. Givord, H. S. Li, and J. Voiron, Physica B **149**, 345 (1988).
[53] N. H. Duc, T. D. Hien, D. Givord, J. J. M. Franse, and F. R. de Boer, J. Magn. Magn. Mater. **124**, 305 (1993).
[54] B.-P. Hu, H.-S. Li, J. P. Gavigan, and J. M. D. Coey, J. Phys.: Condens. Matter **1**, 755 (1989).
[55] G. G. Lonzarich and L. Taillefer, J. Phys. C: Solid State Phys. **18**, 4339 (1985).
[56] B. Kirchner, W. Weber, and J. Voigtländer, J. Phys.: Condens. Matter **4**, 8097 (1992).
[57] Y. Kakehashi and P. Fulde, Phys. Rev. B **32**, 1595 (1985).
[58] J. Hubbard, Phys. Rev. B **19**, 2626 (1979).
[59] H. Hasegawa, J. Phys. Soc. Japan **46**, 1504 (1979).
[60] B. L.. Gyorffy, A. J. Pindor, J. Staunton, G. M. Stocks, and H. Winter, J. Phys. F **15**, 1337 (1985).
[61] J. Staunton and B. L. Gyorffy, Phys. Rev. Lett. **69**, 371 (1992).
[62] Y. Kakehashi, Phys. Rev. B **34**, 3243 (1986).
[63] R. E. Prange and V. Korenman, Phys. Rev. B **19**, 4691 (1979).
[64] K.-H. Müller, D. Eckert, P. A. P. Wendhausen, A. Handstein, S. Wirth, and M. Wolf, IEEE Trans. Magn., in press (1994).
[65] M. Brouha and K. H. J. Buschow, J. Appl. Phys. **44**, 1813 (1973).
[66] Q.-N. Qi, R. Skomski and J. M. D. Coey, J. Phys.: Condens. Matter **6**, 3245 (1994).

[67] R. Coehoorn, Phys. Rev. B **39**, 13072 (1989).
[68] J. F. Herbst, J. J. Croat, R. W. Lee, and W. B. Yelon, J. Appl. Phys. **53**, 250 (1982).
[69] J. C. Slater, Rev. Mod. Phys. **25**, 199 (1953).
[70] D. R. Haman, Phys. Rev. Lett. **23**, 95 (1969).
[71] W. E. Evenson, J. R. Schrieffer, and S. Q. Wang, J. Appl. Phys. **41**, 1199 (1970).
[72] R. Dederichs, in *24. IFF-Ferienkurs*, IFF, Jülich, 1993, ch. 27.
[73] S. H. Liu, Phys. Rev. B **15**, 4281 (1977).
[74] Y. Kakehashi, Phys. Rev. B **43**, 10820 (1991).
[75] Y. Kakehashi, Phys. Rev. B **47**, 3185 (1993).
[76] V. Heine, Phys. Rev. **153**, 673 (1967).
[77] M. Brouha, K. H. J. Buschow, and A. R. Miedema, IEEE Trans. Magn. **10**, 182 (1974).
[78] N. D. Lang and H. Ehrenreich, Phys. Rev. **168**, 605 (1968).
[79] S. Jaakkola, S. Parviainen, and S. Penttilä, J. Phys. F: Met. Phys. **13**, 491 (1983).
[80] D. M. Edwards and E. P. Wohlfarth, Proc. R. Soc. **A 303**, 127 (1968).
[81] J. M. D. Coey, "Intermetallic Compounds and Crystal-Field Interaction", in: *Science and Technology of Nanostructured Materials*, G. C. Hadjipanayis and G. A. Prinz, eds., Plenum Press, New York, 1991, p. 439.
[82] M. D. Kuz'min, Phys. Rev. B **46**, 8219 (1992).
[83] K. N. R, Taylor and M. I. Darby, *Physics of Rare-Earth Solids*, Chapmán & Hall, London, 1972.
[84] D. J. Newman and B. Ng, Rep. Prog. Phys. **52**, 699 (1989).
[85] M. T. Hutchings, Solid State Phys. **16**, 227 (1964).
[86] A. J. Freeman and R. E. Watson, Phys. Rev. **127**, 2058 (1962).
[87] R. Skomski, M. D. Kuz'min, and J. M. D. Coey, J. Appl. Phys. **73**, 6934 (1993).
[88] H.-S. Li and J. M. D. Coey, J. Magn. Magn. Mater. **115**, 152 (1992).
[89] D. Fruchart and S. Miraglia, J. Appl. Phys. **69**, 5578 (1991).
[90] R. Skomski , Phil. Mag. B **69**, in press (1994).
[91] M. Richter, P. M. Oppeneer, H. Eschrig and B. Johansson, Phys. Rev. B **46**, 13919 (1992).
[92] X.-F. Zhong and W. Y. Ching, Phys. Rev. B **39**, 12018 (1989).
[93] K. Hummler and M. Fähnle, Phys. Rev. B **45**, 3161 (1992).
[94] N. H. March, Adv. Phys. **6**, 1 (1957).
[95] R. Skomski, in *Proceedings of the 8th International Symposium on Magnetic Anisotropy and Coercivity in RE-TM Alloys*, University of Birmingham, in press, (1994).
[96] H.-S. Li and J. M. Cadogan, in *Proceedings of the 7th International Symposium on Magnetic Anisotropy and Coercivity in RE-TM Alloys*, University of Western Australia, Perth, 1992, p. 185.
[97] K. H. J. Buschow, Rep. Prog. Phys. **54**, 1123 (1991).
[98] S. G. Brush, Rev. Mod. Phys. **39**, 883 (1967).
[99] K. H. Fischer and J. A. Hertz, *Spin Glasses*, University Press, Cambridge, 1991.

Chapter 24

SPIN REORIENTATION TRANSITIONS IN INTERMETALLIC COMPOUNDS WITH INTERSTITIAL INCLUSIONS

J. BARTOLOME
Instituto de Ciencia de Materiales de Aragón
CSIC-Universidad de Zaragoza
Facultad de Ciencias
E-50009 Zaragoza
Spain

ABSTRACT. Spin reorientation transitions (SRT) appear in many of the new intermetallic compounds as the temperature varies because of the competition between the rare earth and the transition metal sublattice anisotropies. The possible phase transitions are described in terms of both phenomenological and microscopic models. Some simple approximations are introduced to simplify the analysis of the phase diagrams. The modification and induction of the SRT when H, C, or N atoms occupy interstitial sites are reviewed in the $R_2Fe_{14}Z$; Z=B and C and the R_2Fe_{17} series of compounds. Special attention is given to the interpretation of some anomalies detected by a.c. susceptibility which have been erroneously classified as due to SRT in the past.

1. Introduction

The $R_2Fe_{14}B$ and the R_2Fe_{17} compounds are susceptible to charging with interstitial atoms, such as H, C, or N. Such inclusions modify the magnetic properties of the parent compounds, sometimes in a dramatic fashion, even changing the easy magnetization direction. Some of them undergo a spin reorientation transition, SRT, because of competing anisotropies of the magnetic sublattices, thus the charging with interstitial atoms may, and often does, give rise to variations in these transitions. This chapter will describe the mechanisms producing the SRTs and the effect of interstitials on the transitions.

2. Spin Reorientation Transition

A spin reorientation transition is a phase transition which consists of a change in the easy magnetization direction, EMD. Irrespective of whether this change takes place in an abrupt, first order, or continuous, second order, fashion, it implies a qualitative change in symmetry which is always abrupt. If the temperature is changed, while all other external thermodynamic parameters are maintained constant, the SRT occurs at a transition temperature, T_S. Such a SRT is called spontaneous. There are two levels in the description of the SRT, a macroscopic phenomenological level, and a microscopic level.

2.1. MODELS

The phenomenological model is based on the description of the magnetic anisotropy as a series expansion of the anisotropy free-energy in terms of symmetry adapted functions of the angles determining the direction of the magnetization. For tetragonal symmetry the expression is,

F. Grandjean et al. (eds.), Interstitial Intermetallic Alloys, 599–616.

$$F_a(\theta,\varphi) = K_1 \sin^2\theta + (K_2 + K_2' \sin 4\varphi)\sin^4\theta + (K_3 + K_3' \sin 4\varphi)\sin^6\theta , \qquad (1)$$

whereas for hexagonal symmetry it is

$$F_a(\theta,\varphi) = K_1 \sin^2\theta + K_2 \sin^4\theta + (K_3 + K_3' \sin 6\varphi)\sin^6\theta , \qquad (2)$$

where θ and φ are the spherical angles of the magnetization with respect to the crystallographic axes. The relative values of the anisotropy constants, K_i and K_i', result in different orientations of the magnetization which minimize the anisotropy energy. To illustrate this, we show the different cases which may appear in the tetragonal case. For a single magnetic domain in a single crystal of tetragonal symmetry the free energy is

$$F = F_a + F_H, \qquad (3)$$

where F_a has been described above and

$$F_H = - \mathbf{M_S}.\mathbf{H} = - M_S\, H\left(\sin\theta \sin\alpha \cos(\varphi - \beta) + \cos\theta \cos\alpha\right) \qquad (4)$$

is the Zeeman energy for the field, H, applied in the direction determined by the spherical angles, α and β. The vast majority of the SRT's take place at constant φ. Imposing this condition on the free energy (1), yields,

$$F_a(\theta) = K_1 \sin^2\theta + \overline{K}_2 \sin^4\theta + \overline{K}_3 \sin^6\theta , \qquad (5)$$

where $\overline{K}_2 = K_2 \pm K'_2$ and $\overline{K}_3 = K_3 \pm K'_3$, with "+" for the minimum at $\varphi = 0$ and "–" at $\varphi = \pi/2$. The anisotropy of a hexagonal magnet can be transformed in a similar way. The equilibrium position of the system, is obtained from $(\partial F/\partial\theta)_\varphi = 0$.

The minimum of the free energy can occur at $\theta = 0$, $\pi/2$, or θ_o where $\theta_o = \arcsin\sqrt{R}$, with

$$R = \frac{-\overline{K}_2 + \left[\overline{K}_2^2 - 3 K_1\overline{K}_3\right]^{1/2}}{3\overline{K}_3} . \qquad (6)$$

The possible phases have to fulfill the following conditions, if $\overline{K}_2^2 > 3 K_1\overline{K}_3$ and $0 < R < 1$ the orientation is canted ($\theta \neq 0$). If $\overline{K}_2^2 < 3 K_1\overline{K}_3$, $R \geq 1$ or $R < 0$ then the system is either axial ($\theta = 0$) if $\overline{K}_2 + \overline{K}_3 > K_1$ or basal ($\theta = \pi/2$) if $\overline{K}_2 + \overline{K}_3 < K_1$.

Because the anisotropy constants vary with temperature, the easy direction of magnetization may reorient from axial to basal, axial to canted, or canted to basal. We shall see that the first anisotropy constant, K_1 plays a very important role in many of the SRTs. As an example, a single crystal of $DyFe_{11}Ti$ has been studied by means of magnetization (Hu et al., 1990a), magnetic torque, and susceptibility techniques (García et al., 1993) to ascertain the character of the two SRT present in this compound. Upon cooling below room temperature, a second-order SRT takes place at T_{S1}= 192 K. The actual transition temperature may be deduced by taking into account that the SRT may be considered a Landau type of second-order transition with θ as an order parameter, in which case one expects that below T_S, $\theta^2 \propto (T - T_{S1})$. Indeed, this behavior is fulfilled by $DyFe_{11}Ti$, as is seen in Fig. 1. The intersection with the abscissa of the extrapolated linear dependence gives the best estimate of T_{S1}. The transition is due to K_1 approaching zero at T_{S1}, but, because in this case $K_2' > 0$, the EMD cants away from the c-axis but is always contained in the (110) planes. The angle θ increases continuously to $\theta = 42^o$ as the temperature is lowered to T_{S2} = 69K. At this temperature a first-order SRT takes place and the $\theta = 42^o$

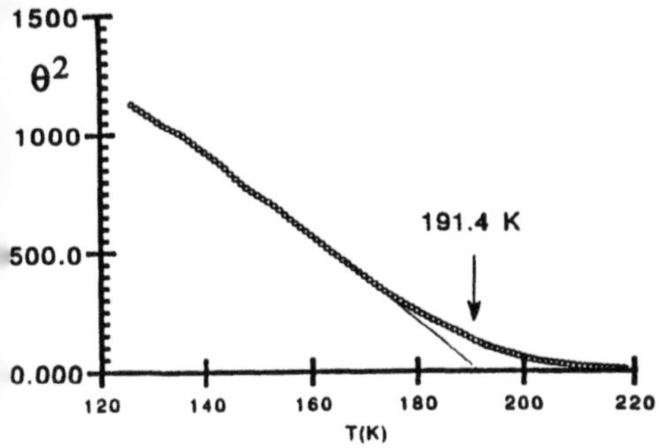

Figure 1. θ^2 versus T of the second-order SRT in $DyFe_{11}Ti$, demonstrating the Landau-like behavior of the EMD canting angle, (García et al., 1993).

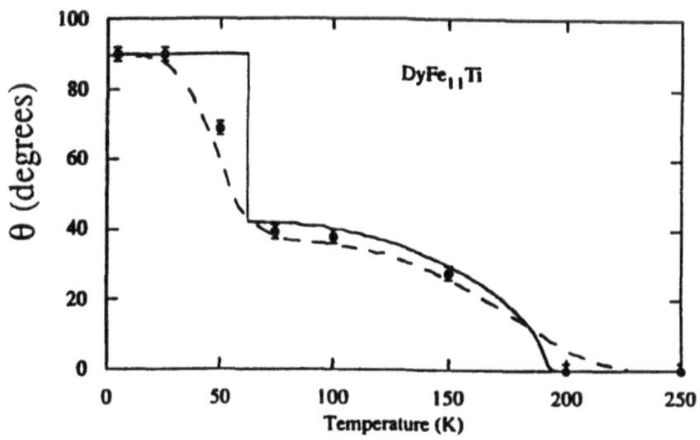

Figure 2. Canting angle of the EMD of $DyFe_{11}Ti$, as a function of temperature. T_{S1} = 192 K (second-order SRT), T_{S2} = 69 K (first-order SRT). (- -) deduced from a.c. susceptibility measurements, (•) from torque measurements, (–) theoretical predictions. (García et al., 1993)

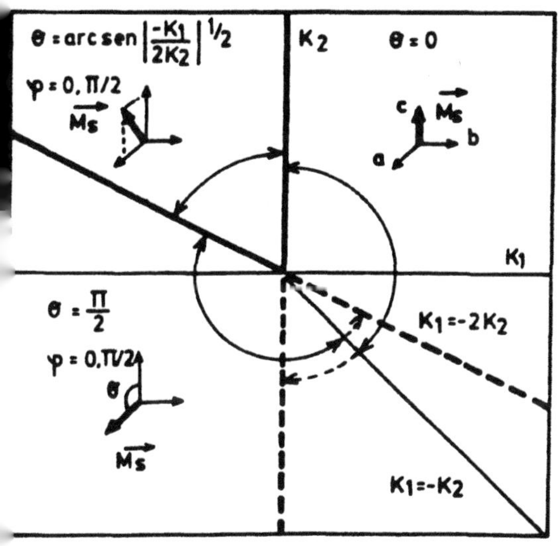

Figure 3. Magnetic phase diagram of a system with cylindrical symmetry, derived from expression [7].Thick lines represent second-order, and dashed lines represent first-order SRTs. Between the two dashed lines there is a metastable region.

phase coexists with the low-temperature $\theta = 90^o$ phase over a wide temperature region. K_1 remains negative throughout the T_{S2}, Fig. 2 (Algarabel et al., 1994).

At high temperature, when $\overline{K}_3$ is negligible, the analysis is much simpler. Expression (6) tends to

$$\theta_o = \arcsin\sqrt{\frac{-\overline{K}_1}{2\overline{K}_2}}, \tag{7}$$

for $\overline{K}_3 \rightarrow 0$, where we shall drop the bar for simplicity. For $K_2 > 0$, two second-order SRT take plac; axial to angular when $K_1 = 0$ and angular to basal when $K_1 + 2K_2 = 0$. For $K_2 < 0$, one first-order SRT takes place when $K_1 + K_2 = 0$. A $\theta = 90^o$ metastable phase exists in the region between $K_1 + K_2 = 0$ and $K_1 + 2K_2 = 0$ and a second $\theta = 0^o$ metastable phase appears between $K_1 + K_2 = 0$ and $K_1 = 0$, see Fig. 3 (Smit and de Wijn, 1959).

Although the phenomenological model is helpful to describe the SRT phenomena, it does not explain why K_i vary with temperature. To achieve this a microscopic model is required. It gives a good account of the SRT temperatures, magnetization curves, and thermodynamic functions. The types of the interactions involved have been described in Chapter 22 in this book. We shall apply in the following a molecular field, MF, model so that the anisotropy may be well described in a single-ion scheme.

The exchange interaction between the rare-earth, R, sublattice and the Fe sublattice is described by the Hamiltonian,

$$\begin{aligned} H_{ex}^{R} &= 2(g_J - 1)\mu_B \mathbf{H}_{ex} . \mathbf{J} \\ &= -2(g_J - 1)\mu_B n_{RFe} M_{Fe}(T)\left(J_x \sin\theta \cos\varphi + J_y \sin\theta \sin\varphi + J_z \cos\theta\right), \end{aligned} \tag{8}$$

where $\mathbf{H}_{ex}$ is the exchange field acting on the R ion with J and g_J, thus describing the magnetic interaction between the 4f and 3d spins. The temperature dependence of $\mathbf{H}_{ex}$ is taken as proportional to the Fe sublattice magnetization, which, in turn, is derived from the measurements of the isostructural Y, La, or Lu compounds scaled to T_C for the particular compound. J_x, J_y, and J_z are the operators of the Cartesian components of the R angular momentum. This Hamiltonian is added to the crystal field Hamiltonian,

$$H_{CF} = \sum_{n,m} \theta_n \langle r^n \rangle A_n^m O_n^m = \sum_{n,m} B_n^m O_n^m , \tag{9}$$

where O_n^m are the Stevens operator equivalents, and θ_n the Stevens constants (Stevens, 1952). An alternative notation also used is $\theta_2 = \alpha_j$. It results in the Hamiltonian $H^R(i) = H_{CF}^R(i) + H_{EX}^R(i)$, where i runs over the different rare-earth sites. The matrix elements, $\langle J,J_z | H^R(i) | J,J_{z'} \rangle$, are calculated and, after diagonalization of the resulting complex Hermitian matrix, we obtain the eigenvalues and eigenvectors. This is done for each T, θ, and φ. The total free energy of the system is given by,

$$F(T,\theta,\varphi) = -kT \sum_i \ln Z(i) + F_{Fe}(\theta), \tag{10}$$

where Z(i) is the canonical partition function and the second term is the Fe sublattice anistropy energy, $F_{Fe} = p_{Fe} K_1 \sin^2\theta$, where p_{Fe} is the number of Fe atoms per formula unit. The equilibrium direction, θ_m, φ_m of the Fe sublattice magnetization is determined for each T as the angles that minimize the total free energy. By substituting the values of θ_m and φ_m into expression (10), we obtain the thermodynamic free energy, F(T), and from it the magnetic and thermodynamic functions may be derived. In this way we have explained quantitatively the first-order and second-order SRT's in $R_2Fe_{14}B$, where R = Nd, Ho, Er, and Tm (Piqué et al., 1994) see Fig. 4a, b, c, and d.

This method is quite rewarding if one knows all the parameters because one may account for the magnetic properties of the compounds. But it is not too convenient when trying to obtain a qualitative understanding of the effect of the interstitial atoms on the different parameters involved, which is one of the main goals of this chapter.

2.2. SOME APPROXIMATIONS

The SRT temperatures are easy to measure and provide a good probe of the effect of the interstitial atoms. The comparison of the measured T_S with the dependence of the predicted values on the exchange and crystal field parameters supplies, in some cases, information on how the interstitial atoms modify these parameters.

The transition temperature, T_S, is obtained in theory, from the condition imposed on the anisotropy constants by the type of transition, in accordance with the description made in Section 2.1. In the transitions present in the $R_2Fe_{14}B$ compounds the condition for a second-order SRT is that $K_1(T_S) = 0$, and for a-first order SRT that $K_1(T_S) + K_2(T_S) = 0$. In the second case the transition temperatures are high and one may apply the approximation that $K_2(T)$ is negligible. As a consequence, if the parameters are known, in both cases the T_S may be found by solving the equation,

$$K_1^{Fe}(T_S) = -K_1^{R}(T_S), \tag{11}$$

because $K_1 = K_1^{Fe} + K_1^{R}$. Conversely, one may obtain information on the parameters if T_S is known. $K_1^{Fe}(T)$ is derived from experiment on isostructural compounds with a non-magnetic rare earth. As mentioned above, a proper temperature scaling is necessary to match the Curie temperature.

The anisotropy constant due to the rare earth sublattice may be derived, in a linear theory, applying the canonical ensemble statistics, one obtains for the tetragonal case (Rudowicz 1985; Lindgard and Danielsen 1975),

$$K_1^R(T) = -N_R\left[\frac{3}{2}B_2^o\langle O_2^o\rangle + 5B_4^o\langle O_{40}\rangle + \frac{21}{2}B_6^o\langle O_6^o\rangle\right], \tag{12}$$

$$K_2^R(T) = \frac{7}{8}N_R\left[5\,B_4^o\langle O_4^o\rangle + 27\,B_6^o\langle O_6^o\rangle\right],\; K'^R_2(T) = \frac{1}{8}N_R\left[B_4^{4c}\langle O_4^o\rangle + 5\,B_6^{4c}\langle O_6^o\rangle\right],$$

$$K_3^R(T) = -\frac{231}{16}N_R B_6^o\langle O_6^o\rangle,\; K'^R_3(T) = -\frac{11}{16}N_R B_6^{4c}\langle O_6^o\rangle,$$

where N_R are the number of atoms in the rare earth sublattice and $\langle O_m^o\rangle$ represents the thermal average of the Stevens operator (Parker, 87). Though this approximation has the advantage of being applicable in a rather broad temperature range, we have found it cumbersome to use because the parameter dependence is imbedded in the unavoidable statistical computer calculation.

A different approximation may be used (Kuz'min 1992) when T_S is high enough in temperature such that, near T_S, the ratio $\Delta_{ex}/kT << 1$, where $\Delta_{ex} = 2|g_J - 1|\mu_B B_{ex}$ is the exchange splitting between two successive energy levels. In this case, all other anisotropy constants are of higher order in 1/T and thus can be neglected. The expression for the second-order anisotropy constant for the rare earth atoms is

$$K_1^R = \frac{1}{20}J(J+1)(2J-1)(2J+3)B_2^o\left(\frac{\Delta_{ex}}{kT}\right)^2 + O(T^{-4}), \tag{13}$$

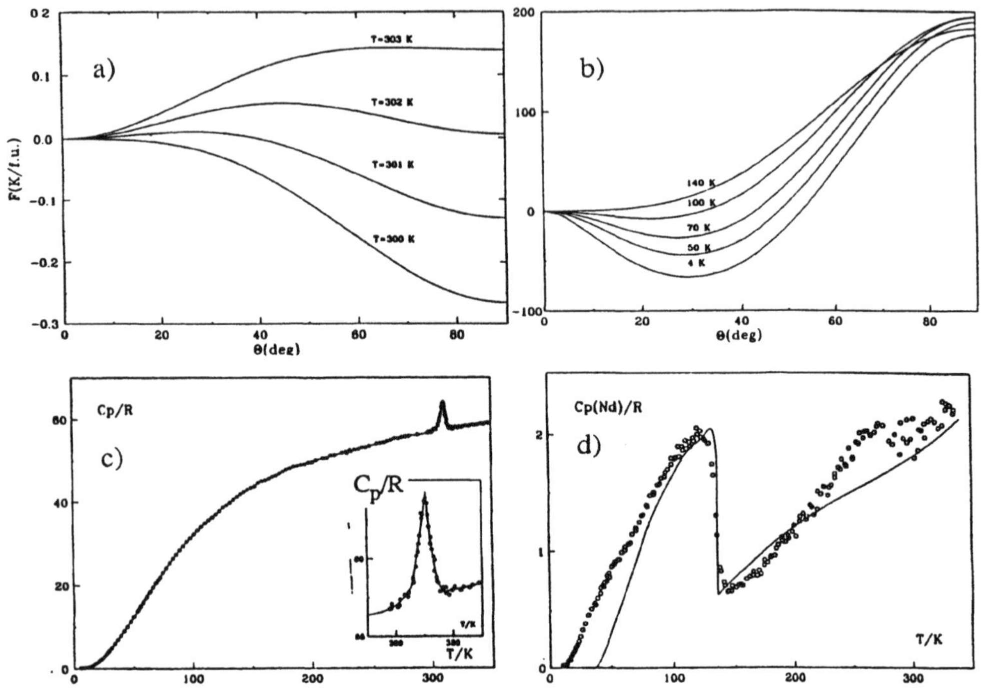

Figure 4. Free energy curves of $R_2Fe_{14}B$ versus canting angle for different temperatures: a) R=Er, with two simultaneous minima (first-order SRT) b) R = Nd, with only one minimum (second order SRT). c) heat capacity data of $Er_2Fe_{14}B$. d) heat capacity data of $Nd_2Fe_{14}B$, (full line) theoretical prediction. (Piqué et al., 1994).

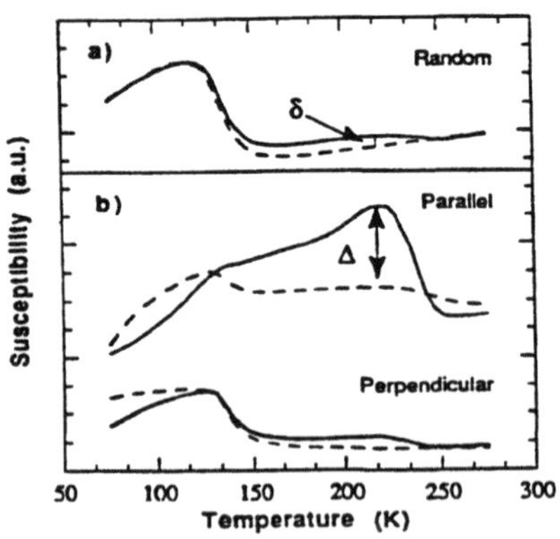

Figure 5. a.c. susceptibility of $Nd_2Fe_{14}B$, a) non-oriented samples (- -) as-quenched, and (–) annealed. b) Oriented samples, (–) with and (- -) without anomaly, measured parallel and perpendicular to the EMD. (Bartolomé et al., 1994)

as was deduced by Kuz'min et al. (1994). By substitution in expression [11] we obtain

$$B_2^o = \frac{20\, K_{Fe}}{J(J+1)(2J-1)(2J+3)} \left(\frac{k\, T_{SR}}{\Delta_{ex}}\right)^2 . \tag{14}$$

As a check for the simple high temperature expansion approach, we have considered the parent compounds, $Er_2Fe_{14}B$ and $Tm_2Fe_{14}B$, which have first-order transitions at the relatively high temperatures of T_S= 317 and 308K, respectively. Using the exchange parameters $n_{ErFe} = 137\mu_o$ and $n_{TmFe} = 135\mu_o$ (Givord et al., 1988) and the value of the Fe sublattice magnetization, $\mu_o M_{Fe}$ = 1.28T at $T = T_S = 0.57T_C$ (reduced by a factor of 0.86 as compared to $\mu_o M_{Fe}$ = 1.49T at T = 0K (Givord et al., 1984)) we get Δ_{ex}/k = 47.1 and 38.7K for $Er_2Fe_{14}B$ and $Tm_2Fe_{14}B$, respectively. The B_2^0 values thus found for the two parent compounds, $B_2^0(0)$ = 0.531 and 1.79K, respectively, practically coincide with the values given by Givord et al. (1988) and Cadogan et al. (1988), averaged over the two rare earth sites, an agreement which confirms the validity of the high temperature approximation in these cases.

At low temperatures another approximation may be applied to evaluate K_1^R for the case $kT << 2\Delta_{ex} + \Delta_1 + \Delta_2$, where Δ_1 is the crystal field contribution to the energy gap between the lowest two energy levels when θ=0, and Δ_2 is the same for the next two succesive levels. In Kuz'min et al. (1994) the following expressions are given for K_1^R,

$$K_1^R = \frac{2J-1}{4}\frac{\Delta_{ex}\Delta_2}{\Delta_{ex}+\Delta_2} + \left(\frac{J}{2}\frac{\Delta_{ex}\Delta_1}{\Delta_{ex}+\Delta_1} - \frac{2J-1}{4}\frac{\Delta_{ex}\Delta_2}{\Delta_{ex}+\Delta_2}\right)\tanh\left(\frac{\Delta_{ex}+\Delta_1}{2\,k\,T}\right), \tag{15}$$

with

$$\Delta_1 = -3\,(2J-1)\,B_2^o - 10\,(2J-1)(2J-2)(2J-3)\,B_4^o - \frac{21}{2}(2J-1)(2J-2)(2J-3)(2J-4)(2J-5)\,B_6^o$$

and (16)

$$\Delta_2 = -3\,(2J-3)\,B_2^o - 10\,(2J-2)(2J-3)(2J-10)\,B_4^o - \frac{21}{2}(2J-2)(2J-3)(2J-4)(2J-5)(2J-21)\,B_6^o\,,$$

This formalism neglects all crystal field parameters, B_n^m, with m ≠ 0. When checking this method with the second-order SRTs which occur in $Nd_2Fe_{14}B$ and $Ho_2Fe_{14}B$, one finds T_S = 117 and 57K, respectively. The respective values found by exact diagonalization of the complete Hamiltonian are T_S = 122 and 57K. The agreement can be considered as good, which justifies the approximations made in this section. The discrepancy with the corresponding experimental values, T_S = 135 and 63K (Piqué et al. 1994) requires a revision of the crystal field and exchange parameters proposed by Givord et al. (1988). To overcome this problem, we rescaled $B_2^0(0)$ for the two compounds so as to reproduce our experimental values of T_S = 135K for $Nd_2Fe_{14}B$ and 55K for $Ho_2Fe_{14}B$ (Kuz'min et al., 1994). This approximation, however, has the drawback that three parameters, B_2^0, B_4^0, and B_6^0, are involved in one equation, so one obtains just a relation between them.

2.3 NON-SRT ANOMALIES

A word of caution should be given at this stage about a controversy as to whether the type of anomaly which appears in the magnetic ac susceptibility is a real SRT, as reported repeatedly, or is due to an extrinsic mechanism. These anomalies have been detected in many rare-earth transition-metal intermetallic compounds, e.g., in $R_2Fe_{14}B$ (Grössinger et al., 1986, Kou et al., 1991b), $R_2Co_{14}B$ (Kou et al., 1991b), $R_2Fe_{14}C$ (Kou et al., 1991a), $R_2Fe_{14}BH_x$ (Obbade et al., 1991), R_2Fe_{17} (Miraglia et al., 1991), R_2Co_{17} (Kou et al., 1992b), $RFe_{11}Ti$ (Kou et al., 1993), and $RFe_{10}Mo_2$ (Christides et al., 1993).

From the phenomenology reported in the cited papers a few general statements may be made and illustrated by the curve of the in-phase susceptibility, χ', of $Nd_2Fe_{14}B$ for random and oriented powders (Fig. 5a and b). (a) The anomaly of χ' is rounded and asymmetric in temperature, has a maximum at T~220K for $Nd_2Fe_{14}B$ and a height which depends, for a given sample, on the exciting a.c. field and frequency, shows thermal hysteresis, and is inhibited by an applied external field. The quadrature component, χ'', shows a maximum coincident in temperature. (b) For a given composition the anomaly may appear irrespective of the microstructure, i.e., for $Nd_2Fe_{14}B$ it has been observed in a single crystal (Chen et al., 1992), in as-cast and annealed ingots (Kou et al., 1991b), or in a powder sample. But, interestingly, it has not been observed in other ingots and powders of different origin or batch (Lázaro et al., 1992a). (c) The anomaly appears irrespective of whether the rare earth is magnetic or non-magnetic (Lázaro et al., 1992a), or the transition metal is Fe or Co (Kou et al., 1991b). (d) Careful measurements performed on oriented powders embedded in epoxy have shown the anomaly to be highly anisotropic. The anomaly is best observed in the powder alignment direction coincident with the spontaneous magnetization, Fig. 5b (Lázaro et al., 1992a). Indeed, it becomes the most remarkable anomaly in that direction, higher than the one due to the SRT in $Nd_2Fe_{14}B$. (e) In the cases in which the materials contained interstitial H, N, or C atoms, the anomaly was modified in height and position (Lazaro et al., 1992b).

There exist different interpretations for the origin of the anomaly, an intrinsic origin, which relates it to the temperature variation of the anisotropy constants, a pseudo-first order magnetization process transition, a pseudo-FOMP (Kou et al., 1991b), other SRT (Obbade et al., 1991), or of extrinsic origin as in the presence of Fe containing particles (Lázaro et al., 1992a), caused by domain wall motion (Kou et al., 1993), or spin-glass-like behavior (Christides et al., 1993).

Our own magnetic susceptibility experiments on this type of sample have proven (Bartolomé et al., 1994) that the thermal annealing of the sample causes the anomaly and modifies its temperature and height. This allows one to classify the anomaly as of extrinsic origin, and thus it cannot be due to an exotic SRT or FOMP. By modifying the stoichiometry and performing scanning electron microscopy we could also rule out impurities, or superparamagnetism of nanometric particles as mechanisms causing the anomaly. However, the anomaly may or may not appear in the process of casting, in an uncontrolled manner. To check whether these conclusions are more general than in $Nd_2Fe_{14}B$ we have checked that all compounds which are reported to present the anomaly, had been annealed for days or weeks at temperatures between 800 and 1000°C. This is consistent with our conclusions.

The fact that the anomaly is magnetically anisotropic, has a definite temperature range, is modified by interstitials, and is coupled to a relaxation process, indicates that the anomalies result from domain wall motion with their mobility depending on their coupling to defects or, perhaps, to spurious H atoms.

3. Modification of the SRT's upon insertion of interstitial atoms

It is possible to identify several ways in which hydrogenation may influence the anisotropy. These include the variation of the Fe sublattice contribution to the first anisotropy constant, K_1, the variation of the R-Fe exchange interaction, and the variation of the crystal field acting on the rare-earth electronic states.

3.1 $R_2Fe_{14}ZH_x$ SERIES WITH Z = B AND C

In the $R_2Fe_{14}B$ parent compounds the Fe sublattice magnetization has axial anisotropy so that for $\alpha_J = 0$ (Y, La, Lu, and Gd) there are no SRT's. Besides, the A_2^0 crystal field coefficient of the rare earth is positive, therefore, for $\alpha_J < 0$ (Pr, Tb, Dy, Nd, and Ho) the crystal field reinforces the Fe axial anistropy and no SRT should take place. However, due to higher order anisotropy terms, an axial to canted SRT appears in the Nd and Ho compounds. Finally, for $\alpha_J > 0$ (Sm, Er, Tm, and Yb) the crystal field promotes basal anisotropy in opposition to the Fe sublattice and a first order axial to basal SRT occurs in these compounds as temperature is lowered, see Fig. 6.

In a recent paper by Kuz'min et al. (1994) the variations of T_S with hydrogenation has been discussed for both the first-order SRT, which take place in $Er_2Fe_{14}BH_x$ and $Tm_2Fe_{14}BH_x$, and the second-order transitions in $Ho_2Fe_{14}BH_x$ and $Nd_2Fe_{14}BH_x$. The experimental results are given in Fig. 7, together with the parallel results for the $R_2Fe_{14}CH_x$ counterparts. From inspection it is evident that the borides and carbides show similar behavior. In the Er and Tm compounds $T_S(x)$ have shallow minima at $x \approx 1$, whereas in the Nd compounds maxima are observed at about the same hydrogen content, and T_S increases monotonically with x in both Ho systems.

The scheme proposed in the previous paragraph is applicable here. The Fe sublattice contribution, K_1^{Fe}, may be estimated from the $K_1^{Fe}(T)$ dependence, determined experimentally for the $Y_2Fe_{14}BH_x$ compounds (Bartashevitch and Andreev, 1989), in which the anomalous maximum in K_1^{Fe}, present in the parent compound, was observed. As mentioned in a previous chapter, $K_1^{Fe}(T)$ has a plateau at low temperatures, and a flat maximum in the 250 to 350K region, a behavior which remains upon hydrogenation. In addition, the value of K_1^{Fe} decreases almost linearly with hydrogenation (Pareti et al., 1988). In this work it is proposed that $K_1^{Fe} \approx 0$ for x = 4, however, in our own magnetization experiments we have found that in the range $3 < x < 5$, K_1^{Fe} is very small, but never zero. Combining both results, one may propose an empirical dependence on T and x as follows,

$$K_1^{Fe}(T,x) = K_1^{Fe}(T,0)(1 - x/4.5), \tag{17}$$

where $K_1^{Fe}(T,0)$ is the anisotropy constant of the parent compound at that temperature.

The R-Fe interaction seems to be less important. From the analysis of the Curie temperatures for $Er_2Fe_{14}BH_x$ and $Lu_2Fe_{14}BH_x$ (Fruchart et al., 1988b) one may derive the expression,

$$n_{ErFe}` \sqrt{T_C^{Er}\left(T_C^{Er} - T_C^{Lu}\right)}, \tag{18}$$

which remains constant, within experimental error, under hydrogenation. Likewise, the ^{161}Dy hyperfine field in $Dy_2Fe_{14}BH_x$ was found to be practically independent of x (Ferreira et al., 1985). From the analysis of the Curie constants in other $R_2Fe_{14}B$ compounds, n_{RFe} seems to increase by 1 to 5% for $0 < x < 1$ and decrease for higher x (Fruchart 1988a). In any case, it seems justified to apply the approximation that the exchange field acting on the rare earth, $H_{ex} = - n_{RFe}M_{Fe}(T)$, depends on temperature but not on x. It is interesting to note that a similar conclusion was reached from the variation of T_S in $(Er_xR_{1-x})_2Fe_{14}B$ under hydrostatic pressure (Ibarra et al., 1992).

The modification of the crystal field parameters may now be deduced. For Er and Tm equation (14) may be used, because the high temperature approximation is applicable, in which the linear dependence on x of $K_1^{Fe}(x)$, is applicable. The $B_2^0(x)$ dependence may then be obtained and is represented in Fig. 8. It is interesting to note that when normalized to unity at x = 0, the dependence is nearly the same for $Er_2Fe_{14}BH_x$ and $Tm_2Fe_{14}BH_x$; it

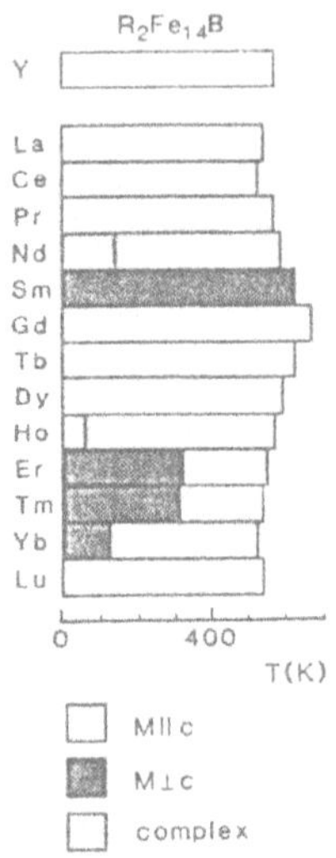

Figure 6. Bar diagrams schematizing the different phases present in the $R_2Fe_{14}B$ series.

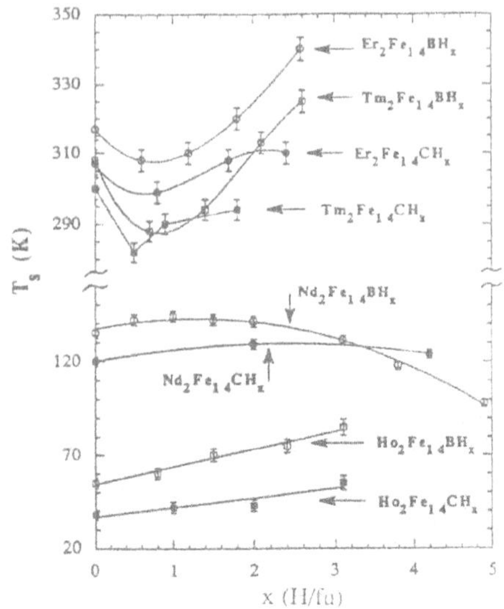

Figure 7. T_S versus concentration of H (x) in $R_2Fe_{14}BH_x$ and $R_2Fe_{14}CH_x$. (Kuz'min et al. 1994).

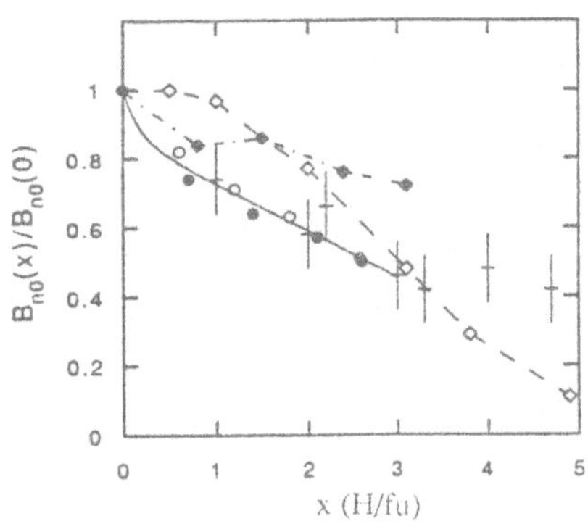

Figure 8. Dependence of the CF parameters in $R_2Fe_{14}BH_x$ on x (normalised to unity at x=0): circles — B_2^o for R=Er (○) and Tm (●), — (+) ^{161}Dy Mössbauer data , — B_4^o, B_6^o CF for R=Nd (◇) and Ho (◆) (Kuz'min et al. 1994)

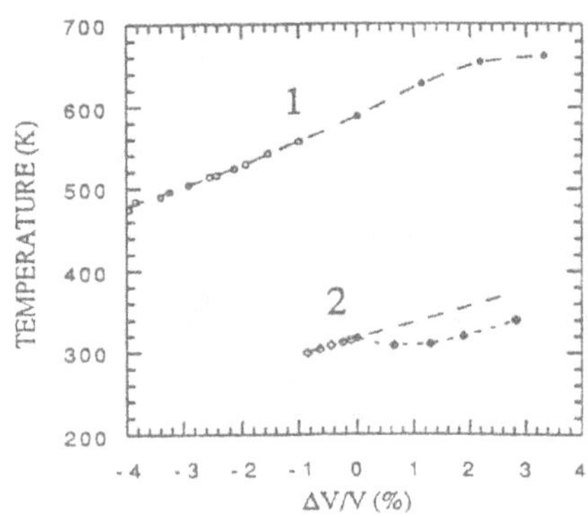

Figure 9. 1) Curie temperature and 2) T_S for $Er_2Fe_{14}B$, (○,◇) from measurements under pressure, (●,◆) from hydrogenated samples. (Kuz'min et al. 1994)

decreases rapidly for low x and more slowly and almost linearly for $x > 0.5$. For the Ho and Nd compounds, the SRT is known to depend on the higher-order anisotropy constants. Their variation with x may be estimated from the low temperature approximation given by equation (15). Now we consider constant Δ_{ex}, and K_1^{Fe}, which is linear in x and B_2^0, follows the dependence observed for the Er and Tm compounds. A further assumption, that B_4^0 and B_6^0 depend similarly on x, needs to be included. By doing so one finds the dependence shown in Fig. 8. Thus a general conclusion is that all crystal field parameters decrease in these borides and carbides.

In the case of B_2^0, a corroboration of this conclusion results from ^{161}Dy Mössbauer spectral data (Ferreira et al., 1985). The quadrupole interaction was found to increase upon hydrogenation, an increase which corresponds to a decrease of the lattice contribution, a contribution which has the opposite sign and is proportional to B_2^0 (Sánchez et al., 1986). The data of Ferreira et al. (1985), subtracted from the dysprosium valence term, e^2qQ = 140 mm/s, and normalized to unity at $x = 0$, (the crosses with error bars in Fig. 8), follow the same trend as B_2^0 in $Er_2Fe_{14}BH_x$ and $Tm_2Fe_{14}BH_x$. In this dependence, two stages can be distinguished, an initial non-linear stage to x~1, and a linear stage. Comparison of our data on T_S in $Er_2Fe_{14}BH_y$ with those for $Er_2Fe_{14}B$ under hydrostatic pressure (Ibarra et al., 1992) enables us to attribute the initial, non-linear, stage to the charge-transfer effects and the following, linear, stage to lattice dilation, see Fig. 9, in which both sets of data, curve 2, are plotted against the relative volume change, $\Delta V/V$. The charge transfer has an influence mainly on the crystal field acting on the rare earth, whereas the dilation affects equally the crystal field and the iron sublattice; hence K_{Fe} is linear in x. In agreement with these ideas, the Curie point of $Nd_2Fe_{14}B$, as a function of $\Delta V/V$, shows no anomaly as it passes from contraction under pressure to expansion under hydrogenation, see Fig. 9, curve 1.

The higher-order crystal field parameters in $Ho_2Fe_{14}BH_y$ seem to behave in principle similarly to $B_2^0(x)$, whereas in $Nd_2Fe_{14}BH_y$ their behavior is distinct in that they change little for $x \lesssim 1$ and decrease approximately linearly at higher x. This difference reflects the fact that the sequence of filling the interstitial sites in $R_2Fe_{14}B$ by hydrogen is different for light and heavy rare-earths elements. When R is a light rare earth, hydrogen preferentially fills the 16k sites having two R and two Fe nearest neighbours, whereas for a heavy rare earth the preference is given to the 8j sites with three R and one Fe nearest neighbours (Obbade et al., 1991). This explains the stronger influence, in the latter case, of the interstitial hydrogen on the rare-earth crystal field. This effect is more pronounced for small x; for large x the filling becomes more uniform due to saturation of the preferential site. In view of these findings, the $B_2^0(x)$ dependence for $Nd_2Fe_{14}BH_x$ may be expected to have a negative curvature at small x. The SRT's occur in the related R = Er, Tm, Ho, and Nd compounds, no doubt because they take place in the parent compounds. There are some problems with the non-SRT anomalies detected in their magnetic susceptibility (Section 2.2), on one hand, and with hydrogen induced SRT's in other members of the series, on the other hand. Upon hydrogenation the non-SRT anomaly decreases in temperature in a systematic way, see Fig. 10 (Lázaro et al., 1992b) and the mechanism explaining these anomalies remains the same as in the parent compound. Namely, the magnetic domain walls vary their mobility with temperature due to their coupling to defects and the temperature dependence of the anisotropy constant. So, beyond doubt, these anomalies are not due to SRT's.

A different problem is the reported induction of SRT's upon H uptake in the $Pr_2Fe_{14}BH$ compound (Pourarian et al., 1986). Indeed, an anomaly is detected in the hydrided material as well as in $Dy_2Fe_{14}BH_{4.5}$ (Zhang et al., 1988) and $Gd_2Fe_{14}BH_x$ (Algarabel et al., 1988), see Fig. 11. To clarify this point, we carried out systematic measurements on all members of the series and found that, in all hydrogenated boride and carbide materials

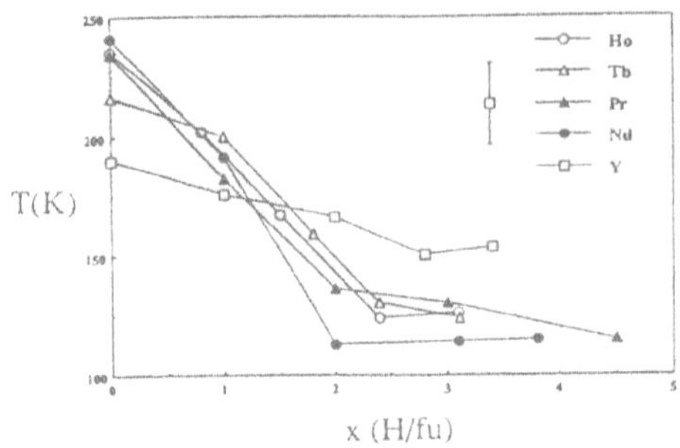

Figure 10. Dependence of the temperature of the anomalous peak on the degree of hydrogenation in $R_2Fe_{14}BH_x$. (Lazaro et al., 1992b)

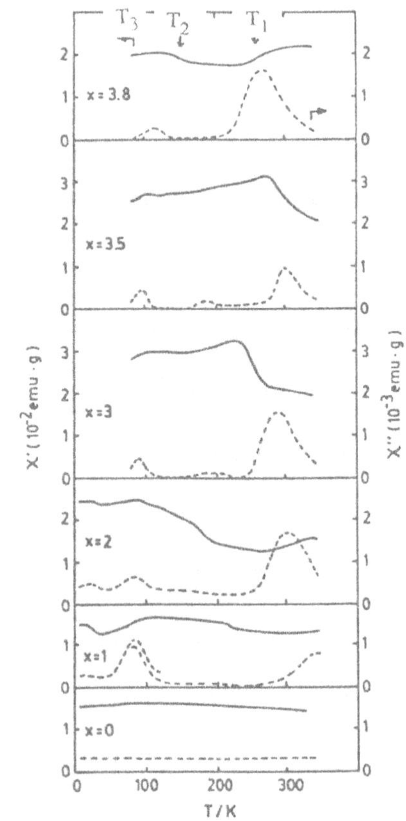

Figure 11. A.c. Magnetic susceptibility of $Gd_2Fe_{14}BH_x$ powders. (–) χ', (- -) χ'' (Algarabel et al. 1988)

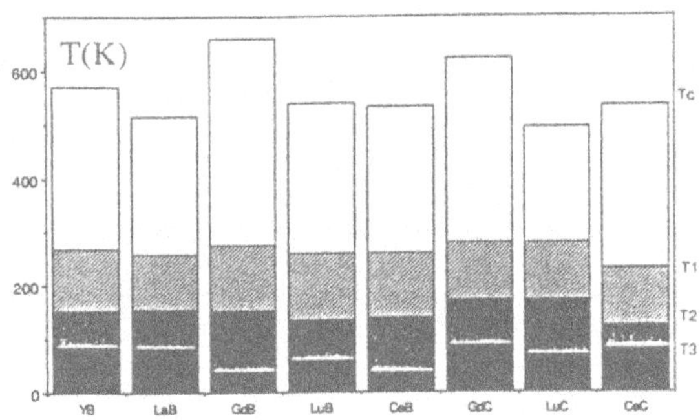

Figure 12. Block diagram showing the four regions of magnetic behavior for $R_2Fe_{14}BH_{max}$ with X=B,C for $\alpha_j=0$. T_C is the Curie temperature, T_1, T_2 and T_3 are the inflexion points present in $\chi(T)$. (Lazaro et al., 1991)

measured a large shoulder is observed in the region of 250 to 300K in the susceptibility measured parallel to the preferential direction of the oriented samples. Naming the inflexion point of this anomaly as T_1 in Fig. 12, the similarity is apparent in the bar diagram of the anomalies found for the non magnetic rare-earth compounds and for those with $\alpha_j = 0$. We think this anomaly is caused by the change with temperature of the mobility of the domain walls coupled to defects.

Another common hydrogen induced effect is a plateau-like anomaly, with inflexion points, T_2 and T_3, as the anomaly appears and disappears upon cooling, respectively. They range from T_2 to T_3, but their width and height depend upon the compound and cannot be related simply to the maximum hydrogen uptake. In addition, T_1, T_2, and T_3 decrease with increasing hydrogen content. In Lázaro et al. (1991) it was shown that this anomaly appears for any value of α_J and even for the non-magnetic rare earth compounds. For this reason, it was concluded that this anomaly is caused by either a reorientation of the Fe sublattice or, more probably, by an α–Fe phase present in all the samples. Today we think it is due to the aforementioned variation in the domain wall mobility as temperature is varied (García et al., 1994a).

3.2 $R_2Fe_{14}BN_x$

Following the success in the formation of intermetallic nitrides with improved magnetic characteristics (Coey and Sun, 1990), nitrogenation of the $R_2Fe_{14}B$ compounds was also attampted with the goal of finding new ways to improve these compounds. In the first work in this direction (Yang et al., 1991, Kou et al., 1992a, Zhang et al., 1992), it was reported that about one atom per formula unit of nitrogen entered into the lattice, producing an increase in T_C and a slight decrease in the anisotropy. In particular, it was reported that, for $Nd_2Fe_{14}BN_{1-\delta}$, the SRT present in the parent compound also appeared at a slightly lower temperature. However, Barret et al., (1993) later carried out a systematic X-ray powder diffraction, scanning microscopy, and magnetic study, and concluded that, at least for R = Y, the existence of the $Y_2Fe_{14}BN_x$ compound could not be confirmed. In fact, they conclude that, in the different nitrogenation batches, there was decomposition of the starting alloy after nitrogenation into other nitrides and borides. For this reason, we refrain from going deeper into these compounds.

3.3. $R_2Fe_{17}H_x$

As described in a previous chapter, for the R_2Fe_{17} parent compounds the Fe sublattice magnetization has strong basal anisotropy, so that for $\alpha_J = 0$ (Y, La, Lu, and Gd), there is no SRT. In addition, the A_n^m crystal field coefficients of the rare earth atoms have opposite signs than in the $R_2Fe_{14}B$ compounds, thus, for $\alpha_J < 0$ (Pr, Tb, Dy, Nd, and Ho) the crystal field reinforces the Fe basal anisotropy and no SRT takes place. Finally, for $\alpha_J > 0$ (Sm, Er, Tm, and Yb) the crystal field promotes axial anisotropy in opposition to the Fe sublattice. However, only in Tm_2Fe_{17} does a SRT occur at $T_S = 72K$ (Gubbens et al., 1973).

The R_2Fe_{17} compounds may be charged with hydrogen quite effectively, however, its effect is to diminish the magnetic anisotropy, as is indicated by Mössbauer spectral studies of ^{155}Gd in $Gd_2Fe_{17}H_x$ (Isnard et al., 1994), while the planar anisotropy of the Fe sublattice diminishes slightly, as evidenced by the reduction of the ^{57}Fe hyperfine field, especially on the 6g site (Qi et al., 1992). In spite of the variation in anisotropies, the net effect of hydrogenation seems to be the reduction of the crystal field anisotropy for all rare earths beyond any capacity to compete with the Fe sublattice, because no SRT has been found either in $Tm_2Fe_{17}H_{3.2}$ or in any other member of the series (García, 1994b).

3.4. $R_2Fe_{17}Z_x$ WITH Z = C, N

The insertion of C or N in the R_2Fe_{17} structure has the opposite effect to the insertion of H; the Fe moments increase and the Fe sublattice planar anisotropy also increases, as detected by Mössbauer experiments (Coey and Sun, 1990). However, the effect on the crystal field anisotropy on the rare earth atoms is spectacular.

For $Sm_2Fe_{17}C_x$, the inclusion of C with x as small as 0.1 allows the axial rare earth anisotropy to overcome the Fe planar anisotropy and, at low temperature an axial phase is observed (Wang and Hadjipanayis, 1991). For $x > 0.5$, $Sm_2Fe_{17}C_x$ becomes an axial compound at all temperatures (Grössinger et al., 1991, Ding and Rosenberg, 1991), and thus no SRT is observed. For Z = N the charged material seems to be well described as a mixture of pure and totally charged $Sm_2Fe_{17}N_{3-\delta}$, because the penetration depth of N is reduced, so it has been impossible to look for SRT in homogeneous systems with intermediate concentrations. The latter composition also becomes axial at all temperatures and, thus, no SRT is present. It is interesting to mention the work of Lu et al., (1992) who studied the crossover of axial to basal anisotropy as a function of substitution of Sm by Y in $(Sm_{1-x}Y_x)_2Fe_{17}N_{3-\delta}$. They concluded that, up to x = 0.8, the system is axial, while only for $Y_2Fe_{17}N_{3-\delta}$ is planar anisotropy found, see Fig. 13. This proves the dominance of the axial anisotropy of the Sm atom, enhanced by N uptake, in this type of compound (Lu et al., 1992).

In the case of $Er_2Fe_{17}C_x$, the Er axial anisotropy is able to overcome the Fe basal anisotropy when x = 0.8, and a SRT from a low-temperature axial to a high temperature basal phase appears at T_S = 83K. For higher rates of C insertion, T_S increases as the C concentration increase, see Fig. 14, (Kou et al., 1991c). This is to be expected because the second order crystal field coefficient A_2^0 becomes more negative by a factor of four to five when one C per formula unit is inserted. Thus the progressive insertion of C will also yield a progressive increase of A_2^0 and, as a consequence, an increase of T_S. For the same reasons as given above, $Er_2Fe_{17}N$ shows a SRT from axial to basal anisotropy, at T_S = 110K, as would be expected for an A_2^0 value which is even more negative than for $Er_2Fe_{17}C$ (Gubbens et al., 1991a), see Fig. 15. In the literature there is a certain degree of scatter in the SRT temperature, undoubtedly originating in the degree of indetermination of the actual N content; one finds T_S = 150K from M(T) measurements, with x = 2.7 (Liu et al., 1991), and T_S = 120K, for the same value of x, from a.c. magnetic susceptibility measurements (Hu et al., 1990b).

In the case of $Tm_2Fe_{17}C_x$, for $0 < x < 1.4$, the SRT present in the parent compound is enhanced in T_S, increasing from 73 to 230K (Grössinger et al., 1991), again due to the strong increase in $|A_2^0|$ brought about by the C insertion (Gubbens et al., 1991b), see Fig. 16. Within the same scheme one may understand that, for $Tm_2Fe_{17}N_{3-\delta}$ ($x \approx 3$), T_S is reported to have increased to 200K by Hu et al. (1990b) or to 225K by Liu et al. (1991).

Concluding, we have used a simple scheme of competition of sublattice anisotropies to study the presence of SRTs in two series of compounds in which H, C, and N may become interstitial atoms. These two series have average second order crystal field parameters of opposite sign, thus illustrating how the magnetic anisotropy depends crucially on the A_n^m parameters.

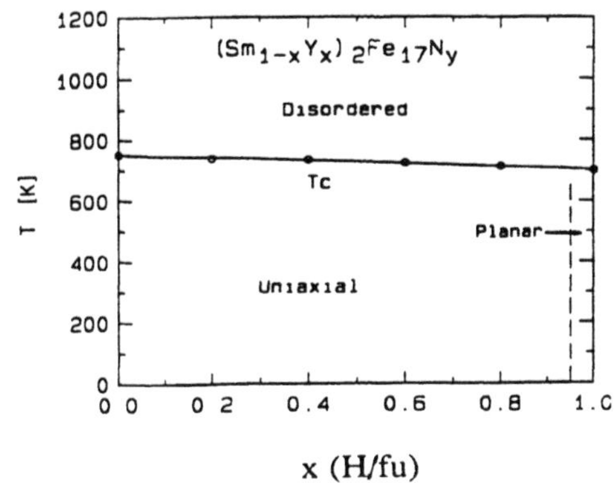

Figure 13. The spin phase diagram for the $(Sm_{1-x}Y_x)_2Fe_{17}N$ system. (Lu et al., 1992)

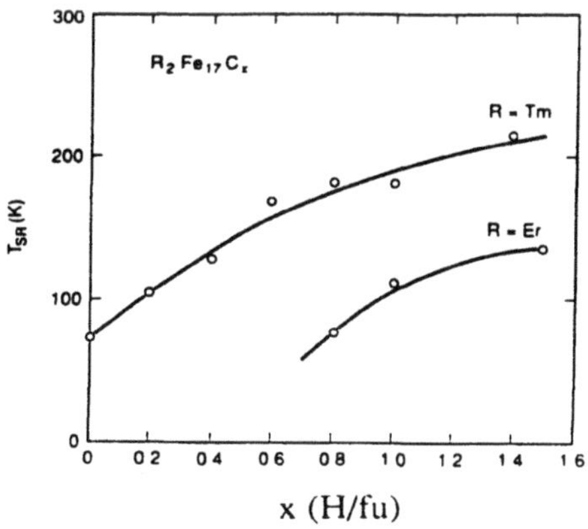

Figure 14. The dependence of T_S on carbon content in $Tm_2Fe_{17}C_x$ and $Er_2Fe_{17}C_x$. (Kou et al., 1991c and Grössinger et al., 1991)

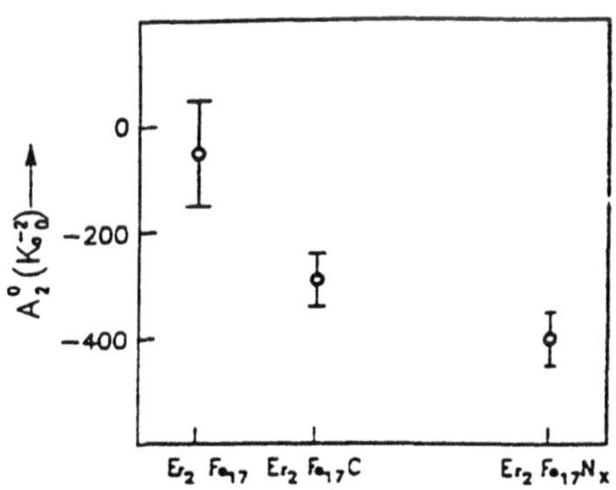

Figure 15. Second-order crystal field coefficient, A_2^0 derived from ^{166}Er Mössbauer spectra of Er_2Fe_{17}, $Er_2Fe_{17}C$ and $Er_2Fe_{17}N_x$. (Gubbens et al., 1991a)

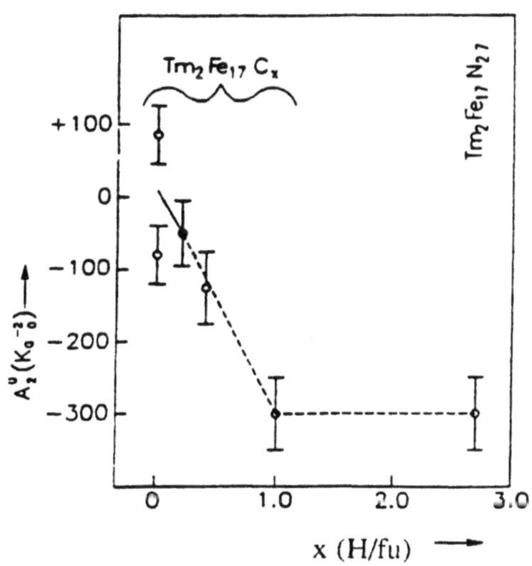

Figure 16. Second-order crystal field coefficient, A_{20} derived from ^{169}Tm Mössbauer spectra of for Tm_2Fe_{17}, $Tm_2Fe_{17}C_x$ and $Tm_2Fe_{17}N_x$. (Gubbens et al., 1991b)

Acknowledgements

Extensive discussions over many years with D. Fruchart are acknowledged. The discussions and aid of L. M. García and M. D. Kuz'min are gratefully acknowledged. The LEA "MANES" covers the Grenoble-Zaragoza cooperation. The CEE CEAM III project has allowed a very intensive exchange of information during this work. This work is also supported by the CICYT project MAT93-0240-C04-04.

References

Algarabel A.P., J.I. Arnaudas, J. Bartolomé, J. Chaboy, A. del Moral, J.M. Fernandez, M.R. Ibarra, C. Marquina, R. Navarro, and C. Rillo (1988) "CEAM report". Ed. I.V. Mitchell (Elsevier, New York) p. 240.

Algarabel P.A., M.R. Ibarra, J. Bartolomé, L.M. García, and M.D. Kuz'min (1994) To be published in J. Magn. Magn. Mater.

Bartolomé J. in "Supermagnets, Hard Magnetic Materials," (1990) Eds. G.J. Long and F. Grandjean (Kluwer) p. 261.

Bartolomé J., L.M. García, F.J. Lázaro, Y. Grincourt, L.G. de la Fuente, C. de Francisco, J.M. Muñoz, and D. Fruchart (1994) IEEE Trans. Magn. in press.

Barret R., D. Fruchart, J.L. Soubeyroux, R. Ferre, R. Fruchart, and A. Stergiou (1993) J. Less-Common Metals 142, 187-194.

Bartashevitch M.I. and A.V. Andreev (1989) Sov. Phys. JETP 69, 1192-1195.

Belov K.P., A.K. Zvezdin, A.M. Kadomtseva, and R.Z. Levitin (1976) Sov. Phys. Usp. 19, 574.

Buschow K.H.J. (1991) Rep. Prog. Phys. 54, 1123-1213.

Cadogan J.M., J.P. Gavigan, D. Givord, and H.S. Li (1988) J. Phys. F 18, 779.

Chen D.X, V. Skumryev, and H. Kronmüller (1992) Phys. Rev. B 46, 3496.

Christides C., A. Kostikas, G. Zouganelis, V. Psyharis, X.C. Kou, and R. Grössinger (1993) Phys. Rev. B 47, 11220-11229.

Coey J.M.D and H. Sun (1990) J. Magn. Magn. Mater. 87, L251-L254.

Ding J. and M. Rosenberg (1991) J. Less Common Metals 168, 335.

Ferreira L.P., R. Guillien, P. Vulliet, A. Yaouanc, D. Fruchart, P. Wolfers, P. L'Héritier, and R. Fruchart (1985) J. Magn. Magn. Mater. 53, 145-152.

Fruchart D., L. Pontonnier, F. Vaillant, J. Bartolomé, J.M. Fernandez, J.A. Puertolas, C. Rillo, J.R. Regnard, A. Yaouanc, R. Fruchart, and P. L'Héritier (1988a) IEEE Trans. Magn. 24, 1641.

Fruchart D., P. Wolfers, S. Miraglia, L. Pontonnier, F. Vaillant, H. Vincent, D. Le Roux, A. Yaouanc, P. Dalmas de Reotier, P. l'Héritier, and R. Fruchart (1988b) "CEAM report" Ed. I.V. Mitchell (Elsevier, New York) p. 214.

García L.M., J. Bartolomé, P.A. Algarabel, M.R. Ibarra, and M.D. Kuz'min (1993) J. Appl. Phys. 73, 5906.

García L.M., J. Bartolomé, F.J. Lázaro, C. de Francisco, J.M. Muñoz, and D. Fruchart (1994b) ICM'94, Warsaw.

García L.M. (1994b) private communication.

Givord D., H.S. Li., and Perrier de la Bâthie (1984) Solid State Comm. 51, 857.

Givord D., H.S. Li, J.M. Cadogan, J.M.D. Coey, J.P. Gavigan, O. Yamada, H. Maruyama, M. Sagawa, and S. Hirosawa (1988) J. Appl. Phys. 63, 3713.

Grössinger R., X.K. Sun, R. Eibler, K.H.J. Buschow, and H.R. Kirchmayr (1986) J. Magn. Magn. Mat. 58, 55-60.

Grössinger R., X.C. Kou, T.H. Jacobs, and K.H.J. Buschow (1991) J. Appl. Phys. 69, 5596.

Gubbens P.C.M. and K.H.J. Buschow (1973) J. Appl. Phys. 44, 3739.
Gubbens P.C.M., A.A. Molenaar, G.J. Boender, A.M. van der Kraan, T.H. Jacobs, and K.H.J. Buschow (1991a) J. Magn. Magn. Mat. 97, 69-72.
Gubbens P.C.M., A.A. Molenaar, T.H. Jacobs, and K.H.J. Buschow (1991b) J. Less-Common Metals 176, 115-121.
Hu H., H. Li, J.M.D. Coey, and J.M. Gavigan (1990a) Phys. Rev. B 41, 2221.
Hu H., H. Li, H. Sun, J.F. Lawler, and J.M.D. Coey (1990b) Solid State Comm. 76, 587-590.
Ibarra M.R., Z. Arnold, P.A. Algarabel, L. Morellon, and J. Kamarad (1992) J. Phys.: Condens. Matter 4, 9721.
Isnard O., P. Vuillet, A. Blaise, J.P. Sanchez, S. Miraglia, and D. Fruchart (1994) J. Magn. Magn. Mater. 131, 83-89.
Kou X.C., R. Grössinger, and H.R. Kirchmayr (1991a) J. Appl. Phys. 70, 6372-6374.
Kou X.C., R. Grössinger, T.H. Jacobs, and K.H.J. Buschow (1991c) Physica B, 168, 181.
Kou X.C and R. Grössinger (1991b) J. Magn. Magn. Mat. 95, 184-194.
Kou X.C., T.S. Zhao, R. Grössinger, H.R. Kirchmayr, X. Li, and F.R. de Boer (1992a) Phys. Rev. B 46, 11204-11207.
Kou X.C., T.S. Zhao, R. Grössinger, and F.R. de Boer (1992b) Phys. Rev. B 46, 6225.
Kou X.C., T.S. Zhao, R. Grössinger, H.R. Kirchmayr, X. Li, and F.R. de Boer (1993) Phys. Rev. B 47, 3231-3242.
Kuz'min M.D. (1992) Phys. Rev. B 46, 8219.
Kuz'min M.D., L.M. García, I. Plaza, J. Bartolomé, D. Fruchart, and K.H.J. Buschow (1994) ICM 94. Warsaw.
Lázaro F.J., L.M. García, F. Luis, C. Rillo, J. Bartolomé, D. Fruchart, O. Isnard, S. Miraglia, S. Obbade, and K.H.J. Buschow (1991) J. Magn. Magn. Mat. 101, 372-374.
Lázaro F.J., L.M. García, J. Bartolomé, S. Miraglia, and D. Fruchart (1992a) "Studies of Magnetic Properties of Fine Particles and Their Relevance to Material Science," J.L. Dormann and D. Fiorani, Eds., Elsevier, Amsterdam, p. 423.
Lázaro F.J., L.M. García, J. Bartolomé, D. Fruchart, and S. Miraglia. (1992b) J. Magn. Magn. Mat. 114, 261-265.
Lindgard P.A. and O. Danielsen (1975) Phys. Rev. B 11, 351.
Liu J.P., K. Bakker, F.R. de Boer, T.H. Jacobs, D.B. de Mooij, and K.H.J. Buschow (1991) J. Less-Common Metals 170, 109-119.
Lu Y., O. Tegus, Q.A. Li, N. Tang, M.J. Yu, R.W. Zhao, J.P. Kuang, F.M. Yang, G.F. Zhou, X. Li, and F.R. de Boer (1992) Physica B 177, 243-246.
Miraglia S., J.L. Soubeyroux, C. Kolbeck, O. Isnard, D. Fruchart, and M. Guillot (1991) J. Less-Common Metals 171, 51-61.
Obbade S., S. Miraglia, P. Wolfers, J.L. Soubeyroux, D. Fruchart, F. Lera, C. Rillo, B. Melaman, and G. le Caër (1991) J. Less-Common Metals 171, 71.
Pareti L., O. Moze, D. Fruchart, P. L'Héritier, and A. Yaouanc (1988) J. Less-Common Metals 142, 187-194.
Parker F.T. (1987) J. Appl. Phys. 61, 2606.
Piqué M., R. Burriel and J. Bartolomé (1994) to be published.
Pourarian F., M.Q. Huang, and W.E. Wallace (1986) J. Less-Common Metals 120, 63.
Qi Qi-nian, H. Sun, R. Skomski, and J.M.D. Coey (1992) Phys. Rev. 45, 12278-12286.
Rudowicz C (1985) J. Phys. C: Solid State Phys. 18, 1415.
Sanchez J.P., J.M. Friedt, A. Vasquez, P. L'Héritier, and R. Fruchart (1986) Solid State Comm. 57, 309.
Smit J. and H.P.J. de Wijn (1959) "Ferrites", Philips Technical Library.

Stevens K.W.H. (1952) Proc. Phys. Soc. A65, 209.
Wang Y.Z. and G.C. Hadjipanayis (1990) J. Magn. Magn. Mat. 71, 203.
Yang Y., X. Zhang, L. Kong, Q. Pan, Y. Hou, S. Huang, and L. Yang (1991) J. Less-Common Metals 170, 37-44.
Zhang L.Y., J.P. Pourarian, and W.E. Wallace (1988) J. Magn. Magn. Mat. 71, 203.
Zhang X., Q. Pan, S. Ge, Y. Yang, J. Yang, Y. Ding, B. Zhang, Ch. Ye, and L. Jin (1992) Solid State Comm. 83, 231-234.

Chapter 25

ROLE OF INTERSTITIAL ALLOYS IN THE PERMANENT MAGNET INDUSTRY

K.H.J. Buschow
Philips Research Laboratories
5600 JA Eindhoven, The Netherlands

1. Introduction

There are three main aspects in which interstitial atoms play a role in permanent magnet materials. The role of the interstitial atoms in these three cases can be described as particle size modification, microstructure modification and intrinsic property modification. In the first two cases the interstitial atoms are hydrogen atoms, and their role is only a transient one. In the third case the interstitial atoms are nitrogen or carbon atoms and their permanent presence in the interstitial positions of compounds is essential.

During the cominution process of cast $SmCo_5$ and $Nd_2Fe_{14}B$ ingots one may profit from the volume expansion associated with the hydrogen absorption which itself presents an effective approach to particle size reduction. A further beneficial influence of the H atoms, in the so called hydrogen decrepitation process (HD), is the modification of the mechanical properties of the interstitial solid solution with respect to the parent material. After having served as modifior during the production process the hydrogen is eventually removed.

In the so-called HDDR process (Hydrogen Decrepitation Desorption Recombination) the transient role of the interstitial hydrogen atoms is of quite a different nature. Thermodynamic properties, such as the formation enthalpy of the ternary hydride and the activation energy for atomic motion, are now the relevant parameters. Also in this case the hydrogen is removed after having served to modify the microstructure.

It is well known that the intrinsic properties of intermetallic compounds can be drastically modified by up-take of interstitial atoms. As described in several previous reviews (Buschow et al. 1982; Buschow 1991; Li and Coey 1991) these modifications may lead to enhancements as well as reductions of Curie temperature, magnetization and magnetocrystalline anisotropy, depending on the nature of the parent compound as well as on the nature of the interstitial atom. In some cases these modifications take a form in which the interstitial solid solutions can be regarded as novel permanent magnet

F. Grandjean et al. (eds.), Interstitial Intermetallic Alloys, 617–632.

materials in their own right, independent of the (unfavourable) properties of the corresponding parent materials. A prerequisite for permanent magnet applications is that these interstial solid solutions have a sufficiently high thermal stability in a limited temperature range, also above room temperature. This restricts these solid solutions to ternary nitrides and carbides.

It follows from the discussion given above that the physical principles involved in the three categories of interstitial solid solutions that play a role in permanent magnet materials are completely different. For this reason they will be discussed in three separate sections.

2. The Hydrogen Decrepitation Proces

Ternary hydrides of the type $R_2Fe_{14}BH_x$ and $SmCo_5H_x$ can be prepared by bringing powdered material into an atmosphere of H_2 gas.

These interstitial hydrides exist because the reaction enthalpy (ΔH_f) of the transformation

$$Nd_2Fe_{14}B + \tfrac{1}{2}\, x\, H_2 \rightarrow Nd_2Fe_{14}BH_x \tag{1}$$

is sufficiently negative to give an equilibrium hydrogen pressure, p $[H_2]$, given by the van 't Hoff relation ($x \approx 6$),

$$\ln p[H_2] = \Delta H_f/3RT - \Delta S_f/3R \tag{2}$$

of ca. 1 bar even at elevated temperatures.

The volume increase accompanying the absorption of hydrogen gas leads to pulverization of even large lumps of the cast material which, when combined with attritor milling or jet milling, is an effective means of producing fine Nd-Fe-B particles (Dalmas de Réotier et al. 1985, Harris et al. 1985). In this HD (hydrogen decrepitation) technique the cast, coarse grained Nd-Fe-B alloy is readily broken up into a relatively fine powder simply by exposure to hydrogen at ca. 1 bar at room temperature. The absorption of a total of about 0.4 $^w/_o$ hydrogen, first by the neodymium rich intergranular material and then by the $Nd_2Fe_{14}B$ matrix phase, results in quite an effective particle size reduction of the cast material. After milling, the powder is aligned, pressed, and sintered. This sintering process is preferably performed in vacuum. It removes all the hydrogen from the sample and leads to the formation of a fully dense body with a fine grain size (McGuiness et al. 1989). Results of detailed studies made by several investigators (Harris et al. 1985, 1987; Williams et al. 1991) showed that the desorption of hydrogen during the sintering process involves three stages. First, hydrogen is released by the $Nd_2Fe_{14}B$ matrix phase at about 200°C and subsequently by the neodymium-rich intergranular material in two stages at about 250 and

600°C. These latter two temperatures have to be associated with the Nd trihydride that had formed from the Nd metal present in the original intergranular material. At 250°C this trihydride decomposes into H_2 gas and Nd dihydride. The dihydride decomposes eventually into H_2 gas and Nd metal at about 600°C, as may be seen from Fig. 1.

It is of importance for the attainment of good permanent magnetic properties such as high coercivity, that the main phase, $Nd_2Fe_{14}BH_x$, be given sufficient opportunity for hydrogen desorption before heating to higher temperature ($T \geq 600°C$). If the desorption is incomplete the ternary hydride, $Nd_2Fe_{14}BH_x$, will disproportionate into the binary Nd hydride, Fe, and Fe_2B (as will be discussed in more detail in the next section). The Nd hydride will desorb H_2 gas and give rise to Nd metal which will not affect the coercivity. But the precipitation of Fe and Fe_2B within the main phase may lead to the formation of nucleation centres for domain walls and hamper the development of large coercivities. Generally one may expect, however, that only small precipitates will disappear during sintering by reacting with the Nd metal to form again $Nd_2Fe_{14}B$.

The HD process is well established and used worldwide by many companies for the production of sintered Nd-Fe-B magnets. The main advantages of the HD process are the following (Harris 1992) :

- Difficulties encountered with breaking-up the ingot by conventional means (associated with the presence of free iron and the concomitant larger mechanical strength) are avoided by the use of hydrogen.

- A preponderance of intergranular failure increases the proportion of single crystal particles. This preponderance occurs because the intergranular material is Nd-rich, absorbs more hydrogen, and becomes more brittle than the main phase.

- The HD particles are very friable and very amenable to further reduction in size by attritor or jet milling. This means, as a consequence, that substantial cost reduction may occur due to the improved milling characteristics.

- The lower oxidizability of the HD material results in less oxygen pick-up during milling.

- The aligned compacts exhibit virtually zero remanent magnetism which entails easier handling.

- The desorption of hydrogen from the green compacts during heating produces a non-oxidizing environment.

. The very clean surfaces are ideal for subsequent sintering and there is a significant reduction in the sintering temperature when compared with that of the conventional powder.

. The sintered magnets have smaller grain sizes compared with those made from conventional material, thus exhibiting higher mechanical strength and offering the opportunity to reach higher coercivities.

. Generally the HD process allows control of the particle size, shape, and distribution by control of the HD conditions.

3. The Hydrogen Decrepitation Desorption Recombination Process

In order to understand the HDDR process it is necessary to compare the two competing reactions,

$$Nd_2Fe_{14}B + \tfrac{1}{2}\, x\, H_2 \rightarrow Nd_2Fe_{14}BH_x \qquad (3)$$

and

$$Nd_2Fe_{14}B + 2.7\, H_2 \rightarrow 2\, NdH_{2.7} + 12\, Fe + Fe_2B. \qquad (4)$$

The reaction enthalpy, expressed per mole H_2, is more negative for reaction (4) than for reaction (3). However, a higher activation energy is required for reaction (4) to proceed. The reason for this is the phase separation of $Nd_2Fe_{14}BH_x$ into grains consisting of $NdH_{2.7}$, Fe_2B, and grains of Fe metal. For the phase separation to take place, long-range diffusion of the metal atoms is required, and such metal atom diffusion is known to be almost absent at temperatures below about 600°C. In fact, it is this limited metal atom diffusion which makes it possible to prepare ternary R-3d-H hydrides, the latter being metastable with respect to binary R-H hydrides and the 3d metal. In the case of $Nd_2Fe_{14}B$ this means that the binary hydride, along with Fe_2B and Fe metal, will be formed when the charging process is performed at too high a temperature or when the $Nd_2Fe_{14}B$ powder is heated too long under hydrogen gas. This degradation or disproportionation reaction starts at about 650°C.

The nature of the disproportionation reaction of $Nd_2Fe_{14}BH_x$ into $NdH_{2.7}$, Fe_2B, and Fe metal was confirmed by magnetic measurements and X-ray diffraction (McGuiness et al. 1990a).

Subsequent heating of the reaction products in vacuum leads to desorption of hydrogen gas from $NdH_{2.7}$ and the formation of Nd metal. The Nd metal then recombines with the Fe_2B and Fe to form fine grained $Nd_2Fe_{14}B$. The result of the whole process is to alter the microstructure of the $Nd_2Fe_{14}B$. In the original cast alloy the $Nd_2Fe_{14}B$ grains are fairly large because they have been formed by nucleation and growth from the liquid

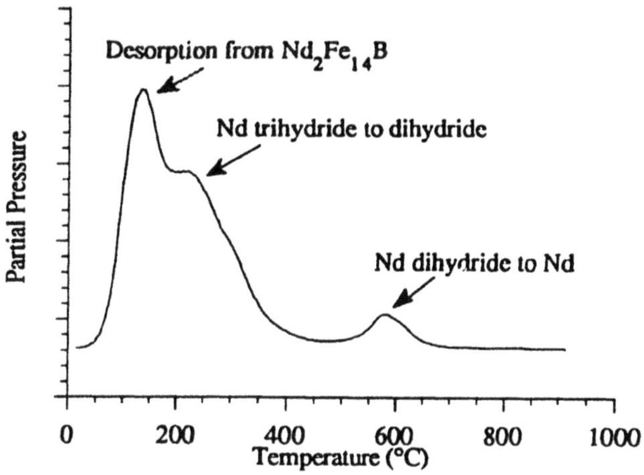

Figure 1. Mass spectroscopy trace of the hydrogen desorption process of hydrogenated $Nd_2Fe_{14}B$ powder (after Harris 1992).

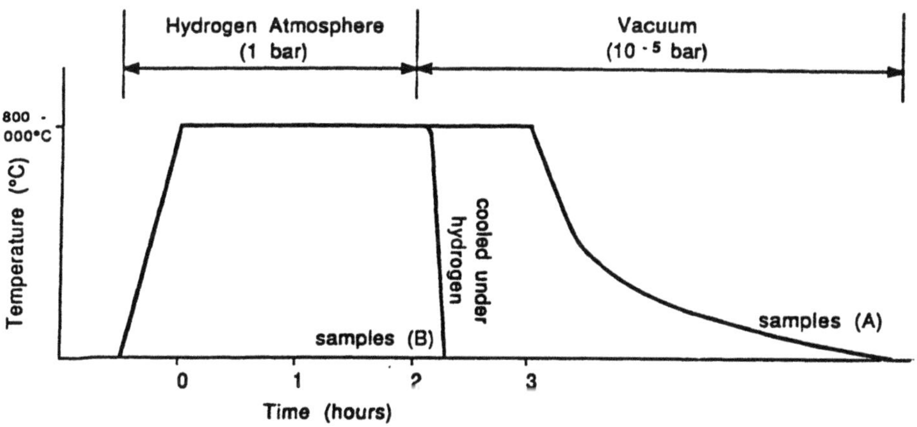

Figure 2. The hydrogen decrepitation process (HDD). Sample A consists of fine-grained $Nd_2Fe_{14}B$ with high coercivity. By contrast, sample B consists of a mixture of $NdH_{2.6}$, bcc Fe, Fe_2B, and some $NdFe_4B_4$. After Harris and McGuiness (1990).

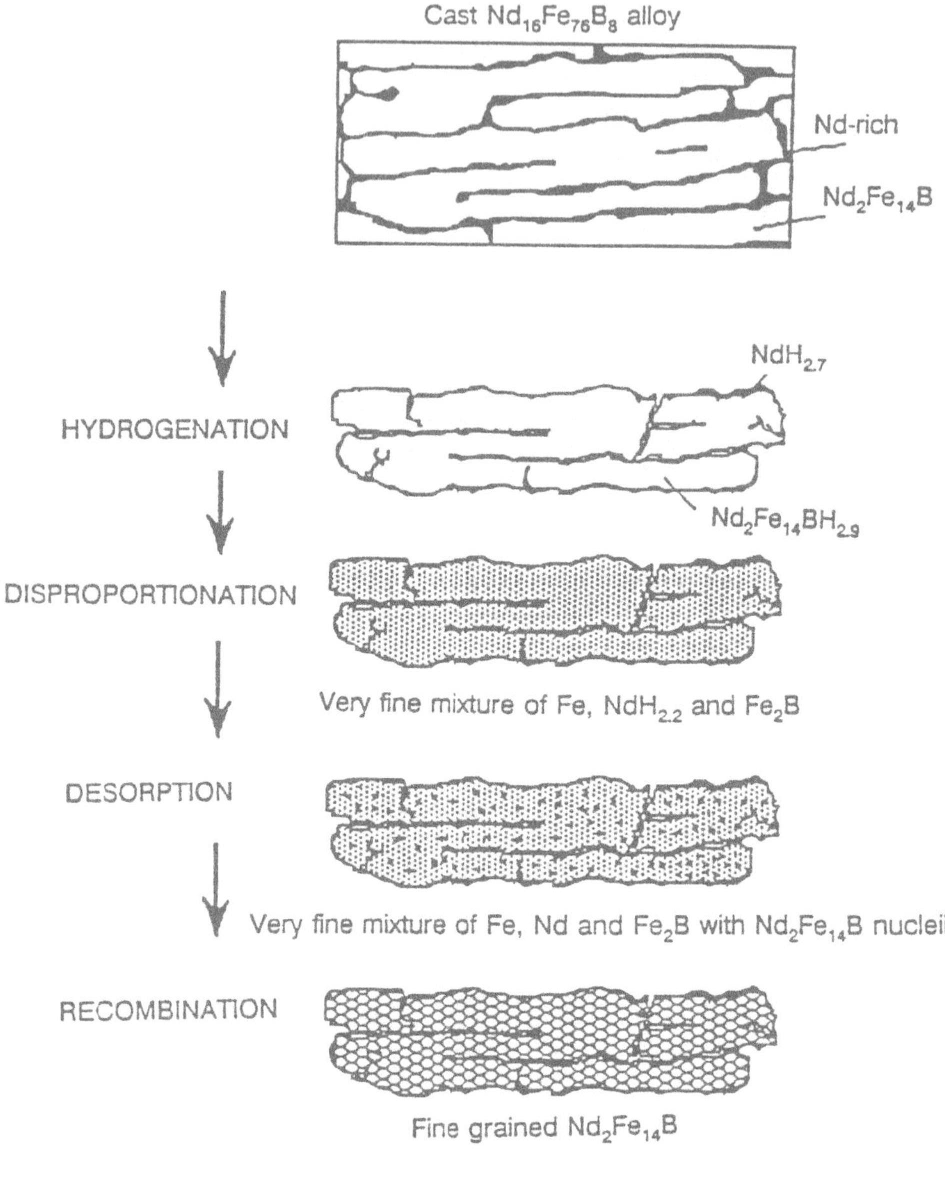

Figure 3. Schematic representation of the modification of the microstructure of cast $Nd_2Fe_{14}B$ type alloys by means of the HDDR process (after Harris 1993).

phase at high temperatures. In contrast, the recombination of the finely divided Nd, Fe_2B, and Fe is a solid state reaction in which the nucleation and growth rates are much lower, leading to much finer grains ($< 1\ \mu m$).

Schematically the HDDR process is shown in Figs. 2 and 3. The advantage of the HDDR process is that fairly large coercivities can be reached in Nd-Fe-B alloys without the requirement of having Nd-rich material at the grain boundaries (Nakayama et al. 1991a). This means that sufficiently high coercivities can be obtained in alloys of a composition close to the stoichiometric composition (Zhang et al. 1991).

It has been observed by Harris and McGuiness (1990) and McGuiness et al. (1990b) that the HDDR grains are very amenable to anomalous grain growth and the formation of relatively large facetted grains. These large grains would be expected to be deleterious to the coercivity. This behaviour is considered to be the origin of the optimal processing temperature reported by Takeshita and Nakayama (1989) and by McGuiness et al. (1990a). It is illustrated for the $Nd_{16}Fe_{76}B_8$ alloy in Fig. 4. The poor coercivities at the lower temperatures have been interpreted in terms of the presence of uncombined iron. The decrease in coercivity at high temperatures was ascribed to grain growth. Zhang et al. (1991) have examined the growth of such grains in near stoichiometric material and have shown that there is a fairly sharp cut-off temperature below which such grains can be avoided. This behaviour is illustrated in Fig. 5.

It has been mentioned by Harris (1992) that the HDDR process can also be used as a very effective means to homogenize Nd-Fe-B type alloys. The manufacture of high-BH_{max} magnets requires high remanence and hence alloy compositions close to the stoichiometric composition. Cast alloys of such compositions invariably have large grains of primary Fe inside the $Nd_2Fe_{14}B$ grains, because of the peritectic formation of the latter phase. The primary Fe crystals are deleterious to the coercivity so that normally a very long, high-temperature annealing treatment is required to remove the primary Fe crystals and to produce single-phase material. However, homogeneous material can be achieved rapidly if the alloy is subjected to the HDDR process. This indicates that there is an enhancement of the solid state diffusion processes in the disproportionated state and quite large Fe dendrites can be removed by quite short annealing treatments (Harris, 1992). If one assumes that mass transport proceeds primarily by intergranular diffusion, the enhancement mentioned is plausible in view of the enormous increase in intergranular diffusion paths and the strongly reduced activation energy for atomic motion in the disproportionated state.

Another important feature of the HDDR process is that it yields coercive powders than can be used in bonded magnets. Here we recall that Nd-Fe-B powders obtained by other types of particle size reduction, for instance by milling, do not exhibit sufficient coercivity. The HDDR powders prepared in this way may be very close to the stoichiometric composition but they are

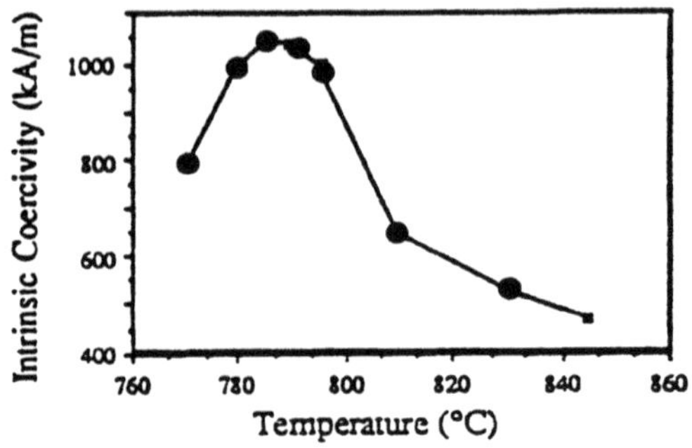

Figure 4. Variation of the coercivity with vacuum degassing temperature for $Nd_{16}Fe_{76}B_8$ alloys (after Harris and McGuiness, 1990).

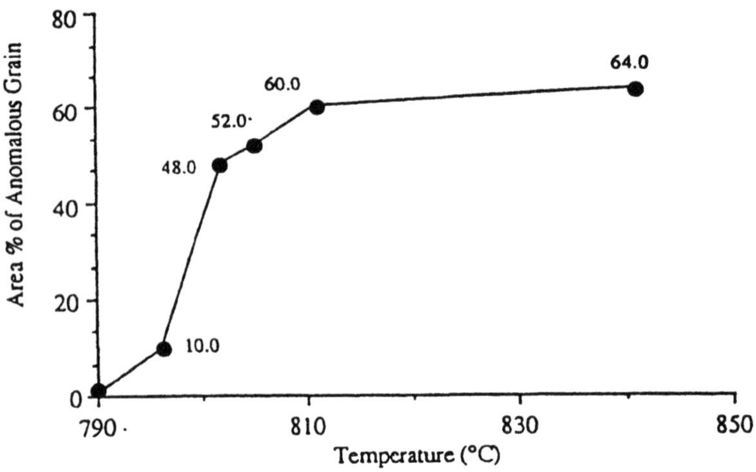

Figure 5. Dependence of the percentage of large facetted HDDR grains in $Nd_{2.1}Fe_{14}B$ on vacuum degassing temperature (after Harris 1993).

magnetically isotropic, meaning that their remanence and maximum energy product are comparatively low. Takeshita and Nakayama (1990) discovered that various additives such as Zr, Hf, and Ga are very effective in producing anisotropic HDDR powder, the amount of additive required being surprisingly low (for instance $Nd_{12.5}Fe_{69.9}Co_{11.5}B_6Zr_{0.1}$). Microscopic investigations described by Harris [13] revealed that large facetted HDDR grains had formed within the original as-cast grain of the alloy. These facetted grains have a common orientation, which is probably the same as that of the original grain. The anisotropic nature of HDDR powders of alloys like $Nd_{12.5}Fe_{75.9}Co_{11.5}B_8Zr_{0.1}$ can thus be understood if one assumes that the HDDR grains have nucleated and grown within an original as-cast grain region from submicron grains, the latter having a common orientation [13]. Evidently the effect of the additive is to bring about nucleation centres for nucleation and growth of the HDDR grains, the latter grains having kept the orientation of the original cast grain. This memory effect is quite interesting and needs further discussion.

It has been mentioned by Nakayama and Takeshita (1991b) that one possible origin of the memory effect might be an incomplete transformation during charging. This would leave a small amount of the $Nd_2Fe_{14}B$ phase dispersed in the degradation products, NdH_2, α-Fe, and Fe_2B. The small untransformed $Nd_2Fe_{14}B$ particles, that still have the orientation of the original cast grains, might then serve as nucleation centers for the HDDR grains.

In order to understand why the addition of small amounts of Zr promotes the occurrence of small amounts of untransformed $Nd_2Fe_{14}B$ one has to consider the following. EXAFS experiments made on $Nd_{2-x}Zr_xFe_{14}B$ by Capehart et al. (1993) showed that the Zr atoms substitute into the Nd sites, meaning that locally one has $Zr_2Fe_{14}B$ (although the latter compound does not form as a true ternary phase). In analogy with reaction 1, one may write

$$Zr_2Fe_{14}B + \tfrac{1}{2}\, x\, H_2 \rightarrow Zr_2Fe_{14}BH_x. \quad (5)$$

It can be shown by model calculations (Buschow et al. 1982) that the hydride formation enthalpy of $Zr_2Fe_{14}BH_x$ is less negative by approximately 30 kJ/(mol H_2) compared to $Nd_2Fe_{14}BH_x$. If we use this result in eq. 2 and take into consideration that ΔS_f for $Zr_2Fe_{14}BH_x$ will be roughly the same as for $Nd_2Fe_{14}BH_x$ (ΔS_f is mainly the entropy of the H_2 gas which is lost upon absorption by the solid) one finds that the equilibrium hydrogen pressure of $Zr_2Fe_{14}BH_x$ would be more than 10^5 higher than for $Nd_2Fe_{14}B$ under comparable circumstances.

It seems plausible therefore that no hydrogen absorption will take place during the HDDR process in those regions of the $Nd_{2-x}Zr_xFe_{14}B$ grains where the Zr concentration is relatively high. The latter regions may be fairly small because the Zr concentrations used are comparatively small, x ≈ 0.01. But this is of no importance when serving as nucleation centers in the recombination process.

Another additive that has proven to be effective in leading to an anisotropic HDDR powder is gallium. Several experiments described in more detail in the review of Buschow (1991) have shown that Ga substitutes into several of the Fe sites in $Nd_2Fe_{14}B$. Model calculations made for $Nd_2Fe_{14-x}Ga_xB$ on the basis of thermodynamic data of de Boer et al. (1988), show that in this case no hydrogen absorption is expected in regions comparatively rich in Ga. These regions can therefore become effective as nucleation centers in the recombination process.

The main characteristics of the HDDR process can be summarized as follows:

- The grain size of $Nd_2Fe_{14}B$ after completion of the HDDR process is about an order of magnitude smaller than the particle size that can be obtained by conventional cominution methods.
- The fine HDDR grains occur in agglomerates, the overall shapes and sizes of which correspond to the HD particles.
- The process leads to fairly high values of the coercivity. The latter exhibits a maximum with processing temperature which can be attributed to the presence of free iron present at lower temperatures and to rapid grain growth at higher temperatures. Above a critical temperature large facetted grains are formed.
- Under appropriate conditions the HDDR process can be applied to solid blocks of the ingot without decrepitation.
- The HDDR process can be used as a very effective means of homogenizing alloys of near stoichiometric compositions that contain α-Fe dendrites.
- Additions of very small amounts of certain elements, such as Zr, Hf, or Ga, result in the formation of an anisotropic HDDR powder. To achieve full alignment of the anisotropic powder, individual, multi-grained HDDR particles must only contain material from a single cast grain. This means that, in practice, a collection of aligned original grains is required for magnets with high remanences.
- Isotropic and anisotropic resin bonded magnets can be manufactured from HDDR powder. The degree of anisotropy can be increased by die-upset forging.

For a more detailed description of the HDDR process the reader is referred to the original papers of Harris (1992) and Takeshita and Nakayama (1993). The demagnetization curve of an anisotropic bonded magnet manufactured from anisotropic powders of $Nd_2(FeCo)_{14}B$-type alloys, with Zr and Ga additives obtained by the latter authors, is shown in Fig. 6. The energy product of the bonded magnets is 72 kJ/m^3 (18 MGOe). Higher energy products of 136 kJ/m^3 or 34 MGOe are reached in hot pressed anisotropic

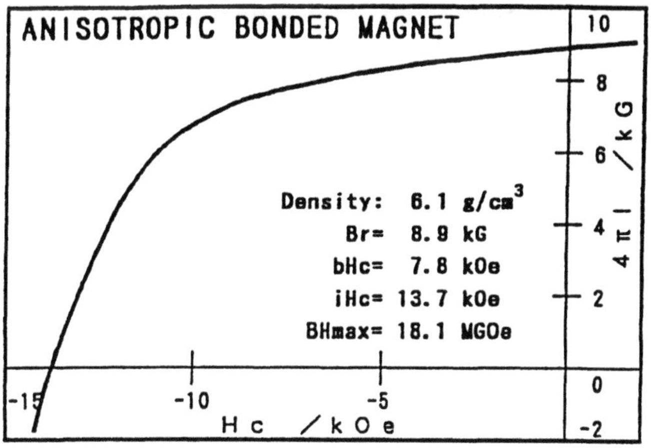

Figure 6. Demagnetization curve of an anisotropic magnet made from HDDR powder of $Nd_2(Fe,Co)_{14}B$ based alloys containing small amounts of Zr and Ga (after Takeshita and Nakayama, 1993).

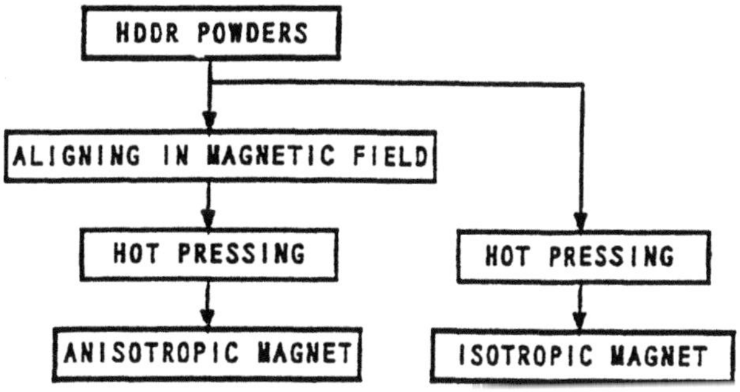

Figure 7. Flow charge for the manufacturing of hot pressed magnets based on HDDR powders (after Takeshita and Nakayama, 1993).

magnets made from HDDR powders. The corresponding manufacturing route is shown in Fig. 7.

4. Interstitial Solutions of C and N in R_2Fe_{17} Compounds

R_2Fe_{17}, of all R-Fe compounds, has the highest saturation magnetizations but, at the same time, the lowest Curie temperatures. Also their relatively low magneto-crystalline anisotropy makes the R_2Fe_{17} compounds less attractive as permanent magnet materials. Considerable improvements with respect to Curie temperature and anisotropy can be obtained, however, by combining these materials with carbon or nitrogen, giving rise to ternary cmpounds of the composition $R_2Fe_{17}C_x$ or $R_2Fe_{17}N_x$ with $0 \leq x \leq 3$.

The ternary carbides, $R_2Fe_{17}C_x$, can be prepared by standard alloying techniques, such as by arc melting rare earths, Fe, and C. It is also possible to first prepare the R_2Fe_{17} compounds and then subject the finely powdered material to hydrocarbon gasses at elevated temperatures. The ternary nitrides, $R_2Fe_{14}N_x$, can be prepared by heating powdered R_2Fe_{17} in a nitrogen or NH_3 atmosphere.

The C and N atoms occupy interstitial positions in the crystal structure of the R_2Fe_{17} compounds. In the hexagonal and rhombohedral modification of the R_2Fe_{17} compounds the preferred interstitial positions are similar. The formal composition is $R_2Fe_{17}C_3$ or $R_2Fe_{17}N_3$ if the interstitial site preferred by the C or N atoms is fully occupied.

The concentration dependence of the Curie temperature generally behaves as follows : For small x values the Curie temperature increases with carbon concentration more rapidly than for larger x values. However, there is a linear behaviour when T_c is plotted versus the corresponding unit cell volumes (Buschow 1991).

The Curie temperature enhancement can be correlated with earlier investigations of R_2Fe_{17}, which have shown that their Curie temperatures decrease strongly under applied pressure. The value of $\Gamma = d\ln T_c/d \ln V$, obtained by combining dT_c/dP with compressibility measurements is 13 for R_2Fe_{17} compounds. This value compares favourably with the Γ values corresponding to the straight lines in plots of ΔT_c versus ΔV. These results indicate that the T_c enhancement is primarily a volume effect, in the carbides as well as in the nitrides.

First indications that the C uptake is accompanied by drastic changes in the rare earth sublattice anisotropy were obtained from ^{169}Tm Mössbauer spectroscopy (Gubbens et al. 1989). From the concentration dependence of the electric field gradient V_{zz}, measured via the quadrupole splitting in various ^{169}Tm Mössbauer spectra of $Tm_2Fe_{17}C_x$, it was concluded that the second order crystal field parameter, A_2°, tends to more negative values with x.

The lowest order expression for the anisotropy constant K_1, see the chapter "Introduction to hard magnetic materials" in this volume, can be written as

$$K_1 = -3/2\, \alpha_J < r^2 > N_R\, (2\, J^2 - J)\, A_2^{\circ} \tag{6}$$

The observation that A_2° tends to more negative values means that K_1 tends to more positive values with increasing x for all compounds in which the R component has a positive value of α_J. As a consequence, the tendency of the R sublattice to align the moments along the c-axis becomes stronger with increasing C content for R elements with $\alpha_J > 0$, i.e. for $Sm_2Fe_{17}C_x$, $Er_2Fe_{17}C_x$, $Tm_2Fe_{17}C_x$, and $Yb_2Fe_{17}C_x$. There is good experimental evidence that the rare earth sublattice anisotropy ($M \parallel c$) is strong enough to overcompensate the Fe sublattice anisotropy ($M \perp c$) in $Sm_2Fe_{17}C_x$ even at fairly low carbon concentrations. For instance, the easy magnetization direction in $Sm_2Fe_{17}C_{0.5}$ is parallel to the c-axis at all temperatures below T_c (Kou et al. 1991), in contrast to the behaviour of Sm_2Fe_{17} where the easy magnetization direction is perpendicular to the c-axis at all temperatures. Wang and Hadjipanayis (1991) reported that a concentration as low as x = 0.1 is sufficient to give rise to an easy magnetization direction parallel to the c-axis at low temperatures.

The Curie temperature enhancement found in interstitial solutions of carbon or nitrogen in R_2Fe_{17} is not much different when compounds of the same interstitial concentration are compared. The strong Curie temperature enhancement is accompanied by only a rather modest Fe moment enhancement, as can be derived from high-field measurements made at 4.2 K (Liu et al. 1990). However, the Curie temperature enhancement leads to quite a substantial increase in the magnetization at room temperature (see, for instance, Long et al. 1992) which is of importance for permanent magnet applications. In fact, because the N and C up-take leads to strong enhancements of the Curie temperature and anisotropy, especially for $Sm_2Fe_{17}Z_x$, these materials are currently considered as good candidates for permanent magnet applications. In order to generate coercivities of sufficient magnitude in these materials, it is a prerequisite that the grain sizes be very small. This has led to four major routes to coercive powders, (1) conventional casting followed by powder metallurgy (Fukuno et al. 1992; Rodewald et al. 1993), (2) conventional casting followed by HDDR treatment (Sugimoto et al. 1992a), (3) melt spinning followed by heat treatment, and (4) mechanical alloying followed by heat treatment (Kuhrt et al. 1993). Charging of the material is subsequently performed in N_2, NH_3, or hydrocarbon gasses at temperatures well below the decomposition temperature of the interstitial solutions (into Sm nitride or carbide and α-Fe). The latter decomposition hampers application of the interstitial alloys in the form of dense sintered magnets, although Sugimoto et al. (1992b) report that the decomposition temperature can be increased to some extent by additives. Generally the coercivities of $Sm_2Fe_{17}N_x$ powders have a magnitude high enough for permanent magnet applications (μ_oH_{cJ} = 4.36 T, Kuhrt et al. 1993) and the remanence of the powders also reach high values under carefully controlled processing conditions (B_r = 1.3 T; Kukuno et al. 1992).

Magnet bodies of nitrogenized Sm_2Fe_{17} powders can be prepared in the form of resin bonded magnets with BH_{max} = 17 MGOe, B_r = 9.0 T, and μ_oH_{cB} = 6.5 T (Kobayashi et al., 1992). In order to explore the favourable weak temperature dependence of the coercivity in magnet bodies suitable for high-temperature applications, Rodewald et al. (1993) and Kuhrt et al. (1993) have investigated Sn and Zn bonded magnets. In these cases, however, the remanences were fairly low (B_r < 0.7 T).

Apart from the materials discussed above Fe-based magnetic materials, consisting mainly of $ThMn_{12}$ type compounds, have occasionally attracted attention. A review of the $ThMn_{12}$-type materials has been given by Buschow (1992). The advantage of these materials is their good corrosion resistance, but it is difficult to generate sufficiently high coercivities for application in permanent magnets. Recent developments have used the fact that the sign of the anisotropy in $ThMn_{12}$ type compounds can be reverted by nitrogenation. This makes Nd based compounds eligible for the attainment of hard materials. Mechanical alloyed and nitrogenated powders of the composition $Nd_{10}Fe_{75}V_{15}N_x$ seem to be the most favourable of this group with T_c = 768 K, M_s (295 K) = 131 Am^2/kg and μ_oH_{cJ} = 0.75 T (Hadjipanayis et al. (1992)).

References

Buschow K.H.J. (1991) Repts. Prog. Phys. 54, 1123.

Buschow K.H.J., Bouten P.C.P. and Miedema A.R. (1982) Rep. Prog. Phys. 45, 937.

Capehart T.W., Mishra R.K., and Pinkerton F.E. (1993) J. Appl. Phys. 73, 6476.

Dalmas de Réotier P.D., Fruchart D., Wolfers P., Vulliet P., Yaouanc A., Fruchart R. and L'Héritier P. (1985) J. Phys. (Paris) 46, C6-249.

De Boer F.R., Boom R., Mattens W.C.M., Miedema A.R. and Niessen A.K. (1988) "Cohesion in Metals", North Holland Publ. Amsterdam.

Fukuno A., Ishizaka C., and Yoneyama T., 12th Int. Workshop on R.E. Magnets and their Applications, Canberra 1992, pp. 60.

Grössinger R., Kou X.C., Jacobs T.H., and Buschow K.H.J. (1991) J. Appl. Phys. 69, 5596.

Gubbens P.C.M., Van der Kraan A.M., Jacobs T.H. and Buschow K.H.J. (1989) J. Magn. Magn. Mater. 80, 265.

Harris I.R., Noble C. and Bailey T. (1985) J. Less Common Metals, 106, pp. L1.

Harris I.R., McGuiness P.J., Jones D.G.R. and Abell J.S. (1987) Phys. Scripta T19, 439.

Harris I.R. and McGuiness P.J. (1990) Proc. 11th Int. Workshop on Rare Earth Magnets and their Applications Pittsburgh PA 1990, pp. 29.

Harris I.R. (1992) Proc. 12th Int. Workshop on R.E. Magnets and their Applications, Canberra 1992, pp. 347.

Hadjapanayis G.C., Tang Z.X., Gong W. and Singleton E.W., 7th Int. Symposium on Magnetic Anisotropy and Coercivity in RE-TM Alloys, Canberra 1992, pp. 402.

Koboyashi K., Iriyama T., Imaoka N., Suzuki T., Kato H. and Nakagawa Y., 12th Int. Workshop on R.E. Magnets and their Applications, Canberra 1992, pp. 32.

Kuhrt C., Schnitzke K. and Schultz L., J. Appl. Phys. (MMM'92).

Kukuno A., Ishizaka C. and Yoneyama T., 12th Int. Workshop on R.E. Magnets and their Applications, Canberra 1992, pp. 60.

Liu J.P., Bakker K., de Boer F.R., Jacobs T.H. and Buschow K.H.J. (1990) J. Less-Common Met. 170, 109.

Long G.J., Pringle O.A., Grandjean F. and Buschow K.H.J. (1992) J. Appl. Phys. 72, 4845.

McGuiness P.J., Devlin E.J., Harris I.R., Rozendaal E. and Ormerod J. (1989) J. Mat. Sci. 24, 2541.

McGuiness P.J., Zhang X.J., Yin X.J. and Harris I.R. (1990a) J. Less-Common Metals 162, 359.

McGuiness P.J., Xhang X.J., Forsyth H. and Harris I.R. (1990b), J. Less-Common Metals 162, 379.

Nakayama R., Takeshita T., Hakura M., Kuwano N. and Oki K. (1991a) J. Appl. Phys. 70 (1), pp. 3770.

Nakayama R. and Takeshita T. (1991b) Proceedings of 71st Meeting of Magnetic Society of Japan. "Topics on Rare Earth Permanent Magnets" 24 July, pp. 25.

Rodewald W., Wall, B., Katter M., Veliscescu M. and Schrey P., J. Appl. Phys. (MMM'92).

Sugimoto S., Nakamura H., Okada M. and Homma M., 12th Int. Workshop on R.E. Magnets and their Applications, Canberra 1992a, pp. 372.

Sugimoto S., Nakamura H., Okada M. and Homma M., 12th Int. Workshop on R.E. Magnets and their Applications, Canberra 1992b), pp. 218.

Takeshita, T. and Nakayama R., 10th Int. Workshop on Rare Earth Magnets and their Applications, Kyoto, 1989, pp. 551.

Takeshita T. and Nakayama R. (1990) 11th Int. Workshop on Rare Earth Magnets and their Applications, Pittsburgh, PA, pp. 7.

Takeshita T. and Nakayama R. (1993) Proc. 12th Int. Workshop on R.E. Magnets and their Applications, Canberra 1993, pp. 670.

Wang Y.Z. and Hadjipanayis G.C. (1991) J. Appl. Phys. 69, 5565.

Williams A.J., McGuiness P.J. and Harris I.R. (1991) J. Less-Common Metals 171 No. 1, pp. 149.

Zhang X.J., McGuiness P.J. and Harris I.R. (1991) J. Appl. Phys. 68 (8), pp. 5838.

Chapter 26

CHARACTERIZATION AND MANUFACTURING OF PERMANENT MAGNETS

K.H.J. Buschow
Philips Research Laboratories,
5600 JA Eindhoven
The Netherlands

1. Introduction

Permanent magnets are indispensable in modern technology. They form important components of many electromechanical and electronic components used in domestic and professional appliances. For instance, an average home contains more than fifty of such devices of which at least ten are in a standard family car. Magnetic resonance imaging used as a medical diagnostic tool is an example where large quantities of permanent magnets are used in professional appliances. Permanent magnet materials are furthermore essential in devices for storing energy in a static magnetic field. Major applications involve the conversion of mechanical to electrical energy and vice versa, or the exertion of a force on soft ferromagnetic objects. The applications of magnetic materials in information technology are continuously growing.

In this chapter a description will be given of the manufacturing and application of modern magnetic materials. It will be shown that the underlying technology is of a highly interdisciplinary nature and combines features of crystal chemistry, metallurgy, and solid state physics. The main emphasis in this chapter will be placed on the more fundamental aspects of magnetism, solid state physics, and chemistry. It will be clear that production and application cannot be properly treated without a discussion of some basic characteristics of permanent magnets. In the second section a brief survey will therefore be given of the most common characteristics of permanent magnet materials. For a description of the fundamental properties of the compounds involved the reader is referred to other chapters in this book dealing with crystal field induced anisotropy, and intra- and intersublattice interactions. Various types of manufacturing routes will be treated in the third section.

F. Grandjean et al. (eds.), Interstitial Intermetallic Alloys, 633–652.

2. Basic Characteristics of Permanent Magnet Materials

2.1. THE ENERGY PRODUCT

Permanent magnetic materials are characterized by a broad magnetic hysteresis loop and a concomitant high coercivity. The remanence, B_r, determines the flux density that remains after removal of the magnetizing field and, hence, is a measure of the strength of the magnet. The coercivity, ${}_BH_c$, is a measure of the magnet's resistance to demagnetizing fields. The performance of a magnet is usually specified by its energy product that can be regarded as a figure of merit and is defined as the product of the flux density, B, and the corresponding opposing field, H. The hysteresis loop of a given permanent magnet material is preferably measured on a long cylindrical sample in order to exclude demagnetizing effects.

The energy product of a particular magnet body made of this material can then be derived relatively easily. This is illustrated in Fig. 1 where two different types of magnet materials (A and B) are compared. In the left parts of the figure the second quadrants of the hysteresis loops of the two magnet materials are shown. In the second quadrant the field is opposed to the flux density which may be due to external fields and/or the magnet's own demagnetizing field. Each point on the B-H curve can be taken to represent a working point of a magnet subjected to the corresponding demagnetizing fields. Small demagnetizing fields and working points close to the B axis apply generally to elongated or rod-shaped magnet bodies in their own demagnetizing field. The demagnetizing fields are smaller the larger the length of the rod as compared with its diameter. By contrast, the working points of magnet bodies with flat or disc-like shapes correspond to much larger demagnetizing fields and, hence, are located closer to the horizontal axis. The energy products, BH, for both types of magnet shapes, as given by the surface area of the corresponding BH rectangles, are comparatively low. The energy products corresponding to all points of the B(H) curve are plotted (horizontal scale) as a function of the flux density (vertical scale) in the right parts of Fig. 2. The largest possible value of the energy product for each magnetic material is indicated by $(BH)_{max}$. The two corresponding working points are indicated on the B(H) curves of both magnet materials as a filled and open circle, respectively.

The maximum energy product is one of the most generally used criteria for characterizing the performance of a given permanent magnet material. The magnitude of this product can be shown to be equal to twice the potential energy of the magnetic field outside the magnet divided by the volume of the magnet.

In addition to the maximum energy product there are other criteria that can be used to specify the quality of a permanent magnet material. Of importance in many static applications is the magnitude of the intrinsic

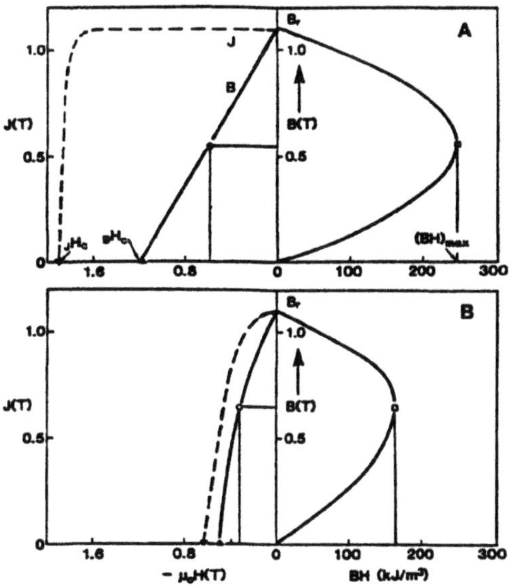

Figure 1. Comparison of the hard magnetic properties of two hard magnetic materials. Left, flux density B (full lines) and magnetic polarization J (broken lines) as a function of the demagnetizing field strength H. Right, the product, BH, (horizontal axis) plotted versus B (vertical axis) for both materials. The working point, corresponding to $(BH)_{max}$, is indicated on the B(H) curve (left point for materials A and B by a filled and open circle, respectively).

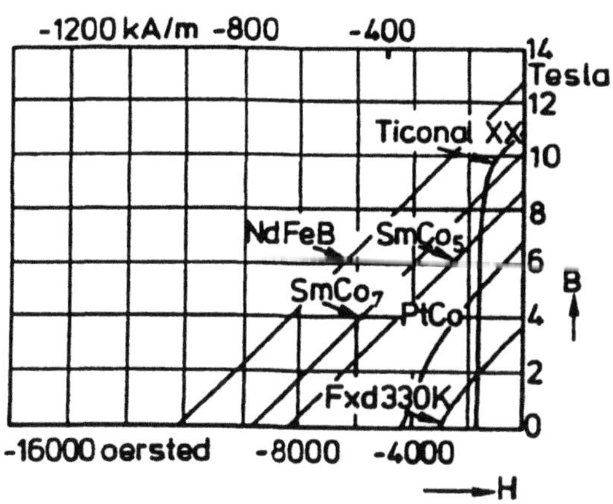

Figure 2. B versus H plots of several permanent magnet materials.

coercivity, $_JH_c$. This may be illustrated again in Fig. 1, which compares the field dependence of the magnetic polarization, J, and the flux density, B, of two different magnet materials that have different hysteresis loops but the same remanence, B_r. It follows from the relation $B = J + \mu_o H = \mu_o M + \mu_o H$, that $_JH_c$ and $_BH_c$ will not differ much when the former value is smaller than the remanence of the permanent magnet material, as is the case for the material in Fig. 1B. In permanent magnet materials based on rare earth compounds, the intrinsic coercive force, $_JH_c$, can become much larger than the remanence. This situation is illustrated by the example shown in Fig. 1A. In this latter case the value of the intrinsic coercivity is considerably in excess of the field corresponding to the $(BH)_{max}$ point and also much in excess of $_BH_c$, at which field the magnetic flux vanishes. In the following it will be assumed that both magnets form part of a magnetic circuit and that they have a shape corresponding to their $(BH)_{max}$ point. When incorporated into a magnetic circuit in which external magnetic fields are present, the magnet material corresponding to B in Fig. 1 is able to tolerate only a relatively small demagnetizing field. For instance, magnetizing fields higher than twice the field corresponding to the $(BH)_{max}$ point will completely demagnetize the magnet body and hence make it useless. By contrast, the magnet material corresponding to A in the same figure is able to tolerate demagnetizing fields more than three times higher than the field at its $(BH)_{max}$ point. It may be seen from the figure that this behaviour originates from the independence of the magnetic polarization, J (broken line), from the negative external and/or internal fields up to a value close to $_JH_c$.

High values of $_JH_c$ can generally be obtained in magnet materials that have a high intrinsic magnetocrystalline anisotropy, as in rare earth compounds. In materials where the hard magnetic properties originate from shape anisotropy, as in alnico type materials, it is not possible to generate large coercivities. The B(H) curves of representative rare earth base magnets can be compared with the B(H) curve of Ticonal XX (alnico type) and some other magnet materials in Fig. 2. It is the presence of large coercivities, in particular that makes the rare earth based magnets suitable for applications requiring flat magnet shapes.

2.2. COERCIVITY

If one applies a steadily increasing magnetic field in a direction opposite to the easy magnetization direction of a perfect single crystal of a magnetic compound, one would expect that all atomic moments reverse their orientation by a process of collinear rotation whenever the applied field becomes equal in size to the anisotropy field, H_A (uniform rotation). Stoner and Wohlfarth (1948) showed that for spheroid particles, in which the major axis coincides with the easy axis as determined by the magnetocrystalline anisotropy, the coercivity is given by

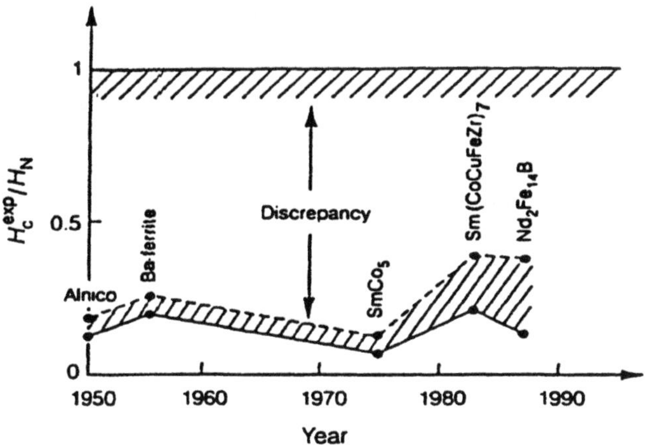

Figure 3. The ratio H_c/H_N of permanent magnets developed during the last four decades (---•---•--- laboratory magnets, ____ commercial magnets). The values of H_N are equal to $2K_1/M_S$. After Kronmüller et al. (1988).

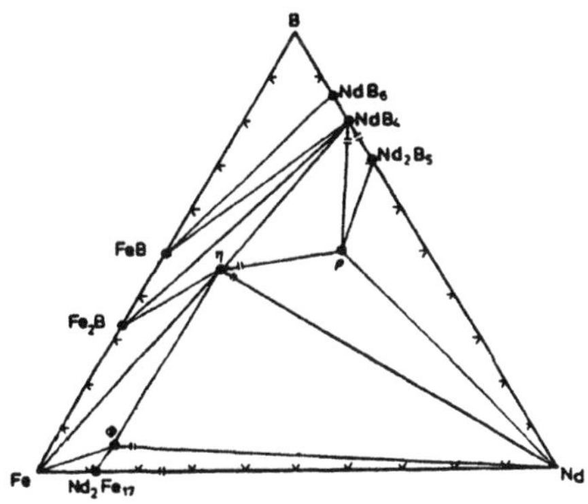

Figure 4. Ternary section of the Nd-Fe-B system at 900°C (Nd-poor alloys) and 600°C (Nd-rich alloys). The three ternary phases are $Nd_2Fe_{14}B$ (ϕ), $Nd_{1.3}Fe_4B_4$ (η) and $Nd_5Fe_2B_6$ (ρ).

$$H_C = 2K_1/M_s-(N - N_{\perp})M_s \tag{1}$$

Coercivities of such large magnitude are, however, seldom encountered in reality. Most permanent magnet materials give rise to magnetization reversal at field strenghts that are only a small fraction (10-15%) of the value of $H_A = 2K_1/M_s$. This behaviour may be illustrated by means of the comparison made in Fig. 3. The reason for this comparatively easy magnetization reversal is the existence of magnetic domain structures. Magnetic particles of sufficiently large size will generally not be uniformly magnetized, but rather will be composed of magnetic domains that are mutually separated by domain walls or Bloch walls. The magnetizations in adjacent domains point in opposite directions in order to reduce the magnetostatic energy. The magnetization in the wall between two domains gradually changes from one preferred magnetization direction to the other. The thickness of the wall is determined by the relative strength of the anisotropy energy and the exchange energy. The former tends to reduce the wall thickness, whereas the latter tends to increase the thickness.

Domain walls, and the corresponding reversed domains, can be nucleated relatively easily near lattice imperfections, where the local values of the exchange and anisotropy fields are sufficiently reduced from the values in the bulk of the material to make a local magnetization reversal possible. This wall nucleation at defects may take place spontaneously or under the influence of an externally applied negative magnetic field. The field required for wall nucleation, commonly referred to as the nucleation field, H_N, is an important parameter for describing the concomitant coercivity, H_c. Non-uniform processes in which the magnetization does not take place by uniform rotation, but by wall nucleation and propagation, are fairly common. These processes dominate in materials with high magnetocrystalline anisotropy. By analogy with eq. (1) an empirical relation of the type

$$H_c = \alpha\, 2K_1/M_s - N_{eff}M_s \tag{2}$$

has proven realistic for describing the nucleation field, H_N, and the concomitant coercivity, H_c. The quantities α and N_{eff} are microstructural parameters, to be discussed in more detail below, which determine the relative importance of the magnetocrystalline anisotropy and the local demagnetizing field, respectively.

In the so-called nucleation-type magnet, the coercivity is mainly determined by difficulties in reverse domain nucleation. Once a wall has been formed adjacent to a nucleated reverse domain it can travel comparatively easily further into the grain. For obtaining high coercivities the wall motion must remain restricted to the affected grain. This means that wall motion must be impeded by grain boundaries, because otherwise a single nucleated wall would lead to magnetization reversal of the entire magnet.

The possibility of wall pinning at grain boundaries is therefore considered to be a prerequisite for nucleation-type magnets.

In the so-called pinning type magnets the situation is completely different. Here wall nucleation can proceed comparatively easily but the Bloch walls cannot travel freely throughout the whole grain. The impeded wall motion is due to magnetic inhomogeneities present in the grains that may act as pinning centers for wall motion. Apart from small changes in magnetization, associated with some wall bending, this pinning will prevent further magnetization reversal. Wall displacement, other than bending, can occur only when the force exerted on the wall becomes sufficiently strong. This is the case when the strength of the external field exceeds the pinning field, H_p, which then determines the coercivity, i.e. $H_c = H_p$. More details regarding the coercivity mechanisms described above can be found in reviews published by Zijlstra (1982), Givord et al. (1990) and Kronmüller et al. (1991).

Permanent magnet materials, such as $Sm(Co,Fe,Cu,Zr)_7$, are pinning-controlled. At high temperatures the alloy consists of a single phase. Heat treatment of the material at lower temperatures leads to the formation of a finely divided precipitate that is able to pin the Bloch walls and produce a high coercivity. In the permanent magnet materials, $Nd_2Fe_{14}B$ and $SmCo_5$, the coercivity is nucleation-controlled, which has important consequences for the manufacturing of these magnets.

It will be clear that any impurity phases able to nucleate reverse domains and walls have to be avoided in the microstructure of nucleation-controlled permanent magnets. An ideal microstructure for nucleation-controlled magnets consists of fine grains of the main phase, for instance $Nd_2Fe_{14}B$, each grain being surrounded by a thin layer of intergranular material. The latter material need not necessarily be single phase but has to be composed of non-magnetic phase. When the nonmagnetic layer is fairly homogeneous the value of α in eq. 2 can become rather large. The presence of small amounts of a magnetic material in the intergranular region can act as wall nucleation centers and destroy the coercivity. By analogy, the grains of the main phase have to be free of any inclusions or precipitates of a magnetic phase. For the manufacture of permanent magnets it is therefore desirable to know the phase relationships of the corresponding systems. For Nd-Fe-B these phase relationships will be briefly discussed in the next section.

2.3. PHASE DIAGRAM INFORMATION

The usefulness of the $Nd_2Fe_{14}B$ type compounds as permanent magnets has initiated quite a number of investigations dealing with the Nd-Fe-B phase diagram. Results obtained in several phase diagram investigations, reviewed by Buschow, (1991) are reproduced in Fig. 4. There are three ternary

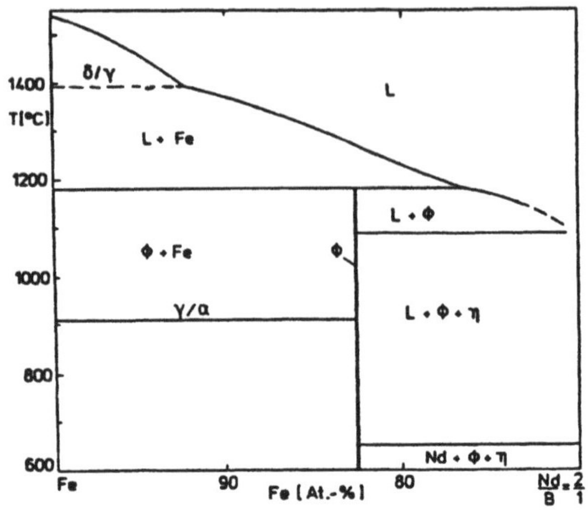

Figure 5. Vertical section in the Nd-Fe-B phase diagram for compositions with a Nd/B ratio of 2 (after Schneider, 1986).

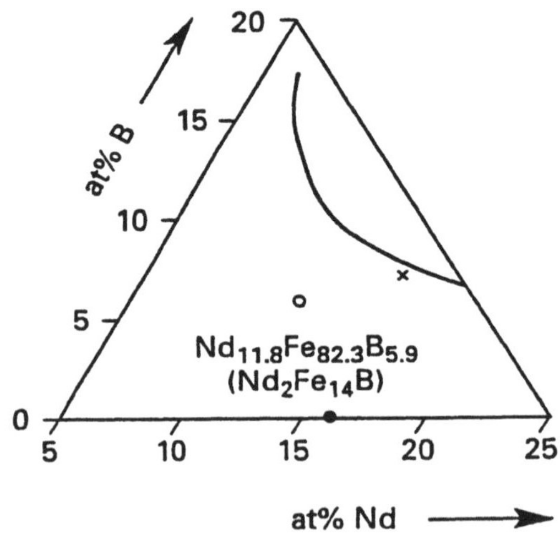

Figure 6. Fe-rich corner of the Nd-Fe-B phase diagram. Primary Fe crystals form on the left side of the curve.

compounds, viz. $Nd_2Fe_{14}B$ (ϕ), $Nd_{1+\epsilon}Fe_4B_4$ (η) and $Nd_5Fe_2B_6$ (ρ). The tie line shown in Fig. 4 between the ϕ phase and Fe is no longer present at temperatures above 900°C. According to the results of Schneider et al. (1986a) a tie line between the phases ϕ and Fe_2B is found instead. These authors also showed that the type of phase transformations observed when Fe-rich liquids are cooled depends strongly on the degree of superheating of the alloys during melting of the constituent elements.

A novel feature that has to be included in the phase diagram shown in Fig. 4 is the occurrence of the binary compound, Nd_5Fe_{17}, first observed by Schneider et al. (1989). It is a stable phase in the Nd-Fe system formed peritectically at 1053 K (Landgraf et al. 1990). Its crystal structure was determined by Moreau et al. (1990). This structure is hexagonal and comprises 7 crystallographic inequivalent Nd sites and 14 inequivalent Fe sites.

A different way of representing the phase relationships in the Fe rich corner of the Nd-Fe-B system is shown in Fig. 5. It is a vertical section passing through the composition $Nd_2Fe_{14}B$. It follows from the results shown in Fig. 5 that, when a molten alloy of the composition $Nd_2Fe_{14}B$ is cooled infinitely slowly, so that equilibrium is reached at each intervening temperature, one would observe the following phases. On cooling from high temperature, the liquidus is traversed for the composition $Nd_2Fe_{14}B$ at about 1280°C. Further cooling then leads to the crystallization of crystals of pure Fe from the melt, their number and magnitude increasing steadily until, at about 1180°C, the peritectic melting point of the $Nd_2Fe_{14}B$ phase is reached. Below this temperature the Fe crystals and the remaining less Fe-rich liquid phase are no longer equilibrium phases and have to transform into $Nd_2Fe_{14}B$. Formation of the latter phase will start on the outside of the Fe crystals which ultimately, together with the liquid phase, will be completely consumed. After sufficient time only the $Nd_2Fe_{14}B$ phase will be left, this being the phase stable at room temperature.

Unfortunately, the $Nd_2Fe_{14}B$ phase formed around the Fe crystals below 1180°C represents a diffusion barrier for the reaction of the Fe in the interior with the less Fe-rich liquid. This means that the peritectic formation of $Nd_2Fe_{14}B$ will be incomplete in a normal casting process, and will leave residues of primary Fe crystals in the $Nd_2Fe_{14}B$ grains, which are in turn surrounded by a relatively Nd-rich liquid that has solidified at the eutectic temperature at about 630°C. As has been discussed in the previous section, the presence of the primary Fe crystals is harmful for the attainment of sufficiently high coercivity. This is one of the reasons that permanent magnet alloys are generally chosen in a concentration range where the crystallization of primary Fe is avoided. These concentrations are found above the solid curve drawn in the diagram of Fig. 6.

3. Manufacturing Routes of NdFeB Type Magnets

The common production routes for NdFeB permanent magnets are schematically represented in Fig. 7. Routes A and B are the well-established powder metallurgical treatments that lead to high-performance magnet bodies. Route B is particularly useful because most rare earth 3d compounds readily absorb large quantities of hydrogen gas at room temperature.

The strong volume increase accompanying the absorption of hydrogen gas entails a pulverization of the material which, when used in combination with attritor milling or jet milling, is an effective means of producing fine Nd-Fe-B particles. In this so-called HD (hydrogen decrepitation) technique the cast, coarse grained, $Nd_{15}Fe_{77}B_8$-type alloy is readily broken up into a relatively fine powder simply by exposure to hydrogen at around 1 bar pressure and room temperature. The absorption of a total of about 0.4 weight percent hydrogen, first by the neodymium rich phase, and then by the $Nd_2Fe_{14}B$ matrix phase, results in the decrepitation process. After milling, the powder is aligned, pressed, and vacuum sintered. This last process removes all the hydrogen from the sample as well as forming a fully dense body with a fine grain size (McGuiness et al. 1989). Results of detailed studies made by several investigators (see the review of Buschow 1991) showed that the desorption of hydrogen during the sintering process involves three stages. First, hydrogen is released by the $Nd_2Fe_{14}B$ matrix phase at about 200°C and subsequently by the neodymium-rich material in two stages at about 250 and 600°C.

A large variety of alloy compositions are generally used as starting materials in routes A and B. Here it should be mentioned that a considerable body of experimental studies has been carried out to improve the intrinsic properties of $Nd_2Fe_{14}B$ by different types of substitutions. A comprehensive description of the results of these studies is given in the review of Buschow (1988). Double substitutions such as Dy for part of the Nd and Co for part of the Fe in $Nd_2Fe_{14}B$ have led to improvements in temperature coefficient of the coercive force and to an increase in the Curie temperature. General drawbacks of most of these substitutions are a diminished magnetization and an increase in the price of the raw material because Dy and Co are more expensive than Nd and Fe, respectively. Other investigations have explored the possibility of changing the microstructure of the sintered magnets by small amounts of additives such as Ga, Al, and Nb. The changes in microstructure due to these additions involve an increase in coercivity without necessarily decreasing the magnetization too much and without increasing the production costs to an undesirable level. This is true for double additives, such as Co and V or Mo (Sagawa et al. 1990, Hirosawa et al. 1990, 1991). Sagawa et al. showed that the prominent improvement of the presence of V is the generation of a microstructure in which the $NdFe_4B_4$ is no longer

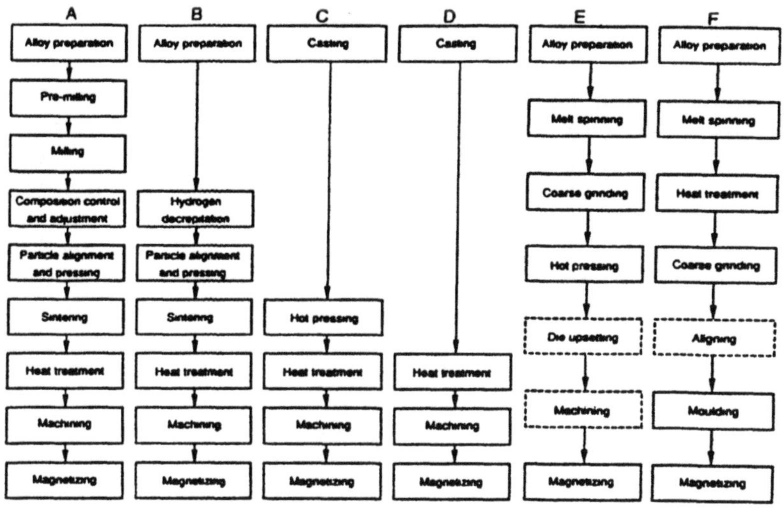

Figure 7. Survey of manufacturing routes in the production of different types of rare-earth permanent magnets.

TABLE 1. Effects of substitution of transition metals (T) in $Nd_2Fe_{14-x}T_xB$ on Curie temperature ΔT_c, and the magnetization, ΔM, and anisotropy field, ΔH_A, at 300 K. The second column indicates the site preference of the T atoms.

T	Site	ΔT_c	ΔM	ΔH_A
Ti		−	−	−
V		−	−	−
Cr	$8j_2$	−	−	−
Mn	$8j_2$	−	−	−
Co	$16k_2$, $8j_1$	+	−	−
Ni	$18k_2$, $8j_2$	+	−	−
Cu		+	−	−
Zr	4f, 4g	−	−	−
Nb		−	−	+
Mo		−		−
Ru		−		−
W		−		
Al	$8j_2$, $16k_2$	−	−	+
Ga	$8j_1$, 4c $16k_2$	−	−	−
Si	$4c(16k_2)$	+	−	+

present in the intergranular region. The beneficial influence on the microstructure stems from the uniform distribution of small $V_{3-x}Fe_xB_2$ precipitates. The presence of these precipitates suppresses excessive growth of the $Nd_2Fe_{14}B$ grains during liquid phase sintering, growth which would have decreased the coercivity. Therefore it is possible to raise the sintering temperature with a concomitant more complete consumation of Nd, Fe, and $NdFe_4B_4$ in the liquid phase to form $Nd_2Fe_{14}B$. This leads to sintered magnets with improved coercivities and corrosion resistance.

The powder metallurgical route is circumvented in route C of Fig. 7 (Nozieres et al. 1988, Akioka et al. 1991). The magnet bodies obtained by this process are of isotropic nature with energy products considerably below those of the magnets obtained via routes A and B. A further simplification is obtained in route D. This method did not lead to useful values of the energy product in the case of NdFeB alloys, but proved very suitable when applied to PrFeB alloys containing additives (Chang et al. 1991). Energy products up to 145 kJ/m^3 were reported.

The melt-spinning process already mentioned in Section 5.1. is employed in route E of Fig. 7. Coarse powders of melt-spun ribbons (Magnequench powder) are commercially available. Hot pressing of the powder at about 750°C leads to full densification without impairing the coercive force as a result of excessive grain growth (Lee, 1985; Lee et al. 1985). Typical values of the energy product are 100 kJ/m^3. But magnets with energy products over 120 kJ/m^3 have also been reported. When the hot pressing is followed by die-upsetting, textured magnet bodies are obtained with their easy magnetization direction parallel to the direction of press. The increase in remanence so obtained is responsible for the fairly large energy products.

Optimized NdFeB powder, obtained by melt spinning, is used in route E of Fig. 7. This powder is incorporated into a polymer binder by injection or compression moulding. The energy product is necessarily small for the corresponding magnets because of the reduced volume fraction of the magnetic material. The main advantage of the bonded magnets is the possibility of giving the magnet bodies the final required shape with close dimensional tolerances.

4. Recent Developments

4.1. Sintered Magnets

As was discussed in the preceding sections, the manufacture of sintered magnets based on $Nd_2Fe_{14}B$ involves starting alloys that are highly off-stoichiometric. The reason for this is the avoidance of primary α-Fe that hampers large coercivities. In addition, the presence of Nd-rich intergranular material magnetically isolates the $Nd_2Fe_{14}B$ grains, which promotes large coercivities. Generally speaking, in order to have a high H_c it is necessary that the particles remain small and magnetically well isolated after sintering. But it is not required, in principle, that the total amount of intergranular material be as high as, for instance, in the standard $Nd_{15}Fe_{77}B_8$ alloy. Better control of the amount and nature of the microstructure offers the possibility of manufacturing sintered magnets with higher remanence, energy product, and coercivity.

Several recent investigations have focused on improving only remanence and energy product because it was realized that the coercivity was more than sufficient in many room temperature applications. Conceptually the steps to be taken to reach this goal look simple, because all that has to be done is to increase the main phase at the cost of the intergranular phase. But the highly peritectic nature of the main phase presents problems which can only partially be solved by changing the phase relationships via additives.

The liquidus projection of the Fe-rich corner of the Nd-Fe-B system is shown in Fig. 6, where the curve defines the concentration region below which primary Fe forms (Schneider et al., 1986). Concentrations close to this line may be used in practice, because particle size reduction of the alloy may bring the small primary Fe crystals in direct contact with the Nd-rich liquid during sintering, causing the disappearance of these crystals. However, when using higher Fe concentration, i.e. concentrations close to the formula composition $Nd_2Fe_{14}B$, the primary Fe crystals are too large and generally will not disappear during sintering. In such cases the α-Fe can be removed before sintering by a heat treatment of the ingots after casting. Homogenized alloys close to the stoichiometric composition have the additional advantage that oxygen pick-up during the powder metallurgical treatment is at a minimum because of the absence of the Nd-rich phase. This greatly facilitates the handling of the corresponding powders.

Particularly high values for the energy product, $(BH)_{max} \approx 400$ kJ/m^3 and remanence, $B_r \approx 1.52$ T, were obtained by Otsuka and Otsuki by using concentrations very close to the stoichiometric composition. The corresponding powder (92 Vol%) was mixed before sintering with a second powder (8 Vol%) of higher Nd content (about 60 wt% Nd). The latter powder acts as a sintering aid. This powder was obtained from alloys prepared by continuous splat cooling and present in amorphous or microcrystalline form. The coercivity of the magnets is comparatively modest ($\mu_0\,{}_JH_c \approx$ 1 T) but higher values for ${}_JH_c$ were obtained when using splat cooled sintering aids containing several types of additives. These additives can be chosen in such a way as to enhance the corrosion resistance of the intergranular phase.

It has, of course, to be realized that the more steps that have to be taken in the manufacturing route, the higher the costs of the permanent magnets. In other words there is a trade-off between higher energy products and higher costs. No substantial cost increase is expected, however, when improving the magnets by means of small amounts of additives. The most common additives are listed in Table I, together with their effect on the Curie temperature, magnetization, and anisotropy field of the main phase. Several of these additives give rise to a maximum or minimum in the concentration dependence of these quantities. A more detailed discussion and references to the original papers can be found in the review of Buschow (1991). In most cases the additives are included in the initial alloy preparation. For additives such as Ti, Zr, and Nb it may be expected that they decrease the extent of the region in Fig. 6 in which primary Fe forms. But the main benefit of all the additives is their strong modification of the intergranular phase. This follows, for instance, because all additives listed in the table have been reported to give rise to an increase in the coercivity while only a few of the additives enhance the corresponding anisotropy field of the main phase (large α in eq. 2). The effect of the additives in the intergranular phase is to form nonmagnetic ternary, pseudobinary, or pseudoternary compounds that can magnetically isolate the grains of the main phase. Another important effect is the growth inhibition of the grains of the main phase during sintering, as was already mentioned above. For improving the corrosion resistance it is important to remove free Nd from the intergranular material. Little effect in this regard is expected from additives such as Mo, W, Nb, Zr, and Ti alone. These additives are therefore always added in combination with other additives.

It may be seen from the table that all additives have the unfavourable property of decreasing the magnetization of the main phase. In fact, the presence in the main phase of the elements added is not at all desirable, with the exception of Co which enhances T_c. This has led several investigators to employ the additives in the form of a second powder that is thoroughly mixed with the powder of the main alloy before compaction and sintering. An example of such a procedure involves the standard alloy, $Nd_{15.5}Fe_{72.5}B_7$, that was thoroughly mixed with fine powder (3 wt %) of $DyGa_2$ before sintering (1h at 1085°C). Results after a sintering and postsintering treatment (90 min at 580°C) lead to a magnet with B_r = 1.2 T and μ_0 $_JH_c$ = 1.3 T.

By means of EPMA (Electron Probe Micro Analysis) it was shown that almost all of the Ga is still contained in the intergranular material after sintering, albeit the composition of the intergranular material is strongly inhomogeneous. After postsintering treatment, the intergranular material became more homogeneous, consisting mainly of Nd and Ga. Most of the Dy and Fe had entered the main phase, presumably by grain growth of the main phase during sintering. The excess Nd originally present in the eutectic phase had combined with the Ga to form a new nonmagnetic intergranular phase of enhanced corrosion resistance.

Similar procedures, using different types of additives were employed by Gandehari (1991) and Sasaki et al. (1991). The additives comprise pure elements, alloys, and various types of intermetallic compounds. When the additive has a melting point higher than the sintering temperature, as for example in $DyGa_2$, it is desirable that the powder particles are very small and are evenly distributed in the powder mixture.

The results described above for the $DyGa_2$ additive show that the situation in the sintered body at the end of the sintering step is always still far from equilibrium. This means that the compositions of the main phase and the intergranular material need not be compatible at the sintering temperature. It is therefore possible to make high performance magnets by sintering a mixture consisting of fine powder of nearly single phase $Nd_2Fe_{14}B$ and very small amounts of a fine powder of an alien alloy regarded suitable as an intergranular material.

Other investigations have focused on improving the high-temperature properties of sintered NdFeB magnets so that they become suitable for electromotive applications at operating temperatures around 200°C, thus replacing the more expensive $SmCo_5$ magnets. The main problem here is the strong temperature coefficient of H_c. Because high remanence is not at a premium in these applications one may increase H_c via a concomitant increase of the anisotropy by substituting Dy and/or Tb for Nd, thereby decreasing the saturation magnetization. But also in the high-temperature NdFeB grade material, a strict control of particle size and the nature of the intergranular material is indispensable.

4.2. Hot-Formed Magnets

As was mentioned in Section 3, the powder metallurgical route is avoided in making hot-formed magnets, route C in Fig. 7. Melt-spun NdFeB alloys have been used as starting materials to produce different grades of magnets, the best results being obtained by die-upsetting. Mishra et al. (1993) have recently been able to improve the properties of die-upset magnets by using melt spun NdFeB ribbons containing small amounts of Co, Ga, and C. The energy product is close to 400 kJ/m³ with B_r = 1.48 T and $\mu_o\ {}_JH_c$ = 1.48 T. This result was achieved mainly through controlled grain growth during densification and die-upsetting, and by keeping the amount of intergranular phase low. Similar results were obtained by Shinoda et al. (1992). These authors also used Co and Ga additives and showed that optimum results with BH_{max} close to 400 kJ/m³, were reached for the composition $Nd_{13.2}Fe_{72.5}Co_{7.5}Ga_{0.8}B_6$.

Hot formed magnets of lower performance, but with considerably lower energy expenditure can be manufactured from Cu containing PrFeB alloys (Akioka et al., 1991). Kwon et al. (1992) used ingots of the composition $Pr_{20}Fe_{74}Cu_2B_4$ to obtain magnets by an upset forging process in which the deformation was carried out in approximately 20 secs. The magnetic alignment during the upset forging was attributed to grain boundary gliding of the plate-like grains (B_r = 1.05 T, μ_o $_JH_c$ = 1.2 T).

Microstructures and magnetic properties of NdFeB ingots, after hot-rolling at 1000°C, were described by Iwamura et al. (1992). The authors stress the relatively low B content required for good performance magnets ($Nd_{15}B_{80}B_5$; BH_{max} = 270 kJ/m^3, B_r = 1.29 T, μ_o $_JH_c$ = 6.6 T).

4.2. Bonded Magnets

Fine powders prepared from cast or heat treated NdFeB alloys do not possess sufficiently high coercivities for application of these powders in resin bonded magnets. However, coercive powders can be obtained by grinding sintered or hot-formed magnets. A more economical route to coercive powders consists of the HDDR process (Takeshita and Nakayama, 1992; Harris, 1992) which will be discussed in more detail elsewhere in this book. This process consists essentially of four steps, hydrogenation of $Nd_2Fe_{14}B$ at low temperatures, decomposition of $Nd_2Fe_{14}BH_x$ into $NdH_{2+\delta}$ + Fe + Fe_2B, desorption of H_2 gas from $NdH_{2+\delta}$, and, finally, recombination of Nd + Fe + Fe_2B into $Nd_2Fe_{14}B$. Because the formation of $Nd_2Fe_{14}B$ in the last step is a solid state reaction, it leads to very fine grains of $Nd_2Fe_{14}B$ with sufficiently large coercivity.

The HDDR powders can be used to manufacture isotropic and anisotropic bonded magnets. The energy product BH_{max} of an anisotropic bonded magnet based on Zr and Ga doped $Nd_2(Fe_2Co)_{14}B$ has been reported to reach values up to 143 kJ/m^3 with B_r = 0.89 T and μ_o $_JH_c$ = 1.37 T (Takeshita and Nakayama, 1992).

4.3. $ThMn_{12}$-type and Fe_3B-type Materials

Apart from the two groups of materials discussed in the two preceding sections, Fe-based magnetic materials consisting mainly of $ThMn_{12}$-type compounds or Fe_3B have occasionally attracted attention. A review of the $ThMn_{12}$-type materials has been given by Buschow (1992). The advantage of these materials is their good corrosion resistance, but it is difficult to generate sufficiently high coercivities for application in permanent magnets. Recent developments in this area comprise permanent magnets based on $ThMn_{12}$ type compounds of the composition $Sm_{1-x}M_x(Fe,Co)_{12}$, where M = Zr, Hf or Bi and $0.1 < x < 0.2$. Values of μ_o $_JH_c$ up to 0.72 T were obtained on sintered magnets (Ohashi, 1992). It is well known that the sign of the anisotropy in $ThMn_{12}$ type compounds can be changed by nitrogenation (see the report of Buschow, 1992, and references cited therein). This makes Nd based compounds suitable for hard magnetic materials. Mechanically alloyed and nitrogenated powders of the composition $Nd_{10}Fe_{75}V_{15}N_x$ seem to be the most favourable of this group with T_c = 768 K, M_s(295 K) = 131 Am^2/kg, and $\mu_o\ {}_JH_c$ = 0.75 T (Hadjapanayis et al., 1992).

The hard magnetic properties of R-Fe-B permanent magnet materials, with Fe_3B as the main phase were, already known in 1987 (Buschow, 1987). They represent the magnet materials with the lowest rare earth content, $Nd_4Fe_{78}B_{18}$, and are prepared by melt spinning followed by a short heat treatment. Remanences are typically 1.2 T with $T_c \approx 800$ K. The main constituent of this material is the metastable tetragonal Fe_3B compound which has a planar type of anisotropy (Coene et al., 1991), the coercivity being mainly due to the presence of small amounts of $R_2Fe_{14}B$ (Eckert et al., 1990). Optimization of the alloy composition and annealing treatment led to coercivities of about 0.4 T (Coehoorn and de Waard, 1990). Although of very good corrosion resistance, the low coercivities have prevented the use of these materials in more general permanent magnet applications. Considerable improvements were made recently by Hirosawa et al. (1992) by means of various types of additives. For melt-spun alloys of the type $Nd_3Dy_2Fe_{70.5}Co_5Ga_1B_{18.5}$, the following values were reported, B_r = 0.98 T, ${}_JH_c$ = 0.6 T. Bonded magnets prepared from these materials are comparatively inexpensive, have a low temperature coefficient of ${}_JH_c$, and are comparatively easy to magnetize, requiring magnetizing fields of only 800 kA/m ($\mu_o H$ = 1 T).

5. References

Akioka K., Kobayashi K., Yamagami T. and Shimoda T., J. Appl. Phys. 69 (1991) 5829.

Barrett R., Ferre R., Fruchart D., Fruchart R., Soubeyroux J.L. and Stergion S., J. Alloys Compds. (1993).

Buschow K.H.J. (1991) Rept. Progr. Phys. 54, 1123.

Buschow K.H.J. (1988) in "Ferromagnetic Materials" vol.4 p.l., North Holland, Amsterdam, E.P. Wohlfarth and K.H.J. Buschow Eds.

Buschow K.H.J., Repts. Prog. Phys. 54 (1991) 1123.

Buschow K.H.J., Repts. Prog. Phys. 45 (1982) 937.

Buschow K.H.J., J. Magn. Magn. Mater. 100 (1992).

Buschow K.H.J., 7th International Workshop on R.E. Magnets and their Application, Bad Soden 1987, pp. 453.

Chang W.C., Paik C.R., Nakamura H., Takahashi N., Sugimoto S., Okada M. and Homma M. (1991) IEEE Trans Mag. MAG.

Christodulou C.N. and Takeshita T., J. Alloys Compds 194 (1993) 31; 194 (1993) 119.

Coehoorn R., and de Waard C., J. Magn. Magn. Mater. 83 (1990) 228.

Coene W., Hakkens F., Coehoorn R., de Mooij D.B., de Waard C., Fidler J. and Grössinger R., J. Magn. Magn. Mater. 96 (1991) 189.

Coey J.M.D. and Hong Sun, J. Magn. Magn. Mater. 87 (1990) L 251.

Eckert D., Müller K.H., Handstein A., Schneider J., Grössinger R. and Krewenka R., IEEE Trans. Magn. MAG-26 (1990) 1834.

Fukuno A., Ishizaka C. and Yoneyama T., 12th Int. Workshop on R.E. Magnets and their Applications, Canberra 1992, pp. 60.

Gandehari M.H., U.S. Patent No. 5.015.304 (May 1991).

Givord D., Lu O., Rossignol M.F., Tenaud P. and Viadieu T. (1990) J. Magn. Magn. Mater. 83, 183-188.

Hadjapanayis G.C., Tang Z.X., Gong W. and Singleton E.W., 7th Int. Symposium on Magnetic Anisotropy and Coercivity in RE-TM Alloys, Canberra 1992, pp. 402.

Harris I.R., Proc. 12th International Workshop on R.E. Magnets and their Applications, Canberra 1992, pp. 347.

Hirosawa S., Tomizawa H., Mino S. and Hamamura A. (1990) IEEE Trans. Magn. MAG-26, 1360-1963.

Hirosawa S., Mino S. and Tomizawa H. (1991) J. Appl. Phys. 69, 5844-5846.

Hirosawa S., Kenkiyo H. and Uehara M., J. Appl. Phys. (MMM'92).

Iwamura E., Yuri T., Nanaki A., Mitani H. and Ytayama K., Proc. 12th Int. Workshop on R.E. Magnets and their Applications, Canberra 1992, pp. 384.

Jurczyk M. and Wallace W.E., J. Magn. Magn. Mater. 59 (1986) L 182.

Jurczyk M., J. Magn. Magn. Mater. 67 (1987) 187.

Kronmüller K. (1991) in Proc. NATO-ASI "Supermagnets, Hard Magnetic Materials" Kluwer Acad. Publ. Dordrecht G.J. Long and F. Grandjean Eds.

Koboyashi K., Iriyama T., Imaoka N., Suzuki T., Kato H. and Nakagawa Y., 12th Int. Workshop on R.E. Magnets and their Applications, Canberra 1992, pp. 32.

Kown H., Bowen P. and Harris I.R., Proc. 12th Int. Workshop on R.E. Magnets and their Applications, Canberra 1992, pp. 705.

Kuhrt C., Schnitzke K. and Schultz L., J. Appl. Phys. (MMM'92).

Kukuno A., Ishizaka C. and Yoneyama T., 12th Int. Workshop on RE Magnets and their Applications, Canberra 1992, pp. 60.

Lee R.W. (1985), Appl. Phys. Lett. 46, 790.

Lee R.W., Brewer E.G. and Schafel N.A. (1985), IEEE Trans. Magn. MAG-21, 1958.

McGuiness P.J., Devlin E.J., Harris I.R., Rozendaal E. and Ormerod J. (1989) J. Mat. Sci., 24, 2541.

Mishra R.K., Panchauwthan V. and Croat J. (1993), J. Appl. Phys. (MMM).

de Mooij D.B. and Buschow K.H.J., J. Less-Common Met. 142 (1988) 349.

Nakayama R. and Takeshita T., J. Alloys Compds. 192 (1993) 259; 192 (1993) 231.

Nozières J.P., Perrier de la Bâthie R. and Gavinet J. (1988) J. Phys. (Paris) 49 C8-667.

Ohashi K., European Patent Appl. No. 0510578A2 (April 1992).

Otsuka T. and Otsuki E., European Patent No. 0261 579 B1.

Rodewald W., Wal B., Katter M., Veliscescu M. and Schrey P., J. Appl. Phys. (MMM'92).

Sagawa M., Tenaud P., Vial F. and Hiraga K. (1990), IEEE Trans. Magn. MAG-26, 1957-9.

Sasaki K., Otsuka T. nd Fujiwara T., European Patent No. 0249 973 B1 (June 1991).

Schneider G., Henig E.Th., Petzow G. and Stadelmaier H.H., Z. Metallkde 77 (1986) 755.

Shimoda T., Akioka K., Kobayashi O. and Yamagami T. (1989) Proc. MMM Conf. Vancouver 1988.

Shinoda M., Iwasaki K., Tamingawa S. and Tokunaga M., Proc. 12th Int. Workshop on R.E. Magnets and their Applications; Canberra 1992, pp.13.

Sugimoto S., Nakamura H., Okada M. and Homina M., 12th Int. Workshop on RE Magnets and their Applications, Canberra 1992, pp. 60.

Sugimoto S., Nakamura H., Okada M. and Homma M., 12th Int. Workshop on RE Magnets and their Applications, Canberra 1992, pp. 218.

Takeshita T. and Nakayama R., Proc. 12th International Workshop on R.E. Magnets and their Applications, Canberra 1992, pp. 670.

Uchida H.H., Uchida H., Yangisawa T., Kise S., Koeninger V., Matsumura Y., Koike U., Kamada K., Kurino T. and Kaneko H., 7th Symposium on Magnetic Anisotropy and Coercivity in RE-TM Alloy, Canberra 1992, pp. 342.

Zijlstra H. (1982) in "Ferromagnetic Materials", Vol. 3, North Holland Publ. Co. Amsterdam 1982, E.P. Wohlfarth Ed.

Chapter 27

PROCESSING AND MICROMAGNETISM OF INTERSTITIAL PERMANENT MAGNETS

R. SKOMSKI and N. M. DEMPSEY
Physics Department
Trinity College (University of Dublin)
Dublin 2, Ireland

Summary. The production of interstitial permanent magnets, in particular bonded $Sm_2Fe_{17}N_3$ magnets, is discussed. The key figure of merit of a permanent magnet is the energy product $(BH)_{max}$, which describes the magnets ability to store energy. Starting from a detailed micromagnetic analysis of the energy product problem we derive processing requirements and show how these requirements can be met. At present, the most promising processing routes are polymer and zinc bonding of oriented $Sm_2Fe_{17}N_3$ powders, but in practice it is difficult to achieve appreciable packing fractions. The final sections are devoted to processing methods which could become important in the near future.

1. Introduction

Interstitially modified permanent magnets materials, in particular the nitride $Sm_2Fe_{17}N_3$, exhibit excellent intrinsic properties [1, 2]. A key figure of merit of permanent magnets is the energy product $(BH)_{max}$, which describes the ability to store magnetostatic energy. In the case of materials with sufficiently-high coercivity, the theoretical energy product is given by $(BH)_{max,th} = 0.25\ \mu_o M_o^2$, where M_o is the saturation magnetization. Comparing the magnetization values of $Sm_2Fe_{17}N_3$ and $Nd_2Fe_{14}B$ we find that $Sm_2Fe_{17}N_3$ with $(BH)_{max,th} = 472$ kJ/m^3 is only slightly inferior to the present room-temperature record holder $Nd_2Fe_{14}B$ with $(BH)_{max,th} = 516$ kJ/m^3 [3, 4], but at about 200 °C, which is a technologically important temperature region, $Sm_2Fe_{17}N_3$ is potentially the best permanent magnet. Moreover, both magnetocrystalline anisotropy and Curie temperature of $Sm_2Fe_{17}N_3$ exceed the respective values for $Nd_2Fe_{14}B$, and it has been possible to realize coercivities as high as 4.36 T [5] in isotropic magnets. On the other hand, the saturation magnetization is proportional to the volume fraction of the magnetic phase, and an appropriate processing is necessary to exploit the intrinsic properties of interstitial nitrides and carbides.

The metastable Sm_2Fe_{17} nitrides and carbides disproportionate at temperatures above about 550 °C, so conventional high-temperature sintering techniques cannot be used to process these materials. Furthermore, the low diffusivity of nitrogen in Sm_2Fe_{17} excludes interstitial modification of powder particles bigger than a few micrometers. This leads to a number of technological problems: Sm_2Fe_{17} pieces have to be milled and sieved in an efficient manner, nitrogenation and bonding have to be conducted in a way which assures a sufficiently high coercivity, and proper alignment and compaction are necessary to utilize the appreciable saturation magnetization $\mu_o M_o \approx 1.54$ T. Finally, we have to compare raw material and processing costs. Both $Sm_2Fe_{17}N_3$ and $Nd_2Fe_{14}B$ are iron rich, and therefore cheaper than $SmCo_5$ and Sm_2Co_{17}, but samarium is more

F. Grandjean et al. (eds.), Interstitial Intermetallic Alloys, 653–671.

expensive than neodymium — it is difficult to compete against $Nd_2Fe_{14}B$ which combines high performance with moderate processing and raw material costs [3]

Here we deal with the processing of high-performance $Sm_2Fe_{17}N_{3-\delta}$ and $Sm_2Fe_{17}C_{3-\delta}$ magnets We investigate the origin of coercivity in $Sm_2Fe_{17}N_3$ particles, describe processing of bonded interstitial magnets, and compare the performance of metal-bonded magnets with that of magnets obtained by other processing methods

2. Micromagnetic Background

Summary The actual performance of a magnet depends on metallurgical microstructure and macroscopic shape The physical description of real structure magnets is based on a free-energy function which includes exchange energy, magnetostatic energy, and magnetocrystalline energy The corresponding non equilibrium equations of state are non linear, and only in a few cases it is possible to obtain analytical solutions such as Bloch walls

2 1 COERCIVITY AND ENERGY PRODUCT

2 1 1 Coercivity Mechanisms Permanent magnetism is a non-equilibrium phenomenon, because the formation of magnetic domains with zero net magnetization is energetically more favourable than the fully aligned state The minimum external magnetic field which destroys an energetically unfavourable magnetic configuration $\mathbf{M}(\mathbf{r}) = M_o\, \mathbf{s}(\mathbf{r})$ is referred to as coercivity There are three main ways to create coercivity in permanent magnets (i) the formation of reversed nuclei is inhibited (nucleation-controlled coercivity), (ii) the expansion of the reversed nucleus is suppressed (pinning-controlled coercivity), or (iii) the particles are small enough to assure coherent rotation (shape anisotropy) In practice, coercivity is based on a suitable metallurgical microstructure involving, for instance, the removal of inhomogenities which facilitate nucleation (Fig 1)

In any case, coercivity presupposes a sufficiently high magnetic anisotropy — else the magnet's magnetization follows any small external field by coherent rotation or the magnet's own demagnetizing field causes the spontaneous demagnetization into a multidomain state Anisotropy in modern permanent magnets is mainly provided by the crystal-field interaction of the rare earth sublattice, while the shape anistropy in materials such as Alnico [6] originates from magnetostatic interaction

2 1 2 Energy product The energy product $(BH)_{max}$ is twice the maximum magnetostatic energy stored by a unit-volume magnet of optimum shape It is measured in kJ/m^3 or, sometimes, in MGOe = $100/4\pi$ kJ/m^3 = 7 958 kJ/m^3 Energy product is determined from the B-H hysteresis loop and equals the area of the largest second-quadrant rectangle which fits under the B-H loop Generally, $(BH)_{max} \leq \mu_o M_r^2/4$ for high-coercivity magnets ($H_c \geq M_r/2$) and $(BH)_{max} \leq \mu_o M_o H_c/2$ for low-coercivity magnets ($H_c \leq M_r/2$) Here $M_r \leq M_o$ is the remanence of the magnet, i e the magnetization after switching off an infinitely large positive field The highest energy products — the "=" limit of energy product inequalities — are obtained in the case of ideally rectangular hysteresis loops In modern rare-earth permanent magnets it is

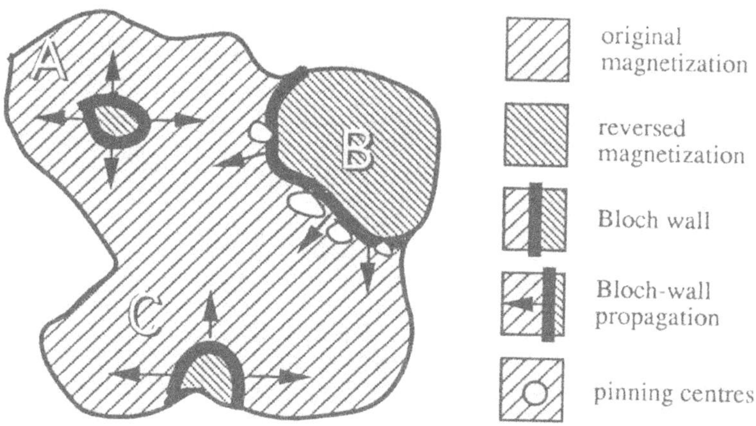

Fig. 1. Coercivity and microstructure.

possible to achieve coercivities larger than $M_o/2$ and the remanent magnetization $M_r \leq M_o$ is the limiting factor. The presence of grain boundary phases, holes in the pressed material, or incomplete alignment reduce M_r and deteriorate the energy product, but nevertheless it has been possible to achieve room-temperature energy products as high as 405 kJ/m^3 in $Nd_2Fe_{14}B$ magnets [4]. Typical energy products of permanent magnetic materials are summarized in Table 1.

Table1. Commercially available permanent magnets [6].

Material	Coercivity mechanism	$\mu_o M_r$ T	$\mu_o H_c$ T	$(BH)_{max}$ kJ/m^3
FeCo (steel)	pinning	1.07	0.02	6
FeCoAlNi (alnico)	shape	1.05	0.06	44
$SrFe_{12}O_{19}$ (ferrite)	nucleation	0.42	0.31	36
$SmCo_5$	nucleation	0.87	0.80	144
$Nd_2Fe_{14}B$	nucleation	1.23	1.11	290

2.2. MICROMAGNETISM

2.2.1. Micromagnetic free energy. The free energy of a given magnetic configuration s(**r**) reads (see e.g.[7-9])

$$F = \int [A\,(\nabla \mathbf{s})^2 - \eta_a - \mu_o M_o \mathbf{H s}]\, d\mathbf{r} \tag{1}$$

with the anisotropy energy density

$$\eta_a = (K_1 + K_2 + K_3)\,(\mathbf{ns})^2 + (K_2 + 3\,K_3)\,(\mathbf{ns})^4 - K_3\,(\mathbf{ns})^6 \tag{2}$$

Here A and **n**(**r**) are the exchange stiffness and the unit vector of the local c-axis direction, respectively. The exchange term $A\,(\nabla \mathbf{s})^2$ describes the quantum-mechanical tendency towards parallel spin alignment, while the uniaxial magnetocrystalline anisotropy originates from the electrostatic crystal-field interaction. In the case of aligned magnets, i.e. $\mathbf{n} = \mathbf{e}_z$, Eq. (2) reduces to the familiar expression $\eta_a = K_1 \sin^2\theta + K_2 \sin^4\theta + K_3 \sin^6\theta$, where θ is the angle between magnetization and c axis. Note that the temperature dependence of the free energy is included in the parameters K_1, A, and M_o.

2.2.2. Internal magnetic field. Magnetostatic interaction, which is similiar to the electrostatic dipole interaction, plays an important role in permanent magnet processing. Ultimately it is responsible for the low energy product of ill-processed magnets, and the macroscopic shape of a magnet is largely determined by magnetostatic requirements.

The interaction energy of two interacting dipoles μ_1 and μ_2 located at $\mathbf{r}_1 = 0$ and $\mathbf{r}_2 = \mathbf{R}$ reads

$$E_{ms} = -\frac{\mu_o}{4\pi}\,\frac{3\,\mu_1\mathbf{R}\,\mu_2\mathbf{R} - \mu_1\mu_2\,R^2}{R^5} \tag{3}$$

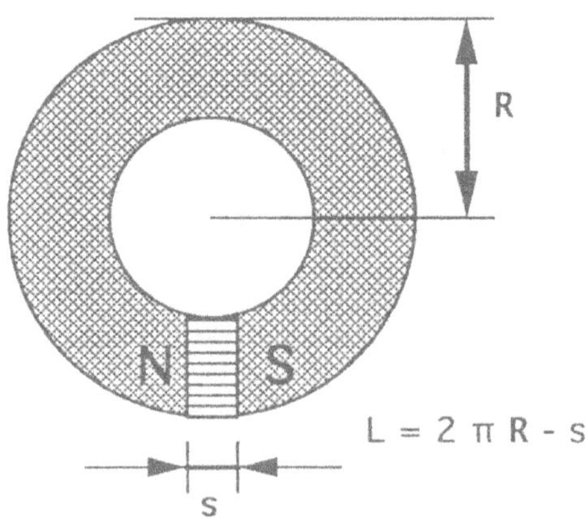

Fig. 2. Magnetic circuit.

and the magnetic field is given by

$$\mathbf{H}(\mathbf{r}) = \frac{1}{4\pi} \int \frac{3\,(\mathbf{R} - \mathbf{r})\,(\mathbf{R} - \mathbf{r})\mu - |\mathbf{R} - \mathbf{r}|^2\,\mu}{|\mathbf{R} - \mathbf{r}|^5}\, d\mathbf{R} \tag{4}$$

Here **a b** means the 'dyadic' product, as opposed to **ab** = **a.b** = $a_x b_x + a_y b_y + a_z b_z$.

A common but not trivial approximation is to replace the magnetostatic interaction by an internal demagnetizing field

$$\mathbf{H}_{int} = - D\,\mathbf{M} \tag{5}$$

where D is the demagnetizing factor. In the case of closed magnetic circuits (Fig. 2) the demagnetization factor is $D = s/(L + s)$. If the coecivity is very low then the magnitude of $\mathbf{H}_{int}$ must be controlled by assuring $D \ll 1$, i.e. $L \gg s$. This the reason for the horse-shoe shape of low-coercivity materials such as steel magnets.

In most cases, the free energy of permanent magnets exhibits two or more metastable equilibrium states $\delta F/\delta \mathbf{s}(\mathbf{r}) = 0$, and tracing the magnetic configuration $\mathbf{s}(\mathbf{r})$ as a function of the external field $\mathbf{H} = H\,\mathbf{e}_z$ is a major task of micromagnetism.

2.2.3. Domains and domain walls. The simplest non-linear $\mathbf{n} = \mathbf{e}_z$ and $\mathbf{H} = 0$ equilibrium solutions of Eq. (1) are Bloch domain walls, which separate $\mathbf{s} = + \mathbf{e}_z$ and $\mathbf{s} = - \mathbf{e}_z$ domains. Based on the ansatz $\mathbf{s}(\mathbf{r}) = \mathbf{e}_z \cos\theta(x) + \mathbf{e}_y \sin\theta(x)$ and the 'tangent' definition [9] of the Bloch-wall width δ_B we obtain by partial integration

$$\delta = \pi \sqrt{\frac{A}{K_1 + K_2 + K_3}} \tag{6}$$

Often the higher-order anisotropy constants are negligible (see Appendix), and is customary to discuss domain-wall properties in terms of [9-10]

$$\delta = \pi \sqrt{\frac{A}{K_1}} \quad \text{(7a)} \qquad \text{and} \qquad \gamma = 4\,\sqrt{AK_1} \quad \text{(7b)}$$

where γ is the domain wall energy. Room-temperature values of A, K_1, δ, and γ are given in Table 2.

The existence of Bloch walls has a number of important consequences:

(i) *Single domain particles.* Domain formation is favourable from the magnetostatic point of view but costs domain-wall energy. If the size of a particle size scales as L, then the condition $\gamma L^2 \approx \mu_0 M_0^2 L^3$ yields a critical single-domain size $L_c \approx \gamma/\mu_0 M_0^2$ below which the formation of domains is energetically unfavourable. A slightly more sophisticated argumentation yields [6, 11] the critical-domain-size radius

$$R_c = 36\,\frac{\sqrt{A\,K_1}}{\mu_0 M_0^2} \tag{8}$$

for spherical particles. Note, however, that the prefactor 36 is sometimes replaced by other estimates.

(ii) *Coherent rotation.* There are two different mechanisms of magnetic reversal: coherent rotation and incoherent rotation. In the case of incoherent rotation, the partly antiparallel orientation of the magnetic regions yields a gain in magnetostatic energy proportional to $\mu_o M_o^2 L^3$, while the spatial variation of the magnetization has to be paid by an exchange energy proportional to $L^3 A/L^2$. This means, incoherent rotation in finite-size particles yields an extra amount of coercivity (Brown's paradox), and below a critical size L_o the exchange contribution dominates and favours coherent rotation [9, 12]. For spherical particles the cross-over radius reads [13]

$$R_o = 5.09 \sqrt{\frac{A}{\mu_o M_o^2}} \qquad (9)$$

Here again, different estimates of the numerical prefactor — based on different nucleation modes — are found in literature [7, 10, 13].

(iii) *Hard-magnetic materials*. Depending on whether the ratio $K_1/\mu_o M_o^2$ is larger or smaller than one, ferromagnets are hard or soft magnetic, respectively. As it can be seen from Eqs. (7) to (9), in hard-magnetic materials the existence of single-domain particles does not automatically imply magnetic reversal by coherent rotation. The critical single-domain size merely determines the ground-state configuration, for instance after thermal demagnetization.

Table 2. Micromagnetic properties of ferromagnetic materials at room temeperature. Numbers deduced from Eqn. (7) are given in italics.

Material	A 10^{-11} J/m	K_1 MJ/m^3	δ nm	γ mJ/m^2	Ref.
Fe	*1.19*	0.047	50	*3.0*	[6]
Fe	0.83	0.047	41	2.5	[10]
Fe	2.50	0.047	72	*4.3*	[14]
Co	*1.29*	0.53	15.5	*10*	[6]
Co	1.03	0.43	15.7	8.2	[10]
Ni	*0.65*	0.0051	110	*0.71*	[6]
Ni	0.34	0.0051	82	0.56	[10]
$BaFe_{12}O_{19}$	0.61	0.325	*13.6*	*5.6*	[6, 15]
$PbFe_{12}O_{19}$	0.51	0.220	*15.1*	*4.2*	[6]
$SmCo_5$	*2.41*	*14.2*	4.1	74	[16]
$SmCo_5$	2.35	17.0	*3.7*	80	[7]
$Nd_2Fe_{14}B$	0.77	4.3	*4.2*	*23*	[7, 14]
$Nd_2Fe_{14}B$	*0.81*	5.0	4.0	*25*	[6]
$Sm_2Fe_{17}N_{3-\delta}$	1.15	8.0	*3.8*	*38*	[11]
$Sm_2Fe_{17}N_{3-\delta}$	4.1[a]	8.3	*7.0*	*74*	[17]
$Sm_2Fe_{17}N_{2.6}$	*0.74*	6.6[a]	*3.3*	28	[18]
$Sm_2Fe_{17}N_{1.0}$	*0.14*	2.2[a]	*4.0*	7[a]	[18]

[a]Estimate.

3. Coercivity in $Sm_2Fe_{17}N_x$ and $Sm_2Fe_{17}C_x$ magnets

Summary. Micro-structured $Sm_2Fe_{17}N_3$ magnets are nucleation-controlled, i.e. their coercivity arises from the absence of bulk and surface inhomogeneities. The mathematical description of nucleation problems leads to eigenvalue problems which are reminescent of Schrödinger's wave equation. In this analogy, the nucleation field is given by the lowest eigenvalue of the equation of state. In practice it is important to remove micron-scaled iron regions, which are harmful to coercivity. Furthermore, partly nitrided grains contain magnetically soft cores which invariably destroy coercivity.

3.1. NUCLEATION

3.1.1. Linearized equation of state. In microstructured magnets such as $Sm_2Fe_{17}N_3$ and $Nd_2Fe_{14}B$ nucleation is the most important coercivity mechanism. High coercivity in these materials presupposes the absence of inhomogenities which may serve as nucleation centres. For the sake of simplicity, let us deal with nucleation in aligned magnets $\mathbf{n}(\mathbf{r}) = \mathbf{e}_z$. Any positive field $\mathbf{H} = + |H|\, \mathbf{e}_z$ stabilizes the fully-magnetized initial state $\mathbf{s}(\mathbf{r}) = + \mathbf{e}_z$, but a sufficiently-strong negative field will induce magnetic reversal. To calculate the nucleation field H_N we can restrict ourselves to small deviations $\mathbf{m}$ from the initial state (nucleation modes). Starting from the identity

$$\mathrm{s} = \sqrt{1 - \mathbf{m}^2}\, \mathbf{e}_z + m_x\, \mathbf{e}_x + m_y\, \mathbf{e}_y \tag{10}$$

with $\mathbf{m} = m_x\, \mathbf{e}_x + m_y\, \mathbf{e}_y$ and $(1 - X)^{1/2} \approx 1 - X/2$ we rewrite the free energy Eq. (1)

$$F = \int \left\{ A\, (\nabla \mathbf{m})^2 + \left(K_1(\mathbf{r}) + \frac{1}{2} \mu_o\, M_o\, H \right) \mathbf{m}^2 \right\} d\mathbf{r} \tag{11}$$

The field $\mathbf{H} = - H_N\, \mathbf{e}_z$ at which the $\mathbf{m} = 0$ free-energy minimum disappears is found by diagonalization of Eq. (11). Putting $H = - H_N$ and $\delta F / \delta \mathbf{m}(\mathbf{r}) = 0$ yields the equation of state

$$- A\, \nabla^2 \mathbf{m} + 2\, K_1(\mathbf{r})\, \mathbf{m} = \mu_o\, M_o\, H_N\, \mathbf{m} \tag{12}$$

where $\mu_o M_o H_N$ is the lowest eigenvalue of the operator $- A\, \nabla^2 + 2\, K_1(\mathbf{r})$.

Eq. (12), which is reminescent of Schrödinger's equation in quantum mechanics, has been solved for a number of cases [7-9, 19-22] and gives a physically reasonable description of macroscopically large permanent magnets.

3.1.2. Brown's paradox. Let us start with the highly idealized homogeneous limit $K_1(\mathbf{r}) = \mathrm{const.}$ In this case the eigenvalue problem Eq. (12) is solved by plane waves $\mathbf{m} = \mathbf{m}_o \exp(i\mathbf{k}\mathbf{r})$, and the lowest eigenvalue corresponds to the $\mathbf{k} = 0$ mode. The result of this consideration is the well-known but nevertheless non-trivial nucleation field $H_N = H_o = 2K_1/\mu_o M_o$. Using $\mu_o M_o = 1.54$ T and $K_1 = 8.89$ MJ/m^3 for $Sm_2Fe_{17}N_3$ [23] we obtain $H_o = 14.5$ T, while observed coercivities are much lower.

The reason for this disagreement, which represents another formulation of Brown's paradox (Section 2.2.3), is the presence of lattice inhomogeneities. Calculation shows that soft impurities reduce the nucleation field if they are larger than the Bloch-wall width δ. Illustratively speaking, comparatively small fields are sufficient to switch large soft regions [20]. The dependence of the nucleation field on the size D of the soft regions is determined by the anisotropy profile. Typical results are $H_N \sim 1/D$ for parabolic K_1 profiles [7] [19] and $H_N \sim 1/D^2$ for soft inclusions in a hard matrix [8, 20]. Some results for random magnets are given in Appendix B.

These results show that the processing of nucleation-controlled permanent magnets requires the removal of all inhomogenities which are larger than the domain-wall width. Possible 'candidates' for coercivity destruction are bulk inhomogenities, soft inclusions, and surface defects, but in practice it is difficult to separate these contributions. Traditional ways to obtain high nucleation fields are high-temperature annealing of the bulk material and restriction to small particle sizes. The latter requirement is helpful to restrict the effect of scattered inhomogeneities to a small volume fraction of the whole magnet.

3.2. THE SOFT-CORE PROBLEM

In the case of partly nitrided $Sm_2Fe_{17}N_x$ grains it is possible to give a clear explanation of the observed low coercivity [24]. As opposed to the strong easy-axis anisotropy of fully modified nitrides and carbides, the Sm_2Fe_{17} parent compound exhibits easy-plane anisotropy ($K_1 < 0$) with nearly no coercivity. However, diffusion kinetics are sluggish at typical modification temperatures (400-500 °C), and due to the R^2 dependence of the necessary diffusion time the core of large grains can remain unmodified even if small particles are fully charged [25].

This 'soft-core' phenomenon is responsible for the low coercivity of incompletely nitrided $Sm_2Fe_{17}N_x$ magnets. Due to the macroscopic size of the soft core the coercivity vanishes if the $K_1 = 0$ at the particle centre [24]. Using experimental diffusion constants [25] and assuming a linear dependence of $K_1(\mathbf{r})$ on $x(\mathbf{r})$ [26] it is predicted that the coercivity of particles, nitrided 5 h at 450 °C, breaks down if the particle radius is larger than 0.75 µm. The experimental break-down radius, obtained by measurements on Zn-bonded powder fractions, is 0.6 µm [24].

4. Processing of $Sm_2Fe_{17}N_3$ Magnets

Summary. The low diffusion of nitrogen in metals excludes the interstitial modification of bulk magnets and various processing routes are employed to prepare Sm_2Fe_{17} material suitable for nitrogenation. The conventional powder route involving mechanical crushing and milling of bulk alloys produces microcrystalline powder suitable for anisotropic applications Nanocrystalline material, isotropic in nature, is produced by mechanical alloying, melt spinning and Hydrogen Disproportionation Desorption Recombination (HDDR). Nitrogen is introduced into these structures upon heating at 400-550 °C in a nitrogen containing atmosphere (e g. N_2, NH_3,H_2/N_2).

4.1. MILLING AND SIEVING

4.1.1. Starting alloy. The starting material is Sm_2Fe_{17} alloy in ingot or powder form. Ingots are produced by both induction and arc melting while reduction diffusion using Sm_2O_3, Fe and Ca yields Sm_2Fe_{17} powder [27]. During induction melting the constituents are melted under an argon atmosphere by induction heating in a crucible which is usually made of alumina, and the molten alloy is poured into cylindrical or rectangular moulds. Another common method is vacuum arc melting where the arc between a tungsten rod and copper crucible melts the alloy. In both cases excess amounts (5-20 wt%) of Sm metal are added to compensate vapourisation losses. To remove secondary α-iron it is necessary to homogenise the alloy at 950-1100 °C for up to 7 days.

The co-reduction process involves the reaction

$$Sm_2O_3 + Fe_2O_3 + 15\,Fe + 6\,Ca \rightarrow Sm_2Fe_{17} + 6\,CaO \quad (13)$$

The advantage of this route is that the reduction of the rare earth oxide is incorporated into the alloy forming step and furthermore, the necessary milling time is reduced.

4.1.2. Milling. Initially, jaw crushing is employed to prepare the ingot material for coarse crushing. The choice of milling devices used to reduce the material to fine powder ranges from laboratory scale motar and roller mills to industrial scale ball, attritor and jet mills. Roller, ball and attritor mills use hardened stainless steel balls and a solvent such as cyclohexane to aid dispersal and reduce oxidation. Wet milling requires an additional drying step, and it is difficult to control the oxygen pick-up. Jet milling is a dry process, where high pressure gas jets smash-up particles into progressively finer powder which is selectively collected.

From the point of view of nitrogenation kinetics, it is desirable to have as small as possible particles, while the rare earth's chemical affinity to oxygen requires a restriction of the surface-to-volume ratio. Experience indicates an optimum particle size of about 3 μm [24, 28]. An important characteristic is the particle size distribution after milling. Large grains exhibit an un-nitrided soft core which leads to zero single-particle coercivity and an unfavourable inflection in the hysteresis loop (Fig. 3). Note that the nitrogenation time scales as the square of the particle size, so the large-size tail of the particle size distribution is particularly critical [25]. A widely used method to avoid these problems is sieving, but from the point of view of industrial-scale processing it is desirable to restrict the number of processing steps as much as possible. It is therefore appropriate to use milling procedures, such as jet milling [29], which yield an intrinsically narrow size distribution.

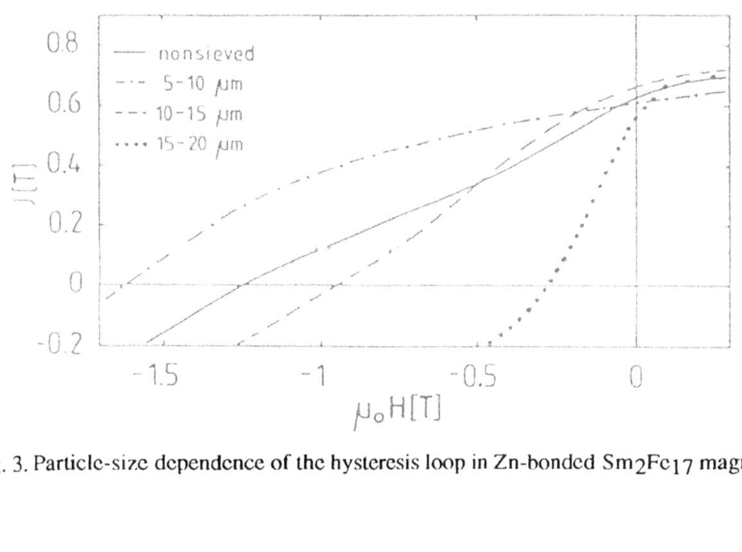

Fig. 3. Particle-size dependence of the hysteresis loop in Zn-bonded Sm_2Fe_{17} magnets [28].

4.2. MECHANICAL ALLOYING, MELT SPINNING AND HDDR

4.2.1. Mechanical alloying. High-energy ball milling of elemental Sm and Fe powder yields an amorphous samarium-iron matrix with embedded nanocrystallites of α–Fe. Suitable annealing is then used to establish a nanocrystalline Sm_2Fe_{17} structure. Mechanical alloying yields isotropic magnets with

$$M_r \leq M_o \int_{\theta=0}^{\pi/2} \cos\theta \sin\theta \, d\theta = \frac{1}{2} M_o \tag{14}$$

and the rather disappointing energy product $(BH)_{max} \leq \mu_o M_o^2/16$, but this disavantage is partly outweighed by the excellent coercivity of mechanically alloyed magnets (see Section 5).

4.2.2. Melt spinning. In the melt-spinning process a jet of molten alloy is fired at a rapidly rotating, water-cooled copper wheel where it solidifies at a cooling rate of about 10^6 K/s. Thrown off the wheel, it forms fine ribbons which are then used to produce bonded magnets. The physical state of the ribbon depends upon the quenching rate, i.e. wheel speed, ejection conditions and melt temperature; grain sizes can be altered by additional annealing treatments. Rapidly quenched ribbons contain a disordered hexagonal $TbCu_7$-type phase (1:9 phase) besides the rhombohedral Sm_2Fe_{17} phase, which destroys coercivity after nitrogenation [30].

4.2.3. HDDR. There are two different reactions of Sm_2Fe_{17} with hydrogen: the gas-solid-solution absorption below about 200 °C and the lattice disproportionation at elevated temperatures (> 550 °C):

$$Sm_2Fe_{17} + (2+\delta)\, H_2 \rightarrow 2\, SmH_{2+\delta} + 17\, Fe \tag{15}$$

where the excess δ arises from the solubility of hydrogen in SmH_2.

HDDR (Hydrogen Disproportionation Desorption Recombination) exploits the formation of samarium dihydride. First, coarse-grained ingot material is heated to form $SmH_{2+\delta}$ and α–Fe according to Eq. (15). Subsequent heating in *vacuum* causes the hydride to decompose, and the thermodynamically unstable intimate mixture of α-Fe and Sm reverts to the equilibrium Sm_2Fe_{17} phase with a favourable fine-grain structure ($D \leq 1$ μm). Note that Sm_2Fe_{17} 'HDDR' is a somewhat unlucky coinage: it must not be confused with $Nd_2Fe_{14}B$ HDDR [31], which has a slightly different meaning (Hydrogen Decrepitation Desorption Recombination).

4.3. NITROGENATION

A detailed description of the nitrogenation kinetics lies beyond the scope of this lecture (see [25, 32]), and we will restrict ourselves to some basic features. The temperature dependece of the nitrogenation reaction is given by the Boltzmann factor $\exp(-E_a/k_bT)$, where E_a=133 kJ/mole is the activation energy for nitrogen diffusion in Sm_2Fe_{17}. Depending on the particle size, typical reaction times vary between a few hours and a few days, while it is suitable to conduct the nitrogenation reaction at temperatures between 400 °C and 450 °C. Note that industrial scale nitrogenation presupposes the control of the exothermic heat of - 57 kJ/mole [25, 32], which is able to cause surface disproportionation and α-Fe precipitation if not properly removed. In fact, it is known that gentle milling after nitrogenation of single crystalline material, isolates potential α-Fe nuclei and improves the coercivity (see e.g. [33]).

Careful hydrogen and ammonia pretreatment can be used to improve the nitrogenation conditions by enhancing the surface activity and causing microcracks, but nitrogenation in pure ammonia easily destroys the intermetallic lattice. The diffusivities of carbon and nitrogen in Sm_2Fe_{17} are comparable [32], so the production of $Sm_2Fe_{17}C_{3-\delta}$ using hydrocarbon gases such as CH_4 is very similar to gas-phase interstitial modification using molecular nitrogen.

5. Bonded Magnets

Summary. Bonding is necessary to produce permanent magnets from $Sm_2Fe_{17}N_3$ powder. Depending on the alignment of the crystallites, there are two main types of bonded magnets: isotropic and anisotropic magnets. Due to the larger remanent magnetisation, magnets with anisotropic texture are superior to those with isotropic alignment. Although some progress has been made with polymer bonding and metal-bonding using low-melting metals such as zinc, the production of bonded nitrides is still in its development stage; the main problem is to achieve alignment in sufficiently dense magnets.

5.1. METAL BONDING

5.1.1. Metallurgical aspects. Liquid phase sintering is a well established processing route in the production of $Nd_2Fe_{14}B$ magnets, where densities of more than 90% are achieved. However, due to the disproportionation behaviour of the $Sm_2Fe_{17}N_3$ it is not possible to produce interstitial magnets by high temperature sintering [1, 34]. Metal-bonding using low-melting metals is one possibility [28, 35-39]; the intention is to wet and separate the $Sm_2Fe_{17}N_3$ grains, much the same as the neodymium-rich grain-boundary phase in Nd-Fe-B magnets does during liquid-phase sintering. At present the best results have been achieved using zinc, though the Sm-Fe-Zn phase diagram does not support this trend: metal bonding is thermodynamically less favourable than, in this particular case, the formation of Sm_2Zn_{17} and bcc iron [40]. Zinc-bonded nitrides are therefore non-equilibrium systems which require a careful heat treatment to optimise coercivity and remanence (Fig. 4). The temperature dependence of the hysteresis loop is shown in Fig. 5.

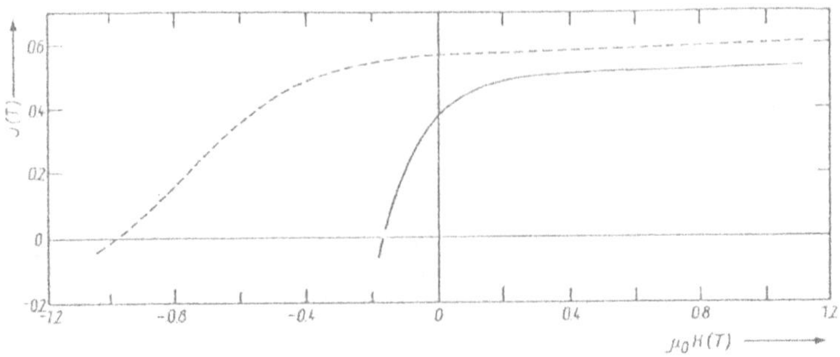

Fig. 4. Typical hysteresis loop of a Zn-bonded magnet for different heat treatments [35].

Zinc has several effects on the microstructure which favour the development of coercivity: (i) it forms the non-magnetic phase Fe_3Zn_7 with harmful free iron found on the surface of nitrided particles, (ii) it tends to smooth the surface of $Sm_2Fe_{17}N_3$ particles by eroding sharp edges and corners which facilitate the nucleation of reverse domains, and (iii) it provides magnetic isolation by forming non-magnetic sheaths coating the individual particles. Note that pressure has been found to improve the zinc coating of $Sm_2Fe_{17}N_3$ grains [5].

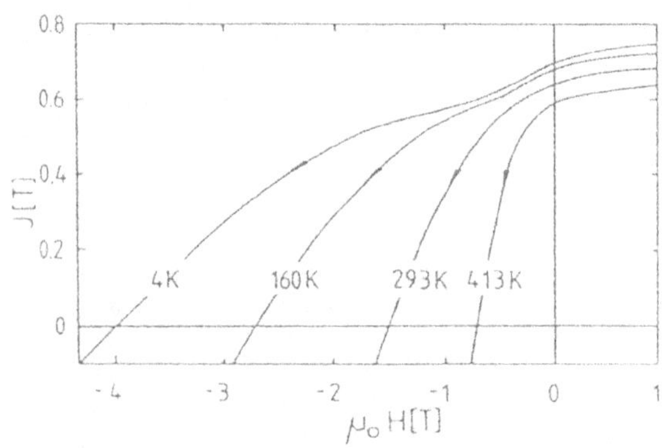

Fig. 5. Temperature dependence of the coercivity in Zn-bonded Sm2Fe17N3 magnets [28].

5.1.2. *Densification.* In a wider sense, the problem of metal bonding consists in achieving a sufficiently dense packing of the aligned hard-magnetic powder. The energy product scales as the square of the remanance M_r, which is, in turn, proportional to the volume fraction f of the magnetic phase. From the maximum volume fraction of randomly packed spheres with no size distribution, f = 0.64 [41], we obtain energy-product reduction factors of $f_2 = 0.41$ and $f_2/4 = 0.10$ for anisotropic and isotropic magnets, respectively. In reality, it is possible to leave the simplistic random-hard-sphere regime, for instance by adding a fraction of somewhat smaller particles which fill the holes of the original random packing to yield $f = 0.72$ [29]. Typical zinc contents are of order 15 wt%; Zn contents much larger than this value dilute the magnetisation and tend to destroy the intermetallic lattice [37, 40]. The temperature dependence of the hysteresis-loop is shown in Fig. 6. Compared to $Sm_2Fe_{17}N_{3-\delta}$, the overall permanent magnetic properties of $Sm_2Fe_{17}C_{3-\delta}$ are slightly inferior (see Appendix). Nevertheless, it has been possible to produce Zn-bonded carbides with a coercivity of $\mu_oH_c = 0.9$ T [42].

5.2. POLYMER BONDING

An alternative method to produce bonded $Sm_2Fe_{17}N_3$ magnets is polymer bonding [38, 44, 45]. The main advantage of polymer bonded magnets is their mechanical behaviour: they are much less brittle than metal bonded or sintered magnets, which makes them suitable for micromechanical applications, for instance in wrist watch motors.

The use of polymers in permanent-magnet processing reflects the versatility of polymeric materials. Epoxy resins are used to prepare laboratory scale magnets for experimental purposes, while simple polymers such as polyvinylchloride $[-CH_2CHCl-]_n$ with plasticizer serve to produce permanent magnets with rubber like properties. Polyamides such as polycaprolactam $[-NH(CH_2)_5CO-]_n$ and its derivatives (nylons) are most suitable to produce high-quality magnets [43] .

Table 3. Magnetic properties of $Sm_2Fe_{17}N_3$ permanent magnets.[a]

H_c T	M_r T	$(BH)_{max}$ kJ/m^3	Processing	Ref.
			Anisotropic	
1.60	0.64	70	zinc-bonded	[28]
1.35	0.77	97	zinc-bonded	[29]
0.88	0.66	67	zinc-bonded	[36]
0.69	0.84	103	resin-bonded	[44]
1.08	0.81	70	resin-bonded	[45]
			Isotropic	
4.36	0.40	30	mechanically alloyed and zinc-bonded	[5]
2.94	0.71	87	resin-bonded	[50]
2.10	0.73	66	melt spun	[30]
2.32	0.61	59	melt-spun carbide[b]	[46]
0.80	0.9	—	HDDR	[47]
1.60	0.75	—	HDDR	[48]

[a]Higher but less reliable $(BH)_{max}$ values are quoted in [51]. [b]The only carbide in this table.

The application temperature of polymer bonded magnets is given by the glass-transition and melting points of the polymer. Many polymers melt at temperatures just above 100 °C, but others, such as polytetrafluoroethylen (teflon) $[-CF_4-]_n$ with T_m=327 °C remain solid at temperatures at which $Sm_2Fe_{17}N_3$ is superior to $Nd_2Fe_{14}B$. On the other hand, low temperatures polymer-bonded magnets become brittle and plasticizers have to be added to suppress the glass-transition temperature.

Using methods such as powder pressing or injection moulding it is comparatively easy to obtain energy products of order 100 kJ/m3, but the achievement of much larger energy products is difficult because it requires the alignment of magnets with a high packing fraction.

6. Outlook

Summary. Until recently, the development of permanent magnetic materials relied nearly entirely on the discovery of new phases with improved magnetization, Curie temperature, and anisotropy. However, the number of suitable iron-rich rare-earth intermetallics is limited, and alternative approaches have to be found to yield a further improvement of permanent magnetic properties. A promising method to achieve this aim is the production of nano-structured two-phase systems.

6.1. REMANENCE ENHANCEMENT

A possible way to improve the comparatively low remanence of isotropic magnets is to utilize exchange coupling [49, 52-54]. Exchange interaction favours magnetic alignment, which can be used to improve the small (remanent) magnetization $M_r < M_o/2$ of isotropic magnets. On the other hand, exchange coupling destroys coercivity, and only the intrinsic surplus anisotropy makes the production of exchange-enhanced magnets possible.

Remanence enhancement becomes more pronounced if nano-structured soft regions with high saturation magnetization are exchange-coupled to a hard isotropic matrix. Examples are nanocrystalline composites $Nd_2Fe_{14}B$/$\underline{Fe_3B}$Fe and $Sm_2Fe_{17}N_3$/Fe produced by melt-spinning [53] and mechanical alloying [49], respectively. Experiment shows that exchange coupling improves the comparatively low remanence $M_r \approx M_o/2$ of the isotropic hard phase [49, 52, 54], but the energy product, though improved with respect to the isotropic single-phase rare-earth material, does not reach the level attained in oriented rare-earth magnets.

6.2. ALIGNED TWO-PHASE MAGNETS

Recent work [8, 22, 60] has shown how it should be possible to substantially increase the energy product in oriented nanostructured two-phase magnets by exploiting exchange coupling between hard and soft regions. The idea behind these systems is to break out of the straitjacket of natural crystal structures by artificially structuring new materials. The concept is similar to that of the 4f-3d intermetallics themselves, but on a different scale, where the atoms are replaced by mesoscopically large soft- and hard-magnetic blocks (Fig. 6). Illustratively speaking, hard regions act as a skeleton which stabilizes the magnetization of the soft phase.

Lowest-order perturbation treatment of Eq. (12), which applies to aligned nano-structed two-phase magnets, predicts a maximum energy product [8, 22, 60]

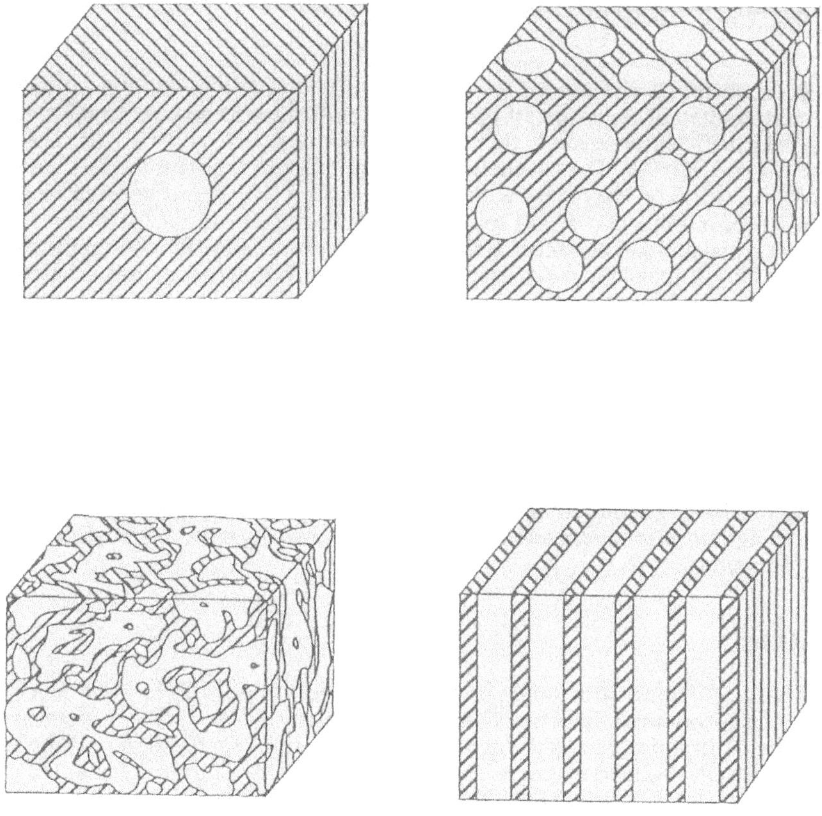

Fig 6 Aligned nanostructured two phase magnets The hard skeleton (dashed area) serves to align soft regions with high magnetizations

$$(BH)_{max} = \frac{1}{4}\,\mu_0 M_s^2 \left(1 - \frac{\mu_0(M_s - M_h)M_s}{2\,K_h}\right) \qquad (16)$$

where the subscripts h and s refer to the the hard and soft phases, respectively. This result is independent of the shape of the hard and soft regions so long as the size of the soft regions is sufficiently small; higher-order corrections to Eq. (16) involve correlation functions $<K_1(\mathbf{r})\,K_1(0)>$. Due to their high magnetization, soft materials such as Fe with $\mu_0 M_s = 2.15$ T or $Fe_{65}Co_{35}$ with $\mu_0 M_s = 2.43$ T [6] could be used to produce $Sm_2Fe_{17}N_3$-containing two-phase magnets with energy products of order 1 MJ/m^3 ('giant energy product'). The factor $1/K_h$ in Eq. (16) indicates the non-applicability of this equation to soft-magnetic materials: giant-energy-product magnetism is based on the strong rare-earth sublattice anisotropy of the hard phase. On the other hand, it is remarkable that these optimum magnets are almost entirely composed of 3d metals, with only 2 wt% samarium.

The practical problem, however, is to realize a structure where the soft regions are sufficiently small to avoid nucleation at small fields while having the hard regions crystographically aligned. A conceivable solution are multilayers consisting of hard and soft magnetic layers, but the advantages of these hypothetical 'megajoule' magnets — very high energy product and low raw material costs — are largely outweighed by the demanding processing requirements. For this reason the use of aligned nano-structured two-phase magnets will be restricted to special applications such as micromechanics or thin-film electronics.

Conclusions

The coercivity of micron-structured permanent magnets is nucleation controlled, so that the processing of oriented $Sm_2Fe_{17}N_3$ magnets requires careful powder processing and bonding. The central problem is the achievement of a sufficiently high packing fraction in polymer and zinc bonded magnets, and it is not possible yet to make detailed market-share predictions for, lets say, PTFE-bonded $Sm_2Fe_{17}N_3$ magnets.

In the case of nanostructured magnets the isotropic character of the material yields a further decrease in energy product, but here the excellent coercivity may open niche applications. Practical progress has been made towards the establishment of Sm-Fe HDDR processing routes, while the production of hypothetical giant-energy-product magnets such as $Sm_2Fe_{17}N_3$/FeCo remains a demanding though realistic challenge.

Acknowledgements

The authors are grateful to J. M. D. Coey, P. A. P. Wendhausen, K. Kobayashi, S. Wirth and K.O'Donnell for help in details. This work forms part of the BRITE/EURAM program of the European Commission.

Appendix: Intrinsic Properties of Permanent Magnets

Table A1. Intrinsic magnetic properties at room temperature.

Material	T_c (°C)	M_o (T)	K_1 (kJ/m^3)	Ref.
Fe	770	2.15	0.047	[14]
Co	1120	1.76	0.53	[52]
Ni	356	0.61	0.0051	[6]
$BaFe_{12}O_{19}$	467	0.478	0.325	[6]
$BaFe_{12}O_{19}$	450	0.470	0.320	[52]
$SrFe_{12}O_{19}$	477	0.478	0.357	[6]
$PbFe_{12}O_{19}$	452	0.402	0.220	[6]
$SmCo_5$	730	1.05	11.9	[52]
$SmCo_5$	747	1.07	17.2	[6]
Sm_2Co_{17}	917	1.20	3.3	[6]
$Nd_2Fe_{14}B$	315	1.61	4.3	[56]
$Nd_2Fe_{14}B$	315	1.59	5.0	[16]
$Sm_2Fe_{17}N_{3-\delta}$	476	1.54	8.9	[8, 23]
$Sm_2Fe_{17}N_{2.9}$	473	1.52	8.9	[57]
$Sm_2Fe_{17}C_{3-\delta}$	395	1.45	7.4	[23]

Table A2. Anisotropy constants of rare-earth intermetallics at and above room temperature.

Material	T (°C)	K_1 (MJ/m^3)	K_2 (MJ/m^3)	M_o (T)	Ref.
$Sm_2Fe_{17}N_{3-\delta}$	RT	8.9	1.46	1.54	[23]
$Sm_2Fe_{17}N_{3-\delta}$	RT	8.1	0.70	1.42	[58]
$Sm_2Fe_{17}N_{2.9}$	RT	8.9	1.90	1.52	[57]
$Sm_2Fe_{17}N_{3-\delta}$	*100*	*6.6*	*0.70*	*1.47*	[23]
$Sm_2Fe_{17}N_{3-\delta}$	*100*	*7.0*	*0.65*	*1.38*	[58]
$Sm_2Fe_{17}N_{3-\delta}$	*180*	*4.7*	*0.25*	*1.36*	[23]
$Sm_2Fe_{17}N_{3-\delta}$	*180*	*5.8*	*0.56*	*1.28*	[58]
$Sm_2Fe_{17}C_{3-\delta}$	RT	7.4	0.74	1.45	[23]
$Sm_2Fe_{17}C_{3-\delta}$	*100*	*5.3*	*0.23*	*1.33*	[23]
$Sm_2Fe_{17}C_{3-\delta}$	*180*	*3.5*	*0.03*	*1.19*	[23]
$Nd_2Fe_{17}N_{2.7}$	RT	- 6.8	0.80	1.60	[57]
$Nd_2Fe_{14}B$	RT	5.0	0.66	1.59	[16]
$Nd_2Fe_{14}B$	RT	4.3	0.65	1.61	[7]
$Sm_2Fe_{14}B$	RT	- 12.0	0.29	1.49	[4]

At room temperature, K_3 = - 0.20 MJ/m^3 for $Nd_2Fe_{14}B$ [7] and, due to $\gamma_J = 0$ for the Sm^{3+} ground state [59], $K_3 \approx 0$ for $Sm_2Fe_{17}N_x$ and $Sm_2Fe_{17}C_x$.

References

[1] J. M. D. Coey and H. Sun, J. Magn. Magn. Mater. **87**, L251 (1990).
[2] J. M. D. Coey, Physica Scripta **T39**, 21 (1991).
[3] J. F. Herbst, Rev. Mod. Phys. **63**, 819 (1991).
[4] M. Sagawa, S. Fujimura, H. Yamamoto, Y. Matsuura, S. Hirosawa, J. Appl. Phys. **57**, 4094 (1985); M. Sagawa, S. Hirosawa, H. Yamamoto, S. Fujimura, and Y. Masuura, Jpn. J. Appl. Phys. **26**, 785 (1987).
[5] C. Kuhrt, K. O'Donnell, M. Katter, J. Wecker, K. Schnitzke, and L. Schultz, Appl. Phys. Lett. **60**, 3316 (1992).
[6] J. E. Evetts, ed., *Concise Encyclopedia of Magnetic and Superconducting Materials*, Pergamon, Oxford, 1992.
[7] H. Kronmüller, "Micromagnetic Background of Hard Magnetic Materials", in: *Supermagnets, Hard Magnetic Materials*, G. J. Long and F. Grandjean, eds., Kluwer, Dordrecht, 1991, p.461.
[8] R. Skomski and J. M. D. Coey, Phys. Rev. B **48**, 15812 (1993).
[9] S. Chikazumi, *Physics of Magnetism*, Wiley, New York, 1964.
[10] D.J. Craik and R. S. Tebble, Rep. Prog. Phys. **24**, 116 (1961).
[11] J.-F. Hu, I. Kleinschroth, R. Reisser, H. Kronmüller, and Sh. Zhou, phys. stat. sol. (a) **138**, 257 (1993).
[12] W. F. Brown, Rev. Mod. Phys. **17**, 15 (1945).
[13] W. F. Brown, Phys. Rev. **105**, 1479 (1957).
[14] H. Kronmüller and T. Schrefl, J. Magn. Magn. Mater. **129**, 66 (1994).
[15] R. S. Tebble and D. J. Craik, *Magnetic Materials*, Wiley, New York, 1969.
[16] K. H. J. Buschow, Mater. Sci. Rep. **1**, 1 (1986).
[17] Y. Otani, A. Moukarika, H. Sun, J. M. D. Coey, E. Devlin, and I. R. Harris, J. Appl. Phys. **69**, 6735 (1991).
[18] T. Mukai and T. Fujimoto, J. Magn. Magn. Mat. **103**, 165 (1992).
[19] H. Kronmüller, phys. stat. sol. (b) **144**, 385 (1987).
[20] R. Skomski, phys. stat. sol. (b) **174**, K77 (1992).
[21] S. Nieber and H. Kronmüller, phys. stat. sol. (b) **153**, 367 (1989).
[22] R. Skomski and J. M. D. Coey, IEEE Trans. Magn. **29**, 2860 (1993).
[23] K.-H. Müller, D. Eckert, P. A. P. Wendhausen, A. Handstein, S. Wirth, and M. Wolf, IEEE Trans. Magn., in press (1994).
[24] R. Skomski, K.-H. Müller, P. A. P. Wendhausen, and J. M. D. Coey, J. Appl. Phys. **73**, 6047 (1993).
[25] R. Skomski and J. M. D. Coey, J. Appl. Phys. **73**, 7602 (1993).
[26] R. Skomski, M. D. Kuz'min, and J. M. D. Coey, J. Appl. Phys. **73**, 6934 (1993).
[27] T. Y. Liu, W. C. Chang, C. J. Chen, T. Y. Chu, and C. D. Wu, IEEE Trans. Magn. **28**, 2593 (1992).
[28] K.-H. Müller, P. A. P. Wendhausen, D. Eckhert, and A. Handstein, Proc. 7th Int. RE-TM Symp., Canberra, 1992, p. 32.
[29] B. Wall, M. Katter, W. Rodewald, and M. Velicescu, IEEE Trans. Magn. 1994 (in press).
[30] M. Katter, J. Wecker, and L. Schultz, J. Appl. Phys. **70**, 3188 (1991).
[31] K. H. J. Buschow, this volume, and references therein.
[32] J. M. D. Coey, R. Skomski, and S. Wirth, IEEE Trans. Magn. **28**, 2332 (1992).
[33] S. Brennan, K. Kobayashi, and J. M. D. Coey, to be published.
[34] R. Skomski and J. M. D. Coey, J. Mater. Eng. Perf. **2**, 241 (1993).

[35] P. A. P. Wendhausen, A. Handstein, P. Nothnagel, D. Eckert, and K.-H. Müller, phys. stat. sol. (a) **127**, K121 (1991).
[36] W. Rodewald, B. Wall, M. Katter, M. Velicescu, and P. Schrey, J. Appl. Phys. **73**, 5899 (1993).
[37] P. A. P. Wendhausen, D. Eckert, A. Handstein, K.-H. Müller, G. Leitner, and R. Skomski, J. Appl. Phys. **73**, (1993) .
[38] W. Rodewald, M. Velicescu, B. Wall, and G. W. Reppel, 12th Int. RE Workshop, Canberra, 1992, p. 191.
[39] Y. Otani, A. Moukarika, H. Sun, J. M. D. Coey, E. Devlin, and I. R. Harris, J. Appl. Phys. **69**, 5584 (1991).
[40] B. Reinsch, H. H. Stadelmaier, and G. Petzow, IEEE Trans. Magn., in press (1993).
[41] D. Stoyan and J. Mecke, *Stochastische Geometrie*, Akademieverlag, Berlin 1983.
[42] P. A. P. Wendhausen, B.-P. Hu, B. Gebel, D. Eckart, A. Handstein, and K.-H. Müller, IEEE Trans. Magn., in press (1994).
[43] K. Kobayashi, private communication (1994).
[44] S. Suzuki and T. Miura, IEEE Trans. Magn. **28**, 994 (1992).
[45] Y.-L. Liu, D.-W. Wang, B.-P. Hu, J.-L. Gao, Q. Song, L. Liu, X.-L. Rao *et al.*, Chin. Sci. Bull. **37**, 289 (1992).
[46] C. Kuhrt, M. Katter, J. Wecker, K. Schnitzke, and L. Schultz, Appl. Phys. Lett. **60**, 2029 (1992).
[47] Sh.-Z. Zhou, J. Yang, M.-C. Zhang, D.-Q. Ma, F.-B. Li, and R. Wang, Proc. 12th Int. Workshop RE Magnets & Appl., Canberra, Australia, 1992, p. 44.
[48] P. A. P. Wendhausen, B. Gebel, N. M. Dempsey, K.-H. Müller, and J. M. D. Coey, to be published.
[49] J. Ding, P. G. McCormick, and R. Street, J. Magn. Magn. Mater. **124**, L1 (1993).
[50] K. Schnitzke, L. Schultz, J. Wecker, and M. Katter, Appl. Phys. Lett. **57**, 2853 (1990).
[51] J. Ding, P. G. McCormick, R. Street, and P. A. I. Smith, Proc. 12th Int. RE Workshop, Canberra, 1992, p. 428.
[52] E. F. Kneller and R. Hawig, IEEE Trans. Magn. **27**, 3588 (1991).
[53] R. Coehoorn, D. B. de Mooij, J. P. W. B. Duchateau, and K. H. J. Buschow, J. Physique **49**, C8-669 (1988).
[54] K.-H. Müller, J. Schneider, A. Handstein, D. Eckhert, P. Nothnagel, and H. R. Kirchmayr, Mat. Sci. Eng. **A133**, 151 (1991).
[56] T. Schrefl and J. Fidler, J. Magn. Magn. Mater. **111**, 105 (1992).
[57] M. Katter, J. Wecker, C. Kuhrt, L. Schultz, X. C. Kou, and R. Grössinger, J. Magn. Magn. Mater. **111** (1992) 293.
[58] J.-F. Hu, X.-C. Kou, H. Kronmüller, and Sh.-Z. Zhou, phys. stat. sol. (a) **134**, 499 (1992).
[59] J. M. D. Coey, "Intermetallic Compounds and Crystal-Field Interaction", in: *Science and Technology of Nanostructured Materials*, G. C. Hadjipanayis and G. A. Prinz, eds., Plenum Press, New York, 1991, p. 439.
[60] J. M. D. Coey and R. Skomski, Physica Scripta **T49**, 315 (1993).

APPENDIX: List of ASI Participants

Juan Bartolomé
Instituto de Ciencias de Materiales de Aragon
Facultad de Ciencias
C. S. I. C.-Univ. de Zaragoza
E-500009 Zaragoza
Spain

Andrea Biscarini
Dipartimento di Fisica
Università di Perugia
Via Pascoli 5
I-06100 Perugia
Italy

David Book
School of Metallurgy
University of Birmingham
Edgbaston
Birmingham, B15 2TT
UK

Aomar Boukraa
c/o Dr. P. Vajda
L. S. I.
Ecole Polytechnique
F-91128 Palaiseau
France

Jan H. V. J. Brabers
Philips Research Laboratories
P. O. Box 80000
NL-5600 JA Eindhoven
The Netherlands

Stephen Brennan
Department of Pure & Applied Physics
Trinity College
University of Dublin
Dublin 2
Ireland

Craig Buckley
Department of Physics
University of Salford
Salford, M5 4WT
UK

K. H. Jürgen Buschow
Van der Waals-Zeeman Laboratorium
University of Amsterdam
Valckenierstraat 65
NL-1018 XE Amsterdam
The Netherlands

Vanyo Christo
Inst. Nuclear Research and Energy
72 Trakia Boulevard
1784 Sofia
Bulgaria

Fermin Cuevas
Dept. Fisica de Materiales, C-IV
Facultad de Ciencias
Univ. Autonoma de Madrid
Cantoblanco
E-28049 Madrid
Spain

Pierre Dantzer
CNRS, URA 446
Université de Paris-Sud
F-91405 Orsay Cedex
France

C. H. de Groot
Philips Research Laboratories
Prof. Holstlaan 4
NL-5656 AA Eindhoven
The Netherlands

Kees de Kort
Philips Research Laboratories
Prof. Holstlaan 4
NL-5656 AA Eindhoven
The Netherlands

Nora Dempsey
Department of Pure & Applied Physics
Trinity College
University of Dublin
Dublin 2
Ireland

Virginie De Pauw
Laboratoire de Microscopie
Electronique
Faculté des Sciences de Rouen
Place Emile Blondel
F-76821 Mont Saint-Aignan Cedex
France

Andrei V. Eremenko
Inst. for Single Crystals
60 Lenin Avenue
Kharkov-1, 310001
Ukrainia

Jose F. Fernandez
Facultad de Ciencias, C-IV
Univ. Autonoma de Madrid
Cantoblanco
E-28049 Madrid
Spain

George Filoti
Experimental Physics VI
Ruhr University
D-44780 Bochum
Germany

Ted B. Flanagan
Department of Chemistry
University of Vermont
Burlington, VT 05405
USA

Andreas Fölzer
Treibacher Chemische Werke AG
Auer v. Welsbachstr. 1
A-9330 Treibach-Althofen
Austria

Luis Miguel Garcia
Instituto de Ciencias de Materiales
de Aragon
Facultad de Ciencias
CSIC-Univ. de Zaragoza
E-50009 Zaragoza
Spain

Danilo Gelli
CNR-ITM
Via Bassini 15
I-20133 Milano
Italy

Fernande Grandjean
Institut de Physique, B5
Université de Liège
B-4000 Sart-Tilman
Belgium

Richard Grant
Department of Physics
Old Dominion University
Norfolk, VA 23529
USA

Michael Hirscher
Max-Planck-Inst. Metallforschung
Heisenbergstrasse 1
D-70569 Stuttgart
Germany

Olivier Isnard
Laboratoire de Cristallographie
BP 166X-CNRS
F-38042 Grenoble Cedex 09
France

Jelena Jaksic
Department of Chemical Engineering
c/o Prof. C. G. Vayenas
University of Patras
GR-26500 Patras
Greece

Sitaram S. Jaswal
Behlen Laboratory of Physics
University of Nebraska
Lincoln, NB 68588-011
USA

Christian Jordy
SAFT Recherche
Route de Nozay
F-91460 Marcoussis
France

Czeslaw Kapusta
Department of Physcis and Astronomy
University of St. Andrews
North Haugh
St. Andrews, Fife KY16 9SS
Scotland, UK

Mohamed Kemali
Physics Department
Joule Laboratory
University of Salford
Salford M5 4WT
England, UK

Markus Kieninger
Max-Planck-Inst. Metallforschung
Seestrasse, 75
D-70174 Stuttgart
Germany

Ibrahim Kilicaslan
Kocaeli Üniversitesi
Teknik Egitim Fakultesi
Kocaeli
Turkey

Michael Kuzmin
ICMA Zaragoza
Dept. Materia Condensada
Facultad de Ciencias
Pl. San Francisco
E-50009 Zaragoza
Spain

Patrick Leblanc
SAFT Recherche
Route de Nozay
F-91460 Marcoussis
France

J. Ping Liu
Van der Waals-Zeeman Laboratorium
Universiteit van Amsterdam
Valckenierstraat 65
NL-1018 XE Amsterdam
The Netherlands

Gary J. Long
Department of Chemistry
University of Missouri-Rolla
Rolla, MO 65401-0249
USA

Jeffrey R. Long
Department of Chemistry
Harvard University
12 Oxford Street
Cambridge, MA 021138
USA

Christof Marte
Max-Planck-Inst. Metallforschung
Seestrasse, 75
D-70174 Stuttgart
Germany

Dave P. Middleton
Philips Research Laboratories
Box WA-12
Prof. Holstlaan 4
NL-5656 AA Eindhoven
The Netherlands

Sanjay R. Mishra
Department of Physics
University of Missouri-Rolla
Rolla, MO 65401
USA

Dieter Nagengast
Hahn-Meitner Institut
Glienickerstrasse 100
D-14109 Berlin
Germany

Peter H. L. Notten
Philips Research Laboratories
P. O. Box 80000
NL-5600 JA Eindhoven
The Netherlands

Emilio Orgaz
Université de Paris-Sud
URA, Bât. 415
F-91405 Orsay-Cedex
France

Antoni T. Pedziwiatr
Institute of Physics
Jagellonian University
Reymonta 4
PL-30059 Krakow
Poland

Carmen Piqué-Rami
Escuela Tecnica Superior
Ingenieros Industriales
Departamento de Fisica
Campus de Viesques
E-33294 Gijon
Spain

Vitali V. Pishko
Institute for Low Temperature
Physics
47 Lenin Avenue
Kharkov-1, 310164
Ukraine

Neculai Plugaru
Institute of Atomic Physics
IFTM
P. O. Box MG-06
Bucharest
Romania

Martin A. Poyser
Department of Pure and Applied
Physics
University of Salford
The Crescent
Salford M5 4WT
UK

Peter C. Riedi
J. F. Allen Laboratory
Dept. Physics and Astronomy
University of St. Andrews
St. Andrew, Fife, KY 16 9SS
Scotland, UK

Javier Rubin
Inst. de Ciencias de Materiales
de Aragon, C. S. I. C.
Facultad de Ciencias
Univ. de Zaragoza
E-50009 Zaragoza
Spain

Umberto Russo
Dept. Inorganic, Metallorganic and
Analytical Chemistry
University of Padua
Via Loredan 4
I-35131 Padova
Italy

Cristo-Laurenzio Saetta
Institute of Materials Science
Demokritos
15310 Ag. Paraskevi
Attiki, Athens
Greece

Ozge Sahin
Dokuz Eylul University
Müh. Fak. Elektrikve Elektronik
Bölümü
TK-35100 Bornova-Izmir
Turkey

Boris Saje
Josef Stefan Institute
39 Jamova
61000 Ljubljana
Slovenia

Herbert R. Schober
Institut für Festkörperforschung
Forschungzentrum Jülich
D-5170 Jülich
Germany

Vladimir M. Shkapa
Ukrainian Academy of Sciences
Institute of Metal Physics
Vernadsky Blvd. 36
252142 Kiev-142
Ukraine

Ralph Skomski
Dept. Pure and Applied Physics
Trinity College
University of Dublin
Dublin 2
Ireland

Lutz Steinbeck
Max-Planck-Society
MPG Research Group
'Electron Systems'
Physics Dept.
TU Dresden
D-01062 Dresden
Germany

Michael A. Todd
Department of Chemistry
Arizona State University
Tempe, AZ 85287-1604
USA

Dimitrios Tsiplakides
Dept. of Chemical Engineering
c/o Prof. C. Vayenas
University of Patras
GR-26500 Patras
Greece

Kurt Wierman
University of Nebraska
116 Brace Labs
Lincoln, NE-68588-0111
USA

William B. Yelon
Research Reactor Center
University of Missouri-Columbia
Columbia, MO 65211
USA

Wei Zhang
Cook Physical Science Building
University of Vermont
Burlington, VT 05405
USA

Tong Zhao
Van der Waals-Zeeman Laboratorium
Universiteit van Amsterdam
Valckenierstraat 65
NL-1018 XE Amsterdam
The Netherlands

Zhi-Gang Zhao
Van der Waals-Zeeman Laboratorium
Universiteit van Amsterdam
Valckenierstraat 65
NL-1018 XE Amsterdam
The Netherlands

Author Index

Subject Index

www.ingramcontent.com/pod-product-compliance
Ingram Content Group UK Ltd.
Pitfield, Milton Keynes, MK11 3LW, UK
UKHW020129290926
15471UKWH00003B/374
9789401102964